Oil Spill Environmental Forensics

Oil Spill Environmental Forensics

Fingerprinting And Source Identification

Zhendi Wang, Ph.D.
ESTD, Environment Canada

Scott A. Stout, Ph.D.
NewFields Environmental Forensics Practice, LLC

AMSTERDAM • BOSTON • HEIDELBERG • LONDON
NEW YORK • OXFORD • PARIS • SAN DIEGO
SAN FRANCISCO • SINGAPORE • SYDNEY • TOKYO

Academic Press is an imprint of Elsevier

Academic Press is an imprint of Elsevier
30 Corporate Drive, Suite 400, Burlington, MA 01803, USA
525 B Street, Suite 1900, San Diego, California 92101-4495, USA
84 Theobald's Road, London WC1X 8RR, UK

This book is printed on acid-free paper. ∞

Copyright © 2007, Elsevier Inc. All rights reserved.

No part of this publication may be reproduced or transmitted in any form or by any means, electronic or mechanical, including photocopy, recording, or any information storage and retrieval system, without permission in writing from the publisher.

Permissions may be sought directly from Elsevier's Science & Technology Rights Department in Oxford, UK: phone: (+44) 1865 843830, fax: (+44) 1865 853333, E-mail: permissions@elsevier.com. You may also complete your request on-line via the Elsevier homepage (http://elsevier.com), by selecting "Support & Contact" then "Copyright and Permission" and then "Obtaining Permissions."

Library of Congress Cataloging-in-Publication Data

British Library Cataloguing-in-Publication Data
A catalogue record for this book is available from the British Library.

ISBN 13: 978-0-12-369523-9
ISBN 10: 0-12-369523-6

For information on all Academic Press publications
visit our Web site at www.books.elsevier.com

Printed in the United States of America
06 07 08 09 10   9 8 7 6 5 4 3 2 1

**Working together to grow
libraries in developing countries**

www.elsevier.com | www.bookaid.org | www.sabre.org

ELSEVIER   BOOK AID International   Sabre Foundation

Biography

Dr. Jan T. Andersson finished his Ph.D. in organic chemistry at the University of Lund, Sweden, in 1976. Following two postdoc years in the USA, he has held faculty positions in Germany and is now professor of analytical chemistry at the University of Münster. Major research interests are polycyclic aromatic sulfur compounds in petroleum and the environment, including high-molecular-weight compounds in fossil materials. The development of analytical methods, from new uses of the atomic emission detector in gas chromatography to novel stationary phases in HPLC, is central to such studies. E-mail: anderss@uni-muenster.de.

Silvana Maria Barbanti is a research scientist focusing on application of biomarkers as parameters of age, source, depositional environment, thermal maturity, and biodegradation to petroleum exploration; and application of diamondoids as indicators of oil cracking and for identification of origins of spilled oils. Ms. Barbanti has an Sc.D. and a master's degree in organic chemistry from Federal University of Rio de Janeiro and a B.S. in pharmacy and biochemistry from University of São Paulo. She has devoted the last 13 years of her life to biomarker research. Dr. Barbanti is a member of the editorial board of *Environmental Forensics*. E-mail: barbanti.ufrj@petrobras.com.br.

Dr. A. Edward Bence has recently retired as senior research advisor from ExxonMobil Upstream Research Company, where his research included compositional kinetics of hydrocarbon generation and the application of petroleum geochemistry to basin exploration, Dr. A. Edward Bence is a former professor of Earth and Space Sciences, from 1968 to 1980, at SUNY Stony Brook, principal investigator NASA Apollo Program, and NSF DSDP and Ocean Crustal Drilling Programs. He has over 150 refereed publications on petroleum geochemistry, planetary basaltic volcanism, electron probe microanalysis, and hydrothermal ore deposits. He earned degrees from the University of Saskatchewan (B.E., geological engineering), the University of Texas at Austin (M.A., geology), and the Massachusetts Institute of Technology (Ph.D., geochemistry). E-mail: tedbence@shaw.ca.

Dr. Paul D. Boehm is group vice president and principal scientist at Exponent, Inc. He has devoted his 30 years of consulting experience on environmental and forensic chemical aspects of aquatic and terrestrial contamination with an emphasis on petroleum and petrochemicals. He has specifically practiced in the areas of environmental forensics, natural resource damage assessment (NRDA), fate and transport of chemicals, contaminated sediments assessments, and environmental impact assessment for new capital projects. Dr. Boehm specializes in chemical baseline issues, reconstruction of historical releases, and technical apportionment of responsibility for petroleum, coal tar, PAH, PCB, and other contamination of aquatic sediments and wildlife. E-mail: pboehm@exponent.com.

Carl E. Brown is a researcher focusing on the detection of oil spills in the marine and terrestrial environments. Dr. Brown has a Ph.D. in physical chemistry from McMaster University and a bachelor of technology degree from Ryerson Polytechnical University. He has been in the oil spill remote sensing field for 14 years and has published over 130 papers and publications in the fields of chemistry and remote sensing. E-mail: carl.brown@ec.gc.ca.

Jan H. Christensen is assistant professor at the Department of Natural Sciences, The Royal Veterinary and Agricultural University, Copenhagen, Denmark. His specialities and primary research interests include oil spills in the marine and terrestrial environment, oil spill identification, transport and transformation of petroleum hydrocarbon mixtures, chemical analysis of complex mixtures of contaminants in the environment (environmental profiling) using modern analytical techniques, development of novel tools for automated chemometric data analysis, and risk assessment of pollutant mixtures. He is responsible for a new bachelor/master course in analytical chemistry and partly responsible for the daily operation of two research-grade analytical instruments: an ion-trap gas chromatograph — mass spectrometer (GC-MS) (Finnigan Polaris/GCQ) and a hybrid quadrupole-orthogonal — time-of-flight mass spectrometer (Q-TOF). Dr. Christensen has started a new laboratory at The Royal Veterinary and Agricultural University, Denmark, and has already authored 36 academic publications, including 13 peer-reviewed articles in highly respected journals within environmental science and analytical chemistry. E-mail: jch@kvl.dk

Per Daling has an M.Sc. in organic analytical chemistry from the Institute of Chemistry, University of Trondheim. He is a senior research scientist at the Marine Environmental Technology Dept. at SINTEF in Trondheim, Norway.

He has 25 years of experience within the field of oil pollution and has been the project manager and scientist responsible for many research programs within

- Weathering and behavior of oil at sea/in ice/on shore
- Oil analysis and oil spill identification (fingerprinting)
- Field and laboratory testing of oil spill dispersants and countermeasure techniques
- Oil spill contingency planning and net environmental benefit analysis

Since 1999, Per S. Daling has been a convenor for the European Committee for Standardisation (CEN) for establishing standardized methodology and guidelines for defensible oil spill identification/fingerprinting. He has authored or co-authored 30 refereed publications and 85 nonrefereed publications (papers and invited presentations). E-mail: per.daling@chem.sintef.no.

Mrs. Maria de Fatima a Guadalupe Meniconi is an environmental geochemist with experience in the petroleum industry since 1986. She has carried out research on chemical fingerprinting of petroleum and derivatives, environmental off-shore production wells monitoring, oil spill assessment and modeling, fisheries safety monitoring, and source identification of hydrocarbons in the environment. She obtained her M.Sc. in chemical engineering from Federal University of Rio de Janeiro (COPPE), Brazil, in 1985, and her M.Phil. in instrumentation and analytical science from University of Manchester Institute of Science and Technology (UMIST), England, in 1999. She is currently a senior research scientist of the Research and Development Center of Petrobras, Rio de Janeiro, Brazil. E-mail: fatimameniconi@petrobras.com.br.

Gregory S. Douglas, Ph.D., is a senior consultant and partner at the NewFields Environmental Forensics Practice and has over 25 years of experience in the field of environmental chemistry. Dr. Douglas received a B.S. in chemical oceanography from the Florida Institute of Technology and M.S. and Ph.D. degrees in chemical oceanography from the Graduate School of Oceanography at the University of Rhode Island. His expertise includes development and application of advanced analytical chemistry methods for the study of the fate and effects of petroleum hydrocarbons in soil, waste, wastewater, and biota. His project experience includes many notable oil spill studies such as the *Exxon Valdez* (USA), *Haven* (Italy), *Trecate Blowout* (Italy), *OSSA II Pipeline* (Bolivia), *M/T Athos* (USA), *North Cape* (USA), *M/V New Carrissa* (USA), *T/V Julie N* (USA), *M/V Sea Empress* (Wales), *M/V Selendang* (USA), and 1991 Gulf War oil spills (Saudi Arabia). Dr. Douglas has also published and presented extensively on analytical methods to identify and reliably monitor the degradation of crude oil in marine sediment and soils. His project support activities include analytical and field program design, analytical method development, oversight of laboratory quality-assurance/quality-control programs, hydrocarbon fingerprinting studies, and litigation support for Natural Resource Damage Assessment (NRDA) programs and Superfund Allocation programs. His e-mail address is gdouglas@newfields.com.

Stephen Emsbo-Mattingly, M.S. (University of Massachusetts), is an environmental scientist with 18 years of analytical chemistry and source identification experience. His area of research includes the composition and changes in the hydrocarbon chemistry of petroleum, tar, and PCB products in the environment. Numerous statistical techniques feature strongly in his work. Mr. Emsbo-Mattingly has authored or co-authored over 50 papers and presentations in the United States and Great Britain. His research has provided a basis for determining environmental liability for both government and industrial clients. He was formerly employed by E3I Environmental Laboratory, META Environmental, and Battelle Memorial Institute before becoming a partner at NewFields Environmental Forensics Practice, LLC, Rockland, Massachusetts. E-mail: smattingly@newfields.com.

Merv Fingas performs research focusing on the behavior and detection of oil spills in the environment. Mr. Fingas has a Ph.D. in environmental physics from McGill University and three master's degrees: one in chemistry, one in business, and another in mathematics, all from the University of Ottawa. He has devoted the last 32 years of his life to spill research and has over 650 papers and publications in the field. Dr. Fingas is a member of several editorial boards including editor-in-chief of the *Journal of Hazardous Materials*. E-mail: merv.fingas@ec.gc.ca.

Glenn S. Frysinger is a professor of chemistry at the United States Coast Guard Academy in New London, CT, USA. He received a B.S. in chemistry (1987) and an M.S. and Ph.D. in experimental physical chemistry (1992) from Rensselaer Polytechnic Institute in Troy, NY, USA. He did postdoctoral work at SRI International in Menlo Park, CA, USA, before joining the chemistry faculty at the USCGA in 1994. He has teamed with Captain Richard Gaines over the last 10 years to develop instrumentation and applications for comprehensive two-dimensional gas chromatography (GC × GC). His focus application areas are forensic and environmental science, including the detection of ignitable liquids in fire debris, oil spill fingerprinting, and detailed petroleum analysis. His e-mail address is Glenn.S.Frysinger@uscg.mil.

Captain Richard B. Gaines has 28 years of active service in the United States Coast Guard. He is currently a professor of chemistry at the United States Coast Guard Academy in New London, CT, USA. In addition to teaching, he also interdicted illicit drugs, responded to oil and hazardous chemical spills, and conducted research and development projects related to the Coast Guard's marine safety mission. He received a B.S. in physical science from the United States Coast Guard Academy in 1978 and an M.S. (1981) and Ph.D. (1998) in analytical chemistry from the University of Connecticut. He has been involved in the development of comprehensive two-dimensional gas chromatography (GC × GC) since its inception. His research, in collaboration with colleague Glenn Frysinger, focuses on environmental and criminal forensics, including detailed analyses of oil spills, arson debris, and petroleum products. His e-mail address is Richard.B.Gaines@uscg.mil.

Asger B. Hansen (MSc) is a senior research scientist at the National Environmental Research Institute (NERI), Denmark. For more than 10 years he has been in charge of NERI's oil spill laboratory and responsible for forensic oil spill identification in Denmark as part of the national marine oil spill response team. He has recently participated in a Nordtest project on revising the Nordtest methodology on oil spill identification. At present, he is a member of the CEN TC/BT TF 120 working group developing CEN guidelines on oil spill identifications. E-mail: aha@dmu.dk.

Dr. Abdelrahman H. Hegazi obtained his B.Sc. in chemistry from Alexandria University, Egypt, and his Ph.D. in 2003 through a scientific channel program between the universities of Alexandria and Münster, Germany. In his thesis he focused on petroleum polycyclic hydrocarbons and sulfur heterocycles in geochemical and environmental investigations. Since 2003, he has been a lecturer at the University of Alexandria with research interests in biomarkers and sulfur heterocycles in petroleum and the environment. E-mail: ahegazih@yahoo.com.

Dr. Alan Jeffrey is a senior geochemist at DPRA/Zymax Forensics in San Luis Obispo, California. Dr. Jeffrey received his Ph.D. in chemical oceanography from Texas A&M University for research using stable isotope ratios to determine the origin of natural gas. He has over 20 years of U.S. and international experience in environmental and petroleum geochemistry. Much of his work has focused on the use of geochemical techniques to solve environmental problems, including sources of spilled hydrocarbon fuels and fugitive methane seeps. E-mail: alanj@zymaxusa.com.

Jeffrey Short is a research chemist at the Auke Bay Laboratory, a division of the Alaska Fisheries Science Center, National Marine Fisheries Service, National Oceanic and Atmospheric Administration (NOAA). Dr. Short has worked on oil pollution issues for over 30 years and has published over 50 research papers on the subject. Dr. Short led several of the initial government studies for the Natural Resource Damage Assessment of the *Exxon Valdez* oil spill and has continued to study the long-term fate and effects of the spill. E-mail: Jeff.Short@noaa.gov.

Paul G. M. Kienhuis (B.Sc.) is an analytical chemist with more than 30 years' experience and works at the Institute of Integral Water Management and Waste Water Treatment (RWS-RIZA, NL). For more than seven years, he has been responsible for the identification of waterborne petroleum and petroleum products from the inland waters of the Netherlands and the Dutch part of the North Sea. At present, he is a member of the CEN TC/BT TF 120 working group developing CEN guidelines on oil spill identifications and is also a member of the ENOSI (European Network for Oil Spill Identification) expert group of the Bonn Agreement (an international agreement by North Sea coastal states). In 2004, together with Dr. G. Dahlmann (BSH, Germany), he started with an annual international ring test for oil spill identification to share and improve knowledge about analytical techniques and limitations in comparing oil samples.

He worked for more than 10 years with LC-MS/MS on polar compounds in surface water and has published several papers in scientific journals dealing with specific methods for screening water samples and developed a library for MS/MS spectra based on chemical ionization and collision-induced dissociation; see www.riza.nl/cicid. E-mail: p.kienhuis@riza.rws.minvenw.nl.

William J. Lehr has a Ph.D. degree in physics from Washington State University. He is currently a senior scientist with the Hazardous Materials Response Division of the National Oceanic and Atmospheric Administration. Dr. Lehr was previously the spill response group leader for the same organization, technical analyst with Jet Propulsion Laboratory, and a professor of applied mathematics at the University of Petroleum and Minerals. Dr. Lehr has also served as an adjunct professor for the World Maritime University and as a consultant for UNESCO.

Dr. Lehr is a recognized expert in the field of hazardous chemical spill modeling with more than 50 peer-reviewed publications. He has served as past guest editor for the journal *Spill Science and Technology* and is currently guest editor for the *Journal of Hazardous Materials*. Dr. Lehr has also served as a reviewer for the National Academy of Science in a study of emulsified fuel spills and as co-chair of the International Oil Weathering Committee. E-mail: Bill.Lehr@noaa.gov.

Mr. Kevin J. McCarthy has over 18 years' experience in the field of environmental forensic chemistry. His specialization is in the area of environmental forensics of nonaqueous phase liquids (NAPL), waters, soils/sediments, and soil- and airborne vapors. He has managed over 200 projects delineating the fate and transport of petroleum-based hydrocarbons and coal-derived wastes in aquatic and terrestrial environments. He has written interpretive reports for over 100 programs, for both legal (outside counsel and in-house counsel) and nonlegal clients. He has participated in developing specialized methodologies for the analysis and characterization of oil and petroleum products. He has authored or co-authored over 50 papers as part of scientific journals and/or books. His expertise includes the use of advanced chemical forensic methodologies to characterize the type(s) of petroleum and non-petroleum materials present and the source relationships among samples and to determine proportional allocations of a broad range of chemicals, including those that comprise crude petroleum, petroleum distillates, solvents, fuels, lubricants, residuals, specialty materials, urban runoff, combustion-related PAHs, coal-derived wastes, and various industrial chemicals. He was a principal research scientist at Battelle Memorial Institute for 15 years and is currently a partner at NewFields Environmental Forensics Practice, LLC, Rockland, Massachusetts. E-mail: kmccarthy@newfields.com.

Robert K. Nelson is a research specialist in the Department of Marine Chemistry and Geochemistry at the Woods Hole Oceanographic Institution in Woods Hole, MA, USA. He has 25 years of chromatographic experience and is a veteran of 16 research cruises. His work focuses on processes by which hydrocarbons and other synthetic contaminants are altered in the environment. In addition to studying weathering processes, he is also interested in developing data visualization technology to aid in the interpretation and communication of comprehensive two-dimensional gas chromatographic data. His e-mail address is RNelson@whoi.edu.

Dr. Ed Owens is a principal with Polaris Applied Sciences, Inc., and is a coastal geologist who has been involved with spill response operations worldwide since 1970. His primary focus is on technical support for shoreline treatment and cleanup. Dr. Owens pioneered the use of aerial videotape surveys to support pre-spill shoreline mapping and for spill documentation and the SCAT process for the description of oiled shorelines. E-mail: ehowens@PolarisAppliedSciences.com.

Dr. Elliot Taylor is an associate and senior consultant with Polaris Applied Sciences, Inc. He is a geological oceanographer who has been involved with oil spill response operations, shoreline surveys and studies, response/contingency planning, and training. Since responding to the 1989 *Exxon Valdez* spill in Alaska, he has been on inland and marine spills and has worked worldwide on major contingency planning efforts. E-mail: ETaylor@PolarisAppliedSciences.com.

Dr. David S. Page is the Charles Weston Pickard Professor of Chemistry and Biochemistry at Bowdoin College in Brunswick, Maine. Since 1975, Dr. Page has published more than 120 peer-reviewed publications, dealing with various aspects of the fate and effects of petroleum and other pollutants on marine environments, including over 30 on the *Exxon Valdez* oil spill. He earned his undergraduate degree in chemistry from Brown University in 1965 and his doctorate in physical chemistry from Purdue University in 1970. E-mail: dpage@bowdoin.edu.

Ms. Heather Parker-Hall is a senior scientist with Polaris Applied Sciences, Inc., and is a physical oceanographer who has been involved with oil spill response, planning, and training for more than 10 years. Prior to becoming a consultant, she was a U.S. National Oceanic and Atmospheric Administration (NOAA) commissioned corps officer and the NOAA Scientific Support Coordinator for California. E-mail: hparker-hall@ Polaris AppliedSciences.com.

Roger Prince is a Senior Research Associate at ExxonMobil Biomedical Sciences, Inc., in New Jersey. Dr. Prince received his B.A. with First Class Honors in biology from the University of York (1971) and his Ph.D. in biochemistry from the University of Bristol (1974), both in the United Kingdom. Prior to joining Exxon in 1983, he was a visiting faculty member at the University of California in Berkeley and on the faculty of the University of Pennsylvania. E-mail: roger.c.prince@ exxonmobil.com.

Christopher M. Reddy is an associate chemist in the Department of Marine Chemistry and Geochemistry at the Woods Hole Oceanographic Institution in Woods Hole, MA, USA. He earned a B.S. in chemistry from Rhode Island College in 1992 and a Ph.D. in chemical oceanography from the Graduate School of Oceanography at the University of Rhode Island in 1997. His work mainly focuses on the source, transport, and fate of organic contaminants in the coastal ocean, with a particular interest in using novel analytical tools such as comprehensive two-dimensional gas chromatography as well as molecular-level isotopic analysis (radiocarbon, stable isotope, and stable chlorine). His e-mail address is CReddy@whoi.edu.

Kristin Rist Sørheim has an M.Sc. in organic analytical chemistry from the Institute of Chemistry, University of Bergen, Norway, with a graduate thesis in organic geochemistry (hydrous pyrolysis experiments for petroleum generation of oil shale).

Since 2003, she has worked as a research scientist in the Marine Environmental Technology Dept. at SINTEF in Trondheim, Norway.

Her main fields of competence and selected programs at SINTEF have been

- analysis of PAH in processed water, emission samples, and pitch from aluminum industry
- oil analysis and oil spill identification (fingerprinting)
- produced water analysis (PAH/NPD, alkylated phenols) in connection with testing different cleaning technologies from the oil industry
- oil spill contingency planning in the Arctic

Since 2005, Kristin has participated in establishing the standardized methodology and guidelines for defensible oil spill identification/fingerprinting — European Committee for Standardisation (CEN). E-mail: Kristin.R.Sorheim@sintef.no.

Debra Simecek-Beatty has been a physical scientist for the National Oceanic and Atmospheric Administration's Hazardous Materials Response and Assessment Division for over 20 years. She has a Master's degree in marine affairs from the University of Washington. During an emergency response, she is responsible for providing estimates of the movement and behavior of the spill. This includes collecting visual observations, remote sensing information, wind and current data, and computer modeling output to form an analysis. In addition, she is responsible for interfacing with local experts (i.e., meteorologist, academia, researchers) in formulating the trajectory analysis. E-mail: Debra.Simecek-Beatty@noaa.ogov.

Kathrine Springman (kspringman@ucdavis.edu) is a research toxicologist at the University of California, Davis, in the Department of Civil & Environmental Engineering. Dr. Springman has focused on multiple stressor interactions, indirect toxic effects, and population-level contaminant effects models. Dr. Springman adapted passive sampler technologies for use in evaluating the sources, fate, and toxicity of complex mixtures, such as oil.

Scott A. Stout, Ph.D., has degrees in oceanography and geology from Florida Tech and

Penn State. He is an organic geochemist with 20 years of petroleum and coal industry experience. He has conducted research on the chemical compositions of crude oil, refined petroleum products, coal, and naturally occurring organic matter and their genesis, distribution, and fate in terrestrial and marine environments. Dr. Stout has authored or co-authored over 100 papers published in scientific journals and books and has provided expert testimony in multiple state and federal courts in the U.S. and overseas. His research has provided a basis for allocating environmental liability for both government and industrial clients. He was formerly employed by Unocal Corporation and Battelle Memorial Institute and is currently a partner at NewFields Environmental Forensics Practice, LLC, Rockland, Massachusetts. Dr. Stout is a member of the American Chemical Society, the European Association of Organic Geochemists, and the American Academy of Forensic Sciences. E-mail: sstout@newfields.com.

Giorgio Tomasi is a Ph.D. student at the Department of Food Science, Quality and Technology group, The Royal Veterinary and Agricultural University, Frederiksberg, Denmark. His primary research interests are application of multilinear models in analytical chemistry, development of algorithms for multiway analysis, and study of alignment methods for chemical signals (obtained with chromatographic techniques as well as NMR spectroscopy). He is the author of eight publications in peer-reviewed journals in the fields of analytical chemistry, environmental science, and computer science. E-mail: giorgio.tomasi@gmail.com.

Allen D. Uhler holds a Ph.D. in chemistry from the University of Maryland and has over 20 years' experience in the field of environmental chemistry. He has published or presented more than 150 treatises on the analysis, occurrence, distribution, and fate of hydrocarbons and persistent anthropogenic industrial chemicals in the environment. Dr. Uhler has conducted assessments of the occurrence, sources, and fate of fugitive petroleum at refineries, offshore oil and gas production platforms, bulk petroleum storage facilities, along petroleum pipelines, and sediments. He has studied the occurrence, behavior, and fate of coal-derived wastes at coke works, former manufactured gas plants, and wood-treating facilities. His experience includes the measurement and environmental chemistry of manmade industrial chemicals including PCB congeners and Aroclors, persistent pesticides, dioxins and furans, metals, and organometallic compounds. D. Uhler was a senior scientist for over 17 years at Battelle Memorial Institute and is currently a partner at NewFields Environmental Forensics Practice, LLC, in Rockland, Massachusetts. E-mail: auhler@newfields.com.

Clifford Walters received Bachelor degrees in both chemistry and biology from Boston University (1976) and a Ph.D. in geochemistry from the University of Maryland (1981). He then entered the private sector to conduct research in petroleum geochemistry; first with Gulf Research & Development (1982), then Sun Exploration & Production Co. (1984), Mobil Research & Development (1988), and finally ExxonMobil Research & Engineering (2000). He currently leads a program investigating the fundamental processes involved in petroleum generation, expulsion, and reservoir alteration, including biodegradation. Cliff has numerous publications in petroleum geochemistry, ranging from theoretical modeling to basinal case studies, and has co-authored the new edition of *The Biomarker Guide*. E-mail: clifford.c.walters@exxonmobil.com.

Dr. Zhendi Wang received degrees in analytical chemistry and environmental chemistry from Peking University and Concordia University. He is a senior research scientist and head of oil spill research of Environment Canada. He has devoted the last 16 years on forensic oil and toxic chemical spill research. His specialties and research interests include development of oil spill fingerprinting and tracing technologies; properties, fate, and behavior of oil and other hazardous organics in the environment; contaminated sediments assessment; oil burn emission and products study; oil biodegradation; and application of modern analytical techniques to oil and chemical spill studies. Dr. Wang has authored or co-authored over 280 publications including peer-reviewed journal papers, invited reviews, book and book chapters, departmental reports, conference proceedings, and other publications. Dr. Wang has received numerous national and international scientific honors. He is currently one of two editors-in-chief of *Environmental Forensics*. E-mail: Zhendi.Wang@ec.gc.ca.

Chun Yang has received a Ph.D. from Nanyang Technological University of Singapore in 2000, a master's degree from Research Centre for Eco-Environmental Sciences of Chinese Academy of Sciences (RCEES-CAS) in 1997, and a bachelor's degree from Beijing Normal University of China in 1992. He specializes in environmental sciences, analytical chemistry, and natural products and has authored or co-authored about 30 academic publications in peer-reviewed journals. Currently, his research in the Emergencies Science and Technology Division of Environment Canada mainly focuses on the chemical fingerprinting and environmental forensics of oil products and other possible spill candidates. E-mail: chun.yang@ec.gc.ga.

Un Hyuk Yim, Ph.D., has degrees in oceanography and environmental chemistry from Seoul National University, Korea. He is an environmental chemist with 10 years of institutional experience. He has conducted research on the environmental occurrence and fate of PAHs and oils in terrestrial and marine environments. Dr. Yim has authored and co-authored over 40 papers published in scientific journals, books, and research reports. His research has provided state-of-the-art technologies in national environmental monitoring and risk assessment of oil spill accidents. He is currently a research scientist at the Korea Ocean Research and Development Institute.

Mohamad Pauzi Zakaria is an associate professor of organic geochemistry at Universiti Putra Malaysia, Selangor, Malaysia. Dr. Zakaria holds a B.S degree from Western Michigan University, Kalamazoo, Michigan. He also has M.S degrees from the Florida Institute of Technology, Melbourne, and the University of Massachusetts, Boston. While at UMass, he conducted his graduate research at Fye Laboratory, Woods Hole Oceanographic Institution, Woods Hole, Massachusetts. He completed his Ph.D. at Tokyo University of Agriculture and Technology, Japan, in organic geochemistry in 2002. His current research interests involve the determination of sources, distribution, and transport pathways of hydrocarbon pollution in the environment (mpauzi@env.upm.edu.my).

Table of Contents

Preface	xlvii
Contributors	xlix

1 Chemical Fingerprinting of Spilled or Discharged Petroleum — Methods and Factors Affecting Petroleum Fingerprints in the Environment 1
- 1.1 Introduction 1
- 1.2 Methods for Chemical Fingerprinting Petroleum 3
  - 1.2.1 Historical Perspective 3
  - 1.2.2 Tier 1 — Chemical Fingerprinting via GC/FID 6
  - 1.2.3 Tier 2 — Chemical Fingerprinting via GC/MS 7
    - 1.2.3.1 Polycyclic Aromatic Hydrocarbons 8
    - 1.2.3.2 Petroleum Biomarkers 9
  - 1.2.4 Quality Assurance and Quality Control 10
    - 1.2.4.1 Quality Control 10
    - 1.2.4.2 Quality Assurance 10
- 1.3 Factors Controlling the Chemical Fingerprints of Spilled or Discharged Petroleum 11
  - 1.3.1 Primary Control — Crude Oil Genesis 13
  - 1.3.2 Secondary Controls — Petroleum Refining 18
    - 1.3.2.1 Gasoline 21
    - 1.3.2.2 Distillate Fuels 22
    - 1.3.2.3 Residual Fuels 27
    - 1.3.2.4 Lubricating Oils 28
    - 1.3.2.5 Oily Waste/Bilge Water Discharges 28
  - 1.3.3 Tertiary Controls — Weathering 29
    - 1.3.3.1 Evaporation 30
    - 1.3.3.2 Dissolution 32
    - 1.3.3.3 Biodegradation 32
    - 1.3.3.4 Photooxidation 34
    - 1.3.3.5 Mousse Formation 34
    - 1.3.3.6 De-Waxing and Wax Enrichment 34
  - 1.3.4 Tertiary Controls—Mixing with "Background" 36
    - 1.3.4.1 What Is "Background"? 36
    - 1.3.4.2 Recognizing and Establishing Background 37
    - 1.3.4.3 Naturally Occurring Background Hydrocarbons 37
      - 1.3.4.3.1 Vascular Plant and Algal Debris. 38
      - 1.3.4.3.2 Particulate Coal and Wood Charcoal. 39
      - 1.3.4.3.3 Natural Oil Seeps. 41
    - 1.3.4.4 Anthropogenic Background Hydrocarbons 41
      - 1.3.4.4.1 Urban and River Runoff 42

1.4	Summary		43
	References		45

2 Spill Site Investigation in Environmental Forensic Investigation — 55
2.1	Introduction		55
2.2	Environmental Site Characterization and Reconnaissance Survey		55
2.3	Site Entry and Safety Issues during the Emergency Response Phase		57
	2.3.1	Management of Safety	57
	2.3.2	Risk Assessment and Characterization	58
	2.3.3	Chemical Toxicity of the Spilled Oil	59
	2.3.4	Working Environment Safety	60
	2.3.5	Personal Protective Equipment (PPE)	60
2.4	Determination of Geographic Boundary and Definition of Different Zones within the Affected Area: 1. Terrestrial Oil Spills		61
2.5	Determination of Geographic Boundary and Definition of Different Zones within the Affected Area: 2. Marine/Coastal Waterborne Oil Spills		62
2.6	Collection of Physical, Ecological, and Environmental Data		63
2.7	Sampling Plan and Design: 1. Spills with Known Source		64
	2.7.1	Water Column Sampling	64
	2.7.2	Oil Source Sampling	65
	2.7.3	Sampling on Land	65
	2.7.4	Sampling Plan Design	65
2.8	Sampling Plan and Design: 2. "Mystery" Spills		66
2.9	Data Management		67
2.10	Conclusions		71
	References		71

3 Petroleum Biomarker Fingerprinting for Oil Spill Characterization and Source Identification — 73
3.1	Introduction		73
3.2	Analytical Methodologies for Petroleum Biomarker Fingerprinting		74
	3.2.1	Petroleum Biomarker Families	74
		3.2.1.1 Acyclic Terpenoids or Isoprenoids	75
		3.2.1.2 Cyclic Terpenoids	75
	3.2.2	Labeling and Nomenclature of Biomarkers	77
		3.2.2.1 Stereoisomers	79
		3.2.2.2 Asymmetric (or Chiral) Carbons and α and β Stereoisomers	80
		3.2.2.3 R and S Stereoisomers of Cyclic Biomarkers	81
	3.2.3	Analysis Methods for Biomarker Fingerprinting	81
	3.2.4	Capillary Gas Chromatography — Mass Spectrometry (GC-MS)	83
		3.2.4.1 Benchtop Quadrupole GC-MS	84
		3.2.4.1.1 Scan Mode	84
		3.2.4.1.2 Selected Ion Monitoring (SIM) Mode	84
		3.2.4.1.3 Example Benchtop GC-MS Conditions (EC Oil Spill Research Laboratory)	85
		3.2.4.1.4 Example Benchtop GC-MS Conditions (Petrobras Geochemistry Laboratory)	85
		3.2.4.2 Triple Quadrupole GC-MS-MS	85
	3.2.5	Mass Spectra and Identification of Biomarkers	86

3.3	Fingerprinting Petroleum Biomarkers		90
	3.3.1	Biomarkers in Crude Oils	90
	3.3.2	Biomarkers in Petroleum Products	99
	3.3.3	Biomarkers in Lubricating Oils	101
	3.3.4	Biomarkers in Oil Fractions with Different Carbon Number Range	104
	3.3.5	Aromatic Steranes in Oils and Petroleum Products	104
	3.3.6	Sesquiterpanes in Oils and Petroleum Products	109
	3.3.7	Diamondoid Compounds in Oils and Lighter Petroleum Products	114
	3.3.8	Application of Biomarker Fingerprintings to Oil Spill Studies	117
	3.3.9	Source-Specific Biomarkers	121
	3.3.10	Using Diagnostic Ratios and Cross-Plots of Biomarkers for Source Identification of Oil Spills	125
		3.3.10.1 Diagnostic Ratios of Biomarkers	126
		3.3.10.2 Cross-Plots of Biomarkers	128
3.4	Effects of Weathering on Biomarker Fingerprinting		130
	3.4.1	Processes Affecting the Fate and Behavior of Spilled Oil	130
	3.4.2	Weathering Effects on Biomarkers Fingerprinting	132
	3.4.3	Biodegradation of Biomarkers in Spilled Oil	133
	3.4.4	Determination of Weathered Percentages Using Biomarkers	134
	3.4.5	Case Study: Source Identification of a Harbor Spill by Forensic Fingerprinting of Biomarkers	134
		3.4.5.1 Product Type-Screening	134
		3.4.5.2 Characterization of Bicyclic Sesquiterpanes	135
		3.4.5.3 Confirmation of Source Identification by Quantitative Evaluation of Alkylated PAHs and Pentacyclic Terpanes and Steranes	135
3.5	Conclusions		138
	References		140

4 Characterization of Polycyclic Aromatic Sulfur Heterocycles for Source Identification   147

4.1	Introduction		147
4.2	Sulfur Compounds in Crude Oil and Petroleum Products		148
4.3	Influence of Refinery Processes on PASH Patterns		150
4.4	Stability of Polycyclic Aromatic Sulfur Heterocycles in the Environment		152
4.5	Petroleum PASH Analysis Techniques		155
	4.5.1	Selective Detection in Gas Chromatography	156
		4.5.1.1 Flame Photometric Detection (FPD)	156
		4.5.1.2 Atomic Emission Detection (AED)	157
		4.5.1.3 Sulfur Chemiluminescence Detection (SCD)	157
		4.5.1.4 Mass-Selective Detection (MSD)	157
	4.5.2	Class Separation of PAH and PASH	157
	4.5.3	Comprehensive Two-Dimensional Gas Chromatography	158
	4.5.4	Quantification of PASH	158
4.6	Petroleum PASH Markers in Environmental Forensic Investigations		159
	4.6.1	PASHs as Source Markers	160
	4.6.2	PASHs as Weathering Markers	162
4.7	Conclusions		164
	References		164

5 Oil Spill Identification by Comprehensive Two-Dimensional Gas Chromatography (GC × GC) — 169
- 5.1 Introduction — 169
  - 5.1.1 The Need for High-Resolution Separations — 169
  - 5.1.2 Multidimensional Methods — 170
- 5.2 Comprehensive Two-Dimensional Gas Chromatography (GC × GC) — 171
  - 5.2.1 Modulation Techniques — 172
  - 5.2.2 Detectors — 172
  - 5.2.3 Data Processing — 173
  - 5.2.4 GC × GC Chromatogram — 174
  - 5.2.5 Peak Identity and Chromatogram Structure — 175
  - 5.2.6 GC × GC Petroleum Applications — 180
- 5.3 Applications of GC × GC to Fingerprint Oil Spills — 181
  - 5.3.1 Mobile Bay Marine Diesel Fuel Spill — 181
  - 5.3.2 West Falmouth No. 2 Fuel Oil Spill — 184
  - 5.3.3 Winsor Cove No. 2 Fuel Oil Spill — 187
  - 5.3.4 Buzzards Bay No. 6 (Bunker C) Spill — 191
  - 5.3.5 Oil Seeps, Santa Barbara, CA, USA — 196
- 5.4 Conclusion — 201
- Acknowledgments — 202
- References — 202

6 Application of Stable Isotope Ratios in Spilled Oil Identification — 207
- 6.1 Introduction — 207
- 6.2 Isotope Ratios and Their Measurement — 207
- 6.3 Bulk Isotope Ratios — 210
- 6.4 Compound-Specific Isotope Analysis (CSIA) — 214
  - 6.4.1 Experimental Considerations — 220
- 6.5 Weathering — 220
- 6.6 Other Isotopes — 224
- 6.7 Conclusions — 224
- References — 225

7 Emerging CEN Methodology for Oil Spill Identification — 229
- 7.1 Introduction — 230
- 7.2 Scope of the CEN Methodology — 231
- 7.3 Strategy for Identifying Oil Spills — 231
- 7.4 Tiered Levels of Analysis and Data Treatment — 233
  - 7.4.1 Decision Chart for Identifying Oil Spills — 233
  - 7.4.2 Visual Characterization and Preparation/Cleanup of Oil Samples — 233
  - 7.4.3 Level 1 — GC/FID Screening — 235
    - 7.4.3.1 Evaluation of Weathering — 236
  - 7.4.4 Level 2 — GC/MS Fingerprinting — 237
    - 7.4.4.1 Diagnostic Ratios from GC/MS Fingerprinting — 237
    - 7.4.4.2 Diagnostic Ratios Derived from Alkylated Polycyclic Aromatic Compounds — 238
    - 7.4.4.3 Diagnostic Ratios Derived from Petroleum Biomarkers — 239
    - 7.4.4.4 Optional Diagnostic Ratios Derived from Sesquiterpanes — 240
  - 7.4.5 Level 3 — Treatment of Results — 245
    - 7.4.5.1 Comparison of Oil Samples Using Diagnostic Ratios — 245

		7.4.5.2	Criteria for Selecting, Eliminating, and Evaluating Diagnostic Ratios	245
		7.4.5.3	Repeatability Limit and Critical Difference	246
		7.4.5.4	Elimination of Diagnostic Ratios Using Signal-to-Noise (S/N) Test	246
		7.4.5.5	Elimination of Diagnostic Ratios Using Duplicate Analyses	248
		7.4.5.6	Optional Comparison of Diagnostic Ratios Using Multivariate Statistics	249
	7.4.6	Final Evaluation and Conclusions		250
7.5	The CEN Methodology in Practice: A Case Study			251
	7.5.1	The Spill Case		251
	7.5.2	GC/FID Screening		251
	7.5.3	GC/MS Fingerprinting		251
	7.5.4	Evaluation and Comparison of Diagnostic Ratios		251
7.6	Summary			254
	Acknowledgment			255
	References			255

8 Advantages of Quantitative Chemical Fingerprinting in Oil Spill Source Identification   257

8.1	Introduction			257
8.2	Qualitative Fingerprinting Methods			258
	8.2.1	Shortcomings of Qualitative Fingerprinting		260
		8.2.1.1	Weathered Oils	260
		8.2.1.2	Genetically Similar Oils	260
		8.2.1.3	Qualitatively Similar Oils	261
		8.2.1.4	Mixing	261
8.3	Quantitative Fingerprinting Methods			263
	8.3.1	Semiquantitative versus Fully Quantitative Methods		263
	8.3.2	Data Generation for Fully Quantitative Fingerprinting		265
		8.3.2.1	Sample Collection	265
		8.3.2.2	Sample Preparation	266
		8.3.2.3	GC/FID Analysis	266
		8.3.2.4	GC/MS Analysis	268
		8.3.2.5	Data Quality	269
	8.3.3	Selection of Diagnostic Indices		269
	8.3.4	Source Identification Protocols for Quantitative Fingerprinting Data		272
8.4	Unraveling Mixed Source Oils Using Quantitative Fingerprinting Data			276
	8.4.1	Two-Component Mixing Models		276
	8.4.2	Case Study 1		277
	8.4.3	Case Study 2		279
		8.4.3.1	Mixing Model Case Study 2	284
8.5	Summary			289
	References			290

9 A Multivariate Approach to Oil Hydrocarbon Fingerprinting and Spill Source Identification   293

9.1	Introduction		293
	9.1.1	Multivariate Methods and Oil Fingerprinting	294

		9.1.2	Integrated Multivariate Oil Fingerprinting (IMOF)	296
	9.2	Sample Preparation and Chemical Analysis		297
		9.2.1	Sample Preparation	297
		9.2.2	Analytical Methods	298
		9.2.3	Fluorescence Spectroscopy	298
		9.2.4	GC-MS	299
		9.2.5	Quality Assurance and Quality Control (QA/QC)	301
	9.3	Data Preprocessing		302
		9.3.1	Partial GC-MS/SIM Chromatograms	303
			9.3.1.1 Baseline Removal	303
			9.3.1.2 Retention Time Alignment	304
			9.3.1.3 Normalization	306
		9.3.2	Diagnostic Ratios	306
		9.3.3	Preprocessing of Fluorescence Spectra	307
	9.4	Multivariate Statistical Data Analysis		308
		9.4.1	Multilinear Models	310
			9.4.1.1 Two-Way Case	310
			9.4.1.2 Higher-Order Arrays	311
		9.4.2	Variable Selection and Scaling	312
	9.5	Data Evaluation		314
		9.5.1	Visual Inspection of Score and Loading Plots	315
		9.5.2	Numerical Comparisons and Statistical Tests	317
	9.6	Conclusions and Perspectives		319
		Acknowledgments		321
		References		322

10 Chemical Heterogeneity of Modern Marine Residual Fuel Oils — 327

10.1	Introduction		327
	10.1.1	Historical Perspective	328
	10.1.2	Production of Heavy Fuel Oils	328
	10.1.3	Marine Fuel Nomenclature and Classification	329
10.2	Forensic Chemistry Considerations		330
	10.2.1	General Chemical Fingerprinting	330
	10.2.2	Samples and Analytical Methods	332
10.3	General Features of Modern Residual Marine Fuel Oils		332
10.4	Molecular Variability among Modern Residual Fuel Oils		336
	10.4.1	Petroleum Biomarkers	336
	10.4.2	Polycyclic Aromatic Hydrocarbons	338
10.5	Distinguishing Heavy Fuel Oils from Crude Oil		343
10.6	Conclusion		346
	References		346

11 Biodegradation of Oil Hydrocarbons and Its Implications for Source Identification — 349

11.1	Introduction		349
11.2	Biochemistry of Petroleum Biodegradation		349
	11.2.1	Aerobic Biodegradation of Hydrocarbons	351
	11.2.2	Anaerobic Biodegradation of Hydrocarbons	355
11.3	Subsurface Biodegradation of Petroleum		357

		11.3.1	The Biodegradation of Hopanes and the Formation of 25-Norhopanes	360
	11.4	Factors Limiting Biodegradation		362
	11.5	Microbial Ecology of Petroleum Biodegradation		365
		11.5.1	The Succession of Microbial Communities	365
		11.5.2	Deep Subsurface Ecology	367
			11.5.2.1 Aerobic Respiration	367
			11.5.2.2 Anaerobic Respiration	368
	11.6	Conclusions; Implications of Biodegradation on Identification		369
		References		370

12 Identification of Hydrocarbons in Biological Samples for Source Determination — 381

- 12.1 Introduction — 381
- 12.2 Determination of the Primary Route of Hydrocarbon Accumulation by Biota — 382
- 12.3 Catabolic Degradation of Hydrocarbons Accumulated by Biota — 387
    - 12.3.1 Catabolic Degradation of PAH — 387
    - 12.3.2 Effects of Catabolism on PAH Accumulation, Persistence, and Depuration — 390
- 12.4 Modes of Toxic Action of Accumulated Hydrocarbons — 393
- 12.5 Case Study: The *Exxon Valdez* Oil Spill — 396
- 12.6 Summary — 398
- References — 398

13 Trajectory Modeling of Marine Oil Spills — 405

- 13.1 Introduction — 405
- 13.2 Forecasting and Hindcasting Oil Spill Movement — 406
- 13.3 Oil Spill Transport — 407
    - 13.3.1 Wind — 409
    - 13.3.2 Currents — 411
    - 13.3.3 Turbulent Diffusion — 413
- 13.4 Evolution of an Oil Spill — 413
    - 13.4.1 Spreading — 413
    - 13.4.2 Oil Weathering — 414
- 13.5 Conclusions and Challenges — 416
- Acknowledgments — 416
- References — 416

14 Oil Spill Remote Sensing: A Forensic Approach — 419

- 14.1 Introduction — 419
- 14.2 Visible Indications of Oil — 420
- 14.3 Optical Sensors — 420
    - 14.3.1 Visible — 420
    - 14.3.2 Infrared — 422
    - 14.3.3 Ultraviolet — 423
    - 14.3.4 Night Vision Cameras — 423
- 14.4 Laser Fluorosensors — 423
- 14.5 Microwave Sensors — 426
    - 14.5.1 Radiometers — 426
    - 14.5.2 Radar — 426
    - 14.5.3 Microwave Scatterometers — 428

14.6	Determination of Slick Thickness	428
	14.6.1 Visual Thickness Indications	428
	14.6.2 Theoretical Approaches	429
	14.6.3 Literature Review of Visual Indications of Oil Slick Thickness	429
	14.6.4 Oil Slick-Thickness Relationships in Remote Sensors	431
	14.6.5 Specific Oil-Thickness Sensors	432
14.7	Acoustic Systems	435
14.8	Satellite Remote Sensing	435
14.9	Detection of Oil under Ice	436
14.10	Real-Time Displays and Printers	438
14.11	Future Trends	438
	References	439

15 Advances in Forensic Techniques for Petroleum Hydrocarbons: The *Exxon Valdez* Experience — 449

15.1	Introduction	449
15.2	Identification of Hydrocarbon Sources in PWS	450
	15.2.1 Multiple Sources of Hydrocarbons	450
	15.2.2 Petrogenic Hydrocarbons	451
	15.2.3 Biogenic Hydrocarbons	453
	15.2.4 Pyrogenic Hydrocarbons	453
15.3	Composition of Exxon Valdez Crude and Its Weathering Products	453
	15.3.1 Bulk Composition and Trace Chemistry	453
	15.3.2 Weathering Trends	457
	15.3.2.1 Data Sources	457
	15.3.2.2 Major Fraction Trends	457
	15.3.2.3 PAH Trends	457
	15.3.2.4 Mass Loss during Weathering	458
15.4	Resolution of Inputs to the Natural Background	458
15.5	Hydrocarbon Source Allocations	462
	15.5.1 Source Allocation Models	462
	15.5.2 Qualitative Allocation Models	464
	15.5.3 Quantitative Models	464
	15.5.3.1 PAH Ratios	464
	15.5.3.2 Statistical Models	465
	15.5.3.3 Statistical Models	466
	15.5.3.3.1 Multivariate Methods — Constrained Least Squares	466
	15.5.3.3.2 Multivariate Analysis — Partial Least Squares	468
	15.5.3.4 Total Organic Carbon (TOC) Constraints on Source Allocations	468
15.6	Allocation of Anthropogenic Sources of PAH	469
15.7	Identification of Hydrocarbons in Biological Samples	470
15.8	Applications of Forensic Methods to Assessments of Oil Bioavailability	477
	15.8.1 PAH Uptake in Biota	477
	15.8.2 Passive Sampling of PAH in Water	481
	15.8.3 Biological Markers	481
15.9	Summary	482

		Acknowledgments	483
		References	483

16 Case Study: Oil Spills in the Strait of Malacca, Malaysia — 489

- 16.1 Strait of Malacca, Malaysia: Introduction — 489
  - 16.1.1 Hydro-Oceanographic Condition of the Strait — 489
  - 16.1.2 Ship Traffic in the Strait of Malacca: Historical and Present — 491
- 16.2 Chronic and Acute Oil Spill Events in the Strait — 491
  - 16.2.1 Contribution of Oil Pollution Sources in Malaysia — 491
- 16.3 Methodology — 492
  - 16.3.1 Sample Collection — 492
  - 16.3.2 Source Petroleum — 492
  - 16.3.3 Tar-Ball Samples — 492
  - 16.3.4 Sediment Samples — 492
  - 16.3.5 Street Dust Samples — 493
  - 16.3.6 Asphalt Samples — 494
  - 16.3.7 Fresh Crankcase Oil — 494
  - 16.3.8 Used Crankcase Oil — 495
  - 16.3.9 Automobile Tire Rubber — 495
  - 16.3.10 Aerosol Samples — 495
- 16.4 Analytical Procedure — 495
  - 16.4.1 Chemicals — 495
  - 16.4.2 Extraction and Fractionation — 496
- 16.5 Instrumental Analysis — 496
  - 16.5.1 Analysis of Alkanes and Hopanes — 496
  - 16.5.2 N-Cyclohexyl-2-Benzothiozolamine (NCBA) — 497
  - 16.5.3 Analysis of PAHs — 497
- 16.6 Establishment and Application of Biomarker Analysis for Source Identification of Oil Pollution Sources in the Strait of Malacca — 498
- 16.7 Case Study 1: Development of the Analytical Method for Oil Pollution Source Identification Using Biomarkers in the Strait of Malacca — 498
  - 16.7.1 Weathering of Tar Balls — 499
- 16.8 The Application of Molecular Markers for Source Identification of Tar-Ball Pollution in Malaysia — 501
- 16.9 Case Study 2: Distribution and Sources of Polycyclic Aromatic Hydrocarbons (PAHs) in Rivers and Estuaries in Malaysia — 502
- 16.10 Conclusions and Future Scenario — 502
- References — 503

17 Evaluation of Hydrocarbon Sources in Guanabara Bay, Brazil — 505

- 17.1 Guanabara Bay and Hydrocarbon Apportioning — 505
  - 17.1.1 Regional Setting — 506
  - 17.1.2 January 2000 Heavy Fuel Oil Spill — 506
- 17.2 Methodology for Hydrocarbon Determination and Source Evaluation — 507
  - 17.2.1 Sampling Design — 507
  - 17.2.2 Chemical Analysis — 508
    - 17.2.2.1 Sediment Sample Extraction — 508
    - 17.2.2.2 Extract Cleanup — 509
    - 17.2.2.3 PAH Analysis of Sediment Samples — 509

		17.2.2.4 Biomarkers	510
	17.2.3	Source Identification Techniques	510
		17.2.3.1 PAH Diagnostic Ratios	510
		17.2.3.2 PAH Multivariate Statistical Analysis	511
		17.2.3.3 Biomarker Diagnostic Ratios	511
17.3	Hydrocarbon Results for Guanabara Bay Sediments		511
	17.3.1	PAH Quantification and Distribution	511
	17.3.2	Hydrocarbon Source Identification	514
		17.3.2.1 PAH Diagnostic Ratios	514
		17.3.2.2 PAH Principal Component Analysis	525
		17.3.2.3 Biomarker Diagnostic Ratios	527
17.4	Conclusions		531
	Acknowledgments		533
	References		534

Index	537

List of Tables

1 Chemical Fingerprinting of Spilled or Discharged Petroleum — Methods and Factors Affecting Petroleum Fingerprints in the Environment
    1-1   Average Estimated Annual Releases (1990–1999) of Petroleum into the Marine Environment in North America and worldwide by Source (in Thousands of Metric Tonnes). Data from National Research Council (2002)   2
    1-2   PAH and Related Compounds or Groups Commonly Targeted in Oil Spill-Related Samples Obtained Using GC/MS-SIM   8
    1-3   Inventory of Petroleum Biomarkers Commonly Used in the Characterization of Spilled Oil Obtained Using GC/MS-SIM   9
    1-4   Inventory of the Factors Controlling the Chemical Fingerprints of Spilled or Discharged Petroleum   11
    1-5   Inventory of Common Biomarker Compound Classes Available for Chemical Fingerprinting of Petroleum in the Environment. The Table Includes the Carbon Boiling Range over Which Each Class Occurs and the Dominant Mass/Charge (m/z) Ratio upon Which They Can Be Identified Using Gas Chromatography-Mass Spectrometry   16
    1-6   Typical Distillation Fractions and Petroleum Products Produced from Crude Oil. (Boiling and Carbon Ranges Are Approximate and Can Vary Slightly with Refiner)   22

2 Spill Site Investigation in Environmental Forensic Investigation
    2-1   Characteristics of Spills on Land and Water (Adapted from Owens, 2002)   56
    2-2   Examples of Factors to be Assessed as Part of a Site Investigation   57
    2-3   Priorities of the Safety Officer during an Oil Spill (Adapted from IPIECA, 2002)   58

3 Petroleum Biomarker Fingerprinting for Oil Spill Characterization and Source Identification
    3-1   Petroleum Biomarkers Frequently used for Forensic Oil Spill Studies   78
    3-2   Common Modifiers and Nomenclatures Used to Modify the Structural Specification of Cyclic Biomarkers   79
    3-3   Surrogate and Internal Standard Used for Oil and Biomarker Fingerprinting   83
    3-4   Characteristic Fragment Ions for Various Biomarkers   92
    3-5   Quantitation results of Major Target Biomarkers in Example Crude Oils and Petroleum Products   98
    3-6   Quantitation Results of Target Biomarkers in the MESA Oil and Its Four Fractions   107
    3-7   Peak Identification of the Triaromatic and Monoaromatic Steranes in the Platform Elly Oil   109
    3-8   Peak Identification of Diamondoid Compounds in Prudhoe Bay Oil   116

3-9	Diagnostic Biomarker Ratios Frequently used for Identification, Correlation, and Differentiation of Spilled Oils	127
3-10	Quantitation Results and Diagnostic Ratios of Sesquiterpanes in Three Oil Samples	137

4 Characterization of Polycyclic Aromatic Sulfur Heterocycles for Source Identification

4-1	Names of Sulfur Compounds in Petroleum as Illustrated in Figure 4-1	149
4-2	Polycyclic Aromatic Sulfur Heterocycles Identified in the Egyptian Crude Oil in Figure 4-2	152

5 Oil Spill Identification by Comprehensive Two-Dimentional Gas Chromatography (GC × GC)

5-1	Peak identities of Compounds Labeled in Figure 5-24	201

6 Application of Stable Isotope Ratios in Spilled Oil Identification

6-1	Stable Isotope and Metal Concentrations in Spilled Oils Near a Tank Farm in Western U.S. Compared with Potential Crude Oil Sources	213
6-2	Sensitivity and Precision for Elements Reported to Date of Compound-Specific Isotope Analysis (CSIA). Data from Schmidt et al. (2004)	214

7 Emerging CEN Methodology for Oil Spill Identification

7-1	Diagnostic Ratios Derived from Selected n-Alkanes and Acyclic Isoprenoids (Diterpanes) Recorded by GC/FID	236
7-2	Recommended Diagnostic Ratios (DR) Derived from Alkylated PACs	237
7-3	Recommended Diagnostic Ratios (DR) Derived from Tri- and Pentacyclic Triterpanes	242
7-4	Recommended Diagnostic Ratios (DR) from Rearranged (Diasteranes) and Regular 14α(H)- and 14β(H)-Steranes	243
7-5	Recommended Diagnostic Ratios (DR) Derived from Triaromatic Steroids	243
7-6	Optional Diagnostic Ratios (DR) Derived from Sesquiterpanes	244
7-7	Example of Comparison of Diagnostic Ratios Based on a Repeatability Limit of 14%	248
7-8	Comparison of Diagnostic Ratios of Acyclic Isoprenoids Using the Repeatability Limit	253
7-9	Comparison of Diagnostic Ratios (in %) of Some Triterpanes Using the Repeatability Limit	253
7-10	Comparison of Diagnostic Ratios (in %) of Spill and Candidate Source Samples Using a Repeatability Limit ($r_{95\%}$) of 14%	254

8 Advantages of Quantitative Chemical Fingerprinting in Oil Spill Identification

8-1	Inventory of Target Saturated Hydrocarbons (SHC), PAHs, and Biomarkers Commonly Analyzed by GC/FID and GC/MS-SIM in Oil Spill Studies	267
8-2	Summary of Typical Data Quality Objectives Used in Quantitative Fingerprinting Studies of Waterborne Oil Spills	269
8-3	List of Common Quantitative Indices Useful in Oil Spill Investigations. This List Is Not Intended to Be All Inclusive	271
8-4	Diagnostic Ratios and Precision-Based Statistics for a Spilled Oil and Two Candidate Source Oils (A and B). All Data are Based Upon the Absolute	

	Concentrations of Individual Compound or Compound Groups Measured by GC/MS as per the Revised Nordtest Method (Daling et al., 2002)	272
8-5	Example Data for Selected Biomarkers for Two Oils Demonstrating How the Absolute Analyte Concentrations are Needed to Calculate Accurate Diagnostic Ratios in Theoretical Mixtures. Ratios Calculated from the Original Ratios, as Might Be Available from Peak Areas (but not Concentrations), Do Not Mix Linearly Due to Concentration Differences in the Two Oils. Therefore, Absolute Concentrations Are Necessary to Develop Accurate Mixing Models	277
8-6	Diagnostic Ratios for the IFO and HFO Source Oils and Field Samples	282

9 A Multivariate Approach to Oil Hydrocarbon Fingerprinting and Spill Source Identification

9-1	List of Mass Fragments (m/z) and Corresponding Compound Groups Analyzed with High Relevance for Oil Hydrocarbon Fingerprinting	300
9-2	Application of a Variable-Outlier Detection Method to 17 Biomarker Ratios	313

10 Chemical Heterogeneity of Marine Heavy Fuel Oils

10-1	ISO Specifications for Marine Residual Fuel Oils	329
10-2	Weight Percentages of Total Hydrocarbons (THC) and Common Isoalkane Ratios for 71 Worldwide IFO 380 Heavy Fuel Oils	331
10-3	Homo- and Heteroatomic PAH Composition for 71 Worldwide IFO 380 Heavy Fuel Oils (Concentrations in µg/kg)	339
10-4	Summary Statistics for Certain PAH Diagnostic Metrics for 71 Worldwide IFO 380 Heavy Fuel Oils (See Table 10-3 for Compound Abbreviations.)	342
10-5	Biomarker-Based Thermal Maturity Parameters Measured in Five Asphalts (S_1–S_5) Manufactured with Vacuum Distillation Residuum Derived from the Parent Crude Oil Source	345

11 Biodegradation of Oil Hydrocarbons and Its Implications for Source Identification

11-1	Comparison of Aerobic and Anaerobic Respiration Reactions	366

13 Trajectory Modeling of Marine Oil Spills

13-1	Grid Resolution of Atmospheric Models Modified from Kalnay (2003)	411
13-2	Estimated Length and Time Scale of Oil Movement Due to Surface Current Transport	411

14 Oil Spill Remote Sensing: A Forensics Approach

14-1	Relationships between Appearance and Oil-Slick Thickness	430
14-2	Relationship between Oil Thickness and Detection Limits in Remote Sensing Instruments	432
14-3	Attributes for Sensor Selection	438
14-4	Sensor Suitability for Various Missions	439

15 Advances in Forensic Techniques for Petroleum Hydrocarbons: The *Exxon Valdez* Experience

15-1A	Major Hydrocarbon Fractions, TOC, and PAH Characteristics of EVC, Weathered EVC, PWS Prespill Benthic Sediments, Oil Residues at Historical Human Activity (HA) Sites, and Potential Background Sources	455

15-1B Saturate and Aromatic Biomarker Indices for EVC, Weathered EVC, PWS Prespill Benthic Sediments, Oil Residues at Historical Human Activity (HA) Sites, and Potential Background Sources ... 456

15-2 Summary of Source Allocation Results for Subtidal Sediments Sampled in Six PWS Embayments in 1999 and 2000 ... 471

15-3 Interpretations of Tissue (Excluding Shellfish) PAH Compositions Reported in the 2004 Version of EVTHD (See Text for Definition of EVTHD) ... 475

16 Case Study: Oil Spills in the Strait of Malacca, Malaysia

16-1 PAHs and alkanes compositions for crude oils and Malaysian tar-ball ... 500

17 Evaluation of Hydrocarbon Sources in Guanabara Bay, Brazil

17-1 Results for the Individual PAH (ng/g dry weight)* of Sediment Samples from Guanabara Bay — Campaign 2000 and Studied Oils ... 512

17-2 Results for the Individual PAH (ng/g dry weight)* of Sediment Samples from Guanabara Bay — Campaign 2003 and Studied Oils ... 514

17-3 Literature Data on PAH Concentration (ng/g dry weight) of Sediments from Various Coastal Sites in the World ... 516

17-4 Results for Biomarkers, Terpanes, and Steranes of Sediment Samples from Guanabara Bay — Campaign 2000 ... 528

17-5 Results for Biomarkers, Terpanes, and Steranes of Sediment Samples from Guanabara Bay — Campaign 2003 the 30-Norhopane Series ... 530

List of Figures

1 Chemical Fingerprinting of Spilled or Discharged Petroleum — Methods and Factors Affecting Petroleum Fingerprints in the Environment

    1-1   Diagram depicting the four factors that affect the chemical fingerprints of petroleum relevant in oil spill investigations.   4

    1-2   Ternary diagram showing the general compositional features of crude oils. After Tissot and Welte (1984). Arrow indicates general compositional change due to biodegradation.   13

    1-3   Examples of aliphatic and aromatic hydrocarbons in crude oils.   14

    1-4   Example of the disparate primary chemical fingerprinting features in crude oils from the North Slope of Alaska and the Niger Delta, Nigeria. (a–b) GC/FID chromatograms, (c–d) partial m/z 191 mass chromatograms, and (e–f) partial m/z 217 mass chromatograms. * = internal standards, left to right: o-terphenyl, 5α-androstane and tetracosane-d_{50}.   17

    1-5   GC/FID chromatograms and partial m/z 191 mass chromatograms for a mystery oil spill and produced crude oil from two nearby production platforms. Differences in the relative abundance of gammacerane were sufficient to distinguish these (and other) genetically related oils from one another. * = internal standards.   19

    1-6   Summary of foreign and domestic petroleum products transported as cargo in U.S. waterways and ports in 2003. Volume calculated from short ton using average density. Data from WCUS (2003).   19

    1-7   Schematic flowchart of a modern complex refinery highlighting the distillation, conversion, and finishing processes used in the production of petroleum products.   21

    1-8   Gas chromatography-flame ionization detection (GC/FID) chromatograms of distinct petroleum products. (A) automotive gasoline, (B) jet fuel A, (C) diesel fuel #2, (D) heavy fuel oil #6, (E) lubricating oil, and (F) petroleum asphalt. # = n-alkane carbon number.   23

    1-9   Partial m/z 191 mass chromatograms for a (A) parent crude oil feedstock and (B) daughter distillate cut showing the progressive loss of tricyclic triterpanes due to distillation effects.   25

    1-10   Partial m/z 123 mass chromatograms showing the distributions of sesquiterpanes in (A) a spilled jet fuel and (B) a candidate source oil. Peak numbers correspond to compounds described by Alexander et al. (1984). The large peak between peaks 10 and 1 is unidentified.   25

    1-11   Chemical fingerprints for two distillate fuel blending stocks. (A) GC/MS TIC for straight-run (virgin) gas oil, (B) GC/MS TIC for light cracked distillate (cycle oil), (C) PAH histogram for straight-run gas oil, and (D) PAH histogram for light cracked distillate. # = n-alkane carbon number, Pr = pristane, Ph = phytane; for PAH analytes, see Table 1-2.   26

1-12 GC/FID chromatograms for four base oils used in the manufacture of lubricating oils. (A) low-viscosity base oil (70N), (B) mediumviscosity base oil (150N), (C) heavy-viscosity base oil (320N), and (D) bright stock (850N). 29

1-13 GC/FID of a bilge oil containing a mixture of different petroleums. (Modified from Emsbo-Mattingly, 2006.) 30

1-14 GC/FID chromatograms showing the progressive loss of compounds from a 34°API crude oil spilled following Hurricane Katrina. (A) slightly evaporated crude oil, (B) moderately evaporated crude oil, and (C) severely evaporated crude oil. 31

1-15 GC/FID chromatograms showing the effect of wax precipitation on the chemical fingerprint of a paraffinic oil. (A) starting oil, (B) wax-depleted fraction within spilled oil 71 hours after the spill, and (C) wax-enriched fraction within the spilled oil 71 hours after the controlled spill experiment. Data from Daling et al. (2002). 35

1-16 GC/FID chromatogram of the TPH extracted from a Louisiana peat containing 2000 mg/kg (dry). 39

1-17 Histograms showing the distribution of PAH in three U.S. coals of varying rank and in modern wood-derived charcoal. For PAH compound abbreviations, see Table 1-2. 40

1-18 GC/FID fingerprints showing the chromatographic character of the UCM "humps" in (A) urban sediment impacted by urban runoff, (B) motor oil, and (C) hydraulic oil. OTP (o-terphenyl) and 5α-androstane are internal standards. 42

1-19 Histograms showing the distributions and concentrations of PAH in an urban sediment from the Elizabeth River (Virginia) containing anthropogenic background PAH and in various urban runoff candidate source components. For compound abbreviations, see Table 1-2. Concentrations refer to mg of PAH per kg of sediment (dry), urban dust (dry), soot (dry), oil, and asphalt (dry). 44

2 Spill Site Investigation in Environmental Forensic Investigation

2-1 Map of tar ball distribution indicating those that matched the source oils on the vessel and those that did not match (adapted from Owens et al., 2002). 69

2-2 Example of chain-of-custody form. 70

3 Petroleum Biomarker Fingerprinting for Oil Spill Characterization and Source Identification

3-1 Molecular structures of example cyclic terpenoid compounds in oil. 76
3-2 Types of isomers. 80
3-3 Oil sample preparation flowchart. 82
3-4 Mass spectra of some common biomarkers used in environmental forensic studies. 87
3-5 The SIM chromatograms at m/z (a) 191 and (b) 217 of a Kuwait crude oil for common terpane and sterane biomarker classes. 91
3-6 Carbon range of common cyclic biomarker classes in crude oil and petroleum products. 93
3-7 GC-FID (left panel) and GC-MS (at m/z 191, right panel) chromatograms of five different oils (light to medium) to illustrate differences in the n-alkane and tri-, tetra-, and pentacyclic terpane distributions between oils. 94

3-8	GC-MS chromatograms at m/z 217 and 218 for five different oils (light to medium) to illustrate differences in sterane distributions between oils.	95
3-9	GC-FID (left panel) and GC-MS (at m/z 191, right panel) chromatograms of five heavy oils from different regions.	96
3-10	GC-MS chromatograms at m/z 217 and 218 of five heavy oils from different regions.	97
3-11	GC-FID (left panel) and GC-MS (at m/z 191, right panel) chromatograms of five common petroleum products (light to heavy) to illustrate differences in the n-alkane and tri-, tetra-, and pentacyclic terpane distributions between oils.	100
3-12	GC-MS chromatograms at m/z 217 and 218 for common petroleum products (light to heavy) to illustrate differences in sterane distributions between oils.	102
3-13	GC-FID (left panel) and GC-MS (at m/z 191, middle panel; m/z 218, right panel) chromatograms of five lubricating oils.	103
3-14	GC-FID chromatograms (left panel) and PAH distributions (right panel) of the MESA oil and its four fractions.	105
3-15	GC-MS chromatograms of the MESA oil and its four fractions at m/z 191 (left panel) and 218 (right panel) to illustrate differences in biomarker distributions.	106
3-16	Peak identification of the triaromatic (m/z 231) and monoaromatic (MA, m/z 253) steranes in the Platform Elly oil.	108
3-17	Mass chromatograms of the TA- (m/z 231) and MA-steranes (253) in the aromatic hydrocarbon fractions of the NIST 1582 oil, a Diesel No. 2 from Korea, IFO-180, and a lubricating oil.	110
3-18	GC-MS (SIM) chromatograms of sesquiterpanes eluting in the n-C_{13} and n-C_{16} range.	112
3-19	*Left panel*: GC-MS chromatograms of sesquiterpanes at m/z 123 for light (API > 35), medium (API: 25–35), and heavy (API < 25) crude oils, including Alaska North Slope (ANS), Arabian Light, Scotia Light oil (Nova Scotia), West Texas, and California API 11. *Right panel*: Comparison of SIM chromatograms of sesquiterpanes at m/z 123 for common petroleum products.	113
3-20	GC-MS chromatograms of diamondoids in Prudhoe Bay oil for peak identification.	115
3-21	GC-MS total ion chromatograms of adamantanes (left) and diamantanes (right) of five representative crude oils including Cold Lake Bitumen, Maya, Mars, Alaska North Slope, and South Louisiana crude oil.	118
3-22	GC-MS total ion chromatograms of adamantanes and diamantanes of five representative refined petroleum products including Jet A, Diesel No. 2, Korea Diesel, Fuel No. 4, and Valvoline-10W-30 lube oil.	119
3-23	Comparison of GC-FID and GC-MS (m/z 191) of three unknown oil samples.	122
3-24	Molecular structures of a selection of "source-specific" biomarkers.	123
3-25	**A**: Cross-plots of the double ratios of Peak 4/Peak 5 versus Peak 3/Peak 5 for over 50 different oils and refined products. The circle indicates related samples from the same origin. **B**: Cross-plots of the double ratios of Peak 4/Peak 5 versus Peak 3/Peak 5 for 11 weathering oil series and 1 diesel weathering series. Each weathering oil series produces a tight cluster.	129
3-26	Extracted ion chromatograms at m/z 123, sesquiterpanes, for three Round Robin samples (right) and their corresponding GC-FID chromatograms for n-alkane analysis (left).	136

3-27 Correlation of diagnostic ratios (normalized to %) of sesquiterpanes between spill sample 2 and suspected source samples 1 (left) and 3 (right) at 95% confidence. All the data (A: left panel) overlap the 1:1 line at 95% confidence, representing a "perfect" match between samples 2 and 1. Conversely, most data points (B: right panel) between samples 2 and 3 do not overlapping the line at 95% confidence, representing a "nonmatch." 138

3-28 Comparison of GC-MS chromatograms of terpanes (m/z 191) and steranes (m/z 218) in three oil samples. 139

4 Characterization of Polycyclic Aromatic Sulfur Heterocycles for Source Identification

4-1 Structures of sulfur compounds in petroleum. For names, see Table 4-1. 149

4-2 Partial extended AED chromatogram of polyaromatic fraction showing the distribution and relative retention times of PASH in an Egyptian crude oil. Peak identifications are listed in Table 4-2. 5FBT, 2FDBT, and 6FBN[2,3-d]T are fluorinated internal standards. (Reprinted from Hegazi et al., 2003, with permission from Elsevier Science.) 151

4-3 GC-FID chromatograms of (a) nondesulfurized diesel fuel and (b) deeply hydrodesulfurized diesel fuel. The asterisk indicates an impurity from the dichloromethane used in the sample workup. (Reprinted from Schade et al., 2002, with permission from Taylor & Francis.) 153

4-4 Kendrick mass defects versus Kendrick nominal mass of vacuum residues: (a) first ligand exchange chromatography (LEC) fraction, (b) second LEC fraction before HDS, (c) first LEC fraction, and (d) second LEC fraction after HDS. The double bond equivalent (DBE) scale given is valid for both panels. (Reprinted from Müller et al., 2005, with permission from the American Chemical Society.) 154

4-5 Gas chromatogram of a diesel (350 ppm sulfur) monitored at m/z 184, which is the mass for DBT. Coinjection with DBT reveals this compound at a retention time of 16.86 minutes. 158

4-6 Mass spectrum of the peak at 16.86 minutes in Figure 4-5 (top) and a mass spectrum of pure DBT (bottom). 159

4-7 Plot of the relative ratios of 2-/3-methyl-DBT to 4-methyl-DBT versus the relative ratios of 1-methyl-DBT to 4-methyl-DBT for 25 different oils. The circles 1 and 2 indicate related samples from origins of North Slope and Terra Nova, respectively. (Reprinted from Wang and Fingas, 1995, with permission from the American Chemical Society.) 160

4-8 Mass fragmentograms of methyl-DBTs (m/z 198) from clams collected in April 1993 (lower trace) and December 1995 (upper trace). 2- and 3-MDBT are biodegraded appreciably faster than 1- and 4-MDBT. (Reprinted from Albaigés et al., 2000, with permission from WIT Press.) 163

4-9 Double-ratio plot of C2/C3 dibenzothiophenes versus C2/C3 phenanthrenes from oil residues from Prince William Sound, Alaska, eight years after the *Exxon Valdez* spill. NSC is the source oil from the *Exxon Valdez*. (Reprinted from Michel and Hayes, 1999, with permission from Elsevier Science.) 163

5 Oil Spill Identification by Comprehensive Two-Dimensional Gas Chromatography

5-1 Conceptual diagram of matching mixture dimensions with separation dimensions to achieve complete separation and an ordered chromatogram.

	(A) Three-dimensional mixture in gray scale, shape, and size; (B) one-dimensional separation in size results in poor separation; (C) two-dimensional separation improves separation and results in a more ordered chromatogram; (D) matching mixture and separation dimensions results in complete separation and ordered chromatogram.	170
5-2	Concepts of comprehensive two-dimensional gas chromatography (GC × GC): (A) first-dimension chromatogram peak containing coeluents; (B) modulator samples first-dimension peaks and injects narrow peak into second dimension; (C) coeluents are separated in the second dimension; (D) serial flame ionization detector data stream is sliced into segments and stacked into an array; (E) array is visualized with a contour. Copyright 2002 from *GC × GC — A New Analytical Tool for Environmental Forensics*, by G.S. Frysinger et al. Reproduced by permission of Taylor & Francis Group, LLC, http://www.taylorand francis.com.	171
5-3	Chromatograms of *Exxon Valdez* cargo oil: (A) One-dimensional gas chromatogram using a nonpolar 100% polydimethylsiloxane stationary phase; (B) GC × GC volatility-by-polarity interpolated color contour plot. The first dimension is a separation using a nonpolar 100% polydimethylsiloxane stationary phase and the second dimension is a separation using a polar 50% phenyl polysilphenylene stationary phase; (C) a small portion of the GC × GC chromatogram visualized as a mountain plot. The mountain plot is excellent for visualizing relative differences among neighboring peaks. See color plate.	174
5-4	A small portion of a volatility-by-polarity GC × GC chromatogram of *Exxon Valdez* cargo oil showing the region containing the triaromatic sterane, hopanes, and sterane biomarkers. GC-MS extracted ion chromatograms (EICs) of these biomarker target ions are overlaid for identification and comparison. Peaks numbered on each EIC are identified in Frysinger and Gaines (2001). The structures of 5α,14α,17α(H)-cholestane (20R) (peak 46), 17α,21β(H)-hopane (peak 58) and cholestane (20S) (peak 69) are included for reference. *x*-axis: minutes; *y*-axis: seconds. Reproduced with permission from Frysinger and Gaines (2001). Copyright 2001 Wiley-VCH.	176
5-5	Partial GC × GC-MS chromatogram of diesel fuel. The *x*-axis is a volatility-based chromatographic separation on a 100% polydimethylsiloxane stationary phase. The *n*-alkanes decane (C_{10}) to tridecane (C_{13}) are identified. The *y*-axis produces polarity-based chromatographic separation on a 14% cyanopropylphenyl methylsiloxane stationary phase. The MS total ion intensity data are displayed as an interpolated color contour. Overlay data: (A–B) GC-MS spectra for specific retention times; (C) GC-MS *m/z* 120 extracted ion chromatogram; (D–H) GC × GC-MS spectra and structures of naphthalene, 1,2,3-trimenthylbenzene, 4-methyldecane, *n*-butylcyclohexane, and decahydronaphthalene, respectively; (I) indan; (J) indene. Copyright 2002 from *GC × GC — A New Analytical Tool for Environmental Forensics*, by G.S. Frysinger et al. Reproduced by permission of Taylor & Francis Group, LLC, http://www.taylorandfrancis.com.	177
5-6	Effect of changing the second-dimension stationary phase on the separation of nonpolar saturates: (A) GC × GC chromatogram of the saturates fraction of a sediment extract using a polar second-dimension 14% cyanopro-pylphenyl polysiloxane stationary phase; (B) GC × GC chromatogram of the same saturates fraction using a chiral γ-cyclodextrin second-dimension	

xxxii   List of Figures

	stationary phase. The expected positions of the n-C_{13} to n-C_{20} alkanes are marked with circles. The region inside the dotted lines is expanded in (C); (C) expanded region showing expected position of n-C_{15} and location of branched alkane bands. Reprinted with permission from Frysinger et al., 2003. Copyright 2003, American Chemical Society. See color plate.	179
5-7	Location of petroleum chemical classes on a volatility-by-polarity GC × GC chromatogram. Nonpolar alkanes are located at the bottom, with one- and multiring aromatics further up the y-axis as their polarity increases. Inset A: naphthalene region showing bands of alkyl-substituted naphthalenes and biphenyls; Inset B: nonpolar region showing high-resolution separations of higher boiling (C_{28}+) acyclic isoprenoids, normal and branched alkanes, and cyclic alkanes. See color plate.	180
5-8	Portion of a volatility-by-polarity GC × GC-FID chromatogram from each of the three samples compared: (a) spill; (b) and (c) potential spill sources. Both axes are in seconds. The boxes contain chemical families and individual compounds used to qualitatively and quantitatively compare two potential source samples with the spill samples. The first column separation shown along the x-axis was accomplished with a nonpolar 5% phenylmethylsiloxane stationary phase. The second column separation shown along the y-axis was accomplished with a polar polyethylene glycol/siloxane copolymer stationary phase. Reprinted with permission from Gaines et al., 1999. Copyright 1999, American Chemical Society. See color plate.	182
5-9	Quantitative comparisons between spill sample and two potential source samples from Figure 5-8. A quality-control (QC) sample, split from the spill, was also quantified. Each panel represents the results of integrating peaks within an individual box shown in Figure 5-8. Each bar represents the total peak or band of peaks integration results normalized to the first peak in the box. The error bar represents an estimated 10% variation calculated from replicate data, but is shown only in the plus direction. Compound classes represented: (a) alkylnaphthalene bands normalized to 2,3,5-trimethylnaphthalene; (b) alkylphenanthrenes normalized to phenanthrene; (c) cycloalkanes normalized to pentadecane; (d) heptadecane, octadecane, pristine, and phytane normalized to each other. Reprinted with permission from Gaines et al., 1999. Copyright 1999, American Chemical Society.	183
5-10	Map of the general area near the 1969 grounding of the barge *Florida*.	185
5-11	Gas chromatograms of Wild Harbor marsh sediment extracts: (A) Nov. 1973; (B) Aug. 2000. Reprinted with permission from Reddy et al., 2002. Copyright 2002, American Chemical Society.	185
5-12	Chromatograms of a Wild Harbor sediment extract, Aug. 2000: (A) Conventional one-dimensional gas chromatogram. The separation was accomplished using a nonpolar polydimethylsiloxane stationary phase; (B) volatility-by-polarity GC × GC chromatogram. The first column separation along the x-axis was accomplished using a nonpolar polydimethylsiloxane stationary phase. The second column separation was accomplished using a polar 14%-cyanopropylphenyl polysiloxane stationary phase. Reprinted with permission from Reddy et al., 2002. Copyright 2002, American Chemical Society. See color plate.	186
5-13	Comparison of Wild Harbor sediment extract with weathered MERL oil: (A) Enlarged volatility-by-polarity GC × GC chromatogram of Wild Harbor extract, Aug. 2000 (as in Figure 5-12); (B) neat MERL oil; (C) 30% by mass	

weathered (WX) MERL oil; (D) 70% by mass weathered (WX) MERL oil. The first column separation along the x-axis was accomplished using a nonpolar polydimethylsiloxane stationary phase. The second column separation was accomplished using a polar 14%-cyanopropylphenyl polysiloxane stationary phase. Weathering experiments were performed in a hood where neat MERL oil was allowed to evaporate until the desired mass loss was achieved. The background is blue. Peak intensity is scaled from white, to red, and then to blue (most intense). Compound abbreviations: (N) naphthalene; (C1N) C_1-naphthalenes; (C2N) C_2-naphthalenes; (C3N) C_3-naphthalenes. Reprinted with permission from Reddy et al., 2002. Copyright 2002, American Chemical Society. See color plate. 188

5-14 Map of the general area near the grounding of the barge *Bouchard 65*. The grounding of the barge *Florida*, discussed previously, is also shown for reference. 189

5-15 Chromatograms of the sediment extract taken from Winsor Cove in 2002 approximately 30 years after being contaminated with No. 2 fuel oil: (A) one-dimensional gas chromatogram showing UCM and the region of the chromatogram visualized in (B). The separation was accomplished using a polydimethylsiloxane stationary phase; (B) volatility-by-polarity GC × GC-FID chromatogram showing the region containing the target compounds. The first-column separation shown along the x-axis was accomplished with a polydimethylsiloxane stationary phase. The second-column separation shown along the y-axis was accomplished with a 50%-phenylpolysilphenylene-siloxane stationary phase. See color plate. 190

5-16 Possible structure of C_5- and C_6-decalins found in degraded No. 2 fuel oil in Winsor Cove approximately 30 years after a spill: (A) 8β(H)-drimane; (B) 8β(H)-homodrimane. 190

5-17 Gas chromatograms of *Bouchard 65* incubation sediment extracts artificially degraded in the laboratory: (A) day 1; (B) day 11; (C) day 46. 191

5-18 Volatility-by-polarity GC × GC $m/z = 123$ extracted ion chromatographs of *Bouchard 65* incubation sediment extracts showing the C_4- to C_6-decalin region: (A) neat *Bouchard 65* fuel oil; (B) day 11; (C) day 36; (D) day 46; (E) day 46 spiked with cis- and trans-1,1,4,4,6-pentamethyldecalin standard; (F) Winsor Cove sediment extract showing the bands of peaks containing C_4-decalins (C4D), C_5-decalins (C5D), C_6-decalins (C6D), and location of 8β(H)-drimane (A) and 8β(H)-homodrimane (B) whose structures are given in Figure 5-16. The first-column separation shown along the x-axis was accomplished with a nonpolar polydimethylsiloxane stationary phase. The second-column separation shown along the y-axis was accomplished with a polar 50%-phenylpolysilphenylene-polysiloxane stationary phase. See color plate. 192

5-19 Map of Buzzards Bay showing the track of the *Bouchard 120* barge on April 27, 2003. Oil-impacted beaches are highlighted in gray around the perimeter of the bay. Nyes Neck (where the oil spill samples discussed in this section were collected) is shown on the inset map. Copyright 2006 from *Tracking the Weathering of an Oil Spill with Comprehensive Two-Dimensional Gas Chromatography*, by Nelson et al. Reproduced by permission of Taylor & Francis Group, LLC., http://www.taylorandfrancis.com. 193

5-20 Conventional GC and GC × GC chromatograms of sample extracts from *Bouchard 120* No. 6 fuel oil-covered rocks collected at various times after the spill: (A) conventional GC chromatogram of sample taken 12 days after the spill. The separation was achieved using a polydimethylsiloxane stationary phase; (B) GC × GC volatility-by-polarity chromatogram of sample taken 12 days after the spill. The first dimension separation along the x-axis was achieved using a nonpolar polydimethylsiloxane stationary phase. The second dimension separation along the y-axis was achieved with a polar 50% phenylpolysilphenylenesiloxane stationary phase; (C) conventional GC chromatogram of sample taken 179 days after the spill; (D) GC × GC volatility-by-polarity chromatogram of sample taken 179 days after the spill; (E) a point-by-point difference chromatogram produced by the subtraction of chromatogram (D) from chromatogram (B) after normalization to the conserved standard 17α(H)-21β(H)-hopane (marked with a star). Copyright 2006 from *Tracking the Weathering of an Oil Spill with Comprehensive Two-Dimensional Gas Chromatography*, by Nelson et al. Reproduced by permission of Taylor & Francis Group, LLC., http://www.taylorandfrancis.com.   194

5-21 Percent losses of select hydrocarbons from *Bouchard 120* No. 6 fuel oil covered rocks collected at Nyes Neck after six months of weathering: (●) n-alkanes; (◆) n-alkylcyclohexanes; (△) branched alkanes. Copyright 2006 from *Tracking the Weathering of an Oil Spill with Comprehensive Two-Dimensional Gas Chromatography*, by Nelson et al. Reproduced by permission of Taylor & Francis Group, LLC., http://www.taylorandfrancis.com.   196

5-22 Volatility-by-polarity GC × GC chromatogram of (A) reservoir crude oil from platform Holly, Santa Barbara Channel, California, and (B) oil stringer emerging through the sea floor collected at the Jackpot seep field, Santa Barbara Channel, California. The first-column separation shown along the x-axis was accomplished with a nonpolar polydimethylsiloxane stationary phase. The second-column separation shown along the y-axis was accomplished with a polar 50%-phenylpolysilphenylene-siloxane stationary phase. A star on each chromatogram indicates the position of the internal standard dodecahydrotriphenylene (DDTP). The distinct difference in n-alkane content between the two samples is evident in this figure. See color plate.   197

5-23 Partial volatility-by-polarity GC × GC chromatogram of the platform Holly produced crude oil showing peaks in the vicinity of n-C_{33} through n-C_{39}. The location of high-molecular-weight isoprenoid alkanes (both acyclic and cyclic), thought to be biomarkers of Archaean membrane lipids, are indicated along with some representative molecular structures. See color plate.   199

5-24 Partial volatility-by-polarity GC × GC chromatogram of the platform Holly reservoir crude oil showing peaks in the vicinity of the sterane and desmethylsterane biomarkers. Identification of labeled peaks is found in Table 5-1. See color plate.   200

6 Application of Stable Isotope Ratios in Spilled Oil Identification

6-1 Carbon isotope ratios in a variety of geochemical materials. (Reprinted with permission from Fuex (1977). Copyright (1977) Elsevier Science.)   209

6-2 Schematic diagram of an instrument set-up to perform compound specific carbon isotope analysis.   209

6-3 Carbon isotope ratios of Exxon Valdez oil and residues, and other tar balls collected in Prince William Sound, Alaska and analyzed by Kvenvolden et al. (1997).   211

6-4	Carbon isotope ratios of saturate and aromatic fractions of oil and tar balls associated with the BP American Trader spill offshore Huntington Beach, California. These are compared with ratios of tar balls north and south of the spill site.	212
6-5	Carbon isotope ratios of individual n-alkanes in a number of oils sourced from fluvio-deltaic (FD), lacustrine (LAC), marine deltaic (MD), and marine carbonate (MC) depositional environments. (Reprinted with permission from Murray et al. (1994). Copyright (1994) Elsevier Science.)	215
6-6	Carbon isotope ratios of individual n-alkanes in asphaltic bitumens from S. Australia (AGSO 934, 935), in a bitumen from N. Australia (AGSO 624), and in oils from Seram, Indonesia (AGSO 324) and Minas, Indonesia (AGSO 436). (Reprinted with permission from Dowling et al. (1995). Copyright (1995) Elsevier Science.)	216
6-7	Gas chromatograms and carbon isotope ratios of individual alkanes in oil from the feathers of birds killed in a major oil spill and from the oil that was the suspected source of the spill. (Reprinted with permission from Mansuy et al. (1997). Copyright (1997) American Chemical Society.)	217
6-8	Bulk carbon isotope ratios of oil from the Erika tanker spill (S1) off the Atlantic coast of France, 1999, and of oil residues collected from the northern (S2–S12) and southern (S13–S18) Atlantic coast. (Reprinted with permission from Mazeas and Budzinski (2002). Copyright (2002) American Chemical Society.)	218
6-9	Gas chromatograms and carbon isotope ratios of individual alkanes in beach tars from the Seychelles Islands. (Reprinted with permission from Philp (2002). Copyright (2002) Academic Press.)	219
6-10	Full scan GC/MS chromatograms of a sediment extract after sequential purification procedures. (A) Al/Si column chromatography (B) High performance liquid chromatography (C) Thin-layer chromatography. (Reprinted with permission from Kim et al. (2005). Copyright (2005) American Chemical Society.)	221
6-11	Carbon isotope ratios of individual naphthalenes and methylnaphthalenes in soils and sediments from McMurdo Station, Antarctica. (Reprinted with permission from Kim et al. (2005). Copyright (2005) American Chemical Society.)	222
6-12	Carbon isotope ratios of individual n-alkane/n-alkene pairs generated by pyrolysing asphaltene fractions of biodegraded oils. These are compared with ratios of individual n-alkanes in the undergraded (initial) oils. (Reprinted with permission from Mansuy et al. (1997). Copyright (1997) American Chemical Society.)	223

7 Emerging CEN Methodology for Oil Spill Identification

7-1	Decision/flowchart of the CEN oil spill identification methodology.	234
7-2	GC/FID chromatogram of a mildly weathered fuel oil displaying the dominating n-alkanes together with the acyclic isoprenoids, pristane and phytane (inserted).	235
7-3	Examples of weathering checks of gas oils by GC/FID. A. Simple overlaying of chromatograms. B. Bar chart comparison of normalized n-alkane distributions.	236
7-4	GC/MS ion fragmentograms of diagnostic alkylated PACs. 4A, methylphenanthrenes (m/z 192); 4B, methyldibenzothiophenes (m/z 198); 4C, C2-phenanthrenes (m/z 206); 4D, C2-dibenzothiophenes (m/z 212);	

	4E, C3-phenanthrenes (*m/z* 220); 4F, C3-dibenzothiophenes (*m/z* 226); 4G, C4-phenanthrenes including retene (*m/z* 234); 4H, methylfluoranthenes/ methylpyrenes/benzofluorenes (*m/z 216*); 4I, C3-chrysenes (*m/z* 270).	239
7-5	Relative distribution of PAC homologues in two different IFO-180 bunker oils (from different refineries).	240
7-6	GC/MS ion fragmentogram of tri- and pentacyclic triterpanes ("hopanes") recorded at *m/z* 191 in the SINTEF standard oil mixture.	241
7-7	GC/MS ion fragmentogram of tetracyclic rearranged (diasteranes) and regular 17α(H)-steranes, and 17β(H)-steranes recorded at *m/z* 217 and *m/z* 218, respectively, in the SINTEF standard oil mixture.	241
7-8	GC/MS ion fragmentogram of triaromatic steroids recorded at *m/z* 231 in the SINTEF standard oil mixture.	242
7-9	GC/MS ion fragmentogram of sesquiterpanes recorded at *m/z* 123 in the SINTEF standard oil mixture.	244
7-10	Protocol/decision chart for selection/elimination of diagnostic ratios.	247
7-11	Signal-to-noise calculation. The distance between the highest and lowest peaks of a part of the noise is used as N (see insert) and the peak height as S. In this example, S/N is 4.5.	249
7-12	PCA score plot of data (diagnostic ratios) from a recent mystery oil spill in the North Sea. Several possible source samples (production samples from nearby offshore fields) together with other production samples were included in the PCA.	250
7-13	GC/FID chromatogram of extracts of (A) candidate source sample (bilge), and (B) spill sample.	252
7-14	Weathering check. Overlay of GC/FID chromatograms of spill sample and candidate source sample.	253

8 Advantages of Quantitative Chemical Fingerprinting in Oil Spill Identification

8-1	Comparison of two oils that exhibit qualitatively similar GC/FID patterns and selected extracted ion profiles (partial) that were interpreted to be "probable matches" by a laboratory conducting ASTM D3328 and D5739. Quantitative analysis showed some measurable differences existed, including higher concentrations of C2- and C3-dibenzothiophenes in oil 2 (see Fig. 8-2). *—internal standards.	262
8-2	Double-ratio plot showing the results of replicate analysis of the two oils shown in Figure 8-1. Despite qualitatively comparable D2 (*m/z* 212), D3, P2 (*m/z* 206), and P3 patterns (e.g., Fig. 8-1) the concentration of D2 and D3 were higher in Oil 2.	263
8-3	Examples of linear regression analysis as per the revised Nordtest oil spill identification method for the data shown in Table 8-4. (A) Example of "non-match" between spill oil and Source A based upon 98% confidence interval. (B) Example of "match" between spill oil and Source B based upon 98% confidence interval.	274
8-4	Revised Nordtest linear regression plots for ten diagnostic ratios (A-J) in the mystery spill oil and two candidate source oils from case study 1. (A) comparison of bilge oil with spilled oil resulting in a non-match, (B) comparison of HFO oil with spilled oil resulting in a non-match, (C) comparison of theoretical mixture of bilge and HFO with spilled oil resulting in a positive match.	278

8-5	Comparison of IFO and HFO source fuels oil fingerprints from a marine oil spill from case study 2. (A) GC/FID chromatogram for IFO, (B) GC/FID chromatogram for HFO, (C) PAH concentration histogram for IFO, (D) PAH concentration histogram for HFO, (E) Partial *m/z* 191 triterpane extracted ion profile for IFO, and (F) Partial *m/z* 191 triterpane extracted ion profile for HFO. H29 = Norhopane, H30 = Hopane * = internal standards.	280
8-6	Comparison of two field samples (mousse and tarball) associated with case study 2 and the vessel containing the fuels shown in Figure 8-5. (A) GC/FID chromatogram for the mousse sample (B) GC/FID chromatogram for the tarball sample, (C) PAH concentration histogram for the mousse sample, (D) PAH concentration histogram for the tarball sample, (E) Partial *m/z* 191 triterpane extracted ion profile for the mousse sample, and (F) Partial *m/z* 191 triterpane extracted ion profile for the tarball sample. H29 = Norhopane, H30 = Hopane * = internal standards.	281
8-7	Revised Nordtest linear regression plots for eleven diagnostic ratios (a-k) in the mousse sample and two candidate source oils (IFO and HFO) from case study 2. (A) Comparison of IFO with the mousse sample resulting in a non-match, (B) comparison of HFO oil with the mousse sample resulting in a non-match.	284
8-8	Revised Nordtest linear regression plots for eleven diagnostic ratios (a-k) in the tarball sample and two candidate source oils (IFO and HFO) from case study 2. (A) Comparison of IFO with the tarball sample resulting in a non-match, (B) comparison of HFO oil with the tarball sample resulting in a non-match.	285
8-9	Plot of the D2/P2 versus percent IFO and HFO from the grounded vessel. This equation was used with the D2/P2 ratios measured in the field samples to calculate the % IFO and % HFO in the field samples. These percentages were then used with the field sample quantitative data to calculate mixed oil source ratios that were used in the revised Nordtest method (e.g., Fig. 8-10).	286
8-10	Revised Nordtest linear regression plots for eleven diagnostic ratios (a-k) in the mousse (A) and tarball (B) samples versus the associated mixed source oils. The parity correlation was improved for both samples when the quantitatively determined mixed source ratios were used. Based on this analysis and considering the additional variability introduced by the mixing model but not reflected in the 98% confidence interval, a source match with the mixed oils is likely.	287
8-11	Revised Nordtest linear regression plots for eleven diagnostic ratios (a-k) in the tarball #2 sample versus the associated mixed source oil. No improvement in parity correlation was observed over the entire range of possible mixtures including the mixed source presented in this figure that was calculated based on the D2/P2 ratio in the sample. This data combined with GC/FID, PAH and biomarker signatures is conclusive evidence that this sample is not related to the oil spill discussed in case #2.	289

9 A Multivariate Approach to Oil Hydrocarbon Fingerprinting and Spill Source Identification

9-1 Illustration of the research strategy behind the IMOF methodology, which is based on four steps: chemical analysis, data preprocessing, multivariate data analysis and data evaluation. The aim of the work in our research groups has

	been to develop rapid, reliable and objective tools for comprehensive analysis and characterization of complex oil hydrocarbon mixtures using the IMOF methodology.	297
9-2	Fluorescence excitation-emission scans of an HFO. The vertical axis and scale bar show the fluorescence intensity. Modified from Christensen et al. (2005b). See color plate.	299
9-3	Partial GC-MS chromatogram of *m/z* 217, which contains tricyclic steranes (eluting between 26 and 34 min) and tetracyclic steranes (eluting between 35 and 42 min).	301
9-4	First derivative of a section of *m/z* 217 for five reference oils and five source oils: (a) before warping and (b) after warping using COW with segment length of 175 data points and a slack of three points. Modified from Christensen et al. (2005d).	303
9-5	Preprocessed fluorescence EEM of the same oil sample as shown in Figure 9-2 after blank subtraction, insertion of missing values, and a small triangle of zeros and excitation/emission correction. The vertical axis and scale bar show the fluorescence intensity. Modified from Christensen et al. (2005b). See color plate.	308
9-6	(a) Weights (RSD_A^{-1}) used for WLS-PCA and (b) the aligned and normalized mean combined chromatogram of unweathered Brent crude oil. The mean was calculated from four replicate samples. Modified from Christensen et al. (2005a).	314
9-7	Score plot (PV2 versus PC4) using WLS-PCA on sections of preprocessed partial GC-MS/SIM chromatograms of *m/z* 217. The PCA model was calculated from the calibration set (61×1231), whereas a reference set (18×1231) of replicate references and test set (22×1231) were calculated by projecting the data onto the loadings. The test set was comprised of 16 *Baltic Carrier* oil spill samples and two spill samples from a Round-Robin exercise analyzed in triplicate (Spill I and Spill II) (Faksness et al., 2002).	316
9-8	Integrated mean-chromatogram (dotted line) and integrated PC4 loadings (solid line) for WLS-PCA in the oil hydrocarbon fingerprinting study in Christensen et al. (2005d). The αα-steranes (αα) and ββ-steranes (ββ) are marked in the plot.	316
9-9	PARAFAC score plot. Factor 3 versus Factor 5. Symbols for LFOs, HFOs, lubes, crude oils, unknown oil samples, replicate reference oils, and the triplicate spill and ship samples are explained in the legend. The 17 replicate references are circled, and arrows mark the position of spill and *Baltic Carrier* ship samples.	318
9-10	Comparison of PARAFAC excitation and emission loading spectra of selected PAHs. Loadings and spectra of individual PAHs have been normalized to ease visual comparison. Furthermore, excitation and emission loadings are shown together in the plots, where excitation loadings appear to the left of the corresponding emission loadings.	318
9-11	Flowchart for the IMOF methodology including the individual oil hydrocarbon fingerprinting methods developed by Christensen et al. (2004, 2005b, 2005c, 2005d).	320

10 Chemical Heterogeneity of Marine Heavy Fuel Oils

10-1	Gas chromatographic "fingerprints" of six IFO-380 HFOs, demonstrating the significant variability in chemical compositions in the sample fuel class. * — internal standards.	334

10-2   Ternary diagram depicting the variability in the bulk hydrocarbon composition among 71 worldwide IFO 380 heavy fuel oil samples. Weight percentages based upon quantitative GC/FID analysis and gravimetric analysis.   335

10-3   Ternary diagrams showing the distribution of C_{27}, C_{28}, and C_{29} regular steranes (5α, 14β, 17β(H) 20S+20R) calculated from absolute concentrations determined from the m/z 218 mass chromatogram for 71 IFO 380 residual fuel oils.   337

10-4   Cross-plots of (A) triterpane-based and (B) sterane-based thermal maturity parameters for 71 IFO 380 fuels. Dashed lines show typical maximum (equilibrium) values for crude oils (from Peters et al., 2005). Triterpane ratios both measured on the m/z 191 mass chromatograms. 14α(H),17α(H) C_{29} steranes measured on the m/z 217 mass chromatogram and 14β(H), 17β(H) C_{29} steranes measured on the m/z 218 mass chromatogram.   337

10-5   Average (±1 standard deviation) concentration of select PAH compounds in 71 worldwide IFO 380 HFOs. See Table 10-3 for compound abbreviations. Inset shows expanded view of 4- to 6-ring PAH.   340

10-6   Percentage C_0–C_4 naphthalenes of the total PAH in 71 worldwide IFO 380 samples. Total PAH represents sum of 54 analytes from Table 10-3.   340

10-7   Histograms comparing distribution and concentration of PAH in (A) low aromatic IFO-380 and (B) high aromatic IFO-380. Gas chromatogram for (C) high aromatic, cracked gas oil blending stock, and (D) a high aromatic IFO-380 containing a prominent cracked gas oil blending stock.   341

10-8   Cross-plots of (A) the range in of the diagnostic source ratios D2/P2 and D3/P3 and (B) the relationship between D2/P2 and total PAH in 71 worldwide IFO 380 heavy fuel oil samples.   343

10-9   Structures of (A) linear 3-ring anth-racene molecule and (B) nonlinear 3-ring phenanthrene molecule.   343

10-10 GC/FID chromatograms for (A) an unweathered IFO 380 fuel and (B) moderately evaporated crude oil and partial m/z 192 mass chromatograms for (C) the unweathered IFO 380 shown in (A) and (D) the moderately evaporated crude oil shown in (B). N — alkyl-naphthalenes, P — alkyl-phenanthrenes; MP — methylphenanthrene, MA — methylanthracene.   344

11 Biodegradation of Oil Hydrocarbons and Its Implications for Source Identification

11-1   Typical reactions catalyzed by hydro-carbon monooxygenases. Octane monooxygenase of *Pseudomonas putida* catalyzes the oxidation of octane to octanol (van Beilen et al., 1994), toluene-3-monooxygenase of *Ralstonia pickettii* catalyzes the oxidation of toluene to 3-cresol (Tao et al., 2004), the cytochrome P450 of *Mycobacterium vanbaalenii* apparently catalyzes the oxidation of benzo[*a*]pyrene to benzo[*a*]pyrene-11,12-epoxide (Moody et al., 2004), and the cyclohexane monooxygenase of *Brachymonas petroleovorans* oxidizes cyclohexane to cyclohexanol (Brzostowicz et al., 2005). Other enzymes catalyze the subterminal oxidation of alkanes (Ludwig et al., 1995), and the oxidation of the 2- or 3-position of toluene (Yeager et al., 1999; Tao et al., 2004).   351

11-2   Typical reactions catalyzed by hydro-carbon dioxygenases. The alkane dioxygenase of an *Acinetobacter* apparently converts alkanes to alkanehydroperoxides (Maeng et al., 1996). Naphthalene dioxygenase of *Pseudomonas putida* oxidizes naphthalene to *cis*-naphthalene-1,2-dihydrodiol (Karlsson et al., 2003).   352

11-3 The opening of aromatic rings by extradiol and intradiol dioxygenases. Examples include the extradiol protocatechuate 4,5-dioxygenase of *Sphingomonas paucimobilis* (Sugimoto et al., 1999) and the intradiol protocatechuate 3, 4-dioxygenase of *Acinetobacter* (Vetting et al., 2000).   353

11-4 The biodegradation of trimethyl, methyl-ethyl, propyl, and isopropyl benzenes by four organisms. *Pseudomonas putida* F1 degrades toluene with a dioxygenase (Cho et al., 2000), while *Ps. putida* mt-2 (Bühler et al., 2000), and *Ps. mendocina* KR (Tao et al., 2004) use monooxygenases, directed at the methyl group and the *para*-position, respectively. *Sphingomonas yanoikuyae* B1 possesses a biphenyl dioxygenase and a xylene monooxygenase (Kim and Zylstra, 1999), and it is certainly possible that all strains contain other hydrocarbon-oxidizing enzymes. These experiments (Prince, V.L., Zylstra, G.J., and Prince, R.C., unpublished) used cells initially grown with vapor phase toluene that were washed and resuspended in minimal medium with 1 µl of gasoline in 10 ml of culture in a 40-ml vial and incubated for 90–113 hr. The traces are for the $m/z = 105$ ion.   354

11-5 Anaerobic hydrocarbon activation by fumarate addition. The denitrifying bacterium HxN1 catalyzes the production of 1-methylpentylsuccinate from hexane (Rabus et al., 2001) and Biegert et al. (1996) were the first to demonstrate the production of benzylsuccinate under anaerobic conditions with *Thauera aromatica*. Similar reactions have been seen under a variety of conditions by a diverse array of organisms (Rabus, 2005).   355

11-6 Anaerobic hydrocarbon activation by carboxylation and dehydration. Carboxylation at the 2-carbon of hexadecane by the sulfate-reducing strain AK-01 was reported by So and Young (1999). Carboxylation at the 3-position was subsequently reported for strain Hxd3 (So et al., 2003). Carboxylation of naphthalene by a sulfate-reducing enrichment culture was reported by Zhang and Young (1997). Recent work on benzene degradation (Chakraborty and Coates, 2005; Ulrich et al., 2005) suggests that a transient hydroxylation precedes carboxylation. While it seems clear that labeled bicarbonate is incorporated into the hydrocarbons, the enzymes responsible have not been characterized. In contrast, ethylbenzene dehydrogenase from *Azoarcus* has been well-characterized (Johnson et al., 2001; Kniemeyer and Heider, 2001); it contains the molybdopterin cofactor.   356

11-7 Anaerobic degradation of hexadecane under methanogenic conditions. A consortium of syntrophic eubacteria and archaea can degrade hexadecane to CH_4, CO_2, and H_2S. [Figure adapted from Parkes (1999) and Zengler et al. (1999)].   357

11-8 Bulk oil properties for a suite of oils (La Luna source) from Eastern Venezuela. Sulfur, nitrogen, nickel, and vanadium increase proportionally with decreasing API gravity. The degree of biodegradation is indicated by the numerical ranking of the Biomarker Biodegradation Scale (Peters et al., 2005) and by descriptive terms used by Wenger et al. (2002).   358

11-9 Gas chromatograms for a suite of selected crude oils from Africa. The oils were generated and expelled from the same source rock under comparable thermal conditions, but indicate quite different amounts of *in situ* biodegradation. Shown are reservoir temperature, API gravity, viscosity, pristane/phytane (Pr/Ph) and $nC_{17}/$ pristane (nC_{17}/Pr) ratios, and mass percentage lost calculated using hopane as a conserved internal standard.

Number labels indicate *n*-alkane carbon number [modified from Wenger et al. (2002)]. 359

11-10 The extent of biodegradation of mature crude oil can be ranked on a scale of 1–10 based on differing resistance of compound classes to microbial attack. Biodegradation is quasi-sequential because some of the more labile compounds in the more resistant compound classes can be attacked prior to complete destruction of less resistant classes [from Peters et al. (2005)]. Arrows indicate where compound classes are first altered (dashed lines), substantially depleted (solid gray), and completely eliminated (black). Sequence of alteration of alkylated PAHs is based on work by Fisher et al. (1998) and Trolio et al. (1999). Qualitative descriptions of the degree of biodegradation from Wenger et al. (2002) reflect changes in oil quality. 361

11-11 Differential biodegradation within compound classes. Modified from Peters et al. (2005). 362

11-12 25-norhopanes have the same optical configuration as hopanes. They differ only by the removal of the methyl group attached to the C-10 carbon. 363

11-13 Mass chromatograms of hopanes (*m/z* 191) and norhopanes (*m/z* 177) from Eastern Venezuela. Ion traces are scaled proportional to their relative response. The 25-norhopanes (indicated by D-carbon number) are believed to originate by loss of a methyl group from C-10 in hopanes. Thus, the single epimer of C_{30} 17α,21β (H)-hopane (top) has been partially altered to C_{29} 25-nor-17α-hopane (bottom), while each of the C_{31}–C_{35} 17α-hopane (22S + 22R) epimers correspond to two C_{30} to C_{34} 25-norhopane epimers. Vertical lines indicate some peaks that yield both *m/z* 191 and 177 ions. Modified from Peters et al. (2005). 363

12 Identification of Hydrocarbons in Biological Samples for Source Identification

12-1 Typical weathering sequence of PAH in Alaska North Slope crude oil. N = naphthalene, F = fluorene, D = dibenzothiophene, P = phenanthrene/anthracene, C = chrysene; numbers indicate substituent alkyl carbon atoms. Weathering increases from top panel (unweathered oil) to bottom panel (moderately weathered oil). Abscissa: proportion of total PAH. 385

12-2 Relative composition of aqueous PAH dissolved from initially unweathered Alaska North Slope oil. Abbreviations and axes as in Figure 12-1. 386

12-3 Relative abundance of PAH and phytane in (A) *Exxon Valdez* oil mousse 11 days following the spill, and (B) in mussels collected six weeks later. Note the PAH weathering evident in the mussels compared with the oil (compare with Figure 12-1), and the prominent phytane in both, indicating ingestion of whole weathered oil droplets by the mussels. Abbreviations: Naph = naphthalene, Menaph = methylnaphthalene, Acenthy = acenaphthylene, Acenthe = acenaphthene, Fluor = fluorene, Dithio = dibenzothiophene, Phenan = phenanthrene, Anthra = anthracene, Fluorant = fluoranthene, Benanth = benzo[*a*]anthracene, Chrys = chrysene, Benzobfl = benzo[*b*]fluoranthene, Benzokfl = benzo[*k*]fluoranthene, Benepyr = benzo[*e*]pyrene, Beneapyr = benzo[*a*]pyrene, Indeno = indeno[1,2,3-*c,d*]pyrene, Dibenz = dibenzo[*a,h*]anthracene, Benzop = benzo[*g,h,i*] perylene (from Short and Harris, 1996b). 387

12-4 Cytochrome P450 1A induction pathway (from Whitlock, 1993; see text in section 12.3.1 for abbreviation). 389

- 12-5 Gill epithelial cells under normal conditions (left), and under CYP 1A induction (right) (from Malins et al., 2004). 389
- 12-6 Catalysis of 7-ethoxyresorufin to resorufin by cytochrome P450 1A (CYP 1A). Resorufin is fluorescent, and its production rate is the basis for the EROD assay for CYP 1A activity. 390
- 12-7 Some effects of oil exposure in herring larvae (photos courtesy of NOAA). 395

13 Trajectory Modeling of Marine Oil Spills
- 13-1 Comparison of a trajectory forecast with oil representing 100 (a) and 1000 (b) particles or Lagrangian elements (LEs). 409
- 13-2 A cross-sectional view of an idealized oil spill in the sea. The arrows below the oil streaks represent the motion of these vortices. The oil is shown collecting in areas with converging surface currents. 414

14 Oil Spill Remote Sensing: A Forensics Approach
- 14-1 Spectra of three fuels with similar physical properties showing the spectral differences in them using a fluorosensor. 425
- 14-2 The return signals from the LURSOT thickness sensor showing a measurement of 9 mm from a light crude oil. Measurement was taken 61 m above ground with aircraft at 200 km/hr. 435

15 Advances in Forensic Techniques for Petroleum Hydrocarbons: The *Exxon Valdez* Experience
- 15-1 PAH compositions of major hydrocarbon source inputs to the marine environment of PWS. A PAH analyte may be a single compound, such as C_0-phenanthrene, or the sum of a group of isomers, such as C_1-phenanthrene, which has four isomers. N1, N2, N3, N4 = C_1, C_2, C_3, C_4-naphthalenes; ACL = acenaphthylene; ACE = acenaphthene; BPH = biphenyl; F, F2, F2, F3 = C_0, C_1, C_2, C_3-fluorenes; P, P1, P2, P3, P4 = C_0, C_1, C_2, C_3, C_4-phenanthrenes/anthracenes; D, D1, D2, D3 = C_0, C_1, C_2, C_3-dibenzothiophenes; FL = fluoranthene; PY = pyrene; FP1 = C_1-fluoranthenes/pyrenes; BaA = benzo(*a*)anthracene; C, C1, C2, C3, C4 = C_0, C_1, C_2, C_3, C_4-chrysenes; BbF = benzo(*b*)fluoranthene; BkF = benzo(*k*)fluoranthene; BeP = benzo(*e*)pyrene; BaP = benzo(*a*)pyrene; Per = perylene; IDP = indeno(1,2,3-cd)pyrene; DBA = dibenzo(*a,h*)anthracene; BgP = benzo(*g,h,i*) perylene. 452
- 15-2 PAH distributions, TPAH and 3–4-ring PAH concentrations, major fraction compositions, and weathering parameters for EVC and four variably weathered PWS shoreline oil samples. The weathering parameter, ω, calculated from the distributions of 14 PAH analytes (Short and Heintz, 1997) becomes indeterminate (IND) at advanced weathering states when one or more analytes are not detected. 454
- 15-3 Pyrogenic index FP/(FP + C24Ph) vs. the ratio of low-molecular-weight (2–3-ring PAH) to total PAH, LPAH/TPAH, for naturally and artificially weathered shoreline oils. LPAH/TPAH decreases with increased weathering. 1989–1990 bioremediation suite oils are from Prince et al. (1994). EM shoreline oil weathering suite from samples collected in 2001–2002. NOAA EVTHD oils from the 2003 version of *Exxon Valdez* Trustees Oil Database on

List of Figures   xliii

	NOAA-NMFS, Auke Bay Laboratory webpage (www.afsc.noaa.gov/abl/oilspill). These include oils that were weathered in the laboratory.	458
15-4	PAH and triaromatic steroid (TAS) compositions (C_2-phenanthrene-normalized) for (A) PWS prespill segment of a benthic core, (B–J) potential source inputs to the natural background. Abbreviations: Np = naphthalenes, Fl = fluorenes, Ph = phenanthrenes, Db = dibenzothiophenes, Ch = chrysenes, Per = perylene, N0 = C_0-naphthalene, N1 = C_1-naphthalene, P0 = C_0-phenanthrene, P1 = C_1-phenanthrene, P2 = C_2-phenanthrene, Db2 = C_2-dibenzothiophene. Adapted with permission from Boehm et al. (2001). Copyright 2001, American Chemical Society.	460
15-5	Triterpane and sterane compositions (C2-phenanthrene-normalized) for (A) PWS benthic core, (B–J) potential source inputs to the natural background. Abbreviations: Ts = $17\alpha(H)$-22,29,30-trisnorhopane, Tm = $18\alpha(H)$-22,29,30-trisnorhopane, 28B = $C_{28,30}$-bisnorhopane, Dpl(ROM) = diplotene, recent organic matter (the Copper River is a major source), 29Nor = norhopane, Hop = C_{30}-hopane, Diaster = diasteranes. Adapted with permission from Boehm et al. (2001). Copyright 2001, American Chemical Society.	461
15-6	C_2-Dibenzothiophenes *versus* C_2-phenanthrenes (D2/P2) relationship of weathered EVC, diesel refined from ANS crude feed stock at a Kenai Refinery, PWS regional background hydrocarbons from deep subtidal cores, and Katalla and Cook Inlet crude oils.	462
15-7	Sediment PAH compositions for a profile from the mid-tide zone to 100-m water depths at a location containing buried EVC residues on the beach, Northwest Bay, Eleanor Island. (Data from S. Rice, personal communication, 2003).	463
15-8	Factor score plot resulting from the PCA analysis of PAH in shoreline and benthic sediments in PWS. Modified after Burns et al. (1997).	465
15-9	Source apportionments for GOA and prespill PWS benthic sediments. KB = Katalla Beach; MGF = Malaspina Glacier Flour. Reprinted with permission from Boehm et al. (2001). Copyright 2001, American Chemical Society.	467
15-10	1999–2000 embayment benthic sediment sampling program with site locations, oiling category, representative sediment PAH profiles, and the results of the CLS source allocations, where ANS = Alaska North Slope oil, Mon = Monterey petroleum, PYR = pyrogenic PAH, and BKG = regional petrogenic background.	470
15-11	PAH and alkane compositions of bald eagle eggshells. (A) Weathered EVC on eggshell collected at Italian Bay on June 8, 1989, EVTHD ID #23363. (B) Relatively fresh diesel on eggshell collected from Hells Hole in eastern PWS outside the spill zone on July 18, 1990, EVTHD ID #20029.	472
15-12	PAH and alkane compositions for sea otter tissues. (A) Heavily weathered EVOS, sea otter skin collected April 12, 1989, EVTHD ID #27757, site not documented. (B) Relatively fresh diesel, sea otter fur collected August 1, 1990, site not documented, EVTHD ID #22851. (C) Procedural artifact, sea otter blood, Ogden Passage, SE Alaska, EVTHD ID #20044.	473
15-13	PAH compositions for Harlequin duck stomach contents (A–B) and corresponding liver tissue (C–D) from the same bird collected by State of Alaska investigators. Adapted, with permission, from *STP 1219 Exxon Valdez Oil Spill: Fate and Effects in Alaskan Water*, copyright ASTM International, 100 Barr Harbor Drive, West Conshohocken, PA 19428.	474

15-14 PAH compositions of tissue exposed to weathered oil in column generator experiments conducted by NOAA/ABL, Juneau, AK. A. Pink salmon eggs after 36 days of exposure to artificially weathered EVC at a gravel loading of 281 ppm oil showing characteristic WSF composition. B. Herring eggs showing mixed oil + WSF composition. 476

15-15 Composition and concentration of PAH in mussel tissues from NOAA oiled mussel bed sites compared with the PAH composition and concentration of underlying sediments for samples collected in 2001. The data are presented as the geometric means of the analyte distributions of replicate samples at each site. Note that the mussel TPAH concentrations are factors of 3 to 300 lower than the TPAH concentrations in the corresponding sediments. 478

15-16 Mean sediment TPAH versus composite clam TPAH for 16 oiled, 3 reference, and 4 HA sites in western PWS. UIZ = upper intertidal zone (+2 m mllw), MIZ = middle intertidal zone (+1–+2 m mllw), LIZ = lower intertidal zone (0–+1 m mllw). 479

15-17 PAH composition of composite clam tissues from two sea otter foraging sites that had been oiled in 1989. 480

15-18 Correlation of rockfish liver ethoxyresorufin 0-deethylase (EROD) activity, a measure of the level of CYP 1A induction, with benthic sediment pyrogenic index (FP)/(FP + C_{24}Ph). Solid symbols = sites oiled in 1989. Open symbols = unoiled reference sites. 482

16 Case Study: Oil Spills in the Strait of Malacca, Malaysia

16-1 Map of Peninsular Malaysia showing the Straits of Malacca. 490

16-2 Oil spill incidents in the Straits of Malacca (Malaysian Marine Department, 2004). 491

16-3 Sampling stations for tar ball samples (solid diamond) in Peninsular Malaysia. 493

16-4 Sampling stations for sediment in Peninsular Malaysia. Solid triangles represent river stations and solid circles represents inshore stations. 494

16-5 Cross-plots for source identification of oils. The circles indicating both categories were established through the analysis of crude oils in Zakaria et al. (2000). 501

17 Evaluation of Hydrocarbon Sources in Guanabara Bay, Brazil

17-1 Social, economical, and physical characteristics of Guanabara Bay. 507

17-2 Oil slick after January 2000 oil spill accident in Guanabara Bay. 508

17-3 Geographical location of the 21 sediment stations in Guanabara Bay sampled in 2000 and 2003. 509

17-4 Predominance of alkylated PAH for intertidal samples from Guanabara Bay 2000 survey in relation to 2003. 517

17-5 Predominance of four- and five-ring PAH for subtidal samples from Guanabara Bay for 2000 and 2003 campaigns. (a) Inside the influence of the oil spill slick; (b) outside the influence of the oil spill slick. 518

17-6 PAH cross-plot of phenanthrene/anthracene versus fluoranthene/pyrene for Guanabara Bay sediments and MF 380, Arabian oil (AL), light oil (Ar), and diesel oil (DM). 520

17-7 PAH cross-plot of anthracene/(anthracene + phenanthrenes) versus fluoranthene/(fluoranthene + pyrene) for Guanabara Bay sediments from

List of Figures xlv

	campaigns 2000 and 2003 and MF 380, Arabian oil (AL), light oil (Ar), and diesel oil (DM).	520
17-8	PAH cross-plot of indeno1,2,3-(cd)pyrene/(indeno1,2,3-(cd)pyrene + benzo(ghi)perylene) versus fluoranthene/(fluoranthene + pyrene) for Guanabara Bay sediments from campaigns 2000 and 2003 and MF 380, Arabian oil (AL), light oil (Ar), and diesel oil (DM).	521
17-9	PAH cross-plot of fluoranthene/(fluoranthene + pyrene) versus phenanthrene + anthracene/(phenanthrene + anthracene + C1phenanthrenes) for Guanabara Bay sediments from campaigns 2000 and 2003 and MF 380, Arabian oil (AL), light oil (Ar), and diesel oil (DM). Interpretive guidelines based upon Yunker et al. (2000).	522
17-10	Plot of the relative ratios Σ (other 3–6-ring PAH)/Σ (5 alkylated PAH series) versus phenanthrene/anthracene for Guanabara Bay sediments from the campaigns 2000 and 2003 and MF 380, Arabian oil (AL), light oil (Ar), and diesel oil (DM). Interpretive guidelines based upon Wang et al. (1999b).	523
17-11	Plot of the relative ratios Σ (other 3–6-ring PAH)/Σ (5 alkylated PAH series) versus fluoranthene/(fluoranthene + pyrene) for Guanabara Bay sediments from the campaigns 2000 and 2003 and MF 380, Arabian oil (AL), light oil (Ar), and diesel oil (DM). Interpretive guidelines from Wang et al. (2002b).	524
17-12	Relative perylene abundance for Guanabara Bay sediments from the campaigns 2000 and 2003.	525
17-13	PCA projections of PAH variables and sediment samples from campaign 2000 and MF 380, Arabian oil (AL), light oil (Ar), and diesel oil (DM).	526
17-14	PCA projections of PAH variables and sediment samples from campaign 2003 and MF 380, Arabian oil (AL), light oil (Ar), and diesel oil (DM).	526
17-15	Terpane and sterane partial mass chromatograms of the spilled marine heavy fuel oil, January 2000.	527
17-16	A plot of C27/C29-steranes $5\alpha,14\beta,17\beta(H),20S$ against C29/C27-steranes $5\alpha,14\alpha,17\alpha(H),20S$ showing the similarity between most of the samples and the correlation between both samples T24 collected in campaigns 2000 and 2003.	531
17-17	A plot of C24-tetracyclic/C26-tricyclic terpanes against C29/(C29 + C30-hopanes) highlighting the correlation between the marine fuel and samples T7 and T9 from the campaign 2000.	531
17-18	A plot of Ts/(Ts + Tm) against 30-norhomohopane/(30-norhomohopane + hopane) showing the correlation between samples and a marine carbonate-sourced crude oil.	531
17-19	A plot of C27/C29-steranes $5\alpha,14\beta,17\beta(H),20S$ against C29/C27-steranes $5\alpha,14\alpha,17\alpha(H),20S$ showing the correlation between most of the samples and a marine carbonate-sourced crude oil.	532
17-20	A plot of C27/C29-steranes $5\alpha,14\beta,17\beta(H),20S$ against C28/C29-steranes $5\alpha,14\beta,17\beta(H),20S$ showing the correlation between the sediment samples and a marine carbonate-sourced crude oil.	532

Preface

Man's widespread reliance upon petroleum as an energy source and chemical feedstock requires the production of vast quantities of crude oil and the subsequent transportation of crude oil and refined products through pipelines and by vessels. Despite continually improving safeguards and increasingly stringent regulations, accidental spills and intentional discharges of petroleum are frequent occurrences in water bodies around the world. Determining the source of "mystery" spills and the extent of impact of both "mystery" and known releases relies largely on environmental forensics — *def.*, the systematic and scientific evaluation of physical, chemical, and historical information for the purpose of developing defensible scientific conclusions relevant to the liability for environmental contamination.

Chemical fingerprinting has played an important role in the rapidly advancing field of environmental forensics of waterborne oil spills. Significant advances in chemical fingerprinting, driven by both the application of petroleum exploration and production geochemistry principles and by advancements in analytical methods and instrumentation, have resulted in the use of fingerprinting in nearly all oil spill investigations worldwide.

The global problem of oil spills, and the global application of environmental forensics in oil spill investigations, warranted a global effort in conveying the state-of-the-science in this book. As such, we've assembled contributions from researchers from Brazil, Canada, Denmark, Egypt, Germany, Japan, Malaysia, Norway, Republic of Korea, The Netherlands, and the United States. These contributions cover both new and emerging chemical fingerprinting technologies and the application and refinement of proven technologies in sufficient detail so as to reasonably represent the current knowledge base of oil spill fingerprinting and source identification. The sequence of chapters spans from an introduction of the methods for and factors affecting chemical fingerprints of petroleum, to oil spill investigation sampling design, to specific chemical fingerprinting features (biomarkers, sulfur-bearing PAHs, and stable isotopes) and instrumentation (GC × GC), data analysis techniques (emerging CEN protocol, quantitative methods, and multivariate analysis), biodegradation affects on and biological uptake of petroleum, fuel chemistry, and non-chemical oil spill identification techniques (transport modeling and remote sensing), before concluding with various case studies. It is hoped that the individual chapters in this book will provide students and scientists with ready access to a comprehensive overview of oil spill fingerprinting and source identification and provide a suitable and up-to-date reference and source of citations for years to come.

We wish to thank the individual efforts of Robert D. Morrison, who helped initiate the publication process, all of the contributing authors and reviewers, and the Elsevier staff for helping bring these contributions to publication.

Zhendi Wang, Environment Canada
Scott A. Stout, NewFields Environmental
Forensics Practice, LLC
June 2006

Contributors

Note: Addresses listed in each group of contributors only refers to the lead author.

A. Edward Bence, David S. Page, and Paul D. Boehm 4480 Ponderosa Drive, Peachland, BC V0H 1X5, Canada (ExxonMobil Upstream Research Co., P.O. Box 2189 Houston, TX 77252-2189) *15 Advances of Forensics Techniques for Petroleum Hydrocarbons: the Exxon Valdez Experience*

Abdelrahman Hegazi and Jan T. Andersson Institut für Anorganische und Analytische Chemie Westfälische Wilhelms-Universität Münster Corrensstrasse 30 D-48149 Münster, Germany *4 Characterization of Polycyclic Aromatic Sulfur Heterocycles for Source Identification*

Alan Jeffrey DPRA 100 San Marcos Blvd., Suite 308, San Marcos, CA 92069 *6 Application of Stable Isotope Ratios in Spilled Oil Identification*

Allen D. Uhler, Scott A. Stout, and Gregory S. Douglas NewFields — Environmental Forensics Practice LLC, 100 Ledgewood Place, Suite 302, Rockland, MA 02370 *10 Chemical Heterogeneity in Marine Heavy Fuel Oils*

Asger B. Hansen, Per S. Daling, Liv-Guri Faksness, Kristin R. Sörheim, Paul Kienhuis, and Rolf Duus National Environmental Research Institute Department of Environmental Chemistry & Microbiology, 399 Frederiksborgvej, P.O. Box 358 DK-4000, Denmark *7 Emerging CEN Methodology for Oil Spill Identification*

Debra Simecek-Beatty and William J. Lehr NOAA — HAZMAT 7600 Sand Point Way, NE Seattle, WA 98115 *13 Trajectory Modeling of Marine Oil Spill*

Edward Owens, Elliott Taylor, and Heather Parker-Hall Polaris Applied Sciences, Inc., #302, 755 Winslow Way East, Bainbridge Island, WA 98110 *2 Spill Site Investigation in Environmental Forensic Investigation*

Fatima Meniconi G. and Silvana M. Barbanti Environmental Geochemist PETROBRAS Research and Development Center (CENPES) Cidade Universitaria, Q.7 — Ilha do Fundao 21949-900, Rio de Janeiro, Brazil *17 Case Study: Evaluation of Hydrocarbon sources in Guanabara Bay, Brazil*

Gregory S. Douglas, Scott A. Stout, Allen D. Uhler, Kevin J. McCarthy, and Stephen D. Emsbo-Mattingly NewFields — Environmental Forensics Practice LLC, 100 Ledgewood Place, Suite 302, Rockland, MA 02370 *8 Advantages of Quantitative Chemical Fingerprinting in Oil Spill Identification*

Jan Christensen and Giorgio Tomasi Department of Natural Sciences, The Royal Veterinary and Agricultural University Thorvaldsensvej 40, 1870 Frederiksberg C, Denmark *9 Multivariate Statistical Methods for Oil Hydrocarbon Fingerprinting and Spill Source Identification*

Jeffrey W. Short and Kathrine R. Springman NOAA Alaska Fisheries Science Center, 11205 Glacier Hwy., Juneau, AK 99801-8626 *12 Identification of Hydrocarbons in Biological Samples for Source Determination*

Merv Fingas and Carl Brown ESTD, Environment Canada, 335 River Road, Ottawa, ON, Canada K1A 0H3 *14 Oil Spill Remote Sensing*

Contributors

Mohamad P. Zakaria and Hideshige Takada Associate Professor, Department of Environmental Sciences Faculty of Environmental Studies, Universiti Putra Malaysia 43400 UPM, Serdang, Selangor Darul Ehsan, Malaysia. Faculty of Agriculture, Tokyo University of Agriculture and Technology, Fuchu, Tokyo 183-8509, Japan *16 Case Study: Oil Spill in the Straits of Malacca, Malaysia*

Richard Gains, Glenn S. Frysinger, Christopher M. Reddy, and Robert K. Nelson Professor Dept. of Science, U.S. Coast Guard Academy, 27 Mohegan Avenue, New London, Connecticut 06320-8101 *5 Oil Spill Source Identification by Comprehensive Two-Dimensional Gas Chromatography (GC × GC)*

Roger C. Prince and Clifford C. Walters ExxonMobil Research and Engineering Co., 1545 Route 22, East Annandale, NJ 08801 *11 Biodegradation of Oil Hydrocarbons and Its Effects to the Source Identification*

Scott A. Stout and Zhendi Wang NewFields — Environmental Forensics Practice LLC, 100 Ledgewood Place, Suite 302, Rockland, MA 02370 *1 Chemical Fingerprinting of Spilled or Discharged Petroleum — Methods and Controlling Factors*

Zhendi Wang, Chun Yang, Merv Fingas, Bruce Hollebone, Un Hyuk Jae Yim, and Ryoung Oh ESTD, Environment Canada, 335 River Road, Ottawa, ON, Canada K1A 0H3 *3 Petroleum Biomarker Fingerprinting for Oil Spill Characterization and Source Identification*

1 Chemical Fingerprinting of Spilled or Discharged Petroleum — Methods and Factors Affecting Petroleum Fingerprints in the Environment

Scott A. Stout[1] and Zhendi Wang[2]

[1] NewFields Environmental Forensics Practice LLC, 100 Ledgewood Place, Suite 302, Rockland, MA 02370.
[2] ESTD, Environment Canada, 335 River Road, Ottawa, ON K1A 0H3.

1.1 Introduction

Oil can enter water by a number of anthropogenic and natural sources. Acute anthropogenic sources of oil can be generally classified as either accidental *oil spills* or intentional *operational discharges*. Despite international antipollution protocols (e.g., Marine Pollution Convention MARPOL 73/78 and its amendments) and improved shipping and handling safeguards (e.g., International Safety Management Code 1998), accidental oil spills and intentional operational discharges from vessels to waterways and the marine environment regularly occur. Table 1-1 shows the estimated masses of petroleum from accidental oil spills and intentional operational discharges in North America and worldwide (NRC, 2003). The primary sources of accidental oil spills are from tanker vessels carrying crude oil or petroleum products, distantly followed by pipeline spills, coastal facility spills, freighter (fuel) spills, and offshore oil production platform spills. The largest sources of intentional operational discharges include discharges from vessels (mostly bilge and other oily discharges), cargo hold washings, and produced water discharges from offshore platforms outside of North America (Table 1-1).

Table 1-1 also shows that chronic anthropogenic sources of oil, such as runoff and atmospheric deposition from terrestrial environments, which are not easily classified as either accidental or intentional, constitute a major source of oil to the marine environment. Since these terrestrial sources are chronic and not easily attributed to a "point source," they collectively constitute a chronic source of anthropogenic "background" hydrocarbons that must be considered in many oil spill investigations. In addition, chronic natural sources of oil (e.g., oil seeps) are thought to exceed all anthropogenic sources combined, contributing up to 2 million tonnes of oil annually (Table 1-1). These naturally occurring oil seeps, as well as other naturally occurring (i.e., biogenic) hydrocarbons, must also be considered in oil spill investigations.

Despite the significant contributions of oil and hydrocarbons to the marine environment from chronic anthropogenic sources, oil seeps, and naturally occurring hydrocarbons, it is the

Table 1-1 Average Estimated Annual Releases (1990–1999) of Petroleum into the Marine Environment in North America and worldwide by Source (in Thousands of Metric Tonnes). Data from National Research Council (2002)

	North America			Worldwide		
	Best Est.	Min.	Max.	Best Est.	Min.	Max.
Accidental Oil Spill Sources						
Tanker/barge vessel spills	5.3	4	6.4	100	93	130
Pipeline spills	1.9	1.7	2.1	12	6.1	37
Coastal facility spills	1.9	1.7	2.2	4.9	2.4	15
Non-tanker (freighter) spills	1.2	1.1	1.4	7.1	6.5	8.8
Oil production platform spills	0.16	0.15	0.18	0.86	0.29	1.4
Intentional Operational Discharge Sources						
Operational discharges (all vessels)	0.22	0.06	0.6	270	90	810
Cargo washings		not allowed		36	18	72
Produced waters from offshore rigs	2.7	2.1	3.7	36	19	58
Recreational marine vessels	5.6	2.2	9		not available	
Atmospheric deposition (offshore rigs)	0.12	0.07	0.45	1.3	0.38	2.6
Atmospheric deposition (vessels)	0.01	trace	0.02	0.4	0.2	1
Jettisoned aircraft fuel	1.5	1	4.4	7.5	5	22
Chronic Terrestrial Sources						
Land-based (river and runoff)	54	2.6	1900	140	6.8	5000
Atmospheric deposition	21	9.1	81	52	23	200
Natural Sources (Seeps)	160	80	240	600	200	2000
Total	**260**	**110**	**2300**	**1300**	**470**	**8300**

acute accidental oil spills and intentional operational discharges that tend to promulgate the greatest need for chemical fingerprinting. These acute releases can require expensive response and cleanup activities and damage marine and terrestrial natural resources; thus, the resulting casualty and punitive financial consequences can be substantial. As such, the source of the spilled oil and the spatial and temporal extent of any damage need to be established confidently so that any resultant liabilities are fairly assessed. Chemical fingerprinting, *def.*, the generation and comparison of diagnostic chemical features among oil samples (i.e., both spill and source oils) and potentially impacted samples (e.g., shorelines, sediments, or biological tissues), can play an important role in assessing this liability and monitoring ecological effects.

The chemical fingerprinting of spilled oils was initiated in the 1970s following an increased environmental awareness and the resulting regulations, such as the Clean Water Act in the U.S. (e.g., Kreider, 1971; Ehrhardt and Blumer, 1972; Duewer et al., 1975; Bentz, 1976, 1978; Albaiges and Albrecht, 1979). Over the past 35 years, the sophistication of the chemical fingerprinting has evolved with the advancement of analytical chemistry methodologies and data interpretation tools — the current state of which is represented in various chapters throughout this book.

The exact role of chemical fingerprinting in an oil spill investigation will vary over time and depending on the circumstances of the release (e.g., Boehm et al., 1997). For example, oil spills or operational discharges that enter surface water can be classified into two basic categories — (1) "mystery" spills and (2) known-source spills. The former typically involves the discovery of a fugitive oil or oily waste at sea, in rivers and harbors, or in other water bodies either in the absence of any known incident or source, or in the presence of multiple source candidates. The latter, as their name implies, involve the release of oil or oily wastes from an identified point source (or at least, a "prime suspect") or a known incident (e.g., accidental discharge during a grounding or collision). In either oil spill scenario there is an opportunity for chemical fingerprinting to answer important questions, which can be used to determine how any liabilities for any cleanup and resource damage are distributed.

In the case of a "mystery" spill, chemical fingerprinting can be used to compare the spilled oil's "fingerprint" to those of any viable candidate source oils in an attempt to unambiguously identify the lone source — or, at least, limit the population of possible sources. These investigations often involve the collection of waterborne oils (i.e., oil slicks) and their comparison to oils collected from various candidate sources or a particular "prime suspect." If no prime suspect is identifiable, based upon other evidence, then a critical component in the success of a chemical fingerprinting investigation of a "mystery" spill is the recognition of and collection of all available viable candidate source oils. Logically, if the population of possible sources is sufficiently comprehensive, there is a better chance of identifying the source. A positive fingerprinting "match" between a mystery spill and a candidate source can sometimes be used alone, or more powerfully in conjunction with other evidence (e.g., trajectory modeling or remote sensing; see Chapters 13 and 14, respectively), to determine responsibility for the mystery spill. Other evidence can be very important when the same oil was present in multiple sources (e.g., a particular lot of fuel or cargo oil carried by multiple vessels and/or terminals within a port). By their nature, investigations of mystery spills usually occur in the short term (days to weeks) following the discovery of the spill or discharge — and typically either solve the mystery at that point, or the spill's source often remains unsolved. Thus, the effects of short-term weathering (e.g., evaporation) also need to be considered in the investigation of mystery spills. Under most scenarios, once a mystery spill is "solved," the need for chemical fingerprinting shifts toward those of a known-source release, described next.

In the case of a known-source release, for example, a grounded and leaking tanker or a production-well blowout, chemical fingerprinting of the spilled oil can be used to establish the spatial and temporal extent of its impact in the environment over the short and long terms. The matrices most often involved in establishing these impacts involve shoreline or benthic sediments and biological tissues, i.e., sites where oil-derived hydrocarbons might accumulate and reside for longer periods of time. Two critical components of this type of chemical fingerprinting investigation are to (1) establish and allocate the contribution of any pre-existing anthropogenic or naturally occurring "background" hydrocarbons from those that are newly introduced by the spilled oil and (2) monitor the changes in spilled oil composition over time (i.e., weathering and natural recovery). Chemical fingerprinting investigations of this type can occur for decades following the release.

The application of chemical fingerprinting in any oil spill or discharge investigation requires an understanding of the factors that collectively contribute to the chemical character of petroleum in the environment — and thereby, its chemical fingerprint. Figure 1-1 shows some of the factors controlling the chemical fingerprints of oil in the environment. Some of these factors influence the chemical fingerprint of petroleum before it is released to the environment, and other factors influence its chemical fingerprint after a release. In this chapter, we review the analytical methodologies used for chemical fingerprinting of waterborne oil spills and discuss the factors that affect the chemical fingerprints of petroleum that are relevant to oil spill investigations. Examples from various oil spill investigations are used throughout in order to demonstrate various aspects of these factors.

1.2 Methods for Chemical Fingerprinting Petroleum

1.2.1 Historical Perspective

The analytical methods available for chemical fingerprinting of spilled oil in the environ-

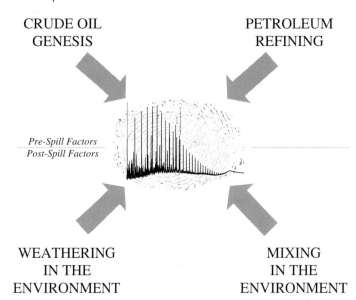

Figure 1-1 Diagram depicting the four factors that affect the chemical fingerprints of petroleum relevant in oil spill investigations.

ment have evolved considerably and basically have paralleled those used by petroleum geochemists in exploration and production applications. In the early 1970s, packed-column gas chromatography and ultraviolet or infrared spectroscopy yielded basic, semiquantitative fingerprints of spilled oils (e.g., Ehrhardt and Blumer, 1972; Bentz, 1976). The application of these methods relied primarily on relatively gross evaluation of the boiling range or n-alkane profiles and degree of weathering of spilled oils. These methods were later supplemented by the development of capillary column gas chromatography (e.g., Rasmussen, 1975) often coupled with low-resolution mass spectrometry (GC/MS) for more detailed molecular characterization of spilled oil (e.g., Albaiges and Albrecht, 1979). The increasingly widespread use of low-resolution GC/MS methods provided a means of measuring the more highly specific compounds that occurred at lower concentrations in oils, such as PAH (e.g., Teal et al., 1978; Grahl-Nielsen et al., 1978) and biomarkers (Seifert and Moldowan, 1978; Seifert et al., 1979), which are still targeted analytes in oil spill studies today. (See ahead and in other chapters throughout this book.) However, over the past 25 years the advances in the use of GC/MS in oil spill investigations have relied on (1) increases in the target analytes of interest, including an increasing array of "diagnostic" compounds or compound groups, (2) improved sensitivity of the analyses (due to increasingly higher degrees of instrumental sensitivity and application of selected ion monitoring data acquisition), and (3) stringent quality assurances (Douglas and Uhler, 1993; Wang and Fingas, 1995a, 1997; Boehm et al., 1997).

While spectroscopic methods have improved and continue to be used in some oil spill investigations (e.g., Staniloae et al., 2001), over the last 30 years, capillary gas chromatography (GC) clearly has proven most effective in the chemical fingerprinting of spilled oils. The primary reason for this is that capillary GC, performed using various column types, can resolve many, but clearly not all (e.g., see Chapter 5 and ahead), of the 100s to 1000s of compounds in petroleum permitting their detection and identification by a variety of detectors. Chemical fingerprinting using GC has relied on a variety of detectors, ranging from nonspecific detectors such as flame ionization detectors (GC/FID) to highly selective mass spectrometers (GC/MS). Other compound-specific detectors such as flame photometric detectors (FPD), chemiluminescence detectors, and atomic emission detectors have been utilized less frequently.

The GC fingerprinting methods most widely used today are based on modifications of the U.S. EPA's codified series of GC techniques intended to measure industrial chemicals (U.S. EPA, 1997). The two modified EPA methods of greatest utility in oil spill fingerprinting studies are (1) method 8015B, *Non-Halogenated Organics Using GC/FID*, and (2) method 8270, *Semi-Volatile Organic Compounds by Gas Chromatography/Mass Spectrometry (GC/MS)*. These standard EPA GC methods are, indeed, excellent base techniques for the chemical fingerprinting of spilled oil, candidate source oils, and potentially impacted environmental media. However, because these methods were developed with the intention of measuring specific chemicals of concern (i.e., the so-called priority pollutant chemicals), fundamental restrictions to the standard EPA methods limit their usefulness in forensic oil spill investigations (Douglas and Uhler, 1993). Of the more than 160 EPA priority pollutant volatile and semivolatile organic compounds, only 20 are petroleum-type hydrocarbons that would potentially be useful in oil spill investigations. This list includes benzene, toluene, ethyl-benzene, xylenes, naphthalene, acenaphthylene, fluorene, anthracene, phenanthrene, fluoranthene, chrysene, pyrene, benzo[a]anthracene, benzo[b]fluoranthene, benzo[a]pyrene, indeno[1,2,3-c,d]pyrene, dibenzo[a,h]anthracene, and benzo[g,h,i]perylene (Sauer and Boehm, 1995). Only half of these compounds are found in significant concentrations in petroleum- or coal-derived products and wastes. Even together, these chemicals alone are of virtually no value in chemical fingerprinting of spilled petroleum because they offer little or no specificity in differentiation between different oils or between oils and other sources of hydrocarbons in the environment. As such, the principal modification to the standard EPA methods involves the expansion of the target analytes to include a greater number of compounds commonly found in petroleums (see ahead).

Modifications of EPA methods 8015B (GC/FID) and 8270C (GC/MS) have been incorporated in multiple oil spill characterization protocols used over the past 30 years. The first of these was 1974's ASTM D3328, *Standard Test Methods for Comparison of Waterborne Petroleum Oils by Gas Chromatography*. This method was based upon GC/FID analysis of spilled oils and their qualitative comparison to potential sources and/or impacted samples. ASTM D3328 has been repeatedly modified, the most current version being updated in 2000 (ASTM, 2000a). The Nordic countries collectively developed a protocol that employed both GC/FID and GC/MS fingerprinting techniques that were first introduced in 1983 (Nordtest, 1983) and subsequently revised in 1991 (Nordtest, 1991). The introduction of the GC/MS fingerprinting component by Nordtest (1991) was, in 1995, followed by the 1995 introduction of ASTM D5739, *Oil Spill Identification by Gas Chromatography and Positive Ion Electron Impact Low-Resolution Mass Spectrometry*. This method was most recently updated in 2000 (ASTM, 2000b). Both the ASTM and Nordtest protocols relied upon qualitative or semiquantitative (i.e., normalized peak areas) comparisons among GC/FID chromatograms and GC/MS extracted ion profiles in comparing a spilled oil to potential sources and to potentially impacted samples (e.g., sediments or biota). Concurrent with the activities of the ASTM and Nordtest organizations, other organizations were developing quantitative fingerprinting protocols in the course of some well-studied oil spill investigations, e.g., *Amoco Cadiz* (e.g., Page et al., 1988), 1991 Gulf War (Michel et al., 1993), or *Exxon Valdez* (e.g., Page et al., 1995). As such, quantitative oil spill fingerprinting protocols in which a large number of target analytes (well beyond those in EPA methods) were quantitatively measured in spilled oils, candidate sources, and in potentially impacted samples, were widely adopted in recent years. Among the analytes most frequently measured in these investigations were *n*-alkanes, acyclic isoprenoids, polycyclic aromatic hydrocarbons

(PAHs), alkylated PAHs and sulfur-containing PAHs, and triterpane and sterane biomarkers (e.g., Douglas and Uhler, 1993; Sauer and Boehm, 1995; Wang et al., 1999a, 2000; Wang and Fingas, 2003; see also Chapters 7 and 8, herein). The success of these studies in fulfilling the roles of chemical fingerprinting in both mystery spills and known-source spills (Section 1.1) has led to the use of quantitative or semiquantitative GC/FID and GC/MS in a logical, "tiered" fashion (Wang et al., 1999a; Stout et al., 2001; Daling et al., 2002; Wang and Fingas, 2003; see also Chapters 7 and 8). An important advantage of these quantitative or semiquantitative chemical fingerprinting methods is the ability to analyze the resulting data using a variety of statistical or numerical methods, rather than rely upon qualitative interpretations of chromatographic patterns (e.g., see Chapters 8 and 9 for more details).

Tier 1 typically is the GC/FID analysis of samples — spilled oils, potential sources, or potentially impacted samples and Tier 2 is the GC/MS analysis of those samples which, after Tier 1, still require additional chemical characterization in order to answer the specific questions at hand. Finally, if necessary, Tier 3 involves the statistical or multivariate analysis of Tiers 1 and 2's data to answer specific questions, e.g., "Is there a positive match to the mystery spill oil" or "what, if any, proportion of hydrocarbons in this sediment is attributable to the spill oil?" Obviously, answers to such questions must consider the precision of the quantitative data, the effects of weathering of any spilled oil, and the potential for mixtures.

1.2.2 Tier 1 — Chemical Fingerprinting via GC/FID

Tier 1 analysis of oil spill-related samples involves a GC/FID based on a modified EPA Method 8015B. The qualitative results available from this tier of analysis include the overall chemical fingerprint of an oil or of extractable hydrocarbons in some matrix (e.g., sediments, soils, or biota). Various methods exist for the extraction, cleanup, and/or fractionation of water, sediment, soil, and tissue extracts so that spilled oil in these matrices can be better characterized either through the removal of external interferences (e.g., biogenic compounds or elemental sulfur) or the concentration of targeted compounds (e.g., compound class or group separation). The reader is directed to sample extraction and preparation procedures described in Sauer and Boehm (1995).

The most appropriate capillary column for oil spill investigations is narrow bore (0.25-mm i.d.) fused silica coated with a 0.25-μm 100% methyl-silicone cross-linked stationary phase. Minimum column length for oil spill investigations is 30 meters, although longer columns, e.g., 60 meters, will afford even better resolution at the expense of longer run times. The keys to the successful application of this method are (1) optimal operation of the GC inlet (McCarthy and Uhler, 1994), which will minimize mass discrimination of heavier hydrocarbons (e.g., >C_{30}), and (2) utilization of a very slow GC oven temperature program, to facilitate optimal resolution of close-eluting compounds. GC oven programs for high-quality Tier 1 fingerprints typically have initial temperatures of 30° to 35°C, and oven ramp rates of 6°C/min or less and total run times of about 60 minutes. It is possible to use shorter run times (higher ramp rates) depending upon the specific character of the oil spill under investigation and questions being addressed.

GC/FID fingerprints will reveal the boiling (carbon) range and, hence, the general hydrocarbon composition of a spilled oil or spill-related samples. Even qualitative chromatographic fingerprints are useful because almost all hydrocarbon assemblages — crude oil, petroleum fuels, petroleum lubricant, and their combustion and waste products — have distinctive chromatographic signatures (e.g., Wang and Fingas, 1997; Stout et al., 2002a). These qualitative differences are the basis of the ASTM D3328 oil spill fingerprinting method. Although weathering and/or mixing (described later in this chapter) can greatly influence the appearance of the chromatographic signature of spilled petroleums, more often than not, distinctive features can be retained in the Tier 1 GC/FID chromatogram. Depending on the extent of weathering, such features may include the ratios between or among resolved compounds that have similar environmental behaviors (e.g., pristane/phytane ratios or carbon preference indices among higher-boiling n-alkanes). Even the profile of the unresolved complex mixture (UCM "hump"), a common feature of weathered petroleum, can be distinctive in some circumstances, e.g., perhaps distinguishing an environmentally evaporated petroleum from a refined/distilled petroleum or background hydrocarbons. Despite what can be revealed by qualitative GC/FID analysis (e.g., overall boiling range, the degree of weathering, or likelihood of mixing), Tier 1 analysis can involve quantitation of any individual compounds that can be chromatographically separated and measured, e.g., the n-alkanes and selected acyclic isoprenoids (e.g., pristane and phytane). The absolute concentrations of these compounds can be quantitatively measured through the use of internal or external standards and appropriate instrument calibration and quality-control measures (see Section 1.2.4), or relative concentrations measured using peak areas.

Strategically, the results of the Tier 1 GC/FID analysis dictate the next step(s) in an oil spill fingerprinting investigation. If, for example, a mystery crude oil spill is under investigation and a potential source candidate is determined to consist of diesel fuel, there is no obvious need to analyze that diesel fuel further, since GC/FID is sufficient to exclude it as a source of the mystery spill. If, however, the Tier 1 results are equivocal in some way, additional analysis via Tier 2 is likely warranted — again, depending upon the specific question(s) at hand. This type of flexibility is inherent in the tiered chemical fingerprinting approach.

1.2.3 Tier 2 — Chemical Fingerprinting via GC/MS

Although waterborne oil spills can sometimes involve volatile petroleums, such as natural gas condensate or automotive gasoline, the impact of such spills on the aquatic and sedimentary environments generally is short lived due to the propensity for evaporation. Fingerprinting of volatile petroleums generally relies upon GC/MS based on modification of EPA Method 8260, sometimes referred to as "PIANO" analysis (Uhler et al., 2002). Most waterborne oil spills involve petroleums containing semi- to minimally volatile constituents that can persist longer in the environment and are amenable to analysis via GC/MS based on a modified EPA Method 8270C.

The operating procedures and GC/MS conditions for optimal detection, separation, and measurement of these semi- to minimally volatile target compounds are somewhat different from those presented in EPA Method 8270 (Stout et al., 2002; Douglas et al., 2004). First, GC and MS operating conditions need to be optimized for separation and sensitivity. At a minimum, 30-m, 0.25-mm i.d., 0.25-µm film thickness 5% phenyl-95% methyl-silicone (or equivalent) capillary column should be used for separation. Optimal operation of the GC should be established to minimize mass discrimination of heavier hydrocarbons (e.g., $>C_{30}$) during sample injection (McCarthy and Uhler, 1994). A slow GC oven temperature program should be used to ensure adequate resolution of closely eluting compounds. Appropriate GC oven profiles have initial temperatures of 30° to 35°C and oven ramp rates of 6°C/min or less. Total GC run times of 1 to 2 hours are typical for these analyses. The MS is commonly operated in the selected ion monitoring (SIM) mode, which greatly enhances the sensitivity and lower detection limits for quantitative analysis. (See also Chapter 3 for a more detailed discussion of GC/MS analysis in oil spill investigations.)

The most common groups of targeted analytes via GC/MS analysis of spill oils, candidate sources, and potentially impacted samples are the (1) homo- and hetero-atomic polycyclic aromatic hydrocarbons (PAHs) and (2) petroleum biomarkers. Other compounds of potentially diagnostic value, e.g., diamondoids, n-alkylcyclohexanes, macrocyclic alkanes, etc., can, of course, be included in any oil spill investigation where necessary. The utility of the PAHs and biomarkers lies in the combination of their specificity to different hydrocarbon sources and their generally reduced susceptibility to weathering on environmental time scales (e.g., see Chapter 11). In crude oil and most petroleum fuels and lubricants, the PAHs and biomarkers are present at much lower concentrations than the more abundant normal, branched, and (nonbiomarker) cyclic alkanes that can easily be detected by Tier 1 GC/FID analysis. Once dispersed in environmental media, the concentrations of PAH and biomarkers are often in the parts-per-million to low parts-per-billion range. In order to detect and quantify these hydrocarbons in sediments and biota potentially impacted by spilled oil, these analyses must be conducted under closely monitored analytical and quality-control protocols (Douglas et al., 2004).

1.2.3.1 Polycyclic Aromatic Hydrocarbons

Table 1-2 presents a list of target homo- and hetero-atomic PAHs that are widely used in Tier 2 analysis of oil spill-related samples. This compilation includes parent and alkyl-substituted PAH with long demonstrated utility in environmental forensics investigations (Boehm and Farrington, 1984; Colombo et al., 1989; Sauer and Uhler, 1994/1995). The list also includes the naphthobenzothiophene homologues, which have been less commonly used in oil spill investigations.

The chromatographic retention times for each target compound or compound group

Table 1-2 PAH and Related Compounds or Groups Commonly Targeted in Oil Spill-Related Samples Obtained Using GC/MS-SIM

Target PAH/PAH Groups	Key	Quant. Ion (m/z)	RF	Target PAH/PAH Groups	Key	Quant. Ion (m/z)	RF
Naphthalene	N0	128	N0	Pyrene	PY	202	PY
C1-Naphthalenes	N1	142	N0	C1-Fluoranthenes/Pyrenes	FP1	216	FL
C2-Naphthalenes	N2	156	N0	C2-Fluoranthenes/Pyrenes	FP2	230	FL
C3-Naphthalenes	N3	170	N0	C3-Fluoranthenes/Pyrenes	FP3	244	FL
C4-Naphthalenes	N4	184	N0	C4-Fluoranthenes/Pyrenes	FP4	258	FL
Acenaphthene	ACE	154	ACE	Naphthobenzothiophenes	NT0	234	NT0
Acenaphthylene	ACY	152	ACY	C1-Naphthobenzothiophenes	NT1	248	NT0
Biphenyl	BPHN	154	BPHN	C2-Naphthobenzothiophenes	NT2	262	NT0
Dibenzofuran	DBF	168	DBF	C3-Naphthobenzothiophenes	NT3	276	NT0
Fluorene	F0	166	F0	C4 Naphthobenzothiophenes	NT4	290	NT0
C1-Fluorenes	F1	180	F0	Benzo[a]anthracene	BaA	228	BaA
C2-Fluorenes	F2	194	F0	Chrysene	C0	228	C0
C3-Fluorenes	F3	208	F0	C1-Chrysenes/benzanthracenes	C1	242	C0
Dibenzothiophene	D0	184	D0	C2-Chrysenes/benzanthracenes	C2	256	C0
C1-Dibenzothiophenes	D1	198	D	C3-Chrysenes/benzanthracenes	C3	270	C0
C2-Dibenzothiophenes	D2	212	D	C4-Chrysenes/benzanthracenes	C4	284	C0
C3-Dibenzothiophenes	D3	226	D	Benzo[b]fluoranthene	BbF	252	BBF
C4-Dibenzothiophenes	D4	240	D	Benzo[j,k]fluoranthene	BjkF	252	BKJF
Anthracene	AN	178	AN	Benzo[a]fluoranthene	BaF	252	BAF
Phenanthrene	P0	178	P0	Benzo[a]pyrene	BaP	252	BAP
C1-Phenanthrenes/Anthracenes	P1	192	P0	Benzo[e]pyrene	BeP	252	BEP
C2-Phenanthrenes/Anthracenes	P2	206	P0	Perylene	PER	252	PER
C3-Phenanthrenes/Anthracenes	P3	220	P0	Indeno[1,2,3-c,d]pyrene	IND	276	IND
C4-Phenanthrenes/Anthracenes	P4	234	P0	Dibenzo[a,h]anthracene	DA	278	DA
Fluoranthene	FL	202	FL	Benzo[g,h,i]perylene	GHI	276	GHI

RF-response factor.

must be established under the appropriate GC/MS operating conditions. The diagnostic and confirmatory ions for each target analyte must be verified from GC/MS analyses of authentic standards, whenever possible. Since only a handful of authentic standards exists for the literally hundreds of C_1 to C_4 alkylated isomers of PAH, characteristic ions, retention time windows, and homologue group patterns must be constructed based on carefully documented PAH and alkyl PAH chromatographic retention indices (Lee et al., 1979), and then redocumented by the analysis of a well-characterized reference petroleum product such as Alaska North Slope (ANS) crude oil or Alberta Sweet Mix Blend (ASMB) (Wang et al., 1994a). Alkyl homologues of PAH are quantified using the straight baseline integration method versus relative response factors (RRFs) assigned from the parent PAH compound (Madsen, 1991). Although the alkylated PAH homologous groups can be quantified using the RRF of the respective parent PAH compounds, it is preferable to obtain RRFs directly from alkylated PAHs (if they are commercially available) for further improvement of PAH quantitation accuracy. For example, the RRFs obtained from 1-methyl-naphthalene, 2-methyl-naphthalene, 2,6-dimethyl-naphthalene, 2,3,5-trimethyl-naphthalene, and 1-methyl-phenanthrene can be used for quantitation of 1-methyl-naphthalene, 2-methyl-naphthalene, C_2-naphthalenes, C_3-naphthalenes, and C_1-phenanthrenes, respectively; and the RRFs of 2,3,5-trimethyl-naphthalene and 1-methyl-phenanthrene can be used for quantitation of C_4-naphthalenes, and C_2-, C_3-, and C_4-phenanthrenes in oil, respectively (Wang et al., 1994a).

1.2.3.2 Petroleum Biomarkers

In Tier 2 GC/MS analysis, the concentrations and/or relative distribution of biomarkers — e.g., acyclic isoprenoids, regular and rearranged steranes, bicyclic, tricyclic, tetracyclic, and pentacyclic terpanes (Table 1-3) — can be measured by taking advantage of the knowledge of characteristic ion fragments and well-documented retention indices for each class of compounds. The analysis of biomarkers is thoroughly described in Chapter 3 and is

Table 1-3 Inventory of Petroleum Biomarkers Commonly Used in the Characterization of Spilled Oil Obtained Using GC/MS-SIM

Triterpanes		Steranes
C_{19} Tricyclic terpane	$17\alpha(H),21\beta(H)$-30-norhopane	C_{20} $5\alpha(H),14\alpha(H),17\alpha(H)$-sterane
C_{20} Tricyclic terpane	$18\alpha(H),21\beta(H)$-30-norneohopane ($C_{29}T_s$)	C_{21} $5\alpha(H),14\beta(H),17\beta(H)$-sterane
C_{21} Tricyclic terpane	$17\alpha(H)$-diahopane (X)	C_{22} $5\alpha(H),14\beta(H),17\beta(H)$-sterane
C_{22} Tricyclic terpane	$17\beta(H),21\alpha(H)$-30-normoretane	$13\beta(H),17\alpha(H)$-diacholestane (20S)
C_{23} Tricyclic terpane	$18\alpha(H)$ & $18\beta(H)$-oleanane	$13\beta(H),17\alpha(H)$-diacholestane (20R)
C_{24} Tricyclic terpane	$17\alpha(H),21\beta(H)$-hopane	$13\beta(H),17\alpha(H)$-diastigmastane (20S)
C_{25} Tricyclic terpanes (a & b)	$17\alpha(H)$-30-nor-29-homohopane	$13\beta(H),17\alpha(H)$-diastigmastane (20R)
C_{26} Tricyclic terpanes (a & b)	$17\beta(H),21\alpha(H)$-moretane	
C_{24} Tetracyclic terpane	22S-$17\alpha(H),21\beta(H)$-30-homohopane	$5\alpha(H),14\alpha(H),17\alpha(H)$-cholestane (20S)
C_{28} Tricyclic terpane (a)	22R-$17\alpha(H),21\beta(H)$-30-homohopane	$5\alpha(H),14\beta(H),17\beta(H)$-cholestane (20R)
C_{28} Tricyclic terpane (b)	Gammacerane	$5\alpha(H),14\beta(H),17\beta(H)$-cholestane (20S)
C_{29} Tricyclic terpane (a)	22S-$17\alpha(H),21\beta(H)$-30-bishomohopane	$5\alpha(H),14\alpha(H),17\alpha(H)$-cholestane (20R)
C_{29} Tricyclic terpane (b)	22R-$17\alpha(H),21\beta(H)$-30-bishomohopane	$5\alpha(H),14\beta(H),17\beta(H),24$-methylcholestane (20R)
$18\alpha(H)$-22,29,30-trisnorhopane (T_s)	22S-$17\alpha(H),21\beta(H)$-30-trishomohopane	$5\alpha(H),14\beta(H),17\beta(H),24$-methylcholestane (20S)
$17\alpha(H)$-22,29,30-trisnorhopane (T_m)	22R-$17\alpha(H),21\beta(H)$-30-trishomohopane	$5\alpha(H),14\alpha(H),17\alpha(H),24$-methylcholestane (20R)
C_{30} Tricyclic terpane (a)	22S-$17\alpha(H),21\beta(H)$-30-tetrakishomohopane	$5\alpha(H),14\alpha(H),17\alpha(H),24$-ethylcholestane (20S)
C_{30} Tricyclic terpane (b)	22R-$17\alpha(H),21\beta(H)$-30-tetrakishomohopane	$5\alpha(H),14\beta H),17\beta H),24$-ethylcholestane (20R)
$17\alpha(H),18\alpha(H),21(\beta)H$-28,30-bisnorhopane	22S-$17\alpha(H),21\beta(H)$-30-pentakishomohopane	$5\alpha(H),14\beta(H),17\beta(H),24$-ethylcholestane (20S)
$17\alpha(H),21\beta(H)$-25-norhopane	22R-$17\alpha(H),21\beta(H)$-30-pentakishomohopane	$5\alpha(H),14\alpha(H),17\alpha(H),24$-ethylcholestane (20R)

not recounted here. It is notable, however, that the GC/MS analysis of biomarkers can be carried out simultaneously with the analysis of PAH described above. Two types of calibration standards are appropriate for determination of biomarker concentrations. The first are retention time marker compounds [e.g., 5β(H)-cholane], against which measured biomarker retention times can be compared (Sauer and Boehm, 1995). The second are discrete biomarker chemicals, available commercially, that can be used to generate response factors for individual compounds and compounds of similar structure.

In some instances where conventional Tier 1 GC/FID and Tier 2 GC/MS analyses achieve equivocal results, additional chemical detail may be necessary to answer specific questions of some oil spill investigations. In recent years, two such analytical methods have been applied to oil spill investigations, namely, (1) compound-specific stable isotope analysis (CSIA) and (2) two-dimensional GC analysis (GC × GC). Both of these methods are discussed in subsequent chapters in this book and are not discussed further herein.

1.2.4 Quality Assurance and Quality Control

Like any high-quality laboratory work, analytical chemistry measurements carried out in support of oil spill investigations must be conducted within the framework of a robust quality assurance (QA) and quality-control (QC) program. Beyond the obvious, a practical reason a laboratory should operate under a well-defined QA/QC program is the need for the utmost confidence in the quality of the data because of the likelihood that the data will be used in litigious inquiries.

1.2.4.1 Quality Control

The quality of chemical fingerprinting data is an essential component of any environmental forensic investigation. In most litigious situations the data quality must be defended (long) before the interpretations of those data. Different laboratories employ different degrees of quality control (QC) governing the sample handling, sample preparation, sample analysis, and reporting of data. The practicalities of running a commercial laboratory often compete with QC of the data. Fingerprinting data to be used in a given oil spill investigation should be analyzed in exclusive analytical batches, on the same analytical instrument, and, to the extent possible, within the same analytical sequence. Whenever practical the data analyses (peak integrations) for a given batch of samples should be conducted by a single GC/MS analyst with experience in petroleum analysis and hydrocarbon pattern recognition. These steps, combined with the calibrations (initial and continuing), appropriate QC samples (procedural blanks, laboratory control samples, reference oils), and replicates (duplicates or triplicates), each contribute to overall data quality.

1.2.4.2 Quality Assurance

A QA program assures laboratory management and project investigators that documented standards for the quality for facilities, equipment, personnel training, and work performance are being attained, and if not, to identify and report the areas that need improvement to meet those standards. A laboratory's QA program should be described in the laboratory's Quality Management Plan (QMP). The QMP should describe the laboratory's policy for management system reviews, quality-control and data-quality objectives, QA project plans, standard operating procedures, training, procurement of items and services, documentation, computer hardware and software, planning and implementation of project work, assessment and response, and corrective action and continuous improvement. A QA system should consist of a minimum of six components, namely,

1. A formal Work/QA Project Plan that describes all work, QA, and quality-control activities associated with a project is developed for each study.
2. Up-to-date standard operating procedures (SOPs) that describe all technical activities conducted by the laboratory.

3. A program to ensure and document that all project personnel are fully trained and qualified to perform project activities before independent activities may begin. Personnel training records should be maintained by the QA Unit and include records of qualifications, prior experience, professional training, and internal training procedures.
4. A documentation and records system that facilitates full sample and data tracking.
5. A quality assessment program for all projects, conducted through management system reviews, technical system audits, performance evaluation samples, data validation, laboratory inspections, and independent data audits. An independent QA Unit within a laboratory should conduct the latter two activities.
6. A continuous improvement program, facilitated through quality assurance audits, a formal corrective action program, and routine, laboratorywide performance assessments and reviews.

In addition, all laboratory deliverables should receive an independent review that includes a quality assurance (QA) and technical component. The QA review should ensure that the data are (1) complete, (2) in compliance with the procedures defined in the QAPP, (3) in compliance with regulatory statutes, and (4) accurate and technically sound. The technical review should ensure that the results and interpretations are (1) objective, (2) defensible among peers, and (3) presented in a manner that conveys the conclusions clearly, often to nonexpert decision makers.

1.3 Factors Controlling the Chemical Fingerprints of Spilled or Discharged Petroleum

Figure 1-1 depicted the four factors that influence the chemical fingerprints of petroleum in the environment that are relevant to oil spill investigations. In this section, these factors are presented and described.

The term "petroleum" collectively refers to the family of materials, naturally occurring and humanmade, that contain complex mixtures of up to tens of thousands of hydrocarbons and nonhydrocarbons (nitrogen-, sulfhur-, oxygen-, and metal-containing compounds; Speight, 1991). Naturally occurring petroleums include gases (e.g., natural gas), liquids (e.g., gas condensates and crude oils), and solids (e.g., solid bitumens and oil shale/sand). The chemistry of naturally occurring petroleums is the collective result of the geologic processes that led to their formation over millions of years, namely (1) the nature of the ancient, organic-rich, source strata, (2) the thermal history of the source strata, and (3) changes brought about during petroleum's migration to and residence in reservoir strata (Tissot and Welte, 1984; Hunt, 1996; Table 1-4).

Table 1-4 Inventory of the Factors Controlling the Chemical Fingerprints of Spilled or Discharged Petroleum

Controlling Factors in Chemical Fingerprints of Petroleum			*Petroleum Types*
Primary **Controls** (Geology)	source rock organic matter type source rock thermal maturity migration effects in-reservoir alteration	*Naturally Occurring*	natural gas/condensates crude oils solid bitumens oil sand/shales
Secondary **Controls** (Refining)	fractionation conversion finishing blending	*Man-Made*	gasoline distillate/residual fuels lubricants waxes/greases/asphalts
Tertiary **Controls** (Weathering & Mixing)	evaporation water-washing biodegradation photo-oxidation mixing with "background"		all fugitive petroleums

These three "primary controls" impart very specific chemical features to naturally occurring petroleums, which will vary markedly between different oil provinces (e.g., coastal California versus Alaska; Kvenvolden et al., 1995). More subtle differences can even exist between oils produced from individual reservoirs within a single oil field (e.g., Kauffman et al., 1990; Peters and Fowler, 2002). As such, the primary controls impart very specific *genetic* features to petroleums that provide the primary basis for the chemical fingerprinting of naturally occurring petroleums in oil spill investigations (Stout et al., 2002; see Section 1.1.1 ahead).

Manmade petroleums are derivatives of naturally occurring petroleums that are produced in the course of petroleum refining. (Minor changes in chemistry may also occur during the production of natural petroleums from the subsurface, e.g., through phase separation, but these are not considered herein.) Manmade petroleums include various fuels (e.g., liquefied petroleum gas, automotive gasoline, and distillate and residual fuel oils), lubricants, and other products (e.g., waxes, greases, and asphalts) and their wastes (e.g., waste oils, bilge oils). Although some of the genetic chemical features of the naturally occurring *parent* petroleum are passed directly to the *daughter* petroleum products, the refining process can impart its own effects on the chemical composition of the resulting manmade petroleums and can, therefore, be considered as a "secondary control" on their chemical fingerprints. For example, distillation and thermal cracking can affect the distributions of some biomarkers and PAH (e.g., Peters et al., 1992; see also Section 1.3.2 ahead). Thus, in total, the chemical compositions of manmade petroleums are effected by the combined affects of the primary and secondary controls (Table 1-4). The chemical specificity imparted by these combined affects provides the basis for the chemical fingerprinting of manmade petroleums in oil spill investigations.

Both naturally occurring and manmade petroleums are transported from areas of their production to areas of their use. Most often it is in the course of this transportation when oil is accidentally spilled or intentionally discharged (Table 1-1). Once petroleum is released into the environment, it immediately is subjected to various physical, chemical, and biological processes that will affect its chemical composition (Jordan and Payne, 1980). These processes are collectively referred to as *weathering*, which can be considered a "tertiary control" on the chemical fingerprinting of petroleum in the environment (Table 1-4). Weathering will affect the chemical composition of fugitive petroleum at widely different rates. For example, evaporation of volatile constituents will occur much more rapidly than the biodegradation of recalcitrant constituents (see Section 1.3.3), and chemical fingerprinting of samples following an oil spill — sometimes collected decades later (e.g., Wang et al., 1994b, 1998; Reddy et al., 2002; Petkewich, 2002; Prince et al., 2002a; Short et al., 2004; see also Chapters 5 and 15 herein) — must consider these longer-term weathering effects. Because of this, both the shorter-term investigations of mystery oil spills and the longer-term investigations of known-source spills must account for the chemical changes to the spilled oil brought about by weathering and focus instead on those features that are un- or minimally affected by weathering (e.g., Douglas et al., 1996; Wang and Fingas, 1995c; Stout et al., 2001; Daling et al., 2002; see also Chapter 7 herein). In fact, in investigations of known-source spills, the effects of weathering are often what chemical fingerprinting is trying to measure — i.e., how has the spilled oil changed over time (e.g., Wang et al., 1994b, 1998).

Concurrent with the weathering, once petroleum is released into the environment it becomes subject to mixing with pre-existing anthropogenic contaminants or naturally occurring materials already present in the environment. These other sources of "background" hydrocarbons can contribute to the collective fingerprint of any floating oils, sediments or soils, waters, and tissues thought to be impacted by the spilled oil (only) and, therefore, need to also be considered as another "tertiary control" on the chemical fingerprint of spilled oil (Table 1-4). If unrecognized, the presence of pre-existing, background hydrocarbons can confound both the identification of a mystery spill's source or the assessment of any long-term effects of a spill on the environment. Even major oil spills from known sources can be confounded by alternative sources of hydrocarbons in the environment. For example, the *Amoco Cadiz* grounding, the Ixtoc-1 blowout, and *Exxon Valdez* groundings each were, in part, confounded by the presence of other anthropogenic and natural hydrocarbons in the spill area (Boehm et al., 1997; Page et al., 1988, 1995). Similarly, smaller oil spills and discharges in ports, which are typically areas with long-term, industrial activities surrounded by urbanized areas, are subject to a multitude of "urban background" hydrocarbons (see Section 1.3.4.4; Stout et al., 2004).

The chemical complexity imparted by the combination of primary, secondary, and tertiary controls on a spilled oil's fingerprint (Table 1-4) provides the oil spill investigator with both the challenges and the means to answer questions surrounding the source and fate of waterborne oil spills or operational discharges in the environment. In the following sections, the bases for these controls on crude oil and petroleum products are described in greater detail (Sections 1.3.1 through 1.3.4).

1.3.1 Primary Control — Crude Oil Genesis

Crude oil is an extremely complex mixture of tens of thousands of individual hydrocarbons and nonhydrocarbons. These compounds range from small, simple, volatile, and distinct compounds (e.g., methane) to extremely large, complex, nonvolatile, colloidally dispersed macromolecules (e.g., asphaltenes). The distribution of these compounds imparts certain physical properties on the oil, and it is these physical properties (e.g., density or viscosity) by which crude oils are generally classified, bought, and sold. These physical properties can be used to grossly classify crude oils as either *conventional* (light or medium) or *heavy*, where conventional oils have an API gravity above 20° or viscosity (at reservoir conditions) below 100 cP (Speight, 1991). Conventional crude oils can be generally classified based upon the predominance of the major hydrocarbon classes — paraffins, naphthenes, and aromatics — as shown in Figure 1-2 and described ahead.

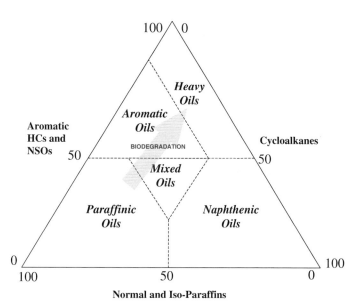

Figure 1-2 Ternary diagram showing the general compositional features of crude oils. After Tissot and Welte (1984). Arrow indicates general compositional change due to biodegradation.

All crude oils contain aliphatic and aromatic hydrocarbons and nonhydrocarbons. The proportions of these three major classes of compounds, and the specific distributions of individual compounds within each class, vary widely among crude oils (Tissot and Welte, 1984). The aliphatic hydrocarbons (sometimes called paraffins or saturates) include multiple compound types, such as normal alkanes (sometimes called normal paraffins), acyclic isoprenoids (sometimes called iso-paraffins), and monocyclic and polycyclic alkanes (collectively called naphthenes), the latter of which include sesquit-, di-, tri-, tetra-, and pentacyclic terpanes and sterane biomarkers (Figure 1-3). The aliphatic hydrocarbons com-

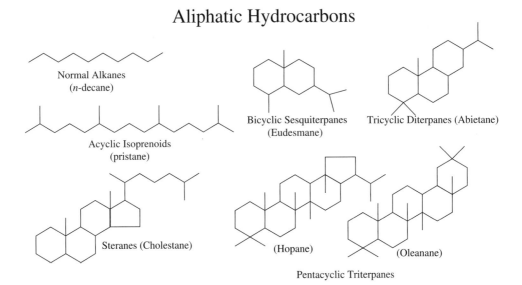

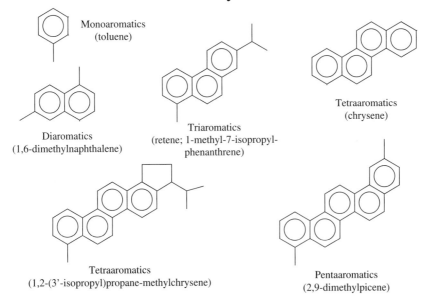

Figure 1-3 Examples of aliphatic and aromatic hydrocarbons in crude oils.

prise the largest weight percentage of most conventional crude oils, and are markedly reduced due to biodegradation in heavy crude oils (Figure 1-2; Speight, 1991; Hunt, 1996). Aromatic hydrocarbons include monoaromatic (BTEX) and polycyclic aromatic hydrocarbons (PAH), the latter of which include various aromatic biomarkers (e.g., PAH with terpane or sterane skeletons; Figure 1-3). Nonhydrocarbons include polars, resins, and asphaltenes. Polars include compounds containing one or more nitrogen, sulfhur, or oxygen atoms (NSOs) that impart a "polarity" to the compounds. Nitrogen-containing compounds include benzocarbazoles, quinolines, and porphyrins, sulfhur-containing compounds include benzo-, dibenzo-, and polynuclear thiophenes, and oxygen-containing compounds include furans, phenols, and acids. Heavy crude oils contain higher percentages of aromatic hydrocarbons, predominantly PAH, and nonhydrocarbons (NSOs) than conventional crude oils (Figure 1-2; Speight, 1991; Hunt, 1996).

Many excellent reviews of crude oil composition are already available (Kinghorn, 1983; Tissot and Welte, 1984; Speight, 1991; Hunt, 1996), which allows this section to focus on those primary molecular features of crude oils that are most useful in oil spill fingerprinting. (Isotopic features of crude oils that are useful in oil spill fingerprinting are discussed in Chapter 6 herein.)

On a molecular level, it can be safely stated that no two crude oils are identical due to the variability in the primary controlling factors leading to their formation (Table 1-4). First, the nature of a crude oil's source rock organic matter type(s) varies widely with geologic age, lithology, and the particular conditions of the ancient depositional environment (e.g., lake, ocean, delta, etc.) in which these source strata accumulated (Tissot and Welte, 1984). Second, the variable kinetic conditions during the generation and expulsion of crude oil from the source strata will impart certain maturity-based features driven by various cracking, isomerization, and aromatization reactions on a crude oil. Third, migration can impart additional changes on the composition of an oil due to geochromatographic or "contamination" effects. Exploration geochemists have long understood the effect that the source rock organic matter type has on the molecular and isotopic composition of crude oil and have used this knowledge to improve crude oil exploration and production (Moldowan et al., 1985). The environmental applications of this knowledge have been increasingly applied to oil spill investigations over the past 35 years or so (see Section 1.3.1 below).

Primary chemical features imparted by the source, maturity, and migration of crude oil are of great value in chemical fingerprinting of both mystery and known-source oil spills. In particular, a variety of petroleum biomarkers offer a very high degree of specificity among crude oils (Table 1-5). Petroleum biomarkers are organic compounds in crude oil whose carbon skeletons can be unequivocally linked to those found in the naturally occurring biochemicals from which they are derived, i.e., biomarkers are "chemical fossils" (Eglinton and Calvin, 1967). The biochemicals that accumulated in ancient sediments are converted into biomarkers during the diagenesis and thermal maturation of the sediments over geologic time, i.e., during their conversion into petroleum source rocks (e.g., Seifert and Moldowan, 1978; Simoneit, 1986). Crude oil expelled from a given source rock sequence will contain biomarkers whose character testifies to the ancient biomass that had originally accumulated in the ancient sediments that later formed the source rock (Didyk et al., 1978; Moldowan et al., 1985; Peters et al., 1986; Volkman, 1988) and its lithology (Rubinstein and Albrecht, 1975). The relative abundances of more and less thermally stable biomarker isomers, or their degree of aromatization, will testify to the thermal maturity of the source rock strata when the oil was expelled (e.g., Seifert and Moldowan, 1978; Mackenzie et al., 1980; Radke, 1988; Farrimond et al., 1998). The migration of crude after its expulsion from the source strata can introduce new compounds "extracted" from carrier beds or reservoir rocks (Curiale, 2002) or experience subtle molecular fractionation (Larter et al., 1996; Matyasik et al., 2000). Finally, after reaching

Table 1-5 Inventory of Common Biomarker Compound Classes Available for Chemical Fingerprinting of Petroleum in the Environment. The Table Includes the Carbon Boiling Range over Which Each Class Occurs and the Dominant Mass/Charge (m/z) Ratio upon Which They Can Be Identified Using Gas Chromatography-Mass Spectrometry

Biomarker Compound Class	Approximate Carbon Boiling Range	Mass Spectral Fragment Ions
n-alkanes	C1 to C45*	m/z 85
acyclic isoprenoids	C12 to C19	m/z 113
bicyclic sesquiterpanes	C13 to C17	m/z 123
diterpanes	C19 to C24	m/z 191
extended tricyclic terpanes	C18 to C26	m/z 191
tetracyclic terpanes	C22 to C23	m/z 191
25-norhopanes (10-desmethylhopanes)	C25 to C33	m/z 177
pentacyclic triterpanes	C26 to C34	m/z 191
diasteranes	C23 to C26	m/z 217
regular steranes	C25 to C29	m/z 217 and 218
C-ring monoaromatic steranes	C23 to C26	m/z 253
triaromatic steranes	C26 to C29	m/z 231

* As measured by conventional (low temperature) gas chromatography.

a reservoir (but prior to its discovery and production) crude oil composition can further change due to water-washing by meteoric water, deasphaltening, or biodegradation in the reservoir (Connan, 1984; Seifert et al., 1984; Palmer, 1993; Cassani and Eglinton, 1991. The latter process, biodegradation, is generally responsible for the vast majority of heavy crude oil deposits worldwide (Tissot and Welte, 1984).

Thus, collectively these primary controls on crude oil (Table 1-4) invoke tremendous variability in the molecular compositions, including the biomarker "fingerprints," of conventional and heavy crude oils (Peters et al., 2005, and refs. therein; see also Chapter 3 herein). Because of this inherent variability, and their relative resistance to biodegradation on environmental time scales (see Chapter 11 herein), petroleum biomarkers have proven particularly useful in crude oil spill investigations (e.g., Kvenvolden et al., 1995; Daling et al., 2002; Page et al., 1995; Volkman et al., 1997; Kaplan et al., 1997, 2001; Stout et al., 2001; Wang and Fingas, 2003; Stout, 2003; Wang et al., 1999b, 2006). The reader is directed to Chapter 3 (herein) to see multiple examples of the biomarker variability among crude oils.

Figure 1-4 demonstrates some disparate primary features for two conventional crude oils. Figures 1-4a and b show the gas chromatography-flame ionization detection (GC/FID) chromatograms for a North Slope crude (NSC) and a Nigerian Bonnie Light crude oil blend, respectively. Both oils contain abundant normal alkanes, but each exhibits a different n-alkane profile and the Nigerian crude exhibits a slight odd-over-even predominance in the $C_{25}+$ normal alkanes. The NSC has a higher nC_{17}/Pr and lower Pr/Ph ratio than the Nigerian crude. Figures 1-4c and d show the partial m/z 191 mass chromatograms in each oil, which reveals a relative abundance of extended tricyclic triterpanes and homohopanes in the NSC, and a relative abundance of oleanane in the Nigerian crude (all versus hopane). The ratios of T_s/T_m vary between the oils, being lower in the NSC. Finally, Figures 1-4e and f show the distributions of dia-(rearranged) and regular steranes, which show a higher proportion of diasteranes and C_{27} regular steranes and a higher C29S/(C29S + C29R) ratio in the NSC than in the Nigerian crude. These differences collectively indicate that these oils' source rocks had varying proportions of marine and terrestrial organic matter and had expelled oil at different heating conditions. These differences in the biomarker distributions are more fully explained in the geochemical literature of these two oil

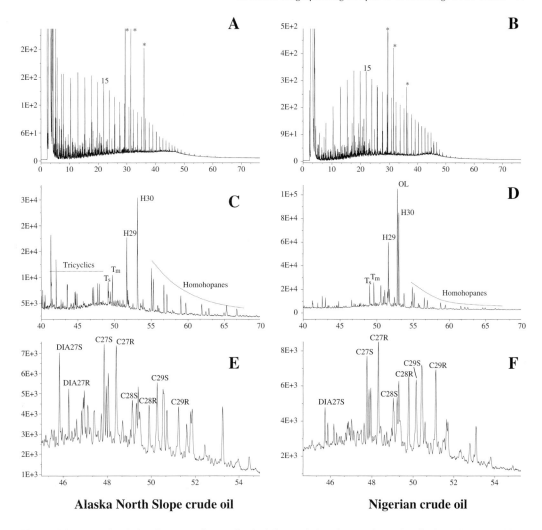

Figure 1-4 Example of the disparate primary chemical fingerprinting features in crude oils from the North Slope of Alaska and the Niger Delta, Nigeria. (a–b) GC/FID chromatograms, (c–d) partial m/z 191 mass chromatograms, and (e–f) partial m/z 217 mass chromatograms. * = internal standards, left to right: o-terphenyl, 5α-androstane and tetracosane-d_{50}.

provinces (e.g., Mackenzie et al., 1985; Ekweozor and Udo, 1988; Ukpabio et al., 1994). However, it should be clear that such differences will provide a strong basis to positively or negatively correlate a spilled crude oil to its candidate sources and to recognize any contribution to potentially impacted sediments or biota. This high degree of chemical disparity between crude oils parallels those recognized in crude oil spill investigations in California and Alaska, where significant primary chemical differences are known to exist between crude oils produced from the Miocene, coastal California basins, the Mesozoic North Slope of Alaska, and natural oil seeps east of Prince William Sound (e.g., Kvenvolden et al., 1995; Bence et al., 1996; and Chapter 15 herein).

The primary chemical differences demonstrated in Figure 1-4 are relatively obvious, given the significant geologic differences between the Mesozoic petroleum system in the North Slope of Alaska versus the Tertiary petroleum system in the Niger Delta. A simple qualitative comparison of these oils' fingerprints (Figure 1-4) can reveal these marked differences. However, crude oils produced from a single petroleum system within a single petroleum province will exhibit far more subtle primary differences than are evident in Figure 1-4. Petroleum geochemists working within a petroleum province recognize these subtle differences, often recognizing "oil families," frequently using multivariate techniques (e.g., Telnaes and Dahl, 1986; Osadetz et al., 1995). Similarly, oil spill investigations in some areas may require precise quantitative (or semiquantitative) comparisons among oils with similar chemical fingerprints in order to facilitate their distinction (see Chapter 8 herein). This situation, for example, as might exist for a mystery oil spill in an offshore producing area where genetically similar crude oils are being produced from different production platforms, has led to the development of numerically or statistically based analysis of crude oils' primary (*genetic*) fingerprinting data in efforts to more quantitatively determine the degree of correlation between spill oils and candidate sources (e.g., Urdal et al., 1986; Stout et al., 2001; Daling et al., 2002; and Chapters 7 and 9 herein). In such cases, the subtle differences between oils produced from individual reservoirs or production zones may reveal important differences.

An example of this type of oil spill investigation is shown in Figure 1-5, which shows the GC/FID chromatograms and partial triterpane (m/z 191) distributions in a mystery spill oil found in an offshore operating area and in two (of the many studied) candidate source oils from nearby crude oil production platforms. All the crude oils produced in the area are "genetically" similar due to the overall consistency in the primary factors described above — source organic matter type, thermal maturity, and migration — within the local petroleum system. The overall similarity of the oils is evident in the highly comparable GC/FID fingerprints of the whole oils. For example, while the spilled oil obviously has been affected by evaporation, each oil is dominated by n-alkanes that exhibit a slight odd-over-even predominance in the $C_{25}+$ range (Figure 1-5). While there is no obvious difference among the oils based upon the GC/FID chromatograms, subtle differences in the primary factors lead to subtle differences in the triterpane distributions in the oils. Close inspection reveals that the relative abundance of gammacerane, a pentacyclic triterpane with a bacterial origin (ten Haven et al., 1989), varies among the oils. This variation eliminated Platform B as a source for the spilled oil (Figure 1-5). In this investigation, all other candidate crude oil sources were similarly eliminated on some subtle basis and Platform A was confirmed to be the likely source of the mystery oil spill.

1.3.2 Secondary Controls — Petroleum Refining

Crude oil is converted into marketable petroleum products through refining. Refined petroleum products are then transported from the refinery to market via trucks, pipelines, and product tankers and barges. The most common petroleum products transported generally reflect their economic value, with automotive gasoline, distillate fuel oils, and residual fuel oils being the most common petroleum products in use. For example, Figure 1-6 shows the approximate volumes of different petroleum products transported as cargo by product tankers and barges in U.S. waterways in 2003 (the last year data are available; WCUS, 2003). These data show that gasoline, distillate fuels, and residual fuels are the most commonly transported petroleum products in U.S. waters. Accidental releases of petroleum product cargo or fuel from vessels, pipelines, and coastal facilities, or intentional discharges of cargo washings or wastes (oily water, slops, and sludges) from vessels can produce both mystery spills and known-source spills. In addition, other sources of fugitive petroleum products in the environment can require investigation as to their source or impact (e.g., jettisoned aircraft fuel; Table 1-1). Thus, although crude oil is the most voluminous petroleum cargo (see above), refined

Chemical Fingerprinting of Spilled or Discharged Petroleum   19

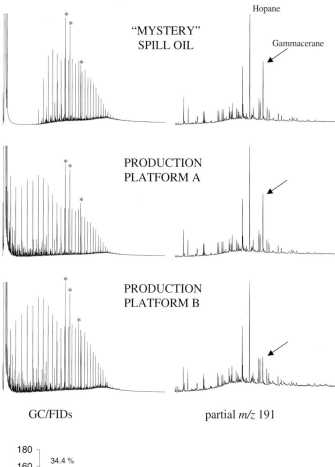

Figure 1-5 GC/FID chromatograms and partial *m/z* 191 mass chromatograms for a mystery oil spill and produced crude oil from two nearby production platforms. Differences in the relative abundance of gammacerane were sufficient to distinguish these (and other) genetically related oils from one another. * = internal standards.

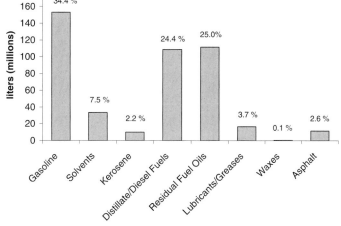

Figure 1-6 Summary of foreign and domestic petroleum products transported as cargo in U.S. waterways and ports in 2003. Volume calculated from short ton using average density. Data from WCUS (2003).

product spills and discharges regularly are the focus of chemical fingerprinting investigations. As such, it is important to recognize the effects of refining on the chemical fingerprints of different petroleum product types.

Refinery chemical engineers concern themselves with the bulk chemical and physical properties of the crude oil feedstock, without specific regard for the molecular details that are of interest in oil spill fingerprinting. The crude oil feedstock processed at any given refinery can include a single "local" crude oil commingled from genetically similar crude oils from within a single basin or a dynamic blend of multiple and genetically dissimilar crude oils produced from different basins, often obtained from different continents. In either case, the commingled production of similar crude oils or the blending of disparate crude oils will result in a crude oil feedstock that exhibits the collective and mass-balanced "primary" chemical (elemental, molecular, and isotopic) features of the individual parent crude oils in the feedstock. As such, the crude oil feedstock at any given refinery at any given time will still exhibit a unique set of primary features (Table 1-4). These features still offer a high degree of chemical specificity that is, at least in part, passed along to the daughter petroleum products or altered to reflect the refining process.

The modern petroleum refinery has evolved from a simple turn-of-the-century facility whose goal was to produce fuel oils for heating and lighting (while discarding then uneconomical gasoline-range hydrocarbons) to modern operations of varying complexity focused on squeezing as much automotive gasoline (the most economically valuable product) out of a barrel of crude oil as feasible. Modern refineries are complex industrial plants where various crude oil feedstocks are utilized to produce not only gasoline, but other economically valuable distillate and residual products. Figure 1-7 shows the complexity that can exist at a modern refinery. For an in-depth understanding of refinery operations, we recommend one of several excellent reference texts on the subject (Gary and Handwerk, 1984; Speight, 1991; Leffler, 2000).

The dominant three processes at any refinery are (1) fractionation, (2) conversion, and (3) finishing (Speight, 1991). Fractionation involves the separation of a crude oil feedstock into various "cuts" that are defined by their boiling points. These cuts, obtained by both atmospheric and vacuum distillation (Figure 1-7), traditionally include light gases (C_1 to C_4 hydrocarbons), straight run naphthas, kerosene, light and heavy gas oils, vacuum distillates, and residual material (bottoms). The approximate boiling and carbon ranges of these distillate cuts are shown in Table 1-6. Few of these distillation cuts are used directly in products (anymore), but rather they are fed to other units in the refinery, either for conversion or for blending into commercial products. Conversion, as the name implies, changes the molecular composition of a distillation cut by cracking larger molecules into smaller ones (thermal or catalytic cracking), sometimes in the presence of excess hydrogen (hydrocracking), stripping hydrogen to convert naphthenes into aromatics or normal paraffins into branched paraffins (catalytic reforming), combining two lower-octane molecules into one higher-octane molecule (alkylation), or converting the chemical structure from one configuration to another (isomerization; Figure 1-7). The finishing process typically involves sweetening and blending of multiple straight-run and converted intermediates into finished petroleum products that meet the required performance and regulatory specifications.

The petroleum products at the right of Figure 1-7 have a range of boiling distributions, which generally increase from top to bottom in the figure. Examples of the chemical fingerprints obtained via Tier 1 GC/FID (Section 1.2.2) of representatives of these different products are shown in Figure 1-8. The carbon (boiling) ranges of these products vary widely and chemical fingerprinting can rely upon these gross differences to distinguish one petroleum product from another.

The detailed chemistry of these various petroleum products depends in part upon the nature of the crude oil feedstock, which, as described above, is highly variable. Furthermore, the detailed product chemistry is a function of the specific operational conditions within a given refinery. Since no two refineries operate in an identical manner, and even a single refinery can change operations daily, the chemical composition of petroleum products on a molecular level is expected to widely vary, while still meeting any required product specifications. This is indeed the case and is why chemical fingerprinting of petroleum

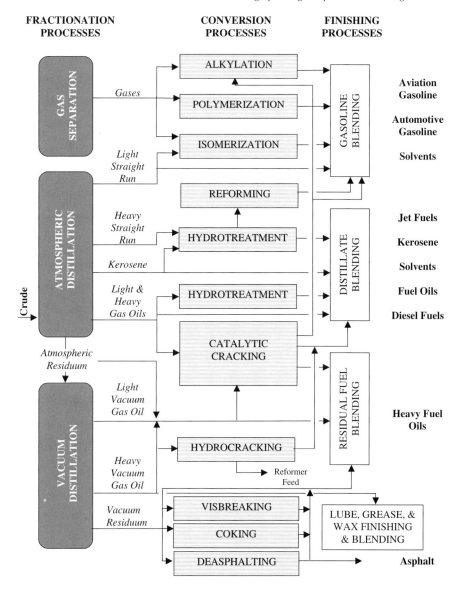

Figure 1-7 Schematic flowchart of a modern complex refinery highlighting the distillation, conversion, and finishing processes used in the production of petroleum products.

products has proven a useful tool in all types of environmental forensic investigations (e.g., Kaplan et al., 1997; Stout et al., 2002a), not just waterborne oil spills.

1.3.2.1 Gasoline

Although automotive gasoline is the most voluminous petroleum product produced and transported, the sources of waterborne gasoline spills and discharges are rarely a "mystery" and their persistence in the surface water environment is relatively short-lived due to their volatility and solubility. As such, the chemical fingerprinting investigations of gasoline spilled onto open water are rare. Occasionally, gasoline spilled onshore into storm water drains can enter waterways through

Table 1-6 Typical Distillation Fractions and Petroleum Products Produced from Crude Oil. (Boiling and Carbon Ranges Are Approximate and Can Vary Slightly with Refiner)

	Boiling Point at 1 atm (°C)	Carbon Range (as n-alkane)
Distillation Fractions		
Gases	−160–0	C1–C4
Light Straight Run	25–90	C5–C6
Heavy Straight Run	85–190	C6–C10
Kerosene	160–275	C9–C15
Atmospheric Gas Oils	250–425	C13–C27
Vacuum Gas Oils	370–575	C22–C45
Vacuum Residuum	>540	>C40
Petroleum Products		
Aviation Gasoline	25–175	C4–C10
Automotive Gasoline	25–215	C4–C12
Volatile Solvents	70–250	C6–C14
Commercial Jet Fuel	80–290	C7–C16
Kerosene (Fuel Oil #1)	200–290	C11–C16
Fuel Oil #2	150–400	C9–C25
Diesel Fuel #2	150–425	C9–C27
Heavy (Residual) Fuel Oils	150–575	C9–C45
Lubricating Oils	270–525	C15–C40

outfalls and require chemical fingerprinting to determine the source. Similarly, small spills during fueling or boat repairs at marinas may have a cumulative effect in some environments and therefore can be investigated using chemical fingerprinting.

Modern automotive gasoline contains hundreds of hydrocarbons and nonhydrocarbons boiling in the C_4 to C_{14} range (Figure 1-8A). The exact number and distribution of these compounds depend on refining processes employed in the production. For example, gasolines blended to contain a high percentage of alkylate or reformate can contain less than half the number of detectable hydrocarbons than in gasolines containing mostly cracked naphthas (Hamilton and Falkiner, 2003). The complexity and composition of automotive gasoline and its chemical fingerprinting in the environment have been recently reviewed (Stout et al., 2006a). This review has shown that modern gasolines are a complex mix of highly refined intermediate blending stocks — such as polymer gas, alkylate, straight-run naphtha, reformate, and cracked naphtha (Figure 1-7) — and various additives (e.g., oxygenates). Historically, a significant proportion of gasoline included a straight-run naphtha blending stock that might retain certain primary features of the crude oil feedstock (e.g., adamantane or C_7 isoparaffin distributions; Stout and Douglas, 2004). However, due to the complicated blending of multiple refinery intermediates in the production of modern gasoline (Figure 1-7), different gasoline grades and brands can exhibit markedly different chemical fingerprints reflecting the blending practice of a particular refiner. Some features in gasoline are more useful than others in distinguishing different gasoline sources. For example, most modern gasolines are blended to include an alkylate blending stock produced during alkylation. The alkylate blending stock is enriched in high-octane trimethylpentane isomers, the abundance and proportions of which can vary depending on the specific nature of the alkylation process and blending practice (Beall et al., 2002). As such, the relatively rare oil spill investigations involving gasolines can still benefit from detailed chemical fingerprinting in an attempt to distinguish the source of waterborne gasoline or to trace its impact in the environment. However, the effects of weathering on waterborne gasoline, which due to its relatively low boiling range (Figure 1-8A), is subject to rapid evaporation that can quickly remove many potentially diagnostic features. In such instances, chemical fingerprinting of the highest boiling fractions of any gasoline residues, e.g., in the C_9+ range may still yield useful results.

1.3.2.2 Distillate Fuels

Distillate fuels, on the other hand, are commonly the focus of waterborne oil spills due to the frequency by which they are transported as cargo (Figure 1-6) and their widespread use as fuels in marine diesel engines. Middle distillate fuel refers to a category of fuels, largely

Chemical Fingerprinting of Spilled or Discharged Petroleum   23

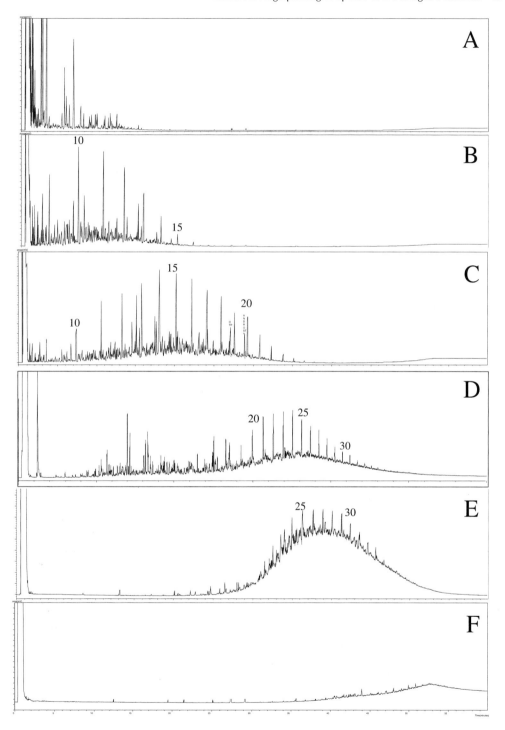

Figure 1-8 Gas chromatography-flame ionization detection (GC/FID) chromatograms of distinct petroleum products. (A) automotive gasoline, (B) jet fuel A, (C) diesel fuel #2, (D) heavy fuel oil #6, (E) lubricating oil, and (F) petroleum asphalt. # = n-alkane carbon number.

classified depending on their intended use. They include civilian and military jet engine fuels, on-road diesel (truck and bus), off-road diesel (diesel-electric locomotives, heavy equipment, and farm machinery), marine diesel engine fuels, nonaviation gas turbine fuels, and domestic and commercial heating fuels. Although each distillate fuel type's specifications are defined by various ASTM standards, the primary distinguishing features of relevance to chemical fingerprinting of the different distillate fuel types relate to their boiling distributions and sulfur contents. In addition, the specific molecular characteristics of a distillate fuel type will depend, in part, on the primary characteristics inherited from the crude oil feedstock.

As their name implies, the production of distillate fuels primarily involves fractionation by distillation. The primary effect of distillation on the different distillation cuts (Table 1-6) is to alter the distributions of compounds depending upon their volatility. An example of this effect appears in Figure 1-9, which shows the distributions of tricyclic triterpane biomarkers in a parent crude oil feedstock and in a middle distillate cut of the feedstock. These tricyclic triterpanes occur in the C_{20} to C_{26} range, i.e., near the higher boiling range of most distillate fuels. As can be seen, there is a gradual reduction in the abundance of tricyclic triterpanes with increasing molecular weight (decreasing volatility) in the distillate cut (Figure 1-9). This type of distillation effect can alter any ratios between compounds with varying volatility (e.g., Peters et al., 1992), particularly those that boil nearer to the end points of the distillate fuel.

Oppositely, molecular features that exist nearer to the middle of any distillate cuts and fuels often are inherited directly from the crude oil feedstock without modification. For the more broad-boiling distillate fuels, these features can include ratios between comparably boiling isoprenoids (pristane/phytane), proportions of methyl-phenanthrene isomers, or distillate range biomarkers such as sesqui- or diterpanes (see Table 1-5 and Chapter 3 herein). Proportions of n-alkanes-to-isoprenoids (nC_{17}/Pr or nC_{18}/Ph) can be altered from the feedstock during blending of distillate blending stock to produce higher cetane fuels, which typically contain a high proportion of n-alkanes (see Figure 1-11 below, in which the nC_{17}/Pr ratio is increased in the cracked distillate relative to the straight-run distillate).

The five methyl-phenanthrene isomers all boil in the C_{20} range and as such can be passed unchanged from the crude oil feedstock into a straight-run distillate. The proportion of these isomers is a primary feature related to the thermal maturity of the source strata, which can be expressed by ratios involving the less stable α-type and more stable β-type isomers (Radke et al., 1982; Radke, 1988). However, conversion refinery processes can alter the distribution of the methyl-phenanthrene isomers — and create new compounds typically absent in the feedstock. Specifically, catalytic cracking can reduce the proportion of the less stable α-type (9-, 4-, and 1-) methyl-phenanthrene isomers and produce methyl-anthracene isomers that are not typically present in crude oil. (An example of this is shown in Chapter 10 herein.)

The sesquiterpanes with a drimane skeleton are "low-boiling" biomarkers that occur in the C_{13} to C_{16} range (Peters et al., 2005). These compounds can provide diagnostic information about the parent crude oil feedstock used in the production of distillate fuels. Under some oil spill circumstances, these biomarkers can distinguish and correlate distillate fuel sources from spilled fuel (Stout et al., 2005, Wang et al., 2005). For example, Figure 1-10 shows the distributions of sesquiterpanes in a spilled jet fuel and a candidate source jet fuel from a nearby storage facility. The patterns of sesquiterpanes were distinct between these fuels, indicating that the candidate source from the storage facility was not the source of the spilled jet fuel.

Another group of compounds that can be useful in the characterization distillate fuels are normal alkylcyclohexanes (as revealed in m/z 83 mass chromatograms). N-alkylcyclohexanes are 6-carbon cyclic hydrocarbons

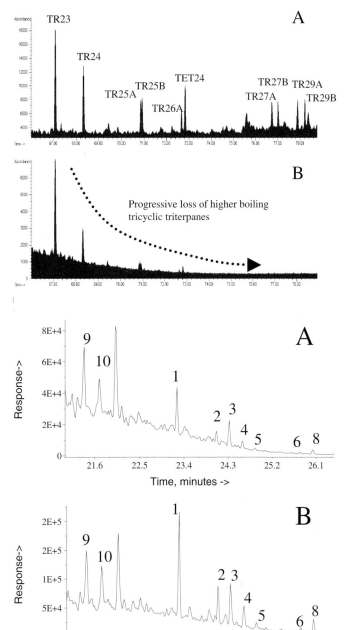

Figure 1-9 Partial *m/z* 191 mass chromatograms for a (A) parent crude oil feedstock and (B) daughter distillate cut showing the progressive loss of tricyclic triterpanes due to distillation effects.

Figure 1-10 Partial *m/z* 123 mass chromatograms showing the distributions of sesquiterpanes in (A) a spilled jet fuel and (B) a candidate source oil. Peak numbers correspond to compounds described by Alexander et al. (1984). The large peak between peaks 10 and 1 is unidentified.

with different numbers of straight-chained or normal "alkyl" side chains — for example, CH-1 (cyclohexane with one methyl group attached), CH-2 (cyclohexane with an ethyl group attached), etc. The distribution of the various CH-1 to CH-18 alkyl-cyclohexanes is indicative of both a petroleum product's nature and type under most weathering states (Kaplan and Galperin, 1997). Thus, as the refinery blends various middle distillate fuels to meet its customer specifications using distinct blending stocks, it will produce distillate fuels with distinctive n-alkylcyclohexane distributions (Song, 2000; Stout et al., 2004, 2006).

Under most circumstances the production of finished distillate fuels that meet modern specifications from a straight-run crude oil distillate(s) is noneconomical. This fact, combined with the availability of many refining intermediate streams (Figure 1-7). Most modern refineries blend cracked intermediated products (e.g., light- and mid-cut cycle oils, visbreaker or coker gas oil, or hydrocrackate) with straight-run distillate products [e.g., light and heavy straight-run (virgin) distillates] in order to produce their distillate fuels (Jewitt et al., 1993). Cracked products can exhibit distinct *secondary* chemical fingerprinting features (Table 1-4) produced during various refinery conversion processes. For example, Figure 1-11 shows the GC/MS total ion chromatograms for a straight-run distillate (Figure 1-11A) and a cracked intermediate distillate (Figure 1-11B) produced from a single crude oil feedstock from the Williston Basin. The straight-run gas oil represents a "slice" of the Williston Basin parent crude oil feedstock and, as such, retains any primary chemical features of the feedstock unaffected by the distillation process (e.g., pristane/phytane ratios, sesquiterpane distributions, and relative abundance of alkylated dibenzothiophenes to alkylated phenanthrenes). The cracked intermediate, however, contains lower relative abundances of isoprenoids (e.g., pristane and phytane relative to n-alkanes) and a three-times higher concentration of PAHs than the straight-run distillate. These differences indicate that excess n-alkanes and PAH apparently were formed, or released from asphaltenes, due to the thermal stresses of the catalytic cracking process. In addition, the distribution of individual PAH groups, as reflected by the relative abundance of alkylated dibenzothiophenes to alkylated phenanthrenes, also has been significantly decreased by the cracking process (Figure 1-11C and D).

Because the sulfur content is an important specification of distillate fuels, the distribution and proportions of sulfur-containing com-

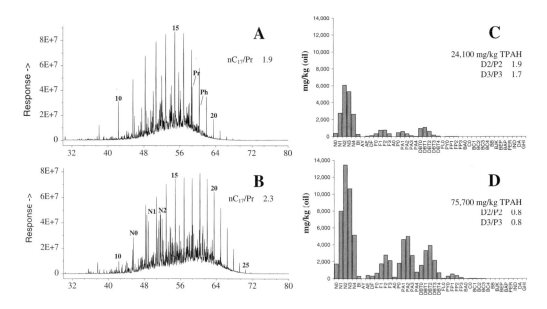

Figure 1-11 Chemical fingerprints for two distillate fuel blending stocks. (A) GC/MS TIC for straight-run (virgin) gas oil, (B) GC/MS TIC for light cracked distillate (cycle oil), (C) PAH histogram for straight-run gas oil, and (D) PAH histogram for light cracked distillate. # = n-alkane carbon number, Pr = pristane, Ph = phytane; for PAH analytes, see Table 1-2.

pounds in middle distillate fuels can prove useful in oil spill investigations of these fuels. The lower sulfur requirements of modern on-road diesel fuel #2 (<50 or 500 ppm) require that refineries employ some form of distillate desulfurization in order to produce low or ultralow sulfur diesel fuels. Most commonly, this requires the deep hydrodesulfurization (HDS) of distillate blending stocks, which reduce sulfur-containing compounds by replacing sulfur with hydrogen (and producing H_2S). Many sulfur-containing compounds, e.g., benzothiophene and sterically hindered alkyl-benzothiophenes like 4-methyldibenzothiophene, 4,6-dimethyldibenzothiophene, and 4-ethyl-6-methyldibenzothiophene, are resistant to HDS (Song, 2000), and therefore the distribution of individual dibenzothiophene isomers can be altered during HDS, which could be useful in some oil spill investigations. (See Chapter 4 herein.)

The sulfur content in distillate fuels is expressed by the relative concentration of alkylated naphthothiophenes, benzothiophenes, or dibenzothiophenes, which are the predominant forms of organic sulfhur within the distillate boiling range (Mossner and Wise, 1999). Most commonly, the relative concentration of the dibenzothiophenes is reflected in the ratio of alkylated dibenzothiophenes to the comparably boiling alkylated phenanthrenes. Both groups of compounds respond similarly in the environment and thereby retain their proportions during environmental weathering (Douglas et al., 1996). Ratios between the two-carbon (C_2) and three-carbon (C_3) alkylated derivatives of dibenzothiophenes and phenanthrenes — D2/P2 and D3/P3 — can help distinguish distillate fuels refined from sweet and sour crude oil feedstocks and/or fuels subject to different degrees of desulfurization (e.g., see Figure 1-11). For example, the relative abundance of alkyl-dibenzothiophenes was a distinguishing feature in identifying the source of a "mystery" diesel fuel spill in the Lachine Canal, Quebec, Canada (Wang et al., 2000). Among other features, including comparable methyl-phenanthrene ratios, these authors showed that the ratios of D2/P2 and D3/P3 were virtually identical between the mystery spill samples and one of the two candidate sources.

As noted above, distillate fuels refined to contain predominantly or exclusively a straight-run distillate component, a difference in the relative abundance of alkyl-dibenzothiophenes to alkyl-phenanthrenes may be identical to that of the crude oil feedstock. This was apparently the case for marine diesel fuels refined from Alaska North Slope (ANS) crude oil versus the *Exxon Valdez* cargo oil, both of which exhibited comparable D2/P2 and D3/P3 ratios (Page et al., 1995). However, the ANS diesel could be distinguished from the cargo oil on the basis of the absence of higher boiling chrysenes in the ANS diesel.

1.3.2.3 Residual Fuels

Residual fuels, as their name implies, are predominantly comprised of the less valuable residues from the various fractionation and conversion processes at a refinery. These fuels are used in marine diesel engines and boilers and in industrial power generation. They require preheating prior to their use due to their high viscosity, relative to distillate fuels, and are often transported by heated barges. As the need for gasoline in distillate fuels increases and refining operations "squeeze" more of these products from crude oil, the nature of the residual hydrocarbon mixtures has become increasingly "heavier" in character. This practice, and the resulting effect on the chemistry residual fuels, is described thoroughly in Chapter 10.

The molecular composition of residual fuels derives, in part, from the molecular composition in the crude oil feedstock. The highest boiling distillation residuum of the feedstock is the primary blending stocks used in the production of residual fuels (Figure 1-7). Molecular features contained within these highest boiling residues largely will be inherited by the residual fuel. These would include the features of most higher-boiling biomarkers such as pentacyclic triterpanes, steranes, aromatic steranes, and porphyrins. However, the effects of heating during distillation, particularly vacuum distillation, can alter various biomarker thermal parameters in residual fuels containing vacuum distillation residuum, i.e., flasher bottoms (Peters et al., 1992; Pieri et al., 1996). Peters et al. (1992) showed that biomarkers with comparable boiling points in a parent crude oil feedstock were variously depleted or enriched compared to daughter refinery streams; e.g., vacuum gas oil or vacuum residuum. Such changes were attributed to either (1) a preferential preservation (i.e., a greater thermal stability) of certain biomarkers or (2) preferential formation (cracking from bound precursors in the oil) of certain biomarkers during heating. Such changes do not preclude the use of biomarker-based ratios in characterizing a spilled residual fuel to its source, but they can confound any attempt to determine the crude oil parent of the spilled fuel. See Chapter 10 for additional information on the compositions of residual fuel oils.

1.3.2.4 Lubricating Oils

Lubricating oils are produced from the finishing (deasphaltening, hydrotreatment, solvent extraction, and dewaxing) of heavy vacuum gas oils (Figure 1-7). The resulting family of base oils is blended to yield different types of lubricating and hydraulic oils that meet various physical- and performance-based properties. Figure 1-12 shows a series of GC/FIDs for different types of base oils. The chemical compositions of the base oils are complex and poorly understood due to the overwhelming predominance of hydrocarbons (straight, branched, cyclic, and aromatic) that cannot be resolved by conventional GC methods. As such, the predominant feature of the GC/FID chemical fingerprints of lubricating oils is the shape of the UCM. Figure 1-12 shows, however, that some base oils contain n-alkanes and biomarkers, the latter of which sometimes can appear as the only resolved peaks in some lubricating oils (e.g., Figure 1-12D). The biomarker distributions in lubricating oils will largely reflect those inherited from the crude oil feedstock and thereby provide a high degree of specificity in the characterization of spilled lubricating oils.

1.3.2.5 Oily Waste/Bilge Water Discharges

Bilge water consists of drippings of various liquids that accumulate in the bottom of a boat. Although bilge water is not comprised of any particular petroleum product, it is worth mentioning here since it most often contains a mixture of petroleum products found onboard a given vessel. As such, the chemical fingerprint of any oily waste contained within the bilge water will reflect the collective features of the fuels, lubricants, and solvents in use on a given vessel. The discharge of oil in bilge water is prohibited in navigable waters of many countries. In order to satisfy the prevailing regulations, most vessels pass bilge water through an oil–water separator to isolate oils until they can be off-loaded at the dock or burned on-board. In addition, large ships install sensors that trigger alarms and the need for immediate corrective actions if the bilge water effluent contains more than 15 parts per million (ppm) of oil and grease. Despite these precautions, poor engine and oil–water separator maintenance are suspected causes of significant chronic petroleum discharges above regulatory concentration limits in coastal waters (Costa, 2001). In addition, the intentional and illegal bypass of the bilge water oil–water separator, unfortunately, is not an uncommon practice.

The chemical fingerprint of oily waste will vary based on the type of materials and waste management practices on-board. Consequently, the chemical composition of hydrocarbons and other chemicals within the waste stream is expected to be highly variable and often reflect the presence of multiple petroleums. Figure 1-13 shows the GC/FID for a bilge oil that exhibits multiple n-alkane and UCM maxima suggesting the presence of middle distillate fuel oil (MFO), heavy fuel oil (HFO), and lubricating oil. Characterization of the oily wastes on-board vessels would employ the same chemical fingerprinting as for the fuels and lubricants used on vessels (described above). Specifically, the distributions of PAHs and biomarkers can be used to characterize oily wastes and their associated discharges.

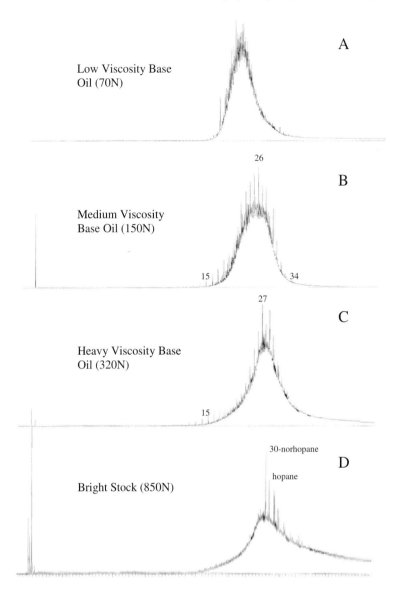

Figure 1-12 GC/FID chromatograms for four base oils used in the manufacture of lubricating oils. (A) low-viscosity base oil (70N), (B) mediumviscosity base oil (150N), (C) heavy-viscosity base oil (320N), and (D) bright stock (850N).

1.3.3 Tertiary Controls — Weathering

In addition to primary and secondary controls on the chemical fingerprints petroleum imposed during petroleum's generation and refining (Sections 1.3.1 and 1.3.2), petroleum released into or onto water immediately is subjected to the collective effects of multiple, naturally occurring chemical, physical, and biological processes, collectively referred to as *weathering*. Weathering will alter the chemical fingerprint of spilled or discharged oil so that its fingerprint may not "match" the original oil.

The most important weathering processes following a release of oil to fresh or marine waters include processes such as spreading, evaporation, dissolution, dispersion into the

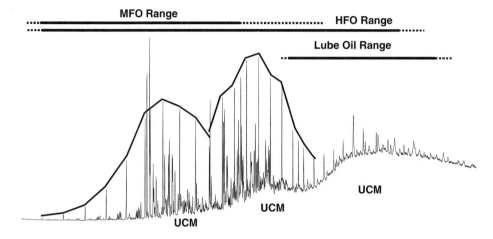

Figure 1-13 GC/FID of a bilge oil containing a mixture of different petroleums. (Modified from Emsbo-Mattingly, 2006.)

water column, formation of water-in-oil emulsions, photochemical oxidation, microbial degradation, adsorption to suspended particulate matter, and stranding on the shore or sedimentation to bottom sediments (Payne and McNabb, 1985; Payne et al., 1987; Mackay and McAuliffe, 1988; Neff, 1990; Wolfe et al., 1994; Wang and Fingas, 1995b). Generally, the mostly physical processes (e.g., dispersion, spreading, emulsification, tar ball formation) do not alone substantially alter the chemical fingerprint of spilled oil. However, these processes largely will affect the rates by which and degrees to which an oil will evaporate, dissolve, or biodegrade — i.e., all processes that will alter a spilled oil's chemical fingerprint. For example, as an oil spreads, the rates of evaporation and dissolution will increase as a greater mass of oil becomes exposed to air and water, respectively, thereby increasing the rates of evaporation and dissolution, respectively. As volatile and soluble constituents preferentially are lost, the physical properties and consequent spreading of the spilled oil are further altered. Thus, the physical behavior and chemical composition of a spilled oil are interrelated and dynamic (Fingas, 1995; Michel et al., 1995).

Extensive reviews on the effects of weathering on spilled oil are found elsewhere (Mackay and McAuliffe, 1988; NRC, 2003). In this section, the effects of weathering on the chemical composition of spilled oils generally are described in terms of their effect on the "chemical fingerprint" of the oil. As such, we focus on the effects of evaporation, dissolution, biodegradation; however, other processes are also considered.

1.3.3.1 Evaporation

Evaporation and dissolution are the primary weathering processes affecting the chemical composition/fingerprint of spilled oil in the hours or days following an oil spill. Although spill-specific conditions (e.g., thickness, mousse formation, wind speed, ambient water/air temperature) will affect the rate of evaporation, under all conditions the propensity for evaporation is ultimately governed by the vapor pressure of individual constituents.

Compounds in petroleum that boil at temperatures below about 270°C, or have vapor pressures greater than about 0.1 mm Hg, tend to evaporate rapidly from the surface of spilled crude oil (Stiver and Mackay, 1984; Fingas, 1995). Included in this category are alkanes from methane to $n\text{-}C_{15}$ and one- and two-ring aromatics ranging from benzene through alkylnaphthalenes (Bobra et al., 1979; Wolfe et al., 1994). The rate of evaporative loss depends on multiple factors, but in general, compounds with boiling points below about 200°C ($\sim C_{12}$) will evaporate within a 24-hour period of a spill (Stiver and Mackay, 1984). The mass reduction attributable to evaporative loss will depend on the composition of the spilled oil, with the percent mass reduction being inversely proportional to the density of the petroleum (e.g., Mackay and McAuliffe, 1988). Under some rare circumstances — e.g., high altitude, 22°C, and turbulent water — evaporative losses can reduce compounds up to C_{36} (Prince et al., 2002b).

The effect of evaporation on the fingerprint of a spilled oil will depend on the oil's original composition. Obviously, spilled gasoline may evaporate completely within a few hours, effectively eliminating its fingerprint, whereas a heavy crude oil will experience minimal losses or change to its fingerprint due to evaporation.

Figure 1-14 shows the effects of evaporation on a 34°API crude oil spilled following the structural failure of an aboveground storage tank (AST) during Hurricane Katrina (Stout et al., 2006b). Figure 1-14A shows the slightly evaporated crude recovered from inside the failed AST two weeks after the failure. It shows a mild loss of hydrocarbons boiling below n-C_{14}. Figure 1-14B shows the residual crude oil recovered from a surface soil

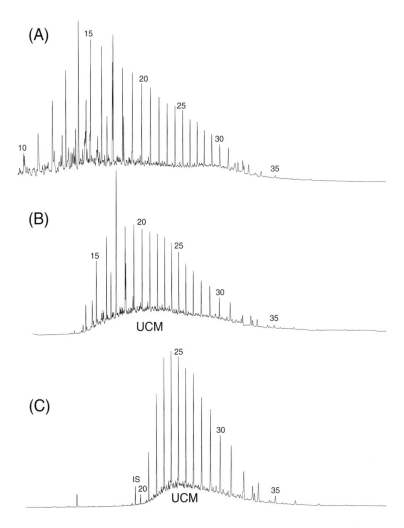

Figure 1-14 GC/FID chromatograms showing the progressive loss of compounds from a 34°API crude oil spilled following Hurricane Katrina. (A) slightly evaporated crude oil, (B) moderately evaporated crude oil, and (C) severely evaporated crude oil.

impacted by the crude oil approximately two weeks after the spill. In this sample, the crude oil has experienced evaporative loss of compounds below n-C_{19} resulting in an increase in the relative mass of the UCM. Figure 1-14C shows the residual crude oil recovered from an oily "bathtub ring" on the side of a building, also about two weeks after the spill. This oily residue has experienced evaporative loss of most all compounds boiling below n-C_{20}.

As demonstrated by the crude oil shown in Figure 1-14, some spilled fuels (e.g., gasoline or kerosene) or high-API gravity crude oils may evaporate completely. Most light- and midweight crude oils will lose between 10 and 50% of their mass to evaporation in the first few days of a spill. Emulsification, the process wherein small water droplets are incorporated into the spilled oil, can substantially retard the rate of evaporation allowing the retention of otherwise volatile compounds in some light- and midweight oils (Fingas, 1995). In the case of heavy crude oils and residual fuel oils, which are already depleted of volatile hydrocarbons, the effects of evaporation generally are lower (Mackay and McAuliffe, 1988). However, evaporation may have a greater effect on certain residual fuel oils that have been blended using a lower-boiling (lower-viscosity) petroleum. In such instances the preferential evaporation of the "diluent" may alter the chemical fingerprinting.

Under most conditions, the effects of evaporation are generally restricted to compounds having vapor pressures above that of approximately n-C_{15}, or to be conservative (and as evident in Figure 1-14), above that of approximately n-C_{20}. As such, chemical fingerprinting of spilled oil residues often must focus on diagnostic features expressed in constituents with vapor pressures lower than about n-C_{20}. In such instances, the effects of evaporation on any diagnostic features (e.g., biomarkers) must be carefully examined.

In general, the effects of evaporative weathering on the oil chemical compositions generally are predictable. Studies of the effects of evaporation on chemical composition changes of a number of oils (Wang et al., 1995b, 2003) at various weathering degrees (from 0 to ~40% loss by mass) reveal that (1) ratios of n-C_{17}/pristane, n-C_{18}/phytane, and pristane/phytane were virtually unaltered, (2) isomeric distributions within C_1-phenanthrenes (4 isomers), C_1-dibenzothiophenes (3 isomers), C_1-fluorenes (3 isomers), and C_1-naphthalenes (2-methyl- and 1-methyl-naphthalene) exhibited great consistency in their relative ratios as the weathering percentages increased, (3) biomarker compounds were concentrated in proportion with the increase of the weathering percentages, and (4) the weathering degree can be readily checked by integration of n-alkanes in the GC-FID chromatograms, as Figure 1-14 shows.

1.3.3.2 Dissolution

Although occurring simultaneously with evaporation, the effects of dissolution are less obvious—and difficult to separate from evaporation. For example, the most soluble constituents in most spilled oils are the monoaromatic hydrocarbons, which are also among the more volatile. Studies have shown that evaporation clearly dominates the losses—for example, as reflected in the lack of preferential removal of benzene over cyclohexane (i.e., compounds with comparable vapor pressures and dissimilar aqueous solubilities; Harrison et al., 1975). It is estimated that dissolution will reduce the mass of a crude oil by 1 to 3 wt% (as opposed to 10 to 50 wt% due to evaporation; Mackay and McAuliffe, 1988). Thus, dissolution is relatively insignificant compared to evaporation (McAuliffe, 1977; Payne et al., 1987; Neff, 1990) and will have little effect on a spilled oil's chemical fingerprint. Understanding the dissolution process, however, may be important in recognizing the impact of spilled oil on biota, which can uptake dissolved hydrocarbons more readily.

1.3.3.3 Biodegradation

The rate of mass loss due to biodegradation is slower than that of evaporation or dissolution.

As such, the effects of biodegradation on a spilled oil's chemical fingerprint are not obvious in the short term following a release. As with the other weathering processes, the effect of biodegradation will depend upon the nature of the spilled oil and the spill conditions.

Following a spill of crude oil or refined petroleum to soils, sediments, or open water, different hydrocarbon classes are degraded simultaneously, but at widely different rates by indigenous microbiota (Atlas et al., 1981; Oudot, 1984; Gough et al., 1992; Prince et al., 2003; see also Chapter 11 herein). Low-molecular-weight n-alkanes with chain lengths of 10 to 22 carbons are biodegraded most rapidly, followed by isoalkanes and higher-molecular-weight n-alkanes, olefins, monoaromatics, PAHs, and, finally, highly condensed cycloalkanes, resins, and asphaltenes. Some sulfur heterocyclics, such as dibenzothiophene and its alkyl homologues, seem to be more resistant than PAHs of similar molecular weight to aerobic microbial degradation (Fedorak and Westlake, 1984; Wang and Fingas, 1995c). However, some strains of aerobic and anaerobic bacteria are able to efficiently degrade dibenzothiophenes (Wu et al., 2002). Compounds that are degraded are typically converted into oxidized compounds that can be further degraded, dissolved, or retained within the oil. Few studies have been directed at understanding that chemical fingerprinting tends to focus mostly on compounds considered resistant to the effects of biodegradation over environmental timescales, namely, PAH and petroleum biomarkers. Though relatively resistant, not all PAH or biomarkers are recalcitrant. Notably, parent PAHs often are biodegraded more rapidly than their alkyl homologues; the rate of biodegradation decreases as the number of alkyl groups on the PAH nucleus increases (Prince et al., 2002a). This is opposite the pattern for photooxidation of PAH (see below); the rate of photooxidation often increases with increasing PAH alkylation (Prince et al., 2003).

Recognizing and monitoring the effects of biodegradation on conventional crude oils and most fuels (kerosene, diesel fuels, fuel oils) are readily reflected in the relative depletion of n-alkanes which is relative to comparably volatile isoprenoids (e.g., n-C_{17}/Pr and n-C_{18}/Ph). However, in the case of heavy crude oil or residual fuels, which can contain little or no n-alkanes by virtue of their formation or refining, the effects of biodegradation are less obvious. For example, heavy crude oils, which (in most cases) have already been biodegraded in an oil reservoir for millions of years, in most cases, are likely to be largely unaffected by microbial biodegradation on environmental timescales. Similarly, residual fuel oils that can be blended using vacuum distillation residuum (Figure 1-7) can contain little or no n-alkanes (see Chapter 10 herein). Nonetheless, given sufficient time, biodegradation can alter even heavy oils and residual fuels, in generally predictable ways. For example, the PAHs in a Bunker C oil, stranded for 22 years on a shoreline in Nova Scotia, Canada (following the 1970 *Arrow* spill), were highly depleted in naphthalene and alkyl-naphthalenes, enriched in C_2- through C_4-phenanthrenes and dibenzothiophenes, and strongly enriched in C_1- through C_3-chrysenes due to weathering, probably primarily by a combination of biodegradation and water-washing (Wang et al., 1994b).

Prince et al. (2003) provide several examples of the effects of biodegradation (and photooxidation) of crude and refined oils following release to the environment. In December 1999, the *Erika* broke up in a storm off the French coast and released about 19,000 tons of a heavy fuel oil, which emulsified and subsequently came ashore (Oudot, 2000). A sample of the cargo oil was collected from the shore at Le Croisic less than four months after the spill and evaluated for loss of hydrocarbons by comparison of analyte concentrations with the concentration of a conservative natural internal marker in the oil, $17\alpha(H)21\beta(H)$-hopane. Almost 20% of the hydrocarbons in the heavy oil had been lost in less than four months (Prince et al., 2003).

Petroleum biomarkers, such as steranes and triterpanes (Table 1-3) are considered to be extremely resistant to microbial degradation (Hughes and Holba, 1987). Because of the persistence of most petroleum-derived steranes and triterpanes, they are used frequently to fingerprint and identify the sources of complex hydrocarbon mixtures in soils and sediments and to quantify rates of biodegradation (Atlas, 1995; Page et al., 1995; Prince et al., 2002a, 2003).

1.3.3.4 Photooxidation

Exposure of spilled crude oil and petroleum products to solar radiation leads to several free radical, photooxygenation reactions that produce a variety of oxygen-containing compounds, including peroxides, aldehydes, ketones, alcohols, carbonyls, sulfoxides, epoxides, and fatty acids (Larson et al., 1977; Thominette and Verdu, 1984a, 1984b; Nicodem et al., 1997). Since these compounds are polar, they are more soluble in water than the parent compounds (Payne and McNabb, 1985; Ehrhardt and Douabul, 1989; Ehrhardt and Burns, 1990; Ehrhardt et al., 1992) and therefore can be preferentially dissolved from the spilled oil. However, photo-sensitive higher-molecular-weight compounds can form cross-links upon photooxidation, thereby becoming increasingly large and insoluble — and the spilled oil increasingly viscous (Thominette and Verdu, 1984a; Daling and Brandvik, 1988). The rate of photooxidation of crude oil is directly related to the intensity of ultraviolet radiation from sunlight, but is not sensitive to ambient temperature (Syndes et al., 2001; Prince et al., 2003).

Photooxidation and biodegradation are the only two natural processes that actually destroy petroleum hydrocarbons and remove them from the environment — in many cases forming the same classes of oxygen-containing products. In many cases the target oil ingredients of the two degradative processes are different. Under suitable conditions, they act together to degrade oil more rapidly than either process working alone (Dutta and Harayama, 2000; Prince et al., 2003).

1.3.3.5 Mousse Formation

Among the various weathering processes, the formation of emulsions (mousse) is the only one that promotes the persistence of oil on the surface of water and on the shore — and concurrently retards the weathering processes. The formation and stability of mousses have been well studied in both the laboratory and field for some time (Payne and Phillips, 1985; Daling and Brandvik, 1988; Durell et al., 1993; Strom-Kristiansen et al., 1993; Ronningsen et al., 1995; Fingas and Fieldhouse, 2003). A stable oil mousse no longer behaves as a well-mixed phase. Emulsions often form a crust of oxidized asphaltene materials that inhibit dissolution, evaporation, photooxidation, and biodegradation of hydrocarbons in the bulk oil phase. As such, the chemical fingerprint of the emulsified oil is preserved relative to a non-emulsified oil. Thus, weathered deposits of mousse, tar, and asphalt pavement on the shore may retain some volatile hydrocarbons and biodegradable compounds for long periods of time (Boehm and Fiest, 1982; Payne and Phillips, 1985; Irvine et al., 1999). If the mousse is exposed to air and sunlight, weathering continues at a slow rate, converting the oil deposit to a tar mat or asphalt pavement that may be extremely persistent and retain many of the primary and secondary features of the spilled oil for long periods of time. Some tar mats and asphalt pavements become brittle with time and wave action may erode tarry particles from the surface, forming tar balls. The "centers" of tar balls can contain virtually unweathered oils, weeks or months after the initial oil spill, thereby providing the potential to generate chemical fingerprints with little effect of biodegradation.

1.3.3.6 De-Waxing and Wax Enrichment

Some oils containing prominent n-alkanes (waxes) are prone to precipitation of the wax when spilled, especially at low temperatures (Strom-Kristiansen et al., 1997). This phenomenon can impart a heterogeneity in the chemical fingerprint of the spilled oil by reducing the relative abundance of n-alkanes in the $C_{20}+$ range in the wax-depleted fraction and increasing their relative abundance in the wax-enriched fraction. An example of this phenomenon is shown in Figure 1-15. In this example, a paraffinic North Sea crude oil (Figure 1-15A) becomes depleted (Figure 1-15B) and enriched (Figure 1-15C) in n-alkanes above n-C_{20} after being spilled under controlled conditions at 3°C for 71 hours (Daling et al., 2002). The effect on the GC/FID fingerprints of the spilled oils is marked and would need to be considered in an oil spill investigation in which the source and spilled oils are compared.

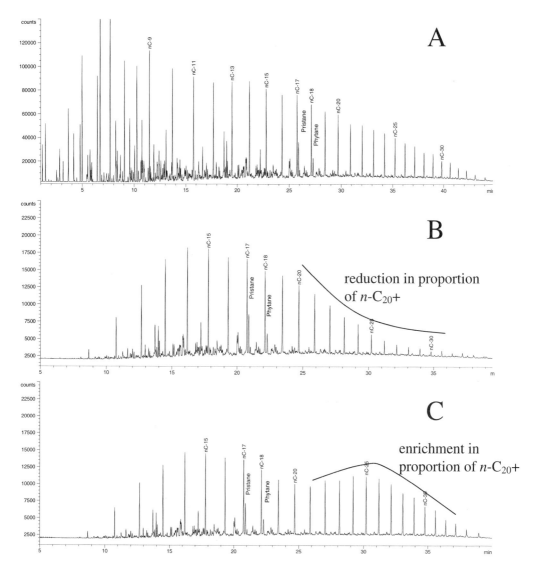

Figure 1-15 GC/FID chromatograms showing the effect of wax precipitation on the chemical fingerprint of a paraffinic oil. (A) starting oil, (B) wax-depleted fraction within spilled oil 71 hours after the spill, and (C) wax-enriched fraction within the spilled oil 71 hours after the controlled spill experiment. Data from Daling et al. (2002).

1.3.4 Tertiary Controls—Mixing with "Background"

Although accidental oil spills or intentional discharges of oil can draw particular attention due to their acute nature, "background" hydrocarbons derived from chronic anthropogenic and natural sources are ubiquitous in modern environments (Table 1-1). The chemical fingerprint of contamination authentically resulting from an accidental oil spill or intentional discharge is often superimposed upon the chemical fingerprint of any pre-existing or "background" chemicals that may have been present before the impact. Recognizing and distinguishing this "background" from any authentic contamination is, therefore, an important component of oil spill investigations. In this section, we review the potential sources of background hydrocarbons in the environment and their potential effect on the chemical fingerprints of spilled or discharged petroleum.

1.3.4.1 What Is "Background"?

"Background" chemicals can take the form of either naturally occurring or anthropogenic chemicals (U.S. EPA, 1989), either of which can be confused with authentic contamination resulting from an oil spill. These two types of background can be defined as follows:

- Naturally occurring background chemicals are inorganic or organic chemicals in the environment that are attributable to natural geological or hydrological characteristics of the area. These chemicals have not been altered by human activity, e.g., natural oil seeps, metals or plant-derived organic matter in soils, forest fire debris, etc.
- Anthropogenic background chemicals are ubiquitous (mostly) synthetic or natural substances that have been released due to human activities but are unrelated to a specific oil spill or discharge. Examples may include agricultural and urban runoff or direct deposition of air pollution particulates.

As implied by these definitions, the issue surrounding background chemicals in oil spill investigations is most often faced in the characterization of soils/sediments and tissues, although water and air can be similarly influenced. In any case, if the contribution of "background" contamination remains unrecognized, the nature and extent of any impact (or even if any impact has occurred at all) can be overestimated. As a result, any environmental risk associated with an oil spill can be overestimated, or unreasonably low cleanup levels may be established. Therefore, every effort should be made to recognize and establish the specific character(s) of the background chemicals and conditions.

The U.S. EPA acknowledges the issue of background contamination in various laws and guidance documents. For example, the Comprehensive Environmental Response, Compensation, and Liability Act includes a provision [CERCLA; 42 USC 9604(a)(3)(A)] that recognizes that elevated background chemicals can render cleanup impractical or impossible and therefore must be assessed. The EPA's *Guidance for Conducting Remedial Investigations and Feasibility Studies Under CERCLA* (U.S. EPA, 1988) includes the requirement that background or reference samples be collected for comparison to impacts associated with site-specific or incident-specific releases. These samples are usually collected in areas where chemicals only attributable to anthropogenic or natural background are expected to occur. Of course, where background conditions exclusively occur is not always apparent ahead of time. Chapter 2 describes the importance of collecting "background" samples in the course of oil spill investigations.

Most of the EPA's guidance for recognizing and distinguishing background contamination arises with respect to inorganic contaminants such as toxic metals. The Office of Solid Waste and Emergency Response has published a forum issue paper that specifically addresses this issue with respect to background metals (U.S. EPA, 1995). The reason for the emphasis on background metals is that they are naturally occurring in all soils and sediments and therefore must always be considered in site assessments involving metal contaminants. However, the issue of background contamination also is crucial in oil spill investigations due to the ubiquitous distribution of anthropogenic and naturally occurring hydrocarbons in nature (Table 1-1).

In this section, the fingerprinting characteristics of some commonly encountered *organic* background chemicals are presented and discussed. The objective is to review the effects these background chemicals can have on the chemical fingerprint of spilled oil.

1.3.4.2 Recognizing and Establishing Background

Two methods are commonly used to assess background contamination, i.e., statistical analysis and chemical fingerprinting. The statistical analysis approach is thoroughly described in the U.S. EPA's issue paper on this topic, which emphasizes the issue with respect to toxic metals (U.S. EPA, 1992). The approach generally relies upon various comparative population statistical methods that are used to establish the range of a chemical's background concentrations (anthropogenic and natural) for a study area, and thereby permit recognition of any impacts above this range. This approach obviously requires a sufficient number of samples from both the background area(s) and the impacted area(s) than can be analyzed statistically. As the number of samples increases, the accuracy and precision of each population's probability density function can be established and compared. However, in most oil spill investigations, the statistical approach is often difficult to achieve since studies can rarely afford a sufficient number of background samples to yield reliable population statistics, and furthermore (as mentioned above), the locations of true background impacts are seldom known *a priori*.

Whereas statistical approaches may be more suited to datasets of toxic metals, which may or may not co-occur with one another, chemical fingerprinting is well suited for datasets involving petroleum. This is because (as described in Section 1.3.1) crude oil and petroleum products are complex mixtures of hundreds of hydrocarbons and nonhydrocarbons, which, while they may all (or mostly) be present in different sources of contamination, are typically assembled in a specific distribution. This allows for their distinction from each other, and from the equally complex mixtures of hydrocarbons and nonhydrocarbons in natural and anthropogenic background materials. This allows for the chemical fingerprinting approach to recognizing and establishing background benefits from the complexity of organic materials and organic-based contaminants. Thus, in most oil spill investigations chemical fingerprinting is the best means of establishing "background" hydrocarbon fingerprints and concentrations.

The chemical fingerprinting approach to recognizing the influence of background chemicals involves either a qualitative visual comparison or a quantitative comparison of diagnostic ratios or distributions within known forms of contamination (e.g., fuels, lubricants, or crude oil) and various types of background (see ahead). Such comparisons can allow an investigator to recognize and thereby account for the influence of "background" hydrocarbons in potentially impacted soils, sediments, and tissues. These influences can then be reconciled with respect to the absolute concentrations attributable to the background to assess the magnitude of any influence attributable to the spilled oil. In the sections that follow, various examples of different forms of natural and anthropogenic organic "background" contamination encountered in oil spill investigations are presented and discussed.

1.3.4.3 Naturally Occurring Background Hydrocarbons

Naturally occurring hydrocarbons can take many forms, including modern plant debris, forest fire debris, bacteria-derived compounds (e.g., sulfur species), crude oil seeps, or sedimentary organic particles (e.g., eroded coal).

The influence of these types of materials typically is greatest in investigations involving the chemical fingerprinting of soils and sediments. For example, most soils and sediments contain some fraction of naturally occurring organic matter. Upon solvent extraction by typical methods (e.g., EPA Method Series 3500), the naturally occurring organic matter in most soils or sediments yields hydrocarbons and nonhydrocarbons within the same boiling range as petroleum. In addition, some individual naturally occurring hydrocarbons also occur within oil (see ahead). Thus, the presence of naturally occurring hydrocarbons can be confused with, or contribute to the concentration of, hydrocarbons derived from spilled or discharged oil. This can result in an overestimate of the concentration of total petroleum hydrocarbons (TPH) or PAH, both of which can erroneously increase the liability for the incident.

During solvent extraction of a soil or sediment in the laboratory (EPA Method Series 3500), any naturally occurring organic matter (OM) that may be present is co-extracted along with any authentic organic contamination present. Thus, the resulting extract contains a combination of soluble natural background and the authentic contamination. Some forms of extractable background contamination can be removed from an extract prior to chemical analysis through various extract "cleanup" procedures (e.g., EPA Method Series 3600). When properly performed, these procedures can eliminate the complications due to the presence of sulfur (EPA 3660), polar compounds (EPA 3610 or 3630), or humic acids, proteins, etc. (EPA 3640). And while such cleanup procedures should routinely be employed in assessing the impacts to soils and sediments, they sometimes are not. Even when they are performed, some background contaminants, e.g., naturally occurring background *hydrocarbons* cannot be separated from authentic *hydrocarbon* contamination. In these instances, chemical fingerprinting is required to recognize the presence of the background hydrocarbons and account for them appropriately. In the following sections, some descriptions of examples of naturally occurring hydrocarbons are discussed.

1.3.4.3.1 Vascular Plant and Algal Debris. The most common naturally occurring organic background material encountered in soils and sediments is due to the presence of vascular (land) plant debris, which is pervasive in most soils and coastal sediments. The extractable component within soils and sediments containing plant debris can be significant, particularly in moist, highly vegetated environments where peat or other organic-rich soil accumulate (or had in the past). For example, Dworian (1997) reported that the concentration of total petroleum hydrocarbons (TPH) as diesel range organics (DRO) in some Alaskan soils was as high as 3600 mg/kg. In such environments, modern plant debris can significantly increase the measured concentrations of total petroleum hydrocarbons (TPH).

The types of compounds extracted from vascular plant debris include various polar compounds (e.g., organic acids and alcohols, sulfur compounds, phenols, tannins) and their diagenetic products. Fortunately, if not removed during cleanup of the extract (e.g., EPA Methods 3610 or 3630) these materials produce "unusual" chemical fingerprints (via gas chromatographic methods such as EPA Method 8015) that can be distinguished from crude oil or petroleum products. However, in some studies these unusual fingerprints are not inspected, and the inflated TPH concentrations reported by the laboratory are accepted blindly. This, of course, can greatly increase what is accepted to be the authentic TPH concentration and aerial extent of any spilled oil.

Even after cleanup of soil/sediment extracts, any plant-derived hydrocarbons (e.g., waxes, terpenes, fats, oils) will persist and potentially contribute to the chemical fingerprint of any spilled oil. Their presence will increase measured TPH concentrations and have a dramatic influence on the appearance and, thus, (mis)interpretation of a potentially impacted sample's chromatographic fingerprint. This circumstance was common in a recent investigation of a crude oil spill following Hurricane

Katrina in Chalmette, Louisiana (Stout et al., 2006b). The massive floodwaters associated with Katrina carried and deposited large masses of peaty debris into the area that, even after alumina cleanup of soil extracts, yielded measurable TPH in surface soils. This TPH could have been mistaken for crude oil spilled during Katrina if not for the inspection of the associated chromatograms. Figure 1-16 shows the GC/FID chromatogram for an alumina-cleaned extract from a peaty soil from a coastal Louisiana bayou. The peat is dominated by a series of odd numbered n-alkanes ranging from $n\text{-}C_{25}$ to $n\text{-}C_{33}$. This odd-over-even pattern is typical of n-alkanes derived from leaf (epicuticular) waxes (Bray and Evans, 1961). GC/MS-SIM analysis of the peat also revealed the presence of anteiso-alkanes in the same carbon ranges. Also present are numerous late-eluting peaks consistent with various plant- and microbial-derived terpenes, including a prominent C_{31} hopene (Figure 1-16). Pristane is absent in this peat, although this isoprenoid is a significant component of some plankton (Blumer et al., 1963), and consequently its presence in coastal sediments is not necessarily due to petroleum. A small UCM hump in the C_{29}–C_{35} range is also evident in this sample. These chromatographic features cannot reasonably be confused with crude oil, even highly weathered crude oil, and thereby demonstrate the importance of fingerprinting in distinguishing low levels of spilled crude oil from allochthonous peat following Hurricane Katrina.

1.3.4.3.2 Particulate Coal and Wood Charcoal.

In many coastal environments, coal is eroded from sedimentary rock outcrops, and minute coal particles are transported and deposited in coastal sediments by rivers. In such areas, coal would be considered a naturally occurring source of background hydrocarbons. Alternatively, many ports formerly hosted industries that utilized coal delivered by vessel or barge. Over the decades of operation particulate coal and coal dust can become widely distributed in sediments. In such areas, coal would be considered as an anthropogenic background material. In either case, particulate coal can occur as a background contaminant that must be distinguished from authentic contamination arising from spilled oil.

The contribution of background hydrocarbons that particulate coal may impart is largely a function of the precursor plant material (e.g., Mesozoic gynmosperm versus Tertiary angiosperm dominant) and the coal's rank (Chaffee et al., 1986). Lower-rank coals (lignite and sub-bituminous) typically yield higher concentrations of nonhydrocarbons during solvent extraction while bituminous coals yield more hydrocarbons (Radke et al., 1980; White and Lee, 1980) with the highest yields resulting from high volatile bituminous coals (Stout et al., 2002b).

The occurrence of coal particles in sediment could confound interpretations surrounding TPH, PAH, and biomarkers in the sample. Figure 1-17 shows the distributions of PAH in three coals of varying rank (obtained from

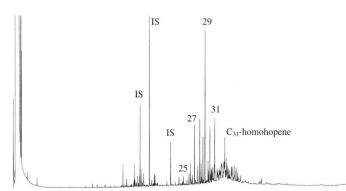

Figure 1-16 GC/FID chromatogram of the TPH extracted from a Louisiana peat containing 2000 mg/kg (dry).

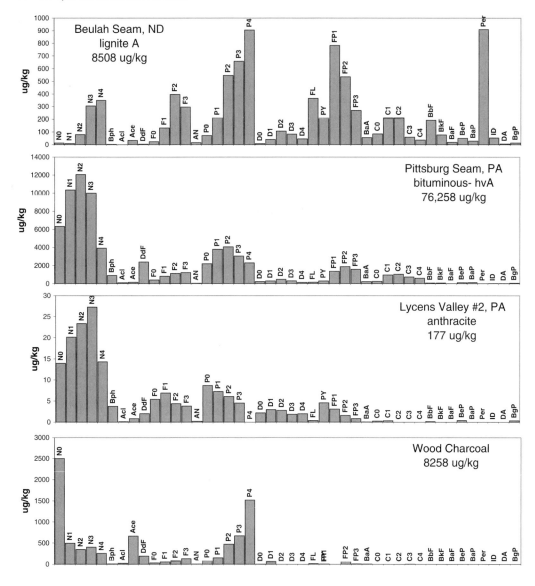

Figure 1-17 Histograms showing the distribution of PAH in three U.S. coals of varying rank and in modern wood-derived charcoal. For PAH compound abbreviations, see Table 1-2.

different U.S. coal fields). The PAH concentrations and distributions vary widely among the coals. The distribution of PAH in the lower-rank coals is dominated by 3- and 4-ring PAH (phenanthrenes P0–P4 and fluoranthene/pyrenes FP1–FP3). Perylene, a commonly recognized naturally occurring PAH (e.g., Silliman et al., 2001), is also abundant. The highly volatile bituminous A coal and the anthracite are both dominated by 2-ring PAH (naphthalenes N0–N4). However, the anthracite, a much higher-rank coal, contains these at far lower concentrations than the bituminous coal. Obviously, the presence of dispersed coal particles in sediments and soils will increase the concentrations of PAH (and TPH), which could be mistakenly identified as spilled oil if the chemical fingerprints were not considered.

In some instances, organic petrographic inspection of the sediment may be necessary to recognize the presence of coal. In fact, the presence of sedimentary coal in Prince William Sound sediments has been argued to have confounded the allocation of the different PAH sources in Prince William Sound following the *Exxon Valdez* oil spill (Short et al., 1999; Boehm et al., 2000a, 2000b; see also Chapter 15 herein).

Finally, debris from forest fires can also occur naturally in many sediments. This debris contains various PAH derived from the combustion/pyrolysis of vegetation (e.g., Laflamme and Hites, 1979; Simoneit and Elias, 2000). This material can also contribute hydrocarbons that, if unrecognized, can affect interpretations and decisions surrounding the potential impact of spilled oil. Figure 1-17 also shows the distribution and concentration of PAH in wood charcoal collected from a recently burned area. The charcoal contained 8258 µg/kg of PAH, which were dominantly comprised of 2- and 3-ring PAH. These occurred in a very different distribution than was observed in the coal samples.

1.3.4.3.3 Natural Oil Seeps. In some oil spill investigations, the occurrence and influence of natural oil seeps must be considered. The reason for this is obvious — *is there any "real" contamination or only background contamination*? Natural oil seeps are a worldwide phenomenon that contributes more petroleum to the environment than all other sources combined (Table 1-1). In coastal southern California the phenomenon is particularly prolific (e.g., Mikolaj et al., 1972; Straughan, 1979) and must always be considered when investigating the origin of forms of coastal contamination, e.g., tar balls (Wang et al., 1993). Chemical fingerprinting studies have shown that some of the hydrocarbons in Prince William Sound sediments also are attributable to long-persistent oil seeps that exist east of Prince William Sound (see Chapter 15 herein). The character of this seeped oil was established through the analysis of radiogenically age-dated sediment cores, which revealed the distinctive fingerprint of the naturally seeped oil in sediments deposited centuries prior to the *Exxon Valdez* oil spill (Burns et al., 1997; Boehm et al., 1997, 2000a).

Of course, in some instances when the suspected spill oil is from the same geologic province as the naturally occurring seeped oil (and therefore shares identical or nearly identical primary chemical features (Section 1.3.1), the ability of chemical fingerprinting to distinguish any spilled oil from any seeped oil is vastly reduced.

There is some expectation that seeped crude oil may exhibit features of a more highly weathered equivalent of locally produced (and typically obtained from deeper production zones) oil. The basis for this is that natural oil seeps often occur when shallower and cooler reservoirs, i.e., those most prone to in-reservoir weathering, are connected to the surface via faults. Alternatively, it is possible that produced (and possibly spilled) oils are derived primarily from deeper reservoir zones where in-reservoir weathering is reduced relative to the natural seeps. In such cases where the naturally seeped oils are already weathered, it may be possible to distinguish spilled crude oil from seeped crude oil — at least in the earlier stages of weathering of the spilled oil. When genetically related seeped and spilled oils from the same petroleum system co-occur in the environment, chemical fingerprinting may be unable to distinguish these two sources unless subtle though measurable variations are present (e.g., as in Figure 1-5). In such instances other forensic tools such as spill trajectory modeling (Chapter 13) or remote sensing (Chapter 14) may prove useful.

1.3.4.4 Anthropogenic Background Hydrocarbons

In port sediments and near urban areas, the anthropogenic background contains PAH and other hydrocarbons that are collectively called "urban background." Urban background consists of hydrocarbons derived from a variety of nonpoint sources such as (1) stormwater runoff, (2) direct deposition (atmospheric fallout) of combustion particles (soot) from vehicle exhaust and factories, (3) surface runoff from proximal roadways, parking lots, and bridges, or (4) discharges from recreational, commercial, and military boat/ship traffic. In many environments, stormwater and river runoff are probably the largest chronic contributors of anthropogenic background PAH to urban and coastal sediments (Eganhouse et al., 1982; Table 1-1). These PAH source materials, which often have been entering a waterway for decades, can impart obvious and recognizable profiles of anthropogenic background PAH to the sediments (e.g., Daskalakis and O'Connor, 1995).

1.3.4.4.1 Urban and River Runoff.

Runoff is the most important source of anthropogenic background hydrocarbons to urban and coastal sediments. It consists of a mixture of organic materials — dust, dirt, particulate matter, soot, solid wastes — that are transported to urban waterways, ports, rivers, or coastal waters and sediments via nonpoint (general runoff) and point (end-of-pipe) sources, often during storm events. In many oil spill investigations involving urban or coastal sediments, the presence of anthropogenic background hydrocarbons derived from runoff must be established in order to assess the potential impact from any authentic contamination derived from spilled or discharged oil.

The presence and composition of the hydrocarbons within runoff have been long recognized (Wade and Quinn, 1979; Eganhouse et al., 1982; Hoffman and Quinn, 1987). Lubricating and other heavy oils contained within urban runoff can be an important source of anthropogenic background hydrocarbons, which must be distinguished from any authentic petroleum contamination. The presence of these (mostly) residual range background hydrocarbons can be generally recognized in GC/FID fingerprints by the presence of an unresolved complex mixture (UCM) that appears as a broad "hump" in the baseline of the fingerprint. The prominence of the UCM is typical of biodegraded petroleum, uncombusted petroleum, lubricating oils, and asphalts, all of which may be components of urban runoff (Gogou et al., 2000). The specific shape of the UCM "hump" can vary in different study areas, but these can generally be distinguished from specific petroleum types. Figure 1-18 shows an example of the UCM hump for an urban runoff-impacted sediment from the Thea Foss Waterway, Tacoma, Washington (Stout et al., 2003). The fingerprints for a motor oil and a hydraulic oil are shown for comparison. Each of these is dominated by a UCM hump, and it is easy to envi-

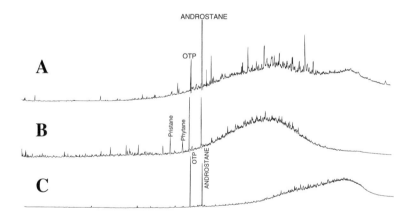

Figure 1-18 GC/FID fingerprints showing the chromatographic character of the UCM "humps" in (A) urban sediment impacted by urban runoff, (B) motor oil, and (C) hydraulic oil. OTP (*o*-terphenyl) and 5α-androstane are internal standards.

sion how these heavy petroleum products, dripped from vehicles and washed from urban road surfaces during storm events, could eventually enter the sediment and produce a broad UCM hump (as appears in Figure 1-17A).

PAHs are also well-known components of urban runoff (Stout et al., 2004). The sources of PAH in urban runoff vary, but the most common sources are (1) soot particles containing combustion-related PAH (principally arising from internal combustion engines, especially diesel based (e.g., Harrison et al., 1996; Marr et al., 1999), (2) street runoff containing traces of used lubricating/hydraulic oils, and (3) illegal or unintentional discharging of waste oil and petroleum products into storm drain systems. The PAHs associated with urban runoff are complex mixtures that tend to be dominated by higher-molecular-weight 4- to 6-ring PAHs (Stout et al., 2004). The prominence of 4- to 6-ring, nonalkylated PAH is indicative of a combustion (pyrolysis) product (Laflamme and Hites, 1978), as is typical in motor exhaust (Westerholm et al., 1988) or wood smoke (Oahn et al., 1999; Simoneit and Elias, 2000). Thus, the PAHs associated with combustion-derived soot particles are likely to contribute the majority of background anthropogenic PAH to urban and coastal sediments.

Figure 1-19 shows the distributions and concentrations of PAH in an urban runoff-influenced sediment from a tributary to the southern branch of the Elizabeth River in Norfolk, Virginia. For comparison, the PAHs in selected components that may be contained within urban runoff are also shown. These include the PAH found in urban dust (SRM 1649a), diesel exhaust soot (SRM 1650), used motor oil (Restek, Inc. Standard), used hydraulic oil, and road asphalt. A strong pyrogenic signature of the PAH is evident in the urban dust and tends to most closely resemble that found in the Elizabeth River sediment. The diesel soot, used motor oil, used hydraulic oil, and road asphalt all exhibit strong petrogenic PAH signatures, although each of these exhibits a distinct distribution. The greater resemblance between the distribution of the PAH in the sediment and in urban dust suggests that the latter (i.e., its equivalent in the Elizabeth River area) is the predominant source of background PAH to the area's sediments. This appears to be the case in spite of the generally low concentration of PAH in urban dust compared to the other contributing sources.

1.4 Summary

Accidental *oil spills* or intentional *operational discharges* of crude oil and petroleum products and wastes are frequent occurrences. Determining the liable source and environmental impact, both spatially and temporally, of these releases relies largely upon chemical fingerprinting. Chemical fingerprinting of petroleum in the environment can be defined as the generation and comparison of diagnostic chemical features among oil samples (i.e., both spill and source oils) and potentially impacted samples (e.g., shorelines, sediments, or biological tissues). The analytical methods available for chemical fingerprinting have evolved over time, with advances in analytical methods generally providing an increasing degree of molecular specificity. Most chemical fingerprinting of oils relies upon gas chromatography (GC) or gas chromatography-mass spectrometry (GC/MS) that can target specific molecular features within petroleum. Because of the chemical complexity of petroleum, which contains hundreds to many thousands of individual compounds, this specificity attainable by GC and GC/MS analysis provides the ability to determine the source and impact of petroleum in the environment. These analyses most often target compounds affording a high degree of specificity between oils, namely, biomarkers and PAHs.

Any determination of the source or impact necessarily relies upon a comparison of the chemical fingerprints of the spilled or discharged petroleum versus those of the candidate source(s) or the potentially impacted samples. Such comparisons, be they qualitative or quantitative, need to consider and account for the factors that collectively affect the chemical fingerprints of petroleum spilled or discharged in the environment. These

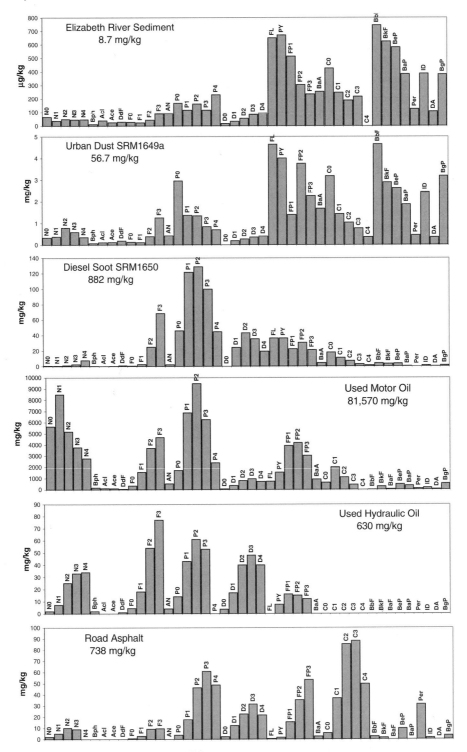

Figure 1-19 Histograms showing the distributions and concentrations of PAH in an urban sediment from the Elizabeth River (Virginia) containing anthropogenic background PAH and in various urban runoff candidate source components. For compound abbreviations, see Table 1-2. Concentrations refer to mg of PAH per kg of sediment (dry), urban dust (dry), soot (dry), oil, and asphalt (dry).

factors can be classified as (1) primary, (2) secondary, and (3) tertiary.

Primary factors include those processes at work during the original generation and accumulation of crude oil over geologic time, and include the (1) character of the ancient organic matter in the source rock strata, (2) the thermal history of the source rock strata, and (3) the processes active during migration and within the crude oil reservoir. Because of these primary factors, crude oils from different geologic provinces and even from different reservoirs in a single oil field can exhibit distinguishable chemical fingerprints, thereby allowing their distinction in oil spill investigations involving crude oil(s).

Secondary factors include those processes used in the refining of crude oil to produce various petroleum products, mostly fuels and lubricants. The modern refining process relies upon various fractionation, conversion, and finishing steps that collectively will determine the chemical fingerprint of the petroleum products produced. Some primary features of the parent crude oil feedstock will be inherited by the daughter petroleum products and other features will be modified by the refining process. The combined effects of primary and secondary processes (i.e., different crude oil feedstocks and different refining processes) can impart different chemical fingerprints for the same petroleum product types, thereby allowing their distinction in oil spill investigations involving a petroleum product(s).

Tertiary factors include the processes that affect the chemical fingerprint of petroleum after it is released into the environment. The two most important tertiary factors are (1) weathering and (2) mixing. It is well established that the processes of weathering will alter the chemical fingerprint of petroleum. For petroleum spilled or discharged to water, the weathering processes of evaporation and biodegradation can have a significant effect on the petroleum's chemical fingerprint, though the effects of other processes on both the chemical fingerprint and the rates of evaporation and biodegradation — dissolution, photooxidation, mousse formation, and dewaxing/wax-enrichment — must also be considered. Another tertiary factor is the effect of mixing of a spilled or discharged oil with any pre-existing, "background" petroleum or other hydrocarbon sources in the environment. "Background" hydrocarbons may be naturally occurring (e.g., modern organic matter, coal or ash, or seeped oil) or anthropogenic (e.g., runoff or atmospheric particulates). The effect of mixing can alter the chemical fingerprint of a petroleum spilled or discharged into the environment, emphasizing the importance of recognizing and characterizing the contribution of "background" when attempting to determine the source or potential impact of spilled petroleum.

Collectively, the effects of these primary, secondary, and tertiary factors must be considered and accounted for in any comparison of chemical fingerprints before any determination as to the source or impact of spilled or discharged petroleum can be established.

References

Albaiges, J. and P. Albrecht. Fingerprinting marine pollutant hydrocarbons by computerized gas chromatography-mass spectrometry. *Int'l. J. Environ. Anal. Chem.*; 1979, **6**, 171–190.

Alexander, R., R.I. Kagi, R.A. Noble, and J.K. Volkman. Identification of some bicyclic alkanes in petroleum. *Org. Geochem.*; 1984, **6**, 63–70.

American Society for Testing and Materials (ASTM). Standard Test Methods for Comparison of Waterborne Petroleum Oils by Gas Chromatography. *ASTM 3328D-00*; 2000a; ASTM International, W. Conshohocken, PA.

American Society for Testing and Materials (ASTM). Oil spill identification by gas chromatography and positive ion electron impact low resolution mass spectrometry. *ASTM D5739–00*; 2000b; ASTM International, W. Conshohocken, PA.

Atlas, R.M. Petroleum biodegradation and oil spill bioremediation. *Mar. Pollut. Bull.*; 1995, **31**, 178–182.

Atlas, R.M., P.D. Boehm, and J.A. Calder. Chemical and biological weathering of oil from the *Amoco Cadiz* spillage within the littoral zone. *Estuar. Coast. and Shelf Sci.*; 1981, **12**, 589–608.

Beall, P.W., S.A. Stout, G.S. Douglas, and A.D. Uhler. On the role of process forensics in the characterization of fugitive gasoline. *Env. Claims J.*; 2002, **14**, 487–506.

Bence, A.E., K.A. Kvenvolden, and M.C. Kennicutt II. Organic geochemistry applied to environmental assessments of Prince William Sound, Alaska, after the *Exxon Valdez* oil spill — a review. *Org. Geochem.*; 1996, **24**, 7–42.

Bentz, A.P. Oil spill identification. *Anal. Chem.*; 1976, **48**, 454A–472A.

Bentz, A.P. Who spilled the oil? *Anal. Chem.*; 1978, **48**, 655A–658A.

Blumer, M., M.M. Mullin, and D.W. Thomas. Pristane in zooplankton. *Science*; 1963, **140**, 974.

Bobra, A.M., D. Mackay, and W.Y. Shiu. Distribution of hydrocarbons among oil, water and vapor phases during oil dispersant toxicity tests. *Bull. Environ. Contam. Toxicol.*; 1979, **23**, 558–565.

Boehm, P.D. and J.W. Farrington. Aspects of the polycyclic aromatic hydrocarbon geochemistry of recent sediments in the Georges Bank region. *Environ. Sci. Technol.*; 1984, **18**, 840–845.

Boehm, P.D., G.S. Douglas, J.S. Brown, D.S. Page, A.E. Bence, W.A. Burns, and P.J. Mankiewicz. Comment on "Natural hydrocarbon background in benthic sediments of Prince William Sound, Alaska: Oil vs. coal." *Environ. Sci. Technol.*; 2000b, **34**, 2064–2065.

Boehm, P.D., G.S. Douglas, W.A. Burns, P.J. Mankiewicz, D.S. Page, and A.E. Bence. Application of petroleum hydrocarbon chemical fingerprinting and allocation techniques after the *Exxon Valdez* oil spill. *Mar. Pollut. Bull.*; 1997, **34**, 599–613.

Boehm, P.D. and D.L. Fiest. Subsurface distributions of petroleum from an offshore well blowout. The Ixtoc I Blowout, Bay of Campeche. *Environ. Sci. Technol.*; 1982, **16**, 67–74.

Boehm, P.D., D.S. Page, W.A. Burns, A.E. Bence, P.J. Mankiewicz, and J.S. Brown. Resolving the origin of the petrogenic hydrocarbon background in Prince William Sound, Alaska. *Environ. Sci. Technol.*; 2000a, **35**, 471–479.

Bray, E.E. and E.D. Evans. Distribution of n-paraffins as a clue to recognition of source beds. *Geochim. Cosmochim. Acta*; 1961, **22**, 2–15.

Burns, W.A., P.J. Mankiewicz, A.E. Bence, D.S. Page, and K.R. Parker. A principal component and least squares method for allocating polycyclic aromatic hydrocarbons in sediment to multiple sources. *Environ. Tox. Chem.*; 1997, **16**, 1119–1131.

Cassani, F. and G. Eglinton. Organic geochemistry of Venezuelan extra-heavy crude oils. 2. Molecular assessment of biodegradation. *Chem. Geol.*; 1991, **91**, 315–333.

Chaffee, A.L., D.S. Hoover, R.B. Johns, and F.K. Schweighardt. Biological markers extractable from coal. In: *Biological Markers in the Sedimentary Record*; Elsevier, Amsterdam, The Netherlands; 1986, 311–345.

Colombo, J.C., E. Pelletier, C. Brochu, and M. Khalil. Determination of hydrocarbon sources using n-alkane and polyaromatic hydrocarbon distribution indexes: Case Study — Rio dc La Plata Estuary, Argentina. *Environ. Sci. Technol.*; 1989, **23**, 888–894.

Connan J. Biodegradation of crude oils in reservoirs. *Adv. Petrol. Geochem.*; 1984, **1**, 299–333.

Curiale, J.A. A review of the occurrences and causes of migration-contamination in crude oil. *Org. Geochem.*; 2002, **33**, 1389–1400.

Daling, P.S. and P.J. Brandvik. A study of the formation and stability of water-in-oil emulsions. In: *Proc. 11th AMOP Technical Seminar, Vancouver, BC, Canada*. Environment Canada, Ottawa, Canada; 1988, 488–499.

Daling, P.S., L.-G. Faksness, A.B. Hansen, and S.A. Stout. Improved and standardized methodology for oil spill fingerprinting. *Environ. Forensics*; 2002, **3**, 263–278.

Daskalakis, K.D. and T.P. O'Connor. Distribution of chemical concentrations in US coastal and estuarine sediment. *Mar. Environ. Res.*; 1995, **40**, 381–398.

Didyk, B.M., B.R.T. Simoneit, S.C. Brassell, and G. Eglinton. Organic geochemical indicators of palaeoenvironmental conditions of sedimentation. *Nature*; 1978, **272**, 216–222.

Douglas, G.S., W.A. Burns, A.E. Bence, D.S. Page, and P.D. Boehm. Optimizing detection limits for the analysis of petroleum hydrocarbons in complex environmental samples. *Environ. Sci. Technol.*; 2004, **38**, 3958–3964.

Douglas, G.S. and A.D. Uhler. Optimizing EPA methods for petroleum-contaminated site assessments. *Environ. Testing and Analysis*; 1993, **5**, 46–53.

Douglas, G.S., A.E. Bence, R.C. Prince, S.J. McMillen, and E.L. Butler. Environmental stability of selected petroleum hydrocarbon source and weathering ratios. *Environ. Sci. Technol.*; 1996, **30**, 2332–2339.

Duewer, D.L., B.R. Kowalski, and T.F. Schatzki. Source identification of oil spills by pattern recognition analysis of natural element composition. *Anal. Chem.*; 1975, **47**, 1573–1783.

Durell, G.S., S.A. Ostazeski, A.D. Uhler, and A.B. Nordvik. Transfer of crude oil weathering technology. Technical Report No. 93–027; Marine Spill Response Corporation, Washington, D.C.; 1993.

Dutta, T.K. and S. Harayama. Fate of crude oil by the combination of photooxidation and biodegradation. *Environ. Sci. Technol.*; 2000, **34**, 1500–1505.

Dworian, P. Interferences by natural organics in diesel analyses. *J. Cold Regions Eng.*; 1997, **1**(2), 71–81.

Eganhouse, R.P., D.L. Blumfield, and I.R. Kaplan. Petroleum hydrocarbons in stormwater runoff and municipal wastes: Input to coastal waters and fate in marine sediments. *Thalassia Jugoslavica.*; 1982, **18**, 411–431.

Eglinton, G. and M. Calvin. Chemical fossils. *Sci. Am.*; 1967, **261**, 32–43.

Ehrhardt, M. and M. Blumer. The source identification of marine hydrocarbons by gas chromatography. *Environ. Pollut.*; 1972, **3**, 179–194.

Ehrhardt, M. and A. Douabul. Dissolved petroleum residues and alkylbenzene photo-oxidation products in the upper Arabian Gulf. *Mar. Chem.*; 1989, **26**, 363–370.

Ehrhardt, M.G. and K.A. Burns. Petroleum-derived dissolved organic compounds concentrated from inshore waters in Bermuda. *J. Exper. Mar. Biol. Ecol.*; 1990, **138**, 35–47.

Ehrhardt, M.G., K.A. Burns, and M.C. Bicego. Sunlight-induced compositional alterations in the seawater-soluble fraction of a crude oil. *Mar. Chem.*; 1992, **37**, 53–64.

Ekweozor, C.M. and O.T. Udo. The oleananes: Origin, maturation and limits of occurrence in Southern Nigeria sedimentary basins. *Org. Geochem.*; 1988, **13**, 131–140.

Emsbo-Mattingly, S.D. Hydrocarbon composition of vessel discharges: Potential sources of petroleum in coastal environments. *Remed. Profess.*; 2005, Dec.

Farrimond, P., A. Taylor, and N. Telnæs. Biomarker maturity parameters: The role of generation and thermal degradation. *Org. Geochem.*; 1998, **29**, 1181–1197.

Fedorak, P.M. and D.W.S. Westlake. Degradation of sulfur heterocycles in Prudhoe Bay crude oil by soil enrichments. *Water, Air, Soil Pollut.*; 1984, **21**, 225–230.

Fingas, M. A literature review of the physics and predictive modeling of oil spill evaporation. *J. Hazard. Mat.*; 1995, **42**, 157–175.

Fingas, M. and B. Fieldhouse. Studies on the formation process of water-in-oil emulsions. *Mar. Pollut. Bull.*; 2003, **47**, 369–396.

Gary, J.H. and G.E. Handwerk. *Petroleum Refining*; Marcel Dekker, New York, NY; 1984.

Gogou, A., I. Bouloubassi, and E.G. Stephanou. Marine organic geochemistry of the Eastern Mediterranean: 1. Aliphatic and polyaromatic hydrocarbons in Cretan Sea surficial sediments. *Mar. Chem.*; 2000, **68**, 265–282.

Gough, M.A., M.M. Rhead, and S.J. Rowland. Biodegradation studies of unresolved complex mixtures of hydrocarbons: Model UCM hydrocarbons and the aliphatic UCM. *Org. Geochem.*; 1992, **18**, 17–22.

Grahl-Nielsen, O., J.T. Staveland, and S. Wilhelmsen. Aromatic hydrocarbons in benthic organisms from coastal areas polluted by Iranian crude oil. *J. Fisheries Res. Board of Canada*; 1978, **35**, 615–623.

Hamilton, B. and R.J. Falkiner. Motor Gasoline. In: G.E. Totten (ed.), *Fuels and Lubricants Handbook: Technology, Properties, Performance, and Testing*; ASTM Manual Series, MNL37WCD; 2003, ASTM International, W. Conshohocken, PA; 61–88.

Harrison, R.M., D.J.T. Smith, and L. Luhana. Source apportionment of atmospheric polycyclic aromatic hydrocarbons collected from an urban location in Birmingham, U.K. *Environ. Sci. Technol.*; 1996, **30**, 825–832.

Harrison, W., M.A. Winnik, P.T.Y. Kwong, and D. Mackay. Disappearance of aromatic and aliphatic components from small sea-surface slicks. *Environ. Sci. Technol.*; 1975, **9**, 231–234.

Hoffman, E.J. and J.G. Quinn. Chronic hydrocarbon discharges into aquatic environments: II — Urban runoff and combined sewer overflows. In: J.H. Vandermeulen and S.E. Hurley (eds.), *Oil in Freshwater: Chemistry, Biology, Countermeasure Technology*; Pergamon Press, New York; 1987, 114–137.

Hughes, W.B. and A.G. Holba. Relationship between crude oil quality and biomarker patterns. *Org. Geochem.*; 1987, **13**, 15–30.

Hunt, J.M. *Petroleum Geochemistry and Geology*, 2nd ed. Freeman Press, USA; 1996.

Irvine, G.V., D.H. Mann, and J.W. Short. Multi-year persistance of oil mousse on high energy beaches distant from *Exxon Valdez* oil source. *Mar. Pollut. Bull.*; 1999, **38**, 572–584.

Jewitt, C.H., S.R. Westbrook, D.L. Ripley, and R.H. Thornton. Fuels for land and marine diesel engines and for nonaviation gas turbines. In: G.V.

Dyroff (ed.), *ASTM Manual on Significance of Tests for Petroleum Products, 6th ed.* Am. Soc. Testing Materials; Phil., PA; 1993.

Jordan, R.E. and J.R. Payne. *Fate and Weathering of Petroleum Spills in the Marine Environment*; Ann Arbor Science Press, Ann Arbor, Michigan; 1980.

Kaplan, I.R. and Y. Galperin. Application of alkylcyclohexane distribution patterns for hydrocarbon fuel identification in environmental samples. In: P.T. Kostecki, E.J. Calabrese, and M. Bonazountas (eds.), *Contaminated Soils*; Amherst Scientific Publishers, Amherst, MA; 1997, **2**, 65–78.

Kaplan, I.R., Y. Galperin, S.-T. Lu, and R.-P. Lee. Forensic environmental geochemistry: Differentiation of fuel-types, their sources and release time. *Org. Geochem.*; 1997, **27**, 289–317.

Kaplan, I.R., S.-T. Lu, H.M. Alimi, and J. MacMurphey. Fingerprinting of high boiling hydrocarbon fuels, asphalts and lubricants. *Environ. Forensics*; 2001, **2**, 231–248.

Kauffman, R.L., A.S. Ahmed, and R.J. Elsinger. Gas chromatography as a development and production tool for fingerprinting oils from individual reservoirs: Applications in the Gulf of Mexico. In: D. Schumaker and B.F. Perkins (eds.), *Proc. 9th Annual Research Conf. of the Soc. Econ. Paleont. Mineral*; New Orleans; 1990, 263–282.

Kinghorn, R.R.F. *An Introduction to the Physics and Chemistry of Petroleum.* J. Wiley & Sons, New York, NY; 1983.

Kreider, R.E. Identification of oil leaks and spills. In: *Proc. 1971 Oil Spill Conf.* API-USEPA-US Coast Guard, Washington, D.C.; 1971, 119–124.

Kvenvolden, K.A., F.D. Hostettler, P.R. Carlson, J.B. Rapp, C.N. Threlkeld, and A. Warden. Ubiquitous tar balls with a California-source signature on the shorelines of Prince William Sound, Alaska. *Environ. Sci. Technol.*; 1995, **29**, 2684–2694.

Laflamme, R.E. and R.A. Hites. The global distribution of polycyclic aromatic hydrocarbons in recent sediments. *Geochim. Cosmochim. Acta*; 1978, **42**, 289–303.

Laflamme, R.E. and R.A. Hites. Tetra- and pentacyclic, naturally occurring, aromatic hydrocarbons in recent sediments. *Geochim. Cosmochim. Acta*; 1979, **42**, 1687–1691.

Larson, R.A., L.L. Hunt, and D.W. Blankenship. Formation of toxic products from a #2 fuel oil by photooxidation. *Environ. Sci. Technol.*; 1977, **11**, 492–496.

Larter, S.R., B.F.J. Bowler, M. Li, M. Chen, D. Brincat, B. Bennett, K. Noke, P. Donohoe, D. Simmons, M. Kohnen, J. Allan, N. Telaes, and I. Horstad. Molecular indicators of secondary oil migration distances. *Nature*; 1996, **383**, 593–597.

Lee, M.L., D.L. Vassilaros, C.M. White, and M. Novotny. Retention indices for programmed-temperature capillary-column gas chromatography of polycyclic aromatic hydrocarbons. *Anal. Chem.*; 1979, **51**, 768–773.

Leffler, W.L. *Petroleum Refining*, 3rd ed. PennWell Corp., Tulsa, OK; 2000.

Mackay, D. and C.D. McAuliffe. Fate of hydrocarbons discharged at sea. *Oil Chem. Pollut.*; 1988, **5**, 1–20.

Mackenzie, A.S., R.L. Patience, and J.R. Maxwell. Molecular parameters of maturation in the Toarcian shales, Paris Basin, France — I. Changes in the configurations of acyclic isoprenoid alkanes, steranes and triterpanes. *Geochim. Cosmochim. Acta*; 1980, **44**, 1709–1721.

Mackenzie, A.S., J. Rullkotter, D.H. Welte, and P. Mankiewicz. Reconstruction of oil formation and accumulation in North Slope, Alaska, using quantitative gas chromatography-mass spectrometry. In: L.B. Magoon and G. Claypool (eds.), *Alaska North Slope Oil/Rock Correlation Study, AAPG Studies in Geology #20*; AAPG, Tulsa, OK; 1985, 319–377.

Madsen, E.L. Determining in situ biodegradation. Facts and challenges. *Environ. Sci. Technol.*; 1991, **25**, 1663–1673.

Marr, L.C., T.W. Kirchstetter, R.A. Harley, A.H. Miguel, S.V. Hering, and S.K. Hammond. Characterization of PAH in motor vehicle fuels and exhaust emissions. *Environ. Sci. Technol.*; 1999, **33**, 3091–3099.

Matyasik, I., A. Steczko, and R.P. Philp. Biodegradation and migrational fractionation of oils from the Eastern Carpathians, Poland. *Org. Geochem.*; 2000, **31**, 1509–1523.

McAuliffe, C.D. Evaporation and solution of C2 to C10 hydrocarbons from crude oils on the sea surface. In: D.Wolfe (ed.), *Fate and Effects of Petroleum Hydrocarbons in Marine Organisms and Ecosystems*; Pergamon Press, NY; 1977, 19–35.

McCarthy, K.J. and R.M. Uhler. Optimizing gas chromatography conditions for improved hydrocarbon analysis; *Proc. 9th Annual Conf. on Contaminated Soils*; Amherst Scientific Publishers, Amherst, MA; 1994.

Michel, J., C.B. Sholz, and B.L. Benggio. Group V fuel oils: Source, behavior, and response issues. In: *Proc. 1995 Oil Spill Conf.* U.S. Coast Guard, American Petroleum Institute, U.S. Environmental Protection Agency, Washington, D.C.; 1995, 559–564.

Michel, J., M.O. Hayes, R.S. Keenan, T.C. Sauer, J.R. Jensen, and S. Narumalani. Contamination of nearshore subtidal sediments of Saudi Arabia from the Gulf War oil spill. *Mar. Pollut. Bull.*; 1993, **27**, 109–116.

Mikolaj, P.G., A.A. Allen, and R.S. Schlueter. Investigation of the nature, extent and fate of natural oil seepage off Southern California. In: *Proc. 4th Annual Offshore Technology Conference*; Dallas, TX; Rpt. No. 1549; 1972.

Moldowan, J.M., W.K. Seifert, and E.J. Gallegos. Relationship between petroleum composition and depositional environment of petroleum source rocks. *Am. Assoc. Petrol. Geol. Bul.*; 1985, **69**, 1255–1268.

Mossner, S.G. and S.A. Wise. Determination of polycyclic aromatic sulfur heterocycles in fossil fuel-related samples. *Anal. Chem.*; 1999, **71**, 58–69.

National Research Council (NRC). *Oil in the Sea III: Inputs, Fates, and Effects*. The National Academy of Science, The National Academies Press, Washington, D.C.; 2003.

Neff, J.M. Composition and fate of petroleum and spill-treating agents in the marine environment. In: J. Geraci and D. St. Aubin (eds.), *Sea Mammals and Oil: Confronting the Risks*; Academic Press, New York; 1990, 1–33.

Nicodem, D.E., M.C.Z. Fernandes, C.L.B. Guedes, and R.J. Correa. Photochemical processes and the environmental impact of petroleum spills. *Biogeochem.*; 1997, **39**, 121–138.

Nordtest. Oil Spill at Sea: Identification, NT Chem 001; Nordtest, Esbo, Finland; 1983.

Nordtest. Oil Spill Identification, NT Chem 001, 2nd ed.; Nordtest, Esbo, Finland; 1991.

Oahn, N.T.K., L.B. Reutergardh, and N.T. Dung. Emission of PAH and particulate matter from domestic combustion of selected fuels. *Environ. Sci. Technol.*; 1999, **33**, 2703–2709.

Osadetz, K.G., L.R. Snowdon, and P.W. Brooks. Oil families in Canadian Williston Basin southwestern Sadkatchewan. *Bull. Can. Petrol. Geol.*; 1995, **42**, 155–177.

Oudot J. Biodegradability of the Erika fuel oil. *Comptes Rendus de l'Academie des Sciences, Serie III, Sci. de la Vie (Life Sciences)*; 2000, **323**, 945–950.

Oudot, J. Rates of microbial degradation of petroleum components as determined by computerized capillary gas chromatography and computerized mass spectrometry. *Mar. Environ. Res.*; 1984, **13**, 277–302.

Page, D.S., J.C. Foster, P.M. Fickett, and E.S. Gilfillan. Identification of petroleum sources in an area impacted by the Amoco *Cadiz* oil spill. *Mar. Pollut. Bull.*; 1988, **19**, 107–115.

Page, D.S., P.D. Boehm, G.S. Douglas, and A.E. Bence. Identification of hydrocarbon sources in the benthic sediments of Prince William Sound and the Gulf of Alaska following the *Exxon Valdez* oil spill. In: P.G. Wells, J.N. Butler, and J.S. Hughes (eds.), *Exxon Valdez Oil Spill: Fate and Effects in Alaskan Waters*; ASTM STP 1219; Am. Soc. Testing Materials, Phil., PA; 1995, 41–83.

Palmer, S.E. Effects of biodegradation and water washing on crude oil composition. In: M.H. Engel and S.A. Macko (eds.), *Advances in Organic Geochemistry — 1993*; Plenum Press, New York; 1993, 511–533.

Payne, J.R. and G.D. McNabb. Weathering of petroleum in the marine environment. *Mar. Technol. Soc.*; 1985, **18**, 1–19.

Payne, J.R. and C.R. Phillips. *Petroleum Spills in the Marine Environment: The Chemistry and Formation of Water-in-Oil Emulsions and Tar Balls*; Lewis Publ., Chelsea, MI; 1985.

Payne, J.R., C.R. Phillips, and W. Hom. Transport and transformations: Water column processes. In: D.F. Bosch and N.N. Rabalais (eds.), *Long Term Environmental Effects of Offshore Oil and Gas Development*; Elsevier Applied Science Publ., London; 1987, 175–231.

Peters, K.E., J.M. Moldowan, M. Schoell, and W.B. Hempkins. Petroleum isotopic and biomarker composition related to source rock organic matter and depositional environment. *Org. Geochem.*; 1986, **10**, 17–27.

Peters, K.E., G.L. Scheuerman, C.Y. Lee, J.M. Moldowan, R.N. Reynolds, and M.M. Pena. Effects of refinery processes on biological markers. *Energy & Fuels*; 1992, **6**, 560–577.

Peters, K.E., C.C. Walters, and J.M. Moldowan. *The Biomarker Guide, 2nd ed*. Cambridge University Press, Cambridge, UK; 2005.

Peters, K.E. and M.G. Fowler. Applications of petroleum geochemistry to exploration and reservoir management. *Org. Geochem.*; 2002, **33**, 5–36.

Petkewich, R. Spilled oil persists after 30 years. *Environ. Sci. Technol.*; 2002, 446.

Pieri, N., F. Jacquot, G. Mille, J.P. Planche, and J. Kister. GC-MS identification of biomarkers in road asphalts and in their parent crude oils. Relationships between crude oil maturity and asphalt reactivity towards weathering. *Org. Geochem.*; 1996, **25**, 51–68.

Prince, R.C., R.M. Garrett, R.E. Bare, M.J. Grossman, T. Townsend, J.M. Suflita, K. Lee, E.H. Owens, G.A. Sergy, J.F. Braddock, J.E. Lindstrom, and R.R. Lessard. The roles of photooxidation and biodegradation in long-term weathering of crude oil and heavy fuel oils. *Spill Sci. Technol. Bull.*; 2003, **8**, 145–156.

Prince, R.C., E.H. Owens, and G.A. Sergy. Weathering of an arctic oil spill after 20 years: The BIOS experiment revisited. *Mar. Pollut. Bull.*; 2002a, **44**, 1236–1242.

Prince, R.C., R.T. Stibrany, J. Hardenstine, G.S. Douglas, and E.H. Owens. Aqueous vapor extraction: A previously unrecognized weathering process affecting oil spills in vigorously aerated water. *Environ. Sci. Technol.*; 2002b, **36**, 2822–2825.

Radke, M. Application of aromatic compounds as maturity indicators in source rocks and crude oils. In: *Advances in Organic Geochemistry — 1986*; Butterworth & Co. Ltd., London; 1988.

Radke, M., R.G. Schaefer, and D. Leythaeuser. Composition of soluble organic matter in coals: Relation to rank and liptinite fluorescence. *Geochim. Cosmochim. Acta*; 1980, **44**, 1787–1800.

Radke, M., H. Willsch, and D. Leythaeuser. Aromatic components of coal: Relation of distribution pattern to rank. *Geochim. Cosmochim. Acta*; 1982, **46**, 1831–1848.

Rasmussen, W.V. Characterization of oil spills by capillary column gas chromatography. *Anal. Chem.*; 1975, **48**, 1562–1566.

Reddy, C.M., T.I. Eglinton, A. Hounshell, H.K. White, L. Xu, R.B. Gaines, and G.S. Frysinger. The West Falmouth oil spill after thirty years: The persistence of petroleum hydrocarbons in marsh sediments. *Environ. Sci. Technol.*; 2002, **36**, 4754–4760.

Ronningsen, H.P., J. Sjoblom, and L. Mingyuan. Water-in-oil emulsions from Norwegian continental shelf 11. Aging of crude oils and its influence on the emulsion stability. *Colloids Surfaces A: Physicochem. and Eng. Aspects*; 1995, **97**, 119–128.

Rubinstein, I. and P. Albrecht. The occurrence of nuclear methylated steranes in a shale. *J. Chem. Soc., Chem. Commun.*; 1975, 957–958.

Sauer, T.C. and P.D. Boehm. Guidance document on hydrocarbon chemistry analytical methods for oil spill assessments. Marine Spill Response Corporation, Tech. Rpt. 95-032, Washington, D.C.; 1995.

Sauer, T.C. and A.D. Uhler. Pollutant source identification and allocation: Advances in hydrocarbon fingerprinting. *Remediation*; 1994/1995; Winter Issue, 25–50.

Seifert, W.K., J.M. Moldowan, and R.W. Jones. Application of biological marker chemistry to petroleum exploration. In: *Proc. 10th World Petroleum Congress, Bucharest, Hungary*; Special Paper SP 8; Heyden & Son Limited, London; 1979, 425–440.

Seifert, W.K. and J.M. Moldowan. Applications of steranes, terpanes and monoaromatics to the maturation, migration and source of crude oils. *Geochim. Cosmochim. Acta*; 1978, **42**, 77–95.

Seifert, W.K., J.M. Moldowan, and G.J. DeMaison. Source correlation of biodegraded oils. *Org. Geochem.*; 1984, **6**, 633–643.

Short, J.W., K.A. Kvenvolden, P.R. Carlson, F.D. Hostettler, R.J. Rosenbauer, and B.A. Wright. Natural hydrocarbon background in benthic sediments of Prince William Sound, Alaska: Oil vs coal. *Environ. Sci. Technol.*; 1999, **33**, 34–42.

Short, J.W., M.R. Lindeberg, P.M. Harris, J.M. Maselko, J.J. Pella, and S.D. Rice. Estimate of oil persisting on the beaches of Prince William Sound 12 years after the *Exxon Valdez* oil spill. *Environ. Sci. Technol.*; 2004, **38**, 1–25.

Silliman, J.E., P.A. Meyers, B.J. Eadie, and J.V. Klump. A hypothesis for the origin of perylene based on its low abundance in sediments of Green Bay, Wisconsin. *Chem. Geo.*; 2001, **177**, 309–322.

Simoneit, B.R.T. Cyclic terpenoids of the geosphere. In: R.B. Johns (ed.), *Biological Markers in the Sedimentary Record*; Elsevier; Amsterdam, The Netherlands; 1986, 43–99.

Simoneit, B.R.T. and V.O. Elias. Organic tracers from biomass burning in atmospheric particulate matter over the ocean. *Mar. Chem.*; 2000, **69**, 301–312.

Song, C. *Introduction to Chemistry of Diesel Fuels*; Taylor and Francis, New York, NY; 2000.

Speight, J.G. *The Chemistry and Technology of Petroleum, 2nd ed.*, Marcel Dekker, Inc., New York, NY; 1991.

Staniloae, D., B. Petrescu, and C. Patroescu. Pattern recognition based software for oil spill identification by gas chromatography and IR spectroscopy. *Environ. Forensics*; 2001, **2**, 363–366.

Stiver, W. and D. Mackay. Evaporation rates of spills and hydrocarbons and petroleum mixtures. *Environ. Sci. Technol.*; 1984, **18**, 834–840.

Stout, S.A. Applications of petroleum fingerprinting in known and suspected pipeline releases — Two case studies. *Appl. Geochem.*; 2003, **18**, 915–926.

Stout, S.A. and G.S. Douglas. Diamondoid hydrocarbons application in the chemical fingerprinting of natural gas condensate and gasoline. *Environ. Forensics*; 2004, **5**, 225–235.

Stout, S.A., A.D. Uhler, and K.J. McCarthy. A strategy and methodology for defensibly correlating spilled oil to source candidates. *Environ. Forensics*; 2001, **2**, 87–98.

Stout, S.A., A.D. Uhler, and K.J. McCarthy. Middle distillate fuel fingerprinting using drimane-based bicyclic sesquiterpanes. *Environ. Forensics*; 2005, **6**, 241–251.

Stout, S.A., A.D. Uhler, and S.D. Emsbo-Mattingly. Characterization of PAH sources in sediments of the Thea Foss/Wheeler Osgood Waterways, Tacoma, Washington. *Soil & Sed. Contam.*; 2003, **12**, 815–834.

Stout, S.A., A.D. Uhler, and S.D. Emsbo-Mattingly. Comparative evaluation of background anthropogenic hydrocarbons in surficial sediments from nine urban waterways. *Environ. Sci. Technol.*; 2004a, **38**, 2987–2994.

Stout, S.A., A.D. Uhler, and K.J. McCarthy. Chemical characterization and sources of distillate fuels in the subsurface, Mandan, North Dakota. *Environ. Forensics*; 2006, **7**(3), 267–282.

Stout, S.A., A.D. Uhler, K.J. McCarthy, and S. Emsbo-Mattingly. Chemical fingerprinting of hydrocarbons. In: B.L. Murphy and R.D. Morrison (eds.), *Introduction to Environmental Forensics*; Academic Press, Boston, MA; 2002a, 137–260.

Stout, S.A., G. Millner, D. Hamlin, and B. Liu. The role of chemical fingerprinting in assessing the impact of a crude oil spill following Hurricane Katrina. *Env. Claims J.*; 2006b, **18**, 169–184.

Stout, S.A., G.S. Douglas, and A.D. Uhler. Automotive gasoline. In: R.D. Morrison and B.L. Murphy (eds.), *Environmental Forensics, Contaminant Specific Guide*; Academic Press, Boston; 2006a, 465–531.

Stout, S.A., S. Emsbo-Mattingly, A.D. Uhler, and K.J. McCarthy. Particulate coal in soils and sediments — Recognition and potential influences on hydrocarbon fingerprinting and concentration. *Contam. Soil Sed. Water*; 2002b, Dec., 12–15.

Straughan, D. Distribution of tar and relationship to changes in intertidal organisms on sandy beaches in Southern California. In: *Proc. 1979 Oil Spill Conference*; Amer. Petrol. Inst., Washington, D.C.; 591–601.

Strom-Kristiansen T., P.S. Daling, and A. Lewis. Weathering properties and dispersability of crude oils transported in U.S. waters. *Technical Report No. 93-032, Marine Spill Response Corporation, Washington, D.C.*; 1993.

Strom-Kristiansen, T., A. Lewis, P.S. Daling, J.N. Hokstad, and I. Singsaas. Weathering and dispersion of naphthenic, asphaltenic, and waxy crude oils. *Proc. 1997 Int'l. Oil Spill Conf.*; Amer. Petrol. Inst., Washington, D.C.; 1997, 631–636.

Syndes, L.K., T.H. Hemmingsen, S. Skare, S.H. Hansen, I.-B. Falk-Petersen, S. Lonning, and K. Ostgaard. Seasonal variations in weathering and toxicity of crude oil on seawater under arctic conditions. *Environ. Sci. Technol.*; 2001, **19**, 1076–1081.

Teal, J.M., K. Burns, and J. Farrington. Analyses of aromatic hydrocarbons in intertidal sediments resulting from two spills of No. 2 fuel oil in Buzzards Bay, Massachusetts. *J. Fisheries Res. Board of Canada*; 1978, **35**, 510–520.

Telnaes, N. and B. Dahl. Oil–oil correlation using multivariate techniques. *Org. Geochem.*; 1986, **10**, 425–432.

ten Haven, H.L., M. Rohmer, J. Rullkotter, and P. Bisseret. Tetrahymanol, the most likely precursor of gammacerane, occurs ubiquitously in marine sediments. *Geochim. Cosmochim. Acta.*; 1989, **53**, 3073–3079.

Thominette, F. and J. Verdu. Photo-oxidative behaviour of crude oils relative to sea pollution. Part I. Comparative study of various crude oils and model systems. *Mar. Chem.*; 1984, **15**, 91–104.

Thominette, F. and J. Verdu. Photo-oxidative behaviour of crude oils relative to sea pollution. Part II. Photo-induced phase separation. *Mar. Chem.*; 1984, **15**, 105–115.

Tissot, B.P. and D.H. Welte. *Petroleum Formation and Occurrence*, 2nd ed. Springer-Verlag, New York, NY; 1984.

Uhler, R.M., E.M. Healey, K.J. McCarthy, A.D. Uhler, and S.A. Stout. Molecular fingerprinting of gasoline by a modified EPA 8260 gas chromatography/mass spectrometry method. *Int'l. J. Environ. Anal. Chem.*; 2002, **83**, 1–20.

Ukpabio, E.J., P.A. Comet, R. Sassen, and J.M. Brooks. Triterpenes in a Nigerian oil. *Org. Geochem.*; 1994, **22**, 323–329.

United States Environmental Protection Agency (U.S. EPA). Guidance for conducting remedial investigations and feasibility studies under CERCLA. Interim Final. EPA/540/G-89/004. Washington, D.C.: Office of Emergency and Remedial Response. 1988.

United States Environmental Protection Agency (U.S. EPA). Risk assessment guidance for Superfund, Vol. I, Human health evaluation manual (Part A). Interim Final. EPA/540/1-89/002. Washington, D.C.: Office of Emergency and Remedial Response. 1989.

United States Environmental Protection Agency (U.S. EPA). Options for addressing high background levels of hazardous substances at CERCLA sites. Draft Final Issue Paper. Washington, D.C.: Office of Emergency and Remedial Response. 1992.

United States Environmental Protection Agency (U.S. EPA). Determination of background concentrations of inorganics in soils and sediments at harzardous waste sites. Engineering Forum Issue Paper. EPA/540/S-96/500. Washington, D.C.: Office of Emergency and Remedial Response. 1995.

United States Environmental Protection Agency (U.S. EPA). Test methods for evaluating solid waste (SW-846). Update III; 1997.

Urdal, K., N.B. Vogt, S.P. Sporstol, R.G. Lichtenthaler, H. Mostad, K. Kolset, S. Nordenson, and K. Esbensen. Classification of weathered crude oils using multimethod chemical analysis, statistical methods and SIMCA pattern recognition. *Mar. Pollut. Bull.*; 1986, **17**, 366–373.

Volkman, J.K. Biological marker compounds as indicators of the depositional environments of petroleum source rocks. In: A.J. Fleets, K. Kelts, and M.R. Talbot (eds.), *Lacustrine Petroleum Source Rocks, Geol. Soc. Spec. Publ. No. 40*; Blackwell Scientific Publishers, Oxford; 1988, 103–122.

Volkman, J.K., A.T. Revill, and A.P. Murray. Applications of biomarkers for identifying sources of natural and pollutant hydrocarbons in aquatic environments. In: R.P. Eganhouse (ed.), *Molecular Markers in Environmental Geochemistry*; *Am. Chem. Soc.*, Washington, D.C.; 1997, 110–132.

Wade, T.L. and J.G. Quinn. Geochemical investigation of hydrocarbons in sediments from mid-Narragansett Bay, Rhode Island. *Org. Geochem.*; 1979, **1**, 157–167.

Wang, Z. and M. Fingas. Differentiation of the source of spilled oil and monitoring of the oil weathering process using gas chromatography-mass spectrometry. *J. Chromatogr. A*; 1995a, **712**, 321–343.

Wang, Z. and M. Fingas. Study of the effects of weathering on the chemical composition of a light crude oil using GC/MS and GC/FID. *J. Microcolumn Separations*; 1995b, **7**, 617–639.

Wang, Z. and M. Fingas. Use of methyldibenzothiophenes as markers for differentiation and source identification of crude and weathered oils. *Environ. Sci. Technol.*; 1995c, **29**, 2842–2849.

Wang, Z. and M. Fingas. Developments in the analysis of petroleum hydrocarbons in oils, petroleum products and oil spill-related environmental samples by gas chromatography. *J. Chromatogr.*; 1997, **774**, 51–78.

Wang, Z. and M. Fingas. Development of oil hydrocarbon fingerprinting and identification techniques. *Mar. Pollut. Bull.*; 2003, **47**, 423–452.

Wang, Z., C. Yang, M. Fingas, B. Hollebone, X. Peng, A.B. Hansen, and J. Christensen. Characterization, weathering, and application of source identification of spilled lighter petroleum products. *Environ. Sci. Technol.*; 2005, **39**, 8700–8707.

Wang, Z., M. Fingas, and D.S. Page. Oil spill identification. *J. Chromat. A*; 1999a, **842**, 369–411.

Wang, Z., M. Fingas, and G. Sergy. Study of 22-year-old *Arrow* oil samples using biomarker compounds by GC/MS. *Environ. Sci. Technol.*; 1994b, **28**, 1733–1746.

Wang, Z., M. Fingas, and K. Li. Fractionation of a light crude oil and identification and quantitation of aliphatic, aromatic, and biomarker compounds by GC-FID and GC-MS, part I. *J. Chromat. Sci.*; 1994a, **32**, 361–382.

Wang, Z., M. Fingas, and L. Sigouin. Characterization and identification of a "mystery" oil spill from Quebec (1999). *J. Chromat. A*; 2001b, **909**, 155–169.

Wang, Z., M. Fingas, and L. Sigouin. Characterization and source identification of an unknown spilled oil using fingerprinting techniques by GC–MS and GC–FID. *LC GC N. Am.*; October 2000, **18**, 1058–1067.

Wang, Z., M. Fingas, M. Landriault, L. Sigouin, B. Castle, D. Hostetter, D. Zhang, and B. Spencer. Identification and linkage of tarballs from the coasts of Vancouver Island and Northern California using GC/MS and isotopic techniques. *J. High Resol. Chromat.*; 1993, **21**, 383–395.

Wang, Z., M. Fingas, S. Blenkinsopp, G. Sergy, M. Landriault, L. Sigouin, and P. Lambert. Study of the 25-year-old Nipisi oil spill: Persistence of oil residues and comparisons between surface and subsurface sediments. *Environ. Sci. Technol.*; 1998, **32**, 2222–2232.

Wang, Z., S.A. Stout, and M. Fingas. Biomarker fingerprinting for spill oil characterization and source identification (Review). *Environ. Forensics*; 2006, **7**(2), 105–146.

Wang, Z.D., B.P. Hollebone, M. Fingas, B. Fieldhouse, and J. Weaver. *Characteristics of spilled oils, fuels, and petroleum products: 1. Composition and properties of selected oils*; U.S. EPA/600/R-03/072, National Exposure Research Laboratory (NERL), EPA, 2003. Also available at the EPA website: www.epa.gov/epahome/recentadditions.htm.

Waterborne Commerce of the United States (WCUS), Part 5 — National Summaries. *U.S. Army Corps of Engineers*; 2003. Also available at the EPA website; also available at the USACE website: http://www.iwr.usace.army.mil/ndc/wcsc/pdf/wcusnatl03.pdf.

Westerholm, R.N., T.E. Alsberg, A.B. Frommelin, M.E. Strandell, U. Rannug, L. Winquist, A. Grigoriadis, and K.-E. Egebäck. Effect of fuel polycyclic aromatic hydrocarbon content on the emissions of polycyclic aromatic hydrocarbons and other mutagenic substances from a gasoline-fueled automobile. *Environ. Sci. Technol.*; 1988, **22**, 925–930.

White, C.M. and M.L. Lee. Identification and geochemical significance of some aromatic components of coal. *Geochim. Cosmochim. Acta*; 1980, **44**, 1825–1832.

Wolfe, D.A., M.J. Hameedi, J.A. Galt, G. Watabayashi, J. Short, C. O'Claire, S. Rice, J. Michel, J.R. Payne, J. Braddock, S. Hanna, and D. Sale. The fate of the oil spilled from the *Exxon Valdez*. *Environ. Sci. Technol.*; 1994, **28**, 561A–568A.

Wu, Q., M.R. Gray, M.A. Pickard, P.M. Fedorak, J.M. Foght. Biocatalytic ring opening of dibenzothiophene and phenanthrene as model substrates dissolved in crude oil. *Petrol. Chem. Div. Preprints No. 615061*; *Am. Chem. Soc.*; 2002, 47.

2 Spill Site Investigation in Environmental Forensic Investigations

E. H. Owens, E. Taylor, and H. A. Parker-Hall

Polaris Applied Sciences, Inc. 12509 130th Lane, NE, Kirkland, WA 98034-7713.

2.1 Introduction

The first step in all forensic investigations surrounding surface oil spills is to characterize the spill site and the affected area to determine the nature and scale of the problem and to identify acute safety issues. Scaling the problem is critical to a forensic investigation and involves essentially defining the source, type, amount, and location of the spilled oil. These elements provide the necessary framework for an assessment of the environmental variables that control the transport and weathering of the oil, and for identification of amplifying evidence on the nature of the spill itself. The identification of safety issues, typically as part of a site safety plan, provides the backbone of all spill-related activities and sets limits on what actions operations personnel and site investigators can and cannot take at the spill site and in the affected area.

There is a fundamental difference between the behavior of oil spilled on land and on water that determines the speed at which spilled oil moves or spreads and the resulting size of the affected area. This difference has a significant influence on the scale and character of the site investigation. Oil spilled on land, except in rare circumstances, flows down slope and typically collects in depressions or against topographic or manmade barriers (Table 2-1). The rate of down-slope movement is a function of the oil viscosity, air/ground temperatures, slope steepness, and the surface condition (roughness, vegetation type, soil type, permeability, etc.). Land surfaces are rarely flat, so that the thickness of layers of oil varies considerably. Oil that reaches creeks, streams, or rivers and oil spilled directly on water is transported and spread by winds and/or surface currents, so that the extent of the affected area, weathering rates, and therefore the scale of the problem, typically increase dramatically.

In this chapter we describe appropriate methods to characterize the spill site in terms of environmental and safety issues, the survey tools that have been developed to define and describe the spill site and the affected area, sampling considerations for spills where the source is known and for "mystery spills," and data management techniques to capture, organize, and present the results of the investigation.

2.2 Environmental Site Characterization and Reconnaissance Survey

An early challenge for spill site investigation is the competing demand and need for emergency response. Initial response activities usually include spill and safety assessment studies, sampling programs, oil containment and treatment, and recovery activities. For any forensics investigation, it is essential to have a clear understanding of the natural site

Table 2-1 Characteristics of Spills on Land and Water (Adapted from Owens, 2002)

Oil on Land	*Oil on Water*
• spilled oil is generally slow-moving or static • the oil collects in depressions or against natural and man-made barriers • usually the size of the affected area is small and it is easy to define the location and amount of surface oil • only light oils spread to form a thin layer; often considerable pooling of oil • weathering slows considerably after approximately 24 hours	• spilled oil is moved by winds and/or currents and often remains in motion for days and sometimes weeks • the size of the affected area increases with time and it can be difficult to locate some or all of the oil; may submerge or sink • oil on the water surface typically spreads to form a very thin layer • weathering and emulsification are dynamic processes that continually alter the physical and chemical properties of the oil

conditions, potential spill sources, factors that influence the area as well as the fate of spilled oil, and safety risks associated with activities and the environment itself. Objectives for site characterization and reconnaissance surveys are to:

- define site features that influence oil fate and persistence,
- delineate areas affected by the spill,
- identify background and incident-specific contributions of oil at the field site,
- describe variations in oil's physical character, concentration, and mode of occurrence in space and time,
- evaluate the variability of oil concentrations and oil penetration depth,
- provide data to help forecast residence time of surface and subsurface oil from knowledge of *in situ* weathering processes,
- provide data of potential use to understand the short- and long-term effects of oiling on resource use.

A first step in site investigation is to establish a scope or scale of the problem: starting with a broad picture and working toward detail. The exact design, size, duration, and scope of the survey program will vary with each spill situation, the environmental conditions, and the operational response. ASTM (American Society for Testing Materials) standards for site investigations of oil and hazardous chemical sites list a number of variables to be considered in a site investigation (ATSM 1995, 1998a, 1998b). Additionally, protocols for study of oiled shorelines are described in detail in PERF (Petroleum Environmental Research Forum) Guidelines (e.g., Owens, 1999).

A site investigation and delineation must take into account steps typical of an ecological risk assessment: identify potential contaminant sources, identify potential receptors that may be affected by exposure to oil, and identify the transport processes that can place oil in contact with receptors (Table 2-2). ASTM Standard D 5745–95 (1999) indicates that site investigation entails assembling existing available information and developing a conceptual site model, including:

- identification of contaminants (oils),
- characterization of background/conditions,
- contaminant (oil) source characterization,
- migration pathway characterization,
- contaminant (oil) mass estimate.

The source(s) of spilled oil may be known, but a site investigation must also clearly identify other potential sources of petroleum hydrocarbons that may pre-exist or naturally occur (e.g., oil seeps, abandoned or active industrial sites, runoff, wrecks, etc.) in the area. Receptors include ecological, human, and economic resources that could be affected by the spill. Transport pathways include surface and subsurface routes. Hence, site investigation and reconnaissance typically must conceive a three-dimensional and temporal model of source and fate of the spilled oil. The type of information to be assessed for site characterization should lead to a clear

Table 2-2 Examples of Factors to be Assessed as Part of a Site Investigation

Attribute	Example — Terrestrial	Examples — Marine
Physical Characteristics	Buildings, structures, landmarks, topography, soils (homogeneity, distribution), debris	Landmarks (points, islands, inlets), topography (nearshore and subtidal), sediments (homogeneity, distribution)
Ecological Use	Vegetation, avifauna, burrows, wildlife, freshwater habitats (lentic, lotic)	Nearshore and intertidal biota (algae, vegetation), epi- and infauna, sessile and mobile organisms, wildlife
Human Use	Industrial, residential, commercial use, public use (parks), transportation (rail, roads, etc.)	Industrial, coastal residential, commercial use, public use (parks), transportation (ports)
Spill Sources	Tanks and pipelines (surface or buried), hoses, tankers (truck, rail), industrial histories, seeps	Vessel traffic, effluent discharges, submerged wrecks, industrial histories, seeps
Surface Pathways	Drainages, surface water (runoff, streams, rivers — direction, speed)	Surface currents, wind, waves and tidal action
Subsurface Pathways	Infiltration (stratigraphy and permeability), groundwater (depth), hydraulic gradient (groundwater speed, direction)	Burial processes, heavier than water oils (initially or after weathering), currents, upwelling, infiltration (stratigraphy and permeability)

understanding of the spilled oil's source and fate. Examples of factors to be assessed are provided in Table 2-2 and in ASTM E-1943 (see Table X2-1, in ASTM, 1998b).

2.3 Site Entry and Safety Issues during the Emergency Response Phase

Scientific and technical support personnel may have an immediate need to rapidly collect ephemeral data and samples during the first response phase as part of a forensic investigation. However, this initial phase of the response is the most uncertain in terms of assessment and characterization of risks to personnel. Scientific and technical staff must be cognizant of any potential safety hazards and of the prescribed safe operating procedures that apply in the area affected by the spilled oil.

One of the paramount concerns of every response operation is for the health and safety of the public and oil spill personnel (MCA, 1998). There are numerous industry standards and governmental guidelines, regulations, and policies regarding human health and safety in the emergency response to spill incidents. The concept of safety guidelines and practices varies from country to country and can fall anywhere along a spectrum of rigid laws and regulations governing safety risks and practices to a case-by-case risk assessment process (IPIECA, 2002). In general, concerns regarding the safety of the work site and working conditions are easier to address and to understand than concerns about exposure to the spilled products themselves (Holliday and Park, 1993).

Key themes in safe site entry following an oil release include site security and management, risk assessment and characterization, chemical toxicity of the spilled material(s), work environment safety, and personal protective equipment (PPE). Personnel entering a spill zone must be trained to appropriate levels in safety protection procedures. Example safety training guidelines are provided in ASTM procedures for hazardous materials and oil spill responders (ASTM, 1986, 2001a, 2001b).

2.3.1 Management of Safety

One of the first appointed roles in emergency spill response management is the safety

Table 2-3 Priorities of the Safety Officer during an Oil Spill (Adapted from IPIECA, 2002)

Safety Management Priorities
1. Site Assessment:
   - Analyze the hazards and document this hazard analysis process
   - Hazard identification
   - Identify appropriate PPE
   - Identify control zones
   - Identify decontamination areas
2. KEY: must use trained, experienced personnel in management of safety issues.
3. Must train all personnel in safety awareness and practices.
4. Develop and implement a site-specific safety plan.
5. Attend daily planning and safety meetings in command center.
6. Correct unsafe acts or conditions as soon as possible.
7. Establish first aid and emergency medical stations and procedures.
8. Conduct safety briefings and communication daily and as needed.

officer(s). Responsibilities assigned to the safety officer are to characterize the nature of existing and dynamic safety hazards, establish and implement safety guidelines, develop a site-specific safety plan, and implement the necessary precautions. This individual, or team of safety personnel, is responsible for monitoring hazardous and unsafe situations and maintaining awareness both of active and developing situations (Table 2-3).

2.3.2 Risk Assessment and Characterization

The safety officer's first step in implementing a safety program is a thorough site assessment and characterization to obtain adequate information on the spill location and the particular character of the spill site, including environmental conditions, to develop specific safe operating procedures (MSRC, 1993). It is best to use standardized procedures or protocols for site assessment, site characterization, and site safety plan development (MSRC, 1993), which can be modified as appropriate to meet the specific conditions of each incident. This comprehensive risk assessment and hazard analysis is conducted immediately after the incident to determine if responders or the general public are in danger. Some of the principal questions that must be answered by this site assessment include (from IPIECA, 2002):

- Is there a risk of explosion or fire?
- Is there a need to monitor for flammable or toxic fumes?
- Is there a need to evacuate people?
- Is the environment safe for people to enter or work in?
- Will oil enter systems that may affect people?
- Is there a need to establish safety or exclusion zones?

Once these key questions have been answered satisfactorily, the safety officer must then determine any risks posed by particular operations or at specific locations within the spill site. These need to be assessed on a case-by-case basis. It is also imperative to document the site risk assessment; many standard forms exist, or one can be created or modified to fit the situation.

The purpose of conducting the initial site risk assessment then is to identify all the potential hazards presented by the incident itself, the work site(s) (including risks inherent to the ambient environmental conditions), and of the response operations. This allows some ability to predict the probability and severity of any potential safety hazard or incident. The response managers, through the safety officer, can then prioritize potential hazards and appropriate guidance and precautions. A common practice is to identify three levels of safety response zones:

HOT — with higher safety risks; stricter PPE requirements; activities include source control, sampling, and cleanup

WARM — general lower personnel exposure; decontamination activities; and

COLD — nonoiled zones; PPE not required for contact with spilled oil, support activities.

Once the safety hazards and their likelihood have been characterized, and appropriate precautions have been determined, safety personnel continue to evaluate their appropriateness and effectiveness as many of the safety risks and potential hazards associated with an oil spill are not static and change with time and/or the response operations. For a continued hazard, there is a common hierarchy of priorities in establishing safety priorities (from IPIECA, 2002):

1. prevent access to the hazard (site control and security),
2. organize the work site(s) so as to reduce exposure to the hazard (safety zones),
3. select and employ appropriate PPE for all personnel in the area.

2.3.3 Chemical Toxicity of the Spilled Oil

Each type of spilled product will have its own inherent chemical properties and associated risks to the safety of the general public and onsite personnel. It is important to remember that these properties or characteristics will change with time as most petroleum products weather (i.e., change their chemical and physical properties when exposed to weather and the natural environment) when spilled, particularly in an open area with exposure to wind, waves, sunlight, etc. Information on the spilled products' properties, such as found in the Material Safety Data Sheets (MSDS), can be used to assess the basic chemical properties and physical characteristics of the spilled product and its hazards to personnel.

The principal health hazards from petroleum products themselves include flammability, the presence of explosive vapors, hydrogen sulfide gas (heavier than air) in sour crudes, exclusion of oxygen, toxicity, and the slippery nature of the oil. The first four of these hazards are generally very short-lived threats, especially for spills in open environments. Except in cases of confined spaces (e.g., trenches, tanks, under docks), the vapor cloud diminishes rapidly, typically within hours after a release.

All of these hazards are a priority when assessing and characterizing the site following a spill and before allowing access to the spill site. Characterization of any of the above explosive or vapor hazards can be accomplished with a comprehensive air monitoring protocol. Air monitoring data are then matched with appropriate safe working exposure limits. For example, guidelines for conducting operations in the presence of potentially explosive vapors are (ExxonMobil, 2002):

- **<10% LEL** — continue response operation. An LEL reading of 10% is equivalent to an airborne concentration of 100,000 ppm, which will require appropriate respiratory protection.
- **≥10% LEL and <25% LEL** — explosion hazard present. Withdraw from area immediately. Proceed with care, especially where there is poor air movement or circulation.
- **>25% LEL** — leave the area quickly and carefully.

Air monitoring must be conducted to define required breathing protection as part of personnel PPE and ensure safe operating conditions. Air and exposure monitoring are conducted using a variety of instruments, including various electronic monitors, Draeger tubes, personal monitors, or passive diffusion monitors.

Although some components of oil are toxic, the actual risk of toxic exposure by responders or the general public to spilled oil is in most cases very low. Spilled petroleum toxic components have a number of pathways of exposure to humans — inhalation of vapors, direct dermal contact, absorption through the skin or eyes, ingestion, or injection (IPIECA, 2002). As with other hazards from spilled products, the risk of a toxic exposure is greatest immediately following a spill, particularly of products containing volatile aromatic compounds such as benzene. Benzene and similar oil compounds are carcinogenic. Safe exposure standards and limits have been defined (see, for example, the NIOSH Pocket Guide to Chemical Hazards, 2005). Air monitoring must be conducted, even though these aromatic components are volatile and generally dissipate rapidly to below prescribed exposure limits, and proper PPE for responders must be prescribed until air monitoring ensures that the risk of exposure to these carcinogens has diminished.

2.3.4 Working Environment Safety

There are numerous characteristics of a spill site that pose hazards to technical personnel. Environmental factors such as spill location, weather conditions, tidal fluctuations, access to cleanup sites, terrain, etc., cannot be controlled. Access points, PPE, and personal security can be controlled, thereby minimizing safety hazards, and site-specific procedures can be used to effectively minimize potential hazards.

Weather is a significant consideration in the safety of response personnel. Response personnel may be exposed to the elements for many hours. Safety hazards can include a range of health issues from hypothermia or frostbite to heat stroke/exhaustion, sunburn, and dehydration. Preventative measures include selecting appropriate clothing, shelter, survival training, selecting appropriate work/rest schedules, and proper communications equipment and weather forecasts (IPIECA, 2002).

Oil spills can occur in a broad range of environments, often within the same spill — e.g., exposed, rocky coasts, ice- and snow-covered tundra or remote mountainous terrain. Safe access and egress must be established for personnel and equipment that accounts for shoreline substrate (cliffs, mangroves, sand, mud, etc.), tidal ranges, riverbank gradients, watercourse flow rate, and depth, and water table characteristics. Indigenous flora and fauna, such as slippery algae, poisonous plants, snakes, alligators, and other dangerous plants and animals must be identified and proper safety precaution briefings given to personnel.

The most common injuries to personnel at spill sites are from slips, trips, and falls. Any surface that encounters spilled oil likely will be slippery. In addition, the work site for the collection of samples may by nature be difficult terrain — rocky coastlines, intertidal zones, cliffs, remote areas, rough sea states, etc. Appropriate PPE and site assessment and sampling equipment can be cumbersome and can make movement or specific actions difficult for scientists or technicians, particularly once clothing or equipment has become oiled.

2.3.5 Personal Protective Equipment (PPE)

Concerns over exposure to spilled contaminants can be addressed by applying broad, conservative standards of PPE and exposure time limits, which could significantly hamper an individual's ability to perform physical activities for prolonged periods, creating a need to balance the risk of exposure to the spilled product with an appropriate, adequate level of PPE (Holliday and Park, 1993). Nonetheless, PPE is essential for each responder to ensure he is able to work safely around chemicals and other materials that may be hazardous to his health.

Decontamination procedures must be established and implemented safely for all personnel and equipment. Technical personnel must be briefed on these procedures, which take place in predesignated locations within the WARM zone. The primary goal of decontamination is to allow personnel and equipment to exit the HOT zone while avoiding cross or secondary contamination. Typically, decontamination procedures are established for both personnel and equipment.

The primary objective of every oil spill response must always be to ensure the safety of the response personnel and the general public. The response management makes a commitment to safety by establishing a safety program led by a safety officer to assess and characterize the potential hazards from the spilled contaminants and the response operations. All personnel in an oil spill site, including scientific and data collection staff, must understand their obligations to work in a safe and responsible manner within the established operating guidelines. Forensic investigation personnel must be adequately trained to a level commensurate with the tasks they will be assigned (ASTM, 2001b; OSHA, 2001).

2.4 Determination of Geographic Boundary and Definition of Different Zones within the Affected Area: 1. Terrestrial Oil Spills

Determination of the geographic boundaries and zones for spills are important to establish safety zones (hot, warm, and cold) for operational concerns, priorities for initial cleanup, and to provide a spatial reference for decision making relative to cleanup, monitoring, and remediation. In terms of a forensics investigation, the boundaries of a terrestrial oil spill site are typically more restricted relative to a waterborne spill primarily because of the limited transport factors in the former: infiltration into soils, retention on vegetation, and possible transport though surface and/or subsurface hydrologic regimes. Delineation of safety zones within terrestrial spills will entail an initial broad reconnaissance followed by detailed site investigation once safety requirements for site entry have been identified and are in place.

An initial assessment of a spill site must consider potential spill sources, which may range from obvious visual surface expression to more difficult subsurface diffusion zones and plumes. Initial delineation of a spill area typically is based on identifying spill source, runoff, and infiltration directions, and delineation of surface and subsurface oil concentrations. Tools for defining the extent of the site under investigation and more in-depth inspection and sampling are

- site aerial photos,
- detailed topographic maps,
- geographic positioning satellite (GPS) receivers (standard or differential),
- survey tie-in to fixed and known reference points (stakes, benchmarks, landmarks, structural features),
- cameras (digital, video, 35 mm).

After establishing safe-site entry requirements and procedures, a site investigation team will

- identify physical aspects and habitats that characterize the site,
- identify potential receptors (sensitive resources),
- identify known or potential spill sources (lines, tanks, tubing),
- delineate surface expressions of spill (visible oil),
- characterize surface and subsurface oil concentrations, distribution, and continuity through pits, trenches, or borings.

A clear record of the above information is best captured through sketches on a field map, accompanied by cross-referenced photographs, samples, and video with narrative. All observations, samples, and photographic evidence should be geographically defined through GPS linkages or surveyed in to fixed features that will allow subsequent assessments to return to the same location. The above information can later be transferred to a geographic information system (GIS) mapping program and associated database for easy reference and subsequent reporting and analysis (see Section 2.9).

A delineated study site should reveal safety zones (hot, warm, cold), oiled areas (concentrations, distribution, surface, and subsurface), nonoiled areas (confirmed), and areas at potential risk of oiling due to identified transport pathways. Because spilled oil is rarely static, site investigation and definition of oiled zones requires additional assessments. The frequency of these assessments depends on how quickly conditions may be expected to change at the site. The use of adopted and standard site-study protocols will help in comparing and interpreting study results for multiple surveys.

2.5 Determination of Geographic Boundary and Definition of Different Zones within the Affected Area: 2. Marine/Coastal Waterborne Oil Spills

Determination of the geographic boundaries and zones for marine and coastal waterborne spills is similar to the process described above for terrestrial spills. A key difference, however, is that waterborne spills typically undergo faster change such that the temporal aspect of site investigation may be more important. For oil spills, an adopted best international practice for characterizing coastal sites and oiling is the Shoreline Cleanup Assessment Technique (SCAT) (Owens, 1999; Owens and Sergy, 2002; ASTM, 1997a, 1997b). Sites along shorelines are identified as specific segments or subsegments, as these are defined by their relative continuity and homogeneity. If shoreline segment maps have not been prepared, the segmentation and site boundaries should consider

- prominent geological features (headlands, streams, etc.),
- changes in shore/sediment types
- changes in oil conditions, or
- habitats.

Once a site has been selected or a shoreline delineated through segmentation, a detailed site study is undertaken. As with terrestrial spills, site investigation will entail delineation of surface oil, subsurface oil (through pits or trenches), and recognition of processes that transported oil to where it was found. Because landmarks may not be as readily available for coastal areas and offshore, GPS (standard and/or differential) and/or field survey techniques (transit, tape and bearing, theodolite, etc.) tied into fixed reference points (particularly underwater) are highly recommended. These benchmarks should serve not only as horizontal controls but also as vertical reference points, particularly where coastal dynamics may change the beach morphology. Tide level can be used as a proxy for a fixed datum; however, apparent tidal level can be affected by winds and groundwater levels. PERF Method 3.2, Monitoring Program Procedures, provides guidelines on tools and methodologies for shoreline surveys and monitoring (Taylor, 1999).

For a recurring or long-term site investigation and monitoring program, transects should be set up across a site, at representative or randomly selected locations, depending on statistical needs of the study program. Periodic surveys are conducted to monitor changes in oil cover, site geomorphology, sediment distribution, types and densities of biota, oil penetration, and other parameters. Study site lengths (measured alongshore) are small enough to obtain adequate resolution and detail on the distribution of oil. PERF Method 3.1, Site Selection and Setup (Taylor, 1999), suggests most study sites would allow for at least 10 across-shore transects spaced no closer than 3 m and no greater than 20 m. Generally, study sites should be in the range of 40 m to 120 m long, and encompass the supratidal to subtidal zones, as appropriate to study requirements.

For offshore areas, definition of geographic locations will vary depending on the spill, and its potential for impact to offshore habitats. Two general categories of offshore study sites are considered: (1) subtidal extension of onshore study sites and (2) offshore areas of potential interest. Taylor (1999) highlighted criteria considered for identifying offshore areas of interest:

- areas underlying oceanographic convergence zones where oil on water is concentrated;
- areas where sediment influx is substantial and represents an important flocculation mechanism;
- areas underlying burned oil or sinking oil; and
- areas in the vicinity of other potential oil sources to the environment (i.e., seeps and wellheads).

The delineated study sites should denote safety zones (hot, warm, cold), oiled areas (concentrations, distribution, surface and subsurface), nonoiled areas (confirmed), and areas at potential risk of oiling due to identified transport pathways. Because spilled oil in the marine environment can be expected to be dynamic, site investigation and definition of oiled zones require frequent assessments. In the initial phases of spill response, these assessments may need to be repeated on the scale of days to weeks.

2.6 Collection of Physical, Ecological, and Environmental Data

The collection of adequate data following an oil spill is integral to characterizing the nature and extent of the release and its potential impacts. Data collected during this time should also consider any forensic questions that might need to be addressed — e.g., the potential source(s) of a "mystery" spill or the establishment of "background" conditions (i.e., pre-existing hydrocarbons in the environment). This section describes an approach to selecting candidate data types that ensure sufficient information is collected, at appropriate temporal and spatial scales, to determine the character, extent, and source(s) of the spill and its potential impact on the environment.

The first set of information to collect is generally related to physical data, such as samples of the spilled material from the release site and, in the case of a known source of oil, from the source. In the case of a "mystery" spill in a marine, lake, or river environment, this would include obtaining samples of slop tank oils, bunkered fuels, and petroleum cargoes from any vessels that recently transited the area. Sample collection may not be a straightforward activity in terms of initially identifying candidate sources and then being able to actually collect a sample. A common situation exists where several or many vessels may have transited the area in which a mystery spill is observed. This was the case for the Dalco Passage oil spill in Puget Sound (2004: Washington, USA), a very busy port area, in which there were "around twenty potential suspect boats" (http://www.epa.gov/oilspill/pdfs/0105update.pdf). In this type of incident, investigators attempt to collect samples from each vessel to try to obtain a match, as well as from potential shore-side sources such as terminals, sewers, rivers and streams, or from sub-sea pipelines. This type of investigation requires the cooperation of suspect parties and may involve the collection of samples from onshore facilities if a vessel has bunkered before leaving a port. Additional challenges to the collection of these vital data are the natural environment itself. Wind, waves, currents, and tides all can have a significant impact on the practicality and feasibility of plans to collect samples of oil, water, or sediments. The nature of the spilled product, whether it floats, submerges, or sinks, also is a factor in sample collection. The spill location, access to the site, and safety considerations play a key role in the ability to collect data to support a forensic investigation. Spills can occur on snow and ice, on frozen or soft, peat-covered tundra, in turbulent rivers, in mangrove forests, on flat, sandy beaches or on steep, rocky cliffs pounded by waves. Spills can be difficult to characterize when they occur on land as pipelines may traverse remote, mountainous regions with rough terrain.

Many types of physical environmental data are ephemeral in nature so that a sampling plan for these data must be designed very early into the spill to capture essential information. The key to the design of an appropriate data collection program to support sampling for a forensic study, particularly in the case of a "mystery spill," is to quickly determine the questions that need to be answered. Example questions are: Where did the oil come from? When did the spill occur? What was the transport pathway? What parameters have affected the weathering of the oil?

Quick identification and collection of the data or observations are required to understand the environmental factors that controlled transport and weathering of the spilled oil and to support the interpretation of the results of sample analyses. Wind speed and direction and weather observations or measurements (temperature and precipitation) may be collected on site or obtained from public service organizations. Current and wave data may be more difficult to obtain and nonsystematic observations may be the only practical option in the early stages of a response. If the study is concerned with transport pathways to determine the origin of a "mystery spill" at sea, in a lake, or on a river, then the collection of meteorological and hydrographic or oceanographic data may be the first priority.

2.7 Sampling Plan and Design: 1. Spills with Known Source

As part of the site characterization of a forensic oil spill investigation, an appropriate sampling plan must be designed and implemented. For spills with a known source, sampling is still integral to determination of the spill pathway and to scaling the nature of the problem in terms of the type, amount, and extent of the spilled material(s). Also integral to this type of spill investigation is the characterization of the pre-existing "background" conditions in the area of the spill. This section describes elements of designing an adequate sampling plan for spills of a known source, with primary focus on the options of discrete water sampling, continuous water sampling, and oil source sampling. Sampling on land is less of a concern as source identification is, in most cases, straightforward. The exceptions occur with spills of light oils, which can penetrate surface soils or sediments and travel through the subsurface, or with underground spills from tanks or pipelines.

Sampling plans are not necessarily lengthy documents. Approved plans are necessary, however, to ensure that the sampling team knows the procedures they are required to follow in order to collect samples that can be analyzed for meaningful and defensible information. Sampling plans should define, at a minimum:

- the type of collection procedure (grab, core, surface, subsurface, etc.),
- the number and location of samples,
- the required procedures for transferring material(s) and the types of container(s) to be used,
- the required procedures for handling, storing, and tracking the sample(s),
- the required procedures for numbering and labeling the sample containers,
- on-site data to be recorded at the time of sampling (GPS coordinates, depth, wind speed and direction, sample team members, etc.),
- site hazards and safety procedures.

2.7.1 Water Column Sampling

Priorities for water column sampling are chosen based on what data or information is required to answer specific questions (Brown, 1999). One method to prioritize water sampling methods or elements is to first evaluate those questions that are time-sensitive. Petroleum products, when spilled into the open environment, undergo rapid physical and chemical alterations when exposed to wind, sunlight, waves, and other natural forces. The greatest rate of change during this weathering process in most cases occurs within the first few hours to the first day after a spill. This ephemeral nature of the oil's physical and chemical properties underscores the importance of initiating sampling as early as possible, adjusting the frequency of sampling to be greatest during the initial hours after a release, and the need to prioritize sampling based upon a rapidly changing time scale. For example, a time-sensitive priority for sampling of an offshore oil slick is to first collect discrete water samples beneath the slick to determine the degree of natural dispersion and/or dissolution of oil components. This can be achieved by continuous water sampling beneath the slick, with attention given to planning for the slick's movement and transport downwind.

Discrete water sample collection and subsequent laboratory analysis provide detailed information on the concentrations of those components of the spilled oil that are of greatest concern, at distinct times after the release. These analyses also provide data for fingerprinting of the oil type and an evaluation of the toxic nature of its components. The potential does exist for contamination by surface oil when collecting subslick samples, so proper sampling procedures and contamination avoidance must be outlined clearly in any sampling plan. Discrete water sample analysis provides integral data to many aspects of the oil spill response and impact assessment and is an essential aspect of all oil spill investigations (Brown, 1999). Samples also should be collected from areas unaffected by the spill in order to provide "background" levels of contamination, which may be a significant factor in an industrialized environment or if there have been prior spills in the region.

2.7.2 Oil Source Sampling

Oil source samples, even in spills in which the source is obvious and known, should always be taken immediately following a release. Laboratory analysis of these samples can provide important information in the detection and investigation of the source. The chemical fingerprint of the source oil obtained from this analysis provides data that are then used to evaluate the potential fate and effects of the oil, both in the short and in the long terms. These detailed, and usually quite accurate, analyses can also shed light onto other background pollutants in the area, potential sources of the oil, and the potential risk to organisms and habitats the oil might encounter.

The approach to setting priorities for oil source sampling is similar to that taken with water column sampling — with a temporal hierarchy based on the ephemeral nature of oil once spilled into the open environment. Below is an example of priorities for several types of oil source samples (adapted from Brown, 1999):

- First priority: collect oil samples from the known or suspected source(s) itself, from the ship's oil tanks, or from the pipeline, well, tank car, or other source. If the source is a ship, take samples from as many segregated tanks as are on the ship, including possible lube oil tanks, oil–water separators, bilges, slops tank(s), fuel oil tanks, and cargo oil tanks, as well as any other petroleum product storage areas.
- Second priority: collect "source" samples of the spilled oil from the spill site itself, from the oil slick, or from a pool of oil on the riverbank or shoreline. Continue to collect samples of the spilled oil at an adequate frequency to characterize its physical and chemical changes as it weathers into such states as mousse or tar balls, if applicable. This sampling can be done concurrently with discrete water column sampling efforts.
- Third priority: sample oil that becomes stranded on shorelines or riverbanks. Continue to collect samples of shoreline oil at intervals adequate to characterize the weathering effects and changes to the oil's physical and chemical properties with time and exposure to the elements.

2.7.3 Sampling on Land

The sampling of oiled sediments, soils, or groundwater on land can be accomplished by surface grab samples, coring, or the installation of groundwater wells, all of which are standard procedures. If multiple potential sources exist, as may be the case in an urban area with a number of active, inactive, or removed underground storage tanks, the sampling design takes into consideration the three-dimensional analysis of potential transport pathways between the location of the oil and the possible sources.

2.7.4 Sampling Plan Design

Discrete water column samples are important for every oil spill, whether the source is known or not and analyses of these samples provide key information on the chemical and physical properties of the oil, its potential toxicity to organisms, its projected persistence in the environment, and other details to aid in spill response decision making. Several elements must be considered when designing a sampling plan, whatever the purpose of the samples

(Brown, 1999). First, there must always be collection of clean, reference samples for comparison. When possible, samples should be collected in triplicate and stored frozen in solvent-washed glass jars.

When choosing the location for sample collection, consideration must be given to the locations directly underneath and in the immediate vicinity of the slick, particularly if sensitive resources are in the area. Within the water column, it is usually the upper 1 meter, or "near surface," that contains the greatest concentration of spilled oil and is usually of most interest for sampling purposes. Selection of specific depth intervals should account for overall water depth and water column mixing conditions (Brown, 1999). In water of 50-m depth or less, under relatively calm conditions, such as in a lake or sheltered coastal bay, mixing is likely confined to the upper few meters and sampling depth intervals at 1 m, 2 m, and possibly 5 m should suffice. Alternatively, for deeper water with greater mixing forces, such as rough weather or seas, discrete water samples may need to be collected from depths deeper than 10 m (Brown, 1999).

Sampling during the early part of a spill must include collection of water from representative background reference sites (i.e., not in the affected or oiled zone). The location of these reference sites can be based upon oil spill trajectories or projected transport pathways of the slick. These projections can identify areas that likely will remain unoiled throughout the spill as well as those areas that are initially unoiled but could later be affected. At each reference site, a minimum of three replicate water collection stations should be established. This replication will facilitate analysis of statistical variability; however, time, equipment, weather, or other constraints may impede this practice (Brown, 1999). Reference samples should be collected at each replication station at each reference site, at the preselected depths and times. Once samples from oiled sites are collected, samples must then be taken as soon as possible at the reference sites. A final important element in the design of a sampling plan is to ensure that sample contamination is avoided during the collection of the water sample, particularly when operating near the slick itself. Specific equipment types and procedures have been developed to assist in the prevention of sample contamination, and due attention should be given this issue in the sampling plan design.

Designing a comprehensive oil source sampling plan is integral in the investigative efforts of an oil spill, and collection of oil from the source is crucial to any forensic program. Oil must be collected from the source tanks, pipeline, well, etc. and from the spill location itself (Brown, 1999). This allows accurate fingerprinting of the spilled oil and helps determine the presence of other oil pollutant sources (i.e., background oil or other sources of hydrocarbons). One set of initial samples must be collected immediately following the event, as close as possible to the source (ship, pipeline, tank, etc.), and then at a given frequency following the release. Samples of floating oil, mousse, tar balls, and beached oil need also to be collected (Brown, 1999).

2.8 Sampling Plan and Design: 2. "Mystery" Spills

The primary goal of sample collection following a "mystery spill," or spill of unknown source, is source identification. The sampling plan is designed to identify potential transport pathways and also to rule out potential spill sources. Most mystery spills occur in marine, lake, or river environments, as it is relatively straightforward to backtrack spills on land. Water column samples must be collected to provide information on likely as well as unlikely sources and on the physical and chemical properties of the spilled material.

An example sampling plan design for a marine mystery spill could include the following elements:

- Investigators first identify and then systematically rule out specific potential sources, including passing vessel traffic in the vicinity; submerged sources such as natural

seeps, subsurface pipelines and wells, and shipwrecks; potential source samples are collected. Oils and wastes contained in a suspected vessel's tanks are sampled with the same thoroughness as are those of a known source (described above). It is important to obtain samples from each of a suspected vessel's tanks since it is possible that no single one of the oils will provide a match to the spilled oil, but a mixture of several oils may (see Chapter 8 herein).
- Samples are collected from oil on water and/or stranded oil in the affected areas and even from affected wildlife.
- Samples are analyzed to determine the type of product and its degree of weathering and to compare oil fingerprinting results with potential source sample analyses.

This information is vital to determine whether the source was a bunker fuel or a cargo from a ship, or whether it could be from a subsurface source such as a pipeline, wreck, or seep. Analyzing the degree of weathering provides information on the relative age or time the oil spent floating on water before it became stranded. This information can then be used, along with historical weather and current information, to calculate a "hindcast," or reverse trajectory, to indicate potential areas of source location (e.g., see Chapter 13 herein). On land, dye tests can be used at a suspected source to determine if the dye follows the same transport pathway and reaches the site of a "mystery" spill. As these clues begin to rule out likely suspect sources, additional information such as from databases of local shipwrecks and vessel transit information from local vessel traffic services can also identify potential source vessels. Further investigation into where and when vessels were in the area can shorten the list of potential suspects to a few, which can then be boarded at their most recent port and samples of their bunkers, bilges, and/or cargoes can be collected and compared to the oil samples taken from the shoreline or affected areas. Should this not provide a match, a closer look must be taken at the submerged sources.

In a case of repeated mystery slicks off San Francisco Bay in 2001, a shipwreck that had been submerged for over 50 years, the M/V *Jacob Luckenbach*, was considered a suspect source early in the investigation. Searches of records for the vessel characteristics revealed that this was a cargo ship operated by the U.S. Navy to transport materials across the Pacific Ocean during the Korean War in 1951 (McGrath et al., 2003). Investigators learned that the Navy fueled such vessels with a particular blend of Navy Fuel Oil (NFO) during that time. A nearby "mothball" fleet of old Navy vessels in San Francisco Bay housed similar vessels of the same era, and an industrious investigator thought to take samples from one of the NFO bunkers to compare with the stranded shoreline oil samples (McGrath et al., 2003). Significant weathering of both samples prevented any conclusive comparisons in this case, but the approach used illustrates how a creative and thorough investigator can uncover the potential source of a "mystery" spill.

Sample designs for mystery spills on land follow similar considerations to those noted in the above. A series of samples may be required to establish a spatial or three-dimensional model of the oiled area and to track the transport pathway to the source. As noted earlier, for coastal or river spills it may be necessary to sample on-land sources, such as sewers or stormwater runoff channels, to investigate potential sources. In the offshore environment where there are multiple operations, such as exploration, production, and gathering or transportation (pipelines), a further sampling challenge occurs when a number of locations or operations have oils derived from similar sources.

2.9 Data Management

Effective and efficient data management begins with the project design so that data are collected, transferred, validated, catalogued, processed, archived, copied, and distributed in a systematic and consistent manner. The most critical elements of data management are to define prior to or at the beginning of the study

1. the parameters that are to be measured or documented,
2. the data collection procedure(s) and method(s) (tools),
3. the format and type of media that are to be used to capture the data,
4. the pathway the data will follow from the field to the final depository, and
5. the quality assurance (QA) and quality-control (QC) procedures.

This discussion summarizes the guidelines developed for the management of scientific data developed by Gundlach and Coogan (1999) and for the application of data management concepts to the documentation of oiling conditions generated by the SCAT method by Lamarche et al. (2005). The QA/QC process is summarized from Chamberlin (1999). These three documents provide a level of detail that can be used to establish the data generation and management protocols in the study design phase.

The use of standardized procedures to collect or measure the raw field data provides a consistent dataset. This approach is best exemplified by the SCAT method to document oiling conditions. This method is based on systematic procedures, standardized terms and definitions for the parameters that are measured or described, and the use of standard paper or electronic forms (Owens and Sergy, 2000). With this approach, multiple field teams generate data that are consistent in space and through time.

The objective of data management is to ensure that the information generated in the field is accurate, reliable, available, and in a suitable and usable format. The types of data that are generated by site investigations typically include both paper documents and digital files that can include field log books, data and chain-of-custody forms, cassette and video tapes, and photographic and digital still or video images. Data management is facilitated if paper documents can be scanned so that the electronic files can be easily stored, copied, and distributed.

The intent of data reduction or processing is to make the raw data available in a suitable format to describe the site or the environmental parameters that have been described or measured. The output can be in the form of tables, diagrams, charts, or maps. As an example, Figure 2-1 summarizes the results of a sampling project related to the grounding of the *New Carissa* and illustrates that a large proportion of the samples did not match any of the source oils on the vessel even near the site of the accident. GIS tools provide a rapid and efficient method to undertake spatial analysis of single or multiple parameters and generate maps that document the geographic variability of selected features. Geographic accuracy is often a critical factor in determining the location of a sample or a specific feature and in the analysis of spatial distributions so that the use of a GPS is now a standard tool that should be included in the study design in order to achieve the necessary level of accuracy for which the data are intended.

The proliferation of digital cameras has benefits and disadvantages for field studies. The ease with which pictures can be taken, and the number of images that can be stored in a camera's internal memory, have led to a very large increase in the number of images that are captured. The documentation and cataloguing of these images can be a very time-consuming process and may not be adequate unless some form of electronic tagging is used that links the images to GPS coordinates. Commercial software is available that enables such links to be applied so than the date, time, latitude, longitude, and other information can be attached to each image. This linkage is a valuable data management tool as it enables images to be stored, catalogued, and retrieved based on location, date, time, photographer's name, etc. A potential problem with digital images is that they can be modified easily and so be made invalid for legal documentation unless strict protocols are followed, for example, as described by Lamarche and Roberts (2004).

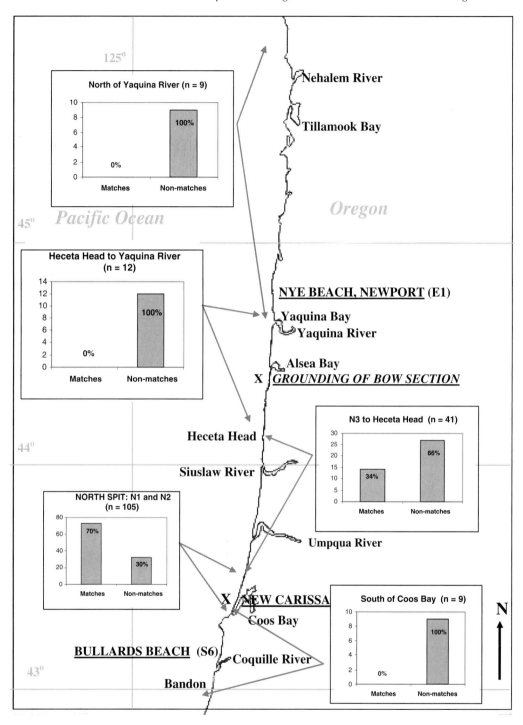

Figure 2-1 Map of tar ball distribution indicating those that matched the source oils on the vessel and those that did not match (adapted from Owens et al., 2002).

Company Name

CHAIN OF CUSTODY RECORD

PROJECT #	PROJECT NAME:
CLIENT	LOCATION:
SAMPLER (SIGNATURE)	

LABORATORY I.D. NUMBER	CLIENT SAMPLE I.D.	DATE	TIME	ANALYSIS REQUESTED	LOCATION / REMARKS

RELINQUISHED BY: (signature)	DATE / TIME	RECEIVED BY: (signature)	DATE / TIME
RELINQUISHED BY: (signature)	DATE / TIME	RECEIVED BY: (signature)	DATE / TIME
RELINQUISHED BY: (signature)	DATE / TIME	RECEIVED BY: (signature)	DATE / TIME

Figure 2-2 Example of chain-of-custody form.

One important element of data management is chain-of-custody (COC) documentation to track the location of samples (Gundlach and Coogan, 1999). Tracking information for every sample that is collected is entered and stored on a standard form (Figure 2-2) that should include, at a minimum,

- the sample ID number,
- the name of the individual who collects the sample,
- the location, date, and time of collection, and boxes for
- the signatures of the individuals that relinquish and receive the sample,
- the date and time that a sample is relinquished and received, and
- the transfer or transportation form (mail, courier, by hand, etc.).

Additional information can be included on the matrix (e.g., oil, water, sediment), preservative used (if any), the type of analysis that is requested, the address to which the sample will be sent, and any appropriate contact information (e.g., names, addresses, and telephone numbers of the client, project manager/chief scientist/team leader, or laboratory).

Quality assurance (QA) and quality-control (QC) procedures ensure that the type, amount, and quality of data are adequate to meet the objectives of the study and are particularly critical if any of the data are to be admissible in court. If data are collected with legal proceedings in mind, a Quality Assurance Project Plan (QAPP) contributes to ensuring that the data will bear critical scrutiny and be scientifically and legally defensible. QA/QC and documentation procedures should be applied from the beginning of the study to ensure that all appropriate data required to meet the study objectives are collected or generated. QA is a management function based on a systematic approach to data collection and documentation linked to the policies or regulations that determine and define the study objectives. QC is the procedure applied to check the quality of the data generated by the field program and covers measurement, documentation, sampling, and analytical techniques. Although these are integral tasks within any oil spill investigation, it is important to remember that they are support services that ensure the quality of the data and, themselves, are not an end product (Robilliard et al., 1991).

2.10 Conclusions

A systematic and well-designed study approach is key to successful forensic oil spill investigations. In the first phase following a spill, or detection of a spill, time is always of the essence, as the situation is dynamic and many data elements and observations integral to the investigation are ephemeral. Key parameters need to be evaluated, measured, and documented as soon as possible. Immediate priorities should be defined so that appropriate samples are collected at necessary locations, times, and frequency to capture critical forensic elements. We can learn from past cases, such as the techniques used to source the "mystery" slicks from the *Luckenbach*, but since every oil spill is different, in a unique environment, a successful study typically involves a degree of creativity that extends beyond simply employing standard investigative and scientific procedures. Creativity and thoroughness are critical in order to adapt to specific spatial and temporal conditions. In this respect, for spills in marine, lake, or river environments, the scientist is not that different from the spill responder in attempting to act quickly and safely in a dynamic and ever-changing situation. The study design should include a QA/QC and a data management plan to ensure a comprehensive, systematic investigation. Although the need to quickly initiate an investigation and comprehensive sampling plan is an inherent and required aspect to all oil spills, an important caveat to note is that rapid actions are not a substitute for quality and carefulness, particularly if study results are not defensible and cannot be used to precisely identify the source.

References

ASTM D5745-95, Developing and Implementing Short-Term Measures or Early Actions for Site Remediation, American Society for Testing and Materials, West Conshohoken, VA, 1999.

ASTM E1739-95, Standard Guide for Risk-Based Corrective Action Applied at Petroleum Release Sites, American Society for Testing and Materials, West Conshohoken, VA, 1995.

ASTM E1912-98, Standard Guide for Accelerated Site Characterization for Confirmed or Suspected Petroleum Releases, American Society for Testing and Materials, West Conshohoken, VA, 2004a.

ASTM E1943-98, Remediation of Ground Water by Natural Attenuation at Petroleum Release Sites, American Society for Testing and Materials, West Conshohoken, VA, 2004b.

ASTM F1011-86, Standard Guide for Developing a Hazardous Materials Training Curriculum for Initial Response Personnel, American Society for Testing and Materials, West Conshohoken, VA, 1986.

ASTM F1656-01, Standard Guide for Health and Safety Training of Oil Spill Responders in the United States, American Society for Testing and Materials, West Conshohoken, VA, 2001b.

ASTM F1686-97, Standard Guide for Surveys to Document and Assess Oiling Conditions on Shorelines, American Society for Testing and Materials, West Conshohoken, VA, 1997a.

ASTM F1687-97(2003) Standard Guide for Terminology and Indices to Describe Oiling Conditions on Shorelines, American Society for Testing and Materials, West Conshohoken, VA, 1997b.

ASTM, F1644-01, Standard Guide for Health and Safety Training of Oil Spill Responders, American Society for Testing and Materials, West Conshohoken, VA, 2001a.

Brown, J., Study Element 2, Water Column and Oil Source Sampling. In S.B. Robertson (ed.), Guidelines for the Scientific Study of Oil Spill Effects. Prepared for PERF (Petroleum Environmental Research Forum) Project 94-10, 1999.

Chamberlin, D., Study Element 13, Quality Assurance/Quality Control. In S.B. Robertson (ed.), Guidelines for the Scientific Study of Oil Spill Effects. Prepared for PERF (Petroleum Environmental Research Forum) Project 94-10, 1999.

ExxonMobil, Oil Spill Response Field Manual. ExxonMobil Research and Engineering Com., Faifax, VA, 2002.

Gundlach, E. and T. Coogan. Study Element 14, Data Management. In S.B. Robertson (ed.), Guidelines for the Scientific Study of Oil Spill Effects. Prepared for PERF (Petroleum Environmental Research Forum) Project 94-10, 1999.

Holliday, M.G. and J.M. Park. Occupational Health Implications of Crude Oil Exposure: Literature Review and Research Needs. Marine Spill Response Corporation, Washington, D.C. MSRC Technical Report Series 93-007, 1993, 55pp.

IPIECA. Oil Spill Responder Safety Guide. International Petroleum Industry Environmental Conservation Association. IPIECA Report Series, Vol. 11. London, UK, 2002.

Lamarche, A. and J.A. Roberts. Framework for the Management of Digital Images Used to Document Oiled Shorelines. *Proceedings 27th Arctic and Marine Oils Spill Program (AMOP) Technical Seminar,* Environment Canada, Ottawa ON, 2004, 235–244.

Lamarche, A., J. Ion, E.H. Owens, and P. Rubec. Providing Successful SCAT Data Management Support during Spill Response. In Proceedings International Oil Spill Conference, American Petroleum Institute, Publication 4686B, Washington DC, 1999, 943–945.

Lamarche, A., E.H. Owens, and G.A. Sergy. Development of a SCAT Data Management Manual. *Proceedings 28th Arctic and Marine Oilspill Programme (AMOP) Technical Seminar*, Environment Canada, Ottawa, ON, 2005, 473–490.

MCGA. Scientific, Technical and Operational Guidance Note-STOp 1/98. www.mcga.gov.uk/c4mcga-stop1_98.pdf, 1998.

McGrath, G., H. Parker-Hall, J. Tarpley, and A. Nack. The Investigation to Identify the SS JACOB LUCKENBACH—Using Technology to Locate a Hidden Source of Oil that Caused Years of Impacts and the Future Implications of Sunken Shipwrecks. In Proceedings, International Oil Spill Conference. American Petroleum Institute, Washington, D.C., 2003, 1219–1224.

MSRC. Worker Health and Safety Workshop, Washington DC (January 7–8, 1993). Marine Spill Response Corporation (MSRC), Washington DC. Technical Report No. 93-012, 1993.

NIOSH (National Institute for Occupational Safety and Health). Pocket Guide to Chemical Hazards, 2005.

OSHA. Training Marine Oil Spill Response Workers under OSHA's Hazardous Waste Operations and Emergency Response Standard. U.S. Department of Labor and the Occupational Safety and Health Administration (OSHA). Washington, DC., 2001.

Owens, E., Study Element 1. Overflights, Photodocumentation, & Shore Description. In S.B. Robertson (ed.), Guidelines for the Scientific Study of Oil Spill Effects. Prepared for PERF (Petroleum Environmental Research Forum) Project 94-10, 1999.

Owens, E.H. and G.A. Sergy. *The SCAT Manual— A Field Guide to the Documentation and Description of Oiled Shorelines (Second Edition)*. Environment Canada, Edmonton AB, 2000.

Owens, E.H., Response Strategies for Spills on Land. *Spill Science and Technology Bulletin,* 2002, vol. 7, no. 3/4, 115–117.

Owens, E.H., G.S. Mauseth, C.A. Martin, A. LaMarche, and J. Brown. Tar ball frequency data and analytical results from a long-term beach monitoring program, *Marine Pollution Bulletin,* 2002, vol. 44, 770–780.

Robertson, S.B. (ed.). Guidelines for the Scientific Study of Oil Spill Effects. Prepared for PERF (Petroleum Environmental Research Forum) Project 94-10, 1999.

Robilliard, G.A., W.H. Desvousges, and R.W. Dinford. Study Element 10, Data Management, QA/QC and Documentation. In Natural Resource Damage Assessment Guidance Manual, Volume 2, Reference Material and Appendices. Report prepared for PERF (Petroleum Environmental Research Forum), 1999.

Taylor, E., Study Element 3, Shoreline and Sediment Sampling. In Guidelines for the Scientific Study of Oil Spill Effects. Robertson, S.B. (ed.), Prepared for PERF (Petroleum Environmental Research Forum) Project 94-10, 1999.

3 Petroleum Biomarker Fingerprinting for Oil Spill Characterization and Source Identification

Zhendi Wang,[1] Chun Yang,[1] Merv Fingas,[1] Bruce Hollebone,[1] Un Hyuk Yim,[2] and Jae Ryoung Oh[2]

[1] Emergencies Science and Technology Division, Environmental Technology Centre, Environment Canada, 335 River Road, Ottawa, Ontario, Canada K1A 0H3.
[2] South Sea Institute, Korea Ocean R & D Institute, 391 Jangmok-Ri, Jangmok-Myon, Geoje-Shi, Kyungnam 656-830, Republic of Korea.

3.1 Introduction

Biological markers, or biomarkers, are one of the most important hydrocarbon groups in petroleum used for chemical fingerprinting. They are complex molecules derived from formerly living organisms. Biomarkers found in crude oils, rocks, and sediments have little or no changes in structures from their parent biochemicals, or so-called biogenic precursors (e.g., terpanoids and steroids), found in living organisms. In comparison with the concentrations of the biogenic precursors in sediments, biomarker concentrations in oil are low, often in the range of several to less than a hundred parts per million (ppm).

Biomarkers are useful for chemical fingerprinting of spilled oils because they retain all or most of the original carbon skeleton of the original natural product, and thereby testify to the specific conditions for oil generation (see Chapter 1 herein). Excellent reviews on the fundamentals of biomarker characterization, their application in petroleum geochemistry, and interpretation of biomarker data for oil exploration and production were published in 1993 (Peters and Moldowan, 1993). A fully updated and expanded edition provides a comprehensive account of the role that biomarker technology plays both in petroleum exploration and in understanding earth history and processes, including environmental applications (Peters et al., 2005). More recently, Wang et al. (2006) have reviewed the environmental applications of biomarker fingerprinting.

Biomarker fingerprinting has historically been used by petroleum geochemists in characterization of oils in terms of (1) oil-to-oil and oil-to-source rock correlation, (2) the type(s) of precursor organic matter present in the source rock, (3) effective ranking of the relative thermal maturity of petroleum, (4) evaluation of migration and the degree of in-reservoir biodegradation based on the loss of n-alkanes, isoprenoids, aromatics, terpanes, and steranes during biodegradation, (5) determination of depositional environmental conditions, and (6) providing information on the age of the source rock for petroleum.

Biomarkers can be detected in low quantities (ppm and sub-ppm level) in the presence of a wide variety of other types of petroleum hydrocarbons by the use of the gas chromatography-mass spectrometry (GC-MS). Relative to other hydrocarbon groups such as alkanes and most aromatic compounds, biomarkers are highly resistant to degradation in the environment (see Chapter 11 herein). Furthermore, due to the wide variety of geological conditions and ages under which oil has formed, every crude oil may exhibit an essentially unique biomarker fingerprint. Therefore, chemical analysis of biomarkers can generate highly specific "source" information of great importance to environmental forensic investigations in terms of determining the source of spilled oil, differentiating and correlating oils, studying the fate and behavior of hydrocarbons in the environment, and monitoring the degradation process and weathering state of oils under a wide variety of environmental conditions. They have also proven useful in identification of petroleum-derived contaminants in the marine and aquatic environments (Stout et al., 2002; Kvenvolden et al., 1995, 2002; Hostettler et al., 1999a; Boehm et al., 1997; Bence et al., 1996; Volkman et al., 1997; Zakaria et al., 2000; Wang et al., 1994a, 1994b, 1999a) and in indicating chronic industrial and urban releases (Stout et al., 1998; Volkman et al., 1992a; Kaplan et al., 1997).

In this chapter we will focus our discussion on a brief description of biomarker chemistry, an overview of analytical methodologies for biomarker separation and analysis, the identification of biomarkers, biomarker distributions in crude oils and various petroleum products, sesquiterpane and diamondoid biomarkers in oils and lighter petroleum products, diagnostic ratios and cross-plots of biomarkers, source-specific biomarkers, weathering effects on oil and biomarker fingerprinting, and an application of biomarkers for oil spill source identification, oil correlation, and differentiation.

3.2 Analytical Methodologies for Petroleum Biomarker Fingerprinting

3.2.1 Petroleum Biomarker Families

Oil consists of complex mixtures of hydrocarbons and nonhydrocarbons that range from small, volatile compounds to large, nonvolatile ones. For example, recently, ultrahigh-resolution Fourier transform ion cyclotron resonance mass spectrometry (Marshall, 2004) revealed that crude oil contains heteroatom-containing (N, O, S) organic components having more than 20,000 distinct elemental compositions ($C_cH_hN_nO_oS_s$). In general, petroleum hydrocarbons are characterized and classified chemically by their structures, including saturates (including straight-chain and branched chain saturates, cycloalkanes, terpanes, and steranes); olefins; aromatics (including the monoaromatic hydrocarbons such as BTEX and other alkyl-substituted benzene compounds, and oil-characteristic alkylated C_0- to C_4-PAH homologous series and other U.S. EPA priority PAHs ranging from 2-ring up through 6-rings); and polar resins (including heterocyclic S, N, and O containing compounds, phenols, acids, alcohols, and monoaromatic steroids) and very high-molecular-weight asphaltenes (Speight, 1999; Berkowitz, 1997).

In 1887, German chemist Otto Wallach determined the structures of several terpenes and discovered that all of them are composed of two or more five-carbon units of isoprene [2-methyl-1,3-butadiene, $CH_2\!\!=\!\!C(CH_3)\!\!-\!\!CH\!\!=\!\!CH_2$]. The isoprene unit maintains its isopentyl structure in a terpene, usually with modification of the isoprene double bonds. The isoprene molecule and the isoprene unit are said to have a "head" (the branched end) and a "tail" (the unbranched ethyl group). Organic chemists and geochemists have long realized that *isoprene* is the basic structural unit of many natural products and all oil biomarker compounds (Peters and Moldowan, 1993; Wade, 2003). Compounds composed of isoprene subunits (that is, obeying the "isoprene rule") are called terpenoids or isoprenoids. The triterpenoids constitute a large diverse group of natural products (Connolly and Hill, 1991). Terpenoids are ubiquitous in microorganisms and in higher and lower plants, and have been characterized to an increasing extent within the animal kingdom. Few have been known for centuries, but in recent decades the level of research and activity in isolating and studying new substances has shown no sign of abating, and the discovery of completely new carbon skeletons among the naturally occurring plant and animal terpenoids is a frequent occurrence.

Terpenoids are grouped according to the number of isoprene units from which they are biogenetically derived, even though some carbons may have been added or lost (Connolly and Hill, 1991). The isoprene rule states that biosynthesis of these compounds occurs by polymerization of appropriately functionized C_5-isoprene subunits. Unlike other biopolymers such as proteins, terpenoids are not readily depolymerized because they are joined together by covalent carbon–carbon bonds. As for the oil-saturated terpenoids, they are generally categorized into families based on the approximate number of isoprene subunits they contain. Terpenoids containing 1 to 8 isoprene subunits are termed as hemi-, mono-, sesqui-, di-, sester-, tri-, and tetra-terpanes. The various oil terpane families are composed of a wide variety of acyclic and cyclic structures (Peters and Moldowan, 1993).

3.2.1.1 Acyclic Terpenoids or Isoprenoids

One of the most important discoveries in petroleum chemistry and organic geochemistry was the detection of a large number of aliphatic isoprenoid hydrocarbons in oils, coals, shales, and dispersed organic materials. The variety of isoprenoid compounds is incomparably large. The linkages between isoprene subunits can be regular (head-to-tail) or irregular (differing in the order of attachment of the isoprene subunits, such as head-to-head or tail-to-tail) linkages. Phytane ($C_{20}H_{42}$), which is one of the most abundant isoprenoids in oil and has been widely used for estimation of the degree of oil biodegradation in the environment, is a typical example of a regular, acyclic isoprenoid consisting of four head-to-tail linked isoprene units. Squalane ($C_{30}H_{62}$) and Botryococcane ($C_{34}H_{70}$) are examples of irregular isoprenoids. Squalane contains six isoprene subunits with one tail-to-tail linkage, while irregular Botryococcane is a highly specific biomarker for lacustrine sedimentation.

Degraded, rearranged, or homologous structures can be categorized into their corresponding parent terpenoid family. The precise number of carbon atoms in a given terpenoid family varies due to differences in source materials, diagenesis, thermal maturity, and in-reservoir biodegradation. For example, pristane ($C_{19}H_{40}$), another isoprenoid compound widely used for environmental oil biodegradation studies, contains one less methylene group (—CH_2—) than phytane ($C_{20}H_{42}$), but it is still classified as an acyclic diterpane. Other examples include pseudohomologous series of regular isoprenoids from C_{15} (farnesane) through C_{16} (trimethyl-C_{13}) and C_{18} (norpristane), which are also quite abundant in oil.

3.2.1.2 Cyclic Terpenoids

The most common cyclic terpenoids in oil are terpanes, steranes (irregular cyclic terpernoid compounds), and aromatic steranes. Although cyclic terpenoids containing almost any number of carbons can occur in theory, only those containing combinations of five or six carbons (cyclopentyl or cyclohexyl) occur commonly in petroleum.

As mentioned above, the terpanes include sesqui- (C_{15}, bicyclic), di- (C_{20}, largely tricyclic), and triterpanes (C_{30}, mainly pentacyclic, and some tricyclic and tetracyclic), which are found in most crude oils. The terpanes comprise several homologous series, including bicyclic, tricyclic, tetracyclic, and pentacyclic compounds. Hopanes are pentacyclic triterpanes commonly containing 27 to 35 carbon atoms in a naphthenic structure composed of four six-membered rings and one five-membered ring (Figure 3-1). Hopanes with the 17α(H), 21β(H)-configuration in the range of C_{27} to C_{35} are characteristic of petroleum because of their large abundance and thermodynamic stability compared to other epimeric (ββ and βα) series.

The four-ringed steranes are a class of biomarkers containing 21 to 30 carbons,

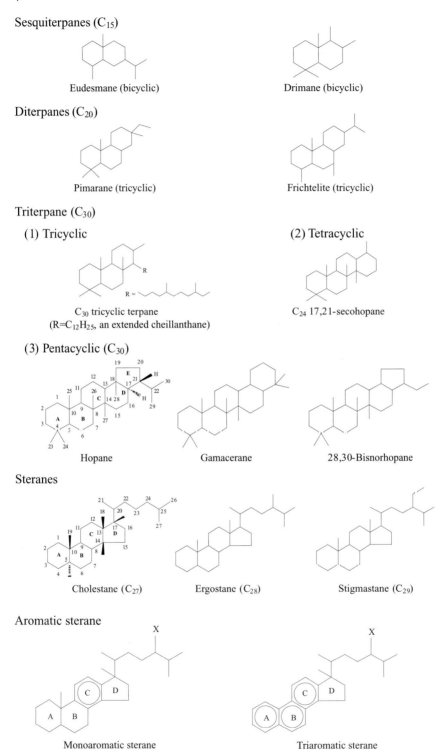

Figure 3-1 Molecular structures of example cyclic terpenoid compounds in oil.

including regular steranes, rearranged diasteranes, and mono- and triaromatic steranes. Among them, the regular C_{27}—C_{28}—C_{29} homologous sterane series (cholestane, ergostane, and stigmastane) are the most common steranes and are useful for chemical fingerprinting because of their high source specificity. These sterane homologue series do not contain an integral number of isoprene subunits, and thus only approximate the isoprene rule. However, they still show some terpenoid character and can be categorized into the corresponding cyclic terpenoid families.

Aromatic steranes are another group of biomarker compounds found in the oil aromatic hydrocarbon fraction. These compounds can also provide valuable information for forensic investigations on oil-to-oil correlation, differentiation, and source identification. The C-ring monoaromatic (MA) steranes are characterized by a series of 20R and 20S C_{27}—C_{28}—C_{29} 5α- and 5β-cholestanes, ergostanes, and stigmastanes. The ABC-ring triaromatic steranes are formed from aromatization of C-ring monoaromatic steranes involving the loss of a methyl group at the A/B ring junction. This fraction is composed mainly of C_{20} and C_{21}, and C_{26}—C_{27}—C_{28} homologous triaromatic steranes. Examples of monoaromatic and triaromatic steranes are also shown in Figure 3-1. As a summary, Table 3-1 lists important biomarker terpane, sterane, and aromatic sterane compounds, used frequently for forensic oil spill studies.

3.2.2 Labeling and Nomenclature of Biomarkers

The chemical structures of terpenoids are more complicated than that of normal alkanes and isoalkanes (Peters and Moldowan, 1993; Morrison et al., 1992; Wade, 2003). The system used for the nomenclature of terpenoids has evolved over many years. For many terpenoid classes, several names have been proposed for the carbon skeleton, but the basic rules of the IUPAC (International Union of Pure and Applied Chemistry) system are used for nomenclature of biomarkers. For example, the acyclic isoprenoids pristane ($C_{19}H_{40}$) and phytane ($C_{20}H_{42}$) are named as 2,6,10,14-tetramethylpentadecane and 2,6,10,14-tetramethylhexadecane, respectively.

Cyclic triterpanes and steranes (Figure 3-1) are labeled according to the following rules (Peters and Moldowan, 1993; Wade, 2003): (1) each carbon atom and the rings in biomarker molecules are labeled systematically. Rings are specified in succession from left to right as the A-ring, B-ring, C-ring, D-ring, and so on. (2) A capital "C" followed immediately by a subscript number refers to the number of carbon atoms in a particular compound (e.g., C_{30} hopane and C_{27} sterane mean that they contain 30 and 27 carbon atoms, respectively). (3) A capital "C" followed by a dash and numbers refers to a particular position within the compound [e.g., C-17 and C-21 in the 17α(H), 21β(H)-hopane is the carbon atoms at positions 17 and 21]. (4) Prefixes are used to indicate the changes to the normal biomarker carbon skeleton, which include the prefixes *nor-*, *seco-*, *neo-*, and others. Table 3-2 summarizes the nomenclature used to modify the structural specification of cyclic biomarkers. The prefix *nor-* is used to indicate loss of carbons from a carbon skeleton. For example, 17α(H), 21β(H)-30-norhopane is identical to C_{30} 17α(H), 21β(H)-hopane except that a methyl group at the C-30 position has been lost from its point of attachment at the C-22 position. Similarly, 25-norhopanes are identical to C_{30} hopane except that a methyl group (at C-25) has been removed from its point of attachment at the C-10 position. If two or three carbons are lost, the prefix "*bisnor-*" or "*trisnor-*" is used, respectively. Thus, 28-, 30-bisnorhopanes have two methyl groups (at C-28 and C-30) removed from their parent C_{30} hopane. The prefix "*seco-*" is used to indicate cleavage of a bond, with the locants for both ends of the broken bond given, e.g., 3,4-secoeudesmane. The 17, 21-secohopane indicates that the bond between carbon number 17 and 21 in the E-ring of C_{30} hopane has been broken, resulting in the formation of a new tetracyclic terpane. The prefix "*homo-*" is used

Table 3-1 Petroleum Biomarkers Frequently used for Forensic Oil Spill Studies

Peak	Compound	Code	Empirical formula	Target ions
	Sesquiterpanes (Bicyclic terpanes)			
	C_{14} sesquiterpanes		$C_{14}H_{26}$	123, 179
	C_{15} sesquiterpanes		$C_{15}H_{28}$	123, 193
	C_{16} sesquiterpanes		$C_{16}H_{30}$	123, 193, 207
	Diamondoids			
	Adamantanes		$C_{10}H_{16}$, alkyl-$C_{10}H_{15}$	136, 135, 149, 163, 177
	Diamantanes		$C_{14}H_{20}$, alkyl-$C_{14}H_{19}$	188, 187, 201, 215, 229
	Terpanes			
1	C_{19} tricyclic terpane	TR19	$C_{19}H_{34}$	191
2	C_{20} tricyclic terpane	TR20	$C_{20}H_{36}$	191
3	C_{21} tricyclic terpane	TR21	$C_{21}H_{38}$	191
4	C_{22} tricyclic terpane	TR22	$C_{22}H_{40}$	191
5	C_{23} tricyclic terpane	TR23	$C_{23}H_{42}$	191
6	C_{24} tricyclic terpane	TR24	$C_{24}H_{44}$	191
7	C_{25} tricyclic terpane (a)	TR25A	$C_{25}H_{46}$	191
8	C_{25} tricyclic terpane (b)	TR25B	$C_{25}H_{46}$	191
9	triplet: C_{24} tetracyclic terpane + C_{26} (S + R) tricyclic terpanes	TET24 + TR26A + TR26B	$C_{24}H_{42}$ + $C_{26}H_{48}$	191
10	C_{28} tricyclic terpane (a)	TR28A	$C_{28}H_{52}$	191
11	C_{28} tricyclic terpane (b)	TR28B	$C_{28}H_{52}$	191
12	C_{29} tricyclic terpane (a)	TR29A	$C_{29}H_{54}$	191
13	C_{29} tricyclic terpane (b)	TR29B	$C_{29}H_{54}$	191
14	Ts: $18\alpha(H),21\beta(H)$-22,29,30-trisnorhopane	Ts	$C_{27}H_{46}$	191
15	$17\alpha(H),18\alpha(H),21\beta(H)$-25,28,30-trisnorhopane	TH27	$C_{27}H_{46}$	191, 177
16	Tm: $17\alpha(H),21\beta(H)$-22,29,30-trisnorhopane	Tm	$C_{27}H_{46}$	191
17	C_{30} tricyclic terpane 1	TR30A	$C_{30}H_{56}$	191
18	C_{30} tricyclic terpane 2	TR30B	$C_{30}H_{56}$	191
19	$17\alpha(H),18\alpha(H),21\beta(H)$-28,30-bisnorhopane	H28	$C_{28}H_{48}$	191, 163
20	$17\alpha(H),21\beta(H)$-25-norhopane	NOR25H	$C_{29}H_{50}$	191, 177
21	$17\alpha(H),21\beta(H)$-30-norhopane	H29	$C_{29}H_{50}$	191
22	$18\alpha(H),21\beta(H)$-30-norneohopane (C_{29}Ts)	C29Ts	$C_{29}H_{50}$	191
23	$17\alpha(H)$-diahopane	DH30	$C_{30}H_{52}$	191,
24	$17\alpha(H),21\beta(H)$-30-norhopane (normoretane)	M29	$C_{29}H_{50}$	191
25	$18\alpha(H)$ and $18\beta(H)$-oleanane	OL	$C_{30}H_{52}$	191, 412
26	$17\alpha(H),21\beta(H)$-hopane	H30	$C_{30}H_{52}$	191
27	$17\alpha(H)$-30-nor-29-homohopane	NOR30H	$C_{30}H_{52}$	191
28	$17\beta(H),21\alpha(H)$-hopane (moretane)	M30	$C_{30}H_{52}$	191
29	$22S$-$17\alpha(H),21\beta(H)$-30 homohopane	H31S	$C_{31}H_{54}$	191
30	$22R$-$17\alpha(H),21\beta(H)$-30-homohopane	H31R	$C_{31}H_{54}$	191
31	Gammacerane	GAM	$C_{30}H_{52}$	191, 412
32	$17\beta(H),21\beta(H)$-hopane	(IS)	(Internal standard)	191
33	$22S$-$17\alpha(H),21\beta(H)$-30,31-bishomohopane	H32S	$C_{32}H_{56}$	191
34	$22R$-$17\alpha(H),21\beta(H)$-30,31-bishomohopane	H32R	$C_{32}H_{56}$	191
35	$22S$-$17\alpha(H),21\beta(H)$-30,31,32-trishomohopane	H33S	$C_{33}H_{58}$	191
36	$22R$-$17\alpha(H),21\beta(H)$-30,31,32-trishomohopane	H33R	$C_{33}H_{58}$	191
37	$22S$-$17\alpha(H),21\beta(H)$-30,31,32,33-tetrakishomohopane	H314S	$C_{34}H_{60}$	191
38	$22R$-$17\alpha(H),21\beta(H)$-30,31,32,33-tetrakishomohopane	H34R	$C_{34}H_{60}$	191
39	$22S$-$17\alpha(H),21\beta(H)$-30,31,32,33,34-pentakishomohopane	H35S	$C_{35}H_{62}$	191
40	$22R$-$17\alpha(H),21\beta(H)$-30,31,32,33,34-pentakishomohopane	H35R	$C_{35}H_{62}$	191
	Steranes			
41	C_{20} $5\alpha(H),14\alpha(H),17\alpha(H)$-sterane	S20	$C_{20}H_{34}$	217 & 218
42	C_{21} $5\alpha(H),14\beta(H),17\beta(H)$-sterane	S21	$C_{21}H_{36}$	217 & 218
43	C_{22} $5\alpha(H),14\beta(H),17\beta(H)$-sterane	S22	$C_{22}H_{38}$	217 & 218
44	C_{27} $20S$-$13\beta(H),17\alpha(H)$-diasterane	DIA27S	$C_{27}H_{48}$	217 & 218, 259
45	C_{27} $20R$-$13\beta(H),17\alpha(H)$-diasterane	DIA27R	$C_{27}H_{48}$	217 & 218, 259
46	C_{27} $20S$-$13\alpha(H),17\beta(H)$-diasterane	DIA27S2	$C_{27}H_{48}$	217 & 218, 259
47	C_{27} $20R$-$13\alpha(H),17\beta(H)$-diasterane	DIA27R2	$C_{27}H_{48}$	217 & 218, 259
48	C_{28} $20S$-$13\beta(H),17\alpha(H)$-diasterane	DIA28S	$C_{28}H_{50}$	217 & 218, 259
49	C_{28} $20R$-$13\beta(H),17\alpha(H)$-diasterane	DIA28R	$C_{28}H_{50}$	217 & 218, 259
50	C_{29} $20S$-$13\beta(H),17\alpha(H)$-diasterane	DIA29S	$C_{29}H_{52}$	217 & 218, 259
51	C_{29} $20R$-$13\alpha(H),17\beta(H)$-diasterane	DIA29R	$C_{29}H_{52}$	217 & 218, 259
52	C_{27} $20S$-$5\alpha(H),14\alpha(H),17\alpha(H)$-cholestane	C27S	$C_{27}H_{48}$	217 & 218
53	C_{27} $20R$-$5\alpha(H),14\beta(H),17\beta(H)$-cholestane	C27$\beta\beta$R	$C_{27}H_{48}$	217 & 218
54	C_{27} $20S$-$5\alpha(H),14\beta(H),17\beta(H)$-cholestane	C27$\beta\beta$S	$C_{27}H_{48}$	217 & 218
55	C_{27} $20R$-$5\alpha(H),14\alpha(H),17\alpha(H)$-cholestane	C27R	$C_{27}H_{48}$	217 & 218
56	C_{28} $20S$-$5\alpha(H),14\alpha(H),17\alpha(H)$-ergostane	C28S	$C_{28}H_{50}$	217 & 218
57	C_{28} $20R$-$5\alpha(H),14\beta(H),17\beta(H)$-ergostane	C28$\beta\beta$R	$C_{28}H_{50}$	217 & 218
58	C_{28} $20S$-$5\alpha(H),14\beta(H),17\beta(H)$-ergostane	C28$\beta\beta$S	$C_{28}H_{50}$	217 & 218
59	C_{28} $20R$-$5\alpha(H),14\alpha(H),17\alpha(H)$-ergostane	C28R	$C_{28}H_{50}$	217 & 218
60	C_{29} $20S$-$5\alpha(H),14\alpha(H),17\alpha(H)$-stigmastane	C29S	$C_{29}H_{52}$	217 & 218
61	C_{29} $20R$-$5\alpha(H),14\beta(H),17\beta(H)$-stigmastane	C29$\beta\beta$R	$C_{29}H_{52}$	217 & 218
62	C_{29} $20S$-$5\alpha(H),14\beta(H),17\beta(H)$-stigmastane	C29$\beta\beta$S	$C_{29}H_{52}$	217 & 218
63	C_{29} $20R$-$5\alpha(H),14\alpha(H),17\alpha(H)$-stigmastane	C29R	$C_{29}H_{52}$	217 & 218
64	C30 steranes	C30S	$C_{30}H_{54}$	217 & 218
	Monoaromatic steranes			253
	Triaromatic steranes			231

Table 3-2 Common Modifiers and Nomenclatures Used to Modify the Structural Specification of Cyclic Biomarkers

Modifier	Description	Example Biomarker
homo-	one additional carbon on the parent molecular structure	C_{31} 17α(H),21β(H)-30-homohopane
bis-, tris-, tetrakis-, pentakis- (also di-, tri-, tetra-, and penta-)	two to five additional carbons on the parent molecular structure	C_{32} 17α,21β-30,31-bishomohopane C_{33} 17α,21β-30,31,32-trishomohopane C_{34} 17α,21β-30,31,32,33-tetrakishomohopane C_{35} 17α,21β-30,31,32,33,34-pentakishomohopane
seco-	cleaved C-C bond	C_{24} 17,21-secohopane (tetracyclic)
nor-	one less carbon on the parent molecular structure	25-norhopane
bisnor-	two less carbons on the parent molecular structure	28,30-bisnorhopane
trisnor-	three less carbons on the parent molecular structure	25,28,30-trisnorhopane
neo-	methyl group shifted from C-18 to C-17 position on hopanes	C_{29}Ts: 30-norneohopane
α	asymmetric carbon in ring with "H" down	17α(H),21β(H)-hopane
β	asymmetric carbon in ring with "H" up	17β(H),21β(H)-hopane
R	asymmetric carbon in acyclic moiety of biomarkers obeying convention in a clockwise direction	C_{27} 20R cholestane
S	asymmetric carbon in acyclic moiety of biomarkers obeying convention in a clockwise direction	C_{27} 20S cholestane

*Modified from Peters and Moldowan (1993).

to indicate addition of a carbon from the parent carbon skeleton, for example, 30-homohopanes are identical to C_{30} hopane except that a methyl group has been added at C-30 position. If two to five carbons are added on the parent molecular structure, the prefixes "*bis-*", "*tris-*", "*tetrakis-*", and "*pentakis-*" are used, respectively. For more information on compound-naming protocols, see *Appendix IV* in the current *Chemical Abstract Index Guide* (CAS, 2002). *Chemical Abstract* uses these prefixes extensively for classes of terpenoids.

3.2.2.1 Stereoisomers

Isomers are different compounds that have the same molecular formula but the atoms are attached in different ways. There are two classes of isomers (Figure 3-2): (1) *constitutional isomers* and (2) *stereoisomers*. Constitutional isomers (or structural isomers) differ in their bonding sequence, and their atoms are connected differently and the number of *constitutional* isomers increases dramatically with the increase of carbon atoms in each compound. For example, there are two constitutional isomers of butane (C_4H_{10}: *n*-butane and isobutane), three isomers of pentane (C_5H_{12}: *n*-pentane, isopentane, and neopentane), respectively, five isomers of hexane, 18 isomers of octane, 75 possible isomers of decane, and 355 possible isomers of eicosane ($C_{20}H_{42}$), respectively.

Stereoisomers are isomers whose atoms are bonded together in the same sequence but differ from each other in the orientation of the atoms in space. Stereoisomers that are mirror images of each other (i.e., differing in the same manner as right and left hands) are called *enantiomers*; while all other stereoisomers, which are not mirror images, are *diaste-*

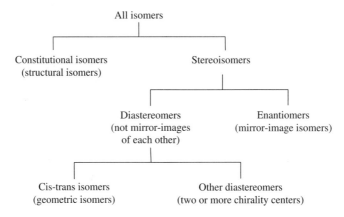

Figure 3-2 Types of isomers.

reomers. The *cis-trans* geometric isomers (such as *cis-* and *trans*-1,2-dimethylcyclopentane) are special types of *diastereomers*. Enantiomer molecules are not superimposable. Many pairs of biomarkers with the same molecular formula (such as 22R and 22S homohopane homologous series in C_{31} to C_{35} range) are *enantiomers*. Differences in special orientation might seem unimportant, but stereoisomers often have remarkably different physical, chemical, and biological properties.

3.2.2.2 Asymmetric (or Chiral) Carbons and α and β Stereoisomers

A carbon atom bonded to four different groups is called *asymmetric* carbon or a *chiral* carbon atom and is often designed by a *. For example, the possible chiral centers for steranes are at C-5, C-14, C-17, C-20, and C-24 (for C_{28} and C_{29} steranes).

As described in Table 3-2, hydrogen atoms attached to an *asymmetric* or *chiral* carbon in a ring structure and are below the plane of the molecule are called α hydrogens, and the bond is drawn with a dashed line and designated as having the α-configuration. Conversely, hydrogen atoms located above the plane of the molecule are called β hydrogens, and the bond is drawn with a wedge bond and designated as having β-configuration. In many common ring systems the α hydrogen atoms found at ring junctions are generally omitted for clarity. For example, in 17α(H), 21β(H)-hopane ($C_{30}H_{52}$, Figure 3-1) the hydrogens at carbon numbers 17 and 21 are down and up; while in 5α(H), 14β(H), 17β(H)-cholestane ($C_{27}H_{48}$, Figure 3-1) the hydrogens attached to carbon numbers 5, 14, and 17 are down, up, and up.

Previously, hopanes were considered to exist as three stereoisomers: 17α(H), 21β(H)-hopane, 17β(H), 21β(H)-hopane, and 17β(H), 21α(H)-hopane (Peters and Moldowan, 1993; Waples and Machihara, 1991). Hopanes in the βα series are also called moretanes. Hopanes with the αβ-configuration in the range of C_{27} to C_{35} are characteristic of petroleum because of their greater thermodynamic stability compared to other epimeric series (ββ and βα). Hopanoids produced by living organisms have generally a ββ-configuration. With increasing maturity the thermodynamically less stable ββ-hopanes are lost or converted to αβ- and βα-hopanes. The ββ series are, generally, not found in petroleum because it is thermally unstable. It was considered that the αα series were not natural products, and it is unlikely that they occur in more than trace levels in petroleum. However, mechanics calculations have shown that the αα-hopanes should be less stable than αβ- and βα-hopanes, but more stable than ββ-hopanes. Recently, Nytoft and Bojesen-Koefoed (2001) found that moderate quantities of 17α(H), 21α(H)-hopanes are present in several sediments and oils. The ratios of C_{30} 17α(H), 21α(H)-hopane to C_{30} 17α(H), 21β(H)-hopane are typically 0.02–0.04 in crude oils and mature sediments, but ratios up to 0.10 have been found in immature sediments.

3.2.2.3 R and S Stereoisomers of Cyclic Biomarkers

Since chiral molecules are not superimposable on their *mirror images*, chirality is a necessary and sufficient condition for existence of enantiomers. Thus, a compound with at least one chiral carbon atom can exist as enantiomers, whereas a compound without chirality cannot exist as enantiomers. The Cahn–Ingold–Prelog convention procedure proposed by R. S. Cahn, C. Ingold, and V. Prelog (Cahn et al., 1966) is the most widely accepted system for naming the *configurations* (the arrangement of atoms that characterizes a particular stereoisomer of chiral centers). Each asymmetric carbon atom is assigned a letter (R) or (S) based on its three-dimensional configuration. To determine the stereoisomeric (R or S) configuration, two steps are involved: (1) following the *sequence rules* (Cahn et al., 1966), the *sequence of priority* is assigned to the four atoms or groups of atoms bonded to the asymmetric carbon atom. In the case of bromochloroiodomethane (CHClBrI), the four atoms attached to the chiral center are all different and priority depends on the atomic number, the atom of higher number having higher priority, thus, the sequence of priority is I, Br, Cl, H. (2) The molecule is oriented so that the group of lowest priority is directed away from the viewer. Subsequently, the remaining groups are arranged. If proceeding the remaining groups in a clockwise direction, that is, from the group of the highest priority to the group of second priority and then to the third, the configuration is specified **R** (Latin: *rectus*, meaning right); if counterclockwise, the configuration is specified **S** (Latin: *sinister*, meaning left). Thus, the compound CHClBrI has two stereoisomers and specified as the R and S enantiomers (their mirror images are nonsuperimposable), respectively.

For cyclic biomarkers, the use of R and S nomenclature is generally restricted to carbon atoms that are not part of a ring, while the use of α versus β nomenclature is used to describe asymmetric configurations at ring carbons. The steranes including the ones most abundant in oils: cholestanes ($C_{27}H_{48}$), ergostanes ($C_{28}H_{50}$), stigmastanes ($C_{29}H_{52}$) can have R- and S-configuration at the acyclic (chain position) carbon atom C-20, resulting in two homologue series with 20R (20R ααα and 20R αββ) and 20S (20S ααα and 20S αββ) configurations. Hopanes with 30 carbons or less show asymmetric centers at C-21 and all ring-juncture carbons including C-5, C-8, C-9, C-10, C-13, C-14, C-17, and C-18. Common homohopanes (C_{31} to C_{35}) have an extended side chain with an additional asymmetric center at C-22, resulting in two homologues with 22R and 22S configurations. These two homologous homohopanes (22R and 22S) can be well separated by GC-MS as well-resolved double peaks, prominent in gas chromatograms. The R and S and α versus β designations are a useful means of describing the relative configuration of biomarker compounds. It should, however, be noted that these designations are determined strictly on the basis of the convention as described by Cahn et al. (1966) without reference to optical rotation.

3.2.3 Analysis Methods for Biomarker Fingerprinting

In the last two decades, a wide variety of instrumental techniques have been developed and used for fingerprinting petroleum hydrocarbons including biomarkers (Wang et al., 1994a, 1995a; ETC Method, 2002; Wang and Fingas, 2003; Uhler et al., 1998–1999; Stout et al., 2002; Dimandja, 2004; Gains et al., 1999; Reddy et al., 2002; Frysinger et al., 2003). A variety of diagnostic ratios, especially ratios of PAH and biomarker compounds, for interpreting chemical data from oil spills have been proposed (Wang et al., 1999a; Stout et al., 2001, 2002; Daling et al., 2002). Many EPA and ASTM methods have been modified (such as the modified EPA Method 8015, 8260, 8270; and the modified ASTM Methods D3328, D5037, and D5739) in recent years to allow flexibility in the deployment of the "standard" analytical methods and to improve specificity

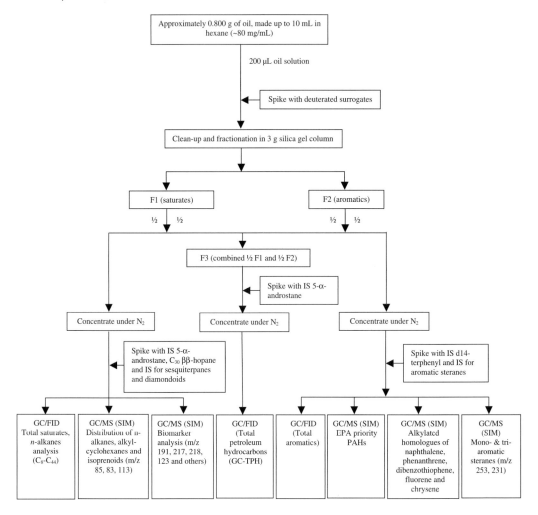

Figure 3-3 Oil sample preparation flowchart.

and sensitivity for measuring spilled oil and petroleum products in soil and water. As an example, Figure 3-3 shows a flowchart of sample preparation procedures used by the Environment Canada Oil Spill Research Laboratory. Table 3-3 summarizes the surrogate and internal standard compounds used for oil and biomarker fingerprinting at the same lab.

Silica gel is used frequently for cleanup and fractionation of oil extracts. The column cleanup procedure used by the Environment Canada Oil Research Laboratory is the following: a chromatographic column with a Teflon stopcock (200 × 10.5 mm i.d.) is plugged with Pyrex glass wool at the bottom, serially rinsed with methanol, hexane, and dichloromethane, and allowed to dry. The column is dry-packed with 3 g of activated silica gel and topped with about 1-cm anhydrous granular sodium sulfate. Columns are then preconditioned using 20 mL of hexane. Just prior to exposure of the sodium sulfate layer to air, appropriate volumes of the oil solutions or concentrated oil extracts are transferred quantitatively to the column (Wang et al., 1994a; ETC Method, 2002). Saturated hydrocarbons are eluted with 12 mL of hexane (Fraction 1, labeled F1). Aromatic hydrocarbons are eluted with 15 mL of hexane : dichloromethane (v/v, 1 : 1, Fraction 2, labeled

Table 3-3 Surrogate and Internal Standard Used for Oil and Biomarker Fingerprinting

Compounds	Chemical Names	Target Ions (m/z)
Surrogates	o-terphenyl (for TPH determination by GC/FID)	
	mixture of d_{10}-acenaphthalene, d_{10}-phenanthrene, d_{12}-benz[a]anthracene, and d_{12}-perylene	164, 188, 240, 264
Internal standards	5-α-androstane (for TPH determination by GC/FID)	
	d_{14}-terphenyl (for quantitation of PAHs)	244
	C_{30} ββ-hopane (for quantitation of terpanes and steranes)	191
	d_3-monoaromatic steranes [5α(H)/5β(H), $C_{21}H_{27}D_3$] (for quantitation of mono- and tri-aromatic steranes)	285
	d_{18}-decahydronaphthalene (cis-) (for quantitation of sesquiterpanes)	156
	d_{16}-adamantane (for quantitation of diamondoids)	152

F2). Saturated biomarkers are eluted with other saturates in F1. Aromatic steranes are eluted in the aromatic fraction, F2. Polar compounds are eluted with 15 mL of methanol (labeled F4). For each sample, half of F1 is used for analysis of the total GC-detectable saturates, n-alkanes and isoprenoids, and biomarker compounds; and half of F2 is used for analysis of alkylated PAH homologues and other EPA priority parent PAHs, and aromatic steranes. The remaining halves of the F1 and F2 are combined into one fraction (Fraction 3, labeled F3) and used for the determination of TPH and UCM. The three fractions are concentrated under a stream of nitrogen to appropriate volumes, spiked with appropriate internal standards, and then adjusted to an accurate pre-injection volume (1.00 mL) for GC-FID and GC-MS analyses.

In accordance of the quality assurance (QA) and quality-control (QC) programs (Page et al., 1995; Douglas et al., 2004; Wang et al., 1999a; Stout et al., 2002; Faksness et al., 2002; EPA, 1997, 1998a, 1998b, 2001; ASTM, 1997a, 1997b), the GC-MS must be calibrated using the terpane standards prior to quantification of the biomarkers in oil. In the Environment Canada Oil Research Laboratory the terpane standards (Table 3-3) include C_{27} 17α(H)-22,29,30-trisnorhopane, C_{29} 17β(H), 21α(H)-30-norhopane, and C_{30} 17β(H), 21α(H)-hopane. The sterane standards include C_{21} 5β(H)-pregnane, C_{22} 20-methyl-5α(H)-pregnane, and the series of C_{27}, C_{28}, and C_{29} steranes. The C_{30} 17β(H), 21β(H)-hopane is used as the internal standard for quantification of tri- to pentacylic biomarkers. The response factors (RRF) are determined relative to the internal standard C_{30} 17β(H), 21β(H)-hopane. In most cases, the average RRF for C_{30} 17β(H), 21α(H)-hopane at m/z 191 are used for quantification of C_{30} 17α(H), 21β(H)-hopane, and other terpanes (in the range of C_{19} to C_{35}). For steranes, the average RRF of C_{29} 20R-ααα-ethylcholestane at m/z 217 relative to the internal standard are used to calculate the concentrations of sterane compounds. The deuterated d_3 monoaromatic steranes [5α(H)/5β(H), $C_{21}H_{27}D_3$] are used as internal standards for quantification of monoaromatic and triaromatic steranes. Certified sesquiterpane standards are not commercially available. The average response factors of cis-decahydronaphthalene ($C_{10}H_{18}$, m/z 138) and 1-methyldecaline ($C_{11}H_{20}$, m/z 152), which have similar molecular structures to those of the sesquiterpanes, relative to the internal standard cis-decahydronaphthalene-d_{18} (m/z 156) were used for quantitation of sesquiterpanes. The d_{16}-adamantane is used as the internal standard for quantification of diamondoid compounds.

3.2.4 Capillary Gas Chromatography — Mass Spectrometry (GC-MS)

GC-MS is the principal instrument used for characterizing biomarkers. Early use of mass chromatograms in organic geochemistry was pioneered at Chevron and led to a stereochemical understanding of steroids and the first practical method of oil fingerprinting based on terpanes and steranes (Seifert, 1977). Today, computerized GC-MS (e.g., benchtop quadrupole GC-MS, high-resolution GC-MS, GC-ion trap MS, and GC-MS-MS) has become the routine technique used in most oil and environmental forensics laboratories to analyze a wide range of petroleum hydrocarbons.

3.2.4.1 Benchtop Quadrupole GC-MS

The quadrupole is the most common mass separator in use today. The benchtop quadrupole GC-MS systems, although lacking the high-resolution capabilities of larger and more expensive magnetic-sector instruments, have sufficient sensitivity and selectivity for most purposes of biomarker analysis. Most benchtop GC-MS use a quadrupole mass filter to separate ions produced from gaseous neutral molecules or species in the ionization chamber. In a high vacuum, ions pass down the lengths of four parallel metal rods to which are applied both a constant voltage and a radiofrequency oscillating voltage. The electric field deflects ions in complex trajectories as they migrate from the ionization chamber toward the detector, allowing only ions with one particular mass-to-charge (m/z) ratio to reach the detector at any instant. Other nonresonant ions collide with the rods and are lost before they reach the detector. By rapidly varying the applied voltages, ions of different masses are selected to reach the detector. A wide range of masses can be recorded in less than 1 second. In this way many mass spectra are taken and stored on a computer as the components of the sample pass from the chromatographic column into the mass spectrometer. Benchtop quadrupole GC-MS can be operated in various modes including (full) scan and selected ion monitoring (SIM).

3.2.4.1.1 Scan Mode. In scan mode, sometimes called *full scan mode*, the mass spectrometer is used to scan (that is, to measure) the entire range of ions generated in the ion source. As the MS detector scans through a predefined mass range (e.g., 50–700 amu), a mass spectrum is generated. Full scan records hundreds of ions per scan (typically, greater than 500 ions per scan are recorded in 3 seconds; the larger the mass range, the fewer scans per second), but with lower sensitivity due to shorter dwell time in comparison with the SIM mode. Each peak that elutes from the GC yields a particular distribution of fragment ion masses. Among these ions generated from the scan, there are always several ions being the most characteristic and diagnostic of the molecule or the compound type, and the most abundant ion in the mass spectrum is called the *base peak*. The magnitude of the total ion current for all mass spectra in an oil sample is generally plotted versus the GC retention time on a total ion chromatogram (TIC) to show a series of peaks that represent relative amounts of components in the sample. Identification and characterization of petroleum hydrocarbons are largely based on the full mass spectral data for structural elucidation, comparison of GC retention data with that of reference standards, recognition of distribution pattern, calculation of retention indexes (RI), and comparison with literature RI values.

3.2.4.1.2 Selected Ion Monitoring (SIM) Mode. In the SIM mode, only a limited number of characteristic ions (for example, the *base peaks* 191, 217, and 218, and those m/z values diagnostic for molecule structural elucidation for target terpanes and steranes) are monitored. For quantification of individual target compounds, the SIM mode is used most frequently, since it shows several advantages in comparison to the scan mode: (1) SIM only records a few selected m/z per scan, resulting in a much longer dwell time for each monitored ion (usually between 25 and 100 milliseconds, depending on the number of m/z selected) than in the scan mode; (2) method detection limits for target analytes are generally lower by almost an order of magnitude than those produced by the full scan GC-MS; (3) the use of the SIM mode is often less noisy and the linear quantification range is increased for trace analytes. As examples, the following briefly describe the analytical GC-MS conditions used by the Environment Canada Oil Spill Research Lab and Petrobras Geochemistry Laboratory (Barbanti, 2004), respectively.

3.2.4.1.3 Example Benchtop GC-MS Conditions (EC Oil Spill Research Laboratory).
Analyses of biomarkers are performed on an Agilent 6890 GC coupled with an Agilent 5973 mass selective detector (MSD). System control and data acquisitions are achieved with the Agilent G1701 BA MSD ChemStation. A 30 m × 0.25 mm i.d. (0.25-μm film thickness) HP-5MS fused-silica capillary column is used. The chromatographic conditions are as follows: carrier gas, helium (1.0 mL/min); injection mode, splitless; injector and detector temperature, 280 and 300°C, respectively. The temperature program employed for biomarkers and alkylated PAHs is 50°C hold for 2 min, then ramp at 6°C/min to 300°C and hold for 20 minutes. Prior to sample analysis, the GC-MS is tuned with perfluorotributylamine (PFTBA). The total run time is 60 minutes.

3.2.4.1.4 Example Benchtop GC-MS Conditions (Petrobras Geochemistry Laboratory).
A 60 m × 0.25 mm i.d. (0.25-μm film thickness) HP-5MS or equivalent 60-m capillary column is used to achieve improved resolution for biomarkers (Barbanti, 2004). The temperature program is as follows: 55°C hold for 2 min, ramp at 20°C/min to 150°C and then 1.5°C to 310°C and hold for 15 minutes. The total run time is 128 minutes.

The 30-m capillary column is used in many environmental forensic labs for most oil spill work. However, the 60-m capillary column with a slow temperature rate and longer running time offers further improved resolution for some paired biomarker isomers, which may not be well resolved by the use of a 30-m column.

3.2.4.2 Triple Quadrupole GC-MS-MS

The combination of two or more MS analyzers, commonly known as MS-MS or tandem mass spectrometry, is a highly specific means of separating mixtures and studying molecular fragments. In the first MS, one ion is isolated and subsequently in the second MS, reactions of that ion are studied further. GC-MS-MS includes linked and de-linked double focusing and triple quadrupole mass spectrometry. Triple quadrupole mass spectrometers are the most common type of tandem mass spectrometers. The first (or parent) and the third (or daughter) quadrupole are MS-1 and MS-2, whereas the second quadrupole in the middle acts as the collision cell. In the collision cell, the transmitted ions formed in the ion source and selected by or passed through MS-1 undergo low-energy collision with an inert gas such as argon. The fragment ions or daughter ions formed in the collision cell are selectively monitored by the daughter quadrupole and recorded using an electron multiplier. Because of the use of three linked quadrupoles, triple quadrupole mass spectrometry allows determination of specific parent–daughter relationships with less interference from other reactions and their related ions and, therefore, increases signal-to-noise ratios and offers improved selectivity for biomarker analysis. Triple quadrupole mass spectrometers can be operated in three GC-MS-MS modes (Linscheid, 2001): (1) precursor (parent) ion scan mode; (2) product (daughter) ion scan mode; and (3) neutral loss scan mode.

In the precursor (parent) ion scan mode, the first quadrupole is scanned and only one or more product ions (daughter ions) are selected and recorded. For example, parent ions of the C_{30}—C_{35} hopanes consist of 412, 426, 440, 454, 468, and 482, respectively. Each of these parent ions produces a major daughter ion at m/z 191 following collision with the inert gas in the collision cell. By the same mechanism, parent ions of common sterane homologous compounds (C_{27}—C_{28}—C_{29}—C_{30} steranes) produce major daughter ions at m/z 217 and 218. Both parent and daughter ions can be selectively monitored to improve signal-to-noise ratio. In comparison with the routine benchtop GC-MS, the GC-MS-MS technique offers a significant refinement for biomarker separation. For example, monitoring the fragment ions at m/z 217 by benchtop GC-MS in SIM mode provides a single mass chromatogram for all steranes (from C_{27} to C_{30}) in an oil sample, many of which co-elute. However, by specifying parent and daughter ions (such as from m/z 386 to m/z 217 for C_{28} steranes), triple quadrupole GC-MS-MS can provide nearly complete separation of an individual sterane family by carbon number. This approach has been successfully used for identification of a biomarker and a biomarker family, product screening, and distribution pattern recognition of biomarkers.

In the product (daughter) ion scan mode, only one precursor (parent) ion is selected and enters the collision cell. The second MS analyzer scans for all product (daughter) ions produced in the collision cell. This type of scan is often used to analyze the fragmentation pattern of a component with specific molecular weight in a complex mixture without interference from any co-eluting compound of different molecular mass.

In the neutral loss scan mode, both analyzers are scanned with the selected mass difference, and for reaction monitoring only one precursor and one product ion species are permitted to travel through MS-1 and MS-2, respectively. This approach can be particularly useful in the search for specific compounds derived from a certain precursor compound.

3.2.5 Mass Spectra and Identification of Biomarkers

Mass spectra produced by GC-MS are one of the most valuable tools for identification of unknown compounds. In addition to molecular formula, the mass spectrum provides structural information of a given molecule. An electron with typical energy of 70 eV (1610 kcal/mol or 6740 kJ/mol) has far more energy than needed to ionize a molecule. Much work on the isolation and identification of individual biomarker components in oils and sediment extracts has been done by petroleum geochemists. The Chevron Biomarker Laboratory developed a *coinjection and mass spectra matching* technique, in combination with other analytical techniques, for provisional identification of unknown biomarker compounds (Peters and Moldowan, 1993).

As an example, Figure 3-4 shows mass spectra for several common petroleum biomarkers used in environmental forensic studies. These figures show that common features of the mass spectra of terpanes, steranes, monoaromatic steranes, and triaromatic steranes: a large parent ion (M^+), an important parent minus a methyl ion ($M^+ - 15$), and a base peak at m/z 191, 217 and 218, 253, and 231, respectively. C_{30} 17β(H), 21β(H)-hopane has, for example, a characteristic parent ion, parent minus methyl ion, and base peak at 412, 397, and 191, respectively. C_{27} 20R ααα-cholastane has a characteristic parent ion, parent minus methyl ion, and base peak at 372, 357, and 217, respectively; while C_{27} 20R αββ-cholastane has a characteristic parent ion, parent minus methyl ion, sterane-characteristic ion, and base peak at 372, 357, 217, and 218, respectively. Adamantane, which is very stable under typical electron impact ionization conditions, has both its base peak and parent ion at m/z 136.

The m/z 191 fragment is often the base peak of mass spectra of cyclic terpanes. It is derived from rings (A+B) of the molecule, but rings (D+E) may also be the source. The m/z 177 fragment is most likely derived from rings (A+B) of triterpane molecules that have lost a methyl group from position 10, that is, 25-norhopanes (Volkman et al., 1983a, 1983b; Grahl-Nielsen and Lygre, 1990). The notable feature of mass spectra for 25-demethylated hopanes is that the m/z 177 fragment has higher intensity than the m/z 191 fragment. Demethylated triterpanes contain different information than the triterpanes and have been suggested as markers for biodegradation (Volkman et al., 1983a). The other triterpanes do also give the m/z 177 fragment upon electron impact in the mass spectrometer, but in lower abundance than the m/z 191 fragment. The fragment is formed by the loss of CH_2 from the m/z 191 fragment and can be seen in all mass spectra of triterpanes. The biomarker

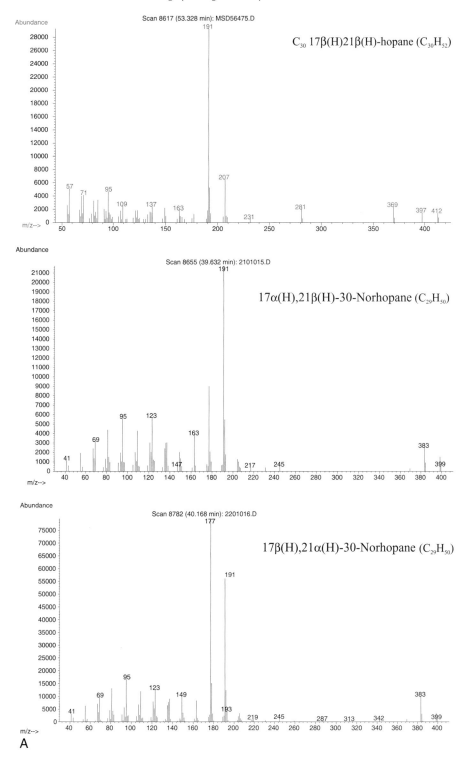

Figure 3-4 Mass spectra of some common biomarkers used in environmental forensic studies.

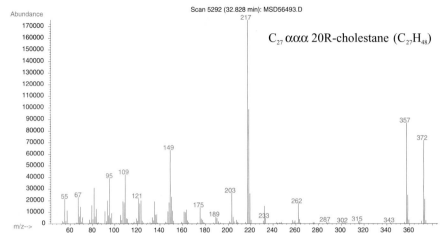

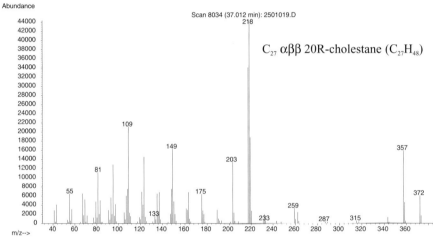

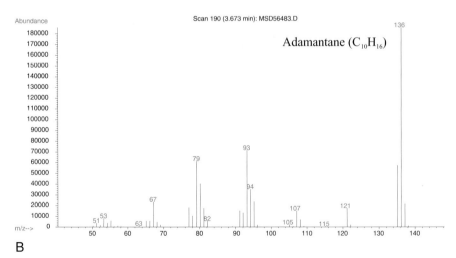

B

Figure 3-4, continued

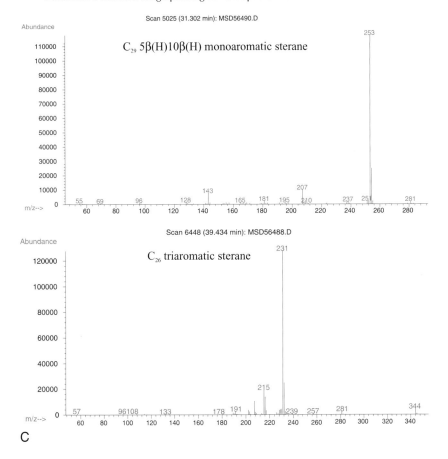

Figure 3-4, continued

C_{29} 18α(H), 21β(H)-30-norneohopane (or C_{29}Ts), which elutes immediately after C_{29} αβ-hopane, has a greater abundance of the m/z 177 ion than of the m/z 191 ion. The proposed mechanism is that the m/z 177 ion is derived from the (D+E) ring fragment and stabilized by the methyl group at position 14 (Gallegos, 1971; Killops and Howell, 1991). Increased stability of the (D+E) ring fragment producing the m/z 177 ion relative to that generating the m/z 191 ion may relate to ring junction configuration or to methyl substitution position in the (D+E) ring. For the similar mechanism, C_{29} 17β(H), 21α(H)-30-norhopane (normoretane) also shows significantly greater abundance of the m/z 177 ion than of the m/z 191 ion [see the mass spectrum of C_{29} 17β(H), 21α(H)-30-norhopane in Figure 3-4].

The m/z 217 and 218 fragment ions are derived from rings (A+B+C) of most 14α(H)- and 14β(H)-steranes. The βαα and ααα steranes have a base peak at m/z 217, while the base peak of αββ steranes is at m/z 218. The relative intensities of m/z 149 to m/z 151 fragment ions in the mass spectra of steranes have been used to distinguish between 5α- and 5β-stereoisomers (Gallegos, 1971). Note that the only significant difference between the mass spectra of 5α- and 5β-epimers is that the m/z 149 fragment is more abundant than the m/z 151 moiety for the 5α-epimer (e.g., 5α-cholestane versus 5β-cholestane, $C_{27}H_{48}$). Furthermore, the GC retention time of the 5β epimer is shorter than that of the 5α isomer. Hence, the stereo configuration of 5α- and 5β-steranes having the same parent ion, a parent ion minus a methyl ion, and a base peak at m/z 217 can be determined from the peak ratio at m/z 149 to m/z 151. If the ratio of the m/z 149 ion to the m/z 151 ion is greater than 1, it is 5α-sterane; otherwise, it is 5β-sterane.

The SIM chromatogram of one ion of given m/z with the GC retention time is often diagnostic of a class of homologous compounds with similar structures but different carbon numbers and isomerism and can be used for identification. As an example, Figure 3-5 shows the SIM chromatograms for common biomarker classes (terpanes at 191 and steranes at 217) in a Kuwait crude oil obtained by using a 60-m column (Barbanti, 2004). Thirty-eight terpanes from C_{19} tricyclic terpane to C_{35} homohopanes (m/z 191) and 19 steranes from C_{21} to C_{29} steranes (m/z 217) in total have been unambiguously identified and characterized in this Kuwait oil (refer to Table 3-1 for peak identity). Paired biomarker isomers (H29 and C29Ts, H30 and NOR30H) and triplet (TET24 + TR26A + TR26B) are well resolved. Less abundant C_{30} steranes can be clearly recognized as well.

Identification of Vegetation Biomarkers. In addition to petroleum biomarkers, oil-contaminated sediment samples may contain modern plant biomarker compounds that represent oxygenated or unsaturated equivalents of biomarkers found in oil. Identification of these biogenic biomarkers of sediment extracts can often provide valuable information about the nature and source of samples. For example, in the Nipisi spill study (Wang et al., 1998a), three unknown vegetation biomarker compounds with significant abundances were detected. They were positively identified as 12-oleanene ($C_{30}H_{50}$, MW = 410.7, RT = 42.27 min), 12-ursene ($C_{30}H_{50}$, MW = 410.7, RT = 42.74 min), and 3-friedelene ($C_{30}H_{50}$, MW = 410.7, RT = 44.26 min). Formation of a six-membered ring E from the baccharane precursor leads to the oleanane group. Oleananes and their derivatives form the largest group of triterpenoids and occur widely in the plant kingdom (Connolly and Hill, 1991). The friedelene-type triterpenoids arise by increasing degrees of backbone rearrangement of the oleanene skeleton. Methyl migration in ring E of the oleanene precursor leads to the ursene skeleton (Connolly and Hill, 1991).

3.3 Fingerprinting Petroleum Biomarkers

Characterization of n-alkanes is achieved using GC-FID and GC-MS at m/z 85, 71, and 57, while characterization of major biomarker groups is achieved using GC-MS at their diagnostic fragment ions (Table 3-4). Various biomarkers occur in different carbon ranges of crude oils (Figure 3-6).

3.3.1 Biomarkers in Crude Oils

Depending on the oil sources and the geological migration conditions, crude oils can have (1) large differences in distribution patterns of the n-alkane and cyclic-alkanes as well as UCM profiles, (2) significantly different relative ratios of isoprenoids to normal alkanes, and (3) large differences in distribution patterns and concentrations of alkylated PAH homologues and biomarkers. For many oils, their GC-MS chromatograms of terpanes at m/z 191 are characterized by the terpane distribution in a wide range from C_{19} to C_{35} often with C_{29} αβ- and C_{30} αβ-pentacyclic hopanes and C_{23} and C_{24} tricyclic terpanes being often the most abundant. As for steranes (at m/z 217 and 218), the dominance of C_{27}, C_{28}, and C_{29} 20S/20R homologues, particularly the epimers of αββ-steranes, among the C_{20} to C_{30} steranes is often apparent. As examples, Figures 3-7 and 3-8 show GC-MS chromatograms at m/z 191, 217, and 218 for five light (API >35) to medium (API: 25–35) crude oils. For comparison, Figures 3-9 and 3-10 present biomarker fingerprints for five heavy oils including California API 15, Sockeye and Platform Elly from California, and Boscan and Orinoco from Venezuela. Table 3-5 summarizes the quantification results of major target biomarkers in these light, medium, and heavy oils.

In addition to composition, the concentrations of biomarkers can vary widely with the

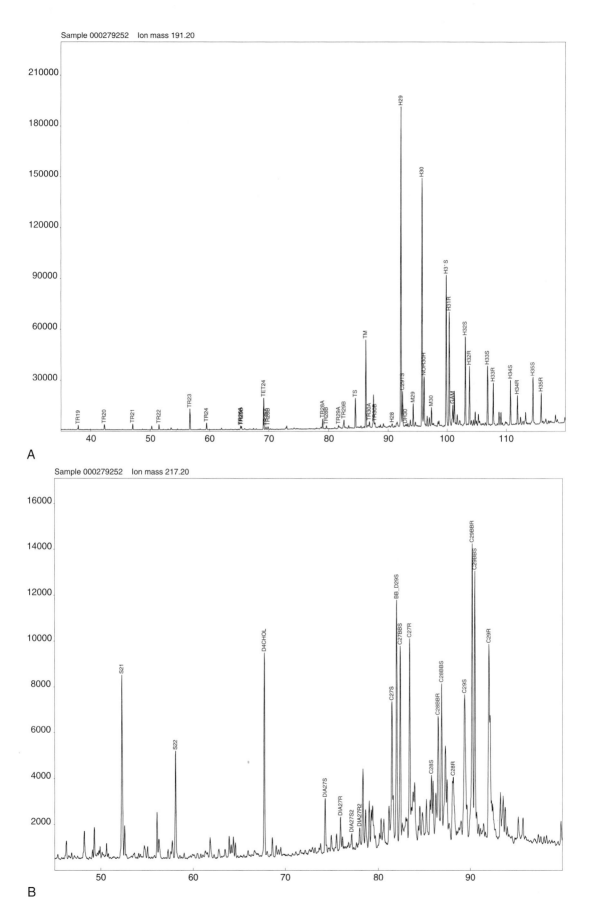

Figure 3-5 The SIM chromatograms at m/z (a) 191 and (b) 217 of a Kuwait crude oil for common terpane and sterane biomarker classes.

Table 3-4 Characteristic Fragment Ions for Various Biomarkers

Biomarkers	Diagnostic Fragment Ions
Acyclic terpenoids	
alkyl-cyclohexanes	83
methyl-alkyl-cyclohexanes	97
isoprenoids	113, 127, 183, M$^+$
Cyclic terpenoids	
sesquiterpanes with drimane structure	123
adamantanes	135, 136, 149, 163, 177, 191
diamantanes	187, 201, 215, 229
tri-, tetra-, penta-cyclic terpanes	191, M$^+$
25-norhopanes	177, 191
28, 30-bisnorhopanes	163, 191
steranes	217, 218
5α(H)-steranes	149, 217, 218
5β(H)-steranes	151, 217, 218
diasteranes	217, 218, 259
methyl-steranes	217, 218, 231, 232
Aromatic steranes	
monoaromatic steranes	253, 267
triaromatic steranes	231, 245

type of depositional environment (noxic/anoxic, freshwater/marine/hypersaline), type of organic matter (e.g., terrigenous origin or marine origin), maturity and biodegradation as well (see Chapter 1 herein). For a given type of organic material, the biomarker concentrations generally decrease with increasing thermal maturity. Very light oils or condensates (e.g., the Scotia Light) typically contain low concentrations of detectable biomarkers. In most cases, characterization of biomarkers should include determination of both absolute concentrations and relative fingerprinting distributions, and should not be just measuring peak ratio alone. This is important because it is possible to have a situation where a source might have a similar biomarker ratio but very different actual amounts of biomarkers. Quantitative determination of biomarkers is also critical in oil spill studies involving recognition and/or allocation of mixtures of different oils (e.g., see Chapter 8).

Figures 3-7 to 3-10 and Table 3-5 qualitatively and quantitatively demonstrate differences in biomarker distributions between 10 oils. Different from most crude oils, the Scotia Light (API = 59) only contains trace amounts of biomarkers [the total concentration of target biomarkers (i.e., terpanes and steranes) is only 29 μg/g oil], far lower than the corresponding values for other crude oils. The Alaska North Slope (ANS) oil contains a wide range of terpanes from C_{20} tricyclic terpane to C_{35} pentacyclic terpanes with the C_{30} αβ hopane as the most abundant, followed by C_{29} αβ hopane. The triplet C_{24} tetracyclic + C_{26} (S + R) tricyclic terpanes are highly abundant as well. In contrast, the Arabian Light, South Louisiana, and Troll oils have terpanes largely located in the C_{27} to C_{35} pentacyclic hopane range, and only contain small amounts of C_{20} to C_{24} tricyclic terpanes. In addition, the abundance of C_{29} αβ hopane is higher than that of C_{30} αβ hopane in Arabian Light crude oil. The steranes are present in all five light to medium crude oils but with different distribution patterns. The characteristic V-shaped C_{27}—C_{28}—C_{29} regular αββ sterane (m/z 218) distribution is clearly demonstrated, which indicates high thermal maturity. The relative abundances of C_{27}—C_{28}—C_{29} steranes in oils reflect the carbon number distribution of the sterols in the precursor organic matter in the source rocks for these oils. In general, a dominance of

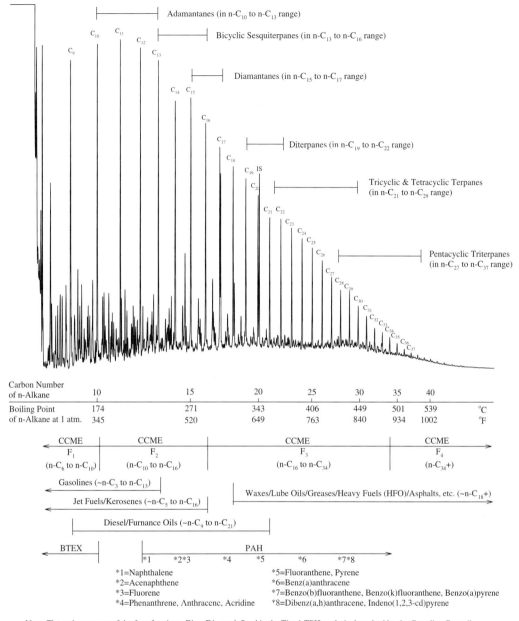

Figure 3-6 Carbon range of common cyclic biomarker classes in crude oil and petroleum products.

C_{27} over C_{29} steranes specifies marine algae organic matter input, while a predominance of C_{29} steranes over C_{27} steranes may indicate a preferential higher plant input (Peters and Moldowan, 1993; Gürgey, 2002). The ANS, South Louisiana, and Troll oils contain higher amounts of diasteranes as well as C_{21} and C_{22} regular steranes. By contrast, the Arabian Light has much lower concentrations of steranes in total (the total of C_{27}—C_{28}—C_{29} αββ steranes is only 110μg/g oil) but displays significantly higher concentration of C_{29} αββ

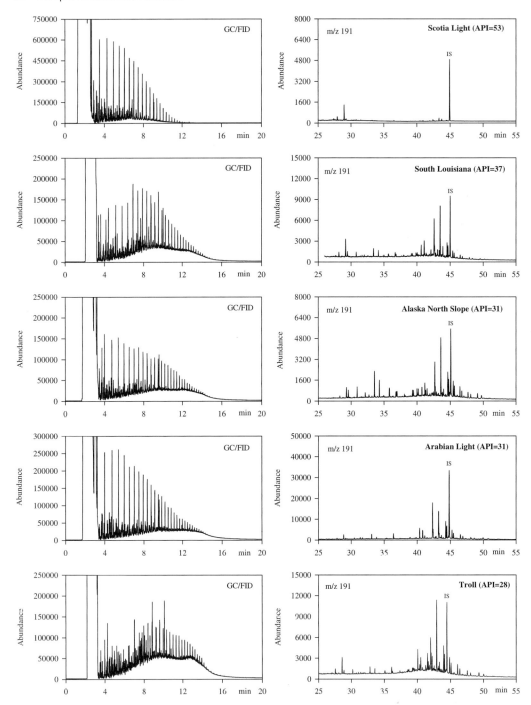

Figure 3-7 GC-FID (left panel) and GC-MS (at *m/z* 191, right panel) chromatograms of five different oils (light to medium) to illustrate differences in the *n*-alkane and tri-, tetra-, and pentacyclic terpane distributions between oils.

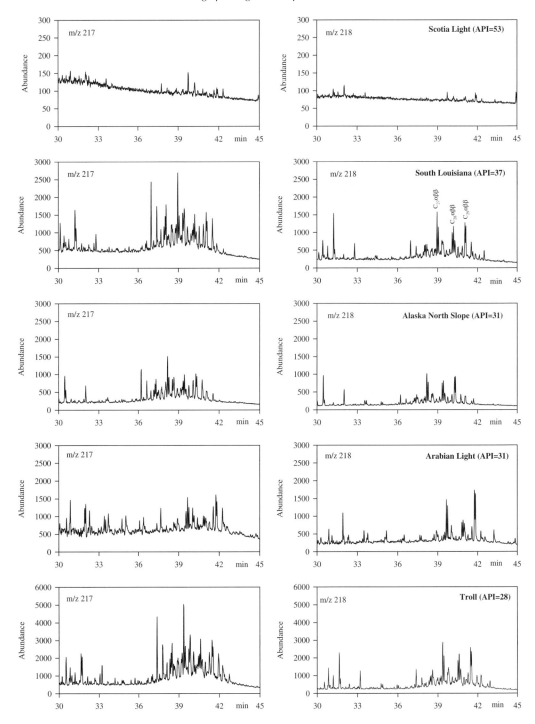

Figure 3-8 GC-MS chromatograms at *m/z* 217 and 218 for five different oils (light to medium) to illustrate differences in sterane distributions between oils.

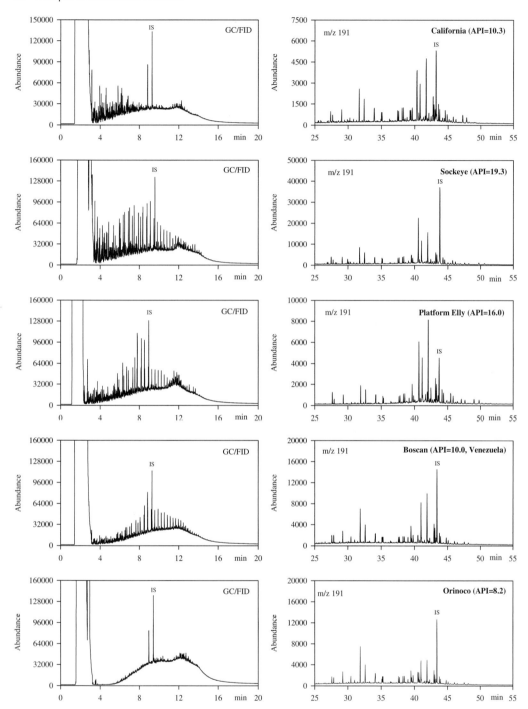

Figure 3-9 GC-FID (left panel) and GC-MS (at *m/z* 191, right panel) chromatograms of five heavy oils from different regions.

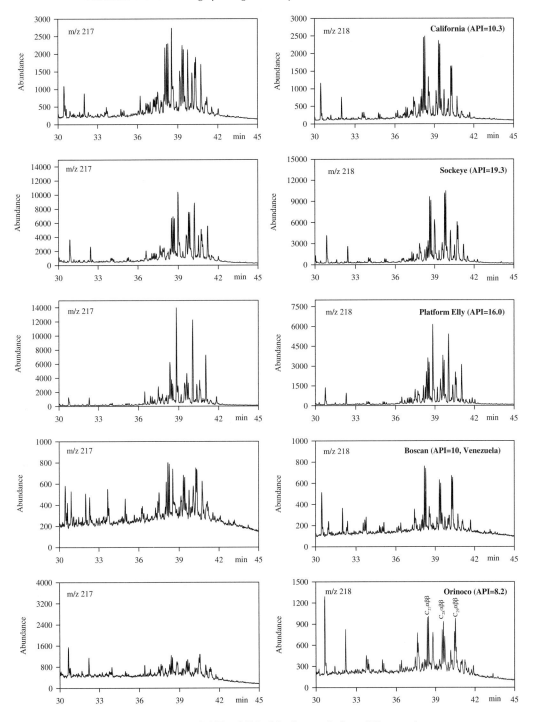

Figure 3-10 GC-MS chromatograms at *m/z* 217 and 218 of five heavy oils from different regions.

Table 3-5 Quantitation results of Major Target Biomarkers in Example Crude Oils and Petroleum Products

Oil Samples	Scotia Light	South Louisiana	Alaska North Slope	Arabian Light	Troll	California (API = 11)	Sockeye	Platform Elly	Used Air Compressor Oil	Valvoline 10W-30 Motor Oil
Biomarker compounds (μg/g oil)										
C_{21}	0.00	9.43	18.7	4.47	7.81	22.5	19.1	20.1	17.1	11.6
C_{22}	0.00	3.53	8.65	4.73	2.96	8.86	17.7	4.32	14.3	15.2
C_{23}	0.87	14.8	49.6	17.7	11.1	56.5	46.2	41.3	86.7	68.2
C_{24}	0.61	10.7	31.6	6.63	9.14	39.3	31.3	33.9	45.0	25.5
$C_{29}\ \alpha\beta$	3.32	74.6	69.3	152	56.6	69.3	61.2	107	190	864
$C_{30}\ \alpha\beta$	5.79	100	112	125	126	109	99.5	216	414	718
C_{31} (S)	1.74	26.4	48.9	79.9	44.3	46.1	38.7	64.6	180	385
C_{31} (R)	1.24	21.5	35.8	65.7	34.5	32.7	40.6	52.5	148	305
C_{32} (S)	0.95	15.2	37.4	48.1	30.4	32.5	27.5	43.0	142	238
C_{32} (R)	0.79	9.94	24.6	29.8	22.0	22.0	18.9	32.2	96.1	164
C_{33} (S)	0.00	8.96	24.2	27.0	26.7	25.1	18.8	35.2	104	140
C_{33} (R)	0.00	5.48	16.1	17.8	16.3	17.6	12.8	28.5	69.5	91.7
C_{34} (S)	0.00	4.65	19.1	14.4	16.4	17.9	8.40	20.0	78.3	77.6
C_{34} (R)	0.00	2.78	11.2	8.80	9.54	11.6	5.70	15.1	43.1	51.6
C_{35} (S)	0.00	3.33	17.7	14.7	12.4	23.0	12.1	22.1	72.5	85.7
C_{35} (R)	0.00	2.27	15.0	7.80	8.73	20.8	9.15	20.9	46.5	47.6
Ts	1.40	20.3	16.2	42.6	34.1	9.08	6.90	13.2	61.9	148
Tm	1.66	29.6	25.2	36.5	23.3	20.7	35.4	55.9	74.8	215
$C_{27}\ \alpha\beta\beta$-steranes	2.84	89.3	124	35.1	172	438	208	649	437	525
$C_{28}\ \alpha\beta\beta$-steranes	2.77	67.4	121	20.1	125	427	260	754	384	363
$C_{29}\ \alpha\beta\beta$-steranes	5.20	89.8	152	55.1	179	289	152	466	761	778
Total	**29.2**	**610**	**979**	**814**	**968**	**1738**	**1129**	**2695**	**3466**	**5318**
Diagnostic ratios										
C_{23}/C_{24}	1.42	1.39	1.58	2.63	1.22	1.44	1.48	1.22	1.93	2.68
$C_{23}/C_{30}\ \alpha\beta$	0.15	0.15	0.45	0.14	0.09	0.52	0.46	0.19	0.21	0.09
$C_{24}/C_{30}\ \alpha\beta$	0.11	0.11	0.28	0.05	0.07	0.36	0.31	0.16	0.11	0.04
$C_{29}\ \alpha\beta/C_{30}\ \alpha\beta$	0.57	0.75	0.62	1.22	0.45	0.64	0.62	0.49	0.46	1.20
$C_{31}(S)/C_{31}(S + R)$	1.40	1.23	1.36	1.22	1.28	1.41	0.95	1.23	1.22	1.26
$C_{32}(S)/C_{32}(S + R)$	1.20	1.53	1.52	1.61	1.38	1.48	1.46	1.33	1.48	1.45
Ts/Tm	0.84	0.69	0.64	1.17	1.46	0.44	0.19	0.24	0.83	0.69
$C_{27}\ \alpha\beta\beta$-steranes/$C_{29}\ \alpha\beta\beta$-steranes	0.55	0.99	0.82	0.64	0.96	1.52	1.37	1.39	0.57	0.67
$C_{30}/(C_{31} + C_{32} + C_{33} + C_{34} + C_{35})$	1.23	0.99	0.45	0.40	0.57	0.44	0.52	0.65	0.42	0.45

steranes than C_{27} αββ cholestane and C_{28} αββ ergostane series (Table 3-5).

The dominance of C_{28} 17α(H), 18α(H), 21β(H)-28,30-bisnorhopane (BHN28) is particularly prominent in California API-11, Sockeye, and Platform Elly (all three oils are from California), and its abundance is even higher than C_{30} and/or C_{29} 17α(H), 21β(H)-hopane (Figure 3-9). A high concentration of C_{28} 17α(H), 18α(H), 21β(H)-28,30-bisnorhopane is typical of petroleum from highly reducing to anoxic depositional environments (Mello et al., 1990). The California API-11 and Platform Elly demonstrate higher concentration of C_{31} to C_{35} homohopanes than the Sockeye oil. Also, the California API-11 has a significantly higher concentration of C_{35} homohopanes (22S + 22R) than C_{34} homohopanes (22S + 22R), further indicating a highly reducing marine environment of deposition with no available free oxygen (Peters and Moldowan, 1993). For the Orinoco Bitumen, C_{23} terpane is the most abundant, followed by the C_{30} and C_{29} hopane; while the Boscan oil demonstrates higher concentrations of C_{29} and C_{30} terpanes than C_{23} terpane. The presence of triplets with different relative distributions is apparent for most heavy oils. Orinoco and Boscan oils have somewhat the V-shaped C_{27}—C_{28}—C_{29} regular αββ sterane (m/z 218) distribution. Three California oils have very high concentrations of steranes (Table 3-5), with a more abundant C_{28} ergostane than C_{27} and C_{29} sterane series. This is also the case for several other heavy California oils including California API-15 and Platform Irene (data not shown here). The high relative levels of C_{28} ergostane may be related to increased diversification of phytoplankton assemblages in the Jurassic and Cretaceous oils.

3.3.2 Biomarkers in Petroleum Products

Petroleum products are refined from crude oils through a variety of refining processes including distillation, cracking, catalytic reforming, isomerization, alkylation, and blending (Olah and Molnar, 1995; Speight, 2002; Simanzhenkov and Idem, 2003). Depending on the chemical composition of their "parent" crude oil feedstocks, varying refining approach and conditions, wide range of applications, regulatory requirements, and economic requirements, refined products can have a wide variety in chemical compositions.

Light distillates are typically products in the C_4 to C_{13} carbon range. They include aviation gas (gasoline-type jet fuel), naphtha, and automotive gasoline. The GC traces of fresh light distillates are featured with dominance of light-end, resolved hydrocarbons and a minimal UCM. Gasoline is a complex mixture of hundreds of different hydrocarbons predominantly in the C_4 to C_{13} boiling range. Additives are often added to gasoline to improve some specific properties and anti-knock properties. The major components of gasoline that are of environmental concern include MTBE, BTEX, C_3-benzenes, and naphthalene. Gasoline and other light distillates do not contain any terpane and sterane biomarker compounds. But, it has been recently reported that gas condensates and some gas-derived nonaqueous phase liquids (NAPLs) contain diamondoid compounds, which can have potential applications in distinguishing natural gas condensate from automotive gasoline (Stout and Douglas, 2004).

Mid-range distillates are typically products in a relatively broader carbon range (C_6 to C_{26}) and include kerosene, aviation jet fuels, and lighter diesel products. Jet fuel is kerosene-based aviation fuel. Jet fuel is used for aviation turbine power units and usually has the same distillation characteristics and flash point as kerosene. Jet fuels are similar in gross composition, and compositional differences are attributable to additives designed to control some fuel parameters such as freeze and pour point characteristics. As Figure 3-11 shows, the chromatogram of a commercial jet fuel (Jet A) is dominated by GC-resolved *n*-alkanes in a narrow range of *n*-C_7 to *n*-C_{18} with maximum around *n*-C_{11} and a well-defined UCM. Diesel fuels were originally straight-run products obtained from the distillation of crude oil. Currently, diesel fuel may also contain varying amounts of selected cracked distillates to increase the available volume. The boiling range of diesel fuel is approximately

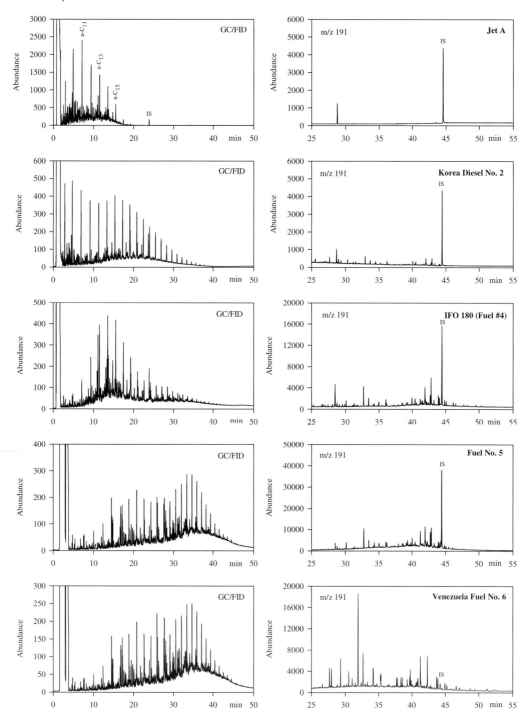

Figure 3-11 GC-FID (left panel) and GC-MS (at *m/z* 191, right panel) chromatograms of five common petroleum products (light to heavy) to illustrate differences in the *n*-alkane and tri-, tetra-, and pentacyclic terpane distributions between oils.

125–380°C. One of the most widely used specifications (ASTM D-975) covers three grades of diesel fuel oils: diesel fuel No. 1, diesel fuel No. 2, and diesel fuel No. 4. The marine fuel specifications have 4 categories of distillate fuels and 15 categories of fuels containing residual components (ASTM D-2069). Diesels consist of hydrocarbons in a carbon range of C_8 to C_{28} and contain high levels of *n*-alkanes, alkyl-cyclohexane, and PAHs. The properties of a given diesel are largely a function of the crude oil feedstock and any blending of various distillate stocks. The GC chromatogram of the Korean diesel fuel No. 2 (Figure 3-11) is dominated by a nearly normal distribution of *n*-alkanes with maxima around *n*-C_{11} to *n*-C_{14}. Also, a central UCM hump is obvious.

Heavy residual fuels. Heavy fuel oils (HFO) are blended products manufactured from residues of various refinery processes. The heavy residual fuels are largely used in marine applications and industrial power generation. Classic heavy fuel types include fuel No. 5 and No. 6 (also known as Bunker C) fuel. For years the term "Bunker C fuel oil" has been widely used to designate the most viscous residual fuels for general land and marine use. The chemical composition of Bunker C (or IFO 380) can vary widely, depending on production oil fields, production years, and processes it has undergone (see Chapter 10 herein for more details). Currently, many Bunker-type fuels are produced by blending residual oils with diesel fuels or other lighter fuels in various ratios to produce residual fuel oil of acceptable viscosity for marine or power plant use. The GC chromatograms of IFO 180, a lighter residual fuel No. 5 (also called Bunker B), and a Bunker C from Venezuela are also shown in Figure 3-11. The differences in the chromatographic profiles, carbon range, shapes of UCM, *n*-alkane, and isoprenoid distributions among these products are obvious. GC-MS chromatograms of *m/z* 191, 217, and 218, for example, jet fuel, diesel, IFO 180, Fuel No. 5 (Bunker B), and Venezuela Fuel No. 6 (Bunker C) are shown in Figures 3-11 and 3-12, respectively.

The differences in the concentrations and relative distributions of tri-, tetra-, and penta-cyclic terpanes and steranes between refined products are apparent. No target terpane and sterane compounds are detected in the Jet A fuel. Generally, most diesels contain none or only a trace of terpanes and steranes. However, the Korean diesel No. 2 demonstrates abundant biomarkers in a much wider carbon number range, indicating that diesels from different manufacturers may have correspondingly varying biomarker fingerprints. The GC-MS chromatogram of terpanes for the Venezuela Fuel No. 6 is characterized by a distribution in a wide range from C_{19} to C_{35} with C_{23} tricyclic triterpanes being the most prominent. As for steranes, the dominance of C_{27}, C_{28}, and C_{29} 20S/20R homologues is apparent. The relative proportion of C_{27}—C_{28}—C_{29} $\alpha\beta\beta$ steranes shows a consistent decrease with increasing carbon number ($C_{27} > C_{28} > C_{29}$). For the IFO-180, the presence of diasteranes is significantly higher than in other heavy fuel oils. As for Fuel No. 5, the dominance of C_{29} sterane peaks, in particular the 20S C_{29} $\alpha\alpha\alpha$ sterane, in SIM chromatogram (*m/z* 217) is pronounced.

3.3.3 Biomarkers in Lubricating Oils

Petroleum-derived lubricating oils are the most commonly used for both automotive and industrial applications. Small-scale spills and contamination by lubricating oil are quite common due to their wide application. Lubricating oils have broad GC profiles in the carbon range of C_{20} to C_{40} with boiling points greater than 340°C. Lubricating oil does not, generally, contain the low boiling fraction of petroleum hydrocarbons. Lubricating oil is largely composed of saturated hydrocarbons, and its GC trace is often dominated by a large UCM with few resolved peaks. In lubricating oil such as hydraulic fluid, for example, the PAH concentrations can be very low, while the biomarker concentrations are, generally, high. As examples, Figure 3-13 shows the GC-FID chromatograms and GC-MS fingerprints of five common lubricating oils.

Significant features of the biomarker distribution in petroleum-derived lubricating oils include the following: (1) biomarkers are

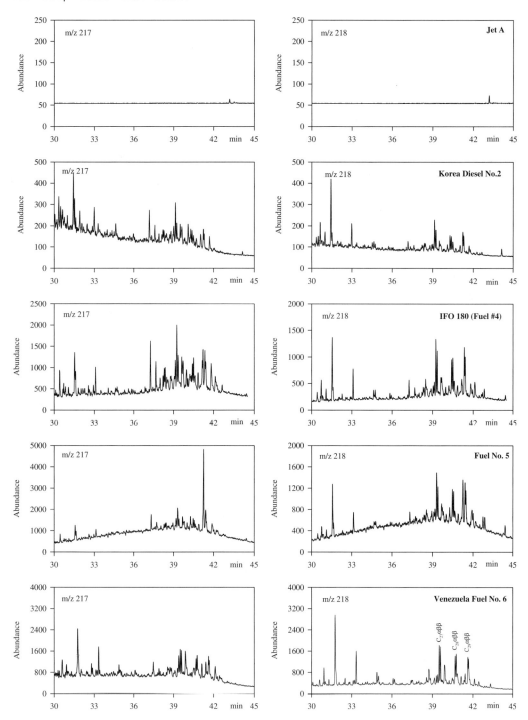

Figure 3-12 GC-MS chromatograms at *m/z* 217 and 218 for common petroleum products (light to heavy) to illustrate differences in sterane distributions between oils.

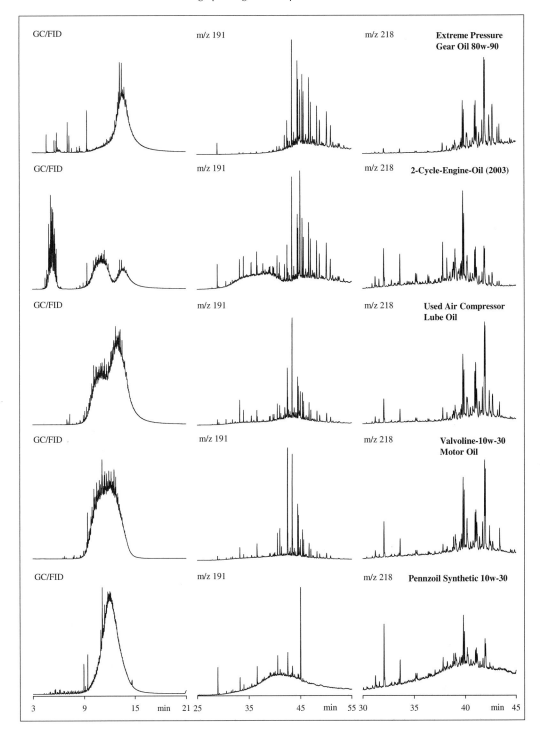

Figure 3-13 GC-FID (left panel) and GC-MS (at *m/z* 191, middle panel; *m/z* 218, right panel) chromatograms of five lubricating oils.

predominantly located in the high carbon number end, because the refining processes have removed low MW biomarkers and concentrated high MW biomarkers from the corresponding crude oil feed stocks; (2) lubricating oils, in general, have high concentrations of target terpane and sterane compounds (2000–6000 µg/g oil) in comparison with most crude oils and petroleum products (Table 3-5); (3) the dominance of characteristic pentacyclic C_{31} to C_{35} homohopanes is particularly prominent; (4) the dominance of C_{27}, C_{28}, and C_{29} 20S/20R homologues is apparent, too; (5) the concentrations of biomarkers are very low in the chemically synthetic lube oil, and the unresolved hump is pronounced.

Lubricating oil contamination through engine exhaust and through leakage and spillage occurs everywhere (Stout et al., 2001; Kaplan et al., 2001). Bieger et al. (1996) have reported the use of terpane biomarker fingerprints of refined oils and motor exhausts to indicate the presence of and trace the origin of diffuse lubricating oil contamination in plankton and sediments around St. John's, Newfoundland, Eastern Canada.

3.3.4 Biomarkers in Oil Fractions with Different Carbon Number Range

A research project to study the relative toxicity of oil fractions to fish (Khan et al., 2004; Clarke et al., 2004) was launched in 2004. The overall aim of this project was to generate useful and relevant data on the mechanism of hydrocarbon toxicity to fish to provide a strong foundation for ecological risk assessment of crude oil. Four oils (Federated crude oil, ANS oil, Scotia Light, and MESA oil) were selected for this research and fractionated into four oil fractions each using a distillation method. The nominal carbon number and boiling point ranges of these fraction are: Fraction 1, C_6 to C_{10}, 50–173°C; Fraction 2, C_{10} to C_{16}, 174–287°C; Fraction 3, C_{17} to C_{34}, 288–481°C; and Fraction 4, >C_{34}, >481°C. Each distillation fraction was quantitatively characterized. Characterization results clearly show that four oil fractions are significantly different from each other in their chemical composition including the carbon range and molecular-weight range, diagnostic ratios of target individual compounds and compound classes, and distribution patterns and profiles of n-alkanes, BTEX and alkyl benzenes, PAHs, and biomarkers. The left panel of Figure 3-14 shows the GC chromatograms of the MESA oil (a medium South American crude, API gravity of 29.7) and four fractions of the oil to illustrate major chemical composition features of each fraction; and the right panel graphically compares quantitative distribution of target alkylated PAHs and other EPA priority PAHs in the oil and its four fractions. Figure 3-15 presents GC-MS chromatograms at m/z 191 and 218 to illustrate differences in biomarker distributions of four fractions. Table 3-6 summarizes the quantitation results of target biomarkers in the MESA oil and its four fractions.

Figure 3-15 clearly reveals that the distribution patterns and profiles of biomarkers in four fractions are significantly different from each other. No biomarkers were present in Fraction 1, and only several smaller biomarker compounds (C_{21} to C_{24} terpanes) with very low abundances were detected in Fraction 2 (2.14 µg/g oil in total). The totals of the target biomarkers were determined to be 2523 and 2045 µg/g oil for Fractions 3 and 4, respectively, much higher than that in the original oil (1771 µg/g oil). Obviously, as the lower-molecular-weight hydrocarbons were removed from the crude oil by distillation, the higher-molecular-weight biomarkers were correspondingly concentrated, resulting in higher concentration of biomarkers in Fractions 3 and 4. In Fraction 4, no biomarkers with a carbon number smaller than C_{27} were detected. The terpanes (pentacyclic hopanes) were confined to a much narrower range of C_{29} to C_{35}, and the C_{29} 20S/20R steranes were significantly more abundant than the C_{27} and C_{28} group steranes.

3.3.5 Aromatic Steranes in Oils and Petroleum Products

Aromatic steranes are another group of biomarker compounds that are highly resistant to

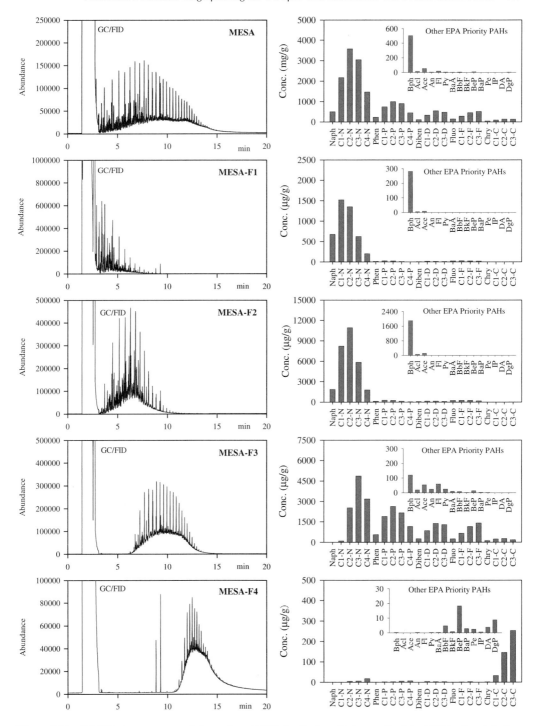

Figure 3-14 GC-FID chromatograms (left panel) and PAH distributions (right panel) of the MESA oil and its four fractions.

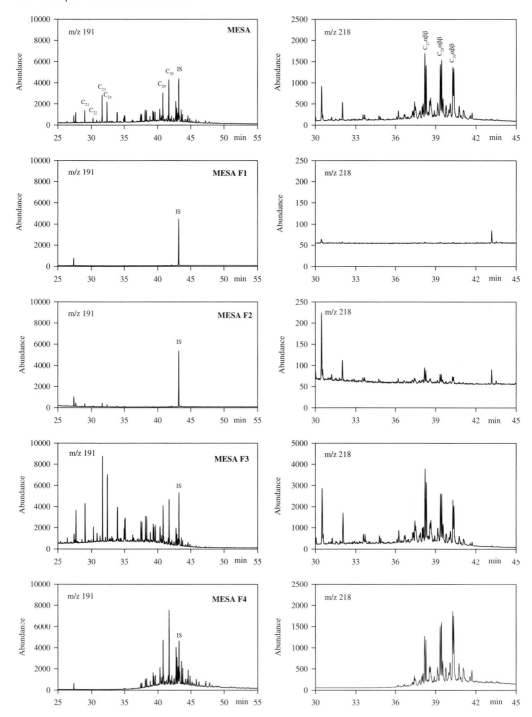

Figure 3-15 GC-MS chromatograms of the MESA oil and its four fractions at *m/z* 191 (left panel) and 218 (right panel) to illustrate differences in biomarker distributions.

Table 3-6 Quantitation Results of Target Biomarkers in the MESA Oil and Its Four Fractions

Oil Samples	MESA	MESA-F1	MESA-F2	MESA-F3	MESA-F4
Biomarker compounds (μg/g oil)					
C_{21}	35.8	0.76	7.73	95.0	0.00
C_{22}	13.2	0.20	2.24	35.4	0.00
C_{23}	77.3	0.76	8.62	211	0.32
C_{24}	59.8	0.42	5.97	161	0.26
$C_{29}\ \alpha\beta$	92.1	0.00	1.02	98.9	150
$C_{30}\ \alpha\beta$	148	0.00	1.62	124	262
C_{31} (S)	65.9	0.00	0.51	44.6	122
C_{31} (R)	46.8	0.00	0.21	31.5	84.2
C_{32} (S)	40.3	0.00	2.35	22.7	72.9
C_{32} (R)	25.4	0.00	0.00	15.2	50.2
C_{33} (S)	26.5	0.00	0.00	13.3	58.0
C_{33} (R)	16.7	0.00	0.00	8.12	38.4
C_{34} (S)	16.9	0.00	0.00	6.97	39.2
C_{34} (R)	9.05	0.00	0.00	3.76	24.8
C_{35} (S)	12.7	0.00	0.00	4.51	34.9
C_{35} (R)	14.0	0.00	0.00	4.13	35.4
Ts	17.9	0.00	0.39	27.4	17.5
Tm	37.6	0.00	0.66	47.4	40.2
$C_{27}\ \alpha\beta\beta$-steranes	366	0.00	6.65	649	236
$C_{28}\ \alpha\beta\beta$-steranes	308	0.00	5.19	506	352
$C_{29}\ \alpha\beta\beta$-steranes	340	0.00	4.06	412	428
Total	1771	2.14	47.2	2523	2045
Diagnostic ratios					
C_{23}/C_{24}	1.29	1.81	1.44	1.31	1.21
$C_{23}/C_{30}\ \alpha\beta$	0.52	NA	5.32	1.70	0.00
$C_{24}/C_{30}\ \alpha\beta$	0.40	NA	3.68	1.29	0.00
$C_{29}\ \alpha\beta/C_{30}\ \alpha\beta$	0.62	NA	0.63	0.79	0.57
$C_{31}(S)/C_{31}(S+R)$	1.41	NA	2.43	1.42	1.44
$C_{32}(S)/C_{32}(S+R)$	1.59	NA	NA	1.50	1.45
Ts/Tm	0.48	NA	0.60	0.58	0.44
$C_{27}\ \alpha\beta\beta$-steranes/$C_{29}\ \alpha\beta\beta$-steranes	1.08	NA	1.64	1.57	0.55
$C_{30}/(C_{31}+C_{32}+C_{33}+C_{34}+C_{35})$	0.54	NA	0.53	0.80	0.47

biodegradation and can be used for oil-to-oil correlation and oil source tracking. Aromatic steranes are monitored using m/z 231 and 253 for triaromatic (TA) and monoaromatic (MA) steranes, respectively. The m/z 231 mass chromatograms of crude oil are characterized by series of 20R and 20S C_{26}-C_{27}-C_{28} triaromatic a steranes (TA-cholestanes, TA-ergostanes, and TA-stigmastanes) plus C_{20} to C_{22} TA-steranes. The m/z 253 mass chromatograms are featured by series of 20R and 20S C_{27}-C_{28}-C_{29} 5β(H) and 5α(H) MA steranes as well as rearranged ring-C 20S and 20R MA-diasteranes. Peak identification of TA- and MA-steranes in the Platform Elly oil is summarized in Figure 3-16 and Table 3-7. As Figure 3-16 shows, all target TA-steranes are well separated under the present GC conditions except that the C_{26} 20R isomer co-elutes with the C_{27} 20S isomer (Peak 5). The structures of rearranged MA steranes have been established as 10-desmethyl 5α- and 5β-methyl (20S and 20R) MA-diasterane isomers (Riolo et al., 1985; Moldowan and Fago, 1986).

To illustrate the differences in TA- and MA-steranes between oils and refined products, Figure 3-17 compares the mass chromatograms of the TA- and MA-steranes in the aromatic hydrocarbon fractions of the NIST 1582 oil and refined products (IFO-180, a lubricating oil, and a Diesel No. 2 from Korea).

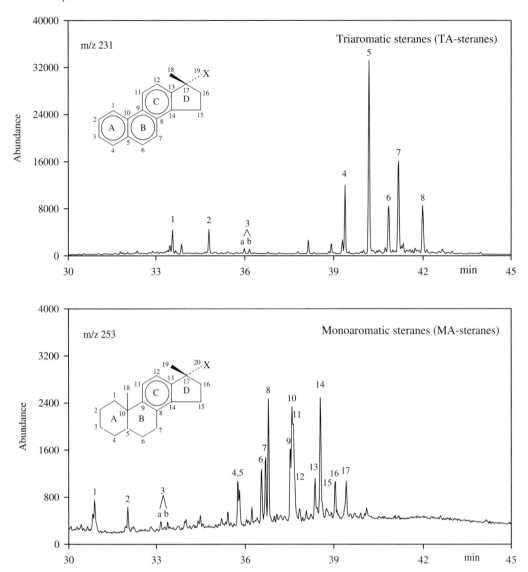

Figure 3-16 Peak identification of the triaromatic (*m/z* 231) and monoaromatic (MA, *m/z* 253) steranes in the Platform Elly oil.

Figure 3-17 shows apparent differences in the relative distributions and absolute concentrations of TA- and MA-steranes between oils and refined products. Generally, triaromatic steranes are much more abundant than monoaromatic steranes for all oils studied. In many lighter oils such as Cook Inlet, Federated, West Texas, and Scotia Light, only trace MA-steranes are detected. This implies that TA-steranes are more valuable marker compounds than MA-steranes for environmental forensic investigations. Unlike most Canadian diesels, the Korean Diesel No. 2 still contains a relatively large quantity of high carbon number TA-steranes. Similarly, lubricating oils do not or only contain trace levels of MA-sterane compounds. Synthetic lubricants should not contain any TA- or MA-sterane compound if they are purely chemically synthesized. However, GC-MS analyses show that

Table 3-7 Peak Identification of the Triaromatic and Monoaromatic Steranes in the Platform Elly Oil

Peak No.	Compounds	Code	Molecular Formula
Triaromatic steranes (TA-steranes, m/z 231)			
1	C_{20} TA-sterane (X = ethyl)	C20TA	$C_{20}H_{20}$
2	C_{21} TA-sterane (X = 2-propyl)	C21TA	$C_{21}H_{22}$
3	C_{22} TA-sterane (X = 2-butyl) (a and b are epimers at C-19)	C22TA	$C_{22}H_{24}$
4	C_{26} TA-chloestane (20S)	SC26TA	$C_{26}H_{32}$
5	C_{26} TA-chloestane (20R) + C_{27} TA-ergostane (20S)	RC26TA + SC27TA	$C_{26}H_{32}$ $C_{27}H_{34}$
6	C_{28} TA-stigmastane (20S)	SC28TA	$C_{28}H_{36}$
7	C_{27} TA-ergostane (20R)	RC27TA	$C_{27}H_{34}$
8	C_{28} TA-stigmastane (20R)	RC28TA	$C_{28}H_{36}$
Monoaromatic steranes (MA-steranes, m/z 253)			
1	C_{21} MA-sterane (X = ethyl)		$C_{21}H_{30}$
2	C_{22} MA-sterane (X = 2-propyl)		$C_{22}H_{32}$
3	C_{23} MA-sterane (X = 2-butyl) (a and b are epimers at C-20)		$C_{23}H_{34}$
4	C_{27} 5β(H) MA-cholestane (20S)		$C_{27}H_{42}$
5	C_{27} MA-diacholestane (20S)		$C_{27}H_{42}$
6	C_{27} 5β(H) MA-cholestane (20R) + C_{27} MA-diacholestane (20R)		$C_{27}H_{42}$
7	C_{27} 5α(H) MA-cholestane (20S)		$C_{27}H_{42}$
8	C_{28} 5β(H) MA-ergostane (20S) + C_{28} MA-diaergostane (20S)		$C_{28}H_{44}$
9	C_{27} 5α(H) MA-cholestane (20R)		$C_{27}H_{42}$
10	C_{28} 5α(H) MA-ergostane (20S)		$C_{28}H_{44}$
11	C_{28} 5β(H) MA-ergostane (20R) + C_{28} MA-diaergostane (20R)		$C_{28}H_{44}$
12	C_{29} 5β(H) MA-stigmastane (20S) + C_{29} MA-diastigmastane (20S)		$C_{29}H_{46}$
13	C_{29} 5α(H) MA-stigmastane (20S)		$C_{29}H_{46}$
14	C_{28} 5α(H) MA-ergostane (20R)		$C_{28}H_{44}$
15	C_{29} 5β(H) MA-stigmastane (20R)		$C_{29}H_{46}$
16	C_{29} 5α(H) MA-stigmastane (20R)		$C_{29}H_{46}$
17	C_{30} 5β(H) MA-sterane (20S)		$C_{30}H_{48}$

TA-steranes are present in the *Synthetic 10W-30* lubricating oil. This fact indicates that this lubricating oil may not be 100% synthesized and may be composed of a portion of petroleum-derived hydrocarbons. Barakat et al. (2002) have recently reported a case study in which oil residues were correlated to a fresh crude oil sample of the Egyptian Western Desert-sourced oil by fingerprinting monoaromatic and triaromatic steranes and by determination and comparison of molecular ratios of the target MA- and TA-sterane compounds.

3.3.6 Sesquiterpanes in Oils and Petroleum Products

Polymethyl-substituted decalins or decahydronaphthalenes (i.e., C_{14}–C_{16} bicyclic alkanes), commonly known as sesquiterpanes, were first reported in 1974 (Bendoraitis, 1974) and later discovered in crude oils of the Loma Novia and Anastasievsko-Troyitskoe deposits (Petrov, 1987). Alexander et al. (1983) identified and confirmed the existence of 8β(H)-drimane and 4β(H)-eudesmane in most Australian oils. Noble (Noble, 1986) identified a series of C_{14} to C_{16} sesquiterpane isomers using synthesized standards and mass spectral studies. Various sesquiterpanes, with the greatest enrichment in condensate, were also identified by Simoneit et al. from fossil resins, sediments, and crude oils (Simoneit et al., 1986), and by Chen and He from a great offshore condensate field of Liaodong Bay, Northern China (Chen and He, 1990).

Bicyclic biomarker sesquiterpanes with the drimane skeleton are ubiquitous components of crude oils and ancient sediments. Most

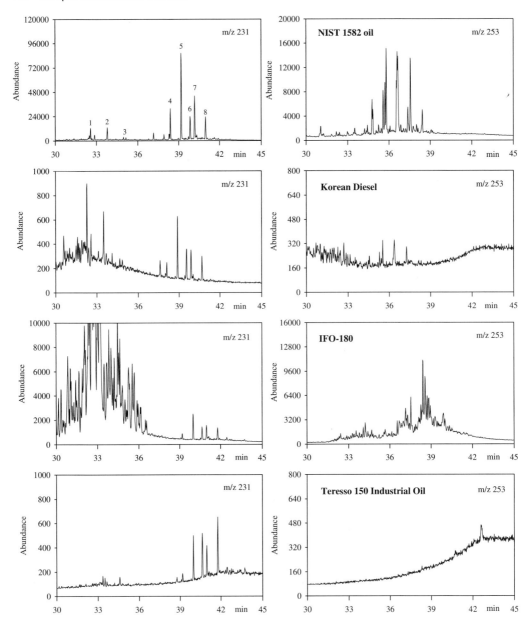

Figure 3-17 Mass chromatograms of the TA- (m/z 231) and MA-steranes (253) in the aromatic hydrocarbon fractions of the NIST 1582 oil, a Diesel No. 2 from Korea, IFO-180, and a lubricating oil.

sesquiterpanes probably originate from higher plants and also from algae or bacteria (Alexander et al., 1984; Philp, 1985; Fan et al., 1991). During the thermal evolution, the relative concentration of C_{14} sesquiterpanes decreases with increasing maturation of organic matters. The concentration of C_{14} bicyclic sesquiterpanes is higher at the immature stage, while those of C_{15} drimane and C_{16} homodrimane are relatively lower. As a result of the dehydroxylation and chemodynamics of their higher-molecular-weight precursors, the concentrations of drimane (C_{15}) and homodrimanes (C_{16}) gradually increase, and the concentrations of C_{14} sesquiterpanes decline (Cheng et al., 1991).

Though biomarker sesquiterpanes have found increasing application in petroleum exploration in recent years, there have been few reports of use of these compounds for forensic oil spill identification (Stout et al., 2005; Wang et al., 2005a). For lighter petroleum products, refining processes have removed most high MW biomarkers from the original crude oil feedstock. Thus, the pentacyclic terpanes and steranes are generally absent or in low abundance in lighter petroleum products (e.g., jet fuels and diesels), while the sesquiterpanes are concentrated in these distillates. The sesquiterpanes with the drimane skeleton are monitored at m/z 123 (a base fragment ion common to all sesquiterpanes). Confirmation ions include 179 (the ion after sesquiterpane $C_{14}H_{26}$ loses $-CH_3$), 193 (the ion after $C_{15}H_{28}$ loses $-CH_3$ and after $C_{16}H_{30}$ loses $-C_2H_5$), and 207 (the ion after $C_{16}H_{30}$ loses $-CH_3$). Examination of GC-MS chromatograms of these characteristic ions of sesquiterpanes provides a highly diagnostic means for correlation, differentiation, and source identification of light to middle-range petroleum products, in comparison with the use of other hydrocarbon groups.

The sesquiterpanes ranging from C_{14} to C_{16} elute between n-C_{13} and n-C_{16} (boiling point 235–287°C) in the SIM chromatogram of the saturated hydrocarbon fraction. Peaks 1 and 2, 3 to 6, and 7 to 10 (Figure 3-18) are identified as C_{14}, C_{15}, and C_{16} sesquiterpanes, respectively. Of 10 identified sesquiterpanes, peaks 5 and 10 are identified to be 8β(H)-drimane and 8β(H)-homodrimane, respectively. GC-MS analyses demonstrate different distribution patterns of sesquiterpanes in crude oils and refined products of different origins. The left panel of Figure 3-19 shows SIM chromatograms of sesquiterpanes at m/z 123 for light (API >35), medium (API: 25–35), and heavy (API <25) crude oils, including Alaska North Slope (ANS), Arabian Light, Scotia Light oil (Nova Scotia), West Texas, and California API 11, while the right panel compares sesquiterpane distributions in common petroleum products from light kerosene to heavy fuel oil.

Ten sesquiterpanes are present in all oils studied. However, distributions and concentrations of sesquiterpanes vary between oils of different origins. Lighter oils ANS, Arabian Light, and Scotia Light have high concentrations of sesquiterpanes, with Peak 10 (C_{16} homodrimane) being the most abundant for the ANS and Arabian Light, and Peak 3 (C_{15} sesquiterpane) for Scotia Light, respectively. The Arabian Light has the lowest concentration of C_{14} sesquiterpanes (Peaks 1 and 2), indicating that this oil is highly mature. On the contrary, the heavy California API 11 oil has the highest concentration of C_{14} sesquiterpane, indicating that this oil is relatively immature.

Sesquiterpanes are absent in light kerosene and heavy lubricating oils. However, refined products IFO-180 and HFO 6303 (Bunker C type) contain high concentrations of sesquiterpanes. It is also noticed that an unknown bicyclic biomarker compound (between Peaks 2 and 3) is abundant in these two products. Jet A is characterized with Peaks 1, 3, and 5 being the most abundant, while Peak 10 is the most abundant in middle-distillates such as diesel due to concentration of lower-MW biomarkers from the original crude oil feedstock. The differences in distribution patterns and concentrations are often apparent between diesels.

Oil spills were reported and sampled on March 17 and 23, 1998, at a sewer outlet flowing into the Lachine Canal in Quebec. Following the accident, a diesel fuel, which was suspected to be the source of the spill, was collected from a reservoir at a pumping station located in Lachine, Quebec. Biomarker fingerprinting of the samples revealed that only trace amounts (<10μg/g oil) of C_{19}–C_{24} tricyclic terpanes, regular C_{20}–C_{22} steranes, and diasteranes were detected. However, the spill samples contained significant amounts of sesquiterpanes. The GC-MS/SIM chromatogram (at m/z 123) and diagnostic ratios of target sesquiterpanes of the spill samples were found to be nearly identical to that of the suspected-source diesel. The only noticeable difference is that, compared to the suspected-source-diesel, the

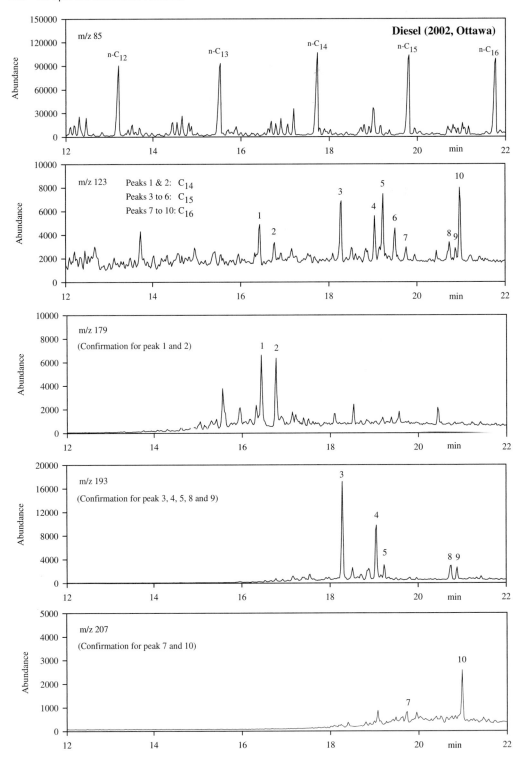

Figure 3-18 GC-MS (SIM) chromatograms of sesquiterpanes eluting in the n-C_{13} and n-C_{16} range.

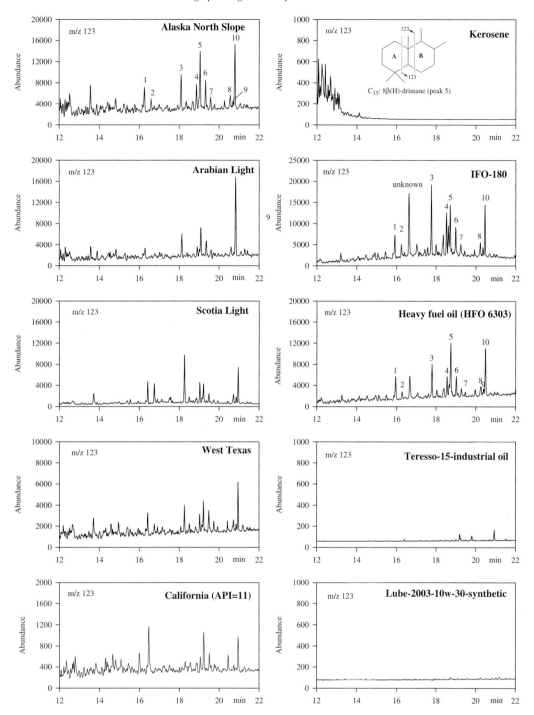

Figure 3-19 *Left panel*: GC-MS chromatograms of sesquiterpanes at *m/z* 123 for light (API > 35), medium (API: 25–35), and heavy (API < 25) crude oils, including Alaska North Slope (ANS), Arabian Light, Scotia Light oil (Nova Scotia), West Texas, and California API 11. *Right panel*: Comparison of SIM chromatograms of sesquiterpanes at *m/z* 123 for common petroleum products.

spilled sample had slightly higher abundances due to weathering. These similarities, in combination with other hydrocarbon quantification results such as bulk hydrocarbon groups, n-alkane distribution, and fingerprints of alkylated PAHs and diagnostic ratios of source-specific PAH compounds (Wang et al., 2000), argued strongly that the suspected diesel collected from the pumping station close to the spill site was the source of the spilled diesel.

3.3.7 Diamondoid Compounds in Oils and Lighter Petroleum Products

The group of *diamondoids* (collective term for adamantane, diamantane, and their alkyl homologous series) is another group of low-boiling cyclic biomarkers of interest to environmental forensics. Diamondoids are rigid, three-dimensionally fused cyclohexane-ring alkane compounds that have a diamondlike cage structure (Chen et al., 1996; Dahl et al., 1999; Grice et al., 2000). They consist of pseudo-homologous series with the general formula, $C_{4n+6}H_{4n+12}$, including adamantane (C_{10}), dia- (C_{14}), tria-, tetra-, and pentamantane ($n = 1-5$, respectively) and higher polymantanes, and their alkylated homologues. Adamantane was first discovered and isolated from a Czechoslovakian petroleum in 1933. Since then, more diamondoids have been found in crude oils (Petrov, 1987; Williams et al., 1986; Wingert, 1992; Grice et al., 2000; Lin and Wilkes, 1995). Adamantane and diamantane found in petroleum are thought to be formed from rearrangements of suitable organic precursors including multiring terpene hydrocarbons under thermal stress with strong Lewis acids (typically clays) acting as catalysts during oil generation (Chen et al., 1996; Dahl et al., 1999). The higher homologues of diamondoids are thought to be formed from lower homologues under extreme temperature and pressure conditions (Grice et al., 2000). The diamond structure endows these molecules with a high kinetic and thermodynamic stability. Laboratory thermal cracking experiments (Dahl et al., 1999) have shown that diamondoids have a higher thermal stability than most other hydrocarbons during thermal cracking of oil; therefore, diamondoids become increasingly enriched in the residual oil or condensate. The increase in methyldiamantane (C_{15}) concentration is directly proportional to the extent of cracking, indicating that under the conditions of the experiments, diamondoids are neither destroyed nor created. Instead, they are conserved and concentrated, and hence can be considered a naturally occurring *internal standard* by which the extent of oil loss can be determined.

Adamantanes and diamantanes elute in the ranges of n-C_{10} and n-C_{13} and n-C_{15} and n-C_{17}, respectively, in the GC-MS chromatogram of a saturated hydrocarbon fraction. Adamantanes are monitored at their characteristic ions at m/z 136 for adamantane, 135 for methyl- and ethyladamantanes, 149 for dimethyladamantanes, 163 trimethyladamantanes, and 179 for tetramethyladamantanes; while diamantanes are monitored at m/z 188, 187, 201, and 215 for diamantane, methyldiamantanes, dimethyl-diamantanes, and trimethyl-diamantanes, respectively. Figure 3-20 shows the GC-MS-SIM chromatograms for analysis of diamondoids in Prudhoe Bay oil. Twenty-six diamondoid compounds were identified, and among these 17 are adamantanes and 9 diamantanes. Peak assignments are presented in Table 3-8. Identification of diamondoid hydrocarbons as based on mass spectra, comparison of GC retention data with reference standards, and calculation of reference index (IR) and comparison with literature RI values (Wingert, 1992; Chen et al., 1996).

Figure 3-20 reveals the following: (1) the differences in concentrations and relative distributions of adamantanes are apparent. 1,3,5,7-tetramethyladamantane (Peak 5) has the lowest concentration among the adamantane series. This is most likely due to the fact that 1,3,5,7-tetramethyladamantane has four methyl groups that could affect each other and cause the molecule structure to be thermally unstable. (2) The group of methyladamantanes contains only two isomers (Peak 2: 1-methyladamantane and Peak 6: 2-methyladamantane)

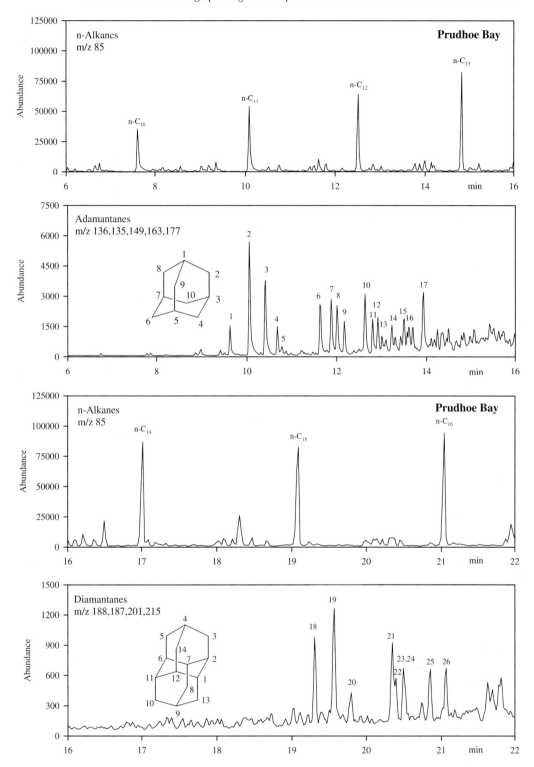

Figure 3-20 GC-MS chromatograms of diamondoids in Prudhoe Bay oil for peak identification.

Table 3-8 Peak Identification of Diamondoid Compounds in Prudhoe Bay Oil

Peak No.	Compounds	Abbreviations	Base Peak	M^+ (m/z)	Formula
Adamantanes					
1	Adamantane	A	136	136	$C_{10}H_{16}$
2	1-Methyladamantane	1-MA	135	150	$C_{11}H_{18}$
3	1,3-Dimethyladamantane	1,3-DMA	149	164	$C_{12}H_{20}$
4	1,3,5-Trimethyladamantane	1,3,5-TMA	163	178	$C_{13}H_{22}$
5	1,3,5,7-Tetramethyladamantane	1,3,5,7-TeMA	177	192	$C_{14}H_{24}$
6	2-Methyladamantane	2-MA	135	150	$C_{11}H_{18}$
7	1,4-Dimethyladamantane, *cis*	1,4-DMA, *cis*	149	164	$C_{12}H_{20}$
8	1,4-Dimethyladamantane, *trans*	1,4-DMA, *trans*	149	164	$C_{12}H_{20}$
9	1,3,6-Trimethyladamantane	1,3,6-TMA	163	178	$C_{13}H_{22}$
10	1,2-Dimethyladamantane	1,2-DMA	149	164	$C_{12}H_{20}$
11	1,3,4-Trimethyladamantane, *cis*	1,3,4-TMA, *cis*	163	178	$C_{13}H_{22}$
12	1,3,4-Trimethyladamantane, *trans*	1,3,4-TMA, *trans*	163	178	$C_{13}H_{22}$
13	1,2,5,7-Tetramethyladamantane	1,2,5,7-TeMA	177	192	$C_{14}H_{24}$
14	1-Ethyladamantane	1-EA	135	164	$C_{12}H_{20}$
15	1-Ethyl-3-methyladamantane	1-E-3-MA	149	178	$C_{13}H_{22}$
16	1-Ethyl-3,5-dimethyladamantane	1-E-3,5-DMA	163	192	$C_{14}H_{24}$
17	2-Ethyladamantane	2-EA	135	164	$C_{12}H_{20}$
Diamantanes					
18	Diamantane	D	188	188	$C_{14}H_{20}$
19	4-Methyldiamantane	4-MD	187	202	$C_{15}H_{22}$
20	4,9-Dimethyldiamantane	4,9-DMD	201	216	$C_{16}H_{24}$
21	1-Methyldiamantane	1-MD	187	202	$C_{15}H_{22}$
22	1,4 & 2,4-Dimethyldiamantane	1,4 & 2,4-DMD	201	216	$C_{16}H_{24}$
23	4,8-Dimethyldiamantane	4,8-DMD	201	216	$C_{16}H_{24}$
24	Trimethyldiamantane	TMD	215	230	$C_{17}H_{26}$
25	3-Methyldiamantane	3-MD	187	202	$C_{15}H_{22}$
26	3,4-Dimethyldiamantane	3,4-DMD	201	216	$C_{16}H_{24}$

due to their structural symmetry. 1-methyladamantanes that have only one methyl group attached to the bridgehead position (that is, at carbon position 1) are the most abundant. Similarly, 4-methyldiamantane (Peak 19), also a bridgehead methylated compound, is the most abundant compound in the diamantane series. The reason is that the methyl substitution in adamantane or diamantane at a bridgehead position (i.e., position of a tertiary carbon in the ring structure) creates a more stable molecule than substitution at a secondary carbon atom (carbon position 2) as the latter produces additional skew-butane repulsions that are not imposed by the bridgehead attachment (Wingert, 1992). Therefore, 1-methyladamantane has a higher thermal stability than 2-methyladamantane. Likewise, 4-methyldiamantane has a higher thermal stability than 1-methyldiamantane (Peak 21) and 3-methyldiamantane (Peak 25). Stable hydrocarbons will gradually increase in relative abundance over the less stable isomeric ones with increasing thermal stress. Hence, the relative distribution of alkyl-substituted diamondoid hydrocarbons may be used for assessing the maturity, especially for highly mature petroleum. Two diamondoid hydrocarbon ratios [methyladamantane index (MAI) and methyldiamantane index (MDI), defined as 1-MA/(1 + 2-MA) and 4-MD/(1 + 3 + 4-MD), respectively] have been developed and used as novel high-maturity indices to evaluate the maturation and evolution of crude oils, and to determine the thermal maturity of thermogenic gas and condensate in several Chinese basins, the maturity of which may be difficult to assess using routine geochemical techniques (Chen et al., 1996). (3) The elution of alkyladamantanes (i.e., the sequence of their boiling points) is quite peculiar. All methyladamantanes substituted at the bridgehead (that is, at position 1) have much lower boiling points than adamantanes with at least one of the methyl groups not situated at the bridgehead (such as 2-methyladamantane, 1,2-dimethyladamantane, 1,4-dimethyladamantane, and 1,3,4-trimethyladamantane). The difference in the boiling points of these adamantanes is so large that 2-methyladamantane (C_{11}) elutes later than 1,3,5,7-tetramethyladamantane (C_{14}).

Experimental results demonstrate that both concentrations and relative distributions of diamondoids vary significantly between crude oils and refined products of different origins. Figure 3-21 diplays the GC-MS chromatograms of adamantanes and diamantanes for five representative crude oils from Cold Lake Bitumen to the South Louisiana oil. Both adamantanes and diamantanes occur in detectable quantities in all of the crude oils studied. Overall, the one-cage adamantanes are much more abundant than two-cage diamantanes. Based on quantitation data, the principal dominant adamantane hydrocarbons are A, 2-MA, 1-MA, 2-EA, 1,2-DMA, and, 1,3-DMA, together accounting for about 50% of all detected adamantanes; and the dominant diamantane compounds are D, 4-MD, 1-MD, 3-MD, and 3,4-DMD. Figure 3-22 compares the GC-MS chromatograms of adamantanes and diamantanes for five representative refined petroleum products including Jet A, two diesel fuels, a fuel No. 4, and a lubricating oil. Adamantanes were found in all fuel oil samples and were detected in most lubricating oils at a trace level. As expected, however, only quite low or no diamantane compounds were detected in light kerosene and heavy-end lubricating oils. Generally, the overall distribution pattern of individual diamondoid compounds in petroleum products is similar to that in crude oils, in which 1-MA and 2-MA, and D and 4-MD, dominated the adamantanes and diamantanes, respectively. The absolute concentrations and distribution patterns of diamondoids differ widely in the petroleum products studied. These differences are attributed to the differences in the crude oil feedstocks used in the production and to the distillation cut point of the petroleum products.

The unique molecular structures of diamondoids imply that their distributions and relative ratios have the potential to differentiate spilled oils, particularly to correlate and differentiate spilled lighter refined products in which high-molecular-weight tri- to pentacyclic biomarkers are present, if at all, in only trace amounts. Although diamondoids have found increasing application in petroleum exploration and refining in recent years, there have been few reports of use of these compounds for forensic oil spill investigations. Recently, Stout and Douglas (2004) reported application of diamondoid hydrocarbons in the chemical fingerprinting of natural gas condensates and gasoline. In this laboratory, diamondoids in over a hundred crude oils and refined products have been quantitatively characterized, and distributions of diamondoids in different oils, oil distillation fractions, and various refined products have been qualitatively and quantitatively compared. A number of diagnostic indices of diamondoids have been developed for forensic oil correlation and differentiation (Wang et al., 2005b).

3.3.8 Application of Biomarker Fingerprintings to Oil Spill Studies

The fingerprints of petroleum biomarkers have been applied to investigations of oil spill accidents (e.g., Barakat et al., 1999; Bence et al., 1996; Kvenvolden et al., 1993; Page et al., 1988; Wang et al., 1994b, 1995b, 1998b; Zakaria et al., 2000, 2001; Stout et al., 2001) and to trace the record of hydrocarbon input to the San Francisco Bay (Hostettler, 1999a). In recent years, biomarkers, together with PAHs and other hydrocarbon characterization results, have been extensively applied to assess the origin of the petrogenic hydrocarbon background in Prince William Sound (PWS) of Alaska, the site of the 1989 *Exxon Valdez* oil spill (i.e., is the petrogenic hydrocarbon background mainly from eroding Tertiary shales

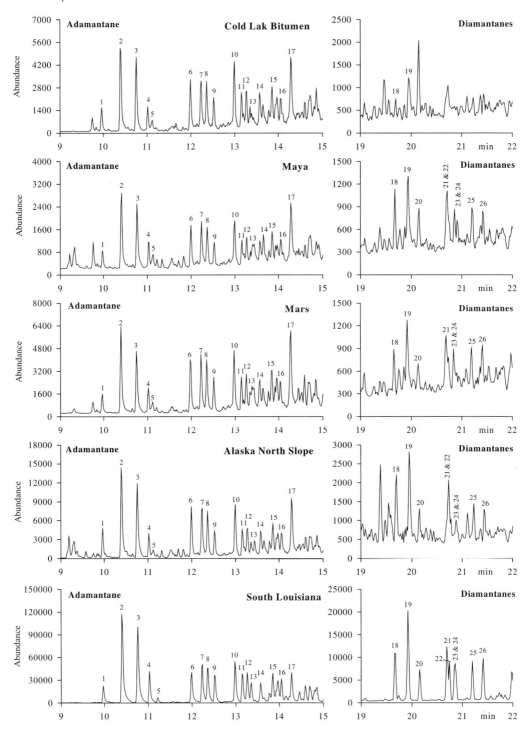

Figure 3-21 GC-MS total ion chromatograms of adamantanes (left) and diamantanes (right) of five representative crude oils including Cold Lake Bitumen, Maya, Mars, Alaska North Slope, and South Louisiana crude oil.

Petroleum Biomarker Fingerprinting for Oil Spill Characterization and Source Identification 119

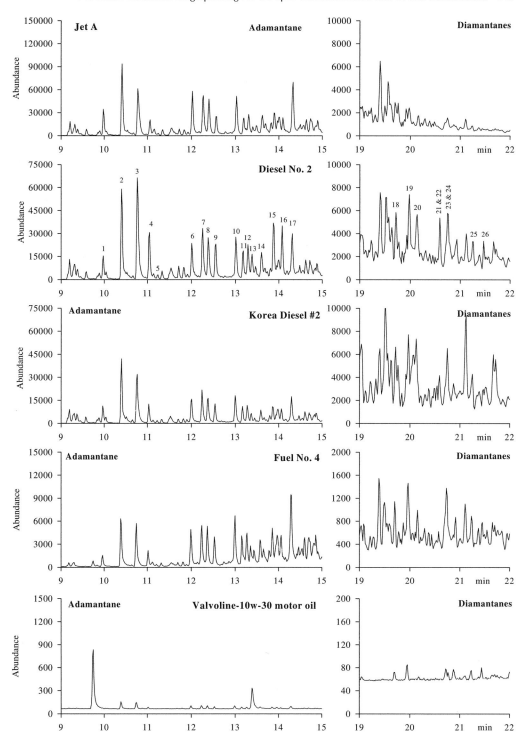

Figure 3-22 GC-MS total ion chromatograms of adamantanes and diamantanes of five representative refined petroleum products including Jet A, Diesel No. 2, Korea Diesel, Fuel No. 4, and Valvoline-10W-30 lube oil.

and residues of natural oil seepage, or mainly from Berling River coals and oil from the Katalla area?) by two groups of scientists (Boehm et al., 2001, 2002; Page et al., 1995, 1996, 2002; Bence et al., 1996; Short et al., 1997, 1999; Hostettler et al., 1999b). (See Chapter 15 herein.)

Examination and comparison of biomarker fingerprinting patterns and profiles are widely used for oil correlation and differentiation in environmental forensic studies. As described above, the distribution pattern and profiles of biomarkers are, in general, different from oil to oil and from oil to refined products of different origins. Various biomarkers can occur in different carbon ranges. Also, concentrations of individual biomarkers could be markedly different. Therefore, qualitative and quantitative comparisons of biomarker distribution are important for spill/source identification: (1) whether target biomarkers detected in spill samples can be found in suspected source candidates; (2) whether the distribution patterns and profiles of biomarkers match; (3) whether the abundances of target biomarkers match; (4) whether there are any "source-specific" or unknown biomarker compounds; (5) whether the diagnostic ratios of major biomarkers match. In most cases, disparity (no matching) of biomarker distribution is strong evidence for lack of correlation between spill sample(s) and suspected source(s). Matching may be an indication of a correlation of spill sample(s) and suspected source(s), but under certain spill scenarios is not necessarily "proof" that samples are from the same source.

Based on analysis of triterpane distribution patterns and determination of two pentacyclic C_{27} triterpanes, Shen (1984) distinguished four Arabian crudes, which in their weathered forms were extremely similar. Volkman et al. (1992b) determined the distribution of various biomarker compounds in a range of aquatic sediment samples to confirm the presence of oil contamination and identify possible oil sources. Among a number of pollution sources, lubricating oils were identified as a major source of hydrocarbon pollution in many estuaries and coastal areas around Australia. Currie et al. (1992) proposed utilization of triterpanes to distinguish tar balls originated from Southeast Asia from those of Australian petroleum sources. Mello et al. (1988) studied the geological and biomarker features of a wide selection of oils from the major Brazilian offshore basins. The study results reveal significant differences in chemical features of various oils, which enable them to be divided into five groups. The diagnostic features used for this classification include the absolute concentrations and distributions of hopanes and steranes, their abundances relative to 4-methylsteranes, and the occurrence and abundance of several specific biomarkers including 18α-oleanane, gammacerane, β-carotane, higher acyclic isoprenoids, 28,30-bisnorhopane, and 25,28,30-trisnorhopane. Barakat et al. (1997) studied the biomarker distribution within five crude oils from the Gulf of Suez, Egypt. The results revealed significant differences in biomarker distribution within the oils, and the oils can be categorized into three groups. Type 1 oils show a high relative abundance of gammacerane indicating a marine saline-source depositional environment. Furthermore, these oils have a predominance of C_{35} over C_{34} 17α(H)-homohopanes. Type 2 oils have an oleanane content of more than 20% of the concentration of C_{30} αβ hopane, indicating they originated from an angiosperm-rich, Tertiary source rock. Type 3 oil has geochemical characteristics intermediate between type 1 and 2 oils. Lu and Kaplan (1992) studied biomarker distribution in natural bitumen extracted from four coals: Rocky Mountain coal (RMC), Australian Gippsland Latrobe Eocene coal (GEC), Australian Gippsland Latrobe Cretaceous coal (GCC), and Texas Wilcox lignite (WL). They found there is a significant difference in the distribution of terpanes among these coals. Whereas pentacyclic triterpanes are dominant in GEC, GCC, and WL, diterpanes strongly predominate in the bitumen of RMC. Furthermore, the composition of triterpanes is also different. For example, tricyclic diterpanes are the only diterpanes present in RMC, whereas tetracyclic and tricyclic diterpanes are both present in GEC and GCC, and tetracyclic diterpanes are most abundant in GEC. However, diterpanes are nearly absent in WL.

Spill Samples Are Different in General Chemical Composition. Wang et al. (1999b) studied oil spilled after a fire broke out at a carpet factory in Acton Vale, Quebec, on June 29, 1998. The GC-FID chromatograms of the spill samples were markedly different from the suspected-source sample. Spill samples were highly weathered (e.g., the *n*-alkanes were nearly completely lost with the abundances of pristane and phytane greatly reduced, and only a hump of UCM was seen in the chromatograms). However, the GC-MS biomarker analysis demonstrated that the distribution pattern of biomarker terpanes and steranes were nearly identical for the highly weathered spill and the relatively fresh suspected-source samples. In addition to the presence of the regular biomarkers from C_{21} to C_{35}, C_{30}-$\beta\alpha$ hopane in high abundance was also observed. It was concluded that the spilled oil was a Bunker C-type fuel, and it matched with the oil in the heat exchange equipment near the boiler, suggesting the oil spilled in the river came from the burned factory.

Spill Samples Are Very Similar in General Chemical Composition. In some cases, unknown oil samples may have very similar bulk chemical composition but markedly different biomarker distribution. On March 28, 2001, three unknown oil samples were received from Montreal for product characterization, correlation, and differentiation. The GC-FID screening results show that the three samples are hydraulic fluid-type products. The samples have very similar GC profiles (Figure 3-23). However, biomarker characterization results demonstrated that samples 1 and 2 are nearly identical in biomarker distribution patterns and concentrations, but sample 3 shows significantly different biomarker distribution from samples 1 and 2 (Figure 3-23). Concentrations of C_{29} and C_{30} hopanes in sample 3 match those in samples 1 and 2. Conversely, concentrations of C_{23} and C_{24}, and the sum of C_{31} to C_{35} homohopanes, are markedly lower and higher than those of the corresponding compounds in samples 1 and 2, respectively. Consequently, the diagnostic ratios of target biomarkers are very similar for samples 1 and 2, but apparently different from sample 3. All these observations point toward the conclusion that samples 1 and 2 are identical, while sample 3 comes from another source (Wang et al., 2002). This case study illustrates that successful forensic investigation will require fingerprinting not only common *n*-alkanes and isoprenoids but also biomarkers and determining their diagnostic ratios of spill samples, in particular for oils and products exhibiting very similar *n*-alkane and isoprenoid distributions.

3.3.9 Source-Specific Biomarkers

Biomarker terpanes with a hopane skeleton and steranes are common constituents of crude oils. However, certain oils may also contain some "source-specific" biomarker compounds including several geologically rarer acyclic alkanes. These biomarkers and their ratios can furthermore provide additional diagnostic information on the types of organic matter that give rise to the crude oil. For example, the geologically rare acyclic alkane botryococcane ($C_{34}H_{70}$) was used to identify a new class of Australian nonmarine crude oils (McKirdy et al., 1986). The presence of botryococcane indicates that the source rock contains remains of the algae *Botryococcus braunii*. The broad platform area of the northern North Sea, including Statford, Gullfacs, Brent, Oseberg, Troll, etc., seems to be specially featured by relative high abundances of C_{28}-*bisnorhopane* (Dahlmann, 2003). Thus, C_{28}-bisnorhopane can be regarded as a "source-specific" parameter. Dahlmann (2003) also found that oils from the Niger Delta (Nigeria) and oils from Africa (in Angola Cabinda and Nemba crudes and in Kongo and Gabon crudes) are characterized by the presence of highly abundant *oleanane* and *gammacerane*, respectively. The presence of 18α(H)-oleanane in benthic sediments in PWS, coupled with its absence in

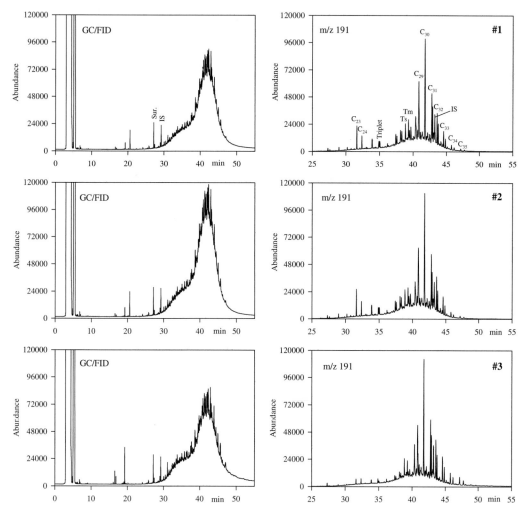

Figure 3-23 Comparison of GC-FID and GC-MS (m/z 191) of three unknown oil samples.

Alaska North Slope crude and specifically in *Exxon Valdez* cargo oil and its residues, confirmed another petrogenic source (Bence et al., 1996; see also Chapter 15). Characterization of 18α(H)-oleanane in oils from the Anaco area and Maturin subbasin, Venezuela, has been used for organic type and age indicator for assessment of the Venezuelan petroleum system (Alberdi and Lopez, 2000). Other "source-specific" petroleum biomarkers (Figure 3-24) include:

1. *C_{30} 17α(H)-diahopane*: C_{30} 17α(H)-diahopane (C_{30}*) elutes right after C_{29} αβ-norhopane and 18α(H), 21β(H)-30-norneohopane (C_{29}Ts) in the m/z 191 mass chromatogram. C_{30} 17α(H)-diahopane has been regarded as a possible terrestrial marker (Moldowan et al., 1991). El-Gayar et al. (2002) characterized seven oils representing the different petroleum-bearing basins in the Western Desert, Egypt. The characterization indicated that Type 2 and Type 3 oils are similar and show relative high pristane/phytane ratios, paucity of C_{30} steranes, and high relative abundance of C_{30}*, suggesting that they probably originated from source rocks containing

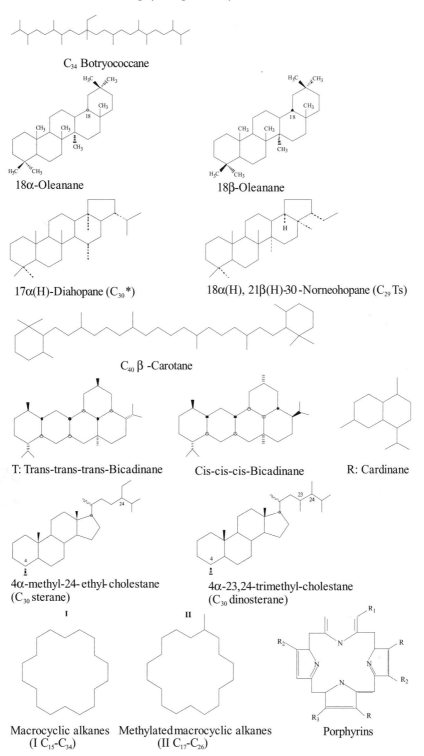

Figure 3-24 Molecular structures of a selection of "source-specific" biomarkers.

significant proportions of higher plant material.

2. *β-carotane*: This compound is a fully saturated C_{40}-dicyclane. It elutes after C_{35} homohopane in the oil saturate fraction. It is highly specific for anoxic, saline, and lacustrine deposition of algal organic matter. It is measured at fragment *m/z* 125 and/or at molecular ion *m/z* 558. β-carotane has been detected in several Chinese oils and the Mississippian Alberta shale. The presence of a significant amount of β-carotane and gammacerane relative to the hopanes has recently been detected, suggesting that the source rocks of the oil from the Liaohe Basin of China were probably deposited in a highly stratified, strongly reducing environment (Wang et al., 1996).

3. *Extended hopanes beyond C_{40}*: A series of side-chain extended 17α(H), 21β(H)-hopanes and 17β(H), 21α(H)-hopanes up to C_{44} have been identified in crude oils and source rock extracts in the Liaohe Basin, Northeast China (Wang et al., 1996). These compounds may be viewed as the representatives of a new class of molecules and may find applications in forensic fingerprinting of unknown spill oils.

4. *Bicadinanes*: Bicadinanes are C_{30}-pentacyclic biomarker compounds and have three configurations, labeled as W (cis-cis-trans-bicadinane), T (trans-trans-trans-bicadinane), and R (cardinane). The mass spectra of bicadinane contain prominent *m/z* 191 and 217 fragments, while peaks can appear in corresponding chromatograms of both hopanes and steranes. All three bicadinanes (W, T, and R) form elute prior to C_{29} hopane in the *m/z* 191 chromatogram. But it can be conveniently monitored with little interference using the *m/z* 412 mass chromatogram. They are highly specific for resinous input from certain higher plants that commonly contributed to source rocks for Tertiary oils from the Far East (van Aarssen et al., 1990). Based on biomarker composition, crude oils from the North, Central, and South Sumatra basins, Indonesia, were classified into three types (Sosrowidjojo et al., 1994), and Group II oils were further distinguished from Group 1 oils by their high abundance of bicadinanes relative to C_{30} hopane on the *m/z* 412 mass chromatogram.

5. *4-methyl steranes*: The 4-methyl steranes can be divided into two major classes: (1) C_{28}–C_{30} analogues of the steranes at positions 4 and 24 (e.g., the C_{30} sterane is 4α-methyl-24-ethylcholestane), and (2) C_{30} dinosteranes (e.g., 4α,23,24-trimethylcholestanes). 4α-Methyl-24-ethylcholestanes often occur in relatively high abundance in Tertiary source rocks and related oils from China (Fu et al., 1992). For example, almost all of the oils from the eastern Pearl River Mouth Basin contain significant amounts of 4-methylsteranes (Zhang et al., 2003). Hu (1991) found that 4-methylsteranes in the range of C_{28}–C_{30} are unusually rich (which comes up to 20–40% of the total steranes) in certain oils from terrestrial facies within the South China Sea. C_{30} 4-methylsterane (M^+ = 414) is particularly abundant among the 4-methylsteranes. Dinosterane has only been reported in petroleum younger than Triassic age (Summons et al., 1992). The presence of dinosterane in relatively high concentrations in asphaltic bitumens from southern Australia (Mckirdy et al., 1994) suggests that their source is no older than mid-Triassic. In a study to re-evaluate the petroleum prospective potential in southeast Australia, Volkman et al. (1992b) examined 10 bitumen samples collected between 1880 and 1915. The high proportions of C_{27} steranes and the presence of C_{30} steranes including dinosteranes suggested that the bitumens were derived from a marine source rock containing mainly marine organic matter.

6. *Macrocyclic alkanes*: Murrisepp et al. (1994) first reported the presence of non-isoprenoidal macrocyclic alkanes in sedimentary material and tentatively identified these cyclic hydrocarbons of the cyclododecane and cyclohexadecane series in the

nonaromatic hydrocarbon fractions of the semicoking oil from an Estonian oil shale. Audino et al. (2001, 2002) have unambiguously identified for the first time a new class of cyclic hydrocarbon biomarker, macrocyclic alkanes and their methylated analogues in a *Botryococcus braunii*-rich sediment (torbanite) of Late Carboniferous age (Audino et al., 2001) and in two Indonesian crude oils (Audino et al., 2002). The compounds consist of a homologous series of macrocyclic alkanes in a wide range from C_{15} to C_{34} and their methylated derivatives (ranging from C_{17} to C_{26}). The distribution of macrocyclic alkanes was measured at the characteristic ion m/z 111. The macrocyclic alkanes appear to be novel markers of *B. braunii* and add to the catalogue of the characteristic hydrocarbons derived from this alga. More importantly, these compounds could be original markers specific to highly resistant algaenan of *B. braunii* in sediments and crude oils.

7. *Porphyrins*: Porphyrins are a special class of N-containing compounds. They are complex derivatives of the basic material porphine. Porphine consists of four pyrrole [(CH=CH)$_2$=NH] units joined by methine, –C=, bridges; the methine bridges establish conjugated linkages between the component pyrrole nuclei, forming a more extended resonance system. Although the resulting structure retains much of the inherent character of the pyrrole components, the larger conjugated system gives increased aromatic character to the porphine molecule. The porphyrin compounds are degradation products of the chlorophyll (photosynthetic pigments of plants and some bacteria). Most of the porphyrin material in crude oils is chelated with metal, of which vanadium is the most important, followed by nickel. Iron and copper-porphyrin chelates may also be present in oil. Porphyrins are not usually considered among the usual nitrogen-containing constituents of petroleum, nor a metallo-containing organic material. Conversely, they are often classified as a unique class of biomarker compounds because they may establish a link between compounds found in the geosphere and their corresponding biological precursors. Crude oils and bitumens contain small amounts of vanadyl and nickel porphyrins. In general, mature, lighter oils contain less of these compounds, whereas heavy oils may contain larger amounts of vanadyl and nickel porphyrins. Chen et al. (1999) have successfully separated nine free petroporphyrin compounds from a Chinese crude oil by reversed-phase HPLC. These were further identified by mass spectrometry as C_{27}E (m/z 408), C_{28}E (m/z 422), C_{29}E (m/z 436), C_{30}E (m/z 450), C_{31}E (m/z 464), C_{29}D (m/z 434), C_{30}D (m/z 448), C_{31}D (m/z 462), and C_{32}D (m/z 476) porphyrins.

The search for source-specific geochemical biomarkers continues to be a fertile area of research for fingerprinting similar sources of petroleum. If an oil shows any additional characteristic compositional features (such as "extra" biomarker peaks), these should of course always be included in the characterization and considered in the identification and correlation. It should be noted, however, that reliable biomarker interpretation is usually based on a whole biomarker distribution chromatogram and a series of biomarker parameters. No single parameter can be exclusively used for unambiguous source identification of unknown spills. Individual unique biomarker parameters only become valuable and meaningful when used together and they agree with other biomarker parameters.

3.3.10 Using Diagnostic Ratios and Cross-Plots of Biomarkers for Source Identification of Oil Spills

Biomarker diagnostic parameters have been long established and are widely used by geochemists for oil correlation, determination of organic input and precursors, depositional environment, assessment of thermal maturity, and evaluation of in-reservoir oil biodegradation (Peters and Moldowan, 1993). Many diagnostic ratios currently used in oil spill studies and environmental forensics originate from the petroleum geochemistry literature.

3.3.10.1 Diagnostic Ratios of Biomarkers

Most biomarkers in spill samples and source oils, in particular those homologous series of biomarkers with similar structure, show little or no changes in their diagnostic ratios. An important benefit of comparing diagnostic ratios of spilled oil and suspected source oils is that concentration effects are minimized. In addition, the use of ratios tends to induce a self-normalizing effect on the data since variations due to the fluctuation of instrument operating conditions day-to-day, operator, and matrix effects are minimized. Therefore, comparison of diagnostic ratios reflects more directly differences of the target biomarker distribution between samples.

Diagnostic ratios can either be calculated from quantitative (i.e., compound concentrations) or semiquantitative data (i.e., peak areas or heights). Diagnostic biomarker ratios frequently used as defensible indices by the environmental chemists for identification, correlation, and differentiation of spilled oils are summarized in Table 3-9. These ratios consist of alkanes, terpanes, steranes, sesquiterpanes, and diamondoids. Ratios are generally defined from (biomarker 1)/(biomarker 2) for simplicity, but can readily be redefined in other forms such as (biomarker 1)/(biomarker 1 + biomarker 2). Selection of diagnostic ratios employed in oil spill studies is mainly based on source-specific variables (e.g., specificity, diversity, and analytical precision). It is important to realize that the suite of diagnostic ratios as listed in Table 3-9 is neither inclusive nor appropriate for all oil spill identification cases. In some spill cases, it may be prudent to include some particularly characteristic ratios. In other situations, the abundance of some biomarkers may be too low to obtain reliable diagnostic ratios. Thus, maintaining flexibility in the selection of diagnostic ratios to be used in specific cases is important.

For diamondoid compounds, numerous diagnostic indices based upon concentrations of target adamantanes have been developed and calculated for the crude oil samples as well as the refined petroleum products. In principle, a large number of diagnostic ratios from 26 identified adamantanes and diamantanes can be produced. However, some ratios are heavily affected by measurement errors due to low peak abundances and poor peak separation; thus, a proper selection of diagnostic ratios of diamondoids is important in order to keep the uncertainties to a minimum and yield reliable results. For this purpose, the diagnostic power (DP) is used for selection of diagnostic ratios (Christensen et al., 2004). DP is defined as the relative standard deviation (RSD_V) of a diagnostic ratio for oils of different origins (~100 oil samples in total) divided by the relative standard deviation (RSD_A) of the same ratios calculated from six measurements of the ESTD reference oil (Prudhoe Bay crude oil, 13.1% weathered). Based on the determined DP values for oil in the Environmental Canada Oil Research Laboratory, diagnostic ratios of 1-MA/2-EA, 1-MA/1,3,4-TMA, 1-MA-1,2-DMA, 1-MA/1,2,5,7-TeMA, 1,3,5,7-TeMA/1,2,5,7-TeMA, 1,3,5-TMA,/1,2,5,7-TeMA, 1,3,5-TMA,/1,3,6-TMA, and 1,4-DMA/1,3,4-TMA with high DP values are selected from more than 50 possible diagnostic ratios as more sensitive and reliable parameters for source correlation and differentiation of oils and petroleum products. It should be noted that the ratios with low DP values, particularly those developed from low abundant diamantanes, are not recommended as reliable distinguishing tools and may be used only as supplementary diagnostic information for certain case studies. Otherwise, higher analytical uncertainties related to these indices could lead to erroneous conclusions for oil source identification.

The triplet ratio, if present, generally varies in oils from different sources and is dependent upon sources, depositional environment, and maturity. The ratio was first used by Kvenvolden et al. (1985) to study a North Slope crude, in which the ratio is ~2. The

Table 3-9 Diagnostic Biomarker Ratios Frequently used for Identification, Correlation, and Differentiation of Spilled Oils

Biomarker classes	Diagnostic ratios	Code
Acyclic isoprenoids	pristane/phytane	pri/phy
	pristane/n-C_{17}	pri/C_{17}
	phytane/n-C_{18}	phy/C_{18}
Terpanes (m/z 191)	C_{21}/C_{23} tricyclic terpane	TR21/TR23
	C_{23}/C_{24} tricyclic terpane,	TR23/TR24
	C_{23} tricyclic terpane/C_{30} αβ hopane	TR23/H30
	C_{24} tricyclic terpane/C_{30} αβ hopane	TR24/H30
	C_{24} tertracyclic/C_{26} tricyclic (S)/C_{26} tricyclic (R) terpane	triplet ratio
	C_{27} 18α,21β-trisnorhopane/C_{27} 17α,21β-trisnorhopane	Ts/Tm
	C_{28} bisnorhopane/C_{30} αβ hopane	H28/H30
	C_{29} αβ-25-norhopane/C_{30} αβ hopane	NOR25H/H30
	C_{29} αβ-30-norhopane/C_{30} αβ hopane	H29/H30
	oleanane/C_{30} αβ hopane	OL/H30
	moretane(C_{30} βα hopane)/C_{30} αβ hopane	M30/H30
	gammacerane/C_{30} αβ hopane	GAM/H30
	tricyclic terpanes (C_{19}–C_{26})/C_{30} αβ hopane	Σ(TR19-TR26)/H30
	C_{31} homohopane (22S)/C_{31} homohopane (22R)	H31S/H31R
	C_{32} bishomohopane (22S)/C_{32} bishomohopane (22R)	H32S/H32R
	C_{33} trishomohopane (22S)/C_{33} trishomohopane (22R)	H33S/H33R
	relative homohopane distribution	H31:H32:H33:H34:H35
	$\Sigma(C_{31} - C_{35})/C_{30}$ αβ hopane	Σ(H31 – H35)/ H30
	homohopane index	H31/Σ(H31 – H35) to H35/Σ(H31 – H35)
Steranes and diasteranes (m/z 217 & 218)	C_{27} 20S-13β(H), 17α(H)-diasterane/ C_{27} 20R-13β(H), 17α(H)-diasterane	DIA 27S/DIA 27R
	relative distribution of regular C_{27}-C_{28}-C_{29} steranes	C27:C28:C29 steranes
	C_{27} αββ/C_{29} αββ steranes (at m/z 218)	C27ββ(S + R)/C29ββ(S + R)
	C_{28} αββ/C_{29} αββ steranes (at m/z 218)	C28ββ(S + R)/C29ββ(S + R)
	C_{27} αββ/(C_{27} αββ + C_{28} αββ + C_{29} αββ) (at m/z 218)	C27ββ/(C27 + C28 + C29)ββ
	C_{28} αββ/(C_{27} αββ + C_{28} αββ + C_{29} αββ) (at m/z 218)	C28ββ/(C27 + C28 + C29)ββ
	C_{29} αββ/(C_{27} αββ + C_{28} αββ + C_{29} αββ) (at m/z 218)	C29ββ/(C27 + C28 + C29)ββ
	C_{27}, C_{28}, and C_{29} ααα/αββ epimers (at m/z 217)	C27αα/C27ββ
		C28αα/C28ββ
		C29αα/C29ββ
	C_{27}, C_{28}, and C_{29} 20S/(20S + 20R) steranes (at m/z 217)	C27 (20S)/C27 (20R)
		C28 (20S)/C28 (20R)
		C29 (20S)/C29 (20R)
	C_{30} sterane index: $C_{30}/(C_{27}$ to $C_{30})$ steranes	C30/(C27 to C30) steranes
	selected diasteranes/regular steranes	
	regular C_{27}-C_{28}-C_{29} steranes/C_{30} αβ-hopanes	C27-C28-C29 steranes/H30
Sesquiterpanes (m/z 123)	relative distribution of sesquiterpanes	
	C_{14} group: Peak 1/Peak 2	P1/P2
	C_{15} group: Peak 3/Peak 5, Peak 4/Peak 5, Peak 6/Peak 5	P3/P5, P4/P5, P6/P5
	C_{16} group: Peak 8/Peak 10	P8/P10
	inter-group: Peak 1/Peak 3, Peak 1/Peak 5,Peak 3/Peak 10, Peak 5/Peak 10	P1/P3, P1/P5, P3/P10, P5/P10
Adamantanes (m/z 135, 149, 163, 177)	methyl adamantane index: 1-MA/(1- + 2-MA)	MAI
	1,4-DMA, cis/1,4-DMA, trans	DMAI
	dimethyl admantane index: 1,3-DMA/(1,3- + 1,4- + 1,2-DMA)	TMAI
	1,3,4-TMA, cis/1,3,4-TMA, trans	
	trimethyl adamantane index: 1,3,4-DMA, cis/(1,3,4-DMA, cis + 1,3,4-DMA, trans)	
	ethyl adamantane index: 1-EA/(1- + 2-EA)	EAI
Diamantanes (m/z 187, 201, 215)	methyl-diamantane index: 4-MD/(1- + 3- + 4-MD)	MDI
	relative distribution of diamantanes: C_0-D:C_1-D:C_2-D:C_3-D	
Triaromatic steranes (m/z 231)	C_{20} TA/(C_{20} TA + C_{21} TA)	
	C_{26} TA (20S)/sum of C_{26} TA (20S) through C_{28} TA (20R)	
	C_{27} TA (20R)/C_{28} TA (20R)	
	C_{28} TA (20R)/C_{28} TA (20S)	
	C_{26} TA (20S)/[C_{26} TA (20S) + C_{28} TA (20S)]	
	C_{28} TA (20S)/[C_{26} TA (20S) + C_{28} TA (20S)]	
Monoaromatic steranes (m/z 253)	C_{27}-C_{28}-C_{29} monoaromatic steranes (MA) distribution.	

*Ratios are defined for simplicity, but can be readily redefined in other forms. For example, the ratio of C_{29} αβ-30-norhopane/C_{30} αβ hopane (H29/H30) can be readily redefined as H29/(H29 + H30) × 100%.

spilled *Exxon Valdez* oil (an Alaska North Slope crude) and its residues also have triplet ratios of ~2. Conversely, many tar balls and residues collected from the shorelines of the Prince William Sound were similar to each other but chemically distinct from the spilled *Exxon Valdez* oil with triplet ratios of ~5. The triplet ratio, combined with other diagnostic biomarker ratios and isotopic compositions, revealed that these non-*Valdez* tar balls originated from California with a likely source being the Monterey Formation (Kvenvolden et al., 1995). During the Arrow oil spill work, the ratio of the most abundant C_{29} to C_{30} hopane as well as C_{23}/C_{24}, Ts/Tm, and $\alpha\beta\beta/(\alpha\beta\beta + \alpha\alpha\alpha)$ of C_{27}, C_{28}, and C_{29} steranes as defined and used by Wang et al. (1994b) as reliable source indicators. Similar approaches, combined with determinations of a number of other "source-specific marker" ratios, were applied to characterize oil samples from the Arctic Baffin Island spill (Wang et al., 1995b), oil on birds (Wang et al., 1997), the 25-year-old wetland Nipisi spills (Wang et al., 1998a), a mystery spill in Quebec (Wang et al., 2001a), and the Detroit River oil spill (Wang et al., 2004). Barakat et al. (2002) have proven the molecular ratios of triaromatic steranes including C_{28}TA 20R/C_{28}TA 20S, C_{27}TA 20R/C_{28}TA 20R, and C_{28}TA 20S/(C_{26}TA 20R + C_{27}TA 20S) were useful source indicators for correlating naturally weathered oil residues in the Egyptian Western Desert to a fresh crude oil sample of the Western Desert-sourced oil.

Use Diagnostic Ratios of Biomarkers in Combination with PAH Ratios for Source Identification. In January/February 1996, a significant number of tar ball incidents occurred along the coasts of Vancouver Island, Washington, Oregon, and California. Samples of the tar balls were collected from the affected beaches and characterized by GC-FID and GC-MS, and further analyzed using a carbon isotopic technique (Wang et al., 1998b). Biomarker characterization revealed that the BC and CA samples have similar diagnostic ratios of most biomarkers, but the CA samples show lower ratios of C_{23}/C_{30} and C_{24}/C_{30} than the BC samples. Only after in combination with characterization results of PAHs and PAH diagnostic ratios, was it defensively concluded that (1) CA/Oregon samples were chemically similar and consistent with the same source of a Bunker-type fuel. (2) BC tar ball samples were chemically similar and consistent with the same source (also Bunker-type fuel). They were similar to the CA/Oregon samples but may have a different source. (3) The spill samples had been highly weathered since release, and the CA samples were more heavily weathered than the BC samples. (4) The source of the tar ball samples was neither ANS nor California Monterey Miocene oil.

In application of diagnostic ratios of biomarkers for spill studies, it is important to acknowledge that regardless of diagnostic parameters used, a basic rule applied to all correlations and differentiations should be

- poorly matching biomarker distribution and/or diagnostic ratios are strong evidence for lack of a correlation between a spill sample(s) and suspected source(s),
- matching may be an indication of a correlation of a spill sample(s) and suspected source(s), but is not necessarily "proof" for identity.

Hence, in order to make more reliable and defensible correlations, the use of a "multi-criteria approach" is often a prerequisite. In a multicriteria approach, the final conclusion is based on analysis and evaluation of the distribution of more than one suite of petroleum compounds (Peters et al., 2005; Stout et al., 2002; Wang et al., 1999a; Christensen et al., 2004; Daling et al., 2002).

3.3.10.2 Cross-Plots of Biomarkers

Cross-plots (i.e., plot of one diagnostic biomarker ratio versus another ratio) are another diagnostic means frequently used in oil geochemistry for oil–oil correlation and determination of oil source and depositional environment (Peters and Moldowan, 1993). Gürgey (2002) analyzed 56 rock and 28 crude oil samples from the sub-salt and supra-salt section of the southern Pre-Caspian Basin.

Based on plots of $C_{24}/C_{26}T$ (C_{24} tetracyclic/C_{26} tricyclic terpanes) versus C_{29}/C_{30} hopane, the author illustrates a clear separation between two populations: Population 1 (1A and 1B) and Population 2. Seifert and Moldowan (1986) applied cross-plots of C_{29} αββ/(αββ+ααα) sterane versus C_{29} 20S/(20S+20R) steranes as a particularly effective measure in describing the thermal maturity of source rocks or oils. Zhang et al. (2003) classified crude oils from the eastern Pearl River Mouth Basin into groups based on cross-plots of relative abundance (at m/z 123) of various isomeric sesquiterpanes versus relative abundances of bicadinanes to C_{30} hopane on the m/z 412 mass chromatogram (bicadinane-T/C_{30}-hopane).

Cross-Plots of Biomarkers for Spill Source Identification. Malaysian coasts are subjected to various threats of petroleum pollution including deliberate and accidental oil spills from various sources. The identification of detailed sources of the oil pollution, therefore, is essential to reduce the oil pollution through effective regulation. Based on chemical evidence that Middle East crude oils were characterized by C_{29} 17α,21β-norhopane and C_{31}–C_{35} homohopanes, whereas these compounds were depleted in South East Asian crude oils, Zakaria et al. (2000, 2001) proposed utility of the cross-plots of C_{29} αβ/C_{30} αβ hopane ratio versus the homohopane index $\Sigma(C_{31}$–$C_{35})/C_{30}$ hopane as key biomarker indicators and successfully distinguished a large number tar ball samples that originated from South East Asian crude oil sources from those of Middle East sources.

Cross-Plots of Sesquiterpane Isomers for Distinguishing Oils and Petroleum Products. Wang et al. (2005a) depict the cross-plots of sesquiterpanes (Peak 4/Peak 5 versus Peak 3/Peak 5) for more than 50 crude oils and refined products (Figure 3-25, left panel). There is large scatter in this set of oils in the cross-plot data: P4:P5 and P3:P5 fall in the

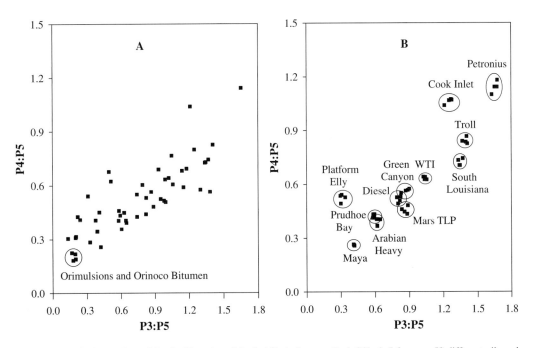

Figure 3-25 **A**: Cross-plots of the double ratios of Peak 4/Peak 5 versus Peak 3/Peak 5 for over 50 different oils and refined products. The circle indicates related samples from the same origin. **B**: Cross-plots of the double ratios of Peak 4/Peak 5 versus Peak 3/Peak 5 for 11 weathering oil series and 1 diesel weathering series. Each weathering oil series produces a tight cluster.

ranges of 0.2–1.2 and 0.1–2.1, respectively. Furthermore, related oils, such as the circle for Orimulsion samples from different batches and for the original Orinoco bitumen, produce tight clusters on the plot. This implies that sesquiterpane ratios, in combination with other fingerprinting data, may be used to discriminate different oils and to identify the source of spill samples. A double ratio plot of P4:P5 versus P3:P5 for 11 weathered crude oil and 1 weathered diesel series is shown in the right panel of Figure 3-25. The four weathered samples for each oil series form tight clusters, indicating that moderate weathering would not be expected to alter sesquiterpane distributions. For example, no depletion of sesquiterpanes, relative to the most abundant Peak 3, was observed for the weathered diesel samples (an Ottawa diesel, 2002) at four weathering percentages of 0, 7.2, 14.2, and 22%.

3.4 Effects of Weathering on Biomarker Fingerprinting

3.4.1 Processes Affecting the Fate and Behavior of Spilled Oil

When oils and petroleum products are released into the environment — water or land — they undergo a series of changes in chemical compositions and physical properties that in combination are termed "weathering." Weathering can strongly influence how oils move and behave in the environment (Jordan and Payne, 1980; Wang et al., 1995c; NRC, 2002). Weathering processes could include evaporation, emulsification, natural dispersion, dissolution, microbial degradation, photooxidation, and other processes (such as sedimentation, adhesion onto the surface of suspended particulate materials, and oil-fine interaction). Each of the weathering processes affects the hydrocarbon family differently. For example, aromatics tend to be more water soluble than aliphatics. Weathering processes occur at very different rates, depending on both the oil type and environmental conditions.

Evaporation. In the short term after an oil spill, evaporation is usually the single most important and dominant weathering process, in particular for the light petroleum products such as gasoline. Evaporation has the greatest effect on the amount of oil remaining on water or land after a spill. In the first few days following a spill, the loss can be up to 70 and 40% of the volume of light crudes and petroleum products, and gasoline can evaporate completely above zero degrees. For heavy or residual oils such as Bunker C oil, the losses due to evaporation comprise only a few percentages of the total volume. The rate at which oil evaporates depends primarily on the oil composition. The more volatile components an oil or fuel contains, the greater the extent and rate of its evaporation. The extent of evaporation is often the most important factor for determining oil properties at a given time after the spill and for changing the behavior of the oil.

Emulsification. Emulsification is the process by which water is dispersed into oil in the form of small droplets. Water droplets can remain in an oil layer in a stable form, and the properties of the emulsified oil are very different from the starting oil. The mechanism of water-in-oil emulsion formation is not yet fully understood, but most likely it starts with sea energy forcing the entry of small water droplets, about 10 to 25 µm in size, into the oil. Emulsions contain about 70% water, and thus, when emulsions are formed, the volume of spilled oil more than triples. In general, water can be present in oil in four ways (Fingas and Fieldhouse, 2003): (1) soluble; (2) unstable emulsion; (3) semi- or meso-stable emulsion; and (4) stable emulsions. Stable emulsions are reddish-brown in color and appear to be nearly solid. These emulsions do not spread and tend to remain in lumps or mats on the sea or shore. It has been noted that when oil forms stable or meso-stable emulsions, the rate of evaporation slows down considerably. Microbial degradation also appears to slow down. The dissolution of soluble components from oil may also cease once emulsification has occurred.

Natural Dispersion. Natural dispersion occurs when fine droplets of oil are transferred into the water column by wave action or turbulence. Small droplets (<20 µm) are relatively stable in water and will remain so for long periods of time. Large droplets tend to rise and larger droplets (>100 µm) will not stay in the water column for more than a few seconds. Natural dispersion is dependent on both the oil type and weather conditions of sea (such as wave action and sea energy). Heavy oils such as Bunker C or a heavy crude will not disperse naturally to any significant extent, whereas light crudes and diesel fuel can disperse significantly. Dispersed oil may also rise to form another surface slick or it may become associated with sediment and be precipitated to the bottom. Dispersant, a chemical spill-treating agent, may be applied to promote the formation of small droplets of oil that "disperse" throughout the top layer of the water column.

Dissolution. Dissolution occurs immediately after the spill. Through the process of dissolution, some of the most soluble components of the oil are lost to the water under the slick. The amount of an individual compound dissolving in the water phase from oil slicks in a given time largely depends on kinetic and equilibrium conditions affected by molecular structure and polarity. In general, (1) the aromatic hydrocarbons are more soluble than aliphatic hydrocarbons, (2) the solubility increases as the degree of alkylation of benzenes and PAHs decrease, (3) the lower-molecular-weight hydrocarbons are more soluble than the high-molecular-weight hydrocarbons in each class of petroleum compounds, and (4) the more polar S-, N-, and O-containing compounds are more soluble than hydrocarbons. Hence BTEX, lighter alkyl-benzene compounds, and PAHs with fewer rings such as naphthalene are particularly susceptible to dissolution or *water-washing*. As only a small amount of oil components actually enters the water column, dissolution does not measurably change the mass balance of the oil. The significance of dissolution is that the soluble aromatic compounds are particularly toxic to fish and other aquatic life. If a spill of oil containing a large amount of soluble aromatic components occurs in shallow water and creates high localized concentrations, then significant numbers of aquatic organisms can be at risk and killed.

Biodegradation. Biodegradation of hydrocarbons by natural populations of microorganisms represents the primary mechanisms by which petroleum and other hydrocarbon pollutants are eliminated from the environment (Prince, 1993; Leahy and Colwell, 1990). The quantitative and qualitative aspects of biodegradation depend on the composition of the microbial community (for example, indigenous bacteria and other microorganisms are often the best adapted and more effective at degrading oil as they are acclimatized to the temperature and other conditions of the area); the type, nature, and amount of oil; and the ambient and seasonal environmental conditions (such as temperature, oxygen, nutrients, salinity, and pH). Petroleum hydrocarbons differ in their susceptibility to microbial attack. Transformations of petroleum hydrocarbons by biodegradation occur stepwise, producing oxidized compounds including alcohols, phenols, aldehydes, and carboxylic acids in sequence by phase 1 and phase 2 metabolic pathways. The compounds may eventually be completely metabolized to carbon dioxide and water, or the polar metabolites may be spread to the surrounding water or accumulated in the residual oil.

Photooxidation. Photooxidation is a potentially significant process in degradation of crude oil spilled at sea, but the effects of photooxidation on the oil composition following oil spills are not yet well understood. In general, photooxidation is considered to be a factor involved in the transformation of crude oil or its products released into the marine environment (Garrett et al., 1998). The photooxidation is dependent on the thickness of the oil slick as well as sun incidence. The photochemical degradation yields a variety of oxidized compounds including alcohols, aldehydes, ketones, and acids, which are more soluble in water than the starting compounds.

Photodegradation affects the oil composition differently than is the case for microbial degradation and can hence complicate the observed weathering patterns for spills in areas with large sun incidence. For most oils, photooxidation is not an important process in terms of changing their fate or mass balance after a spill.

Sedimentation and Oil–Mineral Aggregation. Sedimentation is the process by which oil is deposited on the bottom of the sea or other water body. Once oil is on the bottom, it is usually covered by additional sedimentation and degraded very slowly. Oil–mineral aggregates (OMA) result from interactions among the oil residues, fine mineral particles, and seawater. OMA formation has been identified as an important process that facilitates the natural removal of oil stranded in coastal sediments (Bragg and Owens, 1994; Owens and Lee, 2003). OMA formation is enhanced by physical processes such as wave, energy, tides, or currents. It has recently been noted that oil biodegradation may be enhanced by OMA formation.

3.4.2 Weathering Effects on Biomarkers Fingerprinting

Weathering causes considerable changes in the physical properties and the chemical composition of spilled oil. For severely weathered oils, not only n-alkanes but also branched and cyclo-alkanes are heavily or completely lost, and the UCM becomes pronounced; the BTEX and alkyl benzene compounds can be completely lost, and the PAHs and their alkylated homologous series could also be highly degraded, resulting in the development of a profile in each alkylated PAH family with the distribution of C_0- < C_1- < C_2- < C_3-. Hence it is difficult and often impossible to identify severely weathered oil samples through recognition of n-alkane and PAH fingerprinting patterns. However, the biomarker fingerprinting patterns are often unaltered even for some severely weathered oil samples. Thus, biomarker fingerprints could provide a powerful tool for tracking the source and correlation and differentiation of weathered oils.

The laboratory evaporative weathering (Wang et al., 1995c; Wang and Fingas, 2003; EPA report, 2003) reveals that biomarker terpanes and steranes are not depleted during evaporative weathering; all target biomarker compounds from the C_{19} to C_{35} range are concentrated in proportion with the increase of the weathered percentages; and both terpanes and steranes show a great consistency in the relative ratios of paired biomarker compounds and biomarker compound classes. A number of the laboratory biodegradation studies (Wang et al., 1998c; Blenkinsopp et al., 1996; Swannel et al., 1996; Atlas and Bartha, 1992; Foght et al., 1998) also demonstrate that no sign of alteration in the composition of biomarkers was observed, regardless of the oil type (light, middle, or heavy), incubation times (7, 14, and 28 days), incubation conditions (incubated at 4, 10, 15, and 22°C), with and without the presence of nutrients. The concentrations of terpanes and steranes in the tested oils were consistent, and the diagnostic ratios of paired terpanes and steranes remained constant. For example, the average of the sum of eight target diagnostic biomarker ratios [including C_{23}/C_{24}, Tm/Ts, C_{29}/C_{30}, $C_{32}(S)/C_{32}(R)$, $C_{33}(S)/C_{33}(R)$, C_{23}/C_{30}, C_{24}/C_{30}, and $C_{27}\alpha\beta\beta/C_{29}\alpha\beta\beta$ steranes] from 70 biodegradation samples of the ASMB oil inoculated under various inoculum conditions during 1994 was 8.2 ± 0.2 with relative standard deviations less than 4%. Contrary to the biomarker compounds, n-alkanes, pristine, and phytane were greatly reduced in the positive controls, and n-C_{17}/pristane, n-C_{18}/phytane, and pristane/phytane ratios were significantly altered, indicating degradation of pristane and phytane had also occurred.

Compared to the laboratory-controlled evaporative weathering and biodegradation, the field biodegradation of contaminated petroleum in the environment is generally a long-term and complex process. The study of the 25-year-old Nipisi spill (Wang et al., 1998a) indicates that the surface oil (0–2 cm) has been heavily weathered, evidenced by nearly complete depletion of n-alkanes and isoprenoids and by complete loss of BTEX compounds, striking decreases in the abundances of alkylated naphthalene series, and development of a profile of C_0- < C_1- < C_2- < C_3- in each alkylated PAH group. Conversely, the subsurface residual oil (>30–40 cm) from the same location is still almost unaffected by weathering, with GC chromatographic profiles similar to the reference oil. In contrast to alkane and PAH groups, the biomarker composition of the Nipisi spilled oil is nearly unaffected. The accumulation of terpanes relative to the reference oil during the 25-year period of weathering is apparent, especially for the severely weathered surface sample N2–1A: the concentration of C_{30}-$\alpha\beta$ hopane was approximately 1.8 times that found in the reference oil, and five diagnostic biomarker ratios (C_{23}/C_{24}, Ts/Tm, C_{29}/C_{30}, C_{32} 22S/22R, and C_{33} 22S/22R) were found to be consistent between samples as well.

3.4.3 Biodegradation of Biomarkers in Spilled Oil

Although terpanes and steranes are highly resistant to biodegradation, several studies have shown that they can be degraded to a certain degree under severe weathering conditions (i.e., extensive microbial degradation) (Seifert et al., 1984; Chosson et al., 1991). Based on several geochemical studies, Peters and Moldowan (1993) have created a "quasi-stepwise" sequence for assessing the extent to which biomarkers are degraded in the reservoir. The Arrow (Wang et al., 1994b) and BIOS oil spill studies (Wang et al., 1995b; Prince et al., 2002) have demonstrated degradation of C_{23} and C_{24} tricyclic terpanes. In addition, Tm is degraded faster relative to Ts, even though Ts chromatographically elutes earlier than Tm. In March 1986, sections of peaty mangrove in a tropical ecosystem were polluted by Arabian Light crude oil. Eight years later, Munoz et al. (Munoz et al., 1997) found that isoprenoids were severely degraded and the biomarker distribution altered as well. Norhopanes were found to be the most biodegradation-resistant among the studied terpane and sterane groups, and the C_{30}-$\alpha\beta$ hopane appeared more sensitive to weathering than its higher homologues. Frontera-Suau et al. (2002) examined degradation of petroleum biomarkers using mixed cultures of microorganisms enriched from surface soils at four different hydrocarbon-contaminated sites. They found that these cultures degraded C_{30} 17α, 21β-hopane and the C_{31}–C_{34} extended hopanes in Bonny Light crude oil after 21 days of incubation at 30°C.

Three coastal sites, heavily oiled from the 1974 Metula oil spill in the Strait of Magellan, Chile, were examined in May 1998 to determine the long-term fate and persistence of Metula oil in a marine marsh environment (Wang et al., 2001b). Among the characterized samples, the asphalt pavement samples were the most heavily weathered, evidenced by a complete loss of *n*-alkanes from *n*-C_8 to *n*-C_{41} and by depletion of more than 98% of the alkylated PAHs. Even the most refractory biomarker compounds were affected to varying degrees. Biomarkers showed degradation in the following sequences:

- Biomarkers were altered in the declining order of importance as: diasteranes > C_{27} steranes > tricyclic terpanes > pentacyclic terpanes > norhopanes (C_{29}Ts) ~ C_{29} $\alpha\beta\beta$ steranes.
- Steranes degraded in the order of C_{27} > C_{28} > C_{29} with the stereochemical degradation sequence 20R $\alpha\alpha\alpha$ steranes > 20(R+S) $\alpha\beta\beta$ steranes > 20S $\alpha\alpha\alpha$ steranes.
- Degradation of terpane C_{35} > C_{34} > C_{33} > C_{32} > C_{31} was apparent with a significantly preferential degradation of the 22R epimers over the 22S epimers.
- C_{30}-$\alpha\beta$-hopane appeared more degradable than the 22S epimers of C_{31} and C_{32} homohopanes, but had roughly the same biodegradation rate as the 22R epimers of C_{31} and C_{32} homohopanes and was significantly more resistant to degradation than the 22S and 22R epimers of C_{34} and C_{35} homohopanes.
- C_{29}-18α(H), 21β(H)-30-norneohopane, and C_{29}-$\alpha\beta\beta$ 20R and 20S stigmastanes were

found to be the most degradation-resistant terpane and sterane, respectively, among the studied target biomarkers.

3.4.4 Determination of Weathered Percentages Using Biomarkers

Highly degradation-resistant oil components such as C_{30} αβ hopane or C_{29} αβ norhopane have been applied as conserved "internal standards" for more precise estimation of the weathering degree and extent of the spilled residual oil (Butler et al., 1991; Douglas et al., 1994; Prince et al., 1994; Wang et al., 1995b):

$$P(\%) = (1 - C_s/C_w) \times 100\% \qquad (1)$$

where P is the weathered percentages of the weathered samples, and C_s and C_w are the concentrations of C_{30} αβ-hopane in the source oil and weathered samples, respectively. It should be noted, however, that the weathered percentages can still be underestimated by using C_{30}-αβ-hopane as an internal oil reference for extremely degraded oil samples because C_{30}-αβ-hopane under such circumstances is itself partially depleted, such as in the case of the Metula oil spill (Wang et al., 2001b). However, in most cases, C_{30}-αβ-hopane is the preferred choice and used as "internal standards" for estimating weathered percentages, because C_{30}-αβ-hopane is often the most abundant among C_{19} to C_{35} biomarkers and can thus be quantified more accurately. For lighter refined products, such as diesel samples, which generally do not contain high-molecular-weight terpane and sterane compounds, the bicyclic sesquiterpanes (Wang et al., 2005a) as well as a selection of the more conservative PAHs with a high degree of alkylation such as C_3- or C_4-phenanthrenes can be used as an alternative internal standard for estimating the degree of weathering.

3.4.5 Case Study: Source Identification of a Harbor Spill by Forensic Fingerprinting of Biomarkers

A harbor spill occurred in the Netherlands in 2004. A thick layer of oil (sample 2) was found between a bunker boat and the quay next to the bunker center, and it was suspected that something had gone wrong during bunkering of the vessel. Fuel oils from the bunker boat (sample 1) and the bunker center (sample 3) were collected as suspected sources for comparison with the spill sample. A multi-criterion approach was applied to fingerprint and identify these oil samples and to determine the source of the spill.

3.4.5.1 Product Type-Screening

The samples were type-screened from their GC traces: (1) all have similar GC-FID and GC-MS chromatographic profiles at m/z 83 and 85 for alkyl cyclo-hexanes and n-alkanes, respectively; (2) hydrocarbons ranged between n-C_8 and n-C_{32} with maximal abundances between n-C_{15} to n-C_{17}, and no hydrocarbons heavier than C_{32} were detected; (3) a nearly symmetrical UCM (unresolved complex mixtures of hydrocarbons) of middle-range distillate was apparent; (4) GC-detectable total-petroleum-hydrocarbons (GC-TPH) ranged from 870 to 920 mg/g oil, typical of lighter distillate fuels, significantly higher than most crude oils; (5) total n-alkanes including pristane and phytane were 142, 142, and 145 mg/g oil for the three samples, typical for diesel fuels; (6) all three samples had similar ratios of n-C_{17}/pristane, n-C_{18}/phytane, and pristane/phytane with sample 1 (Bunker boat) being closer to the spill sample 2 than sample 3 (Bunker center); (7) spill sample 2 had been slightly weathered, having considerably lower concentrations of n-C_8, n-C_9, and n-C_{10} than the suspected source samples 1 and 3. All the chromatographic evidence suggests that the spilled oil (sample 2) was a diesel-type fuel and the spill sample was slightly weathered. In this case, two questions remain after the product type-screening: (1) Did these three samples come from the same source? (2) Were the minor differences in chemical composition between samples caused by weathering or mixing with other (pre-existing) contamination? To unambiguously answer these questions, characterization of more than one suite of analytes was performed.

3.4.5.2 Characterization of Bicyclic Sesquiterpanes

Figure 3-26 compares the GC-MS chromatograms of sesquiterpanes at 123 and their corresponding GC-FID chromatograms. Three samples contain significant amounts of sesquiterpanes (Table 3-10). To aid comparison, the diagnostic ratios of paired sesquiterpane isomers with the same carbon number and between groups (with different carbon number) for three samples are presented in Table 3-10 as well. Figure 3-26 and Table 3-10 reveal that (1) sesquiterpanes are extremely abundant in three oil samples. The total concentrations of 10 sesquiterpanes were determined to be as high as 7986, 8255, and 7384 µg/g oil ($n = 3$). (2) Samples 1 and 2 have nearly identical distribution patterns of sesquiterpanes. (3) More importantly, the diagnostic ratios of eight sesquiterpane isomeric pairs were nearly identical for samples 1 and 2 as well. (4) Sample 3 is distinctly different from samples 1 and 2 not only in the diagnostic ratios but also in the concentrations of target sesquiterpanes. In particular, the abundances of Peaks 2, 3, 4, and 8 of sample 3 are much lower than the corresponding peaks of samples 1 and 2. (5) Furthermore, the diagnostic ratios of P3/P5 and P4/P5 for sample 3 are considerably lower than the corresponding ratio values for samples 1 and 2. Conversely, sample 3 has a much higher ratio of P1/P2 than samples 1 and 2. (6) Note that, because of weathering, most probably due to evaporation of the spill sample, the spill sample 2 had slightly higher concentrations of all the observed sesquiterpanes compared to sample 1. Based on the sesquiterpane concentrations, the evaporative mass-loss of sample 2 relative to sample 1 is estimated to be between 4 and 6%.

The diagnostic ratios of sesquiterpanes are compared in double-ratio plots at the 95% confidence limits (Figure 3-27). Specifically, the spill sample (oil sample 2) is compared to both suspected source oil samples 1 and 3, respectively. Based on the criteria described in the revised Nordtest method (Daling et al., 2002), there is a perfectly "positive match" between the spill sample (sample 2) and the spill source candidate (sample 1), while sample 3 is a "nonmatch" to the spill.

3.4.5.3 Confirmation of Source Identification by Quantitative Evaluation of Alkylated PAHs and Pentacyclic Terpanes and Steranes

The source identification by characterizing sesquiterpanes is further validated by quantitative evaluation of five petroleum-characteristic alkylated PAH homologous series (naphthalene, phenanthrene, dibenzothiophene, fluorene, and chrysene) and the pentacyclic biomarker terpanes and steranes. PAH fingerprinting results show that (1) the totals of alkylated PAHs were 24,902, 25,870, and 21,528 µg/g oil for samples 1, 2, and 3, respectively; (2) sample 2 and sample 1 have nearly identical distribution patterns of target alkylated PAHs and other EPA priority PAHs. The pattern for sample 3, however, is noticeably different; (3) diagnostic ratios of target PAH groups and paired PAH isomers are all very similar for all samples, but the ratios for sample 1 and 2 were more similar to each other than to sample 3.

GC-MS analysis found that, in this case, all three oil samples contain detectable amounts of high molecular-weight terpanes and steranes (Table 3-10). The extracted ion chromatograms at m/z 191 and 218 for terpane and sterane characterization are shown in Figure 3-28. The concentrations of target terpanes (C_{21} to C_{31}) and three groups of $\alpha\beta\beta$-steranes (C_{27}, C_{28}, and C_{29}) were determined, and the relative ratios of target biomarker terpanes C_{23}/C_{24}, C_{29}/C_{30}, Ts/Tm, C_{29}-$\alpha\beta$-hopane/C_{30}-$\alpha\beta$-hopane, $C_{31}(22S)/C_{31}(22R)$, and $C_{27}\alpha\beta\beta/C_{29}\alpha\beta\beta$ steranes were also calculated. Terpane and sterane fingerprinting results reveal that (1) only traces of terpanes and steranes were detected in the samples (157, 181, and 101 µg/g oil for samples 1, 2, and 3, respectively),

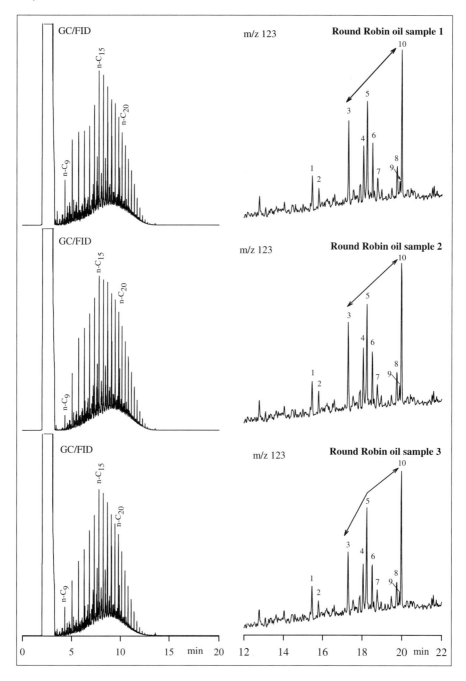

Figure 3-26 Extracted ion chromatograms at *m/z* 123, sesquiterpanes, for three Round Robin samples (right) and their corresponding GC-FID chromatograms for *n*-alkane analysis (left).

Table 3-10 Quantitation Results and Diagnostic Ratios of Sesquiterpanes in Three Oil Samples

Oil Samples	Sample 1	Sample 2	Sample 3
Sesquiterpanes (µg/g oil)			
Peak 1	481 (1.2)*	518 (0.2)	527 (0.6)
Peak 2	334 (0.5)	355 (2.3)	283 (2.6)
Peak 3	1163 (0.9)	1212 (1.4)	965 (2.4)
Peak 4	805 (1.0)	836 (1.4)	666 (2.2)
Peak 5	1349 (2.4)	1392 (1.5)	1370 (1.0)
Peak 6	722 (1.4)	750 (1.2)	658 (2.3)
Peak 7	368 (4.8)	377 (2.4)	384 (4.1)
Peak 8	625 (2.1)	640 (1.4)	507 (2.6)
Peak 9	251 (4.1)	259 (4.6)	220 (4.2)
Peak 10	1889 (0.7)	1916 (0.3)	1803 (1.2)
Total	7986 (0.7)	8255 (0.5)	7384 (0.6)
Sesquiterpane diagnostic ratios			
C_{14} group			
P1:P2	1.44 (1.6)	1.46 (2.3)	1.87 (3.0)
C_{15} group			
P3:P5	0.86 (3.3)	0.87 (2.5)	0.70 (1.5)
P4:P5	0.60 (1.4)	0.60 (2.7)	0.49 (1.5)
P6:P5	0.54 (1.0)	0.54 (2.7)	0.48 (3.2)
C_{16} group			
P8:P10	0.33 (2.7)	0.33 (1.4)	0.28 (1.5)
Intergroup			
P1:P5	0.36 (3.2)	0.37 (1.3)	0.38 (0.6)
P3:P10	0.62 (0.4)	0.63 (1.3)	0.54 (1.9)
P5:P10	0.71 (3.0)	0.73 (1.3)	0.76 (0.8)
Terpanes and steranes (µg/g oil)			
C_{21}	16.2 (2.9)	14.7 (3.6)	15.9 (5.2)
C_{22}	8.47 (2.4)	7.80 (2.7)	7.16 (4.9)
C_{23}	25.7 (0.7)	24.5 (0.9)	22.4 (4.9)
C_{24}	13.1 (1.2)	12.9 (3.7)	11.6 (5.6)
C_{29}	6.61 (1.2)	8.17 (7.0)	3.03 (2.4)
C_{30}	6.39 (0.8)	7.13 (7.0)	2.75 (5.6)
Ts	4.88 (5.0)	6.43 (2.0)	2.14 (2.0)
Tm	4.53 (7.0)	5.67 (2.7)	2.27 (7.3)
$C_{27}\ \alpha\beta\beta$	37.0 (2.4)	48.5 (4.3)	17.8 (3.8)
$C_{28}\ \alpha\beta\beta$	17.1 (3.7)	22.7 (5.4)	8.13 (8.2)
$C_{29}\ \alpha\beta\beta$	17.3 (2.5)	22.1 (6.1)	7.91 (6.4)
Total	157 (1.2)	181 (3.6)	101 (3.7)
Diagnostic ratios of target terpanes and steranes			
C_{21}/C_{22}	1.92 (4.6)	1.89 (1.8)	2.23 (1.1)
C_{23}/C_{24}	1.97 (1.3)	1.90 (3.9)	1.94 (4.2)
C_{23}/C_{30}	4.03 (0.9)	3.44 (6.8)	8.17 (3.6)
C_{24}/C_{30}	2.05 (2.0)	1.81 (4.0)	4.23 (7.5)
C_{29}/C_{30}	1.03 (0.4)	1.15 (0.3)	1.11 (4.4)
Ts/Tm	1.08 (2.0)	1.13 (3.8)	0.95 (8.7)
$C_{27}\ \alpha\beta\beta/C_{29}\ \alpha\beta\beta$	2.13 (3.7)	2.20 (5.3)	2.25 (8.7)

*The concentrations and diagnostic ratios were determined from three measurements. The values in parentheses are relative standard deviation (% RSD) of three measurements.

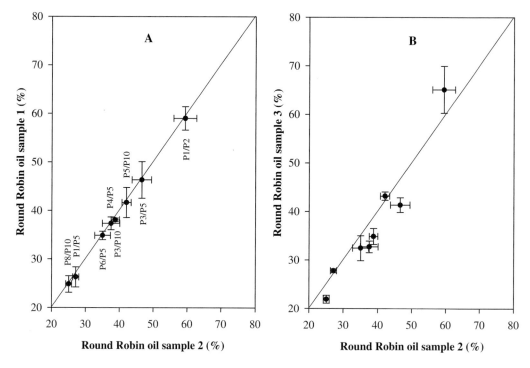

Figure 3-27 Correlation of diagnostic ratios (normalized to %) of sesquiterpanes between spill sample 2 and suspected source samples 1 (left) and 3 (right) at 95% confidence. All the data (A: left panel) overlap the 1:1 line at 95% confidence, representing a "perfect" match between samples 2 and 1. Conversely, most data points (B: right panel) between samples 2 and 3 do not overlapping the line at 95% confidence, representing a "nonmatch."

mostly lower-MW C_{19}–C_{24} terpanes, diasteranes, and C_{27}–C_{29} steranes. No C_{33}–C_{35} pentacyclic hopanes were detected. (2) Samples 2 and 1 have nearly identical terpane and sterane distribution patterns. (3) Sample 3 shows the distribution pattern different from that of samples 1 and 2. The tricyclic terpanes (C_{21} to C_{24}) in sample 3 are similar to samples 1 and 2, but the pentacyclic terpanes (C_{29}–C_{32}) and C_{27}–C_{29} steranes have much lower concentrations than samples 1 and 2. (4) The diagnostic ratios of target hopanes and steranes are similar for samples 1 and 2, while the diagnostic ratios of sample 3, however, are significantly different from either. Clearly, the fingerprinting and quantitation data of PAH and biomarker terpanes and steranes further confirm the conclusion obtained from the fingerprinting results of sesquiterpanes, that is, sample 1 (Bunker boat) is a positive match to the spill sample 2 (spill oil on the water surface), while sample 3 (Bunker center) is a nonmatch to the spill.

The fingerprinting results described above strongly demonstrate that for defensive forensic investigation and unambiguous spill source identification, the use of the "multicriteria" analytical approach must be followed. In many cases, characterization of biomarker and PAH compounds should include determination of both concentrations and diagnostic ratios/relative distributions.

3.5 Conclusions

Biomarkers retain all or most of the original carbon skeleton of the original natural product, and this structural similarity reveals highly specific information about a spilled oil's source than do other compound groups present in oil. Therefore, chemical fingerprinting of source-characteristic and environmentally

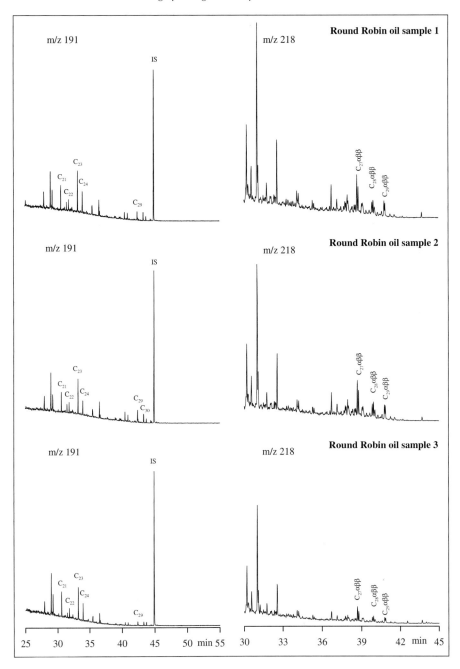

Figure 3-28 Comparison of GC-MS chromatograms of terpanes (*m/z* 191) and steranes (*m/z* 218) in three oil samples.

persistent biomarkers generates information of great importance to environmental forensic investigations in terms of determining the source of spilled oil, differentiating and correlating oils, and monitoring the degradation process and weathering state of oils under a wide variety of conditions. Advancements in spilled oil fingerprinting techniques will continue and these advancements will further enhance the utility and defensibility of oil hydrocarbon fingerprinting and spill source identification.

References

Alberdi, A. and L. Lopez, Biomarker 18α(H)-oleanane, a geochemical tool to assess Venezuelan petroleum systems, *J. South American Earth Sciences*, 2000, **13**, 751–759.

Alexander, R., R. Kagi, and R.A. Noble, Identification of bicyclic sesquiterpanes drimane and eudesmane in petroleum, *J. C. S. Chem. Comm.*, 1983, 226–228.

Alexander, R., R. Kagi, R.A. Noble, and J.K. Volkman, Identification of some bicyclic alkanes in petroleum, In: *Advances in Organic Geochemistry 1983*, P.A. Schenck, J.W. De Leeuw, and G.W.M. Lijmbach (eds.), Pergamon Press, Oxford, 1984.

ASTM Method 3328-90, In: *Annual Book of ASTM Standards, Water (II), Vol. 11.02*, American Society for Testing and Materials, Philadelphia, PA, 1997a.

ASTM Method 5739-95, In: *Annual Book of ASTM Standards, Water (II), Vol. 11.02*, American Society for Testing and Materials, Philadelphia, PA, 1997b.

Atlas, R.M. and R. Bartha, Hydrocarbon biodegradation and oil spill bioremediation, In: *Advances in Microbial Ecology*, K.C. Marshall, (ed.), Plenum Press, New York, Vol. 12, 1992, 287–382.

Audino, M., K. Grice, R. Alexander, and R. Kagi, Macrocyclic-alkanes: A new class of biomarker, *Org. Geochem.*, 2001, **32**, 759–763.

Audino, M., K. Grice, R. Alexander, and R. Kagi, Macrocyclic-alkanes in crude oils from algaenan of *Botryococcus braunii*, *Org. Geochem.*, 2002, **33**, 978–984.

Barakat, A.O., A. Mostafa, M.S. El-Gayar, and J. Rullkotter, Source-dependent biomarker properties of five crude oils from the Gulf of Suez, Egypt, *Org. Geochem.*, 1997, **26**, 441–450.

Barakat, A.O., A.R. Mostafa, J. Rullkotter, and A.R. Hegazi, Application of a multimolecular marker approach to fingerprint petroleum pollution in the marine environment, *Mar. Pollut. Bull.*, 1999, **38**, 535–544.

Barakat, A.O., Y. Qian, M. Kim, and M.C. Kennicutt II, Compositional changes of aromatic steroid hydrocarbons in naturally weathered oil residues in Egyptian Western Desert, *Environmental Forensics*, 2002, **3**, 219–226.

Barbanti, S.M., personal communication, 2004.

Bence, A.E., K.A. Kvenvolden, and M.C. Kennicutt II., Organic geochemistry applied to environmental assessments of Prince William Sound, Alaska, after the *Exxon Valdez* oil spill — a review, *Organic Geochem.*, 1996, **24**, 7–42.

Bendoraitis, J.G., Hydrocarbons of biogenic origin in petroleum: Aromatic triterpanes and bicyclic sesquiterpanes, In: *Advances in Organic Geochemistry 1973*, B. Tissot and F. Bienner (eds.), Editions Technip, Paris, 1974, 209–224.

Berkowitz, N., *Fossil Hydrocarbons: Chemistry and Technology*, Academic Press, San Diego, CA, 1997.

Bieger, T., J. Helou, and T.A. Abrajano, Jr., Petroleum biomarkers as tracers of lubricating oil contamination, *Marine Pollution Bull.*, 1996, **32**, 270–274.

Blenkinsopp, S., Z.D. Wang, J. Foght, D.W.S. Westlake, G. Sergy, M. Fingas, L. Sigouin, and K. Semple, Assessment of the freshwater biodegradation potential of oils commonly transported in Alaska, *Final Report to Alaska Government, ASPS 95–0065*, Environment Canada, Ottawa, 1996.

Boehm, P.D., G.S. Douglas, W.A. Burns, P.J. Mankiewicz, D.S. Page, and A.E. Bence, Application of petroleum hydrocarbon chemical fingerprinting and allocation techniques after the *Exxon Valdez* oil spill, *Mar. Pollut. Bull.*, 1997, **34**, 599–613.

Boehm, P.D., D.S. Page, W.A. Burns, A.E. Bence, P.J. Mankiewicz, and J.S. Brown, Resolving the origin of the petrogenic hydrocarbon background in Prince William Sound, Alaska. *Environ. Sci. Technol.*, 2001, **35**, 471–479.

Boehm, P.D., W.A. Burns, D.S. Page, A.E. Bence, P.J. Mankiewicz, J.S. Brown, and G.S. Douglas, Total organic carbon, an important tool in holistic approach to hydrocarbon source fingerprinting, *Environmental Forensics*, 2002, **3**, 243–250.

Bragg, J.R. and E.H. Owens, Clay-oil flocculation as a natural cleansing process following oil spill: Part 1 — studies of shoreline sediments and residues from past spills, In: *Proceedings of the 17th Arctic and Marine Oil Spill Program (AMOP) Technical Seminar*, Environment Canada, Ottawa, Ontario, 1994, 1–25.

Butler, E.L., G.S. Douglas, W.S. Steinhauter, R.C. Prince, T. Axcel, C.S. Tsu, M.T. Bronson, J.R.

Clark, and J.E. Lindstrom, Hopane, a new chemical tool for measuring oil biodegradation, In: R.E. Hinchee and R.F. Olfenbuttel (eds.), *On-Site Reclamation*, Butterworth-Heinemann, Boston, MA, 1991, 515–521.

Cahn, R.S., C. Ingold, Sir., and V. Prelog, Specification of molecular chirality, *Angewandte Chemie International Ed.*, 1966, **5**, 385–415.

CAS, *Chemical Abstract Index Guide: Appendix IV*, Chemical Abstract Service, Columbus, Ohio, 2002, 175I–327I.

Chen, J. and B. He, The character and genesis of condensate in the north of Liaodong Bay, China, *Organic Geochemistry*, 1990, **6**, 561–567.

Chen, J., J. Fu, G. Sheng, D. Liu, and J. Zhang, Diamondoid hydrocarbon ratios: Novel maturity indices for highly mature crude oils, *Org. Geochem.*, 1996, **25**, 179–190.

Chen, P., Z. Xing, M. Liu, Z. Liao, and D. Huang, Isolation of nine petroporphyrin biomarkers by reversed-phase high-performance liquid chromatography with coupled columns, *J. Chromatogr.*, 1999, **839**, 239–245.

Cheng, K., W. Jin, Z. He, and J. Chen, Application of sesquiterpanes to the study of oil-gas source: The gas-rock correlation in the Qiongdongnan Basin, *J. Southeast Asian Earth Science*, 1991, **5**, 189–195.

Chosson, P., C. Lanau, J. Connan, and D. Dessort, Biodegradation of refractory hydrocarbon biomarkers from petroleum under laboratory conditions, *Nature*, 1991, **351**, 640–642.

Clarke, L.M.J., C.W. Khan, P.V. Hodson, K. Lee, Z.D. Wang, and J. Short, Comparative toxicity of four crude oils to the early life stages of rainbow trout, In: *Proc. 27th Arctic and Marine Oil Spill Program (AMOP) Technical Seminar*, Environment Canada, Ottawa, 2004, 785–792.

Christensen, J.H., A.B. Hansen, J. Mortensen, G. Tomasi, and O. Andersen, Integrated methodology for forensic oil spill identification, *Environ. Sci. Technol.*, 2004, **38**, 2912–2918.

Connolly, J.D. and R.A. Hill, *Dictionary of Terpenoids*, Chapman and Hall, London, 1991.

Currie, T.J., R. Alexander, and R.I. Kagi, Coastal bitumens from Western Australia — long distance transport by ocean currents, *Org. Geochem.*, 1992, **18**, 595–601.

Dahl, J.E., J.M. Moldowan, K.E. Peters, G.E. Claypool, M.A. Rooney, G.E. Michael, M.R. Mello, and M.L. Kohnen, Diamondoid hydrocarbons as indicators of natural oil cracking, *Nature*, 1999, **399**, 54–57.

Dahlmann, G., Characteristic features of different oil types in oil spill identification, Berichte des BSH 31, ISSN 0946–6010, Germany, 2003.

Daling, P.S., L.G. Faksness, A.B. Hansen, and S.A. Stout, Improved and standardized methodology for oil spill fingerprinting. *Environmental Forensics*, 2002, **3**, 263–278.

Dimandja, J.D., GC X GC, *Anal. Chem.*, 2004, **76**, 167A–174A.

Douglas, G.S., R.C. Prince, E.L. Butler, and W.G. Steinhauer, The use of internal chemical indicator in petroleum and refined products to evaluate the extent of biodegradation, In: *Hydrocarbon Bioremediation*, R.E. Hinchee, B.C. Hoeppel, and R.N. Miller (eds.), Lewis Publishers, Boca Raton, FL, 1994, 219–236.

Douglas, G.S., W.A. Burns, A.E. Bence, D.S. Page, and P. Boehm, Optimizing detection limits for the analysis of petroleum hydrocarbons in complex environmental samples, *Environ. Sci. & Technol.*, 2004, **38**, 3958–3964.

El-Gayar, M.S., A.R. Mostafa, A.E. Abdelfattah, and A.O. Barakat, Application of geochemical parameters for classification of crude oils from Egypt into source-related types, *Fuel Processing Technology*, 2002, **79**, 13–28.

EPA, *EPA Test Methods for Evaluating Solid Waste (SW-846)*, Update III, U.S. EPA, Office of Solid Waste and Emergency Response, Washington, DC, 1997.

EPA, *EPA Guidance for Quality Assurance Project Plans*, EPA QA/G-5, U.S. EPA, Washington, DC, 1998a.

EPA, *Guidance for Data Quality Assessment, Practical Method for Data Analysis*, EPA AQ/G-9 QA 97 Updated, U.S. EPA, Washington, DC, 1998b.

EPA, *EPA Requirements for Quality Assurance Project Plans*, EPA QA/R-5, U.S. EPA, Washington, DC, 2001.

EPA Report, *Characteristics of spilled oils, fuels, and petroleum products: 1. Composition and properties of selected oils*, EPA/600/R-03/072, National Exposure Research Laboratory (NERL), EPA, Athens, Georgia. Also, the EPA website: www.epa.gov/epahome/recentadditions.htm, 2003.

ETC Method (updated version), *Analytical Methods for Determination of Oil Components* (by Wang, Z.D.), ETC Method No.: 5.3/1.3/M, Environmental Technology Centre, Environment Canada, Ottawa, Ontario, 2002.

Faksness, L.G., P.S. Daling, and A.B. Hansen, Round Robin Study — Oil Spill Identification, *Environmental Forensics*, 2002, **3**, 279–292.

Fan, P., Y. Qian, and B. Zhang, Characteristics of biomarkers in the recent sediments from Qinghai Lake, Northwest China, *J. Southeast Asian Earth and Science*, 1991, **5**, 113–128.

Fingas, M.F. and B. Fieldhouse, Studies of the formation of water-in-oil emulsions. *Mar. Pollut. Bull.*, 2003, **47**, 369–396.

Foght, J., K. Semple, C. Gauthier, D.W.S. Westlake, S. Blenkinsopp, G. Sergy, Z.D. Wang, and M. Fingas, Development of a standard bacterial consortium for laboratory efficacy testing of commercial freshwater oil spill bioremediation agents, *Environmental Technology*, 1998, **20**, 839–849.

Frontera-Suau, R., F.D. Bost, T.J. Mcdonald, and P.J. Morris, Aerobic biodegradation of hopanes and other biomarkers by crude oil-degrading enrichment cultures, *Environ. Sci. Technol.*, 2002, **36**, 4585–4592.

Frysinger, G., R. Gains, L. Xu, and C.M. Reddy, Resolving the unresolved complex mixture in petroleum-contaminated sediments, *Environ. Sci. Technol.*, 2003, **37**, 1653–1662.

Fu, J., C. Pei, G. Sheng, and D. Liu, A geochemical investigation of crude oils from eastern Pearl River mouth basin, South China, *J. Southeast Asian Earth Science*, 1992, **7**, 271–272.

Garrett, P.M., I.J. Pickerring, C.E. Haith, and R.C. Prince, Photooxidation of crude oils, *Environ. Sci. Technol.*, 1998, **32**, 3719–3723.

Gürgey, K., An attempt to recognize oil populations and potential source rock types in Paleozoic sub- and Mesozoic-Cenozoic supra-salt strata in southern margin of the pre-Caspian basin, Kazakhstan republic, *Org. Geochem.*, 2002, **33**, 723–741.

Gains, R.B., G.S. Frysinger, M.S. Hendrick-Smith, and J.D. Stuart, Oil spill source identification by comprehensive two-dimensional gas chromatography, *Environ. Sci. Technol.*, 1999, **33**, 2106–2112.

Gallegos, E.J., Identification of new steranes, terpanes, and branched paraffins in Great River Shall by combined capillary GC and MS, *Anal. Chem.*, 1971, **43**, 1151–1160.

Garrett, P.M., I.J. Pickerring, C.E. Haith, and R.C. Prince, Photooxidation of crude oils, *Environ. Sci. Technol.*, 1998, **32**, 3719–3723.

Grahl-Nielsen, O. and T. Lygre, Identification of samples of oil related to two spills, *Mar. Pollut. Bull.*, 1990, **21**, 176–183.

Grice, K., R. Alexander, and R.I. Kagi, Diamondoid hydrocarbon ratios as indicators of biodegradation in Australian crude oils, *Organic Geochemistry*, 2000, **31**, 67–73, 2000.

Gürgey, K., An attempt to recognize oil populations and potential source rock types in Paleozoic sub- and Mesozoic-Cenozoic supra-salt strata in southern margin of the pre-Caspian basin, Kazakhstan republic, *Org. Geochem.*, 2002, **33**, 723–741.

Hostettler, F.D., W.E. Pereira, K.A. Kvenvolden, A. Green, S.N. Luoma, C.C. Fuller, and R. Anima, A record of hydrocarbon input to San Francisco Bay as traced by biomarker profiles in surface sediment and sediment cores, *Marine Chemistry*, 1999a, **64**, 115–127.

Hostettler, F.D., R.J. Rosenbauer, K.A. Kvenvolden, PAH refractory index as a source discriminant of hydrocarbon input from crude oil and coal in Prince William Sound, Alaska, *Org. Geochem.*, 1999b, **30**, 873–879.

Hu, G., Geochemical characterization of steranes and terpanes in certain oils from terrestrial facies within South China Sea, *J. Southeast Asian Earth Science*, 1991, **5**, 241–247.

Jordan, R.E. and J.R. Payne, *Fate and Weathering of Petroleum Spills in the Marine Environment: A Literature Review and Synopsis*, Ann Arbor Science Publishers, Ann Arbor, MI, 1980.

Kaplan, I.R., Y. Galperin, S. Lu, and R.P. Lee, Forensic environmental geochemistry differentiation of fuel-types, their sources, and release time, *Org. Geochem.*, 1997, **27**, 289–317.

Kaplan, I.R., S.T. Lu, H.M. Alomi, and J. MacMurphey, Fingerprinting of high boiling hydrocarbon fuels, asphalts and lubricants, *Environmental Forensics*, 2001, **2**, 231–248.

Khan, C.W., L.M.J. Clarke, Z.D. Wang, and B. Hollebone, EROD activity (CYP1A) inducing compounds in fractionated crude oil, In: *Proceedings of the 27th Arctic and Marine Oil Spill Program (AMOP) Technical Seminar*, Environment Canada, Ottawa, 2004, 773–784.

Killops, S.D. and V.J. Howell, Complex series of pentacyclic triterpanes in a lacustrine sourced oil from Korea Bay Basin, *Chem. Geol.*, 1991, **91**, 65–79.

Kvenvolden, K.A., J.B. Rapp, and J.H. Bourell, In: L.B. Magoon and G.E. Claypool (eds.), *Alaska North Slope Oil/Rock Correlation Study*, American Association of Petroleum Geologists Studies in Geology, No. 20, 1985, 593–617.

Kvenvolden, K.A., F.D. Hostettler, J.B. Rapp, and P.R. Carlson, Hydrocarbon in oil residues on beaches of islands of Prince William Sound, Alaska, *Mar. Pollut. Bull.*, 1993, **26**, 24–29.

Kvenvolden, K.A., F.D. Hostettler, P.R. Carlson, J.B. Rapp, C.N. Threlkeld, and A. Warden, Ubiquitous tar balls with a California-source signature on the shorelines of Prince William Sound, Alaska, *Environ. Sci. & Tech.*, 1995, **29**, 2684–2694.

Kvenvolden, K.A., F.D. Hostettler, R.W. Rosenbauer, T.D. Lorenson, W.T. Castle, and S. Sugarman, Hydrocarbons in recent sediment of the Monterey Bay, National Marine Sanctuary, *Marine Geology*, 2002, **181**, 101–113.

Leahy, J.G. and R.R. Colwell, Microbial degradation of hydrocarbons in the environment, *Microbiological Reviews*, 1990, **54**, 305–315.

Lin, R. and Z.A. Wilkes, Natural occurrence of tetramantane, pentamantane and hexamantane in a deep petroleum reservoir, *Fuel*, 1995, **74**, 1512–1521.

Linscheid, M., Mass spectrometry, In: H. Günzler and A. Williams (eds.), *Handbook of Analytical Technique*, WILEY-VCH, Weinheim, Germany, 2001, 579–626.

Lu, S.T. and I.R. Kaplan, Diterpanes, triterpanes, steranes, and aromatic hydrocarbons in natural bitumens and pyrolysates from different coals, Geochimica et Cosmochimica Acta, 1992, **56**, 2761–2788.

Marshall, A.G. and R. Podgers, Petroleomics: The next grand challege for chemical analysis, *Acc. Chem. Res.*, 2004, **37**, 53–59.

Mckirdy, D.M., R.E. Cox, J.K. Volkman, and V.J. Howell, Botryococcane in a new class of Australian non-marine crude oils, *Nature*, 1986, **320**, 57–59.

Mello, M.R., P.C. Gaglianone, S.C. Brassell, and J.R. Maxwell, Geochemical and biological marker assessment of depositional environments using Brazilian offshore oils, *Marine and Petroleum Geology*, 1988, **5**, 205–223.

Mello, M.R., E.A.M. Koutsoukos, M.B. Hart, S.C. Brassell, and J.R. Maxwell, Late cretaceous anoxic events in the Brazilian continental margin, *Organic Geochemistry*, 1990, **14**, 529–542.

Moldowan, J.M. and F.J. Fago, Structure and significance of a novel rearranged monoaromatic steroid hydrocarbon in petroleum, *Geochimica et Cosmochimica Acta*, 1986, **50**, 343–351.

Moldowan, J.M., F.J. Fago, R.M.K. Carlson, D.C. Young, G.V. Duyne, J. Clardy, M. Schoell, C.T. Phillinger, and D.S. Watt, Rearranged hopanes in sediments and petroleum, *Geochimica et Cosmochimica Acta*, 1991, **55**, 3333–3353.

Morrison, R.T., R.N. Boyd, and R.K. Boyd, *Organic Chemistry* (6th ed.), Prentice Hall, Englewood Cliffs, NJ, 1992.

Munoz, D., M. Guiliano, P. Doumenq, F. Jacquot, P. Scherrer, and G. Mille, Long term evolution of petroleum biomarkers in mangrove soil, *Marine Pollution Bulletin*, 1997, **34**, 868–874.

Murrisepp, A.M., K. Urof, M. Liiv, and A. Sumberg, A comparative study of non-aromatic hydrocarbons from kukersite and dictyonema shale semicoking oils, *Oil Shale*, 1994, **11**, 211–216.

NRC, National Research Council, *Oil in the Sea III: Inputs, Fates, and Effects*, The National Academies Press, Washington, DC, 2002.

Noble, R.A., A geochemical study of bicyclic alkanes and diterpenoid hydrocarbons in crude oils, sediments, and coals, *Ph.D. thesis*, Department of Organic Chemistry, University of Western Australia, 1986.

Nytoft, H.P. and J.A. Bojesen-Koefoed, $17\alpha(H)$, $21\alpha(H)$-hopane: Natural and synthetic, Organic Geochemistry, 2001, **32**, 841–856.

Olah, G.A. and Á. Molnar, *Hydrocarbon Chemistry*, Wiley-Interscience, New York, 1995.

Owens, E.H. and K. Lee, Interaction of oil and mineral fines on shorelines: Review and assessment, *Marine Pollution Bulletin*, 2003, **47**, 397–405.

Page, D.S., J.D. Foster, P.M. Fickett, and E.S. Gilfillan, Identification of petroleum sources in an area impacted by the *Amoco Cadiz* oil spill, *Mar. Pollution. Bull.*, 1988, **3**, 107–115.

Page, D.S., P.D. Boehm, G.S. Douglas, and A.E. Bence, Identification of hydrocarbon sources in the benthic sediments of Prince William Sound and the Gulf of Alaska following the *Exxon Valdez* spill, In: P.G. Wells, J.N. Butler, and J.S. Hughes (eds.), *Exxon Valdez Oil Spill: Fate and Effects in Alaska Waters*, ASTM, Philadelphia, PA, 1995, 41–83.

Page, D.S., P.D. Boehm, G.S. Douglas, A.E. Bence, W.A. Burns, and P.J. Mankiewicz, The natural petroleum hydrocarbon background in subtidal sediments of Prince William Sound, Alaska, USA, *Environ. Toxicol. Chem.*, 1996, **15**, 1266–1281.

Page, D.S., A.E. Bence, W.A. Burns, P.D. Boehm, and G.S. Douglas, A holistic approach to hydrocarbon source allocation in the subtidal sediments of Prince William Sound, Alaska,

Embayments, *Environmental Forensics*, 2002, **3**, 331–340.

Peters, K.E., C.C. Walters, and J.W. Moldowan, *The Biomarker Guide (Two Volumes), 2nd Ed*, Cambridge University Press, Cambridge, UK, 2005.

Peters, K.E. and J.W. Moldowan, *The Biomarker Guide: Interpreting Molecular Fossils in Petroleum and Ancient Sediments*, Prentice Hall, Englewood Cliffs, New Jersey, 1993.

Petrov, A.A., *Petroleum Hydrocarbons*, Springer-Verlag, Berlin, Germany, New York, 1987.

Philp, R.P., *Fossil Fuel Biomarkers, Application and Spectra*, Elsevier, New York, 1985.

Prince, R.C., Petroleum spill bioremediation in marine environment, *Crit. Rev. Microbiol*, 1993, **36**, 724–728.

Prince, R.C., D.L. Elmendorf, J.R. Lute, C.S. Hsu, C.E. Haith, J.D. Senius, G.J. Dechert, G.S. Douglas, and E.L. Butler, $17\alpha(H)$, $21\beta(H)$-hopane as a conserved internal marker for estimating the biodegradation of crude oil, *Environ. Sci. Technol.*, 1994, **28**, 142–145.

Prince, R.C., E.H. Owens, and G.A. Sergy, Weathering of an arctic oil spill over 20 years: The BIOS experiment revisited, *Marine Pollution Bulletin*, 2002, **44**, 1236–1242.

Reddy C.M., T.I. Eglinton, A. Hounshell, H.K. White, L. Xu, R.B. Gains, and G.S. Frysinger, The West Falmouth oil spill after thirty years: The persistence of petroleum hydrocarbons in marsh sediments, *Environ. Sci. Technol.*, 2002, **36**, 4754–4760.

Riolo, J. and P. Albrecht, Novel arrangement ring C monoaromatic steroid hydrocarbons in sediments and petroleums, *Tetrahedron Letters*, 1985, **26**, 2701–2704.

Seifert, W.K., Source rock/oil correlations by C27-C30 biological marker hydrocarbons, In: R. Campos and J. Goni (eds.), *Advances in Organic Geochemistry 1974*, Enadimsa, Madrid, 1977, 21–44.

Seifert, W.K., J.M. Moldowan, and G.J. Demaison, Source correlation of biodegraded oils, *Org. Geochem.*, 1984, **6**, 633–643.

Seifert, W.K. and J.M. Moldowan, Use of biological markers in petroleum exploration, In: R.B. Johns (ed.), *Methods in Geochemistry and Geophysics, Vol. 24*. Elsevier, Amsterdam, 1986, 261–290.

Shen, J., Minimization of interferences from weathering effects and use of biomarkers in identification of spilled crude oils by gas chromatography/mass spectrometry, *Anal. Chem.*, 1984, **56**, 214–217.

Short, J.W. and R.A. Heintz, Identification of *Exxon Valdez* oil in sediments and tissues from Prince William Sound and the Northwestern Gulf of Alaska based on a PAH weathering model, *Environ. Sci. Technol.*, 1997, **31**, 2375–2384.

Short, J.W., K.A. Kvenvolden, P.R. Carlson, F.D. Hostettler, R.J. Rosenbauer, and B.A. Wright, Natural hydrocarbon background in benthic sediments of Prince William Sound, Alaska: Oil vs. coal, *Environ. Sci. Technol.*, 1999, **33**, 34–42.

Simanzhenkov, V. and R. Idem, *Crude Oil Chemistry*, Marcel Dekker, Inc., New York, 2003.

Simoneit, B.R.T., J.O. Grimalt, and T.G. Wang, A review on cyclic terpanoids in modern resinous plant debris, amber, coal, and oils, *Annual research report of geochemistry laboratory*, Institute of Geochemistry, Academia Sinica, People's Publishing House, Guizhou, China, 1986.

Sosrowidjojo, I.B., R. Alexander, and R.I. Kagi, The biomarker composition of some crude oils from Sumatra, *Org. Geochem.*, 1994, **21**, 303–312.

Speight, J.G., *The Chemistry and Technology of Petroleum*, Marcel Dekker, Inc., New York, 1999.

Speight, J.G., *Handbook of Petroleum Product Analysis*, Wiley-Interscience, Hoboken, NJ, 2002.

Stout, S.A., A.D. Uhler, T.G. Naymik, and K.J. McCarthy, Environmental forensics: Unraveling site liability, *Environ. Sci. Technol.*, 1998, **32**, 260A–264A.

Stout, S.A., A.D. Uhler, and K.J. McCarthy, A strategy and methodology for defensibly correlating spilled oil to source candidates, *Environmental Forensics*, 2001, **2**, 87–98.

Stout, S.A., A.D. Uhler, K.J. McCarthy, and S. Emsbo-Mattingly, Chapter 6: Chemical fingerprinting of hydrocarbons, In: B.L. Murphy and R.D. Morrison (eds.), *Introduction to Environmental Forensics*, Academic Press, London, 2002, 139–260.

Stout, S.A. and G.S. Douglas, Diamondoid hydrocarbons — application in the chemical fingerprinting of natural gas condensate and gasoline, *Environmental Forensics*, 2004, **5**, 225–235.

Stout, S.A., A.D. Uhler, and K.J. McCarthy, Middle distillate fuel fingerprinting using drimane-based bicyclic sesquiterpanes. *Environmental Forensics*, 2005, **6**, 241–252.

Summons, R.E., J. Thomas, J.R. Maxwell, and C.J. Boreham, Secular and environmental constraints on the occurrence of dinosterane in sediments,

Geochim. Cosmochim. Acta, 1992, **56**, 2437–2444.

Swannel R.P.J., K. Lee, and M. McDonagh, Field evaluation of marine oil spill bioremediation, *Microbiological Reviews*, **60**, 342–365, 1996.

Uhler, A.D., S.A. Stout, and K.J. McCarthy, Increased success of assessments at petroleum sites in 5 steps, *Soil and Groundwater Cleanup*, 1998–1999, Dec/Jan, 13–19.

van Aarssen, B.G.K., H.C. Cox, P. Hoogendoorn, and J.W. de Leeuw, A cadinene biopolymer present in fossil and extract Dammar resins as source for cadinanes and dicadinanes in crude oils from Southeast Asia, *Geochimica et Cosmochimica Acta*, 1990, **54**, 3021–3031.

Volkman, J.K., R. Alexander, R.I. Kagi, and G.W. Woodhouse, Demethylated hopanes in crude oils and their applications in petroleum geochemistry, *Geocim. Cosmochim. Acta*, 1983a, **47**, 785–794.

Volkman, J.K., R. Alexander, R.I. Kagi, and J. Rüllkötter, GC-MS characterization of C27 and C28 triterpanes in sediments and petroleum, *Geocim. Cosmochim. Acta*, 1983b, **47**, 1033–1040.

Volkman, J.K., D.G. Holdsworth, G.P. Neill, and H.J. Bavor, Jr., Identification of natural, anthropogenic and petroleum hydrocarbons in aquatic environments, *Sci. Tol Environ.*, 1992a, **112**, 203–219.

Volkman, J.K., T. O'Leary, R.E. Summons, and M.R. Bendall, Biomarker composition of some asphaltic coastal bitumens from Tasmania, Australia, *Org. Geochem.*, 1992b, **18**, 669–682.

Volkman, J.K., A.T. Revil, and A.P. Murray, Application of biomarkers for identifying sources of natural and pollutant hydrocarbons in aquatic environments, In: R.P. Eganhouse (ed.), *Molecular Markers in Environmental Geochemistry*, American Chemical Society, Washington DC, 1997, 83–99.

Wade, L.G., Jr., *Organic Chemistry* (5th ed.), Prentice Hall, Upper Saddle River, NJ, 2003.

Wang, P., M. Li, and S.R. Larter, Extended hopanes beyond C_{40} in crude oils and source rock extracts from the Liaohe Basin, N.E. China, *Org. Geochem.*, 1996, **24**, 547–551.

Wang, Z.D., M. Fingas, and K. Li, Fractionation of ASMB oil, identification and quantitation of aliphatic, aromatic and biomarker compounds by GC/FID and GC/MSD, *J. Chromatogr. Sci.*, 1994a, **32**, 361–366 (Part I) and 367–382 (Part II).

Wang, Z.D., M. Fingas, and G. Sergy, Study of 22-year-old Arrow oil samples using biomarker compounds by GC/MS, *Environ. Sci. Technol.*, 1994b, **28**, 1733–1746.

Wang, Z.D., M. Fingas, M. Landriault, L. Sigouin, and N. Xu, Identification of alkylbenzenes and direct determination of BTEX and (BTEX + C_3-benzenes) in oils by GC/MS, *Analytical Chemistry*, 1995a, **67**, 3491–3500.

Wang, Z.D., M. Fingas, and G. Sergy, Chemical characterization of crude oil residues from an arctic beach by GC/MS and GC/FID, *Environ. Sci. Technol.*, 1995b, **29**, 2622–2631.

Wang, Z.D. and M. Fingas, Study of the effects of weathering on the chemical composition of a light crude oil using GC/MS and GC/FID, *J. Microcolumn Separations*, 1995c, **7**, 617–639.

Wang, Z.D., M. Fingas, M. Landriault, L. Sigouin, Y. Feng, and J. Mullin, Using systematic and comparative analytical data to identify the source of an unknown oil on contaminated birds, *J. Chromatography*, 1997, **775**, 251–265.

Wang, Z.D., M. Fingas, S. Blenkinsopp, G. Sergy, M. Landriault, and L. Sigouin, Study of the 25-year-old Nipisi oil spill: Persistence of oil residues and comparisons between surface and subsurface sediments, *Environ. Sci. Technol.*, 1998a, **32**, 2222–2232.

Wang, Z.D., M. Fingas, M. Landriault, L. Sigouin, B. Castel, F. Hostetter, D. Zhang, and B. Spencer, Identification and linkage of tarballs from the coasts of Vancouver Island and Northern California using GC/MS and isotopic techniques, *J. High Resolut. Chromatogr.*, 1998b, **21**, 383–395.

Wang, Z.D., M. Fingas, S. Blenkinsopp, G. Sergy, M. Landriault, L. Sigouin, J. Foght, K. Semple, and D.W.S. Westlake, Oil composition changes due to biodegradation and differentiation between these changes to those due to weathering, *J. Chromatography A*, 1998c, **809**, 89–107.

Wang, Z.D., M. Fingas, and D. Page, Oil spill identification, *J. Chromatogr.*, 1999a, **843**, 369–411.

Wang, Z.D., M. Fingas, M. Landriault, L. Sigouin, S. Grenon, and D. Zhang, Source identification of an unknown spilled oil from Quebec (1998) by unique biomarker and diagnostic ratios of "source-specific marker" compounds, *Environmental Technology*, 1999b, **20**, 851–862.

Wang, Z.D., M. Fingas, and L. Sigouin, Characterization and source identification of an unknown spilled oil using fingerprinting techniques by GC-MS and GC-FID, *LC-GC*, 2000, **10**, 1058–1068.

Wang, Z.D., M. Fingas, and L. Sigouin, Characterization and identification of a "mystery" oil spill from Quebec (1999), *J. Chromatogr.*, 2001a, **909**, 155–169.

Wang, Z.D., M. Fingas, E.H. Owens, L. Sigouin, and C.E. Brown, Long-term fate and persistence of the spilled Metula oil in a marine salt marsh environment: Degradation of petroleum biomarkers, *J. Chromatogr.*, 2001b, **926**, 275–190.

Wang, Z.D., M. Fingas, and L. Sigouin, Using multiple criteria for fingerprinting unknown oil samples having very similar chemical composition, *Environmental Forensics*, 2002, **3**, 251–262.

Wang, Z.D. and M. Fingas, Development of oil hydrocarbon fingerprinting and identification techniques, *Mar. Pollut. Bull.*, 2003, **47**, 423–452.

Wang, Z.D., M. Fingas, and P. Lambert, Characterization and identification of Detroit River mystery oil spill (2002), *J. Chromatogr.*, 2004, **1038**, 201–214.

Wang, Z.D., C. Yang, M. Fingas, B. Hollebone, X. Peng, A. Hansen, and J. Christensen, Characterization, weathering, and application of sesquiterpanes to source identification of spilled petroleum products, *Environ. Sci. Technol.*, 2005a, **39**, 8700–8707.

Wang, Z.D., C. Yang, and B. Hollebone, Characterization and application of diamondoids to environmental forensic studies, *Symposium of Environmental Forensics*, Pacifichem 2005, American Chemical Society, Hawaii, 2005b.

Wang, Z.D., S.A. Stout, and M. Fingas, Forensic fingerprinting of biomarkers for oil spill characterization and source identification, *Environmental Forensics*, 2006, **7**(2), in press.

Waples, D.W. and T. Machihara, *Biomarkers for Geologists*, American Association of Petroleum Geologists, AAPG Methods in Exploration Series No. 9, 1991.

Williams, J.A., M. Bjoroy, D.L. Dolcater, and J.C. Winters, Biodegradation in South Texas Eocene oil-effects on aromatics and biomarkers, *Org. Geochem.*, 1986, **10**, 451–462.

Wingert, W.S., GC-MS analysis of diamondoid hydrocarbons in Smackover petroleums, *Fuel*, 1992, **71**, 37–43.

Zakaria, M.P., A. Horinouchi, S. Tsutsumi, H. Takada, S. Tanabe, and A. Ismail, Oil pollution in the Straits of Malacca, Malaysia: Application of molecular markers for source identification, *Environ. Sci. Technol.*, 2000, **34**, 1189–1196.

Zakaria, M.P., T. Okuda, and H. Takada, PAHs and hopanes in stranded tar-balls on the coast of Peninsular Malaysia: Applications of biomarkers for identifying source of oil pollution, *Mar. Pollution. Bull.*, 2001, **12**, 1357–1366.

Zhang, S., D. Liang, Z. Gong, K. Wu, M. Li, F. Song, Z. Song, D. Zhang, and P. Wang, Geochemistry of petroleum system in the eastern Pearl River Mouth Basin: Evidence for mixed oils, *Org. Geochem.*, 2003, **34**, 971–991.

4 Characterization of Polycyclic Aromatic Sulfur Heterocycles for Source Identification

Abdelrahman H. Hegazi[1,2] and Jan T. Andersson[1]

[1]Institute for Inorganic and Analytical Chemistry, University of Münster, Corrensstrasse 30, D-48149 Münster, Germany.
[2]Chemistry Department, Faculty of Science, Alexandria University, Alexandria, Egypt.

4.1 Introduction

The fingerprinting of crude oil spills can be a very complex analytical problem. Not only is the crude oil a highly complex matrix and many of the components of interest are present in low amounts, but an oil in the environment is subject to several weathering factors, such as evaporation and dissolution (physical), biodegradation (biological), and photooxidation (chemical), that often have a major influence on the composition. Furthermore, some of these changes can be dependent on the place and time of the spill. In many cases it is true that the older the pollutant, the more its distinguishing characteristics tend to disappear, but situations are known in which petroleum residues appear after 30 years in concentrations "similar to those observed immediately after the spill" (Reddy et al., 2002), and many biomarkers have proven useful in hydrocarbon fingerprinting and oil spill source identification even in a wide variety of weathering conditions (Stout et al., 2002; Wang and Fingas, 2003).

The most frequent question legal investigators ask of the scientist is to establish the origin of a petroleum spill. Clearly, the chemical analyst can only provide information on the object under investigation based on chemical-analytical data. Linking this information to the source of the pollution and, by implication, the perpetrator, usually needs further evidence that is outside the responsibility of the analytical scientist.

It is reasonable to expect the analyst to state with a certain degree of certainty whether or not a crude oil, finished product, ship bilge, material from tanker washing, or natural seeps is involved. This subject has been studied by a large number of researchers, as is evident from many chapters in this book. To arrive at the most reasonable answer to this question, they have used a multitude of chemical compound classes as markers for different properties, reflecting the complexity of crude oil. Among these are also the polycyclic aromatic sulfur heterocycles (PASHs), which play a major role among the constituents of petroleum and its refined products.

In this chapter we provide recommendations that are directed particularly toward the characterization of PASHs for forensic purposes of any petroleum-derived pollutant that is likely to be encountered mainly in the marine environment and that arrived there through accident or oversight. Obviously, PASHs are only one of several classes of compounds that should be studied for such purposes, as there is no one class of compounds that will be sufficient for answering all the questions of the forensic investigator.

Sulfur compounds in crude oil and petroleum products will be discussed first with the view to showing how varied the compound classes can be in different petroleums and refined products. Next, some changes in PASH patterns when petroleum is treated in refineries are described since this will have an influence on the PASH pattern of petroleum products that may end up in the environment. To understand the importance of PASHs in oil spill source identification, we give an overview on the stability of these compounds in the environment and their reactions there. Analytical methods and techniques used for oil spill fingerprinting have made major advances in recent years, and this development continues. Therefore, in this chapter, we will review the current state of information on PASH analytical techniques. Finally, some case studies in which PASHs were investigated will be described. Our main objective is to clarify how PASHs can be used to help solving problems encountered in oil spill source identification.

4.2 Sulfur Compounds in Crude Oil and Petroleum Products

The petroleum industry has continually been troubled with various problems related to sulfur compounds in petroleum and its products, such as product odor and storage stability, catalyst poisoning, corrosion of processing equipment, and pollution emitted during usage. Thus, it is essential to characterize the structures of sulfur compounds in crude oils and petroleum products. The motivation for this may have technical grounds, for example, to optimize the desulfurization processes, but has the added benefit that the knowledge gained can also be used for other purposes. The discussion ahead will show which kinds of sulfur compounds can be expected and therefore may be targets of an analysis of oil spill residues.

Sulfur is usually the most abundant heteroelement in petroleum. Most of the sulfur present in crude oils is organically bound sulfur while elemental sulfur and hydrogen sulfide usually represent a very minor portion.

The sulfur content is in the range 0.1–3.0% in most crudes (Morrison, 1999) but can reach 8% in the vacuum residue of heavy crudes (Severin and Glinzer, 1984). Organic sulfur compounds in crude oils are distributed over a wide range of molecular structures: aliphatic thiols, mono- and disulfides are sometimes present (Rygle et al., 1984) as well as alkyl phenyl disulfides (Nishioka, 1988), but a large amount occurs in aromatic structures, especially as alkylated thiophene benzologues (Arpino et al., 1987). After distillation, mercaptans, sulfides, and thiophenes are concentrated in the gasoline-range products (Stumpf et al., 1998) while benzothiophenes (BTs), dibenzothiophene (DBT), and alkylated dibenzothiophenes (DBTs) are concentrated in the middle distillate fractions. They may represent up to 70% of the sulfur present in diesel fuel.

The PASH class is by far the most thoroughly investigated of them all, but possibly the sulfides may gain some potential as marker compounds of individual oils since there seems to be quite a diversity of different structures present in fossil material from different sources. A problem could be the ease of their oxidation to sulfoxides in photochemical reactions (Burwood and Speers, 1974) that might lead to a rapid change in their pattern and concentration.

The structures of 1,4,4-trimethyl-4,5,6,7-tetrahydroisothianaphthene and methyl 3,4,5-trimethyl-2-thienyl sulfide (Figure 4-1 and Table 4-1) were established in the kerosene boiling range of Middle East distillates (Birch et al., 1959). Moreover, the sulfides from the petroleum fraction b.p. 200–275° comprised mainly of thiamonocyclanes and included smaller amounts of polycyclic sulfides (Polyakova et al., 1978). The thiamonocyclanes were thiophane homologs. The sulfides from the fraction b.p. 350–450° contained thiamonocyclanes and thiapolycyclanes containing 2–8 rings. The thiapolycyclanes were derivates of thiophane condensed mainly with 5-membered rings. A series of novel, terpenoid-derived polycyclic sulfides were identified by Charrié-Duhaut et al. (2003) in highly desulfurized diesel oils and identified as

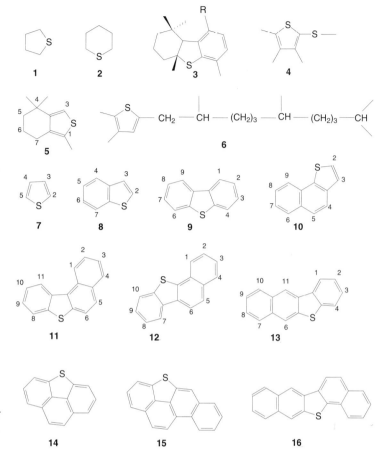

Figure 4-1 Structures of sulfur compounds in petroleum. For names, see Table 4-1.

Table 4-1 Names of Sulfur Compounds in Petroleum as Illustrated in Figure 4-1

Compound	Name
1	Thiophane
2	Thiacyclohexane
3	1,1,4a,6-Tetramethyl-9-alkyl-1,2,3,4,4a, 9b-hexahydrodibenzothiophenes
4	Methyl 3,4,5-trimethyl-2-thienyl sulfide
5	1,4,4-Trimethyl-4,5,6,7-tetrahydroisothianaphthene
6	2,3-Dimethyl-5-(2,6,10-trimethylundecyl)thiophene
7	Thiophene
8	Benzothiophene
9	Dibenzothiophene
10	Naphtho[1,2-*b*]thiophene
11	Benzo[*b*]naphtho[1,2-*d*]thiophene
12	Benzo[*b*]naphtho[2,1-*d*]thiophene
13	Benzo[*b*]naphtho[2,3-*d*]thiophene
14	Phenanthrothiophene
15	Chrysenothiophene
16	Dinaphthothiophene

1,1,4a,6-tetramethyl-9-alkyl-1,2,3,4,4a,9b-hexahydrodibenzothiophenes.

Organo-sulfur compounds may also be present in various other structures. Schmid et al. (1987) and Sinninghe Damsté et al. (1987) describe long-chain dialkylthiacyclopentanes in oils while Payzant et al. (1986) identified terpenoid sulfides in petroleum. Hopanes and steranes are routinely used for petroleum correlation studies but the sulfur-containing compounds do not seem to have been investigated for such purposes. Sinninghe Damsté et al. (1987) and Valisolalao et al. (1984) have tentatively identified steroid and hopanoid thiophenes, respectively, in petroleum. In a later article, Sinninghe Damsté et al. (1989) cited the presence of various highly branched isoprenoid thiophenes in sediments and immature oils. The identification of isoprenoid C20 and

C15 sulfur compounds in Rozel Point Oil has also been described (Sinninghe Damsté and de Leeuw, 1987). These authors reported the occurrence of isoprenoid thiophenes, thiolanes, benzothiophenes, bithiophenes, (thienyl)alkylthiophenes, and thienylthiolanes in this oil and in other oils and sediments. Since isoprenoids often show higher stability in environmental situations, such compounds may be promising for forensic research.

A vast amount of knowledge has accumulated on aromatic sulfur compounds and the following can only give a hint at what is known. Mössner and Wise (1999) identified and quantified naphtho[1,2-*b*]thiophene, dibenzothiophene (DBT), all four methyldibenzothiophenes (MDBTs), three of the four ethyldibenzothiophenes, 15 of the 16 possible dimethyldibenzothiophenes, and 6 of the 28 possible trimethyldibenzothiophenes in a standard reference crude oil sample (SRM 1582), which is a Wilmington crude. They also determined the three isomeric benzo[*b*]naphtho[1,2-*d*]thiophene, benzo[*b*]naphtho[2,1-*d*]thiophene, and benzo-[*b*]naphtho[2,3-*d*]thiophene and 28 of the 30 possible methylbenzonaphthothiophenes (MBNTs) in the same sample. Recently, many of these PASHs were identified in Egyptian crude oils (Hegazi et al., 2003, 2004a). As an illustration of the complexity of the PASH pattern, a partial extended gas chromatogram with the atomic emission detector (AED) in the sulfur-selective mode of an Egyptian crude oil, showing the distribution pattern of BTs and DBTs, is reproduced in Figure 4-2 and an identification of the corresponding peaks is given in Table 4-2. This identification was considerably facilitated due to the availability of reference compounds (PASH, 2006).

In heavy distillates (225–500°C) of Saudi Arabian crude oils (Arabian heavy, light, medium, and extra light), BTs, DBTs, and BNTs are the major PASH types (Farhat et al., 1991). The polar petroleum resins contain many different compounds, such as sulfoxides, sulfides, sulfones, thiophenes, and mixed sulfur-nitrogen heterocycles (Rudzinski and Aminabhavi, 2000; Mitra-Kirtley et al., 1998).

Sulfur compounds in aviation fuels were found to occur mainly as thiols, sulfides, and disulfides while thiophenes and DBTs are minor constituents (Link et al., 2003). Somewhat similar classes of sulfur components in a fluid catalytic cracking (FCC) gasoline were thiols, sulfides, thiophene and alkylthiophenes, tetrahydrothiophene, thiophenols, and benzothiophene (Hatanaka et al., 1997; Cheng et al., 1998). Recent data by Yin and Xia (2004) confirm that thiophene sulfur represents a large fraction of the total sulfur content in FCC gasoline (60 wt. % and more). They detected more than 20 kinds of thiophenes among which some (di- and trimethyl-, ethyl-, ethylmethyl-, di- and triethyl-, *iso*-propyl-, *t*-butyl-) could be identified by GC-MS analysis. Also, they identified cyclo-sulfides in FCC and residue fluid catalytic cracking gasoline produced in China. In an earlier study (Nishioka et al., 1986) 3-ring, 4-ring, and 5-ring PASHs were investigated in a catalytically cracked petroleum vacuum residue. These PASHs are C_{1-5}-DBTs, C_{2-6}-phenanthrothiophenes, C_{1-5}-naphthobenzothiophenes, C_3- and C_4-bis(benzothiophene)s, C_{1-4}-chrysenothiophenes, and C_2-dinaphthothiophene.

4.3 Influence of Refinery Processes on PASH Patterns

Nitrogen and sulfur oxides are a subset of acid rain gases receiving attention throughout the world under agreements aimed at reducing their emission to protect the environment and human health. Current U.S. and E.U. specifications for diesel fuel mandate that the maximum sulfur concentration be 50 ppm, but future values may be as low as 10 ppm (U.S. Environmental Protection Agency, 1999). However, current U.S., Canada, and Europe gasoline sulfur is 30 ppm. Moreover, the introduction of "sulfur-free gasoline" (<10 ppm) has been proposed in Europe for the year 2007 (Nocca et al., 2000).

Those regulations have led to a change in the sulfur compound pattern in refined products that is necessary to recognize since it will be found in the environment in case of spills.

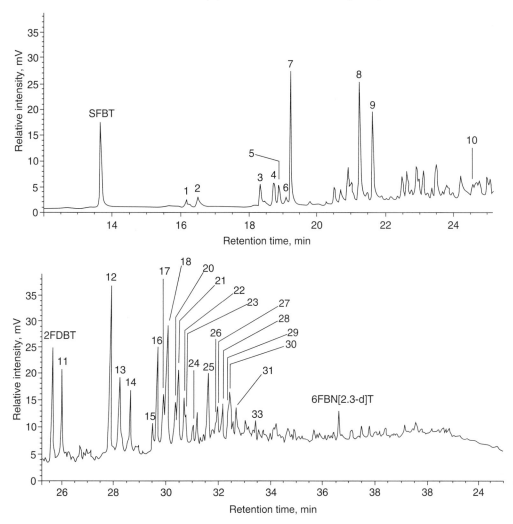

Figure 4-2 Partial extended AED chromatogram of polyaromatic fraction showing the distribution and relative retention times of PASH in an Egyptian crude oil. Peak identifications are listed in Table 4-2. 5FBT, 2FDBT, and 6FBN[2,3-d]T are fluorinated internal standards. (Reprinted from Hegazi et al., 2003, with permission from Elsevier Science.)

The easiest way to limit the amount of sulfur dioxide emitted into the atmosphere is to lower the amount of sulfur in fuels. Conventional physicochemical processes such as hydrodesulfurization (HDS) are effective for the treatment of petroleum. HDS is a catalytic process that uses hydrogen gas to reduce the sulfur in petroleum fractions to hydrogen sulfide, which is readily separated from the fuel.

Under normal conditions for HDS, the most reactive and easiest-to-remove classes of sulfur compounds are thiols, sulfides, and disulfides; compounds such as substituted BTs and DBTs react somewhat more sluggishly (Mössner and Wise, 1999; Lamure-Meille et al., 1995; Shafi and Hutchings, 2000; Knudsen et al., 1999; Stumpf et al., 1998).

Deep hydrodesulfurization of PASHs in light oil was carried out by using Co-Mo/Al$_2$O$_3$ under experimental conditions representative of industrial practice (Kabe et al., 1992). The PASHs were determined by GC-AED and GC-MS. It was found that alkyl-substituted DBTs were the most difficult compounds to desulfu-

Table 4-2 Polycyclic Aromatic Sulfur Heterocycles Identified in the Egyptian Crude Oil in Figure 4-2

Peak	Compound
1	2-Methylbenzothiophene
2	3-Methylbenzothiophene
3	2,7-Dimethylbenzothiophene
4	2,6+3,7+4,7-Dimethylbenzothiophenes
5	4,6-Dimethylbenzothiophene
6	3,5-Dimethylbenzothiophene
7	2,3-Dimethylbenzothiophene
8	2,3,7-Trimethylbenzothiophene
9	2,3,5-Trimethylbenzothiophene
10	2,3,4,7-Tetramethylbenzothiophene
11	Dibenzothiophene
12	4-Methyldibenzothiophene
13	2+3-Methyldibenzothiophenes
14	1-Methyldibenzothiophene
15	4-Ethyldibenzothiophene
16	4,6-Dimethyldibenzothiophene
17	2,4+2,6-Dimethyldibenzothiophenes+ 2-Ethyldibenzothiophene
18	3,6-Dimethyldibenzothiophene
19	2,7+2,8+3,7-Dimethyldibenzothiophenes
20	1,4+1,6+1,8-Dimethyldibenzothiophenes
21	1,3-Dimethyldibenzothiophene
22	3,4-Dimethyldibenzothiophene
23	2,3-Dimethyldibenzothiophene
24	2,4,6-Trimethyldibenzothiophene
25	2,4,8-Trimethyldibenzothiophene
26	2,4,7-Trimethyldibenzothiophene
27	1,4,8-Trimethyldibenzothiophene
28	1,4,7-Trimethyldibenzothiophene
29	2-Methyl-3-ethyldibenzothiophene
30	1,3,7-Trimethyldibenzothiophene
31	2,6-Diethyldibenzothiophene

Source: Reprinted from Hegazi et al., 2003, with permission from Elsevier Science.

rize and 4,6-DMDBT, having substituents in both 4- and 6-positions, remained until the final stage of the reaction (390°C) while alkylbenzothiophenes were completely desulfurized at 350°C. The desulfurization of C_1-BTs results in C_3-benzenes (Depauw and Froment, 1997).

Diesel fuel is an important energy source for industry, a major fuel for ships and vehicles and for building heating. Resembling crude oil's PASH pattern, nondesulfurized diesel shows alkylated BTs and DBTs as the major sulfur-containing compounds (Figure 4-3) (Schade et al., 2002). HDS affects the PASH pattern strongly. Most of the alkylated BTs as well as a large number of the alkylated DBTs (e.g., 1-, 2-, and 3-MDBT) had vanished after a deeper HDS process as seen in Figure 4-3 (Schade et al., 2002), leading to an exceptional prominence of some of the remaining alkylated DBTs, particularly 4MDBT, 4,6DMDBT, and 4E6MDBT. These changes in the pattern of PASHs should help in source identification of spilled refinery products. The HDS reactivities of PASHs are determined by the strength of the C-S bonds, steric hindrance, and electron density on the sulfur atom (Ma et al., 1994). This imparts a reduced desulfurization kinetics to isomers alkylated in the 4-position and thus they are more slowly desulfurized than other alkylated DBTs. More research is needed to establish whether HDS fuels from different processes or different starting crudes retain a common pattern of PASHs, or if individual differences will still be discernible after the HDS. If the latter case turns out to prevail, a tool for source identification may be available.

Recently, high-molecular-weight sulfur containing aromatics were analyzed in a vacuum residue using Fourier transform ion cyclotron resonance mass spectrometry (FTICR-MS) (Müller et al., 2005). The authors reported that HDS affects the distribution patterns of these compounds (Figure 4-4). The amount, as well as class and type, of sulfur compounds is changed during the partial HDS process, which removed primarily compounds with one S atom, whereas those with two S atoms were largely unaffected. Such materials are used as ship fuels and are therefore likely to be found in spills. Although at the moment no restrictions on sulfur in such fuels are legally binding, it is expected that they will be introduced in the future so that again a changed pattern of PASHs can be expected with implications for forensic studies.

4.4 Stability of Polycyclic Aromatic Sulfur Heterocycles in the Environment

As soon as oil is spilled into the marine environment, the processes of spreading, evapora-

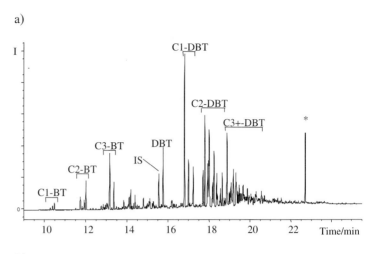

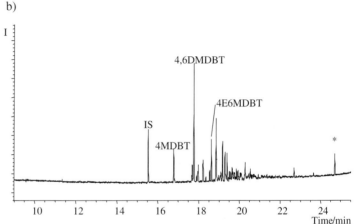

Figure 4-3 GC-FID chromatograms of (a) nondesulfurized diesel fuel and (b) deeply hydrodesulfurized diesel fuel. The asterisk indicates an impurity from the dichloromethane used in the sample workup. (Reprinted from Schade et al., 2002, with permission from Taylor & Francis.)

tion of light ends, emulsification, solubility losses, and microbial and photochemical degradation act to change its composition, making the unambiguous identification of the source of the release and the monitoring of its fate and effect in the environment something of a challenge.

Photooxidation and biodegradation are the two most important factors involved in the chemical transformation of crude oil and its products after a release into the marine environment. Photooxidation affects the aromatic compounds in crude oil and converts them to polar species of higher-aqueous solubility. Natural microbial populations in seawater can biodegrade 28% of crude oil and 36% of photooxidized crude oil within 8 weeks at 20°C (Dutta and Harayama, 2000). For the *Exxon Valdez* accident in Alaska in 1989, the mass balance showed that 50% of the oil was removed through aqueous biodegradation and photolysis and a further 20% through atmospheric photolysis. The remaining 30% were beached, recovered by man or deposited on sediments (Wolfe et al., 1994).

Although seemingly possessing a noticeable degree of stability under environmental conditions, like other organic compounds the PASHs are removed from the environment through different processes, such as microbial degradation (Saftic et al., 1992). Bacteria of the *Pseudomonas* type oxidize methylbenzothiophenes to the sulfoxide if an alkyl substituent is present on the thiophene ring and, if not,

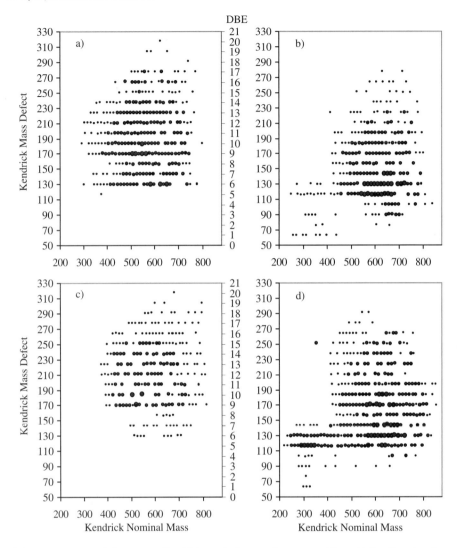

Figure 4-4 Kendrick mass defects versus Kendrick nominal mass of vacuum residues: (a) first ligand exchange chromatography (LEC) fraction, (b) second LEC fraction before HDS, (c) first LEC fraction, and (d) second LEC fraction after HDS. The double bond equivalent (DBE) scale given is valid for both panels. (Reprinted from Müller et al., 2005, with permission from the American Chemical Society.)

as in 7-methylbenzothiophene, to the 2,3-dione with the sulfone as a minor metabolite (Kropp et al., 1994a, b). In some instances, products of other types, namely carboxylic acids, ring cleavage products, and dimers, were noticed.

Dutta and Harayama (2000) noted that complete depletion of alkanes and naphthalene (N) derivatives and less extensive elimination of other components by biodegradation in the order n-alkane > naphthalenes > branched alkanes > fluorenes > phenanthrenes (P) > dibenzothiophenes was the rule. The rate of polycyclic aromatic compound degradation in the environment decreases with increasing ring size and, within a homologous series, decreases with increasing alkylation (Neff, 1979; Douglas et al., 1994; Elmendorf et al., 1994). A study of microbial degradation of organic sulfur compounds in Prudhoe Bay crude oil revealed that the order of susceptibility of the sulfur heterocycles in homologous series was (1) C2-BTs > C3-BTs, (2) DBT > C1-DBTs > C2-DBTs > C3-DBTs (Fedorak and Westlake, 1983, 1984).

Photochemical processes are also effective in the environmental transformation of PASHs. Benzothiophene in aqueous solution is photochemically oxidized to 2-sulfobenzoic acid (Andersson and Bobinger, 1992) and DBT yields 2-sulfobenzoic acid, benzothiophene-2,3-dicarboxylic acid, two isomeric benzothiophenemonocarboxylic acids and an isomer of thiophenetricarboxylic acid (Traulsen et al., 1999). Moreover, monomethylbenzothiophenes are photochemically oxidized through two pathways. One principal reaction pathway involves oxidation of the methyl group(s) to the carboxylic acid(s) via aldehydes, and the other pathway leads, via oxidation of the thiophene ring, to the quinone and then, after ring opening, to 2-sulfobenzoic acids as the ultimate products for all compounds (Andersson and Bobinger, 1996; Bobinger and Andersson, 1998). Opening of the benzo ring is also seen, in which case thiophenealdehydes or -ketones are formed. In a photochemical experiment simulating an oil spill in water, Andersson (1993) found that the sulfur heterocycles are more stable toward photodegradation than the corresponding hydrocarbons.

Even after burning and subsequent weathering (up to 12 days) a petrol sample can be readily identified as such through a sulfur-selective gas chromatogram that also after such severely destructive treatment preserved the monomethylbenzothiophenes. This identification of petrol could not be made based on the universal flame ionization detector (Dynes and Burns, 1987). Diphenyl disulfides and homologues have been found by GC/MS to be indicators of (weathered) gasoline. Since they were not present in creosote or other petroleum derivatives, they can, if present, be used as markers for gasoline (Coulombe, 1995).

Recognition and subsequent correlation of biodegraded and/or water-washed oils with their unaltered counterparts have been long-recognized problems in petroleum geochemistry and oil spill source identification.

Experimental data indicate that sulfur-containing aromatics such as thiophene and 2-ethylthiophene are approximately twice as soluble in water as aromatic hydrocarbons of similar molecular weight (Price, 1976). The data also show that dibenzothiophenes have a higher aqueous solubility compared to that of aromatic hydrocarbons (Palmer, 1984) as is evident from their lower k_{ow} (octanol-water partition coefficient) (Andersson and Schräder, 1999). However, there are indications that PASHs are resistant to water washing (Manowitz and Jeon, 1992) and therefore accumulate in water-washed oils. Although the mentioned results may seem contradictory, many reports in the literature indicate the usefulness of PASHs and their ratios to polycyclic aromatic hydrocarbons (PAHs) in oil spill source identification (see, for example, Wang and Fingas, 1995; Douglas et al., 1996) as discussed ahead.

All those natural effects on a spill mean that before any chemical data are used for a source correlation, a conscientious evaluation of the possibility of changes of the composition of a petroleum sample must be carried out, no matter what the molecular markers used are. If a spill is chemically changed in a selective fashion, for example, in that certain compound classes or isomers are lost at a higher rate than others, it may be that an identification relying heavily on those marker compounds may incorrectly assign the source of the spill. To avoid such misidentifications, several chemical compounds, both within a general group of compounds, such as PASHs, as well as different compound classes should always be used for a source correlation since it is much less likely that they are all affected in a similar way.

4.5 Petroleum PASH Analysis Techniques

Analytical methods for the efficient and unambiguous characterization of petroleum form the basis for any evaluation of the consequences associated with an oil spill. The analytical strategy must be formulated so that it leads to information that can answer questions regarding (1) the identity of the pollutant source, (2) distinguishing spilled oil from background hydrocarbons, (3) evaluating quantifiably the extent of impacted ecosystems, (4) establishing the responsibilities associated with the environmental damage and cleanup operations, and (5) enforcing pollution control laws.

General sampling procedures, including the collection, storage, and transportation of the samples to the laboratory and detailed procedures for recovering oil from different environmental samples, are reviewed in detail by Butt et al. (1986a), and Barman et al. (2000) as well as Wang and Fingas (1997) have reviewed chromatographic techniques for petroleum analysis.

To identify the PASHs in environmental samples, the extract (or oil recovery) must not only be isolated from its particular matrix but requires separation into specific fractions. Numerous procedures for the fractionation of petroleum components have been reported in the literature. Procedures that lead to a fraction that contains the polycyclic aromatic compounds include open column chromatography on various stationary phases (silica, alumina, and silica-alumina combinations), thin layer chromatography, as well as normal-phase high performance liquid chromatography (Wang et al., 1994; Reddy and Quinn, 1999; Jaouen-Madoulet et al., 2000). Solid-phase extraction (SPE) has been used as an alternative technique, with high selectivity, a faster elution profile, and minimization of solvent consumption being among the advantages (Bennett and Larter, 2000; Sauvain et al., 2001; Alzaga et al., 2004).

A question often not addressed in forensic work is that of coelution in gas chromatography. A peak found to appear at the expected retention time is frequently assumed to be the pure compound and its peak height or area is evaluated. As is shown in Section 4.5.1.4, this is not necessarily true. If isomeric compounds are involved, neither sulfur-selective nor mass-selective detection can detect even a massive coelution. For instance, DBT coelutes with naphtho[1,2-*b*]thiophene on nonpolar stationary phases (Schmid and Andersson, 1997). In the standard reference material SRM 1580 Shale Oil the concentration of the naphthothiophene is 46% of that of DBT. If two coeluting compounds have different properties in the environment after a spill but all changes are thought to arise from only one of them, wrong conclusions might be drawn.

4.5.1 Selective Detection in Gas Chromatography

Generally speaking, the fraction containing the polycyclic aromatic compounds can be used for the analysis of both the PAHs and the PASHs if a chromatographic system of a high resolving power and a selective detection system is used. The selective detection is needed since most samples will be too complex for a direct analysis with, e.g., GC-flame ionization detection. Selective detectors include those that show a high selectivity for the element sulfur and mass selective detectors (GC-MS). Unlike in petroleum research, such sulfur-selective detectors have not found much prominence in forensic studies where the mass selective detector is preferentially used, probably because the PASHs can be determined together with many other classes of compounds in one chromatographic run. However, this may not always be advisable (see Section 4.5.1.4). In some cases, a class separation of PAHs and PASHs may present advantages as discussed in Section 4.5.2. It should be stressed that an analytical procedure with as low a detection limit as possible should be used; otherwise false conclusions may easily be drawn regarding the identity of the sample (Douglas et al., 2004).

4.5.1.1 Flame Photometric Detection (FPD)

The aromatic sulfur compounds and their alkylated isomers in crude oil are numerous so the identification and quantification of individual isomers require selective and sensitive methods of detection (Andersson, 2001). Much early research conducted on the identification and application of PASHs was performed using GC-FPD. Although showing a nonlinear response in dependence of the sulfur concentration, the FPD was useful for the identification of heavily biodegraded spill samples (Butt et al., 1986b). A biodegraded crude oil was unrecognizable from its flame ionization detector chromatogram but still readily recognizable from its FPD chromatogram of the sulfur heterocycles resistant to biodegradation. Arpino et al. (1987) determined PASHs in the aromatic fractions, from the cleanup of petroleum and its derived industrial products, by GC-FPD and GC-single ion monitoring (SIM)-MS. GC with dual FID/FPD was used for the detection of PAHs and PASHs in crude oils and sediment samples from North Sea reservoirs (Schou and Myhr, 1988). The same technique together with GC-SIM-MS shows the variation in distribution of PASHs in oils and condensates of different maturity levels (Chakhmakhchev and Suzuki, 1995).

4.5.1.2 Atomic Emission Detection (AED)

The American Society for Testing and Materials (ASTM) method 5623–94 (sulfur compounds in light petroleum liquids by GC and sulfur selective detection) recommends the use of either an AED or a sulfur chemiluminescence detector (SCD) but not the FPD (Quimby et al., 1998). Andersson and Schmid (1995) determined 22 PASHs in a shale oil with GC-AED. The photochemical degradation of monomethylbenzo[*b*]thiophenes and DBTs was studied by GC-AED and GC-MS to elucidate the possible fate of crude oil components after an oil spill (Bobinger and Andersson, 1998; Bobinger et al., 1999). Mössner and Wise (1999) used GC-SIM-MS and GC-AED to determine PASHs in fossil fuel-related samples. Moreover, Schmid and Andersson (1997) examined PASHs quantitatively in three standard reference materials from the National Institute of Standards and Technology (NIST), namely, SRM 1597 coal tar, 1582 crude oil, and 1580 shale oil using GC-AED in the carbon- and sulfur-selective modes. The same technique was used to determine the sulfur compounds in gasoline range petroleum products (Stumpf et al., 1998). Recently, Hegazi et al. (2003, 2004a, b) applied GC-AED to investigate PASHs in Egyptian crude oils and environmentally weathered tar balls.

4.5.1.3 Sulfur Chemiluminescence Detection (SCD)

Although possessing excellent properties for sulfur-selective detection, the SCD has seemingly not received quite the popularity of the other detectors. An example for its use in the determination of various types of sulfur compounds in a crude is given by Andari et al. (1996).

4.5.1.4 Mass-Selective Detection (MSD)

Low-resolution, mass-selective detectors are routinely used with GC for PASH determination but may exhibit an interference that, although known, is not often taken into account. Since DBT and C_4-naphthalenes share the mass 184, they appear together in the chromatogram if this ion is selected, as is illustrated in Figure 4-5. Furthermore, the peak at the correct retention time for DBT, 16.86 min in this case, is not pure but a result of coelution with other substances as is shown by the full spectrum taken at this retention time (Figure 4-6). Any quantification of DBT using single-ion monitoring MS is obviously likely to yield wrong results unless it is shown that the GC peak at the expected retention time is generated only by DBT. Likewise, C_{x+4}-naphthalenes may interfere with the determination of C_x-DBTs, having identical M^+ values. The same is true for the benzothiophenes and benzenes with four more carbon atoms, and benzonaphthothiophenes and phenanthrenes with four more alkyl carbon atoms.

4.5.2 Class Separation of PAH and PASH

If the level of sulfur is low, the selectivity of a sulfur-selective detector such as the AED

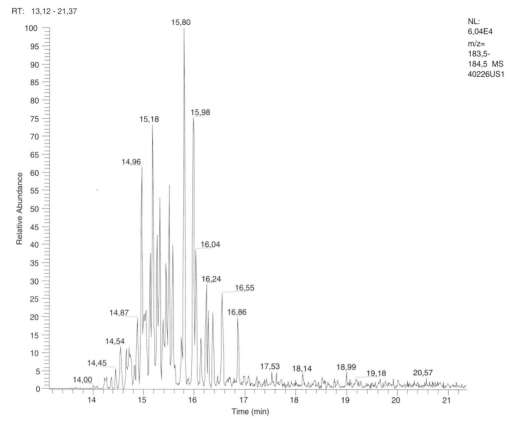

Figure 4-5 Gas chromatogram of a diesel (350 ppm sulfur) monitored at m/z 184, which is the mass for DBT. Coinjection with DBT reveals this compound at a retention time of 16.86 minutes.

(selectivity of sulfur vs. carbon: ca 37,000) may not be high enough to generate a sulfur-selective gas chromatogram without interference from PAHs. Also, as discussed above, the (low-resolution) mass selective detector may include a PAH interference with the PASH determination. In such situations it may be favorable to separate the interfering PAHs from the PASHs, which can then be analyzed using any detector. This procedure can also have the advantage of concentrating the analytes and thus make the analysis more reliable. The only workable method for such a separation is liquid chromatography on palladium(II)-containing stationary phases. A review is available (Andersson, 2001) and more recent work has confirmed the usefulness of the method (Schade et al., 2002; Müller et al., 2005).

4.5.3 Comprehensive Two-Dimensional Gas Chromatography

Fairly recently, comprehensive two-dimensional gas chromatography (GC × GC) with FID has been applied to oil spill source identification (Gaines et al., 1999). PASHs in crude oils have also been separated (Frysinger and Gaines, 2001). The very high chromatographic resolution possible with this technique makes it of interest in forensic studies. Since selective detectors like the AED and MS (van Stee et al., 2003) as well as the SCD (Blomberg et al., 2004) can be used with GC × GC, sulfur compounds may be targets for future applications.

4.5.4 Quantification of PASH

In much forensic work, the absolute quantities of sulfur compounds are not of as great an

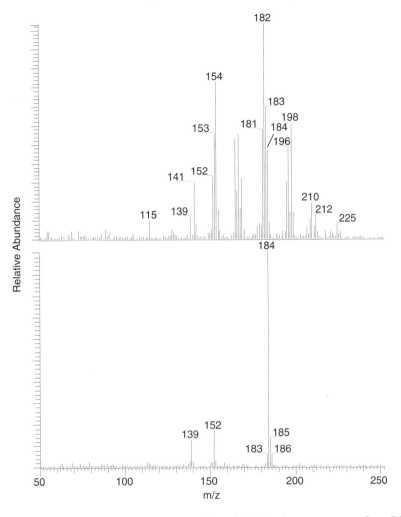

Figure 4-6 Mass spectrum of the peak at 16.86 minutes in Figure 4-5 (top) and a mass spectrum of pure DBT (bottom).

interest as ratios (usually peak heights or areas put in relation to each other) of various compounds. Such ratios can help in establishing the origin of an oil spill through correlation with suspected sources. They can also give information with respect to transformations of the oil in the environment since characteristic changes are known to occur as a result of biotransformation, for example. Ratios are experimentally convenient since a concentration determination presupposes knowledge of the response factor for each compound regardless of detector type (Schmid and Andersson, 1997). Only some detectors show the same molar response for all compounds; if not, a quantification may entail a lot of work and demand the use of pure reference compounds.

4.6 Petroleum PASH Markers in Environmental Forensic Investigations

Compounds used as markers for oil pollution should preferably meet several criteria, some of which demand that markers (1) be present in all crude oils, irrespective of source, (2) be present in large enough amounts that they can be analyzed also in cases of low-level pollution, (3) show a defined degree of stability in the environment, and (4) have no other significant inputs into the environment besides petroleum. Polycyclic aromatic compounds (PACs) are relatively stable and diagnostic constituents of petroleum. Several studies have been carried out using the distribution of the alkylated polycyclic aromatic homologues as environmental fate indicators and source specific markers of oil spills. Among PACs, PASHs fulfill the mentioned criteria to a high degree and therefore their use as organic markers of oil pollution has been suggested. That they have been found in environmental samples many years after an oil spill testifies to their relatively high stability (Friocourt et al., 1982; Reddy et al., 2002).

It is important to keep in mind that PASHs are always used together with several other markers in order to establish forensic evidence. Their occurrence is thus one piece in a chain of chemical information and, as is true for all other markers, they must not be viewed alone. Different markers will provide different pieces of evidence that can be characteristic of the source or the environmental processes that the spill has been subjected to. Since the pattern of the sulfur compounds differs appreciably between different sources, it has a high informational content with respect to source identification. Nonaromatic sulfur compounds have not been used for such purposes, so in this section we will discuss only investigations that show how PASHs can be used to provide important information on an oil spill situation.

4.6.1 PASHs as Source Markers

Ratios of compounds that degrade at the same rates retain the initial oil signature until they can no longer be detected (Boehm et al., 1983). These ratios are called *source ratios* and can be contrasted to ratios that change substantially with weathering and biodegradation, termed *weathering ratios* (Douglas et al., 1996). Alkylated three-ring PAC compounds are quite useful for source identification because of their abundance in petroleum and many of its refined products, their relative concentrations vary among different oils making them source-specific, and they can be quantitatively measured using routine analytical methods (Douglas and Uhler, 1993; Douglas et al., 1994, 1996; Page et al., 1995).

It has been established that the ratio of the three GC peaks from the four methyldibenzothiophenes (2- and 3-MDBT coelute on nonpolar stationary phases) varies strongly with the source of a petroleum (Wang and Fingas, 1995). The ratio of 1-MDBT to 4-MDBT was plotted vs. the ratio of 2/3-MDBT to 4-MDBT for 25 oils and, as shown in Figure 4-7, a satisfactorily wide scattering was obtained, except for oils from the same fields (*ibid.*). This plot can therefore be used to support the suspected origin of a crude oil. An advantage of these compounds is that they have similar volatilities so that evaporative weathering affects them in a consistent manner.

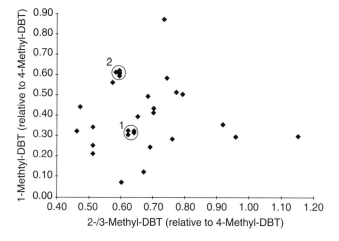

Figure 4-7 Plot of the relative ratios of 2-/3-methyl-DBT to 4-methyl-DBT versus the relative ratios of 1-methyl-DBT to 4-methyl-DBT for 25 different oils. The circles 1 and 2 indicate related samples from origins of North Slope and Terra Nova, respectively. (Reprinted from Wang and Fingas, 1995, with permission from the American Chemical Society.)

A study of composition changes of artificially weathered Alberta Sweet Mixed Blend (ASMB) crude oil indeed indicated that the isomeric distributions within C_1-DBTs, C_1-, C_2-, and C_3-naphthalenes, C_1-phenanthrenes, C_1-fluorenes exhibited great consistency in their relative ratios for ASMB weathered from 0 to 45% (Wang et al., 1998a). Excellent consistency of the relative distribution within the MDBT series was demonstrated not only for artificially weathered oils but also for many short-term field weathering and burned oil samples. It therefore can be used as a method for source identification and differentiation of crude and weathered oils (Wang and Fingas, 1995). The method has distinct advantages: (1) C_1-DBT isomers are found in all crude oils at high enough concentrations and their distribution fingerprints vary significantly; (2) C_1-DBTs are chromatographically well separated with little interference from other compounds (but see the warning in Section 4.5.1.4), so their ratios can be accurately determined; and (3) the relative distributions of C_1-DBTs are subject to little interference from not too severe evaporative weathering.

Wang and Fingas (1997) demonstrated that compared to most crude oils, the Bunker-type fuels have a relative ratio of 2-/3- to 4-MDBT around 1.0, which is unusually high. The relative concentrations of 4-, 2-/3-, and 1-MDBT were determined to be about $1.0:0.92:0.60$ and $1.0:0.92:0.54$ for tarballs from the coasts of Northern California and Vancouver Island of British Columbia, respectively (Wang et al., 1998b). The high 2-/3- to 4-MDBT ratios of these samples together with other source-specific markers indicate that Bunker C was the original source (Wang et al., 1998b).

The relative distribution of alkylated DBTs was used in a study of the PASHs in Egyptian crude oils that indicated a dependence of the PASH distribution patterns on the origin (Hegazi et al., 2003, 2004a). Some oils exhibit a methylbenzothiophene (MBT) pattern, where the ratio 3-MBT/2-MBT is less than unity, whereas it is >1 in other oils. Furthermore, Gulf of Suez oils show a higher relative abundance of 4,6-dimethylbenzothiophene than Western Desert oils. The presence of a V pattern (4-methyl >2-+3-methyl < 1-methyl) in the MDBTs is clear in some oils, while others have a stair-step pattern (4-methyl > 2-+3-methyl > 1-methyl). Moreover, some oils are characterized by their abundance of BTs and a fairly equal distribution of substituted DBTs. Others exhibit low concentrations of BTs and decreasing amounts of C_2- and C_3-DBTs relative to MDBTs. These differences in PASH pattern among crude oils provide a means of resolving multiple oil spill sources.

Groundwater is an important source of potable water. Refined petroleum products and combustion-related products, such as coal tar, are released into soil during spills at industrial facilities, chemical accidents, leakage from hazardous waste landfills, leaking storage and underground fuel tanks, etc. The contamination from these sources will eventually seep into the groundwater, resulting in PAC incorporation into the water. The DBTs are valuable for source correlation because their concentrations reflect the sulfur content of the source, which varies widely between different oils and coal tar (Havenga and Rohwer, 2002). We found that the C1-DBT/C1-P source ratio for groundwater samples is useful for source determination, but the isomers must be present in concentrations $> 0.07\,ng/cm^3$. A double ratio plot of C1-DBT/C1-P source ratio versus C2-N/C1-P weathering ratio could be used as a means of resolving multiple oil spill sources as well as differences in the extent of weathering and degradation from a single source (*ibid.*). This plot could differentiate coal tar- from mineral oil-contaminated groundwater from South Africa.

The higher alkylated an aromatic compound is, the more stable to weathering processes, including biodegradation, it will be. Polycyclic aromatic ratios C_2-DBTs/C_2–phenanthrenes (C2-DBT/C2-P) and C3-DBT/C3-P, which vary among oils having different sulfur contents, have been shown to remain relatively constant as spilled oil weathers and can therefore be used for identification purposes when the residues are heavily degraded (Douglas et al., 1996; Pavlova and Dimov, 2001; Wang and Fingas, 2003). This constancy of C2-DBT/C2-P and C3-DBT/C3-P despite degradation was confirmed in controlled soil biodegradation experiments (Douglas et al., 1996). At up to 70% total petroleum hydrocarbon and 98% total PAC depletion, the ratios remained stable. Therefore, these ratios can be used for moderately degraded oils (30–70%) in both marine and terrestrial environments. DBTs are somewhat more resistant to biodegradation than phenanthrenes so that at higher depletions a trend to slightly higher ratios could be observed. Dutta and Harayama (2000) used a natural seawater for the biodegradation of an Arabian light crude and found that C2-DBT was degraded to 95% while the C_2-phenanthrenes were lost to 100%. Similarly for the C_3-substituted compounds: DBTs were lost to 89%, phenanthrenes to 96%. The photooxidation produced comparative results, e.g., for the C3-derivatives: DBTs were diminished to 81.6%, the phenanthrenes to 100%.

Dibenzothiophenes have played a key role in the environmental assessments conducted following the *Exxon Valdez* oil spill. Their distribution has been used to distinguish *Exxon Valdez* oil and its weathered residues from background petrogenic, pyrogenic, and biogenic compounds in benthic sediments (Bence et al., 1996; see also Chapter 15 herein). The C2-DBT/C2-P and C3-DBT/C3-P ratios were of particular usefulness in conjunction with several other data sources.

Overton et al. (1981) also reported on the utility of C3-DBT/C3-P ratios to identify the source of petroleum residues, in this case Arabian light and Louisiana sweet crudes, after a fire and oil spill. The ratios were also utilized in assessments of the 1991 M/C *Haven* oil spill in Italy to distinguish high-sulfur heavy Iranian cargo crude from a low-sulfur, prespill background (Martinelli et al., 1995).

Stout et al. (2001) utilized a strategy, incorporating statistical analysis of the quantitative chemical data, in order to identify 19 chemical indices (out of 45 evaluated) based on PACs, including sulfur heterocycles, and biomarkers that were both unaffected by weathering and could be precisely measured. The strategy was used in a case study to correlate a spilled heavy fuel oil to 66 candidate sources. Among the diagnostic indices studied were 4MDBT/1MDBT, DBT/P, C2-DBT/C2-P, and C3-DBT/C3-P.

4.6.2 PASHs as Weathering Markers

Such source correlations as described above can only be employed if it is established that the parameters used have not been distorted by alterations in the environment. However, as indicated above, it is well known that aromatic ratios are affected by biodegradation. Not only are, for instance, phenanthrenes degraded faster than DBTs, but, as biodegradation proceeds, the three GC peaks for the MDBTs undergo pronounced changes in their relative ratios (Wang et al., 1998a). 2-/3-MDBTs biodegrade at the fastest rate as shown by the strong decrease of their relative ratio to 4-MDBT, while 1-MDBT is slightly more resistant to biodegradation than 4-MDBT, indicated by an increase of the ratio 1-MDBT:4-MDBT. The same results were obtained for those oil components in clams after the *Aegean Sea* oil spill in Spain (Albaigés et al., 2000) as is evident in Figure 4-8. Hence, this method can be used to indicate the occurrence of biotic degradation of oils, but then, of course, the pattern in the source oil must be known.

A commonly used weathering ratio is defined by C3-DBT/C3-chrysenes (C3-DBT/C3-C) (Douglas et al., 1994). The four-ring chrysenes are less affected by weathering than the three-ring DBTs, and the three alkyl substituents make biodegradation less likely. Again, the corresponding ratios in the source oil must be known so that the change in these indexes can be established.

Plots of a source ratio C3-DBT/C3-P versus a weathering ratio C3-DBT/C3-C provide a means of resolving multiple sources as well as differences in the extent of degradation for samples from a single source (Douglas et al., 1996). These authors used such a plot to distinguish three oil spills; the *Exxon Valdez* (Alaskan North Slope crude oil), M/C *Haven* (heavy Iranian crude oil), and a North Sea production leak (North Sea crude oil). This approach was useful to confirm the finding (Page et al., 1995) of a regional petroleum signature in the deep subtidal sediments that was derived from terrestrial sources of natural hydrocarbons along the northern Gulf of Alaska. The ratios were useful in freshwater deposits also where Athabasca Oil Sands contributed to the PACs (Akre et al., 2004).

Double-ratio plots of C2/C3-DBTs versus C2/C3-phenanthrenes classified 14 oil residues from Prince William Sound, Alaska, eight years after the *Exxon Valdez* spill, according to their weathering degrees (Figure 4-9) (Michel and Hayes, 1999). The oil residues ranged from moderately to extremely weathered. It was found in the same study that the benzonaphthothiophenes had increased in relative abundance and were the dominant PACs in advanced and extreme weathering stages.

The coasts of the Mediterranean city of Alexandria, Egypt, are a major tourist attraction and are vital for fisheries and marine activities. However, they are under constant threat of petroleum pollution in the form of heavy tar loads from several different sources. Furthermore, petroleum contamination from sources such as the oil fields and the Suez-Mediterranean pipeline terminal (SUMED) 27 km west of the city are of concern. A study representing a forensic chemical analysis was carried out to define the liability for the coastal bitumens polluting the beaches of Alexandria (Hegazi et al., 2004b). Tarballs from several locations along the coast were analyzed for their acyclic and polycyclic hydrocarbons as well as sulfur heterocycles using high-resolution GC-AED and GC-MS techniques.

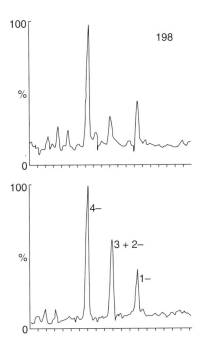

Figure 4-8 Mass fragmentograms of methyl-DBTs (*m/z* 198) from clams collected in April 1993 (lower trace) and December 1995 (upper trace). 2- and 3-MDBT are biodegraded appreciably faster than 1- and 4-MDBT. (Reprinted from Albaigés et al., 2000, with permission from WIT Press.)

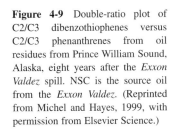

Figure 4-9 Double-ratio plot of C2/C3 dibenzothiophenes versus C2/C3 phenanthrenes from oil residues from Prince William Sound, Alaska, eight years after the *Exxon Valdez* spill. NSC is the source oil from the *Exxon Valdez*. (Reprinted from Michel and Hayes, 1999, with permission from Elsevier Science.)

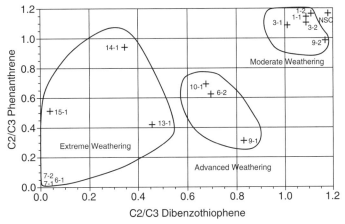

Ratios involving DBT/P, C2-DBT/C2-P, and C3-DBT/C3-P were applied to differentiate between the tarballs according to their sources. The ratios varied from one sample to another, revealing a considerable amount of compositional heterogeneity and suggesting multiple sources of the tarballs.

4.7 Conclusions

The source identification of petroleum spills is a very important subject for both environmentalists and governments and consequently much research has been carried out to facilitate the fingerprinting of crude oil spills. Among the molecular markers finding widespread use, the PASHs, present in all crudes in a variety of derivatives, have been identified as very useful compounds for both source and weathering studies. Despite the fact that they are present in easily measured concentrations using routinely available techniques, some analytical quality considerations must not be overlooked, as stressed in this chapter. With the desulfurization now required in many countries, transportation fuels released into the environment may be too low in sulfur to let the PASHs be an easily measured marker class, but this does not affect crude oils.

References

Akre, C.J., J.V. Headley, F.M. Conly, K.M. Peru, and L.C. Dickson, Spatial patterns of natural polycyclic aromatic hydrocarbons in sediment in the Lower Athabasca River, *J. Environ. Sci. Health*, 2004, A39(5), 1163–1176.

Albaigés, J., C. Porte, D. Pastor, X. Biosca, and M. Solé, The integrated use of chemical and biochemical markers for assessing the effects of the *Aegean Sea* oil spill in the Galicia coast (NW Spain); *Oil and Hydrocarbon Spills II: Modelling, Analysis and Control*, G.R. Rodriguez and C.A. Brebbia (eds.), WIT Press, 2000, 73–84.

Alzaga, R., P. Montuori, L. Ortiz, J.M. Bayona, and J. Albaigés, Fast solid-phase extraction-gas chromatography-mass spectrometry procedure for oil fingerprinting application to the *Prestige* oil spill, *J. Chromatogr. A*, 2004, **1025**, 133–138.

Andari, M.K., H. Behbahani, and A. Stanislaus, Sulfur compound type distribution in naphtha and gas oil fractions of Kuwaiti crude, *Fuel Sci. Technol. Intern.*, 1996, **14**(7), 939–961.

Andersson, J.T., Polycyclic aromatic sulfur heterocycles III. Photochemical stability of the potential oil pollution markers phenanthrenes and dibenzothiophenes, *Chemosphere*, 1993, **27**(11), 2097–2102.

Andersson, J.T., Separation methods in the analysis of polycyclic aromatic sulfur heterocycles, in *Environmental Analysis Handbook of Analytical Separations*, W. Kleiböhmer (ed.), Elsevier Science, Amsterdam, 2001, 3, 75–98.

Andersson, J.T. and B. Schmid, Polycyclic aromatic sulfur heterocycles. IV. Determination of polycyclic aromatic compounds in a shale oil with the atomic emission detector, *J. Chromatogr. A*, 1995, **693**(2), 325–338.

Andersson, J.T. and S. Bobinger, Polycyclic aromatic sulfur heterocycles. II. Photochemical oxidation of benzo[b]thiophene in aqueous solution, *Chemosphere*, 1992, **24**(4), 383–389.

Andersson, J.T. and S. Bobinger, Photochemical degradation of crude oil components: 2-Methyl-, 3-methyl- and 2,3-dimethylbenzothiophene, *Polycycl. Arom. Comp.*, 1996, **11**, 145–151.

Andersson, J.T. and W. Schräder, A method for measuring 1-octanol-water partition coefficients, *Anal. Chem.*, 1999, **71**(16), 3610–3614.

Arpino, P.J., I. Ignatiadis, and G. de Rycke, Sulfur-containing polynuclear aromatic hydrocarbons from petroleum. Examination of their possible statistical formation in sediments, *J. Chromatogr.*, 1987, **390**(2), 329–348.

Barman, B.N., V.L. Cebolla, and L. Membrado, Chromatographic techniques for petroleum and related products, *Critical Rev. Anal. Chem.*, 2000, 30(2&3), 75–120.

Bence, A.E., K.A. Kvenvolden, and M.C. Kennicutt II, Organic geochemistry applied to environmental assessments of Prince William Sound, Alaska, after the *Exxon Valdez* oil spill — a review, *Org. Geochem.*, 1996, **24**(1), 7–42.

Bennett, B. and S.R. Larter, Quantitative separation of aliphatic and aromatic hydrocarbons using silver ion-silica solid-phase extraction, *Anal. Chem.*, 2000, **72**(5), 1039–1044.

Birch, S.F., T.V. Cullum, R.A. Dean, and D.G. Redford, Sulfur compounds in the kerosene boiling range of Middle East distillates. Occurrence of a bicyclic thiophene and a thienyl sulphide, *Tetrahedron*, 1959, **7**, 311–318.

Blomberg, J., T. Riemersma, M. van Zuijlen, and H. Chaabani, Comprehensive two-dimensional gas chromatography coupled with fast sulphur-chemiluminescence detection: Implications of detector electronics, *J. Chromatogr. A*, 2004, **1050**(1), 77–84.

Bobinger, S. and J.T. Andersson, Degradation of the petroleum components monomethylbenzothiophenes on exposure to light, *Chemosphere*, 1998, **36**(12), 2569–2579.

Bobinger, S., F. Traulsen, and J.T. Andersson, Dibenzothiophene in crude oils: Products from the photochemical degradation, *Polycycl. Arom. Comp.*, 1999, **14&15**, 253–263.

Boehm, P.D., D.L. Feist, I. Kaplan, P. Mankiewicz, and G.S. Lewbel, A natural resources damage assessment study: The *Ixtoc I* blowout, in *Proc. 1983 Int. Oil Spill Confe.*, American Petroleum Institute, Washington, DC, 1983, 507–515.

Burwood, R. and G.C. Speers, Photo-oxidation as a factor in the environmental dispersal of crude oil, *Estuarine Coastal Mar. Sci.*, 1974, **2**(2), 117–135.

Butt, J.A., D.F. Duckworth, and S.G. Perry, *Samples: Their collection, storage and transportation; Characterization of spilled oil samples: Purpose, sampling, analysis and interpretation*, John Wiley & Sons, Great Britain, 1986a, 21–41.

Butt, J.A., D.F. Duckworth, and S.G. Perry, *The Analytical Approach to Spill Identification; Characterization of Spilled Oil Samples: Purpose, Sampling, Analysis and Interpretation*, John Wiley & Sons, Great Britain, 1986b, 51–95.

Chakhmakhchev, A. and N. Suzuki, Aromatic sulfur compounds as maturity indicators for petroleum from the Buzuluk depression, Russia, *Org. Geochem.*, 1995, **23**(7), 617–625.

Charrié-Duhaut, A., C. Schaeffer, P. Adam, P. Manuelli, P. Scherrer, and P. Albrecht, Terpenoid-derived sulfides as ultimate organic sulfur compounds in extensively desulfurized fuels, *Angew. Chem.*, 2003, **115**, 4794–4797.

Cheng, W.C., G. Kim, A.W. Peters, X. Zhao, K. Rajagopalan, M.S. Ziebarth, and C.J. Pereira, Environmental fluid catalytic cracking technology, *Catal. Rev.-Sci. Eng.*, 1998, **40**(1/2), 39–79.

Coulombe, R., Chemical markers in weathered gasoline, *J. Forensic Sciences*, 1995, **40**(5), 867–873.

Depauw, G.A. and G.F. Froment, Molecular analysis of the sulphur components in a light cycle oil of a catalytic cracking unit by gas chromatography with mass spectrometric and atomic emission detection, *J. Chromatogr. A*, 1997, **761**, 231–247.

Douglas, G.S. and A.D. Uhler, Optimizing EPA methods for petroleum-contaminated site assessments, *Environ. Test. Anal.*, 1993, **2**(3), 46–53.

Douglas, G.S., R.C. Prince, E.L. Butler, and W.G. Steinhauer, in *Hydrocarbon Bioremediation*, R.E. Hinchee, B.C. Alleman, R.E. Hoeppel, and R.N. Miller (eds.), Lewis Publishers, Ann Arbor, MI, 1994, 219–236.

Douglas, G.S., A.E. Bence, R.C. Prince, S.J. Mcmillen, and E.L. Butler, Environmental stability of selected petroleum hydrocarbon source and weathering ratios, *Environ. Sci. Technol.*, 1996, **30**(7), 2332–2339.

Douglas, G.S., W.A. Burns, A.E. Bence, D.S. Page, and P. Boehm, Optimizing detection limits for the analysis of petroleum hydrocarbons in complex environmental samples, *Environ. Sci. Technol.*, 2004, **38**, 3958–3964.

Dutta, T.K. and S. Harayama, Fate of crude oil by the combination of photooxidation and biodegradation, *Environ. Sci. Technol.*, 2000, **34**, 1500–1505.

Dynes, K. and D.T. Burns, Identification of weathered petrol residues by high-resolution gas chromatography with dual flame ionisation detector-hall electrolytic conductivity detector, *J. Chromatogr.*, 1987, **396**, 183–189.

Elmendorf, D.L., C.E. Haith, G.S. Douglas, and R.C. Prince, in *Bioremediation of Chlorinated and Polycyclic Aromatic Hydrocarbon Compounds*, R.E. Hinchee, A.E. Leeson, L. Semprini, and S.K. Ong (eds.), Lewis Publishers, Ann Arbor, MI, 1994, 188–202.

Farhat, A.M, H. Perzanowski, and S.A. Koreish, Sulfur compounds in high-boiling fractions of Saudi Arabian crude oil, *Fuel Sci. Technol. Intern.*, 1991, **9**(4), 397–424.

Fedorak, P.M. and D.W.S. Westlake, Microbial degradation of organic sulfur compounds in Prudhoe Bay crude oil, *Canadian J. Microbiology*, 1983, **29**(3), 291–296.

Fedorak, P.M. and D.W.S. Westlake, Degradation of sulfur heterocycles in Prudhoe Bay crude oil by soil enrichments, *Water, Air, and Soil Pollution*, 1984, **21**(1–4), 225–230.

Friocourt, M.P., F. Berthou, and D. Picart, Dibenzothiophene derivatives as organic markers of oil pollution, *Toxicol. Environ. Chem.*, 1982, **5**(3&4), 205–215.

Frysinger, G.S. and R.B. Gaines, Separation and identification of petroleum biomarkers by comprehensive two-dimensional gas chromatography, *J. Sep. Sci.*, 2001, **24**(2), 87–96.

Gaines, R.B., G.S. Frysinger, M.S. Hendrick-Smith, and J.D. Stuart, Oil spill source identification by comprehensive two-dimensional gas chromatography, *Environ. Sci. Technol.*, 1999, **33**(12), 2106–2112.

Hatanaka, S., M. Yamada, and O. Sadakane, Hydrodesulfurization of catalytically cracked gasoline. 2. The difference between HDS active site and olefin hydrogenation active site, *Ind. Eng. Chem. Res.*, 1997, **36**(12), 5110–5117.

Havenga, W.J. and E.R. Rohwer, The determination of trace-level PAHs and diagnostic ratios for source identification in water samples using solid-phase microextraction and GC/MS, *Polycycl. Arom. Comp.*, 2002, **22**, 327–338.

Hegazi, A.H., J.T. Andersson, and M.Sh. El-Gayar, Application of gas chromatography with atomic emission detection to the geochemical investigation of polycyclic aromatic sulfur heterocycles in Egyptian crude oils, *Fuel Processing Tech.*, 2003, **85**(1), 1–19.

Hegazi, A.H., J.T. Andersson, M.A. Abu-Elgheit, and M.Sh. El-Gayar, Utilization of GC/AED for investigating the effect of maturity on the distribution of thiophenic compounds in Egyptian crude oils, *Polycycl. Arom. Comp.*, 2004a, **24**(2), 123–134.

Hegazi, A.H., J.T. Andersson, M.A. Abu-Elgheit, and M.Sh. El-Gayar, Source diagnostic and weathering indicators of tar balls utilizing acyclic, polycyclic and S-heterocyclic components, *Chemosphere*, 2004b, **55**, 1053–1065.

Jaouen-Madoulet, A., A. Abarnou, A.M. Le Guellec, V. Loizeau, and F. Leboulenger, Validation of an analytical procedure for polychlorinated biphenyls, coplanar polychlorinated biphenyls and polycyclic aromatic hydrocarbons in environmental samples, *J. Chromatogr. A*, 2000, **886**(1&2), 153–173.

Kabe, T., A. Ishihara, and H. Tajima, Hydrodesulfurization of sulfur-containing polyaromatic compounds in light oil, *Ind. Eng. Chem. Res.*, 1992, **31**(6), 1577–1580.

Knudsen, K.G., B.H. Cooper, and H. Topsoe, Catalyst and process technologies for ultra-low sulfur diesel fuel, *Appl. Catal. A*, 1999, **189**(2), 205–215.

Kropp, K.G., J.A. Goncalves, J.T. Andersson, and P.M. Fedorak, Bacterial transformations of benzothiophene and methylbenzothiophenes, *Environ. Sci. Technol.*, 1994a, **28**(7), 1348–1356.

Kropp, K.G., J.A. Goncalves, J.T. Andersson, and P.M. Fedorak, Microbially mediated formation of benzonaphthothiophenes from benzo[*b*]thiophenes, *Appl. Environ. Microbiol.*, 1994b, **60**(10), 3624–3631.

Lamure-Meille, V., E. Schulz, M. Lemaire, and M. Vrinat, Effect of experimental parameters on the relative reactivity of dibenzothiophene and 4-methyldibenzothiophene, *Appl. Catal. A*, 1995, **131**(1), 143–157.

Link, D.D., J.P. Baltrus, K.S. Rothenberger, P. Zandhuis, D.K. Minus, and R.C. Striebich, Class- and structure-specific separation, analysis, and identification techniques for the characterization of the sulfur compounds of JP-8 aviation fuel, *Energy & Fuels*, 2003, **17**, 1292–1302.

Ma, X., K. Sakanishi, and I. Mochida, Hydrodesulfurization reactivities of various sulfur compounds in diesel fuel, *Ind. Eng. Chem. Res.*, 1994, **33**(2), 218–222.

Manowitz, B. and Y. Jeon, The effects of biodegradation and water washing on sulfur compound speciation in crude oils from Bolivar Coastal Fields, *ACS Division of Petroleum Chemistry Preprints*, 1992, **37**(3), 951–955.

Martinelli, M., A. Luise, E. Tromellini, T.C. Sauer, J.M. Neff, and G.S. Douglas, in *Proc. Inte. Oil Spill Conf.*, Long Beach, CA, API, Washington, DC, 1995, 679.

Michel, J. and M.O. Hayes, Weathering patterns of oil residues eight years after the *Exxon Valdez* oil spill, *Marine Pollution Bulletin*, 1999, **38**(10), 855–863.

Mitra-Kirtley, S., O.C. Mullins, C.Y. Ralston, D. Sellis, and C. Pareis, Determination of sulfur species in asphaltene, resin, and oil fractions of crude oils, *Appl. Spectrosc.*, 1998, **52**(12), 1522–1525.

Mössner, S.G. and S.A. Wise, Determination of polycyclic aromatic sulfur heterocycles in fossil fuel-related samples, *Anal. Chem.*, 1999, **71**, 58–69.

Morrison, R.D., Chemistry and transport of petroleum hydrocarbons, in *Environmental Forensics, Principles & Applications*, CRC Press, Boca Raton, FL, 1999, Chap. 2, 51–90.

Müller, H., W. Schrader, and J.T. Andersson, Characterization of high-molecular-weight sulfur-containing aromatics in vacuum residues using Fourier transform ion cyclotron resonance mass spectrometry, *Anal. Chem.*, 2005, **77**, 2536–2543.

Neff, J.M., *Polycyclic Aromatic Hydrocarbons in the Aquatic Environment: Sources, Fates and*

Biological Effects, Appl. Sci. Publ., London, 1979.

Nishioka, M., Aromatic sulfur compounds other than condensed thiophenes in fossil fuels: Enrichment and identification, *Energy & Fuels*, 1988, **2**(2), 214–219.

Nishioka, M., D.G. Whiting, R.M. Campbell, and M.L. Lee, Supercritical fluid fractionation and detailed characterization of the sulfur heterocycles in a catalytically cracked petroleum vacuum residue, *Anal. Chem.*, 1986, **58**, 2251–2255.

Nocca, J.L., J. Cosyns, Q. Debuisschert, and B. Didillon, *Proc. NPRA Annual Meeting*, Paper AM-00-14, San Antonio, TX, 2000.

Overton, E.B., J.A. McFall, S.W. Mascarella, C.F. Steele, S.A. Antoine, I.R. Politzer, and J.L. Laseter, in *Proc. 1981 Int. Oil Spill Conf.*, American Petroleum Institute, Washington, DC, 1981, 541–546.

Page, D.S., P.D. Boehm, G.S. Douglas, and A.E. Bence, in *Exxon Valdez oil spill: Fate and effects in Alaskan Waters, ASTM Special Technical Publication 1219*, P.G. Wells, J.N. Butler, and J.S. Hughes (eds.), American Society for Testing and Materials, Philadelphia, 1995, 41–83.

Palmer, S.E., Effect of water washing on C_{15+} hydrocarbon fraction of crude oils from Northwest Palawan, Philippines, *AAPG Bulletin*, 1984, **68**(2), 137–149.

PASH, www.pash-standards.de, accessed 2.6.2006.

Pavlova, A. and N. Dimov, Comparison of weathered marine oil spills with the original source of contamination, *Acta Chromatographica*, 2001, **11**, 25–36.

Payzant, J.D., D.S. Montgomery, and O.P. Strausz, Sulfides in petroleum, *Org. Geochem.*, 1986, **9**(6), 357–369.

Polyakova, A.A., G.V. Vasilenko, and L.O. Kogan, Study of the composition and structure of petroleum sulfides using high-resolution mass spectrometry, *Khimiya I Tekhnologiya Topliv I Masel*, 1978, **9**, 58–60.

Price, L.C., Aqueous solubility of petroleum as applied to its origin and primary migration, *AAPG Bulletin*, 1976, **60**, 213–244.

Quimby, B.D., D.A. Grudoski, and V. Giarrocco, Improved measurement of sulfur and nitrogen compounds in refinery liquids using gas chromatography-atomic emission detection, *J. Chromatogr. Sci.*, 1998, **36**(9), 435–443.

Reddy, C.M. and J.G. Quinn, GC-MS analysis of total petroleum hydrocarbons and polycyclic aromatic hydrocarbons in seawater samples after the North Cape oil spill, *Mar. Pollut. Bull.*, 1999, **38**(2), 126–135.

Reddy, C.M., T.I. Eglington, A. Hounshell, H.K. White, L. Xu, R.B. Gaines, and G.S. Frysinger, The West Falmouth oil spill after thirty years: The persistence of petroleum hydrocarbons in marsh sediments, *Environ. Sci. Technol.*, 2002, **36**, 4754–4760.

Rudzinski, W.E. and T.M. Aminabhavi, A review on extraction and identification of crude oil and related products using supercritical fluid technology, *Energy & Fuels*, 2000, **14**(2), 464–475.

Rygle, K.J., G.P. Feulmer, and R.F. Scheideman, Gas chromatographic analysis of mercaptan odorants in liquefied petroleum gas, *J. Chromatogr. Sci.*, 1984, **22**(11), 514–519.

Saftic, S., P.M. Fedorak, and J.T. Andersson, Diones, sulfoxides, and sulfones from the aerobic cometabolism of methylbenzothiophenes by Pseudomonas strain BT1, *Environ. Sci. Technol.*, 1992, **26**(9), 1759–1764.

Sauvain, J.J., T.V. Due, and C.K. Huynh, Development of an analytical method for the simultaneous determination of 15 carcinogenic polycyclic aromatic hydrocarbons and polycyclic aromatic nitrogen heterocyclic compounds. Application to diesel particulates, *Fresenius J. Anal. Chem.*, 2001, **371**(7), 966–974.

Schade, T., B. Roberz, and J.T. Andersson, Polycyclic aromatic sulfur heterocycles in desulfurized diesel fuels and their separation on a novel palladium(II)-complex stationary phase, *Polycycl. Arom. Comp.*, 2002, **22**, 311–320.

Schmid, B. and J.T. Andersson, Critical examination of the quantification of aromatic compounds in three standard reference materials, *Anal. Chem.*, 1997, **69**, 3476–3481.

Schmid, J.C., J. Connan, and P. Albrecht, Occurrence and geochemical significance of long-chain dialkylthiacyclopentanes, *Nature*, 1987, **329**, 54–56.

Schou, L. and M.B. Myhr, Sulfur aromatic compounds as maturity parameters, *Org. Geochem.*, 1988, **13**(1–3), 61–66.

Severin, D. and O. Glinzer, Characterization of heavy crude oils and petroleum residues, *Technip, Paris*, 1984, 19–31.

Shafi, R. and G.J. Hutchings, Hydrodesulfurization of hindered dibenzothiophenes: An overview, *Catal. Today*, 2000, **59**(3–4), 423–442.

Sinninghe Damsté, J.S. and J.W. de Leeuw, The origin and fate of isoprenoid C20 and C15 sulfur compounds in sediments and oils, *Int. J. Environ. Anal. Chem.*, 1987, **28**(1–2), 1–19.

Sinninghe Damsté, J.S., J.W. de Leeuw, A.C.K. Dalen, M.A. de Zeeuw, F. de Lange, I.C. Rijkstra, and P.A. Schenck, The occurrence and identification of series of organic sulfur compounds in oils and sediment extracts. I. A study of Rozel Point Oil (U.S.A.), *Geochim. Cosmochim. Acta*, 1987, **51**(9), 2369–2391.

Sinninghe Damsté, J.S., E.R. Van Koert, Kock-Van Dalen, J.W. de Leeuw, and P.A. Schenck, Characterization of highly branched isoprenoid thiophenes occurring in sediments and immature crude oils, *Org. Geochem.*, 1989, **14**(5), 555–567.

Stout, S.A., A.D. Uhler, and K.J. McCarthy, A strategy and methodology for defensibly correlating spilled oil to source candidates, *Environmental Forensics*, 2001, **2**, 87–98.

Stout, S.A., A.D. Uhler, K.J. McCarthy, and S. Emsbo-Mattingly, Chemical fingerprinting of hydrocarbons, in *Introduction of Environmental Forensics*, B.L. Murphy and R.D. Morrison (eds.), Academic Press, London, 2002, Chap. 6, 139–260.

Stumpf, Á., K. Tolvaj, and M. Juhász, Detailed analysis of sulfur compounds in gasoline range petroleum products with high-resolution gas chromatography-atomic emission detection using group-selective chemical treatment, *J. Chromatogr. A*, 1998, **819**, 67–74.

Traulsen, F., J.T. Andersson, and M.G. Ehrhardt, Acidic and non-acidic products from the photooxidation of the crude oil component dibenzothiophene dissolved in seawater, *Anal. Chim. Acta*, 1999, **392**, 19–28.

U.S. Environmental Protection Agency, Diesel fuel quality, *Advanced Notice of Proposed Rulemaking* EPA 420-F-99-011, Office of Mobile Sources, 1999, Washington, DC.

Valisolalao, J., N. Perakis, B. Chappe, and P. Albrecht, A novel sulfur-containing C35 hopanoid in sediments, *Tetrahedron Lett.*, 1984, **25**(11), 1183–1186.

van Stee, L.L.R., J. Beens, R.J.J. Vreuls, and U.A.T. Brinkman, Comprehensive two-dimensional gas chromatography with atomic emission detection and correlation with mass spectrometric detection: Principles and application in petrochemical analysis, *J. Chromatogr. A*, 2003, **1019**(1–2), 89–99.

Wang, Z.D., M. Fingas, and K. Li, Fractionation of a light crude oil and identification and quantitation of aliphatic, aromatic, and biomarker compounds by GC-FID and GC-MS, Part I, *J. Chromatogr. Sci.*, 1994, **32**(9), 361–366.

Wang, Z.D. and M. Fingas, Use of methyldibenzothiophenes as markers for differentiation and source identification of crude and weathered oils, *Environ. Sci. Technol.*, 1995, **29**(11), 2842–2849.

Wang, Z.D. and M. Fingas, Developments in the analysis of petroleum hydrocarbons in oils, petroleum products and oil-spill-related environmental samples by gas chromatography, *J. Chromatogr. A*, 1997, **774**(1&2), 51–78.

Wang, Z.D., M. Fingas, S. Blenkinsopp, G. Sergy, M. Landriault, L. Sigouin, J. Foght, K. Semple, and D.W.S. Westlake, Comparison of oil composition changes due to biodegradation and physical weathering in different oils, *J. Chromatogr. A*, 1998a, **809**, 89–107.

Wang, Z.D., M. Fingas, M. Landriault, L. Sigouin, B. Castle, D. Hostetter, D. Zhang, and B. Spencer, Identification and linkage of tarballs from the coasts of Vancouver Island and Northern California using GC/MS and isotopic techniques, *J. High Resol. Chromatogr.*, 1998b, **21**(7), 383–395.

Wang, Z.D. and M. Fingas, Development of oil hydrocarbon fingerprinting and identification techniques, *Mar. Pollut. Bull.*, 2003, **47**(9–12), 423–452.

Wolfe, D.A., M.J. Hameedi, J.A. Galt, G. Watabayashi, J. Short, C. O'Claire, S. Rice, J. Michel, J.R. Payne, J. Braddock, S. Hanna, and D. Sale, The fate of the oil spilled from the *Exxon Valdez*, *Environ. Sci. Technol.*, 1994, **28**(13), 560A–568A.

Yin, C. and D. Xia, A study of the distribution of sulfur compounds in gasoline produced in China. Part 3. Identification of individual sulfides and thiophenes, *Fuel*, 2004, **83**, 433–441.

5 Oil Spill Source Identification by Comprehensive Two-Dimensional Gas Chromatography (GC × GC)

Richard B. Gaines,[1] Glenn S. Frysinger,[1] Christopher M. Reddy,[2] and Robert K. Nelson[2]

[1] U.S. Coast Guard Academy, 27 Mohegan Ave., New London, CT, 06320.
[2] Woods Hole Oceanographic Institute, Department of Marine Chemistry and Geochemistry, 360 Woods Hole Rd., Woods Hole, MA 02543.

5.1 Introduction

5.1.1 The Need for High-Resolution Separations

Petroleum is a complex mixture containing thousands of compounds with different chemical functionality including alkanes, one- and multi-ring cycloalkanes, one- and multi-ring aromatics, and compounds containing sulfur, oxygen, and nitrogen. Detailed studies of the source, fate, and transport of petroleum spills in the environment often require high-resolution separations of target compounds and classes of compounds from the rest of the complex mixture. Routine one-dimensional chromatographic methods fall far short of the resolving power required due to the random nature of chromatogram peak spacing. For example, a peak capacity of 100 is needed to separate 100 evenly spaced peaks. Evenly spaced peaks can be expected for simple mixtures such as homologous series of normal alkanes, but if greater chemical diversity is present in the sample, randomly distributed peaks are likely. Statistics show that to separate 98 out of 100 randomly distributed peaks, a peak capacity of 10,000 is required (Davis et al., 1994). Therefore, samples with hundreds or thousands of compounds cannot be resolved using a single chromatographic column.

Giddings (1995) hypothesized that complex mixtures can be described with multiple *dimensions*, such as size, shape and polarity. When multiple dimensions are present in a sample, peaks are randomly distributed in the chromatogram. Complete component separation can be achieved when the separation dimensions match the sample dimensions (provided there is sufficient resolving power in each dimension) and the resulting chromatogram will be ordered.

To illustrate the concept of matching separation dimensions to sample dimensions, consider a mixture of objects having various sizes, shapes, and gray scales, shown in Figure 5-1A. When objects are separated only by size, as in Figure 5-1B, coelutions occur among objects of different shape and gray scale. When objects are separated by two of the three sample dimensions, as in Figure 5-1C, the result is a more ordered chromatogram with less coelution. When the separation dimensions are size and gray scale, the chromatogram exhibits coelution by objects of different shape. When the separation dimensions are size and shape, the chromatogram exhibits coelution by objects of different gray scales. When objects are separated by the same three dimensions as the sample, complete separation results and the chromatogram is ordered. Ordered chromatograms are the result of matching sample and separation dimensionality and represent efficient use of peak capacity.

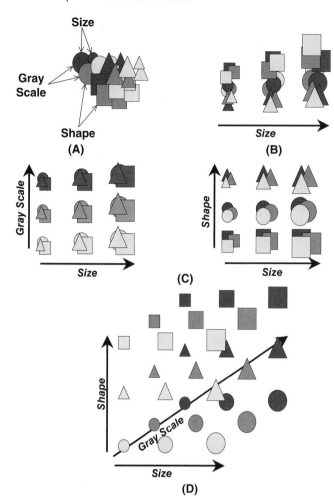

Figure 5-1 Conceptual diagram of matching mixture dimensions with separation dimensions to achieve complete separation and an ordered chromatogram. (A) Three-dimensional mixture in gray scale, shape, and size; (B) one-dimensional separation in size results in poor separation; (C) two-dimensional separation improves separation and results in a more ordered chromatogram; (D) matching mixture and separation dimensions results in complete separation and ordered chromatogram.

5.1.2 Multidimensional Methods

Until recently, multidimensional analyses of petroleum spills were accomplished primarily using gas chromatography — mass spectrometry (GC-MS) systems employing a high-resolution capillary column and a quadrapole mass spectrometer operating in either a full scan or selected-ion-monitoring mode (ASTM, 2000b). Methods such as GC-MS provide information about specific chemical classes found within oil. Although the high-resolution capillary GC column cannot resolve many of the petroleum compounds, the mass spectrometer is able to virtually separate the GC coeluents by mass because components of a specific chemical class often have unique molecular fragmentation patterns. Identification and comparison between compound classes, especially high-molecular-weight biomarkers, produce a more detailed and informative fingerprint (e.g., see Chapter 3 in this volume). GC-MS analyses have certain limitations. GC-MS methods may not always successfully differentiate similar light petroleum distillates. High-resolution GC does not have the resolving power to detect minor component differences, and the high-molecular-weight biomarkers and other compounds with unique fragmentation signatures may be absent in these products. When numerous organic compounds are not fully resolved by the GC, the resulting full-scan mass spectra are the sum of the spectra of all coeluting compounds. This makes identification of potentially new marker compounds by library spectral matching difficult.

5.2 Comprehensive Two-Dimensional Gas Chromatography (GC × GC)

Another approach to a multidimensional separation, called comprehensive two-dimensional gas chromatography (GC × GC), has the potential to revolutionize forensic oil spill analysis. GC × GC is capable of separating an order of magnitude more compounds from complex mixtures than traditional gas chromatography (Bertsch, 2000; Phillips and Beens, 1999; Liu and Lee, 2000). GC × GC has produced chromatograms with thousands of resolved peaks from samples where traditional single-column chromatography resolved fewer than 100 peaks (Venkatramani and Phillips, 1993; Gaines et al., 1999; Frysinger and Gaines, 2001). Increased chromatographic resolution is achieved by using two chromatographic columns with different selectivity coupled together by a modulator. The modulator periodically samples the first column eluent and injects it into the second column. Figure 5-2 illustrates the concepts of comprehensive two-dimensional gas chromatography. The peak in Figure 5-2A represents a typical first-dimension chromatogram peak having a width of about 25 s. In Figure 5-2B, the modulator periodically samples the analyte mass of the first-dimension peak, compresses it, and injects it as a narrow peak into the second column. The vertical dashed lines in the figure indicate the start of each modulation cycle. Figure 5-2C shows how a fast, second-column separation can resolve coeluents present in the first dimension peak. Note that the second-dimension separation is complete before the next modulator injection. In Figure 5-2D, the detector data stream is sliced into segments and packed into a two-dimensional array. Figure 5-2E shows a contour plot created from the two-dimensional array. Peaks are positioned on a two-dimensional retention time

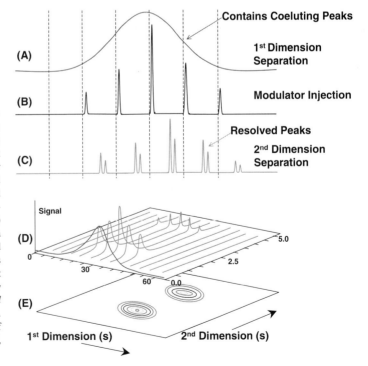

Figure 5-2 Concepts of comprehensive two-dimensional gas chromatography (GC × GC): (A) first-dimension chromatogram peak containing coeluents; (B) modulator samples first-dimension peaks and injects narrow peak into second dimension; (C) coeluents are separated in the second dimension; (D) serial flame ionization detector data stream is sliced into segments and stacked into an array; (E) array is visualized with a contour. Copyright 2002 from *GC × GC — A New Analytical Tool for Environmental Forensics*, by G.S. Frysinger et al. Reproduced by permission of Taylor & Francis Group, LLC, http://www.taylorand francis.com.

plane according to their retention time along each individual retention time axis.

GC × GC is considered *comprehensive* because all of the analyte mass from the first column is transferred to the second column. Thus, all compounds are subjected to two different separation mechanisms. GC × GC is different than two-dimensional "heart-cut" gas chromatography where one zone of analytes eluting from the first column is isolated and subsequently separated in a different column (Bertsch, 1999).

5.2.1 Modulation Techniques

The most critical component of GC × GC is the modulator. The modulator must accomplish three fundamental steps: (1) efficient trapping of the first-column eluent in a short region of the modulator capillary column; (2) rapid release and injection of the trapped compounds onto the head of the second column; and (3) rapid return to trapping conditions. Early commercial modulators employed a section of capillary containing a thick stationary phase to trap first-column eluent, and a rotating slotted heater that when swept across the trapping region, compressed the band, desorbed the trapped analytes, and injected them into the second column (Phillips et al., 1992). Many applications of GC × GC were accomplished using this modulator, including the first application of GC × GC to forensic oil spill identification (Gaines et al., 1999). Modern commercially available modulators such as the KT2004 (Zoex Corporation) employ cryogenic trapping of first-column eluent onto the modulator capillary column by use of a jet of liquid-nitrogen-cooled gas (Gorecki et al., 2004). The trapped analytes are then released into the second column by rapid heating from a dedicated hot jet. Two strategies are employed to prevent breakthrough by first-column eluents while the trap is cooling down. The first strategy employs two pairs of jets that generate two trapping locations. A special sequence of jet operation is used that alternates between the trapping locations to prevent breakthrough. A second strategy uses a loop in the modulator capillary column so that the column falls under the jets twice, thus producing two trapping locations from one cold jet, and avoids the necessity for a special sequence of jet operation. Even though only one set of jets is operating, the loop acts to delay the passage of the eluent released from the upstream trap until the downstream trap is at trapping temperature. When the trapping and releasing temperatures are properly tuned to the analytes, the modulator produces very narrow peaks on the order of 50 milliseconds for a very wide range of compounds, including those normally used to fingerprint crude oil and refined petroleum products (Gaines and Frysinger, 2004). Narrow peaks facilitate fast gas chromatographic separation on the second column necessary for the high peak capacities required to separate complex mixtures such as petroleum. Narrow peaks from spatial band compression during modulation increase signal-to-noise ratio, thus improving the detection and quantification of minor compounds (Liu and Phillips, 1994; Phillips et al., 1999; Lee et al., 2001).

Other GC × GC modulation techniques have been designed and successfully employed, some of which are in various stages of commercialization. For example, GC × GC systems have successfully used moving liquid CO_2 cryogenic modulators (Marriott and Kinghorn, 1999; Beens et al., 2001a), pulsed liquid CO_2 jet modulators (Thermo Electron Corporation) (Beens et al., 2001b), mechanical valve modulators (Bruckner et al., 1998; Seeley et al., 2000; Sinha et al., 2003; Micyus et al., 2005), and stop-flow modulators (Gorecki et al., 2004).

5.2.2 Detectors

GC × GC detectors need to have a fast response. Modulators produce very narrow injections into the second column where fast gas chromatography separates the first column coeluents. Because the chromatography on the second column is also fast, the resulting compound peaks presented to the detector remain very narrow, typically with widths at the base of 80–400 milliseconds depending on the second column retention. To properly sample the narrowest peaks requires a data rate of at least 100 Hz, but a slower data rate of 50 Hz can be used for the wider peaks. Flame ionization detectors are most widely used, but element-specific detectors such as nitrogen chemiluminescence (Wang et al., 2004), sulfur chemiluminescence (Blomberg et al., 2004), and atomic emission (van Stee et al., 2003) show promise to analyze petroleum with possible applications to oil spill fingerprinting.

When a GC × GC is coupled to a mass spectrometer, an additional dimension of information about the sample is available and can be used to identify components separated from the complex mixture. The mass spectrometer can also be used as an additional separation mechanism resulting in a three-dimensional separation. Quadrupole mass spectrometers operating in full scan mode are too slow to properly sample a GC × GC peak unless that peak is broadened (Frysinger and Gaines, 1999). Fast time-of-flight mass spectrometers (TOFMS) that operate with spectral acquisition rates of 100–200 Hz are well-suited for GC × GC and have been used for numerous studies (Dallüge et al., 2003). A GC × GC-TOFMS system is commercially available (LECO Corporation).

5.2.3 Data Processing

GC × GC presents information technology challenges in data handling, visualization, processing, analysis, and reporting due to the quantity and complexity of GC × GC data (Reichenbach et al., 2004). During the early stages of GC × GC development, analysis of GC × GC chromatograms was accomplished by visual means, and peaks were integrated manually. Commercially available software such as Transform (Fortner Software LLC) and Matlab (The Math Works, Inc.) produces interpolated color contour plots from the two-dimensional retention time array. Data are interpolated between contours according to a predefined color palette. The detailed visual nature of the GC × GC plots facilitates effective comparisons between chromatograms. Target peaks are integrated by summing the array data that comprise that peak. This manual integration was made easier due to the improved separation by GC × GC of a target peak from the rest of the sample. This manual approach, while especially useful for visual comparison-based applications such as petroleum fingerprinting, is extremely cumbersome and time-consuming.

Fortunately, new software tools have become available to more easily extract useful information from GC × GC data. GC Image (GCImage LLC) is a software system developed at the University of Nebraska-Lincoln (Reichenbach et al., 2004) that uses advanced information technologies to process and visualize GC × GC data, detect peaks, compare chromatograms, and perform peak deconvolution, pattern recognition and other data mining tasks. ChromaTOF (Leco Corporation) is another software program designed to control Leco's commercially available GC × GC-TOFMS system that has similar functions. Both GCImage and ChromaTOF make effective use of the tremendous amount of data generated when a time-of-flight mass spectrometer is used as a GC × GC detector, including spectral library matching and extracted ion chromatograms. HyperChrom (Thermo Electron Corporation) is a third software program designed to control Thermo Electron's commercially available GC × GC system with flame ionization detection and employs various data processing and visualization capabilities.

GC × GC and GC × GC-TOFMS data structure is well-suited for multivariate quantitative analysis, deconvolution, and pattern matching. New advanced numerical approaches to data mining include chemometric signal deconvolution (Sinha et al., 2004), reduced peak pattern variations that improved template matching (Ni et al., 2005), and enhanced limits of detection using bilinear chemometric analysis (Fraga et al., 2000). Advanced statistical methods will have direct impact on applications such as petroleum fingerprinting by improving current fingerprinting methods and identifying new target compounds useful to fingerprint samples that are not amenable to current methods because they do not contain the appropriate target compounds. For example, analysis of variance (ANOVA)-based feature selection was effectively used to determine chromatogram features important to the pattern recognition of jet fuels (Johnson and Synovec, 2002).

5.2.4 GC × GC Chromatogram

Figure 5-3 shows a chromatographic separation of *Exxon Valdez* cargo oil. Figure 5-3A is a one-dimensional separation using a nonpolar 100% polydimethylsiloxane stationary phase. The carbon range shown is from about C_8 to C_{47} [in part obscured due to (C) inset]. The presence of a hump of unresolved compounds means few compounds have been separated from the mixture. Figure 5-3B is a GC × GC two-dimensional separation visualized as an interpolated color contour. The first dimension separation along the x-axis is on a nonpolar 100% polydimethylsiloxane stationary phase and under the same chromatographic conditions as Figure 5-3A. The second dimension separation along the y-axis is on a polar 50% phenyl polysilphenylene stationary phase. The polarity separation in the second dimension resolved many compounds that coeluted after the first dimension. Compounds that are fully resolved from the mixture are seen as individual spots or peaks on the two-dimensional retention time plane. The back-

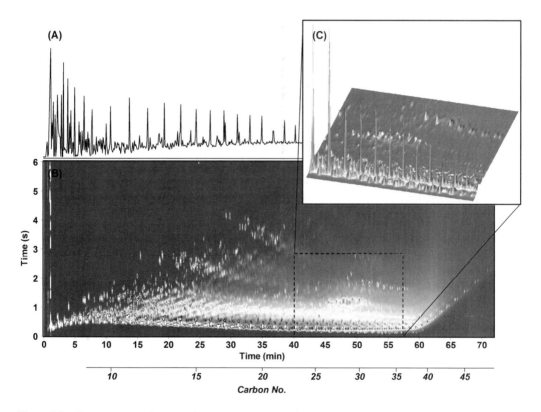

Figure 5-3 Chromatograms of *Exxon Valdez* cargo oil: (A) One-dimensional gas chromatogram using a nonpolar 100% polydimethylsiloxane stationary phase; (B) GC × GC volatility-by-polarity interpolated color contour plot. The first dimension is a separation using a nonpolar 100% polydimethylsiloxane stationary phase and the second dimension is a separation using a polar 50% phenyl polysilphenylene stationary phase; (C) a small portion of the GC × GC chromatogram visualized as a mountain plot. The mountain plot is excellent for visualizing relative differences among neighboring peaks. See color plate.

ground signal is light blue. Peak abundance is colored from white (low) to red (medium) to dark blue (high). To view the low abundance peaks, the highest peaks have been chopped off and this appears as a white portion in the center of the peak.

The crude oil chromatogram has hundreds of resolved peaks, but given the tremendous chemical complexity of the sample, two dimensions of separation are not sufficient to resolve all compounds. This is seen by the many peaks at the bottom of Figure 5-3B that are not fully resolved. It is also possible that some of the individual peaks in the upper portion of the chromatogram have coeluents remaining after the two-dimensional separation as well. Figure 5-3C is a small portion of the GC × GC chromatogram in Figure 5-3B visualized as a mountain plot. Mountain range plots use color and peak height to visualize signal size. Mountain range plots are especially good at revealing small peaks among neighboring larger peaks in GC × GC chromatograms of samples with a wide range of constituent concentrations.

5.2.5 Peak Identity and Chromatogram Structure

Several methods are used to identify the resolved peaks in the GC × GC chromatogram. First, a peak can be identified with a chemical standard. In GC × GC, a match in both the first- and second-dimension retention times provides compound identity with a greater degree of certainty than one-dimensional gas chromatography using a single retention time. A second method to identify peaks in a GC × GC chromatogram is by direct comparison between GC-MS and GC × GC data. A direct comparison is possible if the chromatography columns, conditions, and first-dimension retention times are matched (Frysinger and Gaines, 2001). Figure 5-4 shows a portion of a volatility-by-polarity GC × GC chromatogram of an *Exxon Valdez* cargo oil sample, along with GC-MS extracted ion chromatograms for ions that are diagnostic for sterane, hopanes, and triaromatic sterane biomarker groups. There is excellent retention time and abundance correlation between the GC × GC and GC-MS extracted ion chromatogram peaks. The excellent correlation allows the chemical identities of the extracted ion chromatogram peaks to be transferred to the GC × GC chromatogram band. A third method to identify peaks in the GC × GC chromatogram is to use a mass spectrometer interfaced directly with the GC × GC instrument. Since GC × GC greatly improves the separation of peaks in complex mixtures, a pure compound peak is presented to the mass spectrometer. For example, five GC × GC-MS spectra with corresponding chemical structures are shown in Figures 5-5D to 5-5 H. The mass spectrum of each compound is very clean and is not affected by other compounds that would otherwise coelute and interfere in a GC-MS analysis. The GC × GC-MS spectrum for naphthalene in Figure 5-5D does not have the extraneous ions that degrade the GC-MS spectrum in Figure 5-5A. The GC × GC-MS spectrum for 1,2,3-trimethylbenzene in Figure 5-5E shows how GC × GC-MS can isolate and measure the spectrum for just one of the components contributing to the complex GC-MS spectrum in Figure 5-5B. GC × GC-MS analysis also improves the detection of low-abundance molecular ion peaks. The m/z 156 molecular ion peak for 4-methyldecane and the m/z 140 molecular ion for n-butylcyclohexane are observed in Figure 5-5F and Figure 5-5G, respectively. In a normal GC-MS analysis, these small molecular ion peaks would most likely be lost in the baseline noise. The advantage of GC × GC-MS is that the resulting pure-component spectra can be accurately interpreted and compared to standard mass spectral libraries to determine the chemical identity. Hence, GC × GC-MS analysis can simultaneously provide both the selectivity provided by GC-MS selected ion monitoring and the compound identification provided by a GC-MS full-scan analysis.

A fourth method of peak identification makes use of the ordered chromatograms found in volatility-by-polarity GC × GC separations. This order provides valuable insights

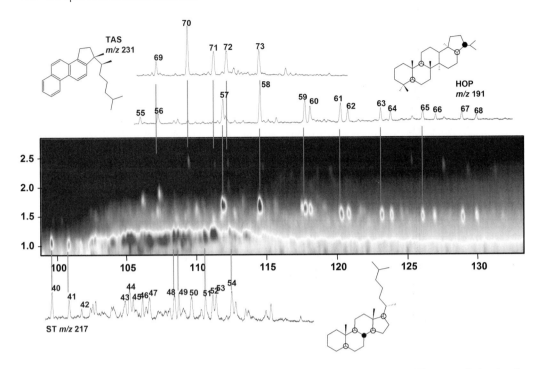

Figure 5-4 A small portion of a volatility-by-polarity GC × GC chromatogram of *Exxon Valdez* cargo oil showing the region containing the triaromatic sterane, hopanes, and sterane biomarkers. GC-MS extracted ion chromatograms (EICs) of these biomarker target ions are overlaid for identification and comparison. Peaks numbered on each EIC are identified in Frysinger and Gaines (2001). The structures of 5α,14α,17α(H)-cholestane (20R) (peak 46), 17α,21β(H)-hopane (peak 58) and cholestane (20S) (peak 69) are included for reference. x-axis: minutes; y-axis: seconds. Reproduced with permission from Frysinger and Gaines (2001). Copyright 2001 Wiley-VCH.

into the properties of unknown species and allows for easy identification of homologous series of compounds within complex mixtures. In Figure 5-5, the first dimension volatility-based separation sorts the complex mixture by boiling point. Since the first dimension is temperature programmed, the homologous series of *n*-alkanes is regularly spaced across the bottom of the two-dimensional retention time plane. The second-dimension separation sorts the complex mixture by functional group from least polar at the bottom to most polar at the top. At the bottom of the GC × GC chromatogram are the nonpolar *n*-alkanes and branched alkanes. Immediately above the alkanes are the one-ring and two-ring cycloalkanes. Above the cycloalkanes are the one-ring and two-ring aromatics. The location of a specific compound in the two-dimensional volatility-by-polarity plane depends primarily on the number of total carbons and the extent to which the chemical structure contributes to molecular polarity.

An important consequence of the first- and second-dimension ordering is that structural isomers are grouped into linear bands. For example, one prominent band in Figure 5-5 is the three-carbon substituted alkylbenzene isomers, which is identified as Figure 5-5C by comparison with the *m/z* 120 extracted ion chromatogram. The structural isomers cover a range of volatility because the boiling points of the isopropyl- and propylbenzene are lower than the ethylmethyl- and trimethylbenzene isomers, but they all have very similar polarity due to the influence of the benzene ring. The last two peaks in the band are indan [peak (I)] and indene [peak (J)], which do not

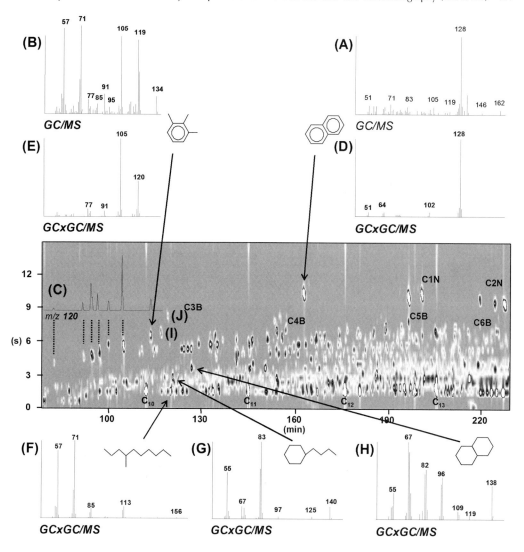

Figure 5-5 Partial GC × GC-MS chromatogram of diesel fuel. The *x*-axis is a volatility-based chromatographic separation on a 100% polydimethylsiloxane stationary phase. The *n*-alkanes decane (C_{10}) to tridecane (C_{13}) are identified. The *y*-axis produces polarity-based chromatographic separation on a 14% cyanopropylphenyl methylsiloxane stationary phase. The MS total ion intensity data are displayed as an interpolated color contour. Overlay data: (A–B) GC-MS spectra for specific retention times; (C) GC-MS *m/z* 120 extracted ion chromatogram; (D–H) GC × GC-MS spectra and structures of naphthalene, 1,2,3-trimethylbenzene, 4-methyldecane, *n*-butylcyclohexane, and decahydronaphthalene, respectively; (I) indan; (J) indene. Copyright 2002 from *GC × GC — A New Analytical Tool for Environmental Forensics*, by G.S. Frysinger et al. Reproduced by permission of Taylor & Francis Group, LLC, http://www.taylorandfrancis.com.

have a corresponding peak in the *m/z* 120 extracted ion chromatogram. Indan and indene peaks would be found in the *m/z* 118 and *m/z* 116 extracted ion chromatograms, respectively. Indan has a cyclopentane ring attached to the benzene, and indene has a cyclopentene ring. In GC × GC, it is logical to extend the family of three-carbon substituted alkylbenzene isomers to include indan and indene because even though their molar masses are different, they both have three carbons attached to the benzene ring in some manner. Most importantly, their volatility and polarity properties place them at the end of that band.

The two-dimensional order also produces other bands in the GC × GC chromatogram. For example, the four-carbon (C4B), five-carbon (C5B), and six-carbon substituted alkylbenzenes (C6B) each form bands to the right of the three-carbon substituted alkylbenzenes (C3B; Figure 5-5). At greater second-dimension retention, the one-carbon naphthalene (C1N) and part of the two-carbon naphthalene (C2N) bands are ordered to the right of the naphthalene peak. The retention time separation between successive bands is about the same as the separation between members of the n-alkane homologous series.

The use of ordered bands to help identify peaks can be facilitated by optimizing the two-dimensional chromatography. In Figure 5-5, the nonpolar alkanes found at the bottom of the volatility-by-polarity GC × GC chromatogram are not very well separated into isomer bands because their polarity difference is very small so their retention time on the second column is the same. Better separation can be achieved using a different second column stationary phase. For example, Figure 5-6(A) shows a portion of a volatility-by-polarity GC × GC separation of the saturates fraction of a sediment sample containing biodegraded petroleum (Frysinger et al., 2003). The line of peaks across the bottom of the chromatogram represents the branched alkanes. The isoprenoid biomarkers norpristane, pristane, and phytane are labeled. The peaks above the branched alkanes are the cycloalkanes, with the position of n-octylcyclohexane shown. Other peaks at about the same second-dimension retention are alkylcyclohexane or alkylcyclopentane isomers. Peaks at greater second-dimension retention are multi-ring cycloalkanes.

The branched alkanes and cycloalkanes in Figure 5-6(A) are not ordered into distinct sloping bands with carbon number spacing. This suggests that the selectivity of the GC × GC separation is not properly tuned to the dimensionality of the mixture. In addition, there is a lot of unused peak capacity (open space) in the two-dimensional retention time plane. A "shape" selective column was better able to separate the branched alkanes and cycloalkanes from one another. Figure 5-6B shows the resulting volatility-by-shape GC × GC chromatogram that contains many separated peaks organized into numerous groups that are spread across the entire second dimension of the GC × GC chromatogram. The branched alkanes, poorly resolved in the GC × GC separation shown in Figure 5-6A, are now separated and organized into numerous diagonal bands (Figure 5-6B). The different bands arise from different alkane substitution patterns. For example, all monomethyl substituted alkanes form one band, and all dimethyl alkanes may form the adjacent band. Circles on the chromatogram indicate the two-dimensional retention time positions of the n-alkanes (n-C_{13} through n-C_{20}). Due to microbial degradation of the sample, only trace peaks remain at each n-alkane position. The number of resolved cycloalkane compounds is also significantly increased in this sample. The cycloalkane compounds are now grouped into bands and spread across the full two-dimensional retention time plane of the gas chromatogram. A small region in Figure 5-6B is expanded in Figure 5-6C to show branched alkane band detail.

Figure 5-7 is a volatility-by-polarity GC × GC chromatogram of a crude oil showing the location of various classes of petroleum constituents that are important to fingerprinting petroleum products. The crude oil spans the carbon number range from about C_8 to C_{47}. A volatility-by-polarity separation of crude oil produces chemical class separation of n-alkanes, branched alkanes, cycloalkanes, and one- and multiring aromatics. Individual chemical classes exhibit banding similar to that shown in Inset A of Figure 5-7. Isomers of alkyl-substituted naphthalenes are grouped into bands of the same total carbons. Bands increase in total carbons going from left to right. Biphenyl is located near the C2-naphthalene band because it has 12 total carbons. Because biphenyl has a slightly different polarity than C2-naphthalene, it is

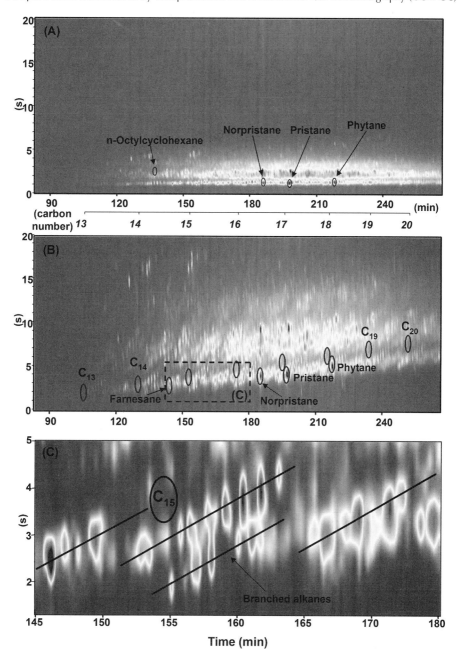

Figure 5-6 Effect of changing the second-dimension stationary phase on the separation of nonpolar saturates: (A) GC × GC chromatogram of the saturates fraction of a sediment extract using a polar second-dimension 14% cyanopropylphenyl polysiloxane stationary phase; (B) GC × GC chromatogram of the same saturates fraction using a chiral γ-cyclodextrin second-dimension stationary phase. The expected positions of the n-C_{13} to n-C_{20} alkanes are marked with circles. The region inside the dotted lines is expanded in (C); (C) expanded region showing expected position of n-C_{15} and location of branched alkane bands. Reprinted with permission from Frysinger et al., 2003. Copyright 2003, American Chemical Society. See color plate.

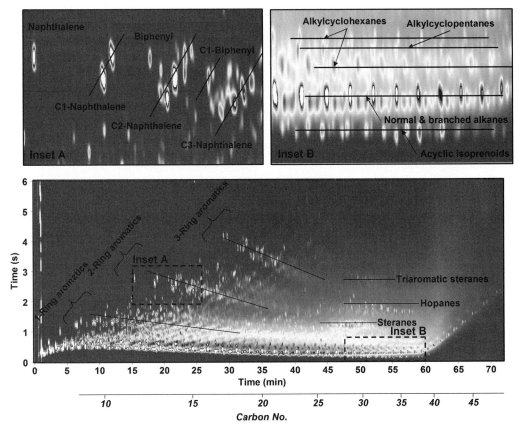

Figure 5-7 Location of petroleum chemical classes on a volatility-by-polarity GC × GC chromatogram. Nonpolar alkanes are located at the bottom, with one- and multiring aromatics further up the *y*-axis as their polarity increases. Inset A: naphthalene region showing bands of alkyl-substituted naphthalenes and biphenyls; Inset B: nonpolar region showing high-resolution separations of higher boiling (C_{28}+) acyclic isoprenoids, normal and branched alkanes, and cyclic alkanes. See color plate.

separated from the C2-naphthalene band. C1-biphenyl has a slightly different polarity than C3-naphthalene, so it is also separated from the C3-naphthalene band. The distance between biphenyl and C1-biphenyl is the same as the distance between the C1- and C2-naphthalene bands. Inset B of Figure 5-7 shows the detailed separation of the higher boiling (C_{28}+) acyclic isoprenoids, normal and branched chain alkanes, and cyclic alkanes achievable with GC × GC. The high-resolution separation of nonpolar and higher boiling compounds makes it possible to investigate new classes of weathering resistant petroleum marker compounds potentially useful for fingerprinting.

5.2.6 GC × GC Petroleum Applications

From the early stages of GC × GC development, petroleum analysis and characterization were logical areas for application. Petroleum samples contain thousands of compounds that are a real challenge to separate by typical one-dimensional chromatographic methods. GC × GC provides the requisite separation power to separate major and minor individual compounds as well as major and minor compound chemical groups (e.g., PIONA: paraffins, isoparaffins, olefins, naphthenes aromatics) from the complex petroleum matrix. For example, group-type analysis has been done for heavy naphtha (Vendeuvre et al., 2005b), naphtha, and other petrochemicals (Vendeuvre et al., 2004; Prazen et al., 2001; Schoenmakers et al., 2000; Blomberg et al., 1997), diesel fuel (Vendeuvre et al., 2005a; Gaines et al., 1999), kerosene (van Deursen et al., 2000), and gasoline (Frysinger and Gaines, 1999). Target compound separation and analysis has been done for biomarkers in crude oil (Frysinger and Gaines, 2001), hydrocarbons (Gaines et al., 1999), nitrogen compounds (Wang et al., 2004), and sulfur compounds (Blomberg et al., 2004) all in diesel fuel, and arson accelerant marker compounds in petrochemicals (Frysinger and Gaines, 2002). This partial list of GC × GC petroleum applications illustrates the enormous potential for GC × GC as a forensic oil spill technique. GC × GC provides a very high-resolution separation with a data output amenable to visual and quantitative comparison between samples using both individual compounds and classes of compounds. Furthermore, GC × GC separations elucidate chemical structure, and therefore new potential target compounds, not usually seen from traditional gas chromatographic separations of petroleum samples.

5.3 Applications of GC × GC to Fingerprint Oil Spills

5.3.1 Mobile Bay Marine Diesel Fuel Spill

The first application of GC × GC to oil spill fingerprinting involved a spill of approximately 100 gallons of marine diesel fuel into a bay near Mobile, Alabama, USA (Gaines et al., 1999). In this example, GC × GC was used to determine qualitative and quantitative similarities and differences between the spill and suspected source samples. The high resolving power of GC × GC separated many hundreds of compounds from the petroleum matrix, making it possible to select individual compounds or groups of compounds for analysis. For this oil spill analysis, several traditional marker classes were used including alkanes, alkylbenzenes, alkylnaphthalenes, and phenanthrenes. One nontraditional class of petroleum constituents found useful was the alkylcyclohexanes. The overall results compared favorably with standard GC fingerprinting methods.

A slightly weathered spill sample (designated Spill) was taken from the surface of the water approximately 24 hr after the spill. Two potential source samples (designated Source 1 and Source 2) were collected from the fuel tanks of two nearby suspect fishing vessels.

Figure 5-8 shows a portion of the volatility-by-polarity chromatogram from each of the three samples analyzed by GC × GC. The spill sample is shown in panel A, one potential source, Source 1, is shown in panel B and one potential source, Source 2, is shown in panel C. The n-alkane range is approximately tridecane (C_{13}) through eicosane (C_{20}). This region contained features that were suitable for fingerprinting of middle distillate fuels. Earlier eluting compounds were not suitable for fingerprinting because of losses due to weathering by evaporation and dissolution while the spilled oil was on the water.

Each sample was compared to the spill qualitatively by visual comparison and quantitatively by integration of selected peaks and bands of peaks. Visual similarities and differences between the spill and source chromatograms were made by comparing the presence or absence of peaks at a particular location on the retention time plane and their relative size. Each source sample exhibits a high degree of similarity with the spill sample. The alkane peaks, found in a line along the bottom of the chromatograms, are very similar, as is the distribution of alkyl-substituted benzene, naphthalene, and phenanthrene compounds found in the aromatic region beginning just above the alkanes (see Figure 5-7 for additional peak locations). The numerous similarities are due to the facts that all three samples are marine diesel fuel and that it is likely the suspect sources fueled from the same fuel source.

Although both suspect source samples are very similar to the spill sample, there are

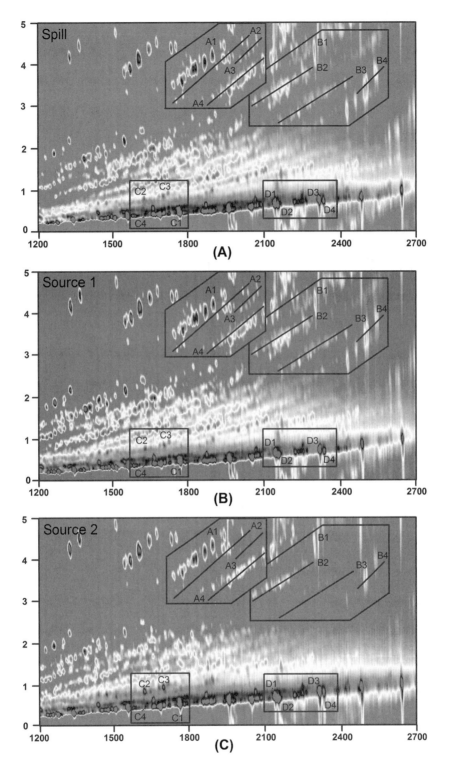

Figure 5-8 Portion of a volatility-by-polarity GC × GC-FID chromatogram from each of the three samples compared: (a) spill; (b) and (c) potential spill sources. Both axes are in seconds. The boxes contain chemical families and individual compounds used to qualitatively and quantitatively compare two potential source samples with the spill samples. The first column separation shown along the x-axis was accomplished with a nonpolar 5% phenylmethylsiloxane stationary phase. The second column separation shown along the y-axis was accomplished with a polar polyethylene glycol/siloxane copolymer stationary phase. Reprinted with permission from Gaines et al., 1999. Copyright 1999, American Chemical Society. See color plate.

numerous small differences that are easily visible in the chromatograms. For example, Source 2 exhibits noticeably fewer peaks in the heavy aromatic region (upper right corner of the chromatograms) than the Spill. Several of these minor chromatogram differences were selected for quantitation. Each chromatogram in Figure 5-8 shows four windows that identify the regions used to quantify differences. Within each window, selected peaks and bands of peaks were integrated and normalized to a specific peak located inside the same window.

Window A contains naphthalene compounds known to be prevalent in petroleum products, are resistant to evaporative weathering, and are useful for fingerprinting. Band A4 contains four carbon substituted naphthalenes (C4N), while bands A2 and A3 contain unidentified compounds of similar volatility and polarity to the alkylnaphthalenes. These bands were normalized to A1, which is 2,3,5-trimethylnaphthalene. The integration results are shown in Figure 5-9(A). The data labeled "quality control" (QC) is a repetitive analysis of the spill sample. The uncertainty in the integration result is 10% (Gaines, 1998) and shown as an error bar on the graph. The bar graph indicates that Source 1 and Spill are similar, while Source 2 and the Spill are different.

Window B in Figure 5-8 contains naphthalene and phenanthrene compounds. The B2 band contains five-carbon-substituted naphthalenes (C5N) and the B3 band contains six-carbon-substituted naphthalenes (C6N). The B4 band contains one-carbon-substituted

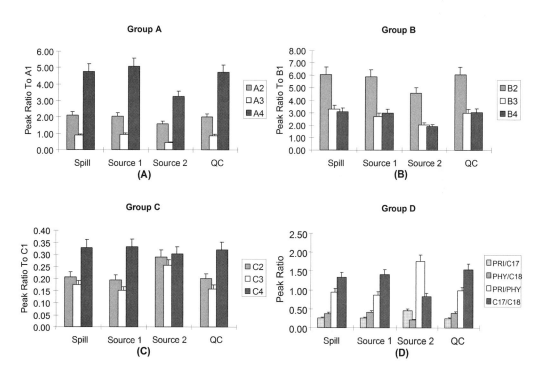

Figure 5-9 Quantitative comparisons between spill sample and two potential source samples from Figure 5-8. A quality-control (QC) sample, split from the spill, was also quantified. Each panel represents the results of integrating peaks within an individual box shown in Figure 5-8. Each bar represents the total peak or band of peaks integration results normalized to the first peak in the box. The error bar represents an estimated 10% variation calculated from replicate data, but is shown only in the plus direction. Compound classes represented: (a) alkylnaphthalene bands normalized to 2,3,5-trimethylnaphthalene; (b) alkylphenanthrenes normalized to phenanthrene; (c) cycloalkanes normalized to pentadecane; (d) heptadecane, octadecane, pristine, and phytane normalized to each other. Reprinted with permission from Gaines et al., 1999. Copyright 1999, American Chemical Society.

phenanthrenes. The bands are integrated and normalized to B1, identified as phenanthrene, and shown in Figure 5-9B. The bar graph indicates that Source 1 and Spill are similar, while Source 2 and the Spill are different.

Window C in Figure 5-8 contains alkane and cycloalkane compounds. The latter of these groups are rarely used in traditional fingerprinting methods, but the differences observed in the GC × GC separation suggest they may be useful in discriminating between similar samples. The peak C1 is n-pentadecane (n-C_{15}), C2 and C3 are tentatively identified by mass spectral library search as two isomers of pentmethyldecahydronaphthalene, and C4 is n-octylcyclohexane. The integration results of peaks C2-C4 are normalized to C1 and shown in Figure 5-9(C).

Window D in Figure 5-8 contains two widely used marker compounds. D2 is pristine and D4 is phytane. D1 is n-heptadecane (n-C_{17}) and D3 is n-octadecane (n-C_{18}). Scientists use four different ratio combinations to discriminate between samples, and these are shown in Figure 5-9(D). The bar graph indicates that Source 1 and Spill are similar, while Source 2 and the Spill are different.

On the basis of both qualitative and quantitative results (Figures 5-8 and 5-9, respectively), Source 1 was determined to be a probable match with the Spill. These results are consistent with the conclusions of the USCG MSL who employed high-resolution gas chromatography (ASTM, 2000a) and GC-MS (ASTM 2000b) standard methods of analyses to determine the source of the spill.

5.3.2 West Falmouth No. 2 Fuel Oil Spill

In this application, GC × GC identified compositional changes in the unresolved complex mixture (UCM) of a No. 2 fuel oil found in sediments some 30 years after being spilled (Reddy et al., 2002). Many chemical classes within the UCM were found to be remarkably persistent including branched alkanes, acyclic isoprenoids, cyclic alkanes, and polynuclear aromatic hydrocarbons (PAHs).

On September 16, 1969, the barge *Florida* went aground near West Falmouth, Mass., and spilled between 650,000 and 700,000 liters of No. 2 fuel oil into Buzzards Bay (Figure 5-10). Strong southwesterly winds mixed the oil into the water column and drove it toward Wild Harbor, located about 1 km north of the spill. Despite the use of oil booms, both subtidal and intertidal areas of Wild Harbor were heavily contaminated with oil. Oil entered the tidal Wild Harbor River, deposited in quiet marsh areas, and sorbed to sediments and grasses at the edge of the river.

The close proximity of Wild Harbor to the Woods Hole Oceanographic Institution made the contaminated area very convenient for the study of long-term fate and effects of petroleum hydrocarbons in the environment, and numerous studies were conducted (Reddy et al., 2002 and refs. therein). One site heavily studied was site M-1, shown in Figure 5-10. Sediment samples analyzed by conventional GC periodically after the spill event showed first a loss of oil compounds in the n-C_{10} to n-C_{13} alkane range due to evaporation and/or water-washing, and then progressive loss of the chromatographically resolved n-alkane peaks by preferential microbial degradation. By 1973, only a baseline hump comprised of an unresolved petroleum compound remained (Reddy et al., 2002).

Marsh sediments collected at the M-1 site in 1989, 20 years after the spill, confirmed the presence of oil residues in Wild Harbor sediments (Teal et al., 1992). In the Teal study, conventional gas chromatograms showed a UCM of petroleum compounds similar to those observed in 1973. Gas chromatography-mass spectrometry analysis showed elevated levels of PAHs in the sediments. The persistence of oil at this site after 20 years was attributed to the heavy contamination of the area by the spill, the high organic carbon content and anoxic conditions in the marsh sediments that hindered microbial degradation, and the presence of a low-energy environment that reduced flushing and water-washing. In particular, the anoxic conditions in the sediment apparently permitted little or no anaerobic degradation of the petroleum.

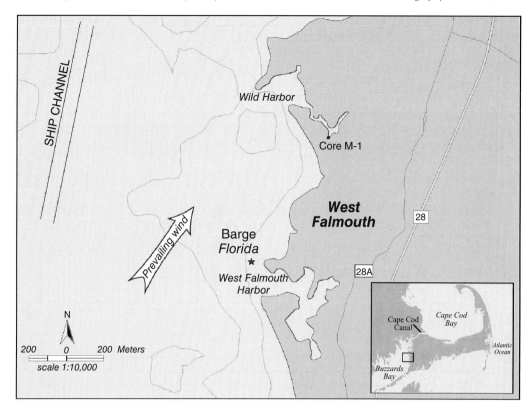

Figure 5-10 Map of the general area near the 1969 grounding of the barge *Florida*.

In 2000, conventional GC of the sediment extract shows little change in the petroleum composition over nearly 30 years since the spill. Figure 5-11 shows a comparison between conventional GC-FID chromatograms of the 1973 and 2000 sediment extracts. The two chromatograms were aligned according to *n*-alkane retention index and exhibit a UCM hump characteristic of degraded petroleum. Unfortunately, these one-dimensional chromatograms provide little information about the chemical composition of the UCM. GC × GC is better suited for analyzing the UCM because of increased resolution and the grouping of compounds by chemical class into bands.

Figure 5-12 shows the one-dimensional gas chromatogram and volatility-by-polarity GC × GC chromatogram of the sediment extract taken at the site in 2000. The GC × GC separation was able to resolve hundreds of

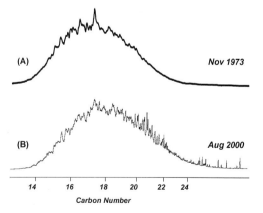

Figure 5-11 Gas chromatograms of Wild Harbor marsh sediment extracts: (A) Nov. 1973; (B) Aug. 2000. Reprinted with permission from Reddy et al., 2002. Copyright 2002, American Chemical Society.

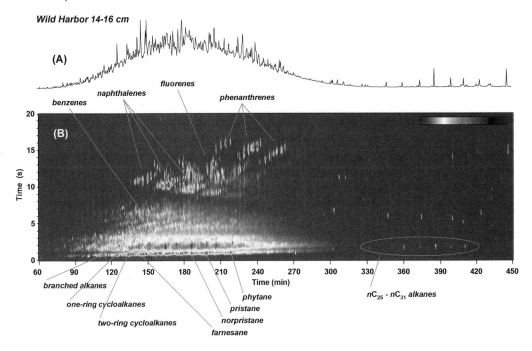

Figure 5-12 Chromatograms of a Wild Harbor sediment extract, Aug. 2000: (A) Conventional one-dimensional gas chromatogram. The separation was accomplished using a nonpolar polydimethylsiloxane stationary phase; (B) volatility-by-polarity GC × GC chromatogram. The first column separation along the *x*-axis was accomplished using a nonpolar polydimethylsiloxane stationary phase. The second column separation was accomplished using a polar 14%-cyanopropylphenyl polysiloxane stationary phase. Reprinted with permission from Reddy et al., 2002. Copyright 2002, American Chemical Society. See color plate.

compounds from the UCM, and analysis of the GC × GC chromatogram reveals several interesting features. The *n*-alkane band normally found in GC × GC chromatograms of unweathered No. 2 fuel oil is no longer present. However, at long retention times, the biogenic (derived from plant wax) $n\text{-}C_{26}$–$n\text{-}C_{33}$ alkanes are present. Despite the loss of the *n*-alkanes, nearly all other compound classes appear to persist in the sediment extract. For example, numerous branched alkanes, including acyclic isoprenoids farnesane, norpristane, pristine, and phytane, are evident in the GC × GC chromatogram. This finding is contrary to studies by Burns and Teal suggesting that branched alkanes at this site were completely degraded within the first seven years after the spill (Burns and Teal, 1979) because they could no longer chromatographically resolve phytane from the UCM background. GC × GC resolved phytane from the background and showed the phytane-to-background ratio was unchanged from the mid-1970s. GC × GC also showed that the UCM contains potentially valuable forensic information conventional GC cannot distinguish from the background.

Because of the susceptibility of the *n*-alkanes to microbial degradation, Volkman et al. (1984) developed a nine-point scale that spans from one (no biodegradation) to nine (extreme biodegradation) that was later applied to refined petroleum products (Burns et al., 2000). (Also see Chapter 11 herein.) The West Falmouth sediments would be ranked between three and four because *n*-alkanes have been removed but alkylcyclohexanes and alkylbenzenes are still present, the acyclic isoprenoids have been reduced but not removed, and the PAHs have been reduced. The high resolving power of GC × GC enabled the rapid determination of microbial degradation of the sample extracts.

GC × GC compound class separations can be very useful for understanding weathering patterns of petroleum and quantifying the influence of evaporation, water-washing, and other processes on residual oil composition. For example, in Figure 5-13 an enlarged portion of the volatility-by-polarity GC × GC chromatogram of the West Falmouth sediment extract is shown along with GC × GC chromatograms of neat Marine Ecosystems Research Laboratory (MERL) No. 2 fuel oil, and MERL oil that has been evaporatively weathered to lose 30% and 70% by mass. When compared to the neat MERL oil, the sediment extract chromatogram has some obvious differences that are due to differences in chemical composition as well as visual scaling. For example, the MERL oil has abundant n-alkanes, so the rest of the image is scaled to the n-alkanes and the rest of the peaks appear small. The sediment extract has no large n-alkane peaks, so smaller peaks are a greater percentage of the mixture, and appear larger than in the MERL oil chromatogram. Therefore, peaks from compounds such as phenanthrenes are more prominent in the sediment extract chromatogram than in the MERL oil chromatogram.

Comparisons between the sediment extract and MERL oil chromatograms in Figure 5-13 show that the n-alkanes and more volatile water soluble (polar) compounds have been preferentially lost, but the mechanism of loss is not easily explained by simple evaporative weathering, water-washing, or biodegradation. For example, the 30% evaporatively weathered MERL oil shows a loss of hydrocarbons less than that of n-C_{10} that is similar to that shown in the sediment extract. However, the 30% weathered MERL oil contains abundant C_1-naphthalene (C1N) and C_2-naphthalene (C2N) compounds that are noticeably absent in the sediment extract chromatogram. Not until the MERL oil is evaporatively weathered to 70% by mass are the C_1- and C_2-naphthalene compounds removed. At the same time, that degree of weathering removed all other hydrocarbons more volatile than n-C_{14} that is not observed in the sediment extract. Thus, it is clear that simple evaporative weathering cannot solely describe the GC × GC pattern observed in the lower-molecular-weight components of the spilled oil found in the Wild Harbor sediment. One contributing mechanism appears to be a preferential loss of the more polar and water-soluble alkylnaphthalenes without the concurrent loss of nonpolar alkanes in the sediment extract. Such a loss may be due to the combined affects of water-washing or preferential biodegradation of more polar compounds.

5.3.3 Winsor Cove No. 2 Fuel Oil Spill

In this application, GC × GC with mass spectrometric detection (GC × GC-TOFMS) found that alkyl decalins (decahydronaphthalenes) are a class of petroleum compounds that are surprisingly persistent in the environment after a spill.

On October 9, 1974, the fuel barge *Bouchard 65* spilled an undetermined amount of No. 2 fuel oil off the west entrance of the Cape Cod Canal. The spilled oil contaminated the Winsor Cove salt marsh located within the Buzzards Bay estuary in Bourne, Mass. (Figure 5-14). To assess the long-term effects of the spilled oil, a core sample of sediment from Winsor Cove was taken in 2001. The sample contained extractable and measurable oil in the upper 4 cm of the sediment.

Analysis of the sediment extract by conventional GC shows a UCM characteristic of a degraded petroleum fuel, shown in Figure 5-15A. When the same sediment extract sample was analyzed by volatility-by-polarity GC × GC-FID, as shown in Figure 5-15B, the absence of normal alkanes and the significant reduction in aromatic hydrocarbons suggest the primary mechanisms of weathering were biodegradation and water-washing (Reddy et al., 2002). The presence of several large peaks in the cyclic alkane regions of the GC × GC chromatogram was noted, and a follow-up study using GC × GC with time-of-flight mass spectrometric detection (GC × GC-TOFMS)

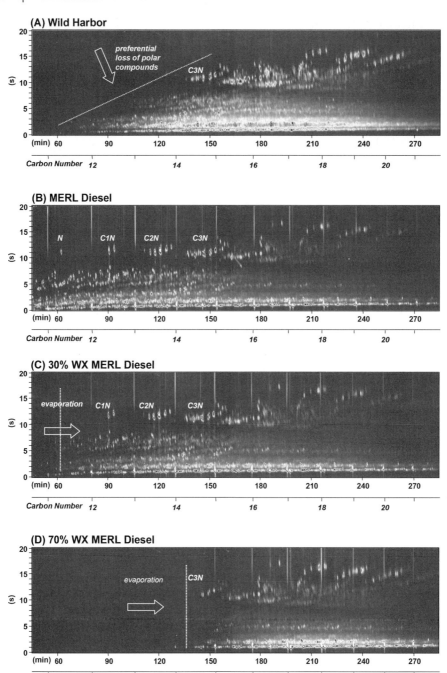

Figure 5-13 Comparison of Wild Harbor sediment extract with weathered MERL oil: (A) Enlarged volatility-by-polarity GC × GC chromatogram of Wild Harbor extract, Aug. 2000 (as in Figure 5-12); (B) neat MERL oil; (C) 30% by mass weathered (WX) MERL oil; (D) 70% by mass weathered (WX) MERL oil. The first column separation along the x-axis was accomplished using a nonpolar polydimethylsiloxane stationary phase. The second column separation was accomplished using a polar 14%-cyanopropylphenyl polysiloxane stationary phase. Weathering experiments were performed in a hood where neat MERL oil was allowed to evaporate until the desired mass loss was achieved. The background is blue. Peak intensity is scaled from white, to red, and then to blue (most intense). Compound abbreviations: (N) naphthalene; (C1N) C_1-naphthalenes; (C2N) C_2-naphthalenes; (C3N) C_3-naphthalenes. Reprinted with permission from Reddy et al., 2002. Copyright 2002, American Chemical Society. See color plate.

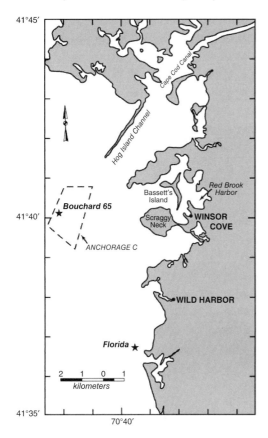

Figure 5-14 Map of the general area near the grounding of the barge *Bouchard 65*. The grounding of the barge *Florida*, discussed previously, is also shown for reference.

determined these compounds to be from the bicyclic sesquiterpanes class of petroleum constituents, most likely pentamethyl- or hexamethyl-substituted decahydronaphthalenes, or hexamethyloctahydroindenes. A comparison of peaks in Figure 5-15(A) with the drimane-based sesquiterpanes identified by Stout et al. (2005) shows a high degree of similarity. Figure 5-16 shows some examples of these compounds, also known as the C_5- and C_6-decalins (Wang et al., 2005). Bicyclic sesquiterpanes such as the C_4- and C_6-decalins have boiling points within the diesel range and are common in oils (Stout et al., 2005; Wang et al., 2005).

The discovery of an archived sample of neat *Bouchard 65* No. 2 fuel oil provided additional opportunities to study the long-term fate of recalcitrant compounds like the C_4- to C_6-decalins. Neat *Bouchard 65* fuel oil was incubated in an aerobic, nutrient-rich environment in the laboratory, and the oil composition was directly compared to the oil found in the sediments at Winsor Cove after 30 years of *in situ* degradation. To illustrate compositional changes in the *Bouchard 65* incubations over time, representative GC-FID chromatograms of the laboratory-degraded oil from three time points are shown in Figure 5-17. Prevalent *n*-alkane and branched alkane peaks apparent in day 1 [Figure 5-17(A)] are significantly smaller by day 11 [Figure 5-17(B)] and nearly indiscernible from the UCM by day 46 [Figure 5-17(C)] of the experiment.

To investigate the effect of weathering and biodegradation on the C_4- to C_6-decalins, volatility-by-polarity GC × GC-TOFMS was used. Figure 5-18 shows GC × GC *m/z* 123 extracted ion chromatograms (EICs) in the region of the target compounds from incubation time 0–46 days and the Winsor Cove sediment extract taken approximately 30 years after the spill. Ion *m/z* 123 is known to be common in the fragmentation pattern of various C_4- to C_6-decalin isomers. The pattern of alkyl decalin peaks in each EIC in Figure 5-18 is similar to that found by Stout et al. (2005) and Wang et al. (2005) using GC-MS, except GC × GC-TOFMS has the advantage of tiling each homologous series of alkyl decalins (C_4-, C_5-, and C_6-decalins) into separate bands. To illustrate this, in Figure 5-18(F), the C_4- to C_6-decalin bands are labeled along with the position of two representative compounds whose structures are shown in Figure 5-16. Peak A is 8β(H)-drimane, a C_5-decalin (C5D). Peak B is 8β(H)-homodrimane, a C_6-decalin (C6D).

Looking at Figures 5-18(A)–(D), there is little change in the distribution and sizes of peaks in the target region as the degradation progresses to 46 days incubation. Figure 5-18(E) shows a day-46 sample spiked with a C_5-decalin standard containing both cis- and trans-1,1,4,4,6-pentamethyldecalin. The standard confirmed the location of the target compounds by producing peaks within the target

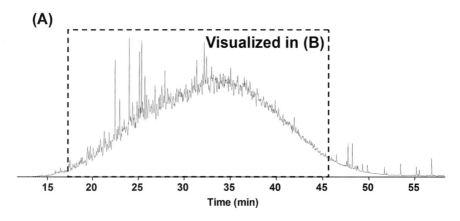

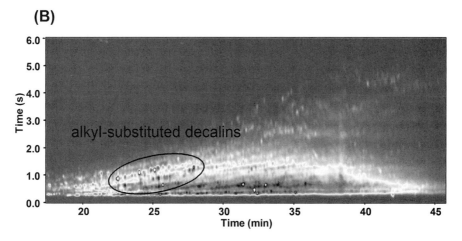

Figure 5-15 Chromatograms of the sediment extract taken from Winsor Cove in 2002 approximately 30 years after being contaminated with No. 2 fuel oil: (A) one-dimensional gas chromatogram showing UCM and the region of the chromatogram visualized in (B). The separation was accomplished using a polydimethylsiloxane stationary phase; (B) volatility-by-polarity GC × GC-FID chromatogram showing the region containing the target compounds. The first-column separation shown along the x-axis was accomplished with a polydimethylsiloxane stationary phase. The second-column separation shown along the y-axis was accomplished with a 50%-phenylpolysilphenylene-siloxane stationary phase. See color plate.

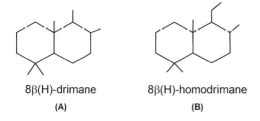

Figure 5-16 Possible structure of C_5- and C_6-decalins found in degraded No. 2 fuel oil in Winsor Cove approximately 30 years after a spill: (A) 8β(H)-drimane; (B) 8β(H)-homodrimane.

compound band in the GC × GC chromatogram. The very large standards peaks made the other peaks seem smaller so they appear slightly different from the same peaks in the other chromatograms. Figure 5-18(F) is the Winsor Cove sediment extract GC × GC chromatogram. The distribution and sizes of peaks in the Winsor Cove GC × GC chromatogram are very similar to that of the day-46 chromatogram, suggesting little if any change in the amounts of these target compounds during weathering or biodegradation conditions encountered by these samples over time.

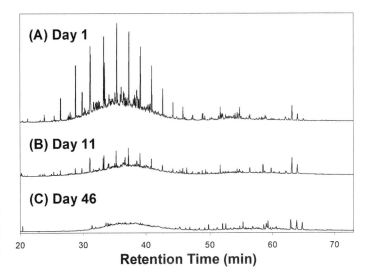

Figure 5-17 Gas chromatograms of *Bouchard 65* incubation sediment extracts artificially degraded in the laboratory: (A) day 1; (B) day 11; (C) day 46.

The recalcitrance of alkyl decalins despite a loss of equally and less volatile compounds in the laboratory degradation experiment as well as that observed in long-term field studies leads to the question of why the alkyl decalins are especially persistent. Structural complexity may limit the availability of these compounds to microorganisms, as the methyl groups surrounding the decalin backbone appear to serve as protection from microbial degradation. The high-resolution separation of these compounds from the UCM by GC × GC and GC × GC-TOFMS may provide additional information relevant to the distribution of these compounds and thus their potential usefulness in fingerprinting methods. For example, in Figure 5-18, the GC × GC m/z 123 ion chromatograms show the separation of alkyl decalin compounds that would have been unresolvable in a GC-MS analysis.

5.3.4 Buzzards Bay No. 6 (Bunker C) Spill

In this application, GC × GC was able to track the compositional changes due to weathering processes of a spilled No. 6 fuel oil over a 6-month period. Along with standard GC × GC analysis, unique data visualization techniques elucidated compositional changes due to evaporation, water-washing, and biodegradation (Nelson et al., 2006).

On April 25, 2003, the petroleum cargo barge *Bouchard 120* struck an underwater ledge and released ~375,000 liters of petroleum into Buzzards Bay, Mass., USA (Figure 5-19). Within 24 hours, helicopter surveys documented a 20-km slick that eventually impacted ~150 km of shoreline along the west, north, and northeast shorelines of Buzzards Bay.

The cargo aboard the *Bouchard 120* was No. 6 fuel oil, also known as Bunker C. This type of oil is a residual fuel that is prepared from the remaining hydrocarbons after the lighter constituents of crude oil have been removed at the refinery (Stout et al., 2002). (See also Chapter 10 herein.) The *Bouchard 120* cargo oil contained compounds with an n-alkane range from n-C_{10} to greater than n-C_{45}; however, most of the GC-detectable mass resides in an elution window between n-C_{14} and n-C_{35}. The *Bouchard 120* cargo also contained abundant alkylated naphthalenes and phenanthrenes that were present in this fuel oil as a result of the addition of a cutting agent used to aid in the transport and delivery of the viscous product.

Figures 5-20(A) and 5-20(C) show traditional gas chromatograms of extracts from

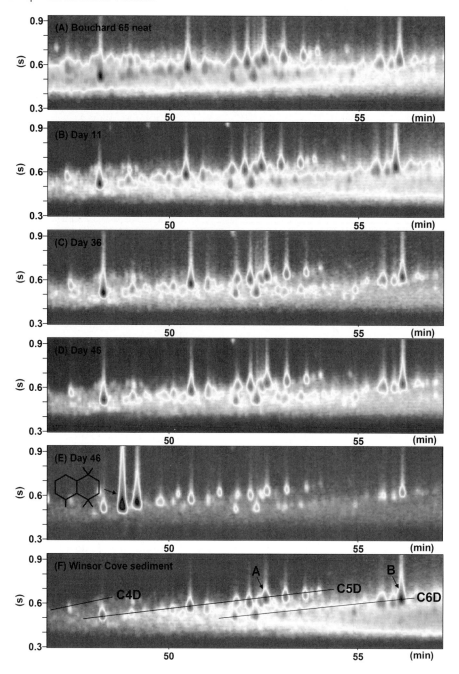

Figure 5-18 Volatility-by-polarity GC × GC $m/z = 123$ extracted ion chromatographs of *Bouchard 65* incubation sediment extracts showing the C_4- to C_6-decalin region: (A) neat *Bouchard 65* fuel oil; (B) day 11; (C) day 36; (D) day 46; (E) day 46 spiked with cis- and trans-1,1,4,4,6-pentamethyldecalin standard; (F) Winsor Cove sediment extract showing the bands of peaks containing C_4-decalins (C4D), C_5-decalins (C5D), C_6-decalins (C6D), and location of 8β(H)-drimane (A) and 8β(H)-homodrimane (B) whose structures are given in Figure 5-16. The first-column separation shown along the *x*-axis was accomplished with a nonpolar polydimethylsiloxane stationary phase. The second-column separation shown along the *y*-axis was accomplished with a polar 50%-phenylpolysilphenylene-polysiloxane stationary phase. See color plate.

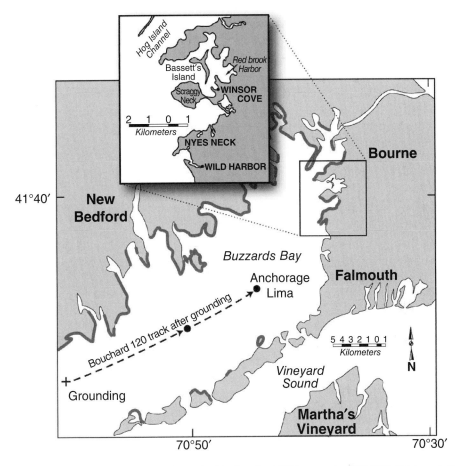

Figure 5-19 Map of Buzzards Bay showing the track of the *Bouchard 120* barge on April 27, 2003. Oil-impacted beaches are highlighted in gray around the perimeter of the bay. Nyes Neck (where the oil spill samples discussed in this section were collected) is shown on the inset map. Copyright 2006 from *Tracking the Weathering of an Oil Spill with Comprehensive Two-Dimensional Gas Chromatography*, by Nelson et al. Reproduced by permission of Taylor & Francis Group, LLC., http://www.taylorandfrancis.com.

oil-covered rocks collected at Nyes Neck on May 9, 2003, and on November 23, 2003, 12 days and 179 days, respectively, after the initial release of No. 6 fuel oil into Buzzards Bay. These dates span the temporal range of samples collected at Nyes Neck. The chromatograms reveal a complex mixture of hydrocarbons that elutes mainly between n-C_{15} and n-C_{37}. Significant differences in these chromatograms are apparent. For example, the n-alkane peaks, which were prominent in the day-12 chromatogram, are not discernable in the day-179 chromatogram. These changes are likely due to the combined weathering effects of biodegradation, water-washing, and evaporation. The end result of nearly six months of weathering is a mixture of petroleum hydrocarbons that is in the form of a broad UCM hump in the conventional gas chromatogram (Figure 5-20C).

The tremendous resolving power of GC × GC and ordered separation of peaks in the GC × GC chromatogram helps identify more detailed compositional changes in the No. 6 fuel oil that occurred during weathering that cannot be seen using traditional gas

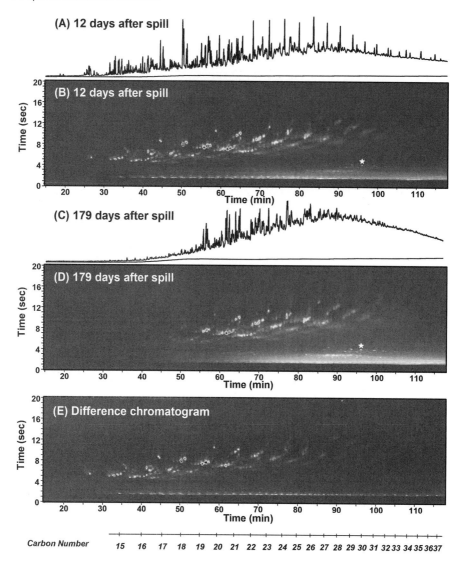

Figure 5-20 Conventional GC and GC × GC chromatograms of sample extracts from *Bouchard 120* No. 6 fuel oil-covered rocks collected at various times after the spill: (A) conventional GC chromatogram of sample taken 12 days after the spill. The separation was achieved using a polydimethylsiloxane stationary phase; (B) GC × GC volatility-by-polarity chromatogram of sample taken 12 days after the spill. The first dimension separation along the *x*-axis was achieved using a nonpolar polydimethylsiloxane stationary phase. The second dimension separation along the *y*-axis was achieved with a polar 50% phenylpolysilphenylenesiloxane stationary phase; (C) conventional GC chromatogram of sample taken 179 days after the spill; (D) GC × GC volatility-by-polarity chromatogram of sample taken 179 days after the spill; (E) a point-by-point difference chromatogram produced by the subtraction of chromatogram (D) from chromatogram (B) after normalization to the conserved standard 17α(H)-21β(H)-hopane (marked with a star). Copyright 2006 from *Tracking the Weathering of an Oil Spill with Comprehensive Two-Dimensional Gas Chromatography*, by Nelson et al. Reproduced by permission of Taylor & Francis Group, LLC., http://www.taylorandfrancis.com.

chromatograms. To illustrate, GC × GC chromatograms for day 12 and day 179 are visualized as interpolated color contour plots in Figures 20(B) and 20(D) respectively. Each chromatogram is normalized and scaled to the conserved biomarker 17α(H)-21β(H)-hopane (Prince et al., 1994) indicated on the plots with a star. Extreme differences in the hydrocarbon composition between day 12 and day 179 are readily apparent throughout the chromatogram. It is possible to identify the chemical classes affected by weathering because petroleum compound classes occupy specific regions of the chromatogram (see Figs. 5-5 and 5-7). For example, most of the alkylated naphthalene and phenanthrene compound peaks eluting before n-C_{18} were lost between day 12 and day 179.

More advanced image processing tools improve data analysis by extracting trends from complex data. In the spill case described here, plots using chromatogram difference, ratio, and addition quantitatively compared point-by-point the difference between two samples that span almost six months of degradation by various weathering processes (Nelson et al., 2006). The differences found, in turn, provide valuable information about the effect of the various weathering processes on the numerous chemical classes found in the petroleum samples. For example, Figure 5-20(E) shows the resulting chromatogram that is generated when the heavily weathered day-179 chromatogram is subtracted from the slightly weathered day-12 chromatogram. The difference chromatograms provide a way to identify compound peaks that have changed in volume relative to 17α(H)-21β(H)-hopane over this six-month weathering time frame. Peaks appearing in the GC × GC difference chromatogram (Fig. 5-20E) are those that lost peak volume through weathering, with prominent peaks indicating a large loss and small peaks indicating a small loss. Over the six-month weathering period, the n-alkanes and the alkylated naphthalenes showed a significant loss of peak volume, while the branched alkanes, located in between the n-alkanes, showed a small loss. The sterane and hopane peaks showed little or no loss as evidenced by a lack of peaks in this region of the difference chromatogram. See Figure 5-7 for the location of these petroleum compound classes in the chromatogram.

To produce good results, peaks for compounds in both chromatograms must have the same first- and second-dimension retention times. Run-to-run retention time reproducibility is generally found acceptable for visual comparisons. Rigorous peak matching needed for more robust quantitative and chemometric comparisons can be facilitated by use of special algorithms (Johnson et al., 2003; Ni et al., 2005). (See also Chapter 9 herein.)

GC × GC chromatogram comparisons such as that employed above identify target chemical classes that are both affected and unaffected during weathering processes. The improved separation by GC × GC of target chemical classes from one another allows the use of a flame ionization detector to quantify the loss of target classes during weathering. Figure 5-21 is a comparison of the relative percent loss for a homologous series of n-alkanes, n-alkylcyclohexanes, and some isoprenoid branched alkanes. The data are presented by retention index because compounds that have the similar retention indices are assumed to have similar tendencies to evaporate. Since these saturates have very low water-solubilities (Eastcott et al., 1988), they all should be affected by water-washing in a comparable manner. The n-alkanes have a near constant loss of 90–100% for retention indices of 1500–2000 and then an ordered dropoff to only 20% loss at retention index of 3600. The n-alkylcyclohexanes only have 30–60% loss in the retention indices of 1500–2000 and a nearly constant loss of 15% from 2100–2600. The differences in weathering loss between these compounds with similar tendencies to evaporate and dissolve may provide a means to discriminate weathering patterns. In this case, these results show how susceptible n-alkanes are to biodegradation relative to n-alkylcyclohexanes once the evaporation component is removed.

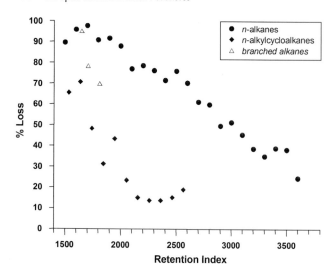

Figure 5-21 Percent losses of select hydrocarbons from *Bouchard 120* No. 6 fuel oil covered rocks collected at Nyes Neck after six months of weathering: (●) *n*-alkanes; (◆) *n*-alkylcyclohexanes; (△) branched alkanes. Copyright 2006 from *Tracking the Weathering of an Oil Spill with Comprehensive Two-Dimensional Gas Chromatography*, by Nelson et al. Reproduced by permission of Taylor & Francis Group, LLC., http://www.taylorandfrancis.com.

5.3.5 Oil Seeps, Santa Barbara, CA, USA

In this final application, GC × GC identifies compositional changes between a reservoir sample of crude oil taken at a Santa Barbara, CA, USA, oil platform and a sample of oil that seeped from a nearby underwater fault zone. The Santa Barbara well sample exhibits unique isoprenoid alkane, hopane, and sterane biomarker chemical signatures that distinguish these oils from other crude oils from around the world.

The offshore Santa Barbara region is known for its very active oil seeps (Hornafius et al., 1999). Carbon-rich marine Miocene sediments provide the source material for the generation of oil in this area (Hornafius et al., 1999; Reed and Kaplan, 1977). Uranium-Thorium (U-Th) measurements of calcite cements around areas where hydrocarbons are believed to have seeped to the surface in the past suggest that hydrocarbons have been moving along faults around the rim of the basin for the past 120,000 and perhaps up to 500,000 years before the present (Boles et al., 2004). Widespread seepage of hydrocarbons occurs along offshore fault zones in the waters of the Santa Barbara basin. Estimates of the amount of hydrocarbons migrating into the environment by way of the natural seeps around Coal Oil Point put the combined gas–liquid hydrocarbon seepage flux at approximately 37 tons per day, with methane emissions comprising 65% of this total (24 tons per day) (Quigley et al., 1996, 1999; Washburn et al., 1996). The volume of gas emanating from the seeps has been fluctuating over time. The fluctuations are most likely linked to daily tidal cycles, oil production at the nearby Holly platform, and seismic activity along the faults in Santa Barbara county (Boles et al., 2001; Leifer et al., 2004).

Modern scientific investigations of the Santa Barbara oil/methane seeps began in earnest during the middle of the 20th century. Some of the earliest GC separations of hydrocarbons upwelling in the Santa Barbara channel date back to 1973 (Mikolaj, 1973). The resolving power of GC coupled to mass spectrometers advanced the study of oil from these seep areas (Kaplan and Reed, 1977; Stuermer et al., 1982). Some of the many chemical structures of compounds extracted from these seep oils were assigned using this approach.

More recent efforts have focused on analyzing seep fluids for the presence and ratios of biomarkers to determine the possible reservoir sources and transport directions of the fluids. Targeted compounds or their ratios typically include trisnorhopane, sterane/hopane ratios, refractory index, C_{28}/C_{29} hopane ratios (Kvenvolden, 2004). Other researchers study-

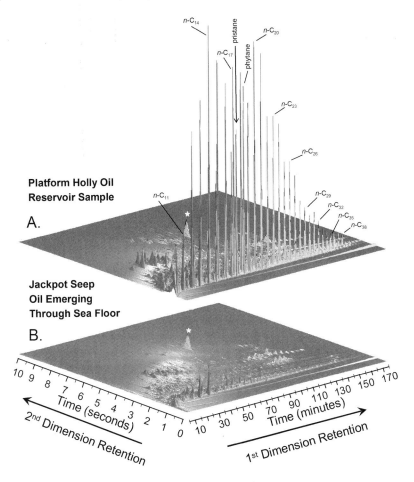

Figure 5-22 Volatility-by-polarity GC × GC chromatogram of (A) reservoir crude oil from platform Holly, Santa Barbara Channel, California, and (B) oil stringer emerging through the sea floor collected at the Jackpot seep field, Santa Barbara Channel, California. The first-column separation shown along the x-axis was accomplished with a nonpolar polydimethylsiloxane stationary phase. The second-column separation shown along the y-axis was accomplished with a polar 50%-phenylpolysilphenylene-siloxane stationary phase. A star on each chromatogram indicates the position of the internal standard dodecahydrotriphenylene (DDTP). The distinct difference in n-alkane content between the two samples is evident in this figure. See color plate.

ing the toxicological effects of seep fluids on marine organisms have measured PAHs (Davis et al., 1981; Spies et al., 1980; Stuermer et al., 1981; Seruto et al., 2005). Geochemical tracing of tarballs from these seeps have been investigated as well (Hostettler et al., 2004). All this work has been limited to traditional GC, which is a technology that is at least two decades old and has provided a limited number of analytes.

To compare the differences in the reservoir composition of crude oil from platform Holly to what is naturally seeping out along coastal Santa Barbara, a series of samples was collected and analyzed by GC × GC. Figure 5-22 shows two volatility-by-polarity GC × GC chromatograms visualized as mountain plots. In (A), the reservoir sample of oil from the platform Holly exhibits the typical inventory of alkane and aromatic constituents. In (B), the sample was from an oil stringer emerging through the sea floor at Jackpot Seep, in close proximity to platform Holly. The star in each chromatogram indicates the position of the internal standard dodecahydrotriphenylene (DDTP) that was added to each sample. Both of the chromatograms were normalized to the peak volume of 17α(H), 21β(H)-hopane, a known conserved biomarker used in monitoring oil weathering processes when crude oil is released into the environment (Prince et al., 1994).

These images reveal the substantial scope and scale of the changes that occur in crude oils due to processes that act on a localized petroleum system, provided that the source oil for the seep and Holly are genetically related. The major differences between the reservoir and seafloor samples appear to occur in the relative abundances of the naphthalene peaks, the branched and isoprenoid alkane peaks, and most noticeably in the normal alkane peaks.

Another feature of Holly oil, like all oils from Monterey shale and carbonate source rocks, is the unique biomarker chemical signatures that distinguish these oils from other types of crude oils worldwide. For example, they may contain several biomarkers derived from Archaea (Orphan et al., 2001). Since their discovery and subsequent classification as a third domain of life on earth (Brock and Freeze, 1969; Brock et al., 1972; Woese and Fox, 1977), members of the domain Archaea are often thought to be nature's "extremists" as they exist in hot springs, deep subsurface sediments, strongly acidic and alkaline springs, volcanic vent systems, extremely saline evaporitic environments, and deep-sea hydrothermal vent systems. Members of the domain Archaea have also been shown to be abundant in picoplankton of open ocean environments as well (DeLong, 1992; DeLong et al., 1998, 1999). However, most Archaea are thermophilic, and many are extremely thermophilic with optimum growth temperatures between 80° and 115°C.

One of the defining characteristics of the Archaea and the source of their usefulness as biomarkers is the unique membrane lipids of this group. The membrane lipids of Archaea are very different than either the eukaryotes or the prokaryotic bacteria. Archaeal lipid membranes form a monolayer not a bilayer as in bacteria and eukaryotes (De Rosa et al., 1980; Kates, 1993; Koga et al., 1993). Archaeal lipid membranes are ether-linked (not ester-linked) to glycerol with C_{20} to C_{40} branched isoprenoid alkane moieties spanning the glycerol units at each end (De Rosa et al., 1980). In the most thermophilic species (the Chrenarchaea), the lipid membranes consist of C_{40} isoprenoid lipids that are thought to reduce the fluidity of the cell membranes at extreme temperatures (De Rosa et al., 1980). The presence of these unique lipid compounds in a geological sample is strong evidence of Archaean activity either in the deep sediments, in the oil reservoirs, or along the path that migrating oil takes as it makes its way through fractures and faults between the reservoirs and the sea floor. Figure 5-23 highlights the region of a GC × GC chromatogram of the Platform Holly produced crude oil where these Archaean biomarkers elute. Along with a suite of acyclic isoprenoid alkanes, trace amounts of two-ring and three-ring cyclic isoprenoid alkanes are present in this sample as well. The peak assignments were made by comparison with a sample rich in cyclic isoprenoid alkanes obtained from argillitic sediments collected in a gold mine in Timmons, Onatario, Canada (data not shown) (Nelson et al., 2005). Biphytane is the largest of the isoprenoid alkanes in this region of the chromatogram. One of the remarkable features of this part of the chromatogram is how well biphytane and the n-C_{35} alkane are resolved from one another.

One of the most striking features of the mid-Miocene Monterey crude oils found at platform Holly is the abundance and distribution of the C_{27}, C_{28}, and C_{29} desmethylsteranes as shown in Figure 5-24. The desmethylsteranes are the most abundant steranes in Monterey Formation-derived crude oils. This is true for many other types of oil as well. Since steranes are derived from sterol precursors, they represent the molecular remnants of eukaryotic life forms from ancient times. In particular, the desmethylsteranes are representative of microalgae and higher plants (Volkman, 2005), diatoms, green algae, red algae, macroalgae, fungi, and yeast (Brocks and Summons, 2004).

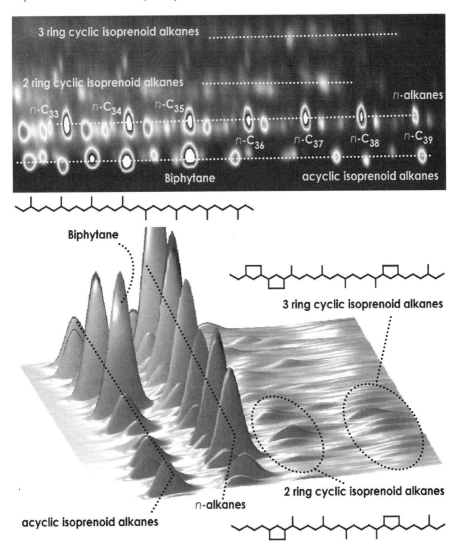

Figure 5-23 Partial volatility-by-polarity GC × GC chromatogram of the platform Holly produced crude oil showing peaks in the vicinity of n-C_{33} through n-C_{39}. The location of high-molecular-weight isoprenoid alkanes (both acyclic and cyclic), thought to be biomarkers of Archaean membrane lipids, are indicated along with some representative molecular structures. See color plate.

Since C_{27}, C_{28}, and C_{29} desmethylsteranes are derived from a number of marine and terrestrial sources, they are not as useful as depositional source indicators as other biomarkers, but they can provide information about the level of maturation of oil reservoirs (Brocks and Summons, 2004). The identities of the desmethylsterane compounds highlighted in Figure 5-24 are given in Table 5-1. For more discussion of biomarker nomenclature, see Chapter 3 herein.

The presence of hopanes in rock or oil samples is a clear indication of ancient sedimentary bacterial activity (Brocks and Summons, 2004). Monterey shale crude oils contain relatively high concentrations of 28,30-bisnorhopane, a hopanoid that is not ubiquitous in crude oils worldwide (Seifert

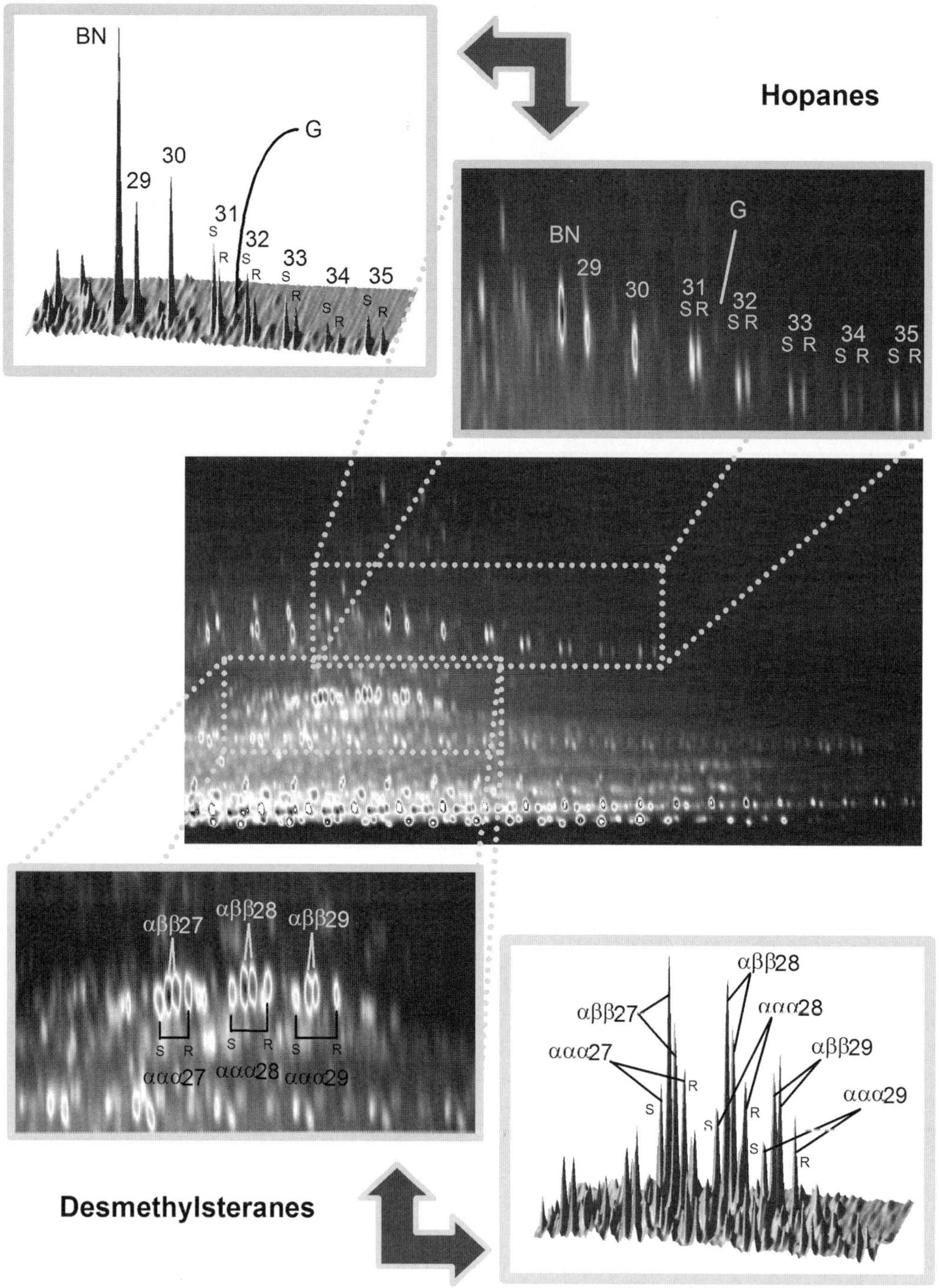

Figure 5-24 Partial volatility-by-polarity GC × GC chromatogram of the platform Holly reservoir crude oil showing peaks in the vicinity of the sterane and desmethylsterane biomarkers. Identification of labeled peaks is found in Table 5-1. See color plate.

Table 5-1 Peak identities of Compounds Labeled in Figure 5-24

Hopane Series Peak Identification	Compound
BN	17α(H),18α(H),21β(H)-28,30-bisnorhopane
29	17α(H),21β(H)-29-norhopane
30	17α(H),21β(H)-hopane
31	S & R epimers of 17α(H),21β(H)-29-homohopane
G	gammacerane
32	S & R epimers of 17α(H),21β(H)-29-bishomohopane
33	S & R epimers of 17α(H),21β(H)-29-trishomohopane
34	S & R epimers of 17α(H),21β(H)-29-tetrakishomohopane
35	S & R epimers of 17α(H),21β(H)-29-pentakishomohopane

Desmethylsterane Peak Identification	Compound
ααα27	5α(H),14α(H),17α(H)-cholestane
αββ27	5α(H),14β(H),17β(H)-cholestane
ααα28	5α(H),14α(H),17α(H)-ergostane
αββ28	5α(H),14β(H),17β(H)-ergostane
ααα29	5α(H),14α(H),17α(H)-stigmastane
αββ29	5α(H),14β(H),17β(H)-stigmastane

et al., 1978; Curiale et al., 1985). The designations 28 and 30 denote the locations in the 28-carbon bisnorhopane where methyl groups are absent when compared to the corresponding 30-carbon hopane molecule. The presence of high concentrations of 28,30-bisnorhopane in sediments that are marine in origin is indicative of a depositional environment that occurs in a well-stratified water column in which the bottom waters are saline in nature and hydrogen sulfide is present (Grantham et al., 1980; Schoell et al., 1992; Brocks and Summons, 2004). A similar depositional environment exists today in the stratified water column of the Black Sea.

Monterey shale oils typically contain 35-carbon homohopanes (17α(H),21β(H)-29-pentakishomohopane, 22S and 22R epimers) that are present in higher concentration than those of the neighboring 34-carbon homohopanes [17α(H),21β(H)-29-tetrakishomohopane 22S and 22R epimers].

Monterey shale derived oils contain gammacerane, a biomarker associated with hypersaline depositional environments (Moldowan et al., 1985; ten Haven et al., 1988). It is likely that gammacerane could not form from its precursor, tetrahymanol, in oxidizing conditions (ten Haven et al., 1989; Sinninghe Damste, 1995). Considered together, the biomarker distribution suggests that the Monterey Formation sediments were deposited in a euxinic marine environment (Peters et al., 2004a, 2004b; Hostettler et al., 2004), the sediments were deposited below a stratified water column, the bottom water was anoxic, saline, and perhaps even hypersaline, and finally the presence of hydrogen sulfide indicates a reducing sedimentary environment. This is a good example of the type of information that can be obtained by examining the distribution of biomarkers in a crude oil sample. The biomarkers present in a crude oil sample can provide a wealth of diagnostic information about the origin of oil.

Just as GC × GC can help resolve the UCM (Frysinger et al., 2003), GC × GC can also shed additional light on the very complicated sterane patterns obtained by conventional GC-MS. This is demonstrated in the lowest panels shown in Figure 5-24, in which the complex sterane pattern is pulled apart. The large number of small, unidentified peaks reveals the complexity in this suite of compounds, yet GC × GC holds promise to start to unravel this complex mixture.

5.4 Conclusion

There are several advantages that a GC × GC separation offers over a conventional GC separation for the analysis of complex mixtures such as petroleum products and spills. The high resolving power of GC × GC gives a more complete picture of the compounds present in petroleum, so it is easier to detect, identify, and quantify individual compounds and families of compounds present in the sample. Peak quantitation in GC × GC can be accomplished using a flame ionization detector. GC often requires a mass spectrometric detector operating as a separation mechanism to help with poorly resolved peaks by using extracted ion data, but this approach requires advanced knowledge of target compounds and compound classes. GC × GC enhances the power of a mass spectrometer detector by presenting to the detector a single-component peak so that the mass spectrum does not contain artifacts that can hinder interpretation and confuse library search algorithms. Each of these advantages is particularly relevant to improved methods of oil spill analysis and source identification. New target compounds or classes of compounds that are useful for fingerprinting can be discovered, identified, and quantified to a degree not attainable with conventional GC and GC-MS methods.

Acknowledgments

The authors acknowledge the following people who contributed to the work described in this chapter. From Woods Hole Oceanographic Institution: Lary Ball, Heather Bischel, George Hampson, Emily Peacock, Desiree Plata, Bruce Tripp, Helen White, and Li Xu; from the University of California at Santa Barbara: David Valentine and George Wardlaw; from the University of Illinois: Fabien Kenig and Todd Ventura. The following organizations provided funding for the work described in this chapter: U.S. Department of Energy, U.S. Environmental Protection Agency, National Science Foundation, U.S. Navy Office of Naval Research, Petroleum Research Fund, The Island Foundation, Robert T. Alexander Trust, and National Institute of Justice.

References

ASTM (American Society for Testing and Materials), *Standard Test Methods for Comparison of Waterborne Petroleum Oils by Gas Chromatography*, D-3328-00, 2000a, W. Conshohocken, PA.

ASTM (American Society for Testing and Materials), *Standard Practice for Oil Spill Identification by Gas Chromatography and Positive Ion Electron Impact Low Resolution Mass Spectrometry*, D-5739-00, 2000b, W. Conshohocken, PA.

Beens, J., J. Dallüge, M. Adahchour, R.J.J. Vreuls, and U.A.Th. Brinkman, Moving cryogenic modulator for the comprehensive two-dimensional gas chromatography (GC × GC) of surface water contaminants, *J. Microcolumn Separation*, 2001a, **13**, 134–140.

Beens, J., M. Adahchour, R.J.J. Vreuls, K. van Altena, and U.A.Th. Brinkman, Simple, nonmoving modulation interface for comprehensive two-dimensional gas chromatography, *J. Chromatography A*, 2001b, **919**, 127–132.

Bertsch, W., Two-dimensional gas chromatography. Concepts, instrumentation, and applications — Part 1: Fundamentals, conventional two-dimensional gas chromatography, selected applications, *J. High Resolution Chromatography*, 1999, **22**, 647–665.

Bertsch, W., Two-dimensional gas chromatography. Concepts, instrumentation, and applications — Part 2: Comprehensive two-dimensional gas chromatography, *J. High Resolution Chromatography*, 2000, **23**, 167–181.

Blomberg, J., P.J. Schoenmakers, J. Beens, and R. Tijssen, Comprehensive two-dimensional gas chromatography (GC × GC) and its applicability to the characterization of complex (petrochemical) mixtures, *J. High Resolution Chromatography*, 1997, **20**, 539–544.

Blomberg, J., T. Riemersma, M. van Zuijlen, and H. Chaabani, Comprehensive two-dimensional gas chromatography coupled with fast sulphur-chemiluminescence detection: implications of detector electronics, *J. Chromatography A*, 2004, **1050**, 77–84.

Boles, J.R., J.F. Clark, I. Leifer, and L. Washburn, Temporal variation in natural methane seep rate due to tides, coal oil point area, California, *J. Geochemical Res.*, 2001, **106**, 27077–27086.

Boles, J.R., P. Eichhubl, G. Garven, and J. Chen, Evolution of a hydrocarbon migration pathway along basin-bounding faults: Evidence from fault cement, *Amer. Assoc. Petroleum Geologists Bull.*, 2004, **88**, 947–970.

Brock, T.D. and H. Freeze, Thermus aquaticus gen. n. and sp. n., a non-sporulating extreme thermophile, *J. Bacteriology*, 1969, **98**, 289–297.

Brock, T.D., K.M. Brock, R.T. Belly, and R.L. Weiss, Solfolobus: A new genus of sulfur-oxidizing bacteria living at low pH and high temperature, *Archeological Microbiology*, 1972, **84**, 54–68.

Brocks, J.J. and R.E. Summons, Sedimentary hydrocarbons, biomarkers for early life, *Treatise on Geochemistry*, H.D. Holland and K.K. Turekian (eds.), 2004, **8**, 63–115.

Bruckner, C.A., B.J. Prazen, and R.E. Synovec, Comprehensive two-dimensional high-speed gas chromatography with chemometric analysis, *Analytical Chem.*, 1998, **70**, 2796–2804.

Burns, K.A. and J.M. Teal, The West Falmouth oil spill: Hydrocarbons in the salt marsh ecosystem, *Estuarine and Coastal Marine Science*, 1979, **8**, 349–360.

Burns, K.A., S. Codi, and N.C. Duke, Gladstone, Australia field studies: Weathering and degradation of hydrocarbons in oiled mangrove and salt marsh sediments with and without the application of an experimental bioremediation protocol, *Marine Pollution Bull.*, 2000, **41**, 392–402.

Curiale, J.A., D. Cameron, and D.V. Davis, Biological marker disribution and significance in oils and rocks of the Monterey formation, California, *Geochimica et Cosmochimica Acta*, 1985, **49**, 271–288.

Dalluge, J., J. Beens, and U.A.Th. Brinkman, Comprehensive two-dimensional gas chromatography: A powerful and versatile analytical tool, *J. Chromatography A*, 2003, **1000**, 69–108.

Davis, J.M., *Statistical Theories of Peak Overlap; Advances in Chromatography*, 1994, P.R. Brown and E. Grushka (eds.), Marcel Decker, New York, **34**, 109–176.

Davis, P.H., T.W. Schultz, and R.B. Spies, Toxicity of Santa Barbara seep oil to starfish embryos: Part 2. The growth bioassay, *Marine Environmental Research*, 1981, **5**, 287–294.

DeLong, E.F., Archaea in coastal marine environments, *Proc. National Acad. Science, USA*, 1992, **89**, 5685–5689.

DeLong, E.F., L.L. King, R. Massana, H. Cittone, A. Murray, C. Schleper, and S.G. Wakeham, Dibiphytanyl ether lipids in nonthermophilic crenarchaeotes, *Applied and Environmental Microbiology*, 1998, **64**, 1133–1138.

DeLong, E.F., L.T. Taylor, T.L. Marsh, and C.M. Preston, Visualization and enumeration of marine planktonic archaea and bacteria by using polyribonucleotide probes and fluorescent in situ hybridization, *Applied and Environmental Microbiology*, 1999, **65**, 5554–5563.

De Rosa, M., E. Esposito, A. Gambacorta, B. Nicolaus, and J.D. Bu'Lock, Effects of temperature on ether lipid composition of Caldariella acidophila, *Phytochemistry*, 1980, **19**, 827–831.

Eastcott, L., W.Y. Shiu, and D. Mackay, Environmentally relevant physical-chemical properties of hydrocarbons: A review of data and development of simple correlations, *Oil and Chemical Pollution*, 1988, **4**, 191–216.

Fraga, C.G., B.J. Prazen, and R.E. Synovec, Enhancing the limit of detection for comprehensive two-dimensional gas chromatography (GC × GC) using bilinear chemometric analysis, *J. High Resolution Chromatography*, 2000, **23**, 215–224.

Frysinger, G.S. and R.B. Gaines, Comprehensive two-dimensional gas chromatography with mass spectrometric detection (GC × GC-MS) applied to the analysis of petroleum, *J. High Resolution Chromatography*, 1999, **22**, 251–255.

Frysinger, G.S. and R.B. Gaines, Separation and identification of petroleum biomarkers by comprehensive two-dimensional gas chromatography, *J. Separation Science*, 2001, **24**, 87–96.

Frysinger, G.S. and R.B. Gaines, Forensic analysis of ignitable liquids in fire debris by comprehensive two-dimensional gas chromatography, *J. Forensic Science*, 2002, **47**, 471–482.

Frysinger, G.S., R.B. Gaines, and C.M. Reddy, GC × GC — A new tool for environmental forensics, *Environmental Forensics*, 2002, **3**, 27–34.

Frysinger, G.S., R.B. Gaines, L. Xu, and C.M. Reddy, Resolving the unresolved complex mixture in petroleum-contaminated sediments, *Environmental Science and Technology*, 2003, **37**, 1653–1662.

Gaines, R.B., G.S. Frysinger, M.S. Hendrick-Smith, and J.D. Stuart, Oil spill source identification by comprehensive two-dimensional gas chromatography, *Environmental Science and Technology*, 1999, **33**(12), 2106–2112.

Gaines, R.B. and G.S. Frysinger, Temperature requirements for thermal modulation in comprehensive two-dimensional gas chromatography, *J. Separation Science*, 2004, **27**, 380–388.

Giddings, J.C., Sample dimensionality: A predictor of order-disorder in component peak distribution in multidimensional systems, *J. Chromatography A*, 1995, **703**, 3–15.

Gorecki, T., J. Harynuk, and O. Panic, The evolution of comprehensive two-dimensional gas chromatography (GC × GC), *J. Separation Science*, 2004, **27**, 359–379.

Grantham, P.J., J. Posthuma, and K. DeGroot, Variation and significance of the C27 and C28 triterpane content of a North Sea core and various North Sea crude oils, *Advances in Organic*

Geochemistry, A.G. Douglas and J.R. Maxwell (eds.), 1980, Pergamon, Oxford, 29–38.

Hornafius, J.S., D.C. Quigley, and B.P. Luyendyk, The world's most spectacular marine hydrocarbon seeps (Coal Point, Santa Barbara Channel, California): Quantification of emissions, *J. Geophysical Research*, 1999, **104**, 20703–20711.

Hostettler, F.D., R.J. Rosenbauer, T.D. Lorenson, and J. Dougherty, Geochemical characterization of tarballs on beaches along the California coast. Part I — Shallow seepage impacting the Santa Barbara Channel Islands, Santa Cruz, Santa Rosa, and San Miguel, *Organic Geochemistry*, 2004, **35**, 725–746.

Johnson, K.J. and R.E. Synovec, Pattern recognition of jet fuels: Comprehensive GC × GC with ANOVA-based feature selection and principal component analysis, *Chemometrics and Intelligent Laboratory Systems*, 2002, **60**, 225–237.

Johnson, K.J., B.W. Wright, K.H. Jarman, and R.E. Synovec, High-speed peak matching algorithm for retention time alignment of gas chromatographic data for chemometric analysis, *J. Chromatography A*, 2003, **996**, 141–155.

Kaplan, I.R. and W.E. Reed, Chemistry of marine petroleum seeps in relation to exploration and pollution, *Proc. Annual Offshore Tech. Conf.*, 1977, **9**, 425–434.

Kates, M., *The Biochemistry of Archaea (Archaebacteria)*, Chap. 9: Membrane Lipids of Archaea, 1993, Elsevier Science Publishers B.V., 261–295.

Koga, Y., M. Nishihara, H. Morii, and M. Akagawa-Matsushita, Ether polar lipids of methanogenic bacteria: Structures, comparative aspects, and biosyntheses, *Microbiological Reviews*, 1993, **57**, 164–182.

Kvenvolden, K.A. and F.D. Hostettler, Geochemistry of coastal tarballs in southern California — a tribute to I.R. Kaplan, *Special Publication — The Geochemical Society*, **9** (Geochemical Investigations in Earth and Space Science), 2004, 197–209.

Lee, A.L., K.B. Bartle, and A.C. Lewis, A model of peak amplitude enhancement in orthogonal two-dimensional gas chromatography, *Analytical Chemistry*, 2001, **73**, 1330–1335.

Leifer, I., J.R. Boles, B.P. Luyendyk, and J.F. Clark, Transient discharges from marine hydrocarbon seeps: Spatial and temporal variability, *Environmental Geology*, 2004, **46**, 1038–1052.

Liu, Z. and J.B. Phillips, Sensitivity and detection limit enhancement of gas chromatographic detection by thermal modulation, *J. Microcolumn Separation*, 1994, **6**, 229–235.

Liu, Z. and M.L. Lee, Comprehensive two-dimensional separations using microcolumns, *J. Microcolumn Separations*, 2000, **12**, 241–254.

Marriott, P. and R. Kinghorn, Cryogenic solute manipulation in gas chromatography — the longitudinal modulation approach, *Trends in Analytical Chemistry*, 1999, **18**, 114–125.

Micyus, N.J., J.D. McCurry, and J.V. Seeley, Analysis of aromatic compounds in gasoline with flow-switching comprehensive two-dimensional gas chromatography, *J. Chromatography A*, 2005, **1086**, 115–121.

Mikolaj, P.G., Composition of oil from the region of new hydrocarbon upwelling in the Santa Barbara channel, Government Report Announcement (U.S.), 1973, **73**, 137.

Moldowan, J.M., W.K. Seifert, and E.J. Gallegos, Relationship between petroleum composition and depositional environment of petroleum source rocks, *Amer. Assoc. Petroleum Geologists Bull.*, 1985, **69**, 1255–1268.

Nelson, R.K., G.S. Frysinger, G.T. Ventura, R.B. Gaines, C.M. Reddy, and F. Kenig, Using comprehensive two-dimensional gas chromatography to obtain direct evidence of the presence of archaean biomarkers from 2.7 billion year old sediments, 28th International Symposium on Capillary Chromatography and Electrophoresis, Las Vegas, NV, May 2005.

Nelson, R.K., B.S. Kile, D.L. Plata, S.P. Sylva, L. Xu, C.M. Reddy, R.B. Gaines, G.S. Frysinger, and S.E. Reichenbach, Tracking the weathering of an oil spill with comprehensive two-dimensional gas chromatography, *Environmental Forensics*, 2006, **7**, 33–44.

Ni, M., S.E. Reichenbach, A. Visvanathan, J. TerMaat, and E.B. Ledford, Jr. Peak pattern variations related to comprehensive two-dimensional gas chromatography acquisition, *J. Chromatography A*, 2005, **1086**, 165–170.

Orphan, V.J., K.-U. Hinrichs, W. Ussler III, C.K. Paull, L.T. Taylor, S.P. Sylva, J.M. Hayes, and E.F. DeLong, Comparative analysis of methane-oxidizing Archaea and sulfate-reducing bacteria in anoxic marine sediments, *Applied and Environmental Microbiology*, 2001, **67**, 1922–1934.

Peters, K.E., C.C. Walters, and J.M. Moldowan, *The Biomarker Guide. Vol. 1: Biomarkers and Isotopes in the Environment and Human History*, 2nd ed., 2004a, Cambridge University Press, Cambridge, pp 12, 58–64, 80, 252, 308.

Peters, K.E., C.C. Walters, and J.M. Moldowan, *The Biomarker Guide. Vol. 2: Biomarkers and Isotopes in the Petroleum Exploration and Earth History*, 2nd ed., 2004b, Cambridge University Press, Cambridge, pp 497–520, 524–526, 608–619, 625–631.

Phillips, J.B., R.B. Gaines, J. Blomberg, F.W.M. van der Wielen, J.-M. Dimandja, V. Green, J. Granger, D. Patterson, L. Racovalis, H.-J. de Geus, J. de Boer, P. Haglund, J. Lipsky, V. Sinha, and E.B. Ledford, A robust thermal modulator for comprehensive two-dimensional gas chromatography, *J. High Resolution Chromatography*, 1992, **22**, 3–10.

Phillips, J.B. and J. Beens, Comprehensive two-dimensional gas chromatography: A hyphenated method with strong coupling between the two dimensions, *J. Chromatography A*, 1999, **856**, 331–347.

Prazen, B.J., K.J. Johnson, A. Weber, and R.E. Synovec, Two-dimensional gas chromatography and trilinear partial least squares for the quantitative analysis of aromatic and naphthene content in naphtha, *Analytical Chemistry*, 2001, **73**, 5677–5682.

Prince, R.C., D.L. Elmendorf, J.R. Lute, C.S. Hsu, C.E. Haith, J.D. Senius, G.J. Dechert, G.S. Douglas, and E.L. Butler, $17\alpha(H),21\beta(H)$-hopane as a conserved internal marker for estimating the biodegradation of crude oil, *Environmental Science and Technology*, 1994, **38**, 142–145.

Quigley, D., J.S. Hornafius, B.P. Luyendyk, R.D. Francis, and E.C. Bartsch, Temporal variations in the spatial distribution of natural marine hydrocarbon seeps in the Northern Santa Barbara Channel, California, *EOS, Trans., Amer. Geophysical Union, Electronic Supplement*, 1996, **77**, F419.

Quigley, D.C., J.S. Hornafius, B.P. Luyendyk, R.D. Francis, J. Clark, and L. Washburn, Decrease in natural marine hydrocarbon seepage near Coal Oil Point, California, associated with offshore oil production, *Geology*, 1999, **27**, 1047–1050.

Reed, W.E. and I.R. Kaplan, The chemistry of marine petroleum seeps, *J. Geochemical Exploration*, 1977, **7**, 255–293.

Reddy, C.M., T.I. Eglinton, A. Hounshell, H.K. White, L. Xu, R.B. Gaines, and G.S. Frysinger, The West Falmouth oil spill after thirty years: The persistence of petroleum hydrocarbons in marsh sediments, *Environmental Science and Technology*, 2002, **36**, 4754–4760.

Reichenbach, S.E., M. Ni, V. Kottapalli, and A. Visvanathan, Information technologies for comprehensive two-dimensional gas chromatography, *Chemometrics and Intelligent Laboratory Systems*, 2004, **71**, 107–120.

Schoell, M., A.M. McCaffrey, F.J. Fago, and J.M. Moldowan, Carbon isotopic composition of 28,30-bisnorhopanes and other biological markers in a Monterey crude oil, *Geochimica et Cosmochimica Acta*, 1992, **56**, 1391–1399.

Schoenmakers, P.J., J.L.L.M. Oomen, J. Blomberg, W. Genuit, and G. van Velzen, Comparison of comprehensive two-dimensional gas chromatography and gas chromatography — mass spectrometry for the characterization of complex hydrocarbon mixtures, *J. Chromatography A*, 2000, **892**, 29–46.

Seeley, J.V., F. Kramp, and C.J. Hicks, Comprehensive two-dimensional gas chromatography via differential flow modulation, *Analytical Chemistry*, 2000, **72**, 4346–4352.

Seifert, W.K., J.M. Moldowan, G.W. Smith, and E.V. Whitehead, First proof of structure of a C28-pentacyclic triterpane in petroleum, *Nature*, 1978, **271**, 436–437.

Seruto, C., Y. Sapozhnikova, and D. Schlenk, Evaluation of the relationships between biochemical endpoints of PAH exposure and physiological endpoints of reproduction in male California Halibut (*Paralichthys californicus*) exposed to sediments from a natural oil seep, *Marine Environmental Research*, 2005, **60**, 454–465.

Sinha, A.E., B.J. Prazen, C.G. Fraga, and R.E. Synovec, Valve-based comprehensive two-dimensional gas chromatography with time-of-flight mass spectrometric detection: Instrumentation and figures-of-merit, *J. Chromatography A*, 2003, **1019**, 79–87.

Sinha, A.E., J.L. Hope, B.J. Prazen, C.G. Fraga, E.J. Nilsson, and R.E. Synovec, Multivariate selectivity as a metric for evaluating comprehensive two-dimensional gas chromatography-time-of-flight mass spectrometry subjected to chemometric peak deconvolution, *J. Chromatography A*, 2004, **1056**, 145–154.

Sinninghe Damste, J.S., F. Kenig, M.P. Koopmans, J. Koster, S. Schouten, J.M. Hayes, and J.W. de Leeuw, Evidence for gammacerane as an indicator of water column stratification, *Geochimica et Cosmochimica Acta*, 1995, **59**, 1895–1900.

Spies, R.B., P.H. Davis, and D.H. Stuermer, Ecology of a petroleum seep off the California coast, *Marine Environmental Pollution*, R. Geyer (ed.), 1980, Elsevier, Amsterdam, 229–263.

Stout, S.A., A.D. Uhler, K.J. McCarthy, and S. Emsbo-Mattingly, Chemical Fingerprinting of Hydrocarbons, *Introduction to Environmental Forensics*, B.L. Murphy and R.D. Morrison (eds.), 2002, Academic Press, San Diego, 135–260.

Stout, S.A., A.D. Uhler, and K.J. McCarthy, Middle distillate fingerprinting using drimane-based bicyclic sesquiterpanes, *Environmental Forensics*, 2005, **6**, 241–251.

Stuermer, D.H., R.B. Spies, and P.H. Davis, Toxicity of Santa Barbara seep oil to starfish embryos: Part 1. Hydrocarbon composition of test solutions and field samples, *Marine Environmental Research*, 1981, **5**, 275–286.

Stuermer, D.H., R.B. Spies, P.H. Davis, D.J. Ng, C.J. Morris, and S. Neal, The hydrocarbons in the Isla Vista marine seep environment, *Marine Chemistry*, 1982, **11**, 413–426.

Teal, J.M., J.W. Farrington, K.A. Burns, J.J. Stegeman, B.W. Tripp, B. Woodin, and C. Phinney, The West Falmouth oil spill after 20 years: Fate of fuel oil compounds and effects on animals, *Marine Pollution Bulletin*, 1992, **24**, 607–614.

ten Haven, H.L., J.W. de Leeuw, J.S. Sinninghe Damste, Schenck, S.E. Palmer, and J.E. Zomberge, Application of biological markers in the recognition of palaeo-hypersaline environments, *Lacustrine Petroleum Source Rock*, K. Kelts, A.J. Fleet, and M.R. Talbot (eds.), 1988, Blackwell Publishing, Oxford, **40**, 123–130.

ten Haven, H.L., M. Rohmer, J. Rullkotter, and P. Bisseret, Tetrahymanol, the most likely precursor of gammacerane, occurs ubiquitously in marine sediments, *Geochimica et Cosmochimica Acta*, 1989, **53**, 3073–3079.

van Deursen, M., J. Beens, J. Reijenga, P. Lipman, and C. Cramers, Group-type identification of oil samples using comprehensive two-dimensional gas chromatography coupled to a time-of-flight mass spectrometer (GC × GC-TOF), 2000, **23**, 507–510.

van Stee, L.L.P., J. Beens, R.J.J. Vreuls, and U.A.Th Brinkman, Comprehensive two-dimensional gas chromatography with atomic emission detection and correlation with mass spectrometric detection: Principles and application to petrochemical analysis, *J. Chromatography A*, 2003, **1019**, 89–99.

Vendeuvre, C., F. Bertoncini, L. Duval, J.-L. Duplan, D. Thiebaut, and M.-C. Hennion, Comparison of conventional gas chromatography and comprehensive two-dimensional gas chromatography for the detailed analysis of petrochemical samples, *J. Chromatography A*, 2004, **1056**, 155–162.

Vendeuvre, C., R. Ruiz-Guerro, F. Bertoncini, L. Duval, D. Thiebaut, and M.-C. Hennion, Characterisation of middle-distillates by comprehensive two-dimensional gas chromatography (GC × GC): A powerful alternative for performing various standard analysis of middle-distillates, *J. Chromatography A*, 2005a, **1086**, 21–28.

Vendeuvre, C., F. Bertoncini, D. Espinat, D. Thiebaut, and M.-C. Hennion, Multidimensional gas chromatography for the detailed PIONA analysis of heavy naphtha: Hyphenation of an olefin trap to comprehensive two-dimensional gas chromatography, *J. Chromatography A*, 2005b, **1090**, 116–125.

Venkatramani, C.J. and J.B. Phillips, Comprehensive two-dimensional gas chromatography applied to the analysis of complex mixtures, *J. Microcolumn Separations*, 1993, **5**, 511–516.

Volkman, J.K., R. Alexander, R.I. Kagi, S.J. Rowland, and P.N. Sheppard, Biodegradation of aromatic compounds in crude oils from the Barrow Sub-basin of Western Australia, *Organic Geochemistry*, 1984, **6**, 619–632.

Volkman, J.K., Sterols and other triterpenoids: Source specificity and evolution of biosynthetic pathways, *Organic Geochemistry*, 2005, **36**, 139–159.

Wang, F.C-Y., W.K. Robbins, and M.A. Greaney, Speciation of nitrogen-containing compounds in diesel fuel by comprehensive two-dimensional gas chromatography, *J. Separation Science*, 2004, **27**, 468–472.

Wang, Z., C. Yang, M. Fingas, B. Hollebone, X. Peng, A.B. Hansen, and J.H. Christensen, Characterization, weathering, and application of sesquiterpanes to source identification of spilled lighter petroleum products, *Environmental Science and Technology*, 2005, **39**, 8700–8707.

Washburn, L., J.S. Hornafius, B.P. Luyendyk, J.F. Clark, D. Quigley, and D.F. Francis, Dispersal of hydrocarbon gas plumes in the Northern Santa Barbara Channel, CA; *EOS, Trans., Amer. Geophysical Union, Electronic Supplement*, 1996, **77**, F419.

Woese, C.R. and G.E. Fox, Phylogenetic structure of the prokaryotic domain: The primary kingdoms, *Proc. National Academy of Sciences*, 1977, **74**, 5088–5090.

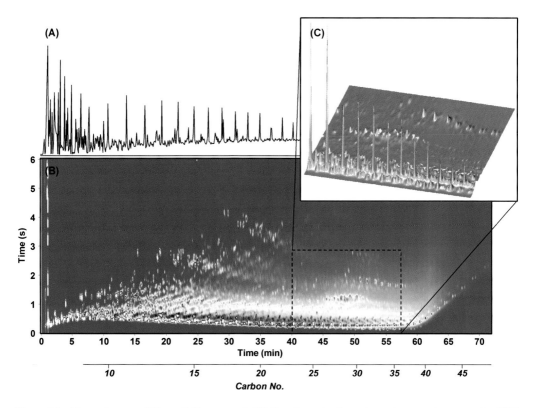

Figure 5-3 Chromatograms of *Exxon Valdez* cargo oil: (A) One-dimensional gas chromatogram using a nonpolar 100% polydimethylsiloxane stationary phase; (B) GC × GC volatility-by-polarity interpolated color contour plot. The first dimension is a separation using a nonpolar 100% polydimethylsiloxane stationary phase and the second dimension is a separation using a polar 50% phenyl polysilphenylene stationary phase; (C) a small portion of the GC × GC chromatogram visualized as a mountain plot. The mountain plot is excellent for visualizing relative differences among neighboring peaks.

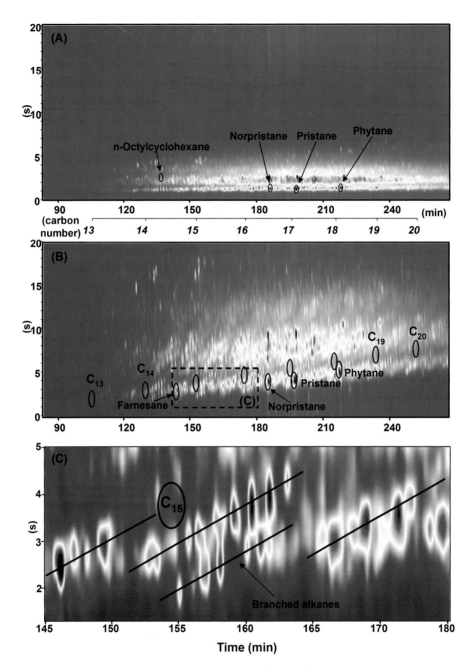

Figure 5-6 Effect of changing the second-dimension stationary phase on the separation of nonpolar saturates: (A) GC × GC chromatogram of the saturates fraction of a sediment extract using a polar second-dimension 14% cyanopropylphenyl polysiloxane stationary phase; (B) GC × GC chromatogram of the same saturates fraction using a chiral γ-cyclodextrin second-dimension stationary phase. The expected positions of the n-C_{13} to n-C_{20} alkanes are marked with circles. The region inside the dotted lines is expanded in (C); (C) expanded region showing expected position of n-C_{15} and location of branched alkane bands. Reprinted with permission from Frysinger et al., 2003. Copyright 2003, American Chemical Society.

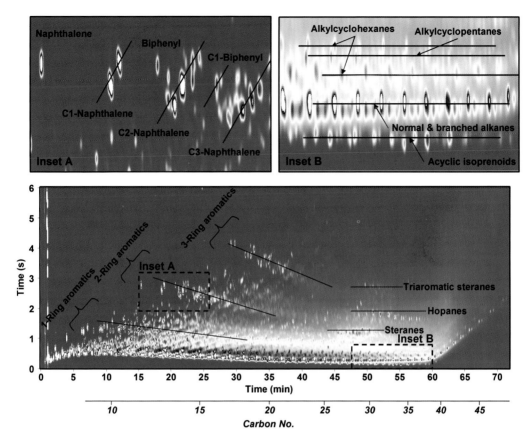

Figure 5-7 Location of petroleum chemical classes on a volatility-by-polarity GC × GC chromatogram. Nonpolar alkanes are located at the bottom, with one- and multiring aromatics further up the y-axis as their polarity increases. Inset A: naphthalene region showing bands of alkyl-substituted naphthalenes and biphenyls; Inset B: nonpolar region showing high-resolution separations of higher boiling (C_{28}+) acyclic isoprenoids, normal and branched alkanes, and cyclic alkanes.

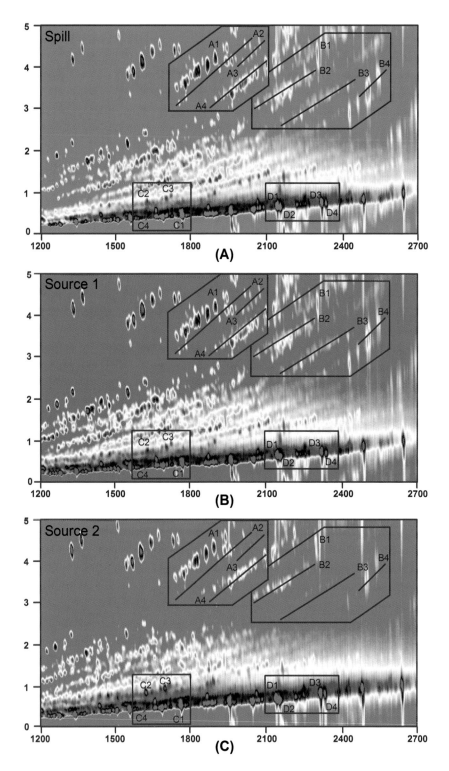

Figure 5-8 Portion of a volatility-by-polarity GC × GC-FID chromatogram from each of the three samples compared: (a) spill; (b) and (c) potential spill sources. Both axes are in seconds. The boxes contain chemical families and individual compounds used to qualitatively and quantitatively compare two potential source samples with the spill samples. The first column separation shown along the x-axis was accomplished with a nonpolar 5% phenylmethylsiloxane stationary phase. The second column separation shown along the y-axis was accomplished with a polar polyethylene glycol/siloxane copolymer stationary phase. Reprinted with permission from Gaines et al., 1999. Copyright 1999, American Chemical Society.

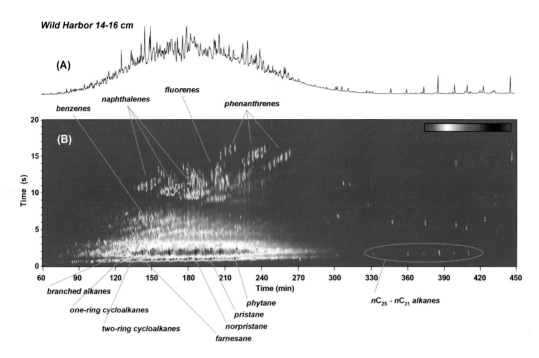

Figure 5-12 Chromatograms of a Wild Harbor sediment extract, Aug. 2000: (A) Conventional one-dimensional gas chromatogram. The separation was accomplished using a nonpolar polydimethylsiloxane stationary phase; (B) volatility-by-polarity GC × GC chromatogram. The first column separation along the x-axis was accomplished using a nonpolar polydimethylsiloxane stationary phase. The second column separation was accomplished using a polar 14%-cyanopropylphenyl polysiloxane stationary phase. Reprinted with permission from Reddy et al., 2002. Copyright 2002, American Chemical Society.

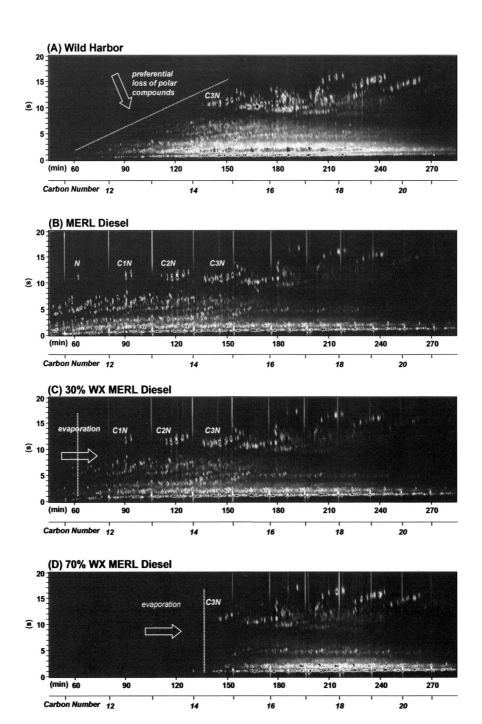

Figure 5-13 Comparison of Wild Harbor sediment extract with weathered MERL oil: (A) Enlarged volatility-by-polarity GC × GC chromatogram of Wild Harbor extract, Aug. 2000 (as in Figure 5-12); (B) neat MERL oil; (C) 30% by mass weathered (WX) MERL oil; (D) 70% by mass weathered (WX) MERL oil. The first column separation along the x-axis was accomplished using a nonpolar polydimethylsiloxane stationary phase. The second column separation was accomplished using a polar 14%-cyanopropylphenyl polysiloxane stationary phase. Weathering experiments were performed in a hood where neat MERL oil was allowed to evaporate until the desired mass loss was achieved. The background is blue. Peak intensity is scaled from white, to red, and then to blue (most intense). Compound abbreviations: (N) naphthalene; (C1N) C_1-naphthalenes; (C2N) C_2-naphthalenes; (C3N) C_3-naphthalenes. Reprinted with permission from Reddy et al., 2002. Copyright 2002, American Chemical Society.

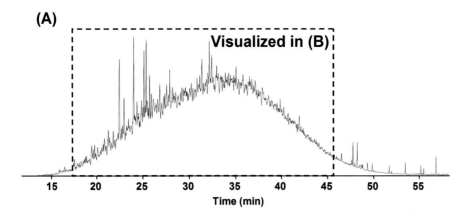

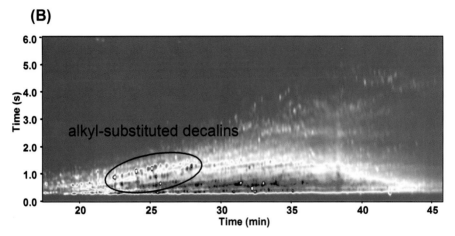

Figure 5-15 Chromatograms of the sediment extract taken from Winsor Cove in 2002 approximately 30 years after being contaminated with No. 2 fuel oil: (A) one-dimensional gas chromatogram showing UCM and the region of the chromatogram visualized in (B). The separation was accomplished using a polydimethylsiloxane stationary phase; (B) volatility-by-polarity GC × GC-FID chromatogram showing the region containing the target compounds. The first-column separation shown along the x-axis was accomplished with a polydimethylsiloxane stationary phase. The second-column separation shown along the y-axis was accomplished with a 50%-phenylpolysilphenylene-siloxane stationary phase.

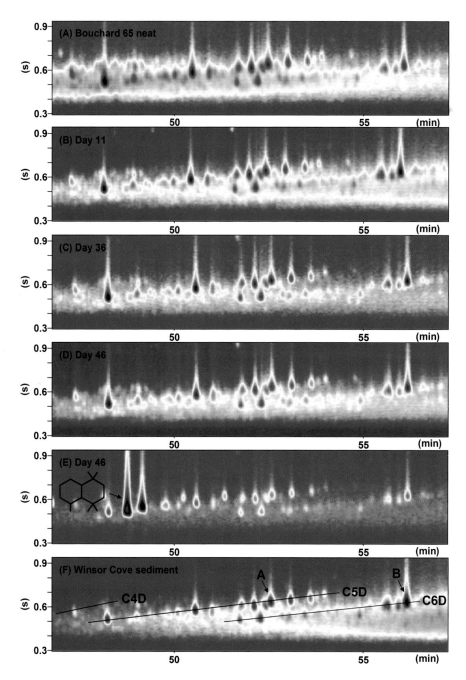

Figure 5-18 Volatility-by-polarity GC × GC $m/z = 123$ extracted ion chromatographs of *Bouchard 65* incubation sediment extracts showing the C_4- to C_6-decalin region: (A) neat *Bouchard 65* fuel oil; (B) day 11; (C) day 36; (D) day 46; (E) day 46 spiked with cis- and trans-1,1,4,4,6-pentamethyldecalin standard; (F) Winsor Cove sediment extract showing the bands of peaks containing C_4-decalins (C4D), C_5-decalins (C5D), C_6-decalins (C6D), and location of 8β(H)-drimane (A) and 8β(H)-homodrimane (B) whose structures are given in Figure 5-16. The first-column separation shown along the *x*-axis was accomplished with a nonpolar polydimethylsiloxane stationary phase. The second-column separation shown along the *y*-axis was accomplished with a polar 50%-phenylpolysilphenylene-polysiloxane stationary phase.

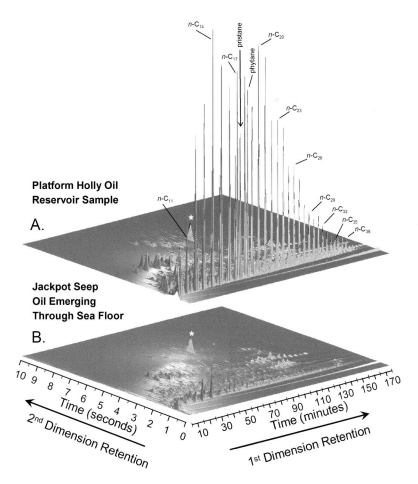

Figure 5-22 Volatility-by-polarity GC × GC chromatogram of (A) reservoir crude oil from platform Holly, Santa Barbara Channel, California, and (B) oil stringer emerging through the sea floor collected at the Jackpot seep field, Santa Barbara Channel, California. The first-column separation shown along the x-axis was accomplished with a nonpolar polydimethylsiloxane stationary phase. The second-column separation shown along the y-axis was accomplished with a polar 50%-phenylpolysilphenylene-siloxane stationary phase. A star on each chromatogram indicates the position of the internal standard dodecahydrotriphenylene (DDTP). The distinct difference in n-alkane content between the two samples is evident in this figure.

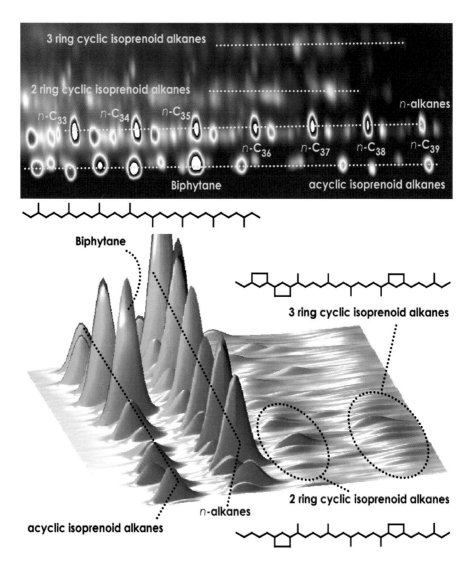

Figure 5-23 Partial volatility-by-polarity GC × GC chromatogram of the platform Holly produced crude oil showing peaks in the vicinity of n-C_{33} through n-C_{39}. The location of high-molecular-weight isoprenoid alkanes (both acyclic and cyclic), thought to be biomarkers of Archaean membrane lipids, are indicated along with some representative molecular structures.

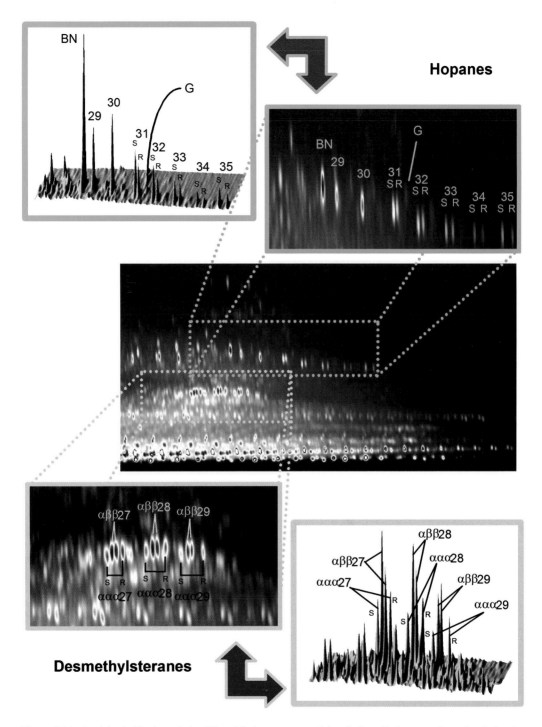

Figure 5-24 Partial volatility-by-polarity GC × GC chromatogram of the platform Holly reservoir crude oil showing peaks in the vicinity of the sterane and desmethylsterane biomarkers. Identification of labeled peaks is found in Table 5-1.

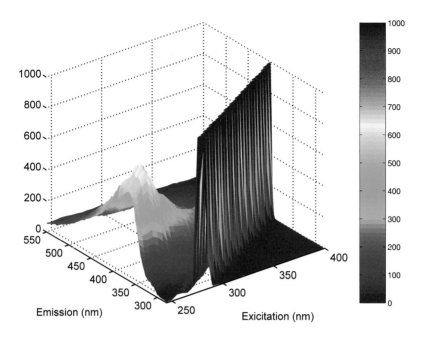

Figure 9-2 Fluorescence excitation-emission scans of an HFO. The vertical axis and scale bar show the fluorescence intensity. Modified from Christensen et al. (2005b).

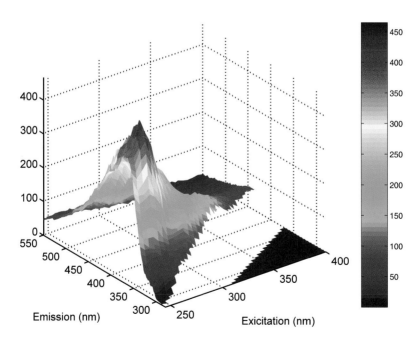

Figure 9-5 Preprocessed fluorescence EEM of the same oil sample as shown in Figure 9-2 after blank subtraction, insertion of missing values, and a small triangle of zeros and excitation/emission correction. The vertical axis and scale bar show the fluorescence intensity. Modified from Christensen et al. (2005b).

6 Application of Stable Isotope Ratios in Spilled Oil Identification

Alan W.A. Jeffrey

DPRA/Zymax, 71 Zaca Lane, Suite 110, San Luis Obispo, CA 93401.

6.1 Introduction

The characterization of spilled oil, slicks, and tars, and their correlation with potential sources, can often be readily accomplished by chemical techniques, as discussed in several chapters. Gas chromatographic techniques such as GC/FID and GC/MS have been used to produce hydrocarbon and other molecular distributions or fingerprints for comparison with the distributions in the sources. Stable isotope ratio measurements provide an additional technique for comparison of spilled petroleum with sources. In a correlation study, this is a useful, nonchemical technique that can complement chemical techniques to support the interpretation.

Stable isotope ratios assume an increased importance where weathering has altered the molecular composition of the spilled oil. Evaporation removes the more volatile chemicals at a rate that depends on the spill conditions and the initial composition of the spilled material. Thus, surface spills in an area where the material can quickly spread, such as from oil tankers at sea, can result in a rapid loss of light hydrocarbons from spilled oil. For a light crude oil, containing a large percentage of volatile hydrocarbons, this can produce a dramatic change in the molecular composition. For lubricating oils and heavy fuel oils, such as Bunker C, which contain much lower proportions of volatile hydrocarbons, evaporation may result in only minor changes in the molecular compositions. Oil spilled into water can lose the more water-soluble chemicals by dissolution into the underlying water.

Aromatic hydrocarbons, particularly benzene, toluene, ethylbenzene, and the xylenes, are more soluble than other compounds in oil (McAuliffe, 1966) and can be rapidly lost from relatively thin layers of product. Microbes in the environment can utilize petroleum hydrocarbons as a carbon source. Because they preferentially degrade certain hydrocarbons, such as *n*-alkanes (Chapelle, 2001), microbial biodegradation results in a change in the composition of the original petroleum. Chapters 3, 11, and 13 provide more detailed discussion of these processes.

Later in this chapter it will be shown that, under most circumstances, stable isotope ratios are changed much less by environmental alteration than are molecular compositions. In some cases, environmental alteration changes the molecular composition so severely that any meaningful correlation between a spilled oil and a possible source is not possible. In these circumstances, stable isotope ratio measurements may provide the only reliable way to correlate the spilled oil with the source. The importance of stable isotope ratio data in a correlation study increases with the degree of weathering. Most of the studies discussed in this chapter concern oil spilled in the sea, which is a particularly harsh environment. The application of stable isotope ratios in the wider area of environmental forensics has been reviewed by Philp (2002).

6.2 Isotope Ratios and Their Measurement

The major elements in petroleum are carbon and hydrogen, and the stable isotope ratios of these elements have been the most intensively studied. Petroleum may also contain sulfur, nitrogen, and oxygen, and stable isotope ratios of these elements may provide additional information to correlate spilled material with sources.

The basis of the stable isotope ratio method is that each of these elements exists in nature in more than one isotopic form. Carbon exists in three isotopic forms: ^{12}C, with 6 protons and 6 neutrons, accounts for about 99% of carbon; ^{13}C, with 6 protons and 7 neutrons, accounts for about 1% of carbon; and ^{14}C, with 6 protons and 8 neutrons, is unstable and decomposes. The hydrogen isotopes of interest are ^{1}H (99.98%) and ^{2}H (deuterium, 0.02%). The isotopes of the other elements in petroleum that find use as stable isotope ratios are nitrogen (^{14}N, 99.6%, and ^{15}N, 0.36%), oxygen (^{16}O, 99.8%, and ^{18}O, 0.02%), and sulfur (^{32}S, 95%, and ^{34}S, 4%). The ratio of the isotopes of an element is not the same in all naturally occurring compounds. There are small variations caused by the different atomic weights of the isotopes. Thus, ^{13}C reacts the same as ^{12}C in chemical transformations, but the heavier ^{13}C can be discriminated against in processes where weight is important, such as evaporation and diffusion. The heavier isotope also forms slightly stronger bonds with other atoms, and when this bond is broken in chemical or enzymatic reactions, slightly less of the ^{13}C—X bonds are broken. If all the C—X bonds are broken, the ratio of $^{12}C/^{13}C$ in the starting material is the same as in the product. But if the reaction does not go to completion, the product may be enriched in ^{12}C, and the starting material enriched in ^{13}C. This accounts for many of the differences in the stable isotope ratios in natural materials. A more detailed discussion of stable isotope geochemistry is given in Hoefs (1997).

Stable isotope ratios are conventionally referenced to an internationally recognized standard, and are expressed in the δ notation, where, for carbon,

$$\delta^{13}C = (R_{sample}/R_{standard} - 1) \times 1000, \text{ and}$$
$$R = {}^{13}C/{}^{12}C.$$

Units are per mil (‰). The standard, by definition, has a δ value of 0, and samples may have positive or negative δ values depending on whether the sample is enriched or depleted in the heavier isotope. More positive δ values are commonly referred to as being isotopically heavier, and more negative δ values are referred to as isotopically lighter. The international standard for carbon is Pee Dee Belemnite (PDB) (Craig, 1957). This sedimentary carbonate lies at the heavy end of the naturally occurring carbon range, so most materials have negative δ values. The range of $\delta^{13}C$ values in naturally occurring materials (Figure 6-1) is almost 100‰, but in petroleum it is only about 15‰. The international standard for hydrogen is Standard Mean Ocean Water (SMOW) (Hagemann et al., 1970).

The elements important in the stable isotope analysis of oils and tars (H, C, N, O, and S) have to be isolated from the material and converted to a gas that can be introduced into an isotope ratio mass spectrometer (IRMS). The gaseous forms of these elements that are commonly used are H_2 (H), CO_2 (C, O), CO (O), N_2 (N), and SO_2 (S). For example, the stable isotopic forms of CO_2, $^{13}CO_2$ and $^{12}CO_2$, are converted in the ion source (Figure 6-2) to the positively charged ions m/e 45 and 44, respectively, which are separated in a stable magnetic field, and the intensities of the ions measured to provide the $^{13}C/^{12}C$ ratio.

Where a sample consists of a mixture of compounds, e.g., n-alkanes in an oil sample, bulk stable isotope ratios can be measured on the total element content in a sample. For example, the $\delta^{13}C$ value of an oil sample is measured by combusting the entire sample in an oxidizing furnace to convert the bulk carbon to CO_2. Alternatively, the $\delta^{13}C$ value of an individual n-alkane in the oil is measured in compound specific isotope analysis (CSIA), by isolating and combusting each of the individual n-alkanes in the oil. A schematic of the CSIA equipment is shown in Figure 6-2.

The first stable isotope ratio instruments were equipped with dual inlets, one for the sample gas, the other for the standard. The sample and standard were analyzed sequentially multiple

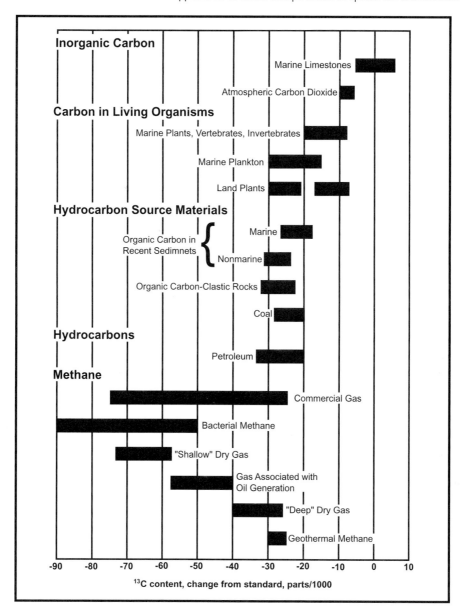

Figure 6-1 Carbon isotope ratios in a variety of geochemical materials. (Reprinted with permission from Fuex (1977). Copyright (1977) Elsevier Science.)

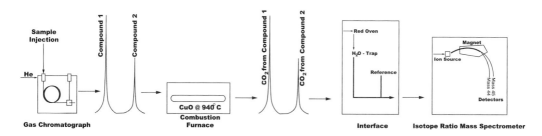

Figure 6-2 Schematic diagram of an instrument set-up to perform compound specific carbon isotope analysis.

times. The sample was converted, in a separate high vacuum line, to an inlet gas, which was then transferred to the inlet system of the IRMS. Analysis of the isotope ratios of an individual compound in a mixture would require isolation of the compounds by preparative GC, conversion of the minute amounts of each compound to the inlet gas, and transfer of the equally minute amounts of gas to the IRMS inlet system. This operation is laborious for a multicomponent mixture, and fraught with many opportunities for isotopic fractionation. The development of continuous-flow isotope ratio instruments greatly enhanced the ability to measure the isotope ratios of individual compounds in a mixture.

In continuous-flow instruments, a carrier gas, helium, continuously passes through the mass spectrometer, similar to analytical GC/MS instruments. This allows the conversion to inlet gases and other preparative techniques to be performed online and in tandem with the isotope ratio measurement. The application of continuous-flow isotope ratio techniques has been reviewed by Lichtfouse (2000). Preparative instruments such as oxidative and reductive element analyzers enable bulk isotope ratio measurements to be performed rapidly online. A GC placed in front of a combustion/pyrolysis furnace, as in Figure 6-2, allows an individual compound in a mixture to be separated, converted to the appropriate inlet gas, and their isotope ratios measured in one operation. The chromatographic process may introduce isotopic fractionation within the peak that represents a particular compound. So the entire peak has to be converted to the inlet gas. Coelution with another compound will obviously complicate the measurement. These experimental difficulties will be discussed in the section on CSIA. Reference gases cannot be introduced in a separate inlet, as in duel inlet instruments, and are added after the conversion stage and analyzed as sample peaks.

6.3 Bulk Isotope Ratios

One of the first reviews of the application of carbon isotope ratios in geochemistry and oil exploration was provided by Fuex (1977), who compiled the bulk carbon isotope ratios of a wide range of geochemical materials in the classic graph shown in Figure 6-1. The factors that controlled the carbon isotope ratio of petroleum hydrocarbons were identified as the ratio of the source organic matter, the isotopic fractionation involved in the formation of the hydrocarbon, and the isotopic fractionation after formation. Fuex considered both fractionations to be unimportant to high-molecular-weight petroleum hydrocarbons, which were influenced primarily by the source material. He stressed that stable isotope ratios be used in conjunction with other geochemical information.

Several studies of bulk isotope ratios in oils have identified some general isotopic relationships that have application to oil spill identification. Silverman and Epstein (1958) distinguished several Tertiary oils from marine and nonmarine source rocks based on carbon isotope ratios. Sofer (1984) extended this by measuring the carbon isotope ratios of the saturate and aromatic fractions of 339 oils. When these values were plotted against each other, oils sourced from terrigenous organic matter could be distinguished from oils sourced from marine organic matter.

There is a general trend toward heavier carbon isotope ratios with decreasing age of petroleum (Stahl, 1977). More recently, Andrusevich et al. (1998) measured the carbon isotope ratios of 514 crude oils spanning a wide age range from Precambrian to Tertiary. Again there was a general trend toward heavier values in younger oils with abrupt isotopic changes at certain discrete time intervals. The trend was attributed to the diversity of phytoplankton preserved in the source rock of the oils.

One of the first uses of stable isotope ratios to identify petroleum-derived pollution was reported by Calder and Parker (1968) in industrial areas in east Texas. At a petrochemical plant, they measured the carbon isotope ratios of dissolved and particulate carbon in the intake and outflow waters. In the outflow ponds, the $\delta^{13}C$ values of dissolved organic carbon was much lighter (−26 to −39‰) than the intake water (−17 to −19‰), which was attributed to contributions from isotopically light petroleum-derived chemicals. In the Houston Ship Channel, which receives contributions from refineries and other industrial activities, light isotope ratios were measured at many locations, reflecting pollution from petroleum-derived chemicals.

Hartman and Hammond (1981) measured the carbon and sulfur isotope ratios and sulfur concentrations of the asphaltene fractions of beach tars deposited near Los Angeles. The asphaltene fraction was used because it was considered to be least affected by weathering. The carbon and sulfur isotope ratios fell within the range of the ratios in local oil seeps (δ^{13}C = −22.5 to −23.2‰; δ^{34}S = +8 to +15‰) and oil from offshore oil wells and were different from crude oils transported into the area by tankers. The source of the majority of the tars was found to be natural seepage from Coal Oil Point, 150 km to the northwest. During the spring, summer, and fall, seeps are transported on ocean currents toward Los Angeles. A smaller number of the beach tars were derived from natural oil seeps in Santa Monica Bay. The authors considered the isotope ratios to be useful in identifying the sources of the beach tars even after 2–4 weeks of weathering.

Macko and Parker (1983) measured carbon and nitrogen isotope ratios of aliphatic, aromatic, and polar fractions of two large oil spills in the northwestern Gulf of Mexico and a number of beach tars along the south Texas coast. Isotopic signatures similar to the oil spills were identified, along with 16 other distinct isotopic signatures in the beach tars. The authors concluded that the multiple signatures reflected multiple sources of spilled oil from oil exploration and transport. The *Exxon Valdez* tanker spill in 1989 deposited a large amount of oil and tarballs along the shore of Prince William Sound in Alaska. Kvenvolden et al. (1993, 1995) analyzed many oil residues that had similar carbon isotope ratios and molecular distributions to the *Exxon Valdez* oil. However, as shown in Figure 6-3, other

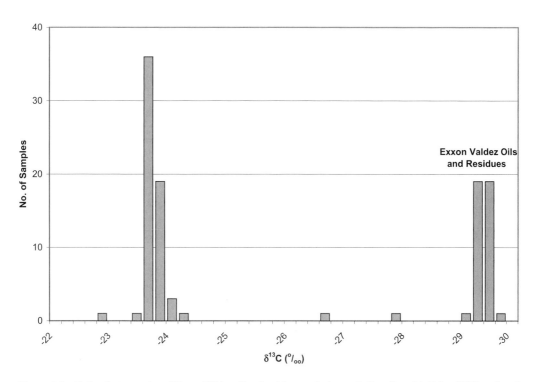

Figure 6-3 Carbon isotope ratios of Exxon Valdez oil and residues, and other tar balls collected in Prince William Sound, Alaska and analyzed by Kvenvolden et al. (1997).

tarballs from the area showed very different characteristics, including $\delta^{13}C$ values averaging $-23.7 \pm 0.2‰$, as compared to the *Exxon Valdez* oil ($-29.4 \pm 0.1‰$). The carbon isotope ratios and biomarker distributions identified the source of the isotopically heavy oil residues as the Southern California Monterey Formation. Kvenvolden et al. (1998) documented the wide use in Alaska of California Monterey oil products, including fuel oils, asphalt, lubricants, and tars, prior to 1970 and the availability of North Slope Alaska oil. The source of the isotopically heavy tarballs was attributed to fuel oil and asphalt stored in the town of Valdez, and spilled during the 1964 Alaskan earthquake. North Slope Alaska oil was the most likely source of lighter carbon ($-29.3 \pm 0.2‰$) in modern pavement and runways.

The opposite situation was demonstrated in a study of an oil spill from the tanker *BP American Trader* in 1990 off Huntington Beach in Southern California (Global Geochemistry Corp., 1990). The oil spilled from the tanker formed a slick that was carried onshore and formed tarballs that could be linked visually with the slick. Tarballs are common along Southern California beaches, and the U.S. Coast Guard, which investigated the spill, was concerned that the extent of the contamination from the spill be delineated scientifically. Biomarker and carbon isotope analyses were performed on oil from the tanker, the oil slick off Huntington Beach, tarballs linked to the slick, and tarballs north and south of the contamination. The carbon isotope ratios of the saturate and aromatic fractions of the samples are compared in Figure 6-4. Tarballs away from the contamination had carbon isotope ratios similar to those measured by Hartman and Hammond (1981), in the range of Southern California Monterey oils. Tanker oil, slick oil, and tarballs associated with the spill had carbon isotope ratios similar to North Slope

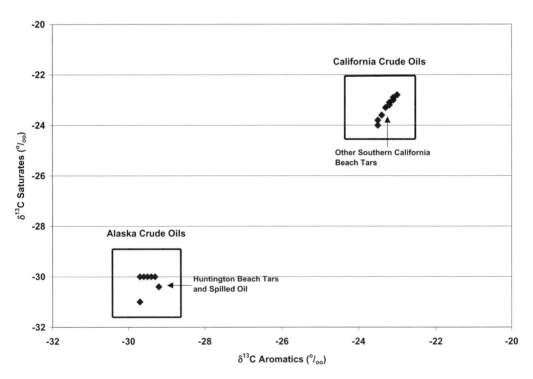

Figure 6-4 Carbon isotope ratios of saturate and aromatic fractions of oil and tar balls associated with the BP American Trader spill offshore Huntington Beach, California. These are compared with ratios of tar balls north and south of the spill site.

Alaska oil, which was the cargo carried by the *BP American Trader*.

Wang et al. (1998) analyzed tarballs along the west coast of North America, from northern California to Vancouver Island, British Columbia, by GC/MS and bulk carbon isotope ratios. The $\delta^{13}C$ values of tarballs from northern California and Oregon had a very narrow isotopic range (−26.8 ± 0.1‰), which is different from the two largest sources of oil on the West Coast, Southern California Monterey and North Slope Alaska. The source of these tarballs was believed to be bunker C fuel oil from a passing tanker.

Carbon and sulfur isotope ratios were used to help identify the sources of crude oil spilled near a tank farm in the western U.S. (Zymax Forensics, 2003). Oil was discovered in several monitoring wells near the tank farm, which had stored crude oil from a number of oil fields in the western U.S. and Canada. Table 6-1 shows the isotope ratios of product samples from four monitoring wells and two seeps, and a sludge sample from an excavation pit. These were compared with six potential source oils. Six of the seven spill samples had very similar $\delta^{13}C$ values (−29.0 to −29.2‰), which correlated with oils 3 and 4. The sulfur isotope ratios of three of the products were similar to oils 3 and 4; the two seep samples were isotopically heavier. Stable isotope ratios could not distinguish between oils 3 and 4. However, nickel and vanadium concentrations allowed oil 3 to be identified as the most likely source of six of the spill samples. Product sample 3 had different carbon and sulfur isotope ratios from the other spill samples, and these ratios correlated with oils 2 and 6. Nickel and vanadium concentrations again allowed oil 6 to be identified as the source of this product. This illustrates the importance of having other geochemical data available when the stable isotope ratios by themselves are inconclusive.

Bulk isotope ratios of oil fractions (saturate, aromatic, polar, and asphaltene) found early use in oil exploration (Stahl, 1977). Northam (1985) increased the number of fractions for isotopic measurement by distilling oil into 11 fractions based on boiling point. Chung et al. (1994) separated 69 oils into 12 fractions, again based on boiling point. When $\delta^{13}C$ was plotted against fraction number, the slope of the $\delta^{13}C$-fraction line varied in oils from different source environments. The CSIA approach, which is discussed in the next section, has replaced these time-consuming techniques.

Table 6-1 Stable Isotope and Metal Concentrations in Spilled Oils Near a Tank Farm in Western U.S. Compared with Potential Crude Oil Sources

Sample ID Spill Products	Carbon Isotope Ratio ‰	Sulfur Isotope Ratio ‰	Vanadium mg/kg	Nickel mg/kg
Product 1	−29.0	6.3	4.1	1.5
Product 2	−29.0	9.4	4.5	1.7
Product 3	−29.9	5.7	7.6	2.8
Product 4	−29.0	9.4	4.9	1.5
Pit sample	−29.2	9.5	6.4	3.0
Seep 1	−29.1	11.0	4.8	1.6
Seep 2	−29.0	10.0	5.4	1.7
Oils				
Oil 1	−28.1	7.1	1.2	<1
Oil 2	−30.1	6.4	22.0	5.5
Oil 3	−28.8	9.1	6.9	2.8
Oil 4	−29.2	9.4	17.0	6.9
Condensate 5	−24.1	24.7	<1	<1
Oil 6	−30.1	6.9	8.3	3.5

6.4 Compound-Specific Isotope Analysis (CSIA)

The first paper discussing the CSIA method was published by Mathews and Hayes (1978), who separated and measured the carbon isotope ratios of methyl esters of C_7–C_{13} acids and amino acid esters. More recently, CSIA principles and technical aspects have been reviewed by Meier-Augustine (1999), who described its applications to areas such as the authenticity of flavors, fragrances, wine, fruit juice, honey, vegetable oils, and drugs. Applications of CSIA in the environmental sciences have been reviewed by Schmidt et al. (2004). A schematic illustration of a CSIA setup is shown in Figure 6-2. Up to the present time, CSIA has been applied to four elements, H, C, N, and O, as shown in Table 6-2. Sulfur isotope ratios have been measured by continuous-flow techniques, but there has been no CSIA application as yet due to the low abundance of sulfur in most organic compounds. In oils, the most obvious applications of CSIA are to C and H isotope ratios. N- and O-containing compounds constitute a very minor component of most oils. S-containing compounds are more important constituents of some oils and are often concentrated by environmental alteration of spilled oil. When CSIA methods are developed for S-containing compounds, they may be very useful in characterizing certain oil spills.

As with other areas of petroleum geochemistry, CSIA of crude oils was developed as a tool in petroleum exploration. This provided a way to isotopically compare individual hydrocarbons in oils with those in potential source rocks. It could also be used to compare hydrocarbons in oils from different locations to determine if they were genetically related.

The carbon isotope ratio of bulk petroleum is inherited from the source organic matter, but is influenced by maturity and by physical and chemical alteration after generation. Isotope ratios of individual hydrocarbons in petroleum are subject to the same controls. But, in addition, they may reflect heterogeneity within the source organic matter itself. Thus, low-molecular-weight n-alkanes may reflect algal or bacterial contributions; high-molecular n-alkanes may originate in plant waxes or from degradation of complex algal and plant molecules. These contributions influence both the $\delta^{13}C$ values of the n-alkanes and the slope of the $\delta^{13}C$-alkane relationship.

An illustration of the carbon isotope distribution of n-alkanes in a number of Tertiary oils formed in different depositional environments is shown in Figure 6-5 (Murray et al., 1994). The lightest oils included fluvio-deltaic oils in the range -25 to $-37‰$; the heaviest were marine carbonate oils in the range -17 to $-22‰$. An interesting feature was the difference in the slope of the $\delta^{13}C$-alkane line with increasing n-alkane size. In fluvio-deltaic oils, the slope was steep, with n-alkanes larger and the C_{30} being much lighter isotopically; in most lacustrine, marine deltaic, and marine carbonate oils, the slope was much shallower or flat. This provides a way of screening possible sources for spilled oil or beach tars of unknown origin, especially in areas of heavy

Table 6-2 Sensitivity and Precision for Elements Reported to Date of Compound-Specific Isotope Analysis (CSIA). Data from Schmidt et al. (2004)

Stable Isotopes	Detection Limits (Nmole Element on Column)	Precision (‰)	CSIA Commercially Available Since
D/H	8–10	5	1998
$^{13}C/^{12}C$	1	0.2	1988
$^{15}N/^{14}N$	0.8–1.5	0.5	1989
$^{18}O/^{16}O$	5	0.8	1998
$^{34}S/^{32}S$	n.a.	n.a.	—

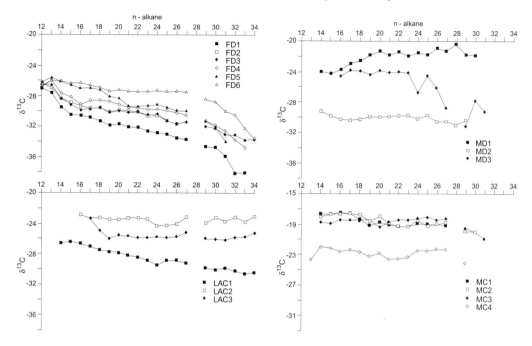

Figure 6-5 Carbon isotope ratios of individual n-alkanes in a number of oils sourced from fluvio-deltaic (FD), lacustrine (LAC), marine deltaic (MD), and marine carbonate (MC) depositional environments. (Reprinted with permission from Murray et al. (1994). Copyright (1994) Elsevier Science.)

tanker traffic carrying oils from different geographical regions.

In one of the first studies to utilize CSIA to identify oil pollution, Ishiwatari et al. (1994) and Uzaki et al. (1993) measured the carbon isotope ratios of C_{27}–C_{33} n-alkanes in a sediment core from Tokyo Bay. In the deeper section of the core, odd- and even-numbered n-alkanes had similar $\delta^{13}C$ values, in the range −29.6 to −32.9‰, which were attributed to higher plant waxes. In the shallower section, dated as post-1965, the isotopic range was shifted toward heavier values (−28.2 to −31.5‰). More significant was that even-numbered n-alkanes were heavier than odd-numbered n-alkanes. The heavier even-numbered n-alkanes were attributed to oil pollution, and the authors noted that pollution in Tokyo Bay increased significantly in the period 1960–1965.

In a comprehensive study of bitumens or beach tars along the northern and southern coasts of Australia, Dowling et al. (1995) used carbon isotope ratios of n-alkanes and other hydrocarbons to help identify the source of the bitumens. The bitumens could be divided into two physical types: a soft, waxy paraffinic type and a hard asphaltic type.

The waxy bitumens occurred along both coasts and had n-alkane carbon isotope ratios that fell within the range of the Minas and Duri oils of the Central Sumatra Basin in Indonesia. Molecular biomarkers in the bitumens also corresponded to those in the Minas and Duri oils, particularly the presence of botryococcane, a marker for oils from lacustrine source rocks. Botryoccane was also markedly isotopically heavier ($\delta^{13}C$ = −10 to −15‰) than the n-alkanes (−25 to −30‰). Carbon isotope ratios of the isoprenoids, pristane and phytane, were slightly heavier than the n-alkanes. Bjorøy et al. (1991) also reported that isoprenoids were generally heavier than n-alkanes in the same oil. The origin of the waxy bitumens in Australia was attributed to natural seeps in Sumatra, coming either from offshore reservoirs, or from onshore reservoirs by river or fault migration to the ocean. The oil then

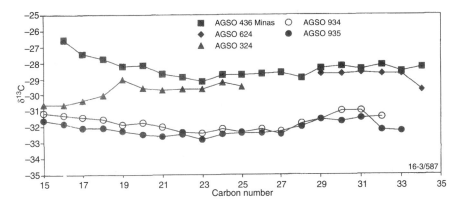

Figure 6-6 Carbon isotope ratios of individual n-alkanes in asphaltic bitumens from S. Australia (AGSO 934, 935), in a bitumen from N. Australia (AGSO 624), and in oils from Seram, Indonesia (AGSO 324) and Minas, Indonesia (AGSO 436). (Reprinted with permission from Dowling et al. (1995). Copyright (1995) Elsevier Science.)

would have been transported south and east on oceanic currents.

The asphaltic bitumens, which were restricted to the southern coast, were isotopically lighter, with $\delta^{13}C$ values in the range −31 to −33‰ (Figure 6-6). Biomarker evidence suggests an origin for these bitumens in clastic marine source rocks. They are believed to arise from natural seeps, but the location was not established.

A bitumen from northern Australia (AGSO 624) exhibited n-alkane carbon isotope ratios very similar to the Minas oil, as shown in Figure 6-6. However, biomarker data identified the bitumen as being derived from a marine carbonate source, which eliminated the lacustrine Minas oil as the source. This illustrates the need to combine stable isotope evidence with molecular evidence to identify the source of spilled oil. In this case, Minas oil, although plausible on a geographical and isotopic basis, was ruled out by the lack of a molecular match. Middle Eastern oil spilled from tankers is one possible source.

Experimental difficulties affecting the CSIA data are illustrated by anomalous results for the carbon isotope ratios of nC_{28} reported by Murray et al. (1994) and omitted from Figure 6-5. This n-alkane was lighter than neighboring alkanes and showed poor reproducibility. This was found to be caused by a non-normal alkane eluting very close to nC_{28}, illustrating the problems that coelution can cause CSIA measurements.

Correlation of spilled light oils, both crude and refined, with potential sources is complicated by the lack of high-molecular-weight biomarkers, such as steranes and triterpanes. C_{15}–C_{16} sesquiterpanes can be useful in some cases, but in my experience, weathering can alter the sesquiterpane distribution, making its correlation with an unweathered potential source inconclusive. CSIA is particularly useful in these cases. Mansuy et al. (1997) showed that three mildly weathered light fuel oils could be correlated with their unweathered source oils with a high degree of confidence, based on carbon isotope ratios of n-alkanes and isoprenoids.

Birds are particularly sensitive to the effects of oil spills, and fouled birds provide a very emotional illustration of the ecological costs of an oil spill. The pathetic sight of oil-soaked birds ignites public indignation and prods the authorities to find the culprit. It is not surprising, therefore, that several studies have focused on identifying the source of oil on bird feathers.

In a light fuel oil spill, which had fouled the feathers of birds, Mansuy et al. (1997) extracted the oil from the feathers to correlate it with a potential source. High-molecular-weight biomarkers were absent, so CSIA provided a unique way to correlate the oils. As

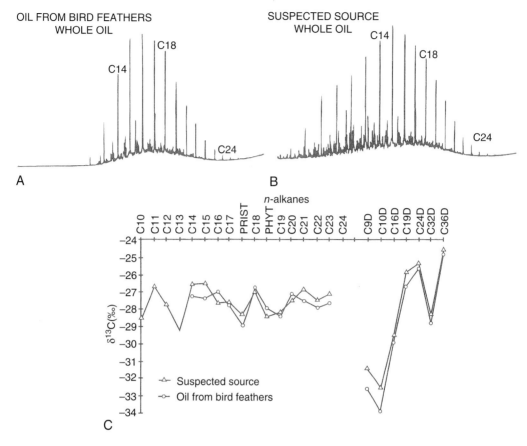

Figure 6-7 Gas chromatograms and carbon isotope ratios of individual alkanes in oil from the feathers of birds killed in a major oil spill and from the oil that was the suspected source of the spill. (Reprinted with permission from Mansuy et al. (1997). Copyright (1997) American Chemical Society.)

shown in Figure 6-7, the oil on the feathers was depleted in the more volatile hydrocarbons compared to the suspected source. However, the similarity between the oils was demonstrated by the carbon isotope ratios of *n*-alkanes and isoprenoids.

Mazeas and Budzinski (2002) used stable carbon isotope ratios, both bulk and compound-specific, to identify the source of oiled bird feathers and oil residues along the Atlantic coast of France after the wreck of the tanker *Erika* in 1999. The tanker was carrying heavy fuel oil, 10,000 tons of which are believed to have spilled. However, other tankers took advantage of the incident to flush their tanks, providing potentially multiple sources for beached oil along the coast. Bulk isotope ratios of oil and tarballs show a significant difference in the samples collected along the northern coast from those collected along the southern coast (Figure 6-8). Northern samples were isotopically very similar to the *Erika* oil. The southern group contained four isotopically heavy samples and two samples with values intermediate between the two groups.

The bulk isotope ratios of oil from fouled bird feathers were found to have large uncertainties caused by interfering compounds. However, CSIA of *n*-alkanes were unaffected by the contamination, and identified the *Erika* oil as the source of the oil on the feathers. In addition, CSIA confirmed the bulk isotope source allocations in the oils and tarballs.

CSIA was also performed on phenanthrene, and methyl, dimethyl, and trimethyl phenanthrenes. This also confirmed the source allocations and showed that the intermediate bulk isotope ratios of the two southern samples in

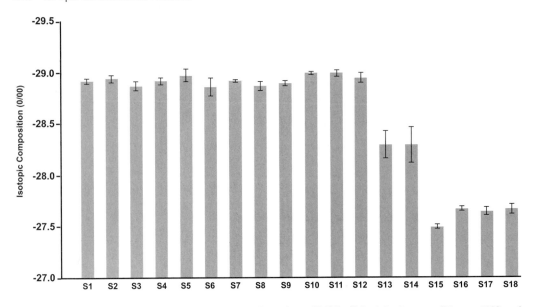

Figure 6-8 Bulk carbon isotope ratios of oil from the Erika tanker spill (S1) off the Atlantic coast of France, 1999, and of oil residues collected from the northern (S2–S12) and southern (S13–S18) Atlantic coast. (Reprinted with permission from Mazeas and Budzinski (2002). Copyright (2002) American Chemical Society.)

Figure 6-8 were not due to mixing of *Erika* oil and the heavier oil, but represented a third source.

The conclusion from the study was that bulk carbon isotope ratio measurement is useful as a screening tool for source identification because it is rapid and requires little sample preparation. However, as illustrated by the oiled bird feathers, contamination can alter the values.

Rogers and Savard (1999) used CSIA to help distinguish naturally occurring hydrocarbons from petroleum pollution along the St. Lawrence River near Quebec City. At sites farthest from industrial activity, $\delta^{13}C$ values of *n*-alkanes were relatively light (−29 to −31‰), with odd and even high-molecular-weight *n*-alkanes showing isotopic differences. At sites near areas of industrial activity, $\delta^{13}C$ values of *n*-alkanes were heavier (−27 to −30‰), with little difference between odd and even *n*-alkanes. The isotopic profiles of *n*-alkanes at some contaminated sites were similar to a bunker fuel ($\delta^{13}C$ = −28 to −29‰) used by ships in the region. The heaviest *n*-alkanes (−25 to −28‰) were found at a contaminated site where bioremediation with fertilizers was in progress. The authors attributed the heavy values to biodegradation.

Philp (2002) studied a number of beach tars from the Seychelles Islands to determine if the source was tanker traffic or natural seeps. Figure 6-9, which shows the gas chromatograms of four tars, indicates a strong similarity in the hydrocarbon distribution in 95M and 95H. Carbon isotope ratios of the high-molecular-weight *n*-alkanes that dominate these samples are also similar. The sterane and terpane distributions were similar and indicated that the tars had been derived from a carbonate source rock and showed similarities to oils from the Middle East. Tars 95M and 95H were, therefore, attributed to releases from passing tankers. The hydrocarbon distributions in 94F and 95F showed similarities to each other, assuming that 94F was a more weathered sample, which would have reduced the maximum around C_{22} seen in 95F. However, the alkane carbon isotope ratios were different, indicating different sources for the two tars. Other information indicated similarities to lacustrine source rocks in the Seychelles area, and the sources were determined to be natural oil seeps.

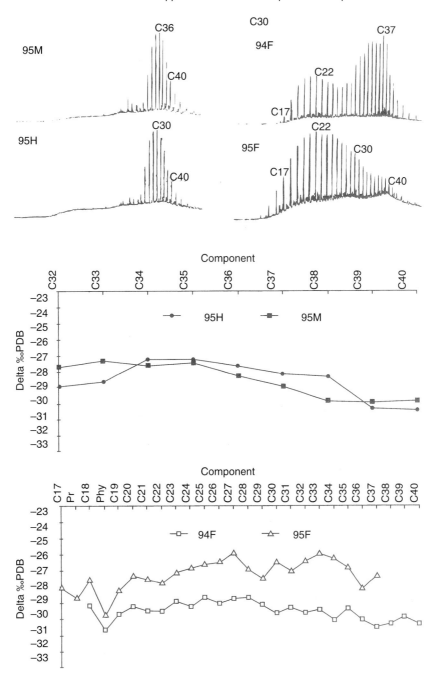

Figure 6-9 Gas chromatograms and carbon isotope ratios of individual alkanes in beach tars from the Seychelles Islands. (Reprinted with permission from Philp (2002). Copyright (2002) Academic Press.)

6.4.1 Experimental Considerations

Sensitivity is a major factor in the application of CSIA, particularly for dissolved compounds such as petroleum hydrocarbons, which are often in very low concentrations in natural waters due to their low solubility. Schmidt et al. (2004) report that commercially available CSIA instruments require at least 1 nmole of carbon or 8 nmole of hydrogen for precise isotopic measurement. Schwarzbauer et al. (2005) were able to measure carbon isotope ratios of anthropogenic contaminants, including aromatic hydrocarbons, reliably on concentrations as low as 50 ng/L. Wang et al. (2004) used semipermeable membrane devices (SPMD) to accumulate petroleum hydrocarbons from river water for CSIA of carbon and hydrogen, with no isotopic fractionation. The SPMDs eliminated the need to process thousands of liters of water required for conventional liquid/liquid and solid phase extraction. CSIA of n-alkane and PAH concentrations as low as <1 ng/L and 3 ng/L, respectively, was performed in water from the Pawtuxet River.

Sensitivity is not a major issue for separate phase oil and tars, which provide abundant carbon and hydrogen for CSIA. The major analytical issue with these materials is co-elution of target hydrocarbons with other compounds, and the change in the target compound's isotope ratio that this introduces, as discussed by Dowling et al. (1995). Elaborate cleanup procedures are often necessary to eliminate potentially coeluting compounds without introducing isotopic fractionation of target compounds.

Kim et al. (2005) isolated and purified PAHs in sediments by performing column chromatography, high-performance liquid chromatography, and thin-layer chromatography sequentially on the dichloromethane extract. The increase in purity with each technique is shown in Figure 6-10. PAH recovery was 80%, and the standard deviation of the $\delta^{13}C$ values of PAH standards purified in this way was <0.4‰. Using this purification procedure, naphthalenes and methylnaphthalenes were isolated from petroleum contaminated soils and offshore sediments at McMurdo Station, Antarctica (Figure 6-11). Soils from the fueling station, helipad, and old oil tank showed similar isotopic distributions, which was attributed to fuel oils. The carbon isotope ratios of soil from the machine shop were more negative and were believed to reflect a wider range of petroleum products, including lubricating oils and waste oils from vehicle repair. Carbon isotope ratios in offshore sediments were even more negative and were attributed to accumulated petroleum product spills from ships and onshore operations at McMurdo Station.

Sohxlet extraction is often the first procedure in isolating oil adsorbed onto solid material. O'Malley et al. (1994) showed this procedure did not cause carbon isotopic fractionation in PAHs.

6.5 Weathering

The resistance of stable isotope ratios to alteration by weathering is considered a major advantage of this technique in correlating spilled oil to possible sources. Many studies support this conclusion. Macko et al. (1981) followed the degradation of crude oil spilled in a salt marsh in Texas under weathering conditions described as harsh. Over a period of 2 years, the $\delta^{13}C$ values of the aliphatic and aromatic fractions of the spilled oil changed by only about 0.5‰ compared to the unweathered oil. The authors considered these changes to be small enough to allow the use of carbon isotope ratios to correlate spilled oils. Several studies have shown that degradation has little effect on carbon isotope ratios of high-molecular-weight compounds (Sun et al., 2005, and references therein). O'Malley et al. (1994) reported that there was no significant isotopic fractionation in 2-5–ring parent PAHs. However, Yanik et al. (2003) reported significant isotopic changes in alkylated naphthalenes and phenanthrenes, but not dibenzothiophenes. Bjorøy et al. (1994) found that evaporation did not cause significant changes in the carbon isotope ratios of C_4–C_{20} hydrocarbons in oils and condensates.

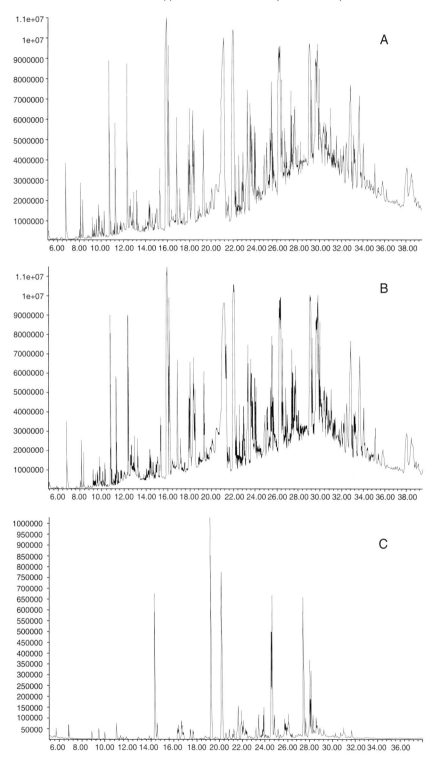

Figure 6-10 Full scan GC/MS chromatograms of a sediment extract after sequential purification procedures. (A) Al/Si column chromatography (B) High performance liquid chromatography (C) Thin-layer chromatography. (Reprinted with permission from Kim et al. (2005). Copyright (2005) American Chemical Society.)

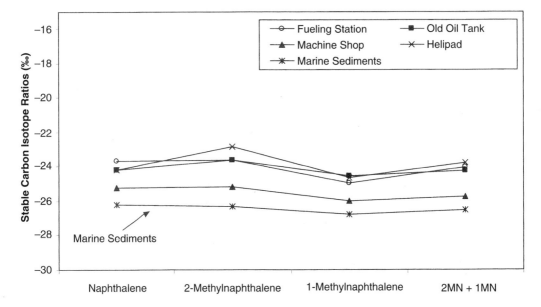

Figure 6-11 Carbon isotope ratios of individual naphthalenes and methylnaphthalenes in soils and sediments from McMurdo Station, Antarctica. (Reprinted with permission from Kim et al. (2005). Copyright (2005) American Chemical Society.)

Neither evaporation nor partitioning into water of jet fuel hydrocarbons (C_{10}–C_{16}) leads to significant isotopic fractionation of the hydrocarbons (Harrington et al., 1999). In addition, Bugna et al. (2004) found that aqueous migration and aerobic degradation of jet fuel hydrocarbons did not result in significant isotopic fractionation. Even most light hydrocarbons in the gasoline range show little carbon isotopic change during water-washing and evaporation (Smallwood et al., 2002). Most of the hydrocarbons in gasoline that show significant isotopic changes are aromatic hydrocarbons, which are not major constituents of most oils.

Mansuy et al. (1997) artificially weathered separate samples of a crude oil by evaporation, water-washing, and biodegradation. After evaporation at room temperature for 4 years, n-alkanes below C_{14} were depleted; after water-washing at room temperature for 38 days, alkanes below C_{15} were depleted; and after biodegradation for 4 months, n-alkanes were completely degraded. The carbon isotope ratios of n-alkanes and isoprenoids in the evaporated and water-washed oils correlate well with the unweathered oil. In the oil biodegraded for 4 months, the lack of alkanes prevented isotopic comparison. After 2 months' biodegradation, however, some n-alkanes and isoprenoids remained, many of which showed a shift to heavier ratios compared to the unweathered oil. This was attributed to interference from the background unresolved complex mixture, which was enhanced in the biodegraded oil. This was shown by isolating n-alkanes in the biodegraded oil by urea adduction. The $\delta^{13}C$ values of the purified n-alkanes in the biodegraded oil corresponded more closely with the values in the unweathered oil.

The complete depletion of n-alkanes in the oil biodegraded for 4 months highlights a major difficulty with CSIA of severely biodegraded oils. To circumvent this, Mansuy et al. (1997) pyrolyzed the asphaltene fractions of the biodegraded oils to generate n-alkanes. The asphaltene fraction is believed to be relatively unaffected by weathering and was used by Hartman and Hammond (1981) to correlate Southern California beach tars with probable source oils based on bulk isotope ratios. Biomarker concentrations in asphaltene pyrolysates are normally low, so correlation is

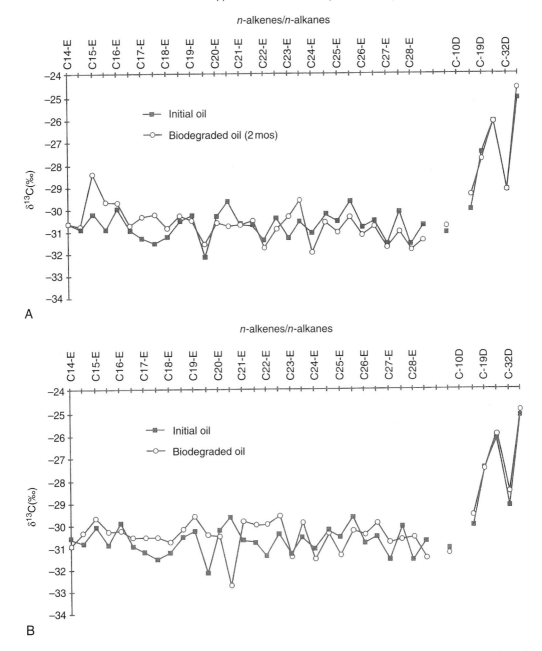

Figure 6-12 Carbon isotope ratios of individual n-alkane/n-alkene pairs generated by pyrolysing asphaltene fractions of biodegraded oils. These are compared with ratios of individual n-alkanes in the undergraded (initial) oils. (Reprinted with permission from Mansuy et al. (1997). Copyright (1997) American Chemical Society.)

limited to n-alkane/n-alkene pairs. As shown in Figure 6-12, there was good carbon isotope correlation between the severely biodegraded oils and the unweathered oil. This study shows that CSIA can be used to correlate even severely biodegraded oils with potential sources. There is evidence that some components of petroleum are changed isotopically by weathering. Palmer (1984) and Sofer (1984) reported a maximum change of 2‰ in the bulk saturate fraction of a marine oil after complete loss of n-alkanes. Philp (2002) also cautions that loss of n-alkanes during biodegradation can result in heavier carbon isotope ratios in the saturate fraction of biodegraded oils.

The evidence to date indicates that weathering does not have a significant effect on the bulk carbon isotope ratios of oils, although biodegradation may have a disproportionate isotopic effect on saturate fractions of oil, shifting isotope ratios to heavier values. The carbon isotope ratios of individual alkanes in oil appear to be resistant to the effects of weathering. This confirms the value of stable isotope ratios, and CSIA in particular, in oil spill studies.

6.6 Other Isotopes

One drawback in the use of carbon isotope ratios to correlate oils, spilled or otherwise, is their relatively narrow natural range of around 15‰. Hydrogen isotope ratios, in contrast, show a much larger isotopic range of up to 160‰ (Li et al., 2001). CSIA of hydrogen isotope ratios has been developed fairly recently (Burgoyne and Hays, 1998; Hilkert et al., 1999) and was applied by Li et al. (2001) to n-alkanes and isoprenoids in a number of oils from western Canada. Isotope ratios of n-alkanes were fairly similar and differed considerably from the isoprenoids, pristane and phytane. This was attributed to the different origins of n-alkanes and isoprenoids during oil formation. The origin of formation water in oil source rocks was also found to have a major influence on the hydrogen isotope ratios. In a study of terrestrially sourced oils from Australia, however, Schimmelmann et al. (2004) found a progressive change in hydrogen isotope ratios with n-alkane size, and little difference between n-alkanes and isoprenoids. The reason for the differences between the two studies is unclear, but it may be related to the depositional environment of the source rocks.

Pond et al. (2002) showed that the hydrogen isotope ratios of n-alkanes in the range C_{15}–C_{18} were shifted to heavier values during biodegradation and could be used to monitor bioremediation of crude oil. High-molecular-weight n-alkanes (C_{19}–C_{27}) were relatively unaffected during biodegradation and could be used for correlation of crude oils. This should also be a useful technique for identifying the source of spilled oil.

Sulfur isotope ratios have found use in bulk isotope studies (Hartman and Hammond, 1981). However, CSIA applications have not been reported. This would seem to be a useful approach in studies of spilled oil, as sulfur is often concentrated in weathered oils. Stable isotope ratios of oxygen and nitrogen, although used extensively in other areas of geochemistry (Brand et al., 1994), have found little use in spilled oil studies. The low concentrations of these elements in petroleum will continue to limit their use.

6.7 Conclusions

Stable isotope ratios have had a long history of use in the identification of the sources of spilled oil and tars. Bulk isotope analysis is relatively quick and low cost and is still useful as a screening technique for these reasons. However, bulk analysis has been largely superceded by the compound-specific isotopic analysis of individual alkanes and aromatic hydrocarbons in oil. By focusing on individual hydrocarbons, CSIA avoids the effects of sample contamination on bulk ratios. Contamination is a potential problem, particularly for carbon, with materials deposited in an environment with multiple sources of extraneous carbon. Advances in the sensitivity of CSIA techniques allow carbon isotope ratios of the low concentrations of petroleum hydrocarbons dissolved in water to be measured reliably. Coelution of target peaks remains a complication of CSIA and has to be monitored to avoid inconsistent results.

The effects of weathering on the carbon isotope ratios of alkanes appear to be minimal, and compound-specific isotope analysis of these hydrocarbons are now widely used to identify spill sources. Hydrogen isotope ratios of high-molecular-weight n-alkanes also appear to be minimally affected by weathering. Compound-specific hydrogen isotope analysis of alkanes offers more resolution than carbon because of the wider δD range, and this technique offers promise for future studies.

Stable isotope ratios may provide the only reliable way of correlating light oils that lack conventional marker compounds like steranes and terpanes. However, many of the studies discussed in this chapter highlight the value of stable isotope ratios as one component in identifying the source of spilled oil.

References

Andrusevich, V.E., M.H. Engel, J.E. Zumberge, and L.A. Brothers, Secular, episodic changes in stable carbon isotope composition of crude oils, *Chemical Geology*, 1998, **152**, 59–72.

Bjorøy, M., K. Hall, P. Gillyon, and J. Jumeau, Carbon isotope variations in *n*-alkanes and isoprenoids of whole oils, *Chemical Geology*, 1991, **93**, 13–20.

Bjorøy, M., P.B. Hall, and R.P. Moe, Variation in the isotopic composition of single components in the C_4–C_{20} fraction of oils and condensates, *Organic Geochemistry*, 1994, **21**, 761–776.

Brand, W.A., R. Tegtmeyer, and A. Hilkert, Compound-specific isotope analysis: Extending toward $^{15}N/^{14}N$ and $^{18}O/^{16}O$, *Organic Geochemistry*, 1994, **21**, 585–594.

Bugna, G.C., J.P. Chanton, C.A. Kelley, T.B. Stauffer, E.L. MacIntyre, and E.L. Libelo, A field test of $\delta^{13}C$ as a tracer of aerobic hydrocarbon degradation, *Organic Geochemistry*, 2004, **35**, 123–125.

Burgoyne, T.W. and J.M. Hayes, Quantitative production of H_2 by pyrolysis of gas chromatographic effluents, *Analytical Chemistry*, 1998, **70**, 5136–5140.

Calder, J.A. and P.L. Parker, Stable carbon isotope ratios as indices of petrochemical pollution of aquatic systems, *Environmental Science and Technology*, 1968, **2**, 535–539.

Chapelle, F.H., *Ground-Water Microbiology and Geochemistry*, 2nd ed., John Wiley & Sons, New York, 2001, 356–357.

Chung, H.M., G.E. Claypool, M.A. Rooney, and R.M. Squires, Source characteristics of marine oils as indicated by carbon isotope ratios of volatile hydrocarbons, *American Assoc. Petroleum Geologists Bull.*, 1994, **78**, 396–408.

Craig, H., Isotopic standards for carbon and oxygen and correction factors for mass-spectrometric analysis of carbon dioxide, *Geochimica Cosmochimica Acta*, 1957, **3**, 133–149.

Dowling, L.M., C.J. Boreham, J.M. Hope, A.P. Murray, and R.E. Summons, Carbon isotope composition of hydrocarbons in ocean transported bitumens from the coastline of Australia, *Organic Geochemistry*, 1995, **23**, 729–737.

Fuex, A.N., The use of stable carbon isotopes in hydrocarbon exploration, *J. Geochemical Exploration*, 1977, **7**, 155–188.

Global Geochemistry Corp., unpublished report, Los Angeles, CA, 1990.

Hagemann, R., G. Nief, and E. Roth, Absolute isotopic scale for deuterium analysis of natural waters. Absolute D/H scale for SMOW, *Tellus*, 1970, **22**, 712–715.

Harrington, R.R., S.R. Poulson, J.I. Drever, P. Colberg, and E.F. Kelley, Carbon isotope systematics of mono-aromatic hydrocarbons: Vaporization and adsorption experiments, *Organic Geochemistry*, 1999, **30**, 765–775.

Hartman, B. and D. Hammond, The use of carbon and sulphur isotopes as a correlation parameter for source identification of beach tars in southern California, *Geochimica Cosmochimica Acta*, 1981, **45**, 309–319.

Hilkert, A.W., C.B. Douthitt, H.J. Schluter, and W.A. Brand, Isotope ratio monitoring gas chromatography/mass spectrometry of *D/H* by high temperature conversion isotope ratio mass spectrometry, *Rapid Communication in Mass Spectrometry*, 1999, **13**, 1226–1230.

Hoefs, J., *Stable Isotope Geochemistry*, 4th ed., Springer, Berlin, 1997.

Ishiwatari, R., M. Uzaki, and K. Yamada, Carbon isotope composition of individual *n*-alkanes in recent sediments, *Organic Geochemistry*, 1994, **21**, 801–808.

Kim, M., M.C. Kennicutt II, and Y. Qian, Polycyclic aromatic hydrocarbon purification procedures for compound specific isotope analysis, *Environmental Science and Technology*, 2005, **39**, 6770–6776.

Kvenvolden, K.A., F.D. Hostettler, J.B. Rapp, and P.R. Carlson, Hydrocarbons in oil residues on beaches of islands of Prince William Sound, Alaska, *Marine Pollution Bull.*, 1993, **26**, 24–29.

Kvenvolden, K.A., F.D. Hostettler, P.R. Carlson, J.B. Rapp, C.N. Threlkeld, and A. Warden, Ubiquitous tarballs with California-source signature on shorelines of Prince William Sound, Alaska, *Environmental Science and Technology*, 1995, **29**, 2684–2694.

Kvenvolden, K.A., P.R. Carlson, A. Warden, and C.N. Threlkeld, Carbon isotopic comparisons of oil products used in the development history of Alaska, *Chemical Geology*, 1998, **152**, 73–84.

Li, M., Y. Huang, M. Obermajer, C. Jiang, L.R. Snowdon, and M.G. Fowler, Hydrogen isotopic compositions of individual alkanes as a new approach to petroleum correlation: Case studies from the Western Canada Sedimentary Basin, *Organic Geochemistry*, 2001, **32**, 1387–1399.

Lichtfouse, E., Compound-specific isotope analysis. Application to archaology, biomedical sciences, biosynthesis, environment, extraterrestrial chemistry, food science, forensic science, humic substances, microbiology, organic geochemistry, soil science and sport, *Rapid Communications in Mass Spectrometry*, 2000, **14**, 1337–1344.

Macko, S.A. and L. Parker, Stable nitrogen and carbon isotope ratios of beach tars on South Texas barrier islands, *Marine Environmental Research*, 1983, **10**, 93–103.

Macko, S.A., L. Parker, and A.V. Botello, Persistence of spilled oil in a Texas salt marsh, *Environmental Pollution Bull.*, 1981, **2**, 119–128.

Mansuy, L., R.P. Philp, and J. Allen, Source identification of oil spills based on the isotopic composition of individual components in weathered oil samples, *Environmental Science and Technology*, 1997, **31**, 3417–3425.

Matthews, D.E. and J.M. Hayes, Isotope-ratio-monitoring gas chromatography-mass spectrometry, *Analytical Chemistry*, 1978, **50**, 1465–1473.

Mazeas, L. and H. Budzinski, Molecular and stable carbon isotopic source identification of oil residues and oiled bird feathers sampled along the Atlantic coast of France after the *Erika* oil spill, *Environmental Science and Technology*, 2002, **36**, 130–137.

McAuliffe, C., Solubility in water of paraffins, cycloparaffins, olefin, acetylene, cycloolefin and aromatic hydrocarbons, *J. Physical Chemistry*, 1966, **70**, 1267–1275.

Meier-Augenstein, W., Applied gas chromatography coupled to isotope ratio mass spectrometry, *J. Chromatography*, 1999, **842**, 351–371.

Murray, A.P., R.E. Summons, C.J. Boreham, and L.M. Dowling, Biomarker and *n*-alkane isotope profiles for Tertiary oils: Relationship to source rock depositional setting, *Organic Geochemistry*, 1994, **22**, 521–542.

Northam, M.A., Correlation of northern North Sea oils: The different facies of their Jurassic source, *Petroleum Geochemistry in Exploration of the Norwegian Shelf*, B.M. Thomas et al. (eds.), Graham and Trotman, London, 1985, 93–99.

O'Malley, V.P., T.A. Abrajano Jr., and J. Hellou, Determination of the $^{13}C/^{12}C$ ratios of individual PAH from environmental samples: Can PAH sources be apportioned, *Organic Geochemistry*, 1994, **21**, 809–822.

Palmer, S.E., Effect of water washing on C_{15} + hydrocarbon fraction of crude oils from northwest Palawan, Philippines, *Amer. Assoc. Petroleum Geologists Bull.*, 1984, **68**, 137–149.

Philp, A.P., Application of stable isotopes and radioisotopes in environmental forensics, *Introduction to Environmental Forensics*, B.L. Murphy and R.D. Morrison (eds.), Academic Press, San Diego, 2002, 99–136.

Pond, K.L., Y. Huang, Y. Wang, and C.F. Kulpa, Hydrogen isotopic composition of individual *n*-alkanes as an intrinsic tracer for bioremediation and source identification of petroleum contamination, *Environmental Science and Technology*, 2002, **36**, 724–728.

Rogers, K.M. and M.M. Savard, Detection of petroleum contamination in river sediments from Quebec City region using GC-IRMS, *Organic Geochemistry*, 1999, **30**, 1559–1569.

Schmidt, T.C., L. Zwank, M. Elsner, M. Berg, R.U. Meckenstock, and S.B. Haderlein, Compound-specific stable isotope analysis of organic contaminants in natural environments: A critical review of the state of the art, prospects, and future challenges, *Analytical and Bioanalytical Chemistry*, 2004, **378**, 283–300.

Schimmelmann, A., A.L. Sessions, C.J. Boreham, D.S. Edwards, G.A. Logan, and R.E. Summons, D/H ratios in terrestrially sourced petroleum systems, *Organic Geochemistry*, 2004, **35**, 1169–1195.

Schwarzbauer, J., L. Dsikowitzky, S. Heim, and R. Littke, Determination of $^{13}C/^{12}C$ ratios of anthropogenic organic contaminants in river water samples by GC-irmMS, *International J. Environmental and Analytical Chemistry*, 2005, **85**, 349–364.

Silverman, S.R. and S. Epstein, Carbon isotopic composition of petroleum and other sedimentary organic materials, *Amer. Assoc. Petroleum Geologists Bull.*, 1958, **42**, 998–1012.

Smallwood, B.J., R.P. Philp, and J.D. Allen, Stable carbon isotopic composition of gasolines determined by isotope ratio monitoring gas chromatography mass spectrometry, *Organic Geochemistry*, 2002, **33**, 149–159.

Sofer, Z., Stable carbon isotope composition of crude oils: Application to source depositional environments and petroleum alteration, *Amer. Assoc. Petroleum Geologists Bull.*, 1984, **68**, 31–49.

Stahl, W.J., Carbon and nitrogen isotopes in hydrocarbon research and exploration, *Chemistry and Geology*, 1977, **20**, 121–149.

Sun, Y., Z. Chen, S. Xu, and P. Cai, Stable carbon and hydrogen isotopic fractionation of individual n-alkanes accompanying biodegradation: Evidence from a group of progressively biodegraded oils, *Organic Geochemistry*, 2005, **36**, 225–238.

Uzaki, M., K. Yamada, and R. Ishiwatari, Carbon isotope evidence for oil-pollution in long chain normal alkanes in Tokyo Bay sediments, *Geochemical Journal*, 1993, **27**, 385–389.

Wang, Z., M. Fingas, M. Landriault, L. Sigouin, B. Castle, D. Hostetter, D. Zhang, and B. Spencer, Identification and linkage of tarballs from the coasts of Vancouver Island and Northern California using GC/MS and isotopic techniques, *J. High Resolution Chromatography*, 1998, **21**, 383–395.

Wang, Y., Y. Huang, J.N. Huckins, and J.D. Petty, Compound-specific carbon and hydrogen isotope analysis of sub-parts per billion level waterborne petroleum hydrocarbons, *Environmental Science and Technology*, 2004, **38**, 3689–3697.

Yanik, P.J., T.H. O'Donnell, S.A. Macko, Y. Qian, and M.C. Kennicutt II, The isotopic compositions of selected crude oil PAHs during biodegradation, *Organic Geochemistry*, 2003, **34**, 291–304.

Zymax Forensics, unpublished report, San Luis Obispo, CA, 2003.

7 Emerging CEN Methodology for Oil Spill Identification

Asger B. Hansen* (NERI, Denmark)

Per S. Daling, Liv-Guri Faksness, and Kristin R. Sörheim (SINTEF Chemistry, Norway)

Paul Kienhuis (RIZA, The Netherlands)

Rolf Duus (Standards Norway, Norway)

*Corresponding author: aha@dmu.dk

Since 1991 the existing Nordtest method on oil spill identification (Nordtest, 1991) has formed an important forensic "platform" in relation to oil spill identification, not only in the Scandinavian countries, but also in other European countries following its recommendation in and adoption to the Bonn Agreement[1] Counter Pollution Manual. Also countries outside Europe have adopted the concept and methodology for oil spill analysis and identification described in this Nordtest method. It basically incorporates a stepwise procedure including initial screening by GC/FID of all samples for characterization and to exclude obviously nonmatching candidate source samples. This step is followed by GC/MS fingerprinting of a few selected samples (spill and potentially matching candidate source samples) recording a suite of key or target petroleum compounds (biomarkers and polycyclic aromatic compounds, PACs). After comparing the GC/MS chromatograms in order to pinpoint possible differences, the conclusion according to the chemical analysis will either be identity (matching chromatograms) or nonidentity (nonmatching chromatograms). When evaluating the chromatograms (both GC/FID and GC/MS), weathering of the oil samples has to be taken into account as the criterion for identity excludes differences arising from weathering.

Experiences over the past 10–15 years with the existing Nordtest (1991) methodology for oil spill identification, however, has shown some need for improvements. For example, the evaluation of chromatograms was generally achieved by qualitative means (i.e., visual comparison) just as the influence of weathering was not always straightforward to interpret, facts that rendered the evaluation and final conclusions to be flawed by subjectivity and hence questionable. Besides, over this time period advances in both analytical and interpretive methods had opened the possibility to obtain more quantitative, objective, and defensible means for verification of the results. Thus in 2000, Nordtest[2] initiated Phase 1 of the

---

[1] The Bonn Agreement is a multilateral agreement by North Sea coastal states, which together with EU will offer: (1) mutual assistance and cooperation in combating pollution, and (2) surveillance as an aid to detecting and combating pollution and to prevent violations of antipollution regulations.

[2] Nordtest is an institution under the Nordic Council of Ministers acting as a joint Nordic body in the field of technical testing and standardization.

project called "Revision of the Nordtest Methodology for Oil Spill Identification" with participation of the national forensic oil spill laboratories in Denmark, Finland, Norway, Sweden, and the Battelle Memorial Institute (Duxbury), USA. Phase 1 (2000–2001) of that project included: (1) a review and assessment of recently published literature on oil spill identification and petroleum geochemistry, and (2) an update of the existing Nordtest (1991) method into a technically more robust and legally defensible oil spill identification methodology by introducing improved laboratory techniques for sample preparation and cleanup, chromatographic analysis, and data evaluation tools. The improved methodology was still based on tiered GC/FID screening and GC/MS fingerprinting procedures, but it also included more objective criteria for evaluation by introducing quantitative diagnostic ratios of key petroleum compounds that could be tested statistically and stricter analytical protocols including QA/QC criteria (Faksness et al., 2002a).

In Phase 2 of the project (2001–2002), the updated Nordtest methodology was evaluated through a Round Robin exercise arranged by SINTEF, Norway, and with the participation of 12 laboratories from 10 countries. Seven oil samples (two artificially weathered "spill" samples and five possible sources) were analyzed according to the recommended analytical protocols of the revised methodology. The Round Robin exercise was a "difficult case," because the two spill samples and three of the suspected sources were qualitatively similar and thereby quite highly correlated to one another. These samples were crude oils from the same oil field in the North Sea, but from different production wells. The Round Robin exercise, however, demonstrated the potential of the updated methodology as a strong technically defensible tool for oil spill identification due to its ability to distinguish between qualitatively similar oils from a spill and potential candidate source samples (Faksness et al., 2002b). Following the Round Robin exercise, the revised methodology has been implemented by several forensic laboratories in Europe and has been used in connection with recent major oil spills, e.g., the *Tricolor* (British Channel, December 14, 2002) and the *Prestige* (off the coast of Galicia, Spain, November 20, 2002).

In 2002, Nordtest proposed the revised methodology (Daling et al., 2002) as a new standard for oil spill identification to the European Committee for Standardization (CEN), which established a task force (CEN BT/TF 120) to evaluate the proposal and eventually prepare a new standard. The standardization process has been split into two work items: (1) sampling and (2) analytical methodology and interpretation of results, involving two working groups that accordingly should produce two guidelines (CEN Technical Reports) on oil spill identification.

7.1 Introduction

The objective of the emerging CEN methodology is to provide a forensic tool for the identification of waterborne oil by comparing samples from spills with those of suspected sources. This methodology, which has been based on a technical revision of the existing Nordtest method (Nordtest, 1991), should be capable of providing both administrative and legal support to the prosecution of an offender ("potential responsible party" — PRP) that has violated national or international regulations by illegally discharging mineral oil into the marine environment. The two working groups under the CEN task force BT/TF 120 have produced two draft guidelines (CEN Technical Reports) that describe the new methodology on oil spill identification — waterborne petroleum and petroleum products accordingly:

- Part 1 — Sampling (CEN, 2005)
- Part 2 — Analytical methodology and interpretation of results (CEN, 2006)

In 2004, RIZA organized a Round Robin (RR) exercise for oil spill identification. The first exercise (RR-2004) dealt with light fuel oil distillates (gas oils). Fifteen laboratories from nine countries participated in the study, and the results of this RR exercise (Kienhuis,

2004) have been taken into account for refining the CEN methodology. A second Round Robin exercise took place in 2005 (RR-2005) and involved bilge water samples. Again more than 10 oil spill laboratories from 9 countries participated (Dahlmann and Kienhuis, 2005), and the emerging CEN methodology now seems to have been implemented by many forensic oil spill laboratories in Europe.

In 2005, SINTEF (Norway) produced a standard oil mixture to facilitate the identification of target compounds and compound groups in oil samples. The mixture is a combination of three crude oils (from Russia, Sicily, and the North Sea) and a heavy bunker oil (IFO-180), and it contains all compounds mentioned in the CEN guidelines. The oil mixture can be obtained from SINTEF, Norway, together with all relevant and integrated chromatograms.

This chapter will describe Part 2 (CEN, 2006) of the emerging CEN methodology that covers the tiered analytical approach, the selected diagnostic target compounds, and the data treatment, whereas sampling techniques and handling of oil samples prior to their arrival at the forensic oil spill laboratory, as described in Part 1 (CEN, 2005), will not be covered here.

7.2 Scope of the CEN Methodology

The scope of the CEN methodology on oil spill identification is to provide a methodology to identify waterborne oils spilled in marine, estuarine, and aquatic environments based on detailed analytical and processing procedures for the comparison of samples from spills with those of suspected sources. When suspected sources are not available, the methodology may be used to characterize the spill as far as possible with respect to oil type and origin.

The methodology is restricted to petroleum and petroleum products containing a significant proportion of petroleum hydrocarbons with boiling points above 200°C. Examples are

- Crude oils
- Light refined products like diesel oils or gas oils
- Heavy refined products like heavy fuel oils, bunker oils, and vaccum residues
- Lubricating oils
- Mixtures of bilge and sludge samples

Still, while the general concepts described in the methodology have a limited applicability for some kerosenes and condensates, it may not be applicable for gasoline.

The described methodology is not intended for oil spilled to groundwater and soil. The chromatograms of oil extracted from soil and groundwater may contain reduced and/or additional compounds compared to the candidate source sample. Including such samples in the methodology would require additional extraction and cleanup methods together with an accounting of which compounds could possibly be reduced and/or which additional peaks could be expected, resulting in a diversion of the final conclusion. Such issues are beyond the scope of the CEN methodology. However, in such cases, including oil spilled to groundwater or soil where appropriate sample preparation and the potential for compound reductions and/or additions are appreciated, a positive match observed according to the CEN methodology would be valid.

7.3 Strategy for Identifying Oil Spills

When an oil spill has been observed, samples should be collected from the current spill and from potential responsible parties, such as suspected ships or other sources, in due course by appropiate authorized personnel. All samples should subsequently be sent either via an authorized "Sampling Coordinator" or directly to the analytical laboratory for forensic characterization and potentially for identification of the spill's source. In the context of the CEN methodology, the process of identifying the source of a spilled oil implies comparison of the chemical composition of the spilled oil with that of candidate source samples.

Conceptually, two results can be achieved in forensic oil spill investigation — "identity" and "nonidentity," depending on whether spill and candidate source samples are "identical" or "nonidentical." "Identical" per se requires all measurable data to match exactly. This criterion is practically and technically impossible to fulfill, and therefore the definition of "identical" is rephrased in operational terms: two samples are identical beyond reasonable doubt if and only if (1) no significant differences in the recorded GC/FID and GC/MS data are observed (i.e., matching chromatograms) or (2) any observed differences are not genuine, but stem from changes introduced in the collected samples after the spill (e.g., due to weathering, contamination, mixing, or degradation). If these criteria are met, the condition of "identity" is achieved. Otherwise, samples are "nonidentical," and "nonidentity" is achieved. The task of looking for genuine significant differences in chemical composition by comparing chromatograms, instead of proving an all-encompassing match, is conceptually more logical and feasible to comply with. According to this definition, it is important to realize that only nonidentity based on genuinely significant differences between samples (i.e., nonmatching chromatograms) can be proved directly (rejection of the null hypothesis), whereas identity always must be based on the process of "ruling out" differences.

Identity must be tested by analyzing and comparing the detailed chemical composition of the selected samples by chemical fingerprinting of a suite of diagnostic or target petroleum compounds. If no or only insignificant differences (i.e., differences of chromatographic peaks being smaller than the analytical variance) are observed, identity could be concluded as being beyond reasonable doubt. On the other hand, if "true" differences (i.e., differences not related to changes in the chemical composition introduced after the spill, e.g., from weathering, etc.) are significant (i.e., larger than the analytical variance observed for the diagnostic compounds), nonidentity could be concluded. In practice, the process of comparing and evaluating chromatograms has been graduated in the CEN methodology by introducing four operational terms (i.e., positive match, probable match, inconclusive, and nonmatch) (cf. Section 7.4.6), to cover the comparison of samples. Only when a *positive match* has been obtained can *identity* be concluded as being beyond reasonable doubt.

In Europe, forensic oil spill identification is performed not only by laboratories that analyze oil samples on a regular basis, but also by laboratories that only compare samples a few times a year. Traditionally, the common practice has been to analyze oil samples qualitatively and compare the chromatograms and ion fragmentograms visually — as per the original Nordtest (1991) protocol. The results of such comparisons depend heavily on the skill and experience of the analyst, and laboratories which rarely analyze oil samples may experience difficulties in reaching sound forensic conclusions. Therefore, the new CEN methodology has introduced the use of diagnostic ratios[3] as an additional and more objective and defensible tool for comparison. The criteria for selecting such ratios take into account the weathering and degradation behavior of the target petroleum compounds and their varying amount and composition in oils of different types and petrogenic origin (Faksness et al., 2002a). To reduce the analytical variance, ratios are preferably generated by using the peak area or height of target compounds, which preferably are recorded by the same m/z value and approximately within the same retention time window. The resulting ratios are successively compared using the repeatability limit ($r_{95\%}$) as the test method. Optionally, diagnostic ratios can also be generated on the basis of chromatographic peaks that have been fully quantified by traditional quantitative chromatographic analysis. Strict QA/QC procedures should be observed when doing fully quantitative analyses.

The use of diagnostic ratios for comparison of oil samples is based on GC/MS data of 29 diagnostic ratios generated from a suite of alkylated polycyclic aromatic compounds

---

[3] Diagnostic ratios (DR) — ratios between the peak height or peak area of single compounds or groups of compounds selected for their diversity in the chemical composition in petroleum and petroleum products and their reported response to weathering and degradation processes.

(PACs) and petroleum biomarkers that are robust against weathering and that have been selected to cover genuine differences in oil samples and oil types. Whereas most of the ratios may be used when crude oils, bunker oils, and bilge samples are involved, only a limited number of ratios may be appropriate for lighter fuel oils (e.g., kerosene, paraffin, diesel, gas oil) because some of the higher-boiling compounds may not be present in such light-end refined products. In spill cases where the spilled oil has only been slightly exposed to weathering, three ratios involving acyclic isoprenoids (i.e., n-C17/pristane, n-C18/phytane, and pristane/phytane ratios) generated from GC/FID data and four ratios involving sesquiterpanes (in the n-C13 to n-C16 retention time window) generated from GC/MS data can also be included for comparison, provided a weathering check shows that these compounds have not been significantly influenced by weathering.

Before integrating (by area or height) the compounds required for generating the selected diagnostic ratios, a visual inspection of the relevant ion fragmentograms should be carried out in order to eliminate those target peaks not present in sufficient amount to fulfill the required signal-to-noise (S/N) criteria and hence not capable of generating robust diagnostic ratios. A visual comparison of the ion fragmentograms is also recommended to exclude obviously different samples and spot highly diagnostic compounds specific for the actual case (e.g., unusual biomarkers). After the comparison and evaluation of the diagnostic ratios using the repeatability limit as criteria, a second visual comparison of the ion fragmentograms of the relevant samples, one by one, should also be carried out to verify (ground truth) the conclusion.

7.4 Tiered Levels of Analysis and Data Treatment

7.4.1 Decision Chart for Identifying Oil Spills

The methodology for identification of oil spills is divided into three tiered levels of analytical procedures and data treatments according the decision or flowchart shown in Figure 7-1.

This operational flowchart guides the reader through the individual steps of the tiered procedures by linking each step of the procedure with the operation performed (analysis/evaluation) at the previous step and its resulting decision until a final conclusion regarding identity can be made. In the flowchart, squared gray boxes refer to operations to be performed, and rounded gray boxes refer to evaluations/conclusions to be made. The flowchart comprises the following steps:

- visual characterization of samples
- sample preparation and cleanup
- GC/FID screening of all samples — visual characterization/classification of oil types — weathering check — calculation of acyclic isoprenoid diagnostic ratios
- GC/MS fingerprinting of selected samples (spill(s) and candidate source(s)) — visual evaluation of chromatograms — calculation of diagnostic ratios — weathering check of PACs (optional)
- Comparison of diagnostic ratios (repeatability limit)
- Conclusion and reporting

7.4.2 Visual Characterization and Preparation/Cleanup of Oil Samples

When arriving at the analytical laboratory, all samples should be carefully examined and described with respect to type (e.g., spill or reference), matrix (e.g., water, sand, feather, etc.), amount, container (glass, plastic, etc.), and general condition. Eventually, all samples should be photographed to assist the visual description and document their condition at arrival at the laboratory. Especially, any sign of jeopardizing or "missing links" regarding the chain-of-custody should be reported.

The applied procedures for forensic identification of spilled oil must strictly observe that any manipulation performed during sample preparation and cleanup can alter its chemical composition and thus weaken the power of

234 Oil Spill Environmental Forensics

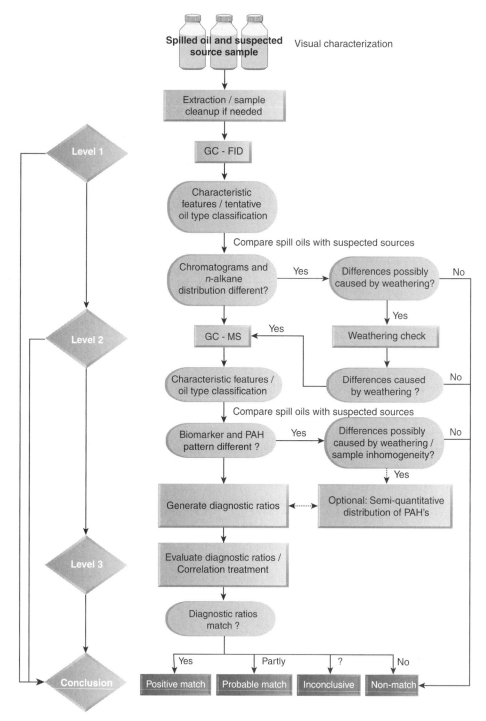

Figure 7-1 Decision/flowchart of the CEN oil spill identification methodology.

evidence. Therefore, sample preparation and cleanup should generally be restricted to a minimum, and in any case precisely stated in the final report. After removal of any pieces of debris (e.g., wood, fabric, feathers, etc.), preparation is generally performed by diluting the sample to appropriate concentrations for GC/FID and GC/MS analysis. For samples contaminated with polar compounds (e.g., bird feathers) or that contain a high amount of heavy components (e.g., heavy bunker oil) a simple cleanup on a silica/alumina column can be used. Descriptions of preparation and cleanup of oil samples are beyond the scope of this chapter, but more detailed procedures for the different sample types can be found in the CEN guidelines (CEN, 2006).

7.4.3 Level 1 — GC/FID Screening

After sample preparation, all samples (i.e., both samples from the oil spill and from suspected sources) are initially characterized by GC/FID screening. This will generally render a descriptive "picture" of the dominating petroleum hydrocarbons in the oil sample (e.g., the overall boiling range and prominence of individual resolved n-alkanes and major isoprenoids as illustrated in Figure 7-2). The GC/FID chromatograms also provide information on the weathering extent of the spilled oil and on any "characteristic features" or "contaminating" components present in the samples. Generally, the n-alkanes are the most characteristic and dominating peaks distributed regularly over the entire retention interval, except for extensively weathered oil samples, such as water samples collected from thin oil films (sheens), highly biodegraded crude oils (biodegradation in the reservoir), and certain refined petroleum products (like lubricating and hydraulic oils).

The GC/FID chromatograms also enable the calculation of diagnostic ratios derived from acyclic isoprenoids (diterpanes) like the n-C17/pristane, n-C18/phytane, and pristane/phytane ratios, unless the spill sample is too weathered, or the amount of these compounds is low relative to the UCM hump ("unresolved complex mixture"), as can often be observed in bunker oils. These ratios can also be indicative of biodegradation of the spilled oil, as they can monitor the effect of microbial degradation at the spill site by the preferential loss of n-alkane hydrocarbons compared to the isoprenoids. The acyclic isoprenoid ratios, however, are easily influenced by weathering and should therefore not be included in any evaluation without a thorough weathering check.

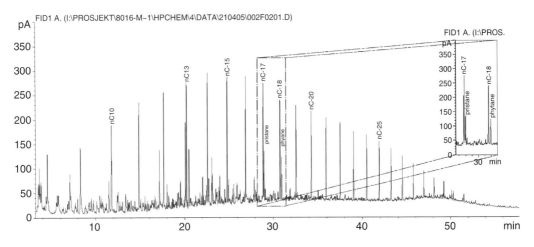

Figure 7-2 GC/FID chromatogram of a mildly weathered fuel oil displaying the dominating n-alkanes together with the acyclic isoprenoids, pristane and phytane (inserted).

7.4.3.1 Evaluation of Weathering

If the chromatograms of a spill and candidate source oil are different and the observed differences could possibly be caused by weathering, a "weathering check" is recommended. Figure 7-3A gives an example of a weathering check simply by overlaying chromatograms of a spill sample and a suspected source (in this case, gas oil). The normalization/manipulation of the chromatograms to a comparable attenuation can easily be done by expanding one of the chromatograms vertically until the peaks in the n-C20 to n-C24 range are of the same height. Alternatively, the weathering check can also be obtained by integrating (by peak height or area) a homologous series (e.g., the n-alkanes) in the GC/FID chromatograms, then normalizing the peaks to nonweathered compounds (e.g., the mean of n-C20 to n-C24), and eventually displaying the normalized peaks in a bar chart (e.g., prepared by Excel^{TM}) for comparison of the spill sample with a nonweathered source sample; see Figure 7-3B for an example of such a normalized n-alkane bar chart of two gas oil samples. The bar chart presentation of the n-alkane distribution in oil samples may also provide information on

Table 7-1 Diagnostic Ratios Derived from Selected n-Alkanes and Acyclic Isoprenoids (Diterpanes) Recorded by GC/FID

Diagnostic Ratio	Definition
DR-nC17/Pri	n-heptadecane/pristane
DR-nC18/Phy	n-octadecane/phytane
DR-Pri/Phy	pristane/phytane

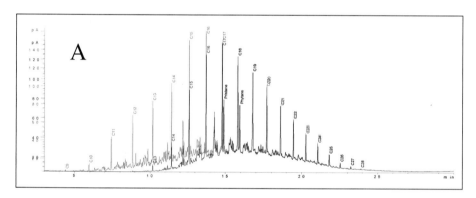

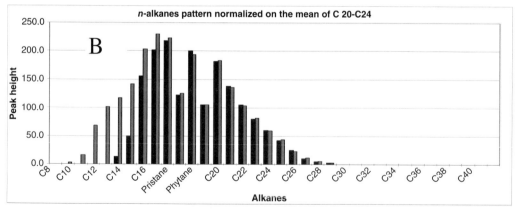

Figure 7-3 Examples of weathering checks of gas oils by GC/FID. A. Simple overlaying of chromatograms. B. Bar chart comparison of normalized n-alkane distributions.

possible wax/paraffin redistribution as a part of the weathering process (Strøm-Kristiansen et al., 1997).

If the comparison of the GC/FID chromatograms of the spill samples with the candidate source samples reveals differences in the hydrocarbon distribution, in the unresolved complex mixture distribution, and/or in the acyclic isoprenoid ratios that obviously are not caused by weathering and that are significantly higher than the analytical variance, then these source samples are nonmatching (i.e., nonidentity is achieved) and should be ruled out and eliminated from additional levels of analysis (cf. Figure 7-1). If, however, there are any doubts about the conclusions, the samples should still be considered as potential sources and analyzed in accordance with Level 2 of the flowchart (Figure 7-1).

7.4.4 Level 2 — GC/MS Fingerprinting

At this level, selected spill samples and those candidate source samples that have not been eliminated by the GC/FID screening (Level 1) are analyzed by tiered GC/MS fingerprinting, generally performed in selected ion-monitoring mode (GC/MS-SIM). The GC/MS analysis at this level is used for characterizing and assessing the content and distributions of a suite of diagnostic and target alkylated PAC and petroleum biomarker analytes, from which recommended diagnostic ratios can successively be generated.

7.4.4.1 Diagnostic Ratios from GC/MS Fingerprinting

Analytical data from PAC and biomarker compounds form the basis for generating a suite of diagnostic ratios. It is generally recommended that applied ratios are based on single compounds recorded at the same m/z value (e.g., m/z 192: DR-2-MPhe/1-MPhe; Table 7-2) to eliminate the mass spectrometer's varying response for different ions. In many cases, however, it may be of diagnostic value also to include ratios based on different m/z groups (e.g., m/z 212/206: DR-C2-Dbt/C2-Phe; Table 7-2) to assess different levels of various compound groups.

For the PACs, diagnostic ratios based on peak areas are generally recommended as most compounds are well-resolved peaks with smooth baselines that are easily integrated for area; besides, some diagnostic ratios are based on whole isomer groups (e.g., DR-C2-Dbt/C2-Phe; cf. Section 7.4.4.2) in which case the ratio has to be based on the total area. For the biomarkers, however, integration of peak heights are generally recommended; diagnostic ratios are often based on peaks that are not well-

Table 7-2 Recommended Diagnostic Ratios (DR) Derived from Alkylated PACs[1]

Ratio Name	Definition	m/z Value
DR-2-MPhe/1-MPhe[2]	2-Methylphenanthrene/1-Methylphenanthrene	192
DR-4-MDbt/1-MDbt	4-Methyldibenzothiophene/1-Methyldibenzothiophene	198
DR-C2-Dbt/C2-Phe	C2-Dibenzothiophenes/C2-Phenanthrenes	212/206
DR-C3-Dbt/C3-Phe	C3-Dibenzothiophenes/C3-Phenanthrenes	226/220
DR-C3-Dbt/C3-Chr	C3-Dibenzothiophenes/C3-Chrysenes	226/270
DR-Retene/C4-Phe	Retene (7-isopropyl-1-methylphenanthrene)/C4-Phenanthrenes	234
DR-BaF/4-MPy	Benzo(a)fluorene/4-Methylpyrene	216
DR-B(b+c)F/4-MPy	Benzo(b+c)fluorene/4-Methylpyrene	216
DR-2-MPy/4-MPy	2-Methylpyrene/4-Methylpyrene	216
DR-1-MPy/4-MPy	1-Methylpyrene/4-Methylpyrene	216

[1] C# here denotes the number of carbon atoms in the PAC alkyl substituent (i.e., C2 could either denote dimethyl or ethyl and C3 either trimethyl or ethyl-methyl).
[2] In cases where the double peaks of the methylphenanthrenes are not properly resolved, this diagnostic ratio could alternatively be generated from the area of the double peaks (i.e., DR-[(3-MPhe + 2-MPhe)/(9/4-MPhe + 1-MPhe)]).

resolved and with noisy baselines, in which case ratios based on peak heights are more robust. When integrating individual peaks, the use of coeluting peaks should be avoided, if possible. If peaks of, for example, C29Ts and C29αβ coelute (cf. Figure 7-6), use either the peak height of the combined peaks or the total area of the combined peaks in the combined ratio (C29Ts + C29αβ)/C30αβ.

Diagnostic ratios could be calculated either as A/B or A/(A + B), where the latter will always give values in the range of 0 to 1. The use of ratios compared to single components generally lowers the analytical variance, and ratios calculated as A/(A + B) result in lower relative standard deviations (RSDs) compared to ratios calculated as A/B. However, while ratios calculated as A/B give constant RSDs independent of the numerical values of A and B, this is not the case for ratios calculated as A/(A + B), where RSDs depend on the actual ratio value (Kienhuis, 2005; Sect. 3.3). Thus, ratios calculated as A/(A + B) are less suitable for comparison, since the confidence interval has to be adjusted to reflect that the RSD depends on the actual value. Generally, it is therefore recommended that diagnostic ratios (DR) between two distinct and well-resolved peaks (A and B) should be based on the ratio formula:

- DR = A/B, or
- DR = 100 × A/B (in %, depending on the preference of the user)

7.4.4.2 Diagnostic Ratios Derived from Alkylated Polycyclic Aromatic Compounds

Petroleum originating from different oil fields and petrogenic provinces generally has sufficient differences in the distribution of alkylated PAC isomers to be of diagnostic value. Several PAC diagnostic ratios have recently been described and applied for oil spill identification (Wang et al., 1999; Weiss et al., 2000; Daling et al., 2002; Stout et al., 2002; Wang and Fingas, 2003). The most common diagnostic groups of alkylated PAC isomers are the phenanthrenes (3-ring) and the heterocyclic dibenzothiophenes (3-ring) that can be found in all sample types mentioned in Section 7.2. However, other alkylated PAC isomer groups can also be of diagnostic value, like the fluorenes (3-ring) and the chrysenes (4-ring). In the CEN methodology, additional ratios of the methylfluoranthenes/methylpyrenes/ benzofluorenes (4-ring) are also included, because compounds within this group may show significant variation between different oils. Thus, in total, 10 diagnostic ratios derived from alkylated PAC isomers are recommended as listed in Table 7-2.

To facilitate and ensure proper identification of the chromatographic peaks used for generating the recommended diagnostic ratios, the CEN methodology includes a series of relevant ion fragmentograms as shown in Figure 7-4A to 7-4I. Of particular importance are the methyl-phenanthrenes (m/z 192), as they may be used to differentiate between crude oils and heavy bunker fuel oils (cf. Figure 7-4A). For example, it has been reported that in crude oils the first pair of peaks (i.e., the 3-methyl- and 2-methylphenanthrenes) are generally less abundant then the second pair of peaks (i.e., the 9-/4- and 1-methylphenanthrenes), while in heavy fuel oil this is reversed (Dahlmann, 2003). The isomer compound cluster recorded by m/z 216 (i.e., the methylfluoranthenes, methylpyrenes, and benzofluorenes) may also provide valuable diagnostic characteristics that have shown to be relatively stable and suitable especially for comparing light fuel oil samples like gas oils. More detailed features of individual oil types encountered in most environmental oil spill situations have recently been characterized and described by Dahlmann (2003).

To provide additional information, a semi-quantitative histogram (e.g., Excel™ bar chart as in Figure 7-5) established from a suite of selected PAC homologues by their relative distribution [individual compounds or compound groups normalized to a nonweathered compound, e.g., C30-17α(H),21β(H)-hopane], may be used as a supplementary diagnostic fingerprint and as a weathering check of the PAC homologues.

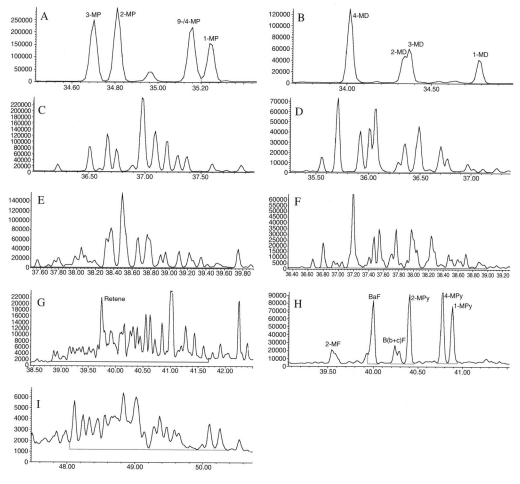

Figure 7-4 GC/MS ion fragmentograms of diagnostic alkylated PACs. 4A, methylphenanthrenes (m/z 192); 4B, methyldibenzothiophenes (m/z 198); 4C, C2-phenanthrenes (m/z 206); 4D, C2-dibenzothiophenes (m/z 212); 4E, C3-phenanthrenes (m/z 220); 4F, C3-dibenzothiophenes (m/z 226); 4G, C4-phenanthrenes including retene (m/z 234); 4H, methylfluoranthenes/methylpyrenes/benzofluorenes (m/z 216); 4I, C3-chrysenes (m/z 270).

7.4.4.3 Diagnostic Ratios Derived from Petroleum Biomarkers

Biomarkers are naturally occurring, ubiquitous, and stable hydrocarbons that are present in crude oils and most petroleum products. They are complex "molecular fossils" derived from once-living organisms. Biomarkers' specificity, diversity, complexity, and relative high resistance to weathering therefore make them extremely useful as diagnostic "markers" in the characterization and differentiation of spilled oils and candidate source oils (Stout et al., 2000; see also Chapter 3 herein). The most common biomarkers used by organic geochemists include sesquiterpanes (e.g., drimanes), triterpanes (e.g., hopanes), diasteranes/steranes (e.g., diacholestanes/cholestanes), and mono- and triaromatic steroids (Peters et al., 2005).

By exploiting the experience gained by petroleum exploration and production geochemistry, combined with the results of an extensive analysis of a large number of oils (Faksness et al., 2002a), a suite of 19 diagnostic biomarker ratios have been selected as

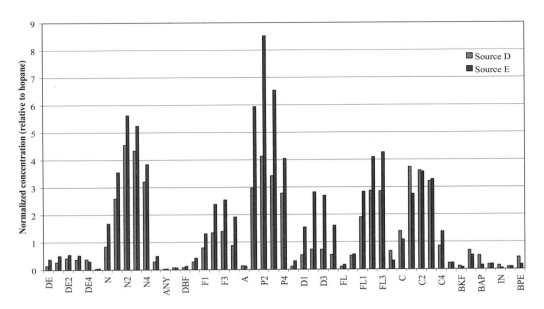

Figure 7-5 Relative distribution of PAC homologues in two different IFO-180 bunker oils (from different refineries).

technically defensible indices to differentiate among qualitative similar oils from spills and available sources.

Examples of ion fragmentograms displaying the target peaks of hopanes and other tri- and pentacyclic triterpanes, rearranged (diasteranes) and regular steranes, and triaromatic steroids are presented in Figures 7-6, 7-7, and 7-8, respectively. Accordingly, definitions of the recommended ratios within each group of biomarkers are listed in Tables 7-3, 7-4, and 7-5, respectively.

Aromatic steroid hydrocarbons (derived from steranes) may also be used as diagnostic compounds suitable for oil spill identification as has recently been described by Barakat et al. (2002) with respect to terrestrial spilled oil. In the CEN guideline, three ratios derived from triaromatic steroids are recommended as diagnostic tools. The monoaromatic steroids, typically recorded by m/z 253, are not recommended here as they may coalesce with higher paraffins (above C19) also recorded at this m/z value on GC/MS instruments with low mass resolution (e.g., quadropoles).

7.4.4.4 Optional Diagnostic Ratios Derived from Sesquiterpanes

Sesquiterpanes may be included as an optional group of biomarkers. Sesquiterpanes include a group of bicyclic (C14–C16 polymethyl-substituted decalins) biomarkers that comprise one of the largest of the terpenoid classes. Thus, sesquiterpanes, including drimane and eudesmane, are common components of crude oils and ancient sediments. In lighter to middle petroleum products like jet fuel and diesel, where refining processes have removed most of the higher-molecular-weight tetracyclic steranes and pentacyclic triterpanes, the lower-molecular-weight bicyclic sesquiterpanes are generally concentrated. In GC/MS chromatograms these compounds may be examined by their characteristic fragment ions (m/z 123, 179, 193, and 207), from where highly diagnostic ratios for correlation, differentiation, and source identification of lighter- to middle-range petroleum products may be acquired (Wang et al., 2005). Like any other low-boiling compounds, however, the sesquiterpanes are

Emerging CEN Methodology for Oil Spill Identification   241

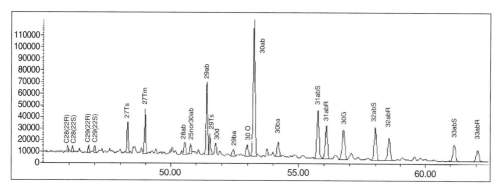

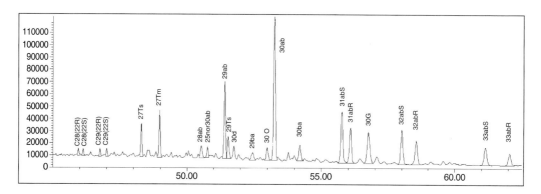

Figure 7-6 GC/MS ion fragmentogram of tri- and pentacyclic triterpanes ("hopanes") recorded at m/z 191 in the SINTEF standard oil mixture.

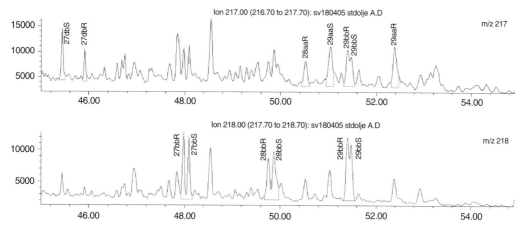

Figure 7-7 GC/MS ion fragmentogram of tetracyclic rearranged (diasteranes) and regular 17α(H)-steranes, and 17β(H)-steranes recorded at m/z 217 and m/z 218, respectively, in the SINTEF standard oil mixture.

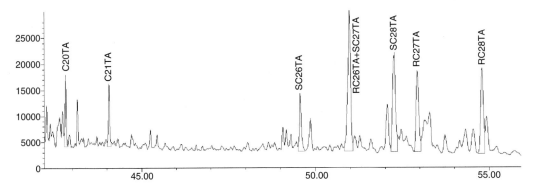

Figure 7-8 GC/MS ion fragmentogram of triaromatic steroids recorded at m/z 231 in the SINTEF standard oil mixture.

Table 7-3 Recommended Diagnostic Ratios (DR) Derived from Tri- and Pentacyclic Triterpanes

Abbreviation	Compound Name	m/z Value
C28 (22S)	C28 tricyclic triterpane (Cheilanthane)	191
C28 (22R)	C28 tricyclic triterpane (Cheilanthane)	191
C29 (22S)	C29 tricyclic triterpane (Cheilanthane)	191
C29 (22R)	C29 tricyclic triterpane (Cheilanthane)	191
C27Ts	C27 18α(H)-22,29,30-trisnorneohopane	191
C27Tm	C27 17α(H)-22,29,30-trisnorhopane	191
C28αβ	C28 17α(H),21β(H)-28,30-bisnorhopane	191
25norC29αβ	C29 17α(H),21β(H)-25-norhopane	191
C29αβ	C29 17α(H),21β(H)-30-norhopane	191
C29Ts	C29 18α(H)-30-norneohopane	191
C30d	C30 15α-methyl-17α(H)-27-norhopane (diahopane)	191
C29βα	C29 17β(H),21α(H)-30-norhopane (normoretane)	191
C30O	C30 18α(H)-oleanane	191
C30αβ	C30 17α(H),21β(H)-hopane	191
C30βα	C30 17β(H),21α-(H)-hopane (moretane)	191
C31αβS	C31 17α(H),21β(H),22S-homohopane	191
C31αβR	C31 17α(H),21β(H),22R-homohopane	191
C30G	C30 Gammacerane	191
C32αβS	C32 17α(H),21β(H),22S-bishomohopane	191
C32αβR	C32 17α(H),21β(H),22R-bishomohopane	191
C33αβS	C33 17α(H),21β(H),22S-trishomohopane	191
C33αβR	C33 17α(H),21β(H),22R-trishomohopane	191

Ratio Name	Definition	Ratio Name	Definition
DR-C28	C28(S+R)/C30αβ	DR-C29αβ	C29αβ/C30αβ
DR-C29	C29(S+R)/C30αβ	DR-C29Ts	C29Ts/C30αβ
DR-(C28+C29)	C28(S+R) + C29(S+R)/C30αβ	DR-C30d	C30d/C30αβ
DR-C27Ts	C27Ts/C27Tm	DR-C30O	C30O/C30αβ
DR-C28αβ	C28αβ/C30αβ	DR-C30G	C30G/C30αβ
DR-25norC30αβ	25norC29αβ/C30αβ		

Table 7-4 Recommended Diagnostic Ratios (DR) from Rearranged (Diasteranes) and Regular 14α(H)- and 14β(H)-Steranes

Abbreviation	Compound Name	m/z Value
C27dbS	C27 13β(H),17α(H),20S — diacholestane (diasterane)	217
C27dbR		217
C28ααR	C27 13β(H),17α(H),20R — diacholestane (diasterane)	217
C29ααS	C28 24-methyl-5α(H),14α(H),17α,20R — cholestane	217
C29ββR	C29 24-ethyl-5α(H),14α(H),17α,20S — cholestane	217
C29ββS	C29 24-ethyl-5α(H),14β(H),17β(H),20R — cholestane	217
C29ααR	C29 24-ethyl-5α(H),14β(H),17β(H),20S — cholestane	217
C27ββR	C29 24-ethyl-5α(H),14α(H),17α(H),20R — cholestane	218
C27ββS	C27 5α(H),14β(H),17β(H),20R — cholestane	218
C28ββR	C27 5α(H),14β(H),17β(H),20S — cholestane	218
C28ββS	C28 24-methyl-5α(H),14β(H),17β(H),20R — cholestane	218
C29ββR	C28 24-methyl-5α(H),14β(H),17β(H),20S — cholestane	218
C29ββS	C29 24-ethyl-5α(H),14β(H),17β(H),20R — cholestane	218
	C29 24-ethyl-5α(H),14β(H),17β(H),20S — cholestane	

Ratio Name	Definition
DR-C29ααS	C29ααS/C29ααR
DR-C29ββ	C29ββ(R+S)/C29αα(S+R)[1]
DR-C27ββSTER	C27ββ(R+S)/[C28ββ(R+S) + C29ββ(R+S)][1]
DR-C28ββSTER	C28ββ(R+S)/[C27ββ(R+S) + C29ββ(R+S)][1]
DR-C29ββSTER	C29ββ(R+S)/[C27ββ(R+S) + C28ββ(R+S)][1]

[1] Although it is generally recommended to use peak heights for generating the biomarker diagnostic ratios, the use of peak areas may in some cases be justified as with the steranes. Thus, as the ββR- and ββS-isomers are often not well-resolved, it is recommended to measure the C27ββ(R+S), C28ββ(R+S), and C29ββ(R+S) double peaks by integrating the whole area and hence use that area for generating the relevant diagnostic ratios as shown in Figure 7-7. For the DR-C29ααS ratio, either peak height or area can be used.

Table 7-5 Recommended Diagnostic Ratios (DR) Derived from Triaromatic Steroids

Abbreviation	Compound Name	m/z Value
C20TA	C20-triaromatic steroid (pregnane derivative)	231
C21TA	C21-triaromatic steroid (homopregnane derivative)	231
SC26TA	C26 20S-triaromatic steroid (cholestane derivative)	231
RC26TA+SC27TA	C26 20R- + C27 20S-triaromatic steroids	231
SC28TA	C28 20S-triaromatic steroid (ethylcholestane derivative)	231
RC27TA	C27 20R-triaromatic steroid (methylcholestane derivative)	231
RC28TA	C28 20R-triaromatic steroid (ethylcholestane derivative)	231

Ratio Name	Definition
DR-C21TA	C21TA/RC28TA
DR-SC26TA	SC26TA/SC28TA
DR-RC27TA	RC27TA/RC28TA

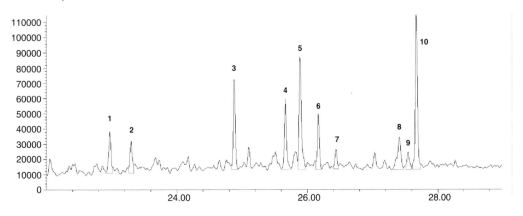

Figure 7-9 GC/MS ion fragmentogram of sesquiterpanes recorded at *m/z* 123 in the SINTEF standard oil mixture.

Table 7-6 Optional Diagnostic Ratios (DR) Derived from Sesquiterpanes

Abbreviation	Compound Name (Peak No.)	m/z Value
SES1	C14H26-sesquiterpane (1)	123
SES2	C14H26-sesquiterpane (2)	123
SES3	C15H28-sesquiterpane (3)	123
SES4	C15H28-sesquiterpane (4)	123
SES5	C15H28-8β(H)-drimane (5)	123
SES6	C15H28-sesquiterpane (6)	123
SES7	C15H28-sesquiterpane (7)	123
SES8	C16H30-sesquiterpane (8)	123
SES9	C16H30-sesquiterpane (9)	123
SES10	C16H30-8β(H)-homodrimane (10)	123

Ratio Name	Definition (Incl. Peak No.)
DR-SES1/SES2	C14-sesquiterpane (1)/C14-sesquiterpane (2)
DR-SES3/SES5	C15-sesquiterpane (3)/C15-8β(H)-drimane (5)
DR-SES4/SES6	C15-sesquiterpane (4)/C15-sesquiterpane (6)
DR-SES5/SES10	C15-8β(H)-drimane (5)/C16-8β(H)-homodrimane (10)

subject to evaporative weathering, a fact that has to be taken into account before using them for diagnostic purposes. Figure 7-9 gives an example of a sesquiterpane GC/MS ion fragmentogram (*m/z* 123) of the SINTEF standard oil mixture, and Table 7-6 lists the identified peaks and recommended diagnostic ratios.

After analyzing the selected samples by GC/MS, the recorded chromatograms should be visually checked for characteristic features or obvious differences, which could possibly eliminate any suspected source sample from the candidate sources. If there is any doubt whether observed differences are due to genuine differences or the effect of weathering and degradation, an additional weathering check of the PAC distribution (cf. Section 7.4.4.2) should be performed. Together with the information obtained from the weathering check of the *n*-alkanes (cf. Section 7.4.3.1), it can now be concluded which compounds are most probably affected and hence not suitable for generation of diagnostic ratios. After the visual evaluation of the GC/MS chromatograms and the weathering check, the diagnostic ratios are generated using only those compounds not or only slightly affected by weathering.

7.4.5 Level 3 — Treatment of Results

7.4.5.1 Comparison of Oil Samples Using Diagnostic Ratios

If two oil samples are identical, their chemical composition is by definition the same apart from those changes introduced after the spill as the result of weathering, contamination, mixing, and degradation (cf. Section 7.3). Accordingly, measured ratios between any pair of compounds should also match — up to a certain analytical related statistical confidence level — in identical samples. Various ratios between individual compounds or group of compounds have been described extensively in the literature and used to compare oil samples. In many cases such ratios may be referred to as "diagnostic" if they possess the quality to discriminate genuinely different oils. This is often the case with ratios derived from biomarkers. Other ratios, like some based on PACs, may or may not possess similar diagnostic significance; especially regarding refined products, refinery processes (e.g., cracking and reforming) may have affected the genuine petrogenic PAC distribution. However, genuinely diagnostic or not, any ratio not influenced by weathering or degradation should still display the same value for identical samples and thus be of "diagnostic" value in a specific spill case and contribute in discriminating and ruling out "nonsource" oils.

It is important to realize, however, that the suite of diagnostic PAC (Table 7-2) and biomarker ratios (Table 7-3 to 7-6) are neither all-inclusive nor appropriate for all oil spill identification cases. In some instances it may be prudent to include a certain characteristic feature of the spilled oil that is recognized as particularly diagnostic. In other situations, the abundance of some compounds necessary for determining the recommended diagnostic ratios are below the recommended S/N ratio. Thus, maintaining flexibility in the selection of diagnostic ratios to be used in a specific spill case is very important.

7.4.5.2 Criteria for Selecting, Eliminating, and Evaluating Diagnostic Ratios

In the CEN methodology, the criteria for selecting diagnostic ratios have generally been based on (1) specificity and diversity, (2) resistance to weathering, and (3) analytical precision/complexity (e.g., ratios calculated using different m/z values are generally not recommended).

Basically, only those ratios that can be measured with low variation should be evaluated for comparing candidate sources to spilled oil (Stout et al., 2000). Peaks with low S/N ratios have an increased variance and should only be used for a visual comparison and not for the comparison of diagnostic ratios. To accommodate both for the limitation in analytical precision and impact of sample heterogeneity, Stout et al. (2000) have suggested a protocol by which candidate diagnostic ratios are evaluated in order to identify those that are most useful for further correlation analysis. The evaluation of the diagnostic ratios was conducted by a simple statistical test (relative standard deviation of triplicate analyses) to identify those ratios that were unaffected by, for example, sample heterogeneity or low analytical precision. However, the use of triplicate analyses of one or some samples in order to decide whether a ratio should be used or not may not be particularly robust and result in highly varying RSDs among the diagnostic ratios.

Another approach was suggested by Faksness et al. (2002a) who applied a Student's t-test on triplicate analyses of one sample and the 95% and 98% confidence levels as critical limits to decide whether two ratios matched or not. Again this approach is not very robust, as the triplicate analyses sometimes may result in very low variances for some ratios, and in the end this may result in very small critical differences (cf. Equation 7-2) and hence non-matching ratios (i.e., false negative or the type I error).

For the CEN methodology, a comparable yet different approach has been applied. It recommends a fixed RSD of 5% for all diagnostic ratios to overcome the variation in critical differences. Before applying this approach, however, the laboratory should validate its analytical method by analyzing a sample at least 7 times to test that it complies with the 5% RSD limit. This limit is applied as a quality criterion, because methods producing higher RSD values should not be used to analyze and compare oil samples in a forensic context. A weakness with this approach, however, could be that small peaks may have relatively high RSDs. Therefore, it is generally recommended to analyze some of the samples from a spill case in duplicate and to compare the ratios of the duplicates. When small peaks result in differences larger than the critical difference, these peaks should only be used for a visual comparison.

To decide whether or not a particular ratio is sufficiently precise and robust to be used for comparison, the CEN methodology describes two consecutive tests to evaluate diagnostic ratios as described by the protocol/decision chart shown in Figure 7-10. The two consecutive tests used for selection and elimination of the diagnostic ratios comprise:

- elimination by means of a signal-to-noise (S/N) tests (Section 7.4.5.4)
- elimination by means of the comparison of the duplicate analyses (Section 7.4.5.5)

7.4.5.3 Repeatability Limit and Critical Difference

To estimate an acceptable difference between two analytical results, the standard deviation of the analysis method is used (ISO, 1994a). Repeatability (**r**) is applied as a test method to compare individual ratios, assuming that the two samples to be compared originate from the same source. Repeatability conditions are met when the samples are analyzed in one series. If the repeatability limit[4] is exceeded, it is

[4] The repeatability limit (**r**) is the difference between two test results; the associated standard deviation is $\sigma\sqrt{2}$. In normal statistical practice, for examining the differences between these two values, the critical difference (CD) used is f times this standard deviation, i.e., $f * \sigma\sqrt{2}$. For a normal distribution at 95% probability level, f is 1.96 and $f * \sqrt{2}$ then is 2.77. As the purpose of this guideline is to give some simple "rules of thumb" to be applied by nonstatisticians when examining the test results, a "rounded" value of 2.8 has been suggested instead of $f * \sqrt{2}$ (ISO, 1994b). Therefore, the repeatability limit $r_{95\%}$ is calculated by multiplying s_r with 2.8.

beyond reasonable doubt that this assumption is not valid and that the samples originate from different sources.

The repeatability limit of validation is based on standard normal distribution. Determining the repeatability standard deviation s_r of an analytical method depends on the quality assurance (QA) system of the laboratory. In general, s_r is calculated by analyzing samples relevant for a method at least seven times in one series when a new or revised method is being implemented and if the method has to be validated. Calculation of the standard deviation or relative standard deviation (RSD) generates **r**.

The repeatability limit $r_{95\%}$ (i.e., 95% confidence level) is calculated by multiplying the fixed RSD (s_r) with a factor of 2.8 by using Eq. (7-1):

$$r_{95\%} = 2.8 * 5\% = 14\% \quad (7\text{-}1)$$

This implies that when samples are analyzed under repeatability conditions, any ratio with an s_r of 5% to be used for the evaluation must not differ more than 14% relatively.

The corresponding critical difference (CD) is calculated by using Eq. (7-2):

$$CD = (\text{mean} * r_{95\%})/100 \quad (7\text{-}2)$$

When the absolute difference between a pair of corresponding ratios of two samples to be compared is lower than the critical difference (CD), based on the repeatability limit $r_{95\%}$, then the comparison gives a positive match; if it is higher, they do not match. This test has to be performed for every diagnostic ratio applied in the evaluation. The repeatability limit $r_{95\%}$ is a test at the 95% confidence level, which implies that 5%, or 1 out of 20 of the pair of ratios slightly above the CD, is acceptable without jeopardizing the conclusion. Table 7-7 shows an example of a ratio comparison.

7.4.5.4 Elimination of Diagnostic Ratios Using Signal-to-Noise (S/N) Test

Only peaks with S/N > 3 to 5 should be used for comparing diagnostic ratios. The S/N criterion is based on the method recommended by IUPAC (Ettre, 1993). N is the peak-to-peak

Emerging CEN Methodology for Oil Spill Identification 247

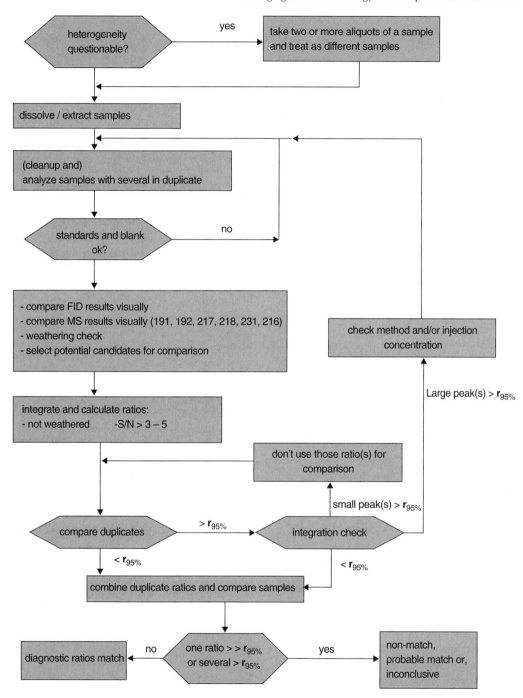

Figure 7-10 Protocol/decision chart for selection/elimination of diagnostic ratios.

Table 7-7 Example of Comparison of Diagnostic Ratios Based on a Repeatability Limit of 14%

Diagnostic Ratios (in %)		Mean (%)	Absolute Difference (%)	Critical Difference (CD, %)[1]	Conclusion
Sample A	Sample B				
48	52	50	52 − 48 = 4	50 ∗ 14/100 = 7	Match
35	45	40	45 − 35 = 10	40 ∗ 14/100 = 5.6	Nonmatch

[1] The critical difference is based on the repeatability limit ($r_{95\%}$) of 14% (at 5% RSD); cf. Eq. 7-2.

value of a part of the noise around a peak and S the peak height. A decision range of 3 to 5 is given, because it may often be difficult to estimate the noise precisely due to the many small peaks present in the chromatographic fine structure of oil samples.

As an example, a part of the *m/z* 191 ion fragmentogram (from C28 22R-tricyclic triterpane to C27 17α(H)-22,29,30-trisnorhopane) of a bunker oil sample is shown in Figure 7-11. Normally, only a part of the baseline around the peak would be used to calculate the noise. A comparison of the noise from 37.6 min–40 min with the noise of another section of the chromatogram [e.g., from 51.5 min to 55 min (see insert)] shows that the noise may not only be caused by the instrument, but that a significant part arises from the very small peaks present in the fine structure of the chromatogram, as mentioned above. In this example, S = 1950 (for the C28 22R-tricyclic triterpane) and N is 450 resulting in S/N = 4.5; therefore, the C28 22R-tricyclic triterpane peak could be either excluded or included in the DR comparison test.

Even more difficult, however, is the integration of groups of isomer compound. For example, in a propeller shaft lubricating oil, many of the methylated PACs mentioned in Table 7-2 are present at a low concentration showing only the highest peaks of an isomer group. In such cases, the isomer pattern should only be used for visual comparison, and in general the S/N > 3 to 5 criterion should be used as a "rule-of-thumb" to facilitate the elimination of small peaks and groups of peaks from the ratio comparison.

7.4.5.5 Elimination of Diagnostic Ratios Using Duplicate Analyses

Duplicate analyses of two aliquots taken from a homogenous sample or the duplicate injections of the same extract will provide information about the actual performance of the analytical system for the samples involved. To improve the value of this test, it is strongly advised to analyze all samples in duplicate in cases with two or three samples. In cases with many samples, at least some of the samples should be analyzed in duplicate.

Ideally, duplicate analyses of an extract should result in identical diagnostic ratios, and observed differences can only be caused by analytical variation. As a result of this test, several options are possible according to the repeatability test (cf. Section 7.4.5.3):

(1) Ratios of large peaks (i.e., with S/N ≫ 5) differ more than 14% (cf. Section 7.4.5.3, Eq. 7-1), then the integration of the peaks involved (retention time and baseline drawing), instrumental conditions, and/or injection concentration should be rechecked and the sample (e.g., after cleanup) preferably re-analyzed.

(2) Ratios of small peaks (i.e., S/N ≈ 5) differ more than 14%, then again the integration of the peaks involved should be rechecked (retention time and baseline drawing); if one or two ratios are still above the critical difference (cf. Section 7.4.5.3, Eq. 7-2), the peaks involved should only be used for a visual comparison, while if several ratios are still above the critical difference, proceed as with large peaks.

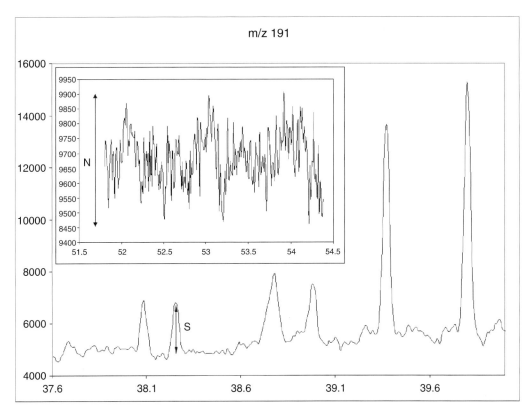

Figure 7-11 Signal-to-noise calculation. The distance between the highest and lowest peaks of a part of the noise is used as N (see insert) and the peak height as S. In this example, S/N is 4.5.

(3) All ratios differ less than 14%, then proceed to the next step of the decision chart (Figure 7-10) and compare the samples involved. Use the mean value of the diagnostic ratios of the samples analyzed in duplicate; the mean value has a lower variance and comes closer to the actual real ratio of the sample.

Generally, duplicate or triplicate analyses make the comparison of diagnostic ratios statistically more robust and reduce the relative standard deviation (RSD) by $\sqrt{n}$. Hence, the RSD of the mean value (m) of duplicate analyses becomes RSD/$\sqrt{2}$. For simplicity, however, it has been decided in this guideline not to reduce the repeatability limit of 14% and not to make a distinction between samples analyzed just once or in duplicate.

7.4.5.6 Optional Comparison of Diagnostic Ratios Using Multivariate Statistics

Multivariate statistical methods like principal component analysis (PCA) can be used as a tool to compare diagnostic ratios between multiple samples (e.g., Stout et al., 2002; Christensen et al., 2004) instead of the purely univariate approach described in previous sections. One advantage of multivariate compared to univariate statistical methods is the ease by which relationships between multiple samples and variables (e.g., diagnostic ratios) can be resolved and visualized by so-called factor scores and loading plots. Additional advantages include noise reduction, obtained by multiple measurements of the same phenomenon (i.e., interrelated variables), and their ability to detect outliers.

The fundamentals of PCA as a data treatment tool in chemical fingerprinting are that it summarizes the information in many correlated diagnostic ratios into a few so-called principal components that are weighted sums of these ratios. Hence, a model with few principal components (e.g., PC1 and PC2) describes the most prominent trends in data. By plotting principal components [e.g., the first principal component (PC1) vs. the second (PC2)], chemical information retained in numerous diagnostic ratios can be compared simultaneously. Oil samples with similar chemical composition will plot closely in score plots, whereas the opposite is the case for dissimilar oils.

The comparison of oil samples by PCA can be further refined by taking into account the uncertainty of the individual diagnostic ratios. This ensures an objective matching of oil spill samples and suspected sources based on all the available information (see Chapter 9, this volume). An example of the application of PCA to an actual oil spill case is shown by the score plot in Figure 7-12. Here the spill sample plots closely to the source sample (Statfjord crude), while nonsource samples (other North Sea crudes) plot more distantly.

7.4.6 Final Evaluation and Conclusions

The final conclusion should be based on a total evaluation using all available data. The comparison of the diagnostic ratios is an important part of the evaluation of the data, however, not all conclusive. It is important to visually inspect all the chromatograms and identify possible characteristic features, and not only to evaluate the measured ratios, before final conclusions are made. In accordance to the flowchart in Figure 7-1, the identification of an oil spill using this new CEN methodology should be concluded with respect to one of the four terms: positive match, probable match, inconclusive, or nonmatch.

A *positive match* states that source and spill samples are identical beyond reasonable doubt. A visual inspection of the chromatograms (GC/FID) and ion fragmentograms (GC/MS) shows only differences which can be

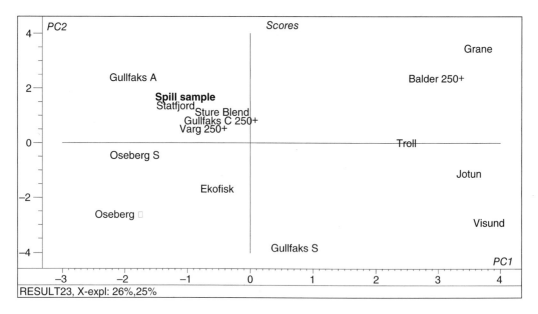

Figure 7-12 PCA score plot of data (diagnostic ratios) from a recent mystery oil spill in the North Sea. Several possible source samples (production samples from nearby offshore fields) together with other production samples were included in the PCA.

explained by weathering/degradation, and all observed differences between diagnostic ratios are below the repeatability limit.

A *nonmatch* applies when differences between chromatograms and diagnostic ratios cannot be explained by weathering/degradation, and when several pairs of ratios are outside the repeatability limit. If only very small differences (close to the repeatability limit) are observed, if contamination is expected, or if just one pair of ratios is clearly outside the repeatability limit, a *probable match* is the obvious alternative conclusion.

If the total amount of oil in a sample is very low, and consequently there is higher analytical variance of the diagnostic peaks, it might result in differences between diagnostic ratios (based on repeated analyses of the same sample) that are higher than the recommended repeatability limit. This would eventually imply the elimination of so many of the diagnostic ratios from further comparisons that it would render the test *inconclusive*.

7.5 The CEN Methodology in Practice: A Case Study

7.5.1 The Spill Case

In an actual spill case from 2004, a waterborne oil slick was discovered, and samples were collected from the spill. According to observations, a specific vessel was suspected of the illegal discharge of oil, and samples were collected from the vessel's bilge tank as candidate source samples. All samples were sent to the analytical laboratory for analysis and eventual identification of the spill.

7.5.2 GC/FID Screening

A spill sample and a bilge sample from the suspected vessel were extracted with dichloromethane, and the extracts dried over sodium sulphate and diluted to appropriate concentrations before being analyzed by GC/FID. Resulting chromatograms of the spill and candidate source samples are shown in Figure 7-13A and B, respectively. Both samples were analyzed in duplicate.

Both samples were characterized as bilge oils according to their GC/FID chromatograms, and no differences were observed apart from those related to the weathering of the spill sample. By overlaying the two chromatograms (see Figure 7-14), the degree of weathering could be evaluated. The spill sample appeared to be affected by evaporation up to about n-C18.

As the spill sample did not seem heavily affected by weathering above n-C18, the acyclic isoprenoid ratios (see Table 7-1) could be suitable for comparison. As is listed in Table 7-8, the comparison of these diagnostic ratios based on a repeatability limit of 14% shows that the absolute differences between the pairs of ratios for the spill and candidate source samples are all well below the repeatability limit; thus, these ratios give a positive match.

7.5.3 GC/MS Fingerprinting

As the GC/FID screening did not reveal significant differences between the spill and candidate source samples, the evaluation was continued to level 2, GC/MS fingerprinting, according to the flowchart in Figure 7-1. The GC/MS analysis included all the m/z values listed in Tables 7-1 to 7-5, while the optional sesquiterpanes (cf. Table 7-6) were not included as these compounds could be affected by weathering. After analyzing the samples and before generating any diagnostic ratios, the ion fragmentograms were compared visually to check for obvious differences between the two samples. As no obvious differences were observed, all peaks relevant for the generation of diagnostic ratios were integrated and the diagnostic ratios calculated.

7.5.4 Evaluation and Comparison of Diagnostic Ratios

According to the flowchart in Figure 7-10, diagnostic ratios were first eliminated on the S/N criteria. Hence, peaks with S/N < 3–5 were not used for generating diagnostic ratios; this included the following triterpane ratios: DR-C28αβ, DR-25norC29αβ, DR-C30O (cf.

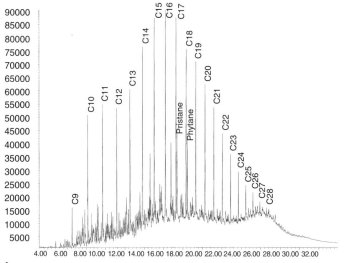

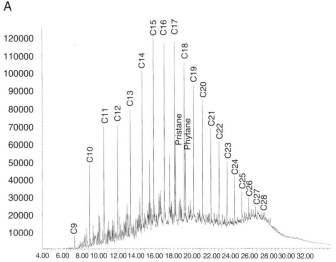

Figure 7-13 GC/FID chromatogram of extracts of (A) candidate source sample (bilge), and (B) spill sample.

Table 7-4), and the three triaromatic steroid ratios: DR-C21TA, DR-SC26TA and DR-RC27TA (cf. Table 7-5). In the next step, ratios were evaluated and eventually eliminated based on duplicate analyses using the repeatability limit. Table 7-9 gives an example on how this selection was achieved for some of the biomarker ratios.

Only one ratio, the DR-C30d, exceeded the repeatability limit of 14%. A closer look at the chromatogram showed that the C30d peak was small, and although it was not eliminated on the S/N criteria, the peak size and the noisy baseline made it difficult to obtain robust values for this ratio, which was therefore eliminated. According to the flowchart in Figure 7-1, Level 3, the remaining ratios were eventually compared on the basis of the repeatability limit, as shown in Table 7-10.

The comparison of 22 diagnostic ratios based on the repeatability limit showed that none of these ratios had absolute differences that exceeded the critical difference and that according to that criteria they all revealed a positive match as was also observed with the three diagnostic ratios derived from acyclic

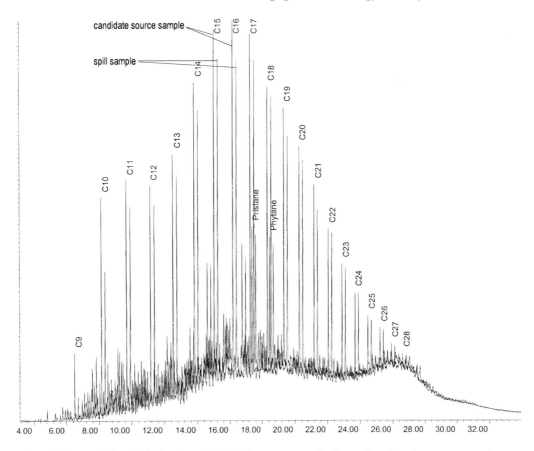

Figure 7-14 Weathering check. Overlay of GC/FID chromatograms of spill sample and candidate source sample.

Table 7-8 Comparison of Diagnostic Ratios of Acyclic Isoprenoids Using the Repeatability Limit

Diagnostic Ratio	Spill Sample[1]	Candidate Source Sample[1]	Mean Value	Absolute Difference	Critical Difference[2]
DR-nC17/Pri	1.94	2.06	2.00	0.12	0.28
DR-nC18/Phy	2.12	2.15	2.13	0.03	0.30
DR-Pri/Phy	1.23	1.22	1.22	0.01	0.17

[1] DR values are means of duplicate analyses.
[2] The critical difference is based on the repeatability limit ($r_{95\%}$) of 14% (at 5% RSD) (cf. Eq. 7-2).

Table 7-9 Comparison of Diagnostic Ratios (in %) of Some Triterpanes Using the Repeatability Limit

Diagnostic Ratio	Spill Sample Duplicate 1	Spill Sample Duplicate 2	Mean Value (%)	Absolute Difference (%)	Critical Difference (%)[1]
DR-C29	115.7	112.9	114.3	2.8	16.0
DR-C27Ts	78.2	77.6	77.9	0.6	10.9
DR-C29Ts	23.2	24.3	23.7	1.1	3.3
DR-C30d	5.0	4.1	4.6	***0.9*** *	***0.6*** *
DR-C30G	8.4	8.6	8.5	0.2	1.2

[1] The critical difference is based on the repeatability limit ($r_{95\%}$) of 14% (at 5% RSD) (cf. Eq. 7-2).
*Exceeded the repeatability limit, see text for description.

Table 7-10 Comparison of Diagnostic Ratios (in %) of Spill and Candidate Source Samples Using a Repeatability Limit ($r_{95\%}$) of 14%

	Spill Sample	Candidate Source Sample	Mean Value	Absolute Difference	Critical Difference (%)[1]
DR-2-MPhe/1-MPhe	112, 81	109, 23	111, 02	3, 58	15, 54
DR-4-MDbt/1-MDbt	364, 39	385, 03	374, 71	20, 64	52, 46
DR-C2-Dbt/C2-Phe	65, 39	61, 91	63, 65	3, 48	8, 91
DR-BaFl/4-MPy	94, 47	98, 43	96, 45	3, 96	13, 50
DR-B(b+c)Fl/4-MPy	32, 81	35, 16	33, 99	2, 35	4, 76
DR-2-MPy/4-MPy	61, 27	66, 98	64, 12	5, 7	8, 98
DR-1-MPy/4-MPy	52, 87	58, 16	55, 52	5, 29	7, 77
DR-C3-Dbt/C3-Phe	86, 15	84, 43	85, 28	1, 72	11, 94
DR-Retene/C4-Phe	4, 93	5, 28	5, 11	0, 35	0, 72
DR-C3-Dbt/C3-Chr	3756, 7	3535, 6	3646, 1	221, 1	510, 5
DR-(C28+C29)	95, 48	92, 20	93, 84	3, 28	13, 14
DR-C28	40, 16	38, 98	39, 57	1, 18	5, 54
DR-C29	55, 32	53, 22	54, 27	2, 09	7, 60
DR-C27Ts	77, 90	78, 66	78, 28	0, 75	10, 96
DR-C29$\alpha\beta$	114, 3	106, 6	110, 5	7, 69	15, 47
DR-C29Ts	23, 73	22, 84	23, 29	0, 89	3, 26
DR-C30G	8, 52	9, 23	8, 87	0, 71	1, 24
DR-C29$\alpha\alpha$S	114, 1	115, 8	115, 0	1, 69	16, 09
DR-C29$\beta\beta$	131, 8	134, 8	133, 3	2, 96	18, 66
DR-C27$\beta\beta$STER	41, 51	41, 79	41, 65	0, 28	5, 83
DR-C28$\beta\beta$STER	27, 07	25, 94	26, 51	1, 13	3, 71
DR-C29$\beta\beta$STER	31, 42	32, 27	31, 84	0, 85	4, 46

[1] The critical difference is based on the repeatability limit ($r_{95\%}$) of 14% (at 5% RSD) (cf. Eq. 7-2).

isoprenoids. Thus, all 25 diagnostic ratios evaluated and compared in this study revealed a positive match. As the final visual comparison (ground truth) of all chromatograms neither revealed any significant differences apart from weathering effects, it could be concluded that the spill and candidate source samples matched positively and that the spill sample was identical to the bilge sample from the suspected vessel beyond reasonable doubt.

7.6 Summary

The emerging CEN methodology for oil spill identification presented here is based on a three-level tiered approach, including (1) GC/FID screening of all involved samples (Level 1); (2) GC/MS fingerprinting of selected spill and candidate source samples (Level 2), from which up to 29 diagnostic ratios of selected PAHs and biomarkers are derived, diagnostic ratios are selected and evaluated on the basis of their analytical variability and changes due to weathering; and (3) correlation of spill and candidate oil samples based on those diagnostic ratios that can be precisely measured and are resistant to weathering effects (Level 3). By statistical treatment of the ratios and an overall assessment of results from all analytical levels, the oil spill identification using this methodology can be concluded with respect to one of four operational and technically defensible terms: positive match, probable match, inconclusive, or nonmatch.

The revised methodology has been implemented by many European forensic oil spill laboratories and as such has been used recently in connection to several oil spill identification cases in, for example, Norway (e.g., Almås and Daling, 2001), Denmark (Hansen et al., 2002), and the Netherlands (e.g., the *Tricolor* incident in the British Channel, and the *Prestige* incident in Spain and France) (Guyomarch, 2002).

The methodology has been demonstrated to be a strong, technically defensible tool capable of differentiating among qualitatively similar oils from a spill and available candidate sources.

Acknowledgment

Nordtest, Standards Norway (SN), and the Norwegian Pollution Control Authorities (SFT) are gratefully acknowledged for co-financing this project.

References

Almås, I.K. and P.S. Daling, *Identification of Oil Spill Observed by "KV Lafjord,"* October 2001, SINTEF report STF66 F01178, 2001 (in Norwegian).

Barakat, A.O., Y. Qian, M. Kim, and M.C. Kennicutt II, Compositional changes of aromatic steroid hydrocarbons in naturally weathered oil residues in the Egyptian western desert, *Environmental Forensics*, 2002, **3**, 219–225.

Christensen, J.H., A.B. Hansen, G. Tomasi, J. Mortensen, and O. Andersen, Integrated methodology for forensic oil spill identification, *Environmental Science & Technology*, 2004, **38**, 2912–2918.

Christensen, J.H. and G. Tomasi, Multivariate statistical methods for petroleum hydrocarbon fingerprinting and spill source identification, Chapter 9, this volume.

Dahlmann, G., Characteristic Features of Different Oil Types in Oil Spill Identification, Berichte des BSH, 2003, **31**, (Available at http://www.bsh.de/de/Produkte/Buecher/Berichte/Bericht31/index.jsp).

Dahlmann, G. and P. Kienhuis, Oil Spill Identification — Round Robin 2005, The Comparison of Four Bilge Samples, RIZA Report No. 2006.043X, 2006.

Daling, P.S., L.-G. Faksness, A.B. Hansen, and S. Stout, Improved and standardized methodology for oil spill fingerprinting, *Environmental Forensics*, 2002, **3**, 263–278.

Ettre, L.S., Nomenclature for chromatography, *IUPAC Recommendations*, 1993, Section 4.3, *Pure & Applied Chemistry*, 1993, **65**(4), 819–872.

CEN, European Committee for Standardization, Oil Spill Identification — Waterborne Petroleum and Petroleum Products — Part 1: Sampling, TC/BT TF 120 WI CSS27004, 2005.

CEN, European Committee for Standardization, Oil Spill Identification — Waterborne Petroleum and Petroleum Products — Part 2: Analytical Methodology and Interpretation of Results, TC/BT TF 120 WI CSS27003, 2006.

Faksness, L.G., H. Weiss, and P.S. Daling, Revision of the Nordtest methodology for oil spill identification — technical report, SINTEF report STF66 A01028, 2002a.

Faksness, L.G., P.S. Daling, and A.B. Hansen, CEN/BT/TF 120 Oil Spill Identification, Summary Report: Round Robin Test Series B; SINTEF report STF66 A02038; 2002b.

Faksness, L.-G., P.S. Daling, A.B. Hansen, P. Kienhuis, and R. Duus, New guidelines for oil spill identification of waterborne petroleum and petroleum products, *Proc 28th Arctic and Marine Oilspill Program (AMOP) Technical Seminar*, Calgary, Canada, June 7–9, 2005, Environment Canada, Ottawa, **1**, 183–202.

Guyomarch, J., Identification du fuel du Prestige-2002, CEDRE Report no. GC.02-15; 2002 (in French).

Hansen, A.B., J. Avnskjold, and C.A. Rasmussen, Application of PAH and biomarker diagnostic ratios in forensic oil spill identification by the revised Nordtest methodology, *Oil and Hydrocarbon Spills III, Modeling, Analysis, and Control*, WIT Press, Southampton, UK, 2002, 59–66.

ISO 5725-2, Accuracy (trueness and precision) of measurement methods and results, Part 2: Basic method for the determination of repeatability and reproducibility of a standard measurement method, International Organization for Standardization, ISO TC69/SC6, 1994a.

ISO 5725-6, Accuracy (trueness and precision) of measurement methods and results, Part 6: Use in practice of accuracy values, International Organization for Standardization, ISO TC69/SC6, 1994b.

Kienhuis, P.G.M., Oil spill identification — Round robin 2004, the comparison of three gas oil samples, RIZA Report No. 2005.042X, 2005.

Nordtest, *Oil Spill Identification*, NT Chem 001, 2nd ed., 1991.

Peters, K.E., C.C. Walters, and J.M. Moldowan, *The Biomarker Guide*, 2nd ed., Cambridge University Press, Cambridge, UK, Vol. 2, *Biomarkers and Isotopes in the Petroleum Exploration and Earth History*, 2005, 475–1155.

Stout, S., W.P. Naples, A.D. Uhler, K.J. McCarthy, L.G. Roberts, and R.M. Uhler, Use of quantitative biomarker analysis in the differentiation and characterization of spilled oil, Paper prepared for the SPE International Conference on Health, Safety, and the Environment, Stavanger, Norway, 26–28 June, 2000.

Stout, S.A., A.D. Uhler, and K.J. McCarthy, A strategy and methodology for defensibly correlating spilled oil to source candidates, *Environmental Forensics*, 2001, **2**, 87–98.

Stout, S.A., A.D. Uhler, K.J. McCarthy, and S. Emsbo-Mattingly, Chemical fingerprinting of hydrocarbons, *Introduction to Environmental Forensics*, B. Murphy and R. D. Morrison (eds.), Academic Press, San Diego, CA, 2002, 137–260.

Strøm-Kristiansen, T., A. Lewis, P.S. Daling, J.N. Hokstad, and I. Singsaas, Weathering and dispersion of naphthenic, asphaltenic and waxy crude oils, *Proc. 1997 Inte. Oil Spill Conf,* 1997, 631–636.

Wang, Z., M. Fingas, and D.S. Page, Oil spill identification, *J. Chromatography A,* 1999, **843**, 369–411.

Wang, Z. and M. F. Fingas, Development of oil hydrocarbon fingerprinting and identification techniques, *Marine Pollution Bull.*, 2003, **47**(9–12), 423–452.

Wang, Z., C. Yang, M. Fingas, B. Hollebone, X. Peng, A.B. Hansen, and J.H. Christensen, Characterization, weathering, and application of sesquiterpanes to source identification of spilled lighter petroleum products, *Environmental Science & Technology*, 2005, **39**, 8700–8707.

Weiss, H.M., A. Wilhelms, N. Mills, J. Scotchmer, P.B. Hall, K. Lind, and T. Brekke, *NIGOGA — The Norwegian Industry Guide to Organic Geochemical Analyses*, Edition 4.0, Published by Norsk Hydro, Statoil, Geolab Nor, SINTEF Petroleum Research, and the Norwegian Petroleum Directorate, 2000, (Available at http://www.npd.no/engelsk/nigoga/nigoga.pdf.)

8 Advantages of Quantitative Chemical Fingerprinting in Oil Spill Source Identification

Gregory S. Douglas, Scott A. Stout, Allen D. Uhler,
Kevin J. McCarthy, and Stephen D. Emsbo-Mattingly

NewFields Environmental Forensics Practice LLC, 100 Ledgewood Place, Suite 302, Rockland, MA 02370

8.1 Introduction

The world economy is dependent on the exploration, production, transportation, and refining of petroleum. The demand for petroleum continues to increase as developing countries move from agricultural- to industrial-based economies. To meet the increased demand for oil over the next two decades, the world oil supply of crude oil will have to increase by over 45 million barrels/day (NRC, 2003), and it is apparent that most of this petroleum will be transported by large tankers (NRC, 2003). Much of the national and international shipping of raw materials and commercial goods in this growing world economy will also be transported by marine vessels. Thus, the risk of accidental discharges of petroleum to the environment — either as crude oil from production operations or as marine fuels from the discharge of maritime cargo vessels — will most certainly increase.

Oil spills can take place in rivers, open waters, and navigable coastal waterways. The natural resources damages (NRD) and liability associated with the release of even a small volume of petroleum warrants a thorough study of the fate of the spilled petroleum. The fundamental elements of almost all oil spill studies includes sufficient chemical characterization so investigators may (1) defensibly determine the source of the oil, (2) distinguish spilled oil from pre-existing background hydrocarbons, (3) quantifiably evaluate the extent of impacted ecosystems (Stout et al., 2001), and (4) reliably monitor oil spill cleanup and remediation.

Detailed chemical analysis of petroleum — often referred to as "chemical fingerprinting" — has played an important role in the identification of oil arising from accidental spills (Wang and Fingas, 1999; Wang et al., 2001; Stout et al., 2001; Page et al., 1995). The results of "chemical fingerprinting" are often supplemented by other lines of evidence (e.g., spill records, operational records, proximity of candidate sources to a slick, and oil slick trajectory modeling) (Lehr et al., 1999) in order to develop a comprehensive conceptual model of a spill event.

The modern chemical fingerprinting analytical methods used today have evolved over the last two decades, largely due to the development and increased sophistication of analytical instrumentation (Boehm et al., 1997; see also Chapter 1 herein). The cornerstone of modern petroleum fingerprinting is high-resolution capillary gas chromatography (GC). The GC technique provides a means to physically separate a complex mixture of hydrocarbons into the individual chemical compounds that can then be detected, identified, and measured by various means. The most appropriate and common means for the detection, identification, and measurement of the individual compounds comprising petroleum include GC in combination with flame ionization detection (FID) or mass spectrometry (MS). Using both of these methods the relative abundance of an individual compound is converted to an electronic signal that is reflected as a peak on the resulting chromatograms. The magnitude of the peak represents the concentration of that compound in the mixture (Douglas et al., 1994). The exponential increase of low-cost computing power and data storage has also provided oil spill investigators with powerful statistical, numerical, and graphical analysis tools that can be used for quantitative chemical fingerprinting. The ability to store large amounts of raw instrument file outputs for reference oils and field samples has also accelerated the development of petroleum product reference libraries rich in information that can be easily accessed and evaluated as new methods and interpretive approaches are developed.

Petroleum is comprised of thousands of individual compounds, many of which can be separated by GC and measured by FID or MS. Thus, GC analysis provides a means of separating hydrocarbon chemicals, producing a distribution of peaks or varying proportions that represent the "fingerprint" of the oil. The GC "fingerprint" of the spilled oil can be compared to the fingerprints for any number of candidate source oils analyzed by the same method(s).

GC fingerprinting of spilled oils, candidate sources, and potentially impacted samples can be conducted qualitatively or quantitatively (see ahead). In turn, any correlations between a spilled oil and its potential sources or potentially impacted samples also can be made qualitatively or quantitatively. Stout et al. (2005) identified the strengths and weaknesses of each approach and concluded that quantitative correlations provide the most reliable and unbiased basis upon which the source or impact of spilled oil can be defensibly determined. One particular advantage of quantitative data is the ability to address the issue of spilled petroleum that is comprised of mixtures, such as (1) mixtures from the commingling of multiple spilled oils from the same or different sources or (2) mixtures of a spilled oil and any pre-existing oil (or other hydrocarbons) in the environment (Stout et al., 2005). If unrecognized, these complications can confound "spill oil-to-source oil" or "spill oil-to-impacted sample" correlations using some correlation statistical methods, such as the revised Nordtest approach (Daling et al., 2002) or the emerging European Committee for Standardization (CEN) protocol (Chapter 7). Mixing of this sort is not uncommon. For example, large marine vessels carry hundreds of thousands of gallons of fuels (diesel marine oil [DMO], intermediate fuel oil [IFO], heavy fuel oil [HFO]), which may be mixed internally within bilges or during fuel blending, or externally after a release due to physical mixing in coastal waters, ports, and harbors. Also, hydrocarbons from anthropogenic and natural sources are ubiquitous in the environment and, therefore, will be available to mix with any spilled oil. In this chapter we discuss the development and field validation of a quantitative chemical fingerprinting mixing model approach that will improve the resolution and accuracy of statistical correlations when mixing — either among multiple spilled oils or between a spilled oil and other, often pre-existing hydrocarbons in the environment — has potentially affected the samples under investigation.

8.2 Qualitative Fingerprinting Methods

The utility of qualitative fingerprinting methods such as ASTM 3328, ASTM 5739, and Nordtest (1991) in oil spill investigations have been previously discussed (Stout et al. [2005]; see also Chapter 1 herein). Qualitative chemical fingerprinting analysis of spilled oil, candidate sources, and background materials can be best described as a visual comparison between various spectroscopic or chromatographic fingerprints. Such comparisons inescapably introduce a degree of subjectivity to the source identification evaluation, which is an undesirable feature of science.

The qualitative approach to chemical fingerprinting of oil spills was formalized in two standards of the American Society for Testing and Materials (ASTM), which are still used today in oil spill investigations in the U.S. ASTM D3328 is a qualitative GC/FID fingerprinting method (which can also include GC-flame photometric detection) that was originally approved in 1990 (ASTM, 1990). This method states,

The matching of oil samples is essentially a profiling technique based on the premise that identical oils give identical chromatograms. Normally, the matching of a spilled oil to a suspect oil can be accomplished by comparison of the chromatograms for each of the oils in a spill case.

ASTM D3328 goes on to state that after considering the effects of weathering,

Normally, a direct comparison of chromatograms will suffice for establishing identity or nonidentity between samples [and] if the chromatograms are the same on the basis of the peak-for-peak matching, there is a high degree of probability that the samples are from the same source.

This protocol defines the match criteria as follows:

Match — like the sample submitted for comparison, that is, the chromatographic pattern is a virtual overlay.
Probable match — the chromatographic pattern is similar to that of the samples submitted for comparison, except (a) for changes that could be attributed to weathering or (b) differences attributable to specific contamination.
Indeterminate — the chromatographic pattern is somewhat similar to that of the sample submitted for comparison, except for certain differences (due to weathering) that it make impossible to ascertain whether the unknown is the same oil heavily weathered, or a totally different oil.
Nonmatch — unlike the samples submitted for comparison.

These criteria introduce considerable subjectivity into the correlation analysis, the degree to which can vary depending upon the data quality and experience of the interpreter. The issue of data quality is very important since ASTM D3328 acknowledges that "no statement is made about either the precision of bias of this test method since the result merely states whether there is conformance to the criteria for comparison specified in the procedure." Ironically, strict interpretation of the "match" criteria (above) would mean that any weathering of the spilled oil, a phenomenon that is virtually guaranteed to occur when oil is spilled onto water, would prohibit ever concluding that any unweathered candidate source can be a "match" for a weathered spilled oil. This method, therefore, provides a relatively weak scientific basis upon which to assign multimillions of dollars of liability for spilled oil.

In 2000, ASTM introduced D5739, an additional procedure that relies upon GC/MS fingerprinting [American Society for Testing and Materials (ASTM), 2000]. The introduction of this procedure tends to acknowledge and substantiate the shortcomings of ASTM D3328. D5739 uses GC/MS to acquire a greater number of more detailed "fingerprints" for a prescribed list of compound groups found in any given oil sample, which undoubtedly provide a more detailed basis for comparing oil spills and their candidate sources. However, like ASTM D3328 (above), ASTM D5739 also relies upon interpretation of these fingerprints as "qualitative comparisons" and "direct visual comparison." The method concludes by stating the procedure is "a means of making qualitative comparisons between petroleum samples; quantitation of the various chemical components is not addressed." Thus, despite acquiring chemical fingerprinting data of increasing specificity using GC/MS, the method's reliance upon "direct visual comparison" of the resulting fingerprints by "placing the EIC's (fingerprints) one over the other," again introduces a significant degree of subjectivity into the source identification process. This method defines its own match criteria, summarized as follows:

Similar — all the fingerprints examined for two oil samples display the same qualitative pattern by visual examination... except those attributable to the precision of the analysis (as determined by analysis of a sample in duplicate) or weathering.

Inconclusive — all the fingerprints for two oil samples display no qualitative differences except one or more slightly greater than the precision of the analysis (as determined by analysis of a sample in duplicate). In this case the questionable oil should be reprepared and re-analyzed if possible and be deemed inconclusive if similar results are obtained.

Dissimilar — some or all of the fingerprints for two oil samples display discrepancies, particularly among the most weathering-resistant target compounds.

These criteria are highly subjective and prone to interpretation differences. Furthermore, since two of the match criteria are based on the sample duplicate precision with no stated minimum performance criteria, it is possible that one laboratory with poor precision will achieve more "matches" than another laboratory with excellent precision. Therefore, the absence of quality control renders this method poorly suited for the definitive determination of liability or guilt in the case of mystery oil spills.

In summary, qualitative chemical "fingerprinting" analysis of spilled oil, candidate sources, and background materials can best be described as a visual comparison between various chromatographic or other forms of chemical fingerprints. Such comparisons inescapably introduce a degree of subjectivity that can be undesirable when subtle differences, environmental weathering, data from multiple sources or vintages, and mixed petroleums are significant factors in the evaluation.

8.2.1 Shortcomings of Qualitative Fingerprinting

For some oil spill investigations, the qualitative approaches of ASTM D3328 and D5739 or Nordtest (1991) may be sufficient to reach defensible conclusions regarding the source of an oil spill. These investigations are generally limited to situations in which there are markedly disparate candidate source oils, only one of which can reasonably match the spilled oil and where extensive weathering of the sample has not occurred. However, most oil spill studies are not this straightforward. At least four particular circumstances in which qualitative fingerprinting can be problematic are spill situations involving (1) significant weathering of the spilled oil, (2) comparison of genetically similar spill and source oils or fuels, (3) qualitatively similar spill and source oils but with varying concentrations, and (4) mixing of spill oils with each other or with pre-existing hydrocarbons in the environment. These situations are briefly described ahead.

8.2.1.1 Weathered Oils

The first of these shortcomings, weathering, is acknowledged by the ASTM D3328 method's match definitions described previously. Specifically, fingerprinting differences attributable to weathering require ASTM D3328 to conclude, at best, a *probable match* exists. Even if qualitative matches are achieved between a spill and source oil using ASTM D5739, the strongest conclusion one may reach is that the oils are "similar." These ASTM-defined conclusions, "probable match" or "similar" are not particularly useful conclusions for litigious situations. Thus, weathering is an inevitable shortcoming in the application of the qualitative ASTM methods or of Nordtest (1991).

8.2.1.2 Genetically Similar Oils

It is common for oils produced, transported, and refined within a given geographic province to share a certain degree of chemical similarity. This stems from the fact that crude oil generated within a specific geologic province will, as a consequence of similar character of the ancient organic matter and similar subsurface heating conditions that produced the crude (Tissot and Welte, 1984), yield crude oils of a related chemical character. Very subtle differences between genetically similar oils from a given oil province, field, or even well, can still be recognized with the appropriate chemical data (e.g., Kaufman et al., 1990; McCaffrey et al., 1996; Nicolle et al., 1997; Hwang et al., 2000). When these crude oils are subsequently refined into specific petroleum products at nearby refineries (e.g., IFO), the resulting petroleum products often will also exhibit genetically similar chemical features "inherited" from the parent crude oil (Peters et al., 1992; see also Chapter 1 herein). Furthermore, as these petroleum products are then distributed throughout the market, multiple handlers (e.g., terminal, pipelines, and tankers) may carry the same fuel. The difference between the characteristics of a fuel used on one vessel versus that used by another may only depend upon the nature of any previous fuel that was present in their tanks at the time of refueling.

In summary, the chemical differences among crude oils or fuels in a geographic area may be very subtle, and therefore go unrecognized using qualitative fingerprinting techniques. This particular shortcoming was the basis for revision of the Nordtest (1991) protocol that was commonly applied to oil spills in the North Sea, where genetically similar crude oils and fuels co-occur, thereby confounding too many oil spill investigations in which only qualitative fingerprinting data were available (Daling, personal communication, 2000).

8.2.1.3 Qualitatively Similar Oils

It is conceivable that some oils may exhibit qualitatively similar fingerprints but differ in the absolute concentrations of different compounds. Consider the situation where the patterns of pentacyclic triterpanes and steranes in two oils are qualitatively similar, but the concentration of triterpanes is much higher in one oil than in the other. This type of difference would go unnoticed in a qualitative fingerprinting protocol such as ASTM D5739 or Nordtest (1991). While this type of situation may seem extraordinary — in part, perhaps because it is not commonly pursued — it can occur. An example from a mystery spill investigation is provided here.

In this case, another laboratory analyzed two oils (a spill and presumed source) via ASTM D3328 and D5730 protocol. A qualitative interpretation of these data concluded that the two oils were a "probable match" in accordance with ASTM D3328 and D5739. Some justification for this conclusion is evident upon inspection of the qualitatively similar GC/FIDs and selected extracted ion plots (EIPs) shown in Figure 8-1. (These data, though representative of the data from the other laboratory, were generated in our laboratory using the quantitative analysis as described in Section 8.3.) The other laboratory had considered the minor differences observed in the GC/FIDs (e.g., disparate UCM profiles, isoprenoid distributions, and Pr/Ph ratios) as being attributable to weathering. Similarly, their qualitative comparisons among the various EIPs obtained via ASTM D5739 concluded the two oils were correlated. While indeed the oils are qualitatively similar, minor differences can be observed (e.g., relative abundance of oleanane). During a qualitative comparison such minor differences might go unnoticed. This is an obvious shortcoming of qualitative interpretations. Of particular importance in this study, however, were the qualitatively similar distributions of C2-dibenzothiophenes (D2) and C2-phenanthrenes (P2) in which both oils showed almost identical m/z 212 and m/z 206 patterns, respectively (Figure 8-1). However, quantitative analysis of the absolute concentrations of the D2 and P2, as well as the D3 and P3, indicated that the two oils contained disparate absolute concentrations (Figure 8-2). This important difference could not be observed upon qualitative comparison of the EIPs patterns using ASTM D5739. Consequently, we concluded that the two oils, though qualitatively similar, were definitively not the same oil.

8.2.1.4 Mixing

In addition to changes brought about by weathering, the chemical fingerprint for a spilled oil will change if it becomes mixed with other oils released concurrently into the environment or

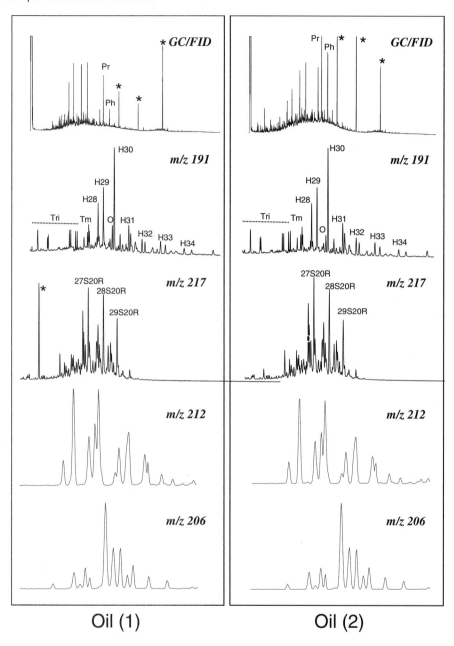

Figure 8-1 Comparison of two oils that exhibit qualitatively similar GC/FID patterns and selected extracted ion profiles (partial) that were interpreted to be "probable matches" by a laboratory conducting ASTM D3328 and D5739. Quantitative analysis showed some measurable differences existed, including higher concentrations of C2- and C3-dibenzothiophenes in oil 2 (see Fig. 8-2). *—internal standards.

with any pre-existing oil (or other hydrocarbons) already present in the environment. For example, large merchant vessels may store hundreds of thousands of gallons of diesel, intermediate, and heavy fuels, which, if released during an accident or illegal discharge, may yield variably "mixed" fingerprints that would not qualitatively match any of the individual bunkered fuels. Qualitative comparison of discrete source samples to the

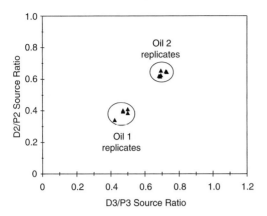

Figure 8-2 Double-ratio plot showing the results of replicate analysis of the two oils shown in Figure 8-1. Despite qualitatively comparable D2 (*m/z* 212), D3, P2 (*m/z* 206), and P3 patterns (e.g., Fig. 8-1) the concentration of D2 and D3 were higher in Oil 2.

spill samples, which are composed of variable mixtures of cargo and/or fuel, would fall short of recognizing the true nature of such a mixture. Similarly, if oil is spilled into an area already containing pre-existing "background" oil or other hydrocarbons, the resulting fingerprint will be qualitatively different from the source oil. Thus, under conditions of potential mixing, qualitative fingerprinting can be easily confounded and result in false negative correlations (e.g., fingerprint identification is a *non-match* or is *dissimilar*).

8.3 Quantitative Fingerprinting Methods

In order to confidently identify the source(s) of weathered, genetically, or qualitatively similar oils, or oil mixtures in environmental samples, some form of quantitative fingerprinting is more likely to yield more defensible results (Stout et al., 2005). The principal reason for this is that recognizing the often subtle differences among spill and source oils cannot reasonably rely upon the qualitative ASTM methods just described. As a result, over the past 15 years there has been an acknowledged need for and resulting advances toward quantification in the chemical fingerprinting of oil spill investigations.

8.3.1 Semiquantitative versus Fully Quantitative Methods

"Quantification" can mean different things to different laboratories. In some environmental laboratories quantitative data refers to the determination of peak areas (or sometimes heights) of compounds within a single chromatogram or extracted ion profile. These data are more accurately considered *semiquantitative* as they do not rely upon the use of internal or external standards or response factors, nor do they result in determination of concentrations of individual target compounds or compound groups. In addition, the quality of peak area or peak height data is frequently not analyzed with a multiconcentration initial calibration that demonstrates the applicable linear range of the target peaks. Continuing calibration standards are also frequently absent, which results in the failure to demonstrate the stability of the instrument over time. Nevertheless, semiquantitative data are still more useful than qualitative data (the latter of which rely entirely upon visual examination of chromatographic patterns). For example, diagnostic ratios can be developed from peak areas and allow statistical comparison between samples, as is described in Chapter 7. However, the use of diagnostic ratios generated from semiquantitative peak height or area responses still cannot address the issue of mixing between oils or between spilled oils and pre-existing hydrocarbons. The reason is that ratios do not mix linearly. A simple example of this is shown in the following data.

	Oil #1	*Oil #2*	*50 : 50 Mix*
Peak A (Response)	1,000	1,000	1,000
Peak B (Response)	5,000	10,000	7,500
Ratio A/B	0.20	0.10	0.13

In this simple example, assume that 1 μL of two oils has been injected and the responses for two peaks (A and B) were measured. The ratios of A/B in oil #1 and #2 are 0.2 and 0.1, respectively. Because the responses (a reflection of their relative concentrations) of peaks A and B vary between the oils, a 50 : 50 mix of these oils yields a ratio of 0.13 and not 0.15, which might be anticipated if the original A/B ratios (0.20 and 0.10) in each oil were mathematically mixed 50:50. It is possible for semi-quantitative peak areas or heights to be normalized to the mass of oil injected into the gas chromatograph — and all chromatographic conditions are kept constant (e.g., split ratios). Under these conditions, the measured responses (but not the ratios) could be used in addressing the issue of mixing.

These potential difficulties are avoided if internal standards and multilevel calibration standards, the latter containing a suite of structurally similar representatives of the targeted compounds, are employed to determine the absolute concentrations of target analytes in the samples studied. This approach is considered fully *quantitative* and is described further ahead.

The investigations following the *Exxon Valdez* oil spill in 1989 first promulgated the use of fully quantitative fingerprinting methods in oil spill investigations (Douglas et al., 1996; Kvenvolden et al., 1995; Page et al., 1995; Bence and Burns, 1995; Boehm et al., 1997). In these studies the absolute concentrations of the target analytes included those from ASTM D5739, as well as other diagnostic analytes derived from the petroleum geochemistry literature (e.g., Peters and Moldowan, 1993; Peters et al., 2005a, 2005b; Radke et al., 1986). These chemical concentrations in oils, sediments, and tissues were measured quantitatively, thereby providing absolute concentrations of diagnostic chemicals (Page et al., 1999; Stout et al., 2001; Daling et al., 2002a; *Federal Register*, 1994). Thus, instead of a qualitative, visual comparison of chromatographic peaks or semiquantitative comparisons of ratios based upon peak heights, tables of the numerical concentration data were developed and used for the calculation of diagnostic ratios (Stout et al., 2001; Page et al., 1995; Douglas et al., 1996; Douglas and Uhler, 1993) and other parameters. These factors allow simultaneous, statistical, or numerical comparisons among many samples and even allocation among disparate sources (Burns et al., 1997). This quantitative approach formed the basis for modern oil spill fingerprinting methods (Boehm et al., 1997).

Briefly, fully quantitative fingerprinting requires that the concentrations of individual and homologue series of diagnostic or environmentally important hydrocarbons be determined using the internal standard method. In this approach, recovery and quantitation internal standards, typically deuterated analogues of PAHs or n-alkanes, are added to the samples in known concentrations prior to GC/FID and GC/MS analysis. The internal standards serve as internal references in quantification of target compounds and as surrogates to judge method performance. Overarching the analytical methods are rigorous quality-control procedures that ensure accuracy and precision, and minimize bias. Typical QC protocols include procedural blanks, replicates, laboratory control samples, and reference oils (Page et al., 1995; Douglas et al., 1994; *Federal Register*, 1994).

Use of these techniques produces precise and accurate numerical concentration data that allow for comparison of spilled oil to candidate source oils using graphical (Boehm et al., 1997), statistical (Daling et al., 2002a; Christensen et al., 2004), or numerical analysis tools (Urdal et al., 1986; Burns et al., 1997; Stout et al., 2001; Mudge, 2002). This type of comparison reduces the subjectivity of the interpreter with greater reliance on data objectivity. These mathematically based approaches to fingerprinting interpretation also permit intercomparison of many oils to each other simultaneously, thereby improving the investigator's ability to identify similarities and/or differences among impacted field samples and suspect sources. The quantitative fingerprinting approach can identify the often subtle chemical relationships that help correlate or decouple samples and candidate sources. It is these subtle, but very important, features that often go unrecognized in qualitative fingerprinting methods (ASTM D3328 and 5739) in which patterns are compared visually.

Concentrations of polycyclic aromatic hydrocarbons (PAHs) and petroleum biomarkers (or biological markers) are particularly useful quantitative measures that benefit oil spill investigations because of their characteristic distributions and environmental recalcitrance (Douglas et al., 1996; Wang et al., 1999; Stout et al., 2001, 2000). Biomarkers are naturally occurring, ubiquitous, and stable hydrocarbons that occur in crude oils and most petroleum products (Peters and Moldowan, 1993). The distribution of biomarkers in oil reflects the "genetic" history of that oil, specifically, the type(s) of precursor ancient organic matter and the thermal history of the rock strata in which the oil was formed. Because of the diversity of organic matter and oil-forming environments, there are varied biomarker patterns among oils found in the many petroleum reservoirs around the world (Peters et al., 1993; see also Chapter 3).

Quantitative PAH and biomarker data form the basis of the oil spill investigation protocols and have been used in many oil spill investigations around the world (MSRC, 1995; Wang and Fingas, 1999; Wang et al., 1999). In virtually all the applications of quantitative oil spill investigations, the data have been evaluated using some kind of graphical or advanced data analysis tool (Urdal et al., 1986; Page et al., 1995; Henry et al., 1997; Burns et al., 1997; Stout et al., 2000, 2001; Mudge, 2002). In the sections that follow, we present an overview of the methods used in the collection of quantitative chemical fingerprinting data and the protocols by which these quantitative data can be applied to oil spill studies.

8.3.2 Data Generation for Fully Quantitative Fingerprinting

8.3.2.1 Sample Collection

The success of any oil spill investigation begins with collection of representative samples, using appropriate techniques (see Chapter 2). Protocols for the collection of the floating oil, mousse, tarballs, sheens, vessel tanks, and port supplier tanks have been described adequately (e.g., Nordtest, 1991). In addition, some natural resource damage assessment (NRDA) guidelines have adequately described sampling of oiled shorelines, sediments, animals, and vegetation (e.g., PERF, 1995).

In the immediate aftermath of an oil spill, it is crucial to identify and sample all the viable candidate sources. Identification of appropriate candidate sources is accomplished by a combination of common sense (e.g., collection of obvious source samples), combined with a thorough record review (e.g., sailing and cargo records, fueling records, etc.) for identification of important, but less obvious, sources. In many cases, oil spill samples are collected by personnel from the investigating agency who often provide split samples to potentially responsible party (PRP) representatives. Ideally, the PRP representative should collect samples of fugitive and source oils in parallel with regulatory investigators.

The typical types of samples collected in a harbor/port spill setting include samples of floating fugitive oil, oil stranded on shorelines and pilings, and oiled animals or water birds. Candidate source samples include representatives of ship cargo and fuel oil (e.g., fuel tank oils, bilge tank oils, oil water separator, settling tank oils, post oil sensor lines, external hull wipes, and drain tank oils from suspect vessels), as well as oil from any nearby port's fuel storage facility. In coastal investigations, oil/sheens near storm drain outfalls should also be collected if present. Often, oil on water is observed as sheen — an ultrathin accumulation of oil floating on water. Because of their nature, sheen samples cannot be collected using simple grab sample techniques. Rather, sheen samples are collected using a 4-inch-diameter, highly porous, TFE-fluorocarbon polymer net. This net was developed in conjunction with the U.S. Coast Guard to capture thin oil sheens in surface water (General Oceanics, 2006). The net is passed through the sheen five times, removed from the net holder ring, and placed into a glass jar for shipment. In our experience, we have found that the Teflon net must be rigorously precleaned with methylene chloride and blank tested prior to use. The precleaned nets should be stored in a clean environment and used within 1–3 months of cleaning. In all cases, appropriate blanks must accompany the samples to the laboratory (e.g., wipe sample and sheen blanks). Samples of free oil, sheen netting, sediments, and other oiled media should be stored at 4°C immediately after collection. Storage of oil samples at ambient temperatures in sealed glass containers will, in most cases, not impact the integrity of the sample for forensic studies.

8.3.2.2 Sample Preparation

Spilled oil/tarball and candidate source oil samples are prepared for analysis by weighing approximately 50 mg of oil into a tared 10-mL volumetric flask, bringing it to volume in dichloromethane (DCM) and removing a 1-mL aliquot for analysis. If a recovered oil sample is suspected to contain entrained water, the water in the extract is removed with anhydrous sodium sulfate prior to taking a 1-mL analytical aliquot. Each analytical batch of authentic samples ($n < 20$) should include appropriate quality-control samples, for example, a procedural blank (PB: 1 ml of DCM), a laboratory control sample (LCS) consisting of 1 mL of DCM spiked with selected hydrocarbons in known concentrations to monitor method accuracy, and one set of triplicate oils (i.e., a single oil prepared three times by drawing separate aliquots and spiking each individually) as a measure of precision and reproducibility of the data. After spiking the 1-ml aliquots of each sample with surrogate internal standard (SIS; o-terphenyl, naphthalene-d_8, phenanthrene-d_{10}, chrysene-d_{10}, or equivalent) and recovery internal standard (RIS; 5α-androstane, acenaphthene-d_{10}, fluorene-d_{10}, benzo[a]pyrene-d_{12}, or equivalent), the samples are split for quantitative GC/FID and GC/MS analysis.

Sheen samples are removed from the sample jar with precleaned tweezers and placed in a widemouth 250-mL jar with a Teflon lined cap for extraction. The sample is spiked with surrogate compounds and serially extracted two times with 50 mL of methylene chloride. The sample container should be rinsed with the first 50 mL of extraction solvent. The combined extract is concentrated to 1 mL by Kuderna Danish/nitrogen evaporation methods, spiked with internal standards, and analyzed.

8.3.2.3 GC/FID Analysis

A high-quality GC/FID analysis provides a hydrocarbon "fingerprint" of the sample, an assessment of the degree of sample weathering, and quantitative measures of important compositional and source-specific chemical compounds. Target analytes commonly measured by GC/FID analysis are listed in Table 8-1. The measured parameters include the total petroleum hydrocarbon (TPH) carbon ranges of n-C_{10} to n-C_{44+}, and n-C_{10} to n-C_{28} (diesel range organics [DRO]) and a variety of individual target hydrocarbons including n-alkanes and isoprenoid compounds.

Gas chromatography-flame ionization detection (GC/FID) should be conducted using a modern capillary gas chromatograph with a splitless injection port. In our work, we use an Agilent 6890 GC. The gas chromatograph should be fitted with a 60 m × 0.32 mm ID, 0.25 μm film thickness, DB-5 capillary column (or equivalent). A suitable GC oven program should be used to facilitate baseline resolution of n-alkane and isoprenoid hydrocarbons (e.g., n-C_{17} versus pristane); the program we favor begins with an initial temperature of 40°C (1 min), followed by a 6°C/min ramp rate to a final temperature of 315°C, followed by a 30-min hold. Ideally, hydrogen should be used as the carrier gas.

Prior to sample analysis, a minimum five-point calibration is performed to demonstrate the linear range of the analysis. The calibration solution is composed of selected aliphatic hydrocarbons within the n-C_9 to n-C_{40} range. Analyte concentrations in the standard solutions range from 1 ng/μL to 200 ng/μL. Target analytes not in the calibration solution may be quantified with the average relative response factor (RRF) of the nearest eluting compound(s). In our work, we assign the following response factors to certain target compounds: RRF of n-C_{14} assigned to 1380 and 1470 (C_{15} isoprenoids); RRF of n-C_{16} is

Table 8-1 Inventory of Target Saturated Hydrocarbons (SHC), PAHs, and Biomarkers Commonly Analyzed by GC/FID and GC/MS-SIM in Oil Spill Studies

Target SHC and PAH	Key	Quant. Ion (m/z)	RF	Target Biomarkers	Key	Quant. Ion (m/z)	RF
SHC — GC/FID		NA	RF_{alk}	**Sesquiterpanes**	SQx	123	Decalin
n-C_8 > n-C_{40}, pristane, phytane	n-Cx	NA	RF_{alk}	**Alkylcyclohexanes**	CH-X	83	Alk
nor-pristane, TPH (C8–C44)		NA	RF_{alk}				
PAH Groups — GC/MS		NA		**Tricyclic Triterpanes — GC/MS**			
Naphthalene	N	128	N	C_{23} Tricyclic triterpane	TC23	191	Hop
C1-Naphthalenes	N1	142	N	C_{24} Tetracyclic triterpane	TC24	191	Hop
C2-Naphthalenes	N2	156	N	C_{25} Tricyclic triterpane	TC25	191	Hop
C3-Naphthalenes	N3	170	N	C_{26} Tricyclic triterpanes	TC26	191	Hop
C4-Naphthalenes	N4	184	N	C_{28} Tricyclic triterpanes	TC28	191	Hop
Acenaphthene	ACE	154	ACE	C_{29} Tricyclic triterpanes	TC29	191	Hop
Acenaphthylene	ACY	152	ACY	C_{30} Tricyclic triterpanes	TC30	191	Hop
Biphenyl	BPHN	154	BPHN				Hop
Dibenzofuran	DBF	168	DBF	**Pentacyclic Triterpanes — GC/MS**			Hop
Fluorene	F0	166	F	$18\alpha(H)$-22,29,30-trisnorhopane (T_s)	27Ts	191	Hop
C1-Fluorenes	F1	180	F	$17\alpha(H)$-22,29,30-trisnorhopane (T_m)	27Tm	191	Hop
C2-Fluorenes	F2	194	F	$17\alpha(H),21\beta(H)$-28,30-bisnorhopane	28ab	191	Hop
C3-Fluorenes	F3	208	F	$17\alpha(H),21\beta(H)$-25-norhopane	25nor	191	Hop
Dibenzothiophene	DBT	184	D	$17\alpha(H),21\beta(H)$-30-norhopane	29ab	191	Hop
C1-Dibenzothiophenes	D1	198	D	$18\alpha(H)$-30-norneohopane ($C_{29}T_s$)	29Ts	191	Hop
C2-Dibenzothiophenes	D2	212	D	$17\beta(H),21\alpha(H)$-normoretane	29ba	191	Hop
C3-Dibenzothiophenes	D3	226	D	$18\alpha(H)$ and $18\beta(H)$ oleanane	30O	191	Hop
C4-Dibenzothiophenes	D4	240	D	$17\alpha(H),21\beta(H)$-hopane (**Hop**)	30ab	191	Hop
Phenanthrene	P0	178	P	$17\beta(H),21\alpha(H)$-moretane	30ba	191	Hop
Anthracene	A0	178	A	22S-$17\alpha(H),21\beta(H)$-30-homohopane	31abS	191	Hop
C1-Phenanthrenes/Anthracenes	P1 or PA1	192	P	22R-$17\alpha(H),21\beta(H)$-30-homohopane	31abR	191	Hop
C2-Phenanthrenes/Anthracenes	P2 or PA2	206	P	Gammacerane	30G	191	Hop
C3-Phenanthrenes/Anthracenes	P3 or PA3	220	P	22S-$17\alpha(H),21\beta(H)$-30-bishomohopane	32abS	191	Hop
C4-Phenanthrenes/Anthracenes	P4 or PA4	234	P	22R-$17\alpha(H),21\beta(H)$-30-bishomohopane	32abR	191	Hop
Fluoranthene	FL	202	FL	22S-$17\alpha(H),21\beta(H)$-30-trishomohopane	33abS	191	Hop
Pyrene	PY	202	PY	22R-$17\alpha(H),21\beta(H)$-30-trishomohopane	33abR	191	Hop
C1-Fluoranthenes/Pyrenes	FP1	216	FL	22S-$17\alpha(H),21\beta(H)$-30-tetrakishomohopane	34abS	191	Hop
C2-Fluoranthenes/Pyrenes	FP2	230	FL	22R-$17\alpha(H),21\beta(H)$-30-tetrakishomohopane	34abR	191	Hop
C3-Fluoranthenes/Pyrenes	FP3	244	FL	22S-$17\alpha(H),21\beta(H)$-30-pentakishomohopane	35abS	191	Hop
C4-Fluoranthenes/Pyrenes	FP4	258	FL	22R-$17\alpha(H),21\beta(H)$-30-pentakishomohopane	35abR	191	Hop
Naphthobenzothiophenes	NBT	234	NBT	**Steranes — GC/MS**			
C1-Naphthobenzothiophenes	NBT1	248	NBT	$13\beta,17\alpha$-diacholestane(20S)	27dbS	217	Chol
C2-Naphthobenzothiophenes	NBT2	262	NBT	$13\beta,17\alpha$-diacholestane(20R)	27dbR	217	Chol
C3-Naphthobenzothiophenes	NBT3	276	NBT	$5\alpha,14\beta,17\beta$-cholestane(20R)	27bbR	218	Chol
C4-Naphthobenzothiophenes	NBT4	290	NBT	$5\alpha,14\beta,17\beta$-cholestane(20S)	27bbS	218	Chol
Benzo[a]anthracene	BAA	228	BAA	$5\alpha,14\alpha,17\alpha$-cholestane(20R)	27aaR	217	Chol
Chrysene	C	228	C	$5\alpha,14\beta,17\beta,24$-methylcholestane(20R)	28bbR	218	Chol
C1-Chrysenes	C1	242	C	$5\alpha,14\beta,17\beta,24$-methylcholestane(20S)	28bbS	218	Chol
C2-Chrysenes	C2	256	C	$5\alpha,14\alpha,17\alpha,24$-methylcholestane(20R)	28aaR	217	Chol
C3-Chrysenes	C3	270	C	$5\alpha,14\alpha,17\alpha,24$-ethylcholestane(20S)	29aaS	217	Chol
C4-Chrysenes	C4	284	C	$5\alpha,14\beta,17\beta,24$-ethylcholestane(20R)	29bbR	218	Chol
Benzo[b]fluoranthene	BBF	252	BBF	$5\alpha,14\beta,17\beta,24$-ethylcholestane(20S)	29bbS	218	Chol
Benzo[j,k]fluoranthene	BKJF	252	BKJF	$5\alpha,14\alpha,17\alpha,24$-ethylcholestane(20R)	29aaR	217	Chol
Benzo[a]fluoranthene	BAF	252	BAF	$5\beta(H)$cholane (**Chol**)	Chol	217/218	
Benzo[a]pyrene	BAP	252	BAP	**Triaromatic Steroids (TAS) — GC/MS**			Chol
Benzo[e]pyrene	BEP	252	BEP	C_{20} — TAS	C20TA	231	Chol
Perylene	PER	252	PER	C_{21} — TAS	C21TA	231	Chol
Indeno[1,2,3-c,d]pyrene	IND	276	IND	$C_{26},20S$ — TAS	SC26TA	231	Chol
Dibenzo[a,h]anthracene	DAH	278	DAH	$C_{26}20R + C_{27},20S$ — TAS	RC26TA SC27TA	231	Chol
Benzo[g,h,i]perylene	BGHI	276	BGHI	$C_{28},20S$ — TAS	SC28TA	231	Chol
Single PAHs				$C_{27},20R$ — TAS	RC27TA	231	Chol
4-Methyldibenzothiophene	4MD	198	D	$C_{28},20R$ — TAS	RC28TA	231	Chol
2/3-Methyldibenzothiophene	2MD	198	D				
1-Methyldibenzothiophene	1MD	198	D	**Decalins**			
3-Methylphenanthrene	3MP	192	P	Decalin	DE	138	DE
2/4-Methylphenanthrene	2MP	192	P	C1-Decalins	DE1	152	DE
2-Methylanthracene	2MA	192	P	C2-Decalins	DE2	166	DE
9-Methylphenanthrene	9MP	192	P	C3-Decalins	DE3	180	DE
1-Methylphenanthrene	1MP	192	P	C4-Decalins	DE4	194	DE
Retene	R	234	P				
Cadalene	CD	198	N				
5-Methylchrysene	MC	242	C				

assigned to 1650 (i.e., nor-pristane). Carbon range TPH analyses are performed by employing the baseline integration technique of Douglas et al. (1994). The "window" for each TPH range is determined from the *n*-alkane calibration standard. All calibration solution compounds that fall within the appropriate carbon range window are used to generate the average RRF for a given carbon range. Areas for surrogate and internal standard compounds are subtracted from the range integration prior to computation of the TPH.

Instrument calibration is assured by analysis of a mid-level calibration check standard after every 10 authentic samples. The check standard's response is compared versus the average RF of the respective analytes contained in the initial calibration. All authentic samples and quality-control samples are bracketed by passing mid-check standards.

8.3.2.4 GC/MS Analysis

Gas chromatography/mass spectrometry (GC/MS) analysis of oil spill samples arguably provides the most important quantitative "fingerprinting" data used in oil spill correlation analysis. We utilize an Agilent 6890 GC interfaced to a Hewlett-Packard 5973 mass selective detector (MSD). The GC should be equipped with a 60 m × 0.25 mm ID, 0.25 μm film thickness, DB-5 capillary column (or equivalent). The GC oven program we use begins at an initial temperature of 40°C with a 1-min hold, followed by a 6°C/min oven ramp to a final temperature of 315°C, followed by a 30-min hold time. Helium is the preferred carrier gas in this quadrapole mass spectrometry analysis.

The GC/MS is calibrated with perfluorotributylamine (PFTBA) at the beginning of each analytical sequence. A minimum 5-point initial calibration consisting of selected target compounds is established to demonstrate the linear range of the analysis (Table 8-1). Analyte concentrations in the standard solutions range from 0.01 ng/μL to 10 ng/μL.

Data acquisition is performed in the select ion monitoring (SIM) mode for optimal sensitivity and selectivity. Quantification of target compounds is performed by the method of internal standards using the average relative response factor of the parent compounds or otherwise representative compounds (Table 8-1) determined from the 5-point initial calibration (*Federal Register*, 1994).

Target compounds measured in this analysis include the polycyclic aromatic hydrocarbons (PAH) and the tricyclic and pentacyclic triterpanes and steranes listed in Table 8-1. Many alkyl PAH isomers are commercially unavailable as analytical standards. Thus, the quantitative response factors for such compounds are based upon that of their respective parent PAH (Table 8-1; *Federal Register*, 1994). Similarly, most target biomarkers are not contained in the initial calibration solution (largely due to the high costs of standards or lack of availability). The RFs for biomarkers are based on those of two representative compounds (Table 8-1). In the case of selected PAH groups, individual PAH isomers are also quantified using RFs derived for the individual isomers contained in the calibration solutions. Among these are four methyl-dibenzothiophene isomers, five methyl-phenanthrene isomers, retene (1-methyl-7-isopropyl-phenanthrene), cadalene (1,6-dimethyl-4-(1-methylethyl)-naphthalene), and 5-methylchrysene.

PAH and biomarker concentrations (as well as alkanes, described above) are calculated by the method of internal standards using the following:

$$C_a = [(A_a/A_i) \times (Amt_i/RF_i) \times D]/V_a$$

where

C_a = concentration of target analyte
A_a = area of quantification ion for target analyte
A_i = area of quantification ion for RIS
Amt_i = amount of RIS added to sample
RF_i = average RF for analyte determined from intial 5-point calibration
D = dilution factor (if applicable)
V_a = sample size (volume or mass)

Biomarker identifications are based upon comparison to selected authentic standards (Chiron

Laboratories), elution patterns in the peer-reviewed literature (e.g., Philp, 1985), and mass spectral interpretation from full-scan GC/MS analyses conducted in the laboratory. Triterpane concentrations are calculated using the RRF of C_{30} 17α(H), 21β(H)-hopane relative to the internal standard d_{12}-chrysene. Sterane concentrations are calculated using the RRF of 5β(H)-cholane relative to the internal standard d_{12}-chrysene.

8.3.2.5 Data Quality

Samples are analyzed in exclusive analytical batches, on the same GC/MS instrument (when possible) and, to the extent possible, within the same analytical sequence. Data analyses (peak integrations) are conducted by a single GC/MS analyst with experience in PAH and biomarker pattern recognition. Each of these steps, combined with the calibrations (initial and continuing), procedural blanks (PB), laboratory control samples (LCS), control oil analysis, surrogate recoveries, and duplicate/triplicate analyses, provides measures of data quality. Typical data quality objectives for these analyses are listed in Table 8-2.

8.3.3 Selection of Diagnostic Indices

Although the comparison of absolute concentrations of target compounds can be used to compare spill and source oils, more often than not diagnostic ratios calculated from these concentrations offer some advantages when comparing samples. One benefit of comparing the diagnostic indices rather than absolute concentrations of spilled oil and suspected source oils or potentially impacted samples is that the analytical variability due to concentration effects is minimized. This technique has been used in other oil spill investigation protocols in which semiquantitative data (e.g., peak heights or areas; Daling et al., 2002a, 2002b;

Table 8-2 Summary of Typical Data Quality Objectives Used in Quantitative Fingerprinting Studies of Waterborne Oil Spills

QC Element or Sample Type	Minimum Frequency	Data Quality Objective/Acceptance Criteria
MS Tuning	Prior to each run sequence using PFTBA	m/e 69: Base Peak (~100,000 counts minimum) m/e 219: 30–60% Base Peak abundance m/e 502: 2–8% Base Peak abundance
Initial Calibration	Prior to every instrument batch sequence or as needed indicated by continuing calibration check	5 point curve, minimum of 5 point. %RSD ≤25% for 90% of analytes and ≤35% for all analytes
Continuing Calibration	Must end analytical sequence, and every 12 field samples or 16 hours, whichever is more frequent	%RSD ≤25% for 90% of analytes. %RSD ≤35% for all analytes
Procedural Blank	Every batch/every 15–20 samples	Less than the reporting limit unless analyte not detected in associated sample(s) or associated sample analyte concentration is >5× blank value
Laboratory Control Sample (LCS)/LCS Duplicate	Every batch/every 15–20 samples	50%–130% recovery SVOCs
Recovery/Surrogate Standards	Every sample	50%–130% recovery for other SVOCs
Internal Standard (IS)	Every sample	50%–200% of the area of the IS in the associated calibration standard SVOC
Instrumental Check	One per initial calibration	80%–120% recovery
Control Oil	One per initial calibration One per sequence for FID	65%–135% recovery PAHs and hopane for biomarkers (for NSC only) alkanes — for chromatogram only

Faksness et al., 2002; Henry et al., 1997) or absolute concentration data are converted to useful quantitative ratios (Burns et al., 1997; Stout et al., 2001; Page et al., 1995). On the other hand, because each index depends on the responses or concentrations of multiple analytes, any errors in these may be increased in the calculation of a ratio. Thus, the precision of the indices is more important than the precision of the concentrations and must be evaluated prior to use in oil spill correlation analysis. It is our experience that in order for a ratio to be useful for interpretive purposes, the analytical precision of a representative oil sample replicate ($n = 3$) must generally be 10% (residual standard deviation) or less. The emerging CEN protocol described in Chapter 7 suggests the use of 14% RSD as a suggested measure of precision.

The data acquisition described earlier provides a set of GC/FID chromatograms, selected extracted ion profiles from GC/MS-SIM, and the tabulated absolute concentration data for each target analyte or analyte group (Table 8-1) identified in the field and QC samples. Evaluations of the biomarker, PAH, total petroleum hydrocarbon (TPH), and selected alkane concentration data are subsequently used to generate a suite of diagnostic indices allowing comparison among the spilled oil, candidate sources, and potentially impacted samples (Table 8-3).

The indices listed in Table 8-3 each serve some diagnostic purpose, but each is not necessarily useful in correlating all types of spilled oil to candidate sources, primarily due to the effects of weathering on some indices (Douglas et al., 1996; Uhler and Emsbo-Mattingly, 2006; Emsbo-Mattingly et al., 2006). For example, the "bulk" ratios of n-C_{18}/n-C_{30} or percent total petroleum hydrocarbons (TPH) in the diesel range (C_{10}–C_{25}/C_{10}–C_{44}) will undoubtedly decrease with evaporative weathering. Alternatively, these same ratios would increase if an HFO had been cut with diesel fuel #2. Other indices may be affected by biological degradation (e.g., n-C_{17}/Pr) or water-washing (e.g., %2-ring PAH/total PAH).

On the other hand, the selected PAH (e.g., C2-dibenzothiphenes/C2-phenanthrenes [D2/P2]) and biomarker indices (e.g., moretane/hopane) are largely considered independent of weathering on an environmental timescale. Therefore, it is proposed that these and other such indices are most suitable for correlating a spilled oil to candidate sources. Among the many potential source ratios, those of greatest value are the ones that clearly differentiate the candidate source oils. Of course, chemical reasonableness in the properties of spilled and candidate source oils should always be considered. In other words, all chemical features (and not only the PAH and biomarker indices) must also support any "positive" correlation based on diagnostic source ratios (Stout et al., 2001).

The impact of oil weathering (e.g., evaporation, solubilization, and biodegradation) on source ratio stability should be considered before it is used in an oil spill investigation (Douglas et al., 1996; Bost et al., 2001; Stout et al., 2001; Daling et al., 2002a; Uhler and Emsbo-Mattingly, 2006). For most waterborne oil spill situations, the spilled oil is primarily influenced by evaporation and volatilization (NRC, 1985); therefore, n-C_{15} and greater hydrocarbons will be less affected. Oil exposed for longer periods of time in the environment will be further altered by biodegradation (NRC, 1985; Prince, 2002). Under severe biodegradation conditions, even the most refractory source ratios can be distorted (Bost et al., 2001); however, these conditions are rarely observed in oil spill events.

Selection of the PAH and biomarker diagnostic indices (e.g., Table 8-3 and Table 8-4) is largely based upon "genetically" significant (i.e., source-specific) variables known to occur among crude oils from different geologic basins. These "genetic" differences are largely carried forward during refining (Peters et al., 1992) and, therefore, can still provide diagnostic information on petroleum products (e.g., a fuel oil). Much of the knowledge of which PAH and biomarker indices are truly "genetically" significant comes from the oil exploration and production geochemical liter-

Table 8-3 List of Common Quantitative Indices Useful in Oil Spill Investigations. This List Is Not Intended to Be All Inclusive

Bulk Indices[1]	PAH Indices[1]	Biomarker Indices[1]
Isoprenoid Indices	**% Ring Number**	**Sterane Indices**
norpristane/pristane	%[2-ring PAH/Σtotal PAH]	%[$C_{27}\beta\alpha$ diasterane (S/S+R)]
n-C_{17}/pristane	%[3-ring PAH/Σtotal PAH]	%[$C_{29}\alpha\alpha\alpha$ steranes (S/S+R)]
n-C_{18}/phytane	%[4- to 6-ring PAH/Σtotal PAH]	%[$C_{29}\alpha\beta\beta$(R+S)/total C_{29} steranes]
Pristane/Phytane	%[S-PAH/Σtotal PAH]	%[$C_{27}\alpha\beta\beta/C_{27}$–$C_{29}$ $\alpha\beta\beta$]
Boil Range Indices	**Methyl-Phenanthrene Indices**	%[$C_{28}\alpha\beta\beta/C_{27}$–$C_{29}$ $\alpha\beta\beta$]
n-C_{18}/n-C_{30}	MPI 1 = 1.5(2MP + 3MP)/(P + 1MP + 9MP)	%[$C_{29}\alpha\beta\beta/C_{27}$–$C_{29}$ $\alpha\beta\beta$]
%[ΣTPHC$_{10}$–C_{25}/ΣTPHC$_{10}$–C_{44}]	MPI 2 = 3(2MP)/(P + 1MP + 9MP)	**Triterpane Indices**
CPI (odd C_{25-33}/(even C_{26-34}) alkanes	MPR = 2MP/1MP	$C_{23}+C_{24}/C_{28}+C_{29}$ tricyclics
	Dibenzothiophene Source Indices	$C_{28}+C_{29}$ tricyclics/Hopane
	MDR = 4MDBT/1MDBT	T_s/Hopane
	DBT/P (Dibenzothiophene/Phenanthrene)	Moretane/Hopane
	D/P	BNH/Hop (28,30-bisnorhopane/Hopane)
	D2/P2	25NH/Hop (25-norhopane/Hopane)
	D3/P3	BNH + 25NH/Hopane
	Naphthobenzothiophene Source Indices	Nor/Hop (Norhopane/Hopane)
	NBT2/C1	Ol/Hop (Oleanane/Hopane)
	NBT3/C2	$C_{29}\alpha\beta/C_{29}T_s$ Hopane
	Retene/Total C4-phenanthrenes	%[C_{31} Hopane (S/S+R)]
	%[Retene/P4]	%[C_{32} Hopane (S/S+R)]
	Chrysene Profile	%[C_{35} Hopanes/ΣC_{30}–C_{35} Hopanes]
	%[C0/Chrysene total]	Hopane/$C_{29}\alpha\alpha\alpha$ 20R Sterane
	%[C1/Chrysene total]	C_{24} tricyclic/C_{26}(S+R) tricyclics
	%[C2/Chrysene total]	Ts/Tm
	%[C3/Chrysene total]	**Triaromatic Sterane (TAS) Source Indices**
	%[C4/Chrysene total]	
	Miscellaneous PAH Source Indices	C_{28},20S TAS/C_{28},20R TAS
	FL/PY (BBF + BKF)/BAP	C_{26},20R+C_{27},20S TAS/C_{27},20R TAS
	BBF+BKF/C C/FP1	Hop/C_{28},20S TAS
	PER/C IP/GHI	
	BBF/BKF PER/BAP	
	BF/FP1 BKF/BAP	
	BBF/BAP BEP/BAP	
	BA/C	

[1] Diagnostic indices may be expressed as A/B or A/(A + B).

ature (Peters and Moldowan, 1993; Peters et al., 2005a, 2005b). These ratios in crude oils are known to be related to (1) the thermal maturity of the source rocks that gave rise to a crude oil, (2) the type of organic matter present in the source rock (e.g., terrestrial versus marine), and/or (3) the degree of weathering that may have occurred in the crude oil reservoir prior to oil production. Understanding the meaning of these ratios, and if they are in any way susceptible to changes due to environmental weathering (evaporation, water-washing, or biodegradation) following an oil spill, certainly needs to be considered (see Chapter 1 herein). However, in this respect, the use of diagnostic ratios based upon biomarker concentrations is particularly useful since these compounds are highly resistant to weathering over environmental timescales (Prince et al., 1994; see also Chapter 11 herein). Finally, although the alkylated PAH compounds are not generally as resistant to weathering as the

Table 8-4 Diagnostic Ratios and Precision-Based Statistics for a Spilled Oil and Two Candidate Source Oils (A and B). All Data are Based Upon the Absolute Concentrations of Individual Compound or Compound Groups Measured by GC/MS as per the Revised Nordtest Method (Daling et al., 2002)

Diagnostic Ratio*	Spill Oil (Rep 2)	Spill Oil (Rep 1)	Spill Oil (Rep 3)	Spill Mean	Std. Dev.	Percent RSD**	95% CI***	98% CI***	Source A	Source B
%27Ts	59	56	57	57	1.5	2.7	3.8	6.1	56	60
%28ab	22	21	18	20	2.1	10.2	5.2	8.4	12	20
%25nor30ab	3	3	2	2	0.5	18.3	1.1	1.8	1	3
%29ab	31	34	31	32	1.7	5.4	4.3	7.0	34	32
%C29Ts	17	18	17	17	0.6	3.3	1.4	2.3	17	18
%30d	9	10	11	10	0.9	8.5	2.1	3.4	13	12
%27dia	46	47	48	47	1.0	2.1	2.5	4.0	48	49
%29aaS	43	49	45	46	3.1	6.7	7.6	12.3	44	44
%29bb	51	53	51	52	1.2	2.2	2.9	4.6	52	49
%27bbSTER	39	38	39	39	0.6	1.5	1.4	2.3	36	40
%28bbSTER	28	30	28	29	1.2	4.0	2.9	4.6	30	27
%29bbSTER	33	34	33	33	0.6	1.7	1.4	2.3	35	33
%TA21	52	50	55	52	2.5	4.8	6.3	10.1	60	54
%TA26	37	36	40	38	2.1	5.5	5.2	8.4	32	39
%TA27	54	56	55	55	1.0	1.8	2.5	4.0	49	56
%D2/P2	31	32	27	30	2.6	8.8	6.6	10.6	17	29
%D3/P3	30	29	25	28	2.6	9.4	6.6	10.6	17	27
%D3/C3	71	77	72	73	3.2	4.4	8.0	12.9	63	75
%2MP/1MP	45	46	47	46	1.0	2.2	2.5	4.0	53	45
%4MD/1MD	76	76	77	76	0.6	0.8	1.4	2.3	85	77

*See Daling et al. (2002) for definition of specific diagnostic ratios.
**Relative percent standard deviation.
***Confidence Interval as per Student's t test (Harris, 1995).

biomarkers, certain ratios of select PAHs have proven to be reliable and stable for fingerprinting purposes, even in the face of significant evaporative and biodegradative weathering (e.g., C2-dibenzothiophenes/C2-phenenathrenes, Page et al., 1995; Douglas et al., 1996; Uhler and Emsbo-Mattingly, 2006). The chemical compounds used in these ratios must have similar physical and chemical properties such that they both weather at the same rates in the environment and retain the original oil source ratio (Overton et al., 1981).

It is important to realize that the indices listed in Table 8-3 should not be considered all-inclusive or appropriate for all oil spill studies. Sometimes it may be prudent to include a certain index that is recognized as particularly diagnostic of the spilled oil. Acquiring the breadth of concentration data (n-alkanes, isoprenoids, PAHs, and biomarkers) described here allows one that flexibility. In addition, the analytical and interpretive methods discussed earlier are not limited only to marine oil spill situations. The basic technical approach and diagnostic indices can also be applied to groundwater/soil investigations where nonaqueous phase liquids (NAPL) and petroleum contaminated soils are present.

8.3.4 Source Identification Protocols for Quantitative Fingerprinting Data

Diagnostic ratios calculated from quantitative or semiquantitative data need to be compared in an unbiased manner. Several methods of increasing sophistication are available for this. First, as we noted earlier, several of the *Exxon Valdez* investigations relied upon quantitative fingerprinting data to reach conclusions regarding the sources of hydrocarbons in the sediments of Prince William Sound. These studies often relied upon simple but diagnostic *x-y* cross-plots of selected diagnostic ratios to demonstrate similarities and differences among a large number of samples (e.g., Bence et al., 1996). Plots relying upon the C2-dibenzothiophenes/C2-phenanthrenes, C3-dibenzothiophenes/C3-phenanthrenes, oleanane/hopane, 28,30-bisnorhopane/hopane, and Ts/Tm ratios were particularly useful distinguishing cargo oil from other hydrocarbon sources in the Sound. A simple example of this approach to comparing samples is shown in Figure 8-2. Double-ratio cross-plots have the advantage of being able to compare a large number of samples simultaneously, but these have the disadvantage of necessarily relying on only two ratios at a time.

Second, more recently, the Nordtest organization developed a revised oil spill identification protocol based on semiquantitative or quantitative chemical fingerprinting that considers as many diagnostic ratios as is appropriate simultaneously (Daling et al., 2002a). The revised Nordtest protocol relies on the linear correlation of diagnostic ratios calculated from semi- or fully quantitative data generated using GC/FID and GC/MS. In this method, the degree of statistical correlation depends on the analytical precision of the diagnostic ratios, as measured by the 95% and 98% confidence intervals derived from the "Student's t" statistical tool (Harris, 1995) determined from replicate (usually triplicate) analyses. Three levels of correlation were suggested in the revised Nordtest protocol:

Positive match — all diagnostic ratios within the 95% confidence interval.

Probable match — all diagnostic ratios within the 98% confidence interval.

Nonmatch — any key diagnostic ratio outside the 98% confidence interval.

An example of the revised Nordtest protocol involved a mystery oil slick discovered in an offshore oil production area containing numerous oil production platforms, pipelines, and barge traffic. Samples of the slick and numerous candidate sources (production streams from different platforms, pipelines, and barges) were collected for analysis. Since the crude oils produced in the area were generated from oil source rock strata containing similar ancient organic matter that had been heated to a similar extent, there was not a great deal of heterogeneity among the candidate oils. In other words, they were genetically similar, which could have confounded an assessment of the slick's source if quantitative data were not available (as described in Section 8.2.1.3).

The mystery spill oil was prepared and analyzed in triplicate via GC/FID and GC/MS as described in Section 8.3.2. Several candidate source oils were also prepared and analyzed in the same manner. Table 8-4 contains the quantitative results for 20 diagnostic ratios determined for the spill oil (in triplicate) and two of the candidate source oils. The relevant statistics for the spill oil (e.g., mean, standard deviation, relative percent standard deviation, and Student's t confidence interval) were calculated according to Daling et al. (2002a). Important among these is the very good analytical precision evident in the %RSD less than 10 for most diagnostic ratios (Table 8-4).

According to the revised Nordtest method's linear regression protocol (Daling et al., 2002a), the diagnostic ratios for the mean spill oil are plotted against the diagnostic ratios for the two candidate source oils on simple *x-y* plots (Figure 8-3). If there was a perfect match between the spill oil and any candidate source oil, each point (representing a particular diagnostic ratio from Table 8-4) would fall perfectly along a straight line. Such perfect matches are rarely observed, inevitably due to some degree of analytical precision error. Two examples among the various candidate source oils analyzed were selected for demonstrating the revised Nordtest's elegantly simple regression approach — Source A and Source B (Figure 8-3). Figure 8-3A clearly shows that Source A has several diagnostic ratios that plot

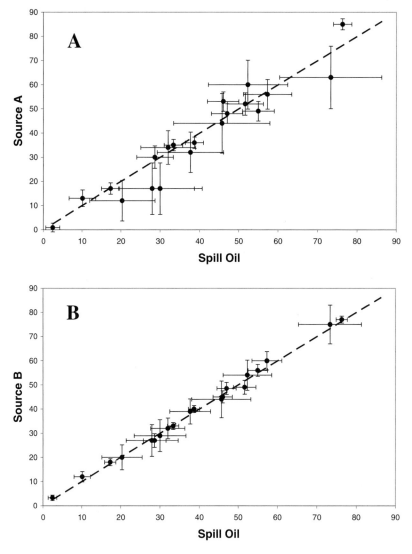

Figure 8-3 Examples of linear regression analysis as per the revised Nordtest oil spill identification method for the data shown in Table 8-4. (A) Example of "non-match" between spill oil and Source A based upon 98% confidence interval. (B) Example of "match" between spill oil and Source B based upon 98% confidence interval.

beyond the 98% CI indicating that, according to the revised Nordtest method's recommended criteria, Source A is a "nonmatch" to the spilled oil. On the other hand, Figure 8-3B shows that Source B has all 20 of the diagnostic ratios within the 95% CI of the spilled oil's ratios indicating that, according to the recommended criteria, Source B is a "positive match" to the spill oil.

Despite the overall genetic similarity among the candidate source oils, the Source B oil was determined to be the only "positive match" among the available source candidates. This provided a strong chemical basis upon which to conclude it was the likely source of the mystery slick. Importantly, there was no opportunity of bias entering into this statistically based interpretation of the quantitative data.

Although the revised Nordtest protocol offers an entirely statistical and objective correlation approach, one potential issue with this approach is that laboratories with exceptional analytical precision for their replicates will have a smaller error and therefore more rigorous match criteria than laboratories with a lower analytical precision. Therefore, it is possible that the less precise laboratory will identify a "positive match" when the more precise lab will not. Another potential issue of the revised Nordtest approach is that because it relies on correlation between only two oils at a time (typically a spill versus a potential source), it can have difficulty dealing with large datasets and with spill samples influenced by mixtures (Stout et al., 2005). The revised Nordtest protocol served as the starting point for the emerging CEN protocol described in Chapter 7.

One issue related to the revised Nordtest protocol is that its simple graphical technique (e.g., Figure 8-3) can become cumbersome if a large number of candidate oils need to be compared to a spilled oil. In such a situation, we develop a correlation matrix based on some standard similarity index, such as the correlation coefficient (r or r^2), based on a statistical comparison of the diagnostic ratios among the oils. A simple ranking of these indices for each candidate source oil versus the spilled oil — from highest to lowest — helps identify those candidate source oils most highly correlated with the spill. The revised Nordtest graphical approach to matching can then focus on these most highly correlated oils in order to determine if they meet appropriate match criteria.

Some additional comments on the use of similarity indices in oil spill correlation studies are warranted. While similarity indices, such as r or r^2, can be used effectively to rank the relative similarity of field samples, it is important to recognize the limitations of correlation analysis for source identification purposes. First, it is important to use quantitative data (e.g., analyte concentrations) with known levels of precision and accuracy. It is particularly helpful to include replicate results and confidence intervals whenever possible. Second, as described, the diagnostic parameters should resist environmental weathering. The indiscriminant use of all analyte data is commonly confounded by nonlinear changes in analyte concentrations due to the preferential weathering of more volatile or biodegradable constituents. Dimensionless concentration ratios of recalcitrant analytes (e.g., biomarkers and 3- to 6-ring PAHs) generally serve well as source indicators in correlation analysis. Third, diagnostic ratios should be equally weighted so that a small number of source indicators with a wide dynamic range do not obscure the statistical importance of indicators with a narrow dynamic range. Specifically, the analyte ratios should be Z-score or range-normalized before running the correlation analysis. Fourth, the selection of levels of significance for correlation coefficients should not be established based on an absolute scale (e.g., "$r^2 > 0.75$ is a source signature match"). Among other factors, the level of significance for the correlation coefficient should take into account the degree of similarity (e.g., genetic similarity) among multiple source area samples and degrees of difference among samples collected from nonimpacted reference areas. Finally, the results of any correlation analysis should consider the potential for mixing. Oil from the same source may accumulate different features as it accumulates hydrocarbons from the ambient environment or from independent releases. Accordingly, the application and results of any simple correlation analysis based on some similarity index must be evaluated for chemical reasonableness and checked on an individual basis using qualitative and quantitative fingerprinting techniques.

An alternative means of analyzing a large suite of quantitative fingerprinting data simultaneously is to employ some multivariate statistical method, such as principal component analysis (PCA) or partial least-squares analysis (e.g., Aboulkassim and Simoneit, 1995; Burns et al., 1997; Stout et al., 2001; Lavine et al., 2001; Mudge, 2002; Christensen, 2002; Christensen et al., 2004, 2005; Li et al., 2004). These methods and their advantages are more fully described in Chapter 9. The primary benefit of these methods is that — as long as the input data consist of data whose precision is known and that is considered source-diagnostic (and not prone to weathering effects) — multivariate analysis of quantitative data can produce objective and confident positive and negative correlations. Obviously, such analyses cannot be performed on qualitative data.

As discussed earlier, correlations based on simple double-ratio cross plots, revised Nordtest-like correlation methods, or even multivariate analyses methods cannot be performed in a vacuum. In other words, diagnostic ratios evaluated by any of these methods need to consider all of the available data before a "positive match" is achieved. Also, keep in mind that a chemical match does not unequivocally mean that the correlated candidate source oil is the only possible source oil. Each spill's circumstances, including any other site information such as location of the spill, water current direction, operational practices, or chemical information such as the degree of weathering and trace biomarker differences, may provide the additional evidence required to isolate the spill source.

8.4 Unraveling Mixed Source Oils Using Quantitative Fingerprinting Data

8.4.1 Two-Component Mixing Models

The one-to-one linear regression protocol described in the revised Nordtest method and exemplified in Figure 8-3, or multivariate analyses, can work extremely well if there is no likelihood of multiple sources mixing to produce a mystery spill. However, there are oil spills in which there is a significant prespill oil component (background) or mixtures of different oils spilled simultaneously that need to be recognized and accommodated. The revised Nordtest protocol (Daling et al., 2002a) acknowledges that "mixed source signals" could be difficult to recognize and that when mixtures are suspected, "a simple linear mixing algorithm could be developed to explain the oil spill." Simple linear mixing of a two-component mixture is straightforward and has been used in oil field studies (e.g., Kaufman et al., 1990; Peters and Fowler, 1992). However, since the diagnostic ratios relied on in the revised Nordtest method do not necessarily respond linearly to mixing (see discussion in Section 8.3.1), the development of any linear mixing algorithm *requires* the use of the absolute concentration of the compound(s) in the algorithm. In other words, the mixing of the diagnostic ratios themselves does not consider that the individual compound concentrations will likely vary between the two mixed oils. For example, Table 8-5 contains a set of data in which the absolute concentrations of two biomarkers (A and B) and two PAHs (C and D) are measured in two oils and the ratios $A/(A + B)$ and $C/(C + D)$ calculated in each oil. Theoretical mixtures of these two oils will yield a set of diagnostic ratios with values intermediate to starting ratios. When the ratios are calculated based on the absolute concentrations of the biomarkers and PAHs, the resulting ratios accurately reflect the ratios in their mixtures. However, suppose the original ratios of the candidate source oil were based upon relative abundances (e.g., peak heights). Mixing the original ratios results in set of diagnostic ratios that do not accurately reflect the mixtures due to the varying concentrations of these biomarkers and PAHs in the oils. Thus, in an oil spill investigation where mixtures of different oils (or oils with background contamination) are under investigation, fully quantitative data are necessary to develop accurate mixing models. The need for concentration data is heightened when the analysis involves data generated by multiple laboratories or data generated over a long period of time. This is a very strong argument for the need to obtain fully quantitative (absolute concentration) data for individual chemicals in oil spill fingerprinting studies.

In other situations, it is possible that the candidate sources collected for a mystery oil spill investigation do not fully represent the population of candidate sources. For example, consider the situation where a vessel's slop tank is empty at the time (after the spill event) samples are being collected, but it was not

Table 8-5 Example Data for Selected Biomarkers for Two Oils Demonstrating How the Absolute Analyte Concentrations are Needed to Calculate Accurate Diagnostic Ratios in Theoretical Mixtures. Ratios Calculated from the Original Ratios, as Might Be Available from Peak Areas (but not Concentrations), Do Not Mix Linearly Due to Concentration Differences in the Two Oils. Therefore, Absolute Concentrations Are Necessary to Develop Accurate Mixing Models

	OIL 2 (mg/kg)	Theoretical Mixtures of OIL 1 and OIL 2									OIL 1 (mg/kg)
(OIL1:OIL2)	(0:100)	(10:90)	(20:80)	(30:70)	(40:60)	(50:50)	(60:40)	(70:30)	(80:20)	(90:10)	(100:0)
Biomarker A	63.9	59.1	54.4	49.7	45.0	40.3	35.5	30.8	26.1	21.4	16.6
Biomarker B	83.3	78.2	73.1	68.1	63.0	58.0	52.9	47.9	42.8	37.8	32.7
PAH C	80.9	74.5	68.2	61.9	55.6	49.2	42.9	36.6	30.3	23.9	17.6
PAH D	403.8	399.4	394.9	390.5	386.0	381.6	377.1	372.7	368.2	363.8	359.3
Diagnostic Ratios in Mixtures Based Upon Absolute Analyte Concentrations											
%A/(A + B)	43.4	43.1	42.7	42.2	41.6	41.0	40.2	39.2	37.9	36.1	33.7
%C/(C + D)	16.7	15.7	14.7	13.7	12.6	11.4	10.2	8.9	7.6	6.2	4.7
Diagnostic Ratios in Mixtures Based Upon Simple Mixtures of Original Source Ratios											
%A/(A + B)	43.4	42.4	41.5	40.5	39.5	38.6	37.6	36.6	35.7	34.7	33.7
%C/(C + D)	16.7	15.5	14.3	13.1	11.9	10.7	9.5	8.3	7.1	5.9	4.7

when the spill occurred. Since the vessel's slop tank might have contained heavy fuel oil (HFO), lubricating oil, or other oils from the vessel, it is possible that some mixture of these oils was present in the slop tank at the time of the spill. Mathematically reconstructing a theoretical suite of possible "slop tank" mixtures — based on the individual oil that might have been present in the slop tank — requires fully quantitative (absolute concentration) data for target PAH and biomarkers in each of the individual end-member candidate oils. We demonstrate this approach in two case studies that involved a mixed source signal.

8.4.2 Case Study 1

In this case study, an oil spill was discovered adjacent to a docked marine crude oil tanker — which may not seem like a mystery. However, quantitative GC/MS analysis of PAH and biomarkers in three candidate source oils collected from the vessel's cargo tanks, HFO fuel tanks, and bilge tanks showed them each to be "nonmatches" to the spilled oil according to the revised Nordtest criteria (Figure 8-4). One of the three candidate source oils — the crude oil cargo — was very different from the spill and could be eliminated for further consideration. However, of the 10 diagnostic ratios utilized (A through J; Figure 8-4), the spilled oil exhibited multiple diagnostic ratios that appeared intermediate between the HFO fuel and bilge oil. For example, inspection of Figure 8-4A shows that the bilge oil exhibited several diagnostic ratios that were higher than observed in the spill oil (e.g., ratios B and F) and another that was lower than in the spill oil (ratio C). Inspection of Figure 8-4B shows that the HFO exhibits diagnostic ratios that were lower than observed in the spill oil (e.g., ratios B and F). Thus, at this point in the investigation, there was a suspicion that the spilled oil might have been derived from a mixture of the bilge and HFO oils, but this needed to be proven.

In order to evaluate the candidacy of a mixed oil release, the PAH and biomarker absolute concentration data were evaluated using a linear mixing algorithm in which the concentrations of each PAH and biomarker analyte were calculated in a series of theoretical, two-component mixtures according to the following equation:

$$(\text{Conc. A} \times \%\text{bilge}) + (\text{Conc. A} \times \%\text{HFO}) \quad (\text{Eq. 8-1})$$

Recalculation of the 10 diagnostic ratios (A through J) based on the concentrations of each ratio's numerator and denominator chemicals

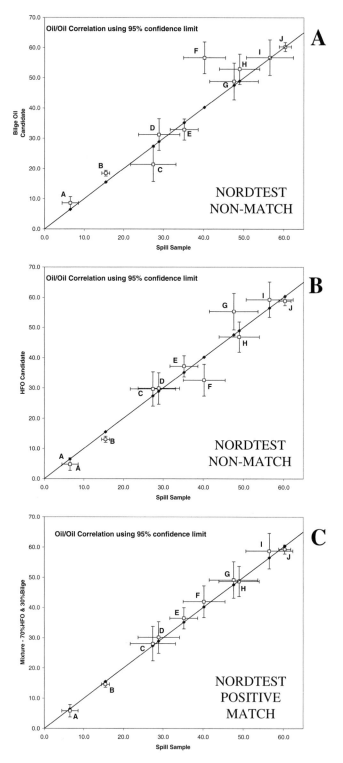

Figure 8-4 Revised Nordtest linear regression plots for ten diagnostic ratios (A-J) in the mystery spill oil and two candidate source oils from case study 1. (A) comparison of bilge oil with spilled oil resulting in a non-match, (B) comparison of HFO oil with spilled oil resulting in a non-match, (C) comparison of theoretical mixture of bilge and HFO with spilled oil resulting in a positive match.

using this two-component mixing algorithm showed that a mixture of 70%HFO and 30%bilge oil provided a "positive match" to the spilled oil according to the revised Nordtest's linear regression protocol, the result of which is shown in Figure 8-4C. Thus, it was reasonably concluded that the suspect vessel had discharged an approximately 70:30 mixture of HFO and bilge oil and was thereby likely responsible for the mystery spill. This conclusion would have been entirely speculative if the absolute concentration data were not available and the two-component mixing model could not be developed.

8.4.3 Case Study 2

Another recent case study evaluated mixed oil signals using quantitative chemical fingerprinting data combined with the revised Nordtest method. In this case, a commercial vessel carrying large quantities of intermediate and heavy fuel oils (IFO and HFO, respectively) and a small amount of marine diesel fuel (MDO) was accidentally grounded. The grounding resulted in IFO and HFO being released to the marine environment. During the subsequent investigation, samples of the IFO and HFO from the vessel as well as field samples potentially containing oil(s) released from the grounded vessel were collected. The objective of the study was to determine if any field samples (mousse and tarballs) collected after the release were "positive matches" to the source oils from the grounded vessel. Although this objective may seem like a simple task, standard interpretive approaches to source identification require careful considerations when mixtures of oils potentially are present (e.g., bilges; Section 8.2.1.4).

Figure 8-5 shows the GC/FID chromatograms, PAH concentration histograms, and tricyclic and pentacyclic triterpane biomarker fingerprints for the IFO and HFO collected from the vessel's fuel tanks after the grounding. The disparate nature of these two residual fuel oils is reflected in their distinct GC/FID chromatograms (Figure 8-5A-B), which can be distinguished well based on a qualitative inspection of the hydrocarbon patterns. The GC/FID chromatographic differences are best expressed by the wider range of n-alkanes and bimodal unresolved complex mixture (UCM, e.g., higher molecular weight) present in the HFO relative to the IFO.

Quantitative chemical fingerprinting analysis revealed a particularly significant difference in the PAH composition of the two fuels, namely the substantially lower C2-dibenzothiophenes/C2-phenanthrenes and C3-dibenzothiophenes/C3-phenanthrenes (D2/P2 and D3/P3, respectively) ratios in the IFO relative to the HFO (Figure 8-5C-D). The environmental stability and source character of these ratios are well established (Douglas et al., 1996, Page et al., 1995), indicating that these fuels contain different levels of sulfur-bearing PAH. In addition, quantitative biomarker analysis showed markedly lower biomarker concentrations of triterpanes in the IFO relative to the HFO, the absence of C28 and C29 tricyclic terpanes in the HFO, and markedly different "triplet" distributions (i.e., C26 tricyclic and C24 tetracyclic terpanes; see inset) and 30-norhopane/hopane signatures (Figure 8-5E-F).

Because of the obvious differences in the GC/FID chromatograms, PAH distributions, and triterpane concentrations/distributions, qualitative fingerprinting would likely be sufficient to distinguish these two source oils from one another, presumably even after moderate weathering has occurred. Thus, it is clear that the two fuels carried by the grounded vessel were chemically distinct from one another.

Figure 8-6 depicts the GC/FID, PAH, and triterpane biomarker fingerprints for two apparently impacted field samples — a mousse sample and a tarball sample — collected after the grounding. The on-site responders confidently linked the mousse sample to the release based on visual observations after the spill. Prior to obtaining the vessel fuel oil samples, the mousse sample was initially used as a representative source sample to identify other mousse and tarball samples in the area. Direct qualitative comparison of the GC/FID

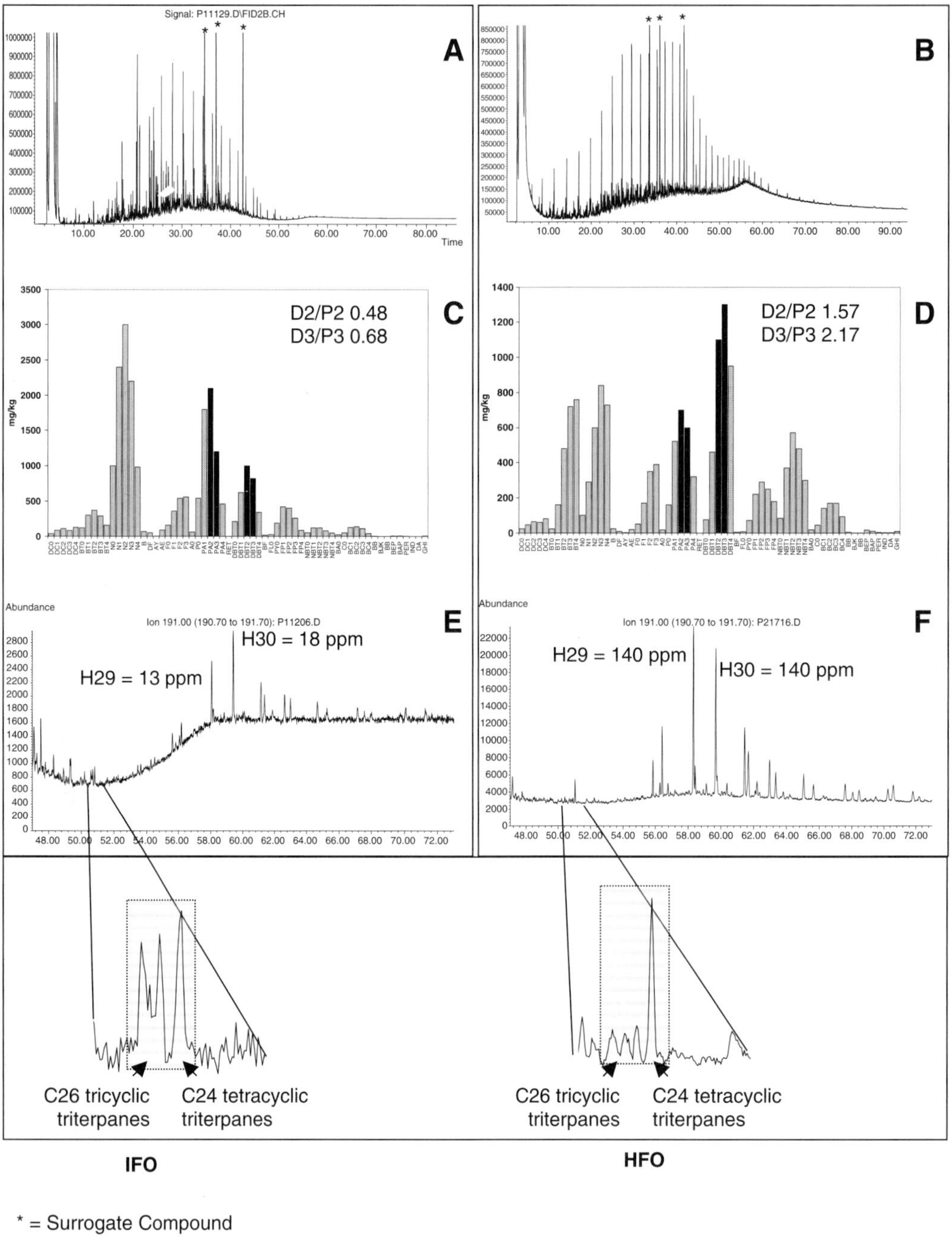

Figure 8-5 Comparison of IFO and HFO source fuels oil fingerprints from a marine oil spill from case study 2. (A) GC/FID chromatogram for IFO, (B) GC/FID chromatogram for HFO, (C) PAH concentration histogram for IFO, (D) PAH concentration histogram for HFO, (E) Partial m/z 191 triterpane extracted ion profile for IFO, and (F) Partial m/z 191 triterpane extracted ion profile for HFO. H29 = Norhopane, H30 = Hopane * = internal standards.

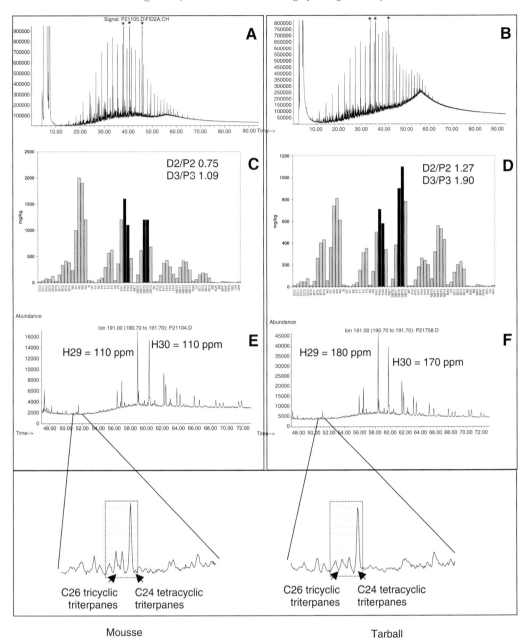

Figure 8-6 Comparison of two field samples (mousse and tarball) associated with case study 2 and the vessel containing the fuels shown in Figure 8-5. (A) GC/FID chromatogram for the mousse sample (B) GC/FID chromatogram for the tarball sample, (C) PAH concentration histogram for the mousse sample, (D) PAH concentration histogram for the tarball sample, (E) Partial m/z 191 triterpane extracted ion profile for the mousse sample, and (F) Partial m/z 191 triterpane extracted ion profile for the tarball sample. H29 = Norhopane, H30 = Hopane * = internal standards.

chromatograms revealed that the two field samples (and others not discussed in this case study) exhibited heterogeneities in the distribution of n-alkanes and in the shape of the UCM hump (Figure 8-6A-B). It would be difficult to say with certainty that these differences were not attributable to variable degrees of weathering of a single oil. For example, evaporative losses of compounds from the HFO (Figure 8-5B) might yield a GC/FID fingerprint comparable to the tarball sample (Figure 8-6B). However, the quantitative PAH data indicated that the mousse and tarball samples were not derived from a single oil. For example, the PAH distributions and D2/P2 and D3/P3 ratios clearly showed that the field samples ($n > 100$) contained varying abundances of sulfur-containing PAHs (Figure 8-6C-D). This result raised some questions as to the actual source(s) of these and other field samples, particularly for samples collected in areas not previously identified as oil-impacted. The questions posed by the on-site responders were (1) are there multiple sources of oil contamination at this site? (2) why are the field samples chemically different from the two source fuel oils? and (3) what tools are available to *confidently* identify spill- and nonspill-related field samples?

Direct qualitative comparison of the two source oils from the grounded vessel (Figure 8-5) and the two field samples (compare Figures 8-5 and 8-6) indicates that neither of the sources would likely match either of the mousse or tarball samples. (This was later confirmed quantitatively using the revised Nordtest method; see ahead.) Although there were general chemical similarities between the two source oils and the field samples (e.g., all were seemingly comprised of a broad-boiling, n-alkane-enriched oil), there were significant chemical differences (e.g., D2/P2 and D3/P3 ratios) that could *not* be explained by simple weathering. Because the grounding and oil spill occurred in an area of heavy commercial shipping traffic, where many of the vessels may have carried generally similar fuels, further investigation as to the other potential source(s) of oil or pre-existing oil in the field samples was warranted.

A suite of the source-specific, diagnostic ratios was developed for these oils (Table 8-6) and was investigated with the objective of determining if either of the vessel's fuel oils had contributed to the oil in the field samples. These ratios were generated from quantitative analysis of the source oils and field samples as described in Section 8.3.2 and were selected

Table 8-6 Diagnostic Ratios for the IFO and HFO Source Oils and Field Samples

Source Ratio	Abbr.	Source Samples		Field Samples		Calculated Mixtures		Field Sample (Unrelated)	
		IFO	HFO	Mousse Sample	Tarball Sample	Mousse Sample Mixture (50:50)	Tarball Sample Mixture (12:88)	Tarball #2 Sample	Tarball #2 (100:−2)
$100 \times [Pr/(Pr+Ph)]$	a	63.89	59.41	58.56	63.06	62.91	60.90	61.96	63.91
$100 \times [odd\ C25–C33/(odd\ C25–C33+even\ C26–C34)]$	b	61.29	57.87	53.31	56.77	58.48	57.97	62.52	61.57
$100 \times [DBT/(DBT+P0)]$	c	28.00	31.91	29.41	33.01	28.93	30.71	15.49	27.98
$100 \times [D2/(D2+P2)]$	d	32.26	61.11	42.86	55.90	42.86	55.52	31.71	31.98
$100 \times [D3/(D3+P3)]$	e	40.59	68.42	52.17	65.48	54.09	64.82	30.16	40.15
$100 \times [Ts/(Ts+Hop)]$	f	16.67	14.63	17.29	16.67	14.87	14.67	15.44	16.96
$100 \times [More/(More+Hop)]$	g	0.00	7.89	6.86	8.11	7.06	7.77	16.36	−1.29
$100 \times [NHop/(NHop+Hop)]$	h	41.94	50.00	50.00	51.43	49.20	49.88	43.90	40.51
$100 \times [C31\ S/(C31\ S+C31\ R)]$	i	56.50	55.38	57.84	56.90	55.52	55.41	67.01	56.65
$100 \times [C32\ S/(C32\ S+C32\ R)]$	j	62.04	58.67	56.25	58.10	59.19	58.75	49.62	62.38
$100 \times [C35\ S+R/((C35\ S+R)+\Sigma C30–C35\ Hopanes)]$	k	10.89	13.48	10.95	12.10	13.14	13.43	0.00	10.57

based on their source-specific properties, their reported resistance to environmental weathering, and their low relative percent difference (RSD < 10% indicates a high degree of analytical precision) for the triplicate analyses of the mousse. Thus, these diagnostic ratios would meet the general criteria for diagnostic ratios (e.g., Stout et al., 2001). The diagnostic ratios (DR) were calculated using the general formula

$$DR = 100 \times [A/(A + B)] \quad \text{(Eq. 8-2)}$$

The diagnostic ratio results range from 0 to 100, per the revised Nordtest (Daling et al., 2002). The diagnostic ratios for the two source oils, mousse, and tarball samples are provided in Table 8-6 — along with some calculated diagnostic ratios for theoretical mixtures of the IFO and HFO sources and a nonspill-related tarball sample (Tarball #2) — that is discussed later in this section.

The diagnostic ratios for each field sample were plotted versus the same ratios for each of the vessel's source oils (Figures 8-7 and 8-8) relative to the parity line, per the revised Nordtest method. The parity line represents the 1:1 source/sample ratio, where identical oils would plot close to or directly on this line. The 98% confidence limits derived from the triplicate analysis of the mousse sample are plotted with each data point (x and y ranges) to reflect the associated analytical variability. Samples where all ratios fall within this confidence level are identified as a probable match using the recommended revised Nordtest criteria discussed in Section 8.3.4.

Analysis of these figures indicates that the sulfur-related diagnostic ratios (e.g., ratio d = [D2/D2 + P2] × 100, ratio e = [D3/D3 + P3] × 100) are clearly different from the IFO and HFO for both the mousse and the tarball samples, respectively (i.e., they plot off the parity line). This was observed earlier in Figures 8-5 and 8-6, where D2/P2 and D3/P3 ratios in the field samples were markedly different than those observed in the field samples. Given that the mousse sample was visually identified by the responder as being spill-related, and the fact that many of the shoreline tarball samples also failed the revised Nordtest criteria, we evaluated the possibility that the field samples were indeed various mixtures of the two source oils.

Several lines of physical and chemical evidence indicated that many of the field samples were likely mixtures of the two source oils. Specifically,

1. Visual observations by on-site responders identified areas of shoreline near the grounded vessel that were clearly related to the spill event (e.g., mousse sample).
2. Qualitative and quantitative chemical analysis of the source oils and field samples suggested that although most of the field samples were a chemical "nonmatch," they appeared to have some genetic biomarker similarities (e.g., C26 tricyclic and C24 tetracyclic terpanes).
3. The range of source ratios observed in the field samples was mostly bounded by the IFO (low) and HFO (high).
4. The primary differences in the revised Nordtest correlation plots appeared to be driven by the sulfur ratios (e.g., D2/D2 + P2), in which the field samples' ratio was intermediate to the two source oils.

Although similarity was readily apparent between the candidate source oils and the spill samples, the revised Nordtest approach was not able to fully unravel the true source of a mixed product that formed the spill sample (e.g., Figure 8-7). Similarly, this approach could not be used to correlate the individual spilled products to tarballs (formed both during the spill and from pre-existing spills) discovered on the nearby coastline. To resolve this issue, a two-component mixing model composed of various proportions of the two source oils was developed using published methods (Page et al., 1995; Burns et al., 1997; Arouri and McKirdy, 2004; Kauffman et al., 1990). Because the sulfur ratios (e.g., D2/D2 + P2) in the source oils and field samples accounted for a majority of the variance in the samples (Figures 8-7 and 8-8), these ratios were used to develop a simple mathematical model to estimate the percentage of each

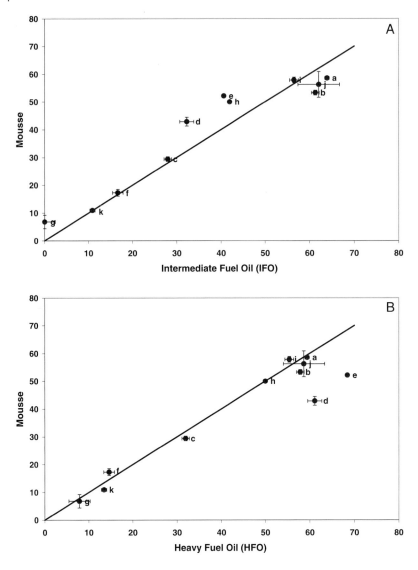

Figure 8-7 Revised Nordtest linear regression plots for eleven diagnostic ratios (a-k) in the mousse sample and two candidate source oils (IFO and HFO) from case study 2. (A) Comparison of IFO with the mousse sample resulting in a non-match, (B) comparison of HFO oil with the mousse sample resulting in a non-match.

source oil within the field sample(s). A new set of diagnostic ratios was then calculated based on the percentage of each source oil in the field sample and used to evaluate the sensitivity of the revised Nordtest method to resolve the identity of the field samples.

8.4.3.1 Mixing Model Case Study 2

Commercial merchant vessels often store three types of fuel; IFO, HFO, and MDO (EPA, 1999). IFO and HFO represent the majority of the vessel's fuel cargo (typically hundreds of thousands of gallons), whereas smaller amounts of MDO are commonly used only to operate auxiliary generators on the vessel. Uhler et al. (2006) (Chapter 10 in this book) document the significant variability in the ratios of D2/P2 and D3/P3 in HFOs. Again, in this case, a significant difference in these ratios between the cargo fuels was recognized (as discussed ealier) and used to develop the spill

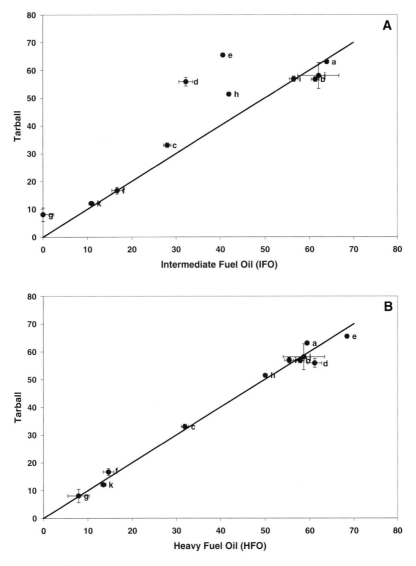

Figure 8-8 Revised Nordtest linear regression plots for eleven diagnostic ratios (a-k) in the tarball sample and two candidate source oils (IFO and HFO) from case study 2. (A) Comparison of IFO with the tarball sample resulting in a non-match, (B) comparison of HFO oil with the tarball sample resulting in a non-match.

mixing model. Rather than simply use the C2-dibenzothiophene concentration in the source oils, source ratios of C2-dibenzothiophenes/C2-phenanthrenes (D2/P2) were used because

- These source ratios vary widely in oils depending on its crude source (Page et al., 1995).
- These source ratios are stable over a wide range of weathering (Douglas et al., 1996).
- These source ratios are less variable than concentrations because they are internally

normalized (to C2-phenanthrene) in the field samples.

The mixing model ratios are calculated from *absolute concentrations* of C2-dibenzothiophenes and C2-phenanthrenes in the source oils. Some error may be introduced because the MDO was not included in the model; however, given the relatively small volume of MDO on the vessel and the lack of a clear MDO signal in the field samples, any bias would be minor. In addition, there was some

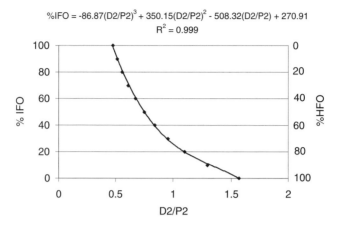

Figure 8-9 Plot of the D2/P2 versus percent IFO and HFO from the grounded vessel. This equation was used with the D2/P2 ratios measured in the field samples to calculate the % IFO and % HFO in the field samples. These percentages were then used with the field sample quantitative data to calculate mixed oil source ratios that were used in the revised Nordtest method (e.g., Fig. 8-10).

information from the on-site responders that the MDO tank had been pumped out. Figure 8-9 is a plot of the D2/P2 ratios versus %IFO in a mixture of the two source oils (IFO and HFO). The equation is expressed by the following third-order polynomial equation:

$$\%\text{IFO} = -86.87(D2/P2)^3 + 350.15(D2/P2)^2 - 508.32(D2/P2) + 270.91 \quad \text{(Eq. 8-3)}$$
$$r^2 = 0.9995$$

The %HFO is simply 100% − %IFO. Once the IFO:HFO proportion for each field sample is determined, a new suite of diagnostic ratios (Table 8-6) is calculated from the absolute concentration data using an equation comparable to Eq. (8-1) (but for the two source oils in Case Study 2).

The best fit for the mixing model in Figure 8-9 is not linear due to the differences in concentrations of the two compound groupings within the source oils (C2-dibenzothiophenes, C3-dibenzothiophenes, Page et al., 1995). This fact re-emphasizes why absolute concentrations are necessary in developing mixing models. For example, the slope of the mixing model curve is flatter with higher percentages of HFO because it takes very little HFO to have a large influence on the D2/P2 ratio in the mixture. Conversely, the slope of the curve is greater with higher IFO concentrations because a given change in D2/P2 requires substantially more IFO relative to the HFO. For the mousse sample, the best calculated relative percentage of source oils based on Eq. (8-3) and the D2/P2 of the sample was a 50:50% mixture of IFO and HFO. For the tarball sample, the calculated relative percentage of source oils was a mixture of 12:80% IFO and HFO. These mixtures were consistent with other chemical factors observed in the sample including UCM distribution, alkane and PAH distribution, tricyclic and tetracyclic triterpanes distribution, and biomarker distributions.

Figure 8-10A and B is a plot of the calculated diagnostic ratios based on the mixing model versus the quantitative mousse (Figure 8-10A) and tarball sample data (Figure 8-10B). The recalculation of the mixed source(s) based on the mixing model significantly improved the fit for both the mousse and tarball samples and strongly indicated that the field samples were indeed variable mixtures of the source oils. For the mousse sample, the improvement was dramatic (Figure 8-7A) and brought the sample very close to the revised Nordtest method's recommended positive match criteria (Figure 8-10A). The primary drivers (as discussed earlier) were the sulfur ratios as exhibited by source ratios c ([DBT/DBT + P0] × 100), d ([D2/D2 + P2] ×

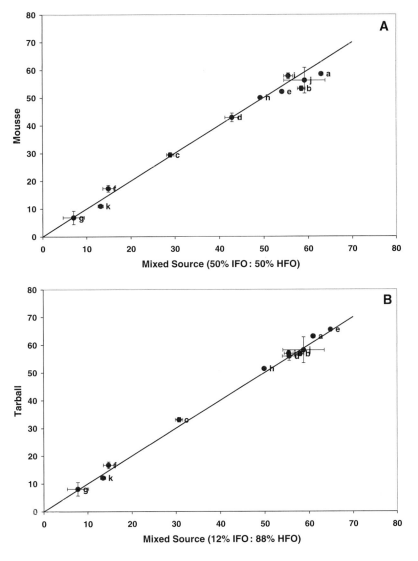

Figure 8-10 Revised Nordtest linear regression plots for eleven diagnostic ratios (a-k) in the mousse (A) and tarball (B) samples versus the associated mixed source oils. The parity correlation was improved for both samples when the quantitatively determined mixed source ratios were used. Based on this analysis and considering the additional variability introduced by the mixing model but not reflected in the 98% confidence interval, a source match with the mixed oils is likely.

100), and e ([D3/D3 + P3] × 100), which were shifted closer to the parity line for a mixture of 50% IFO and 50% HFO. A similar shift closer to the parity line was observed in the tarball sample for a mixture of 12% IFO and 88% HFO. These results are also consistent with the GC/FID, PAH, and biomarker interpretive results. This model was then applied to a suite of oil, mousse, and tarball field samples collected during the ongoing field investigation months after the spill to confidently resolve spill-related oiling from nonspill-related (pre-existing) contamination.

The revised Nordtest method (and the emerging CEN method; Chapter 7) provides a series of specific guidelines to reliably identify mystery oil sources within single component oil spills. However, when mixtures of oils are present, the simple application of this approach may produce a substantial number of false negative results. Under these circumstances, as demonstrated within this case study, the application and source resolution of the revised Nordtest method are greatly enhanced through the development of source oil mixing models. Observation of a significant improvement in parity line correlation between the mixed source and the field sample is a primary indicator of mixing model efficacy (Figure 8-10). Unlike the revised Nordtest approach, however, the error associated with the correlation analysis is not only related to the analytical variability (precision) determined from replicates of a single oil, it also includes the added potential variability introduced by the mixing model itself. This potential variability is a function of how representative the source oils are to the bulk fuel oil, and if all possible sources have been collected from the vessel. This additional error has not been incorporated into the current study although it is expected to increase the 98% confidence intervals associated with the calculated mixtures (e.g., x-axis). Principal component and least-squares methods (Burns et al., 1997) based on all of the source ratios (not simply D2/P2) may substantially reduce the variance associated with the simple two-component mixing model described earlier. These multivariate methods can also more readily accommodate mixtures involving more than two potential sources. In addition to the calculation of mathematical mixes, source oil mixtures could be physically prepared in the laboratory and triplicate analyses performed to more accurately estimate this potential increase in variance. For the purposes of this study, the post-model improvements (Figure 8-10) relative to the sample/mixed-source parity line is strong evidence that the mixed oils are representative of the spilled oils and should be considered, at a minimum, as probable matches based on the revised Nordtest method. In addition to field observations and the qualitative inspection of the hydrocarbon patterns discussed previously, the samples exhibiting a high degree of correlation with the parity line constitute a defensible match with the source oils in this case.

The mixing model approach used in this case study provides an example of a useful method for exploring a range of source signatures derived from binary mixtures of potential source oils. Evaluating a broader range of potential source signatures provides a higher level of confidence for the match/nonmatch conclusion and minimizes the false negatives encountered with simpler correlations of uni-component sources to spill samples composed of mixtures of spilled oils. The examination of an infinite number of mixtures bounded by the sources dramatically increases the accuracy of the fingerprinting method when a nonspill-related sample is identified. Within oil spill response situations, it is the identification of nonspill-related oils that are most problematic to the environmental regulators; therefore, the methods used to reach such a conclusion must be highly defensible and supported by the chemical data.

Figure 8-11 is a mixed-source ratio plot for another field sample from this case study (Tarball #2, Table 8-6), but it was not related to the known oil spill under investigation. The initial revised Nordtest plots relative to the IFO and HFO indicated a clear nonmatch, which was not unexpected given that we had identified the possibility of source oil mixtures at the site. The analysis was then expanded to examine all possible combinations of the source oils, including the best combination identified with the mixing model described above. In all cases there was no combination of the two source oils that could explain the source ratio chemistry observed in the sample. The results presented in Figure 8-11 for Tarball #2 show the disparity between the tarball and the best possible source oil mixture. These results demonstrate that the sample is unrelated to the oil spill and *not* a false negative. In addition to the revised Nordtest results, the conclusion was confirmed by direct examination of the GC/FID, PAH, and biomarker distributions and abundances.

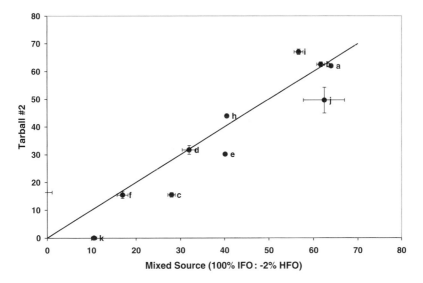

Figure 8-11 Revised Nordtest linear regression plots for eleven diagnostic ratios (a-k) in the tarball #2 sample versus the associated mixed source oil. No improvement in parity correlation was observed over the entire range of possible mixtures including the mixed source presented in this figure that was calculated based on the D2/P2 ratio in the sample. This data combined with GC/FID, PAH and biomarker signatures is conclusive evidence that this sample is not related to the oil spill discussed in case #2.

Based on our experience and as discussed earlier, chemical differences between potential oil spill sources at a site can be relatively small or vastly different. The parity plot analysis discussed earlier may be used to (1) evaluate the potential for source mixtures, (2) rank the degree of correlation between field samples relative to source oils, and (3) define a subset of source oils most likely related to the release. This refined subset of possible sources can then be evaluated in more detail based on the preponderance of the evidence such as the spill location, site history, process chemistry, degree of weathering, and subtle but unique differences in the GC/MS and GC/FID chemical signatures.

8.5 Summary

The chemical fingerprinting approaches available for the identification of oil spill sources and potentially impacted samples fall into two categories: (1) qualitative and (2) quantitative. The qualitative approach relies upon visual comparison of various chromatographic fingerprints and is exemplified by the ASTM D3328 and D5739 methods. The quantitative fingerprinting methodology relies on measurements of the concentrations (relative or absolute) of targeted and diagnostic chemicals, typically PAH and biomarkers, and a subsequent statistical or other numerical analysis of various diagnostic parameters calculated from these concentrations.

The quantitative approach is represented by the revised Nordtest methodology (Daling et al., 2000a), discussed in this chapter, and the emerging CEN methodology (Chapter 7 herein). The quantitiative approach is preferable for most oil spill investigations since the means of interpretation are more objective and robust in the sense that they facilitate numerical comparison of diagnostic details and reduce interpretation bias. The quantitative approaches have a much better opportunity to account for and recognize the effects of weathering, and the subtle but real differences among genetically or otherwise qualitatively similar oils than qualitative fingerprinting methods. However, quantitative approaches such as the revised Nordtest or emerging CEN methods still can be confounded when mixing affects the chemical fingerprints of samples under consideration. Mixing is an important consideration in any oil spill investigation involving multiple source oils or when a spilled oil mixes with any pre-existing hydrocarbons in the ambient environment, both of which will alter the chemical fingerprint of a spilled oil.

To account for the effects of mixing, absolute concentration data are needed for the samples under consideration. The effects of mixing can be mathematically accommodated with numerical mixing models so long as fully quantitative data (i.e., absolute concentration data) are available. When absolute concentration data are available for end-members, a numerical mixing model can predict the character of intermediate mixtures. The use of the revised Nordtest (or emerging CEN) sample-to-sample correlation approach in conjunction with numerical mixing models can improve the chemical comparison between spilled oil(s), candidate source oils, and potentially impacted environmental samples. These numerical mixing models must be based on absolute (concentration) quantitative chemical data since diagnostic ratios calculated from semiquantitative data do not mix linearly due to absolute concentration differences between samples. As with any forensic study, a weight of evidence approach should be used and traditional analysis of all the available chromatographic and analytical data should be carefully examined to ensure that any conclusions regarding correlation between samples is consistent with the mixing model and subsequent statistical analysis.

References

Aboulkassim, T.A.T. and B.R.T. Simoneit, Petroleum hydrocarbon fingerprinting and sediment transport assessed by molecular biomarker and multivariate statistical-analyses in the Eastern harbor of Alexandria, Egypt. *Marine Pollution Bulletin*, 1995, **30**(1), 63–73.

Arouri, K.R. and D.M. McKirdy, The behavior of aromatic hydrocarbons in artificial mixtures in Permian and Jurassic end-member oils: Application to in-reservoir mixing in the Eromanga Basin, Australia. *Organic Geochemistry*, **36**, 105–113.

ASTM, Oil spill identification by gas chromatography and positive ion electron impact low resolution mass spectrometry. ASTM D-5739–00. American Society for Testing and Materials International, W. Conshohocken, PA, 2000.

ASTM, Standard Test Methods for Comparison of Waterborne Petroleum Oils by Gas Chromatography. ASTM D-3328–90. American Society for Testing and Materials International, W. Conshohocken, PA, 1990.

Bence, A.E. and W.A. Burns, Fingerprinting hydrocarbons in the biological resources of the *Exxon Valdez* spill area. In P.G. Wells, J.N. Butler, and J.S. Hughes (eds.), Exxon Valdez *Oil Spill: Fate and Effects in Alaskan Waters*. ASTM STP 1219. American Society for Testing and Materials, W. Conshohocken, PA, 1995, pp. 84–140.

Bence, A.E., K.A. Kvenvolden, and M.C. Kennicutt, Organic geochemistry applied to environmental assessments of Prince William Sound, Alaska, after the *Exxon Valdez* oil spill — a review. *Organic Geochemistry*, 1996, **24**(1), 7–42.

Boehm, P.D., G.S. Douglas, W.A. Burns, P.J. Mankiewicz, D.S. Page, and A.E. Bence, Application of petroleum hydrocarbon chemical fingerprinting and allocation techniques after the *Exxon Valdez* oil spill. *Mar. Pollut. Bull.*, 1997, **34**(8), 599–613.

Bost, F.D., R. Frontera-Suau, T.J. McDonald, K.E. Peters, and P.J. Morris, Aerobic biodegradation of hopanes and norhopanes in Venezuelan crude oil. *Organic Geochemistry*, 2001, **32**, 105–114.

Burns, W.A., P.J. Mankiewicz, A.E. Bence, D.S. Page, and K.R. Parker, A principal component and least squares method for allocating polycyclic aromatic hydrocarbons in sediment to multiple sources. *Environ. Toxicol. Chem.*, 1997, **16**(6), 1119–1131.

Christensen, J.H., Application of multivariate data analysis for assessing the early fate of petrogenic compounds in the marine environment following the *Baltic Carrier* oil spill. *Polycyclic Aromatic Compounds*, 2002, **22**(3–4), 703–714.

Christensen, J.H., A.B. Hansen, G. Tomasi, J. Mortensen, and O. Andersen, Integrated methodology for forensic oil spill identification. *Environ. Sci. Technol.*, 2004, **38**(10), 2912–2918.

Christensen, J.H., G. Tomasi, and A.B. Hansen, Chemical fingerprinting of petroleum biomarkers using time warping and PCA. *Environmental Science & Technology*, 2005, **39**(1), 255–260.

Daling, P.S., L.-G. Faksness, A.B. Hansen, and S.A. Stout, Improved and standardized methodology for oil spill fingerprinting. *Environ. Forensics*, 2002a, **3**, 263–278.

Daling, P.S. and L.-G. Faksness, Laboratory and reporting instructions for the CEN/BT/TF 120 Oil Spill Identification Round Robin test — May 2001. SINTEF Report STF66 A02027, 2002b.

Douglas, G.S., A.E. Bence, R.C. Prince, S.J. McMillen, and E.L. Butler, Environmental stability of selected petroleum hydrocarbon source and weathering ratios. *Environ. Sci. Technol.*, 1996, **30**(7), 2332–2339.

Douglas, G.S., R.C. Prince, E.L. Butler, and W.G. Steinhauer, The use of internal chemical indicators in petroleum and refined products to evaluate the extent of biodegradation. In R.E. Hinchee, B.C. Alleman, R.E. Hoeppel, and R.N. Miller (eds.), *Hydrocarbon Bioremediation*. Ann Arbor, MI, Lewis Publishers, 1994.

Douglas, G.S. and A.D. Uhler, Optimizing EPA methods for petroleum contaminated site assessments. *Env. Test. Anal.*, 1993, **2**, 46–53.

Emsbo-Mattingly, S.D., A.D. Uhler, S.A. Stout, G.S. Douglas, K.J. McCarthy, and A. Coleman, Determining the source of PAHs in sediment. *Land Contamination & R*, 2006, **14**(2).

EPA, In-use marine diesel fuel. Engine programs and Compliance Division, Office of Mobile Sources, U.S. Environmental Protection Agency. EPA420-R-99-027, 1999.

Faksness, L.-G., P.S. Daling, and A.B. Hansen, Round-robin study — oil spill identification. *Environ. Forensics*, 2002, **3**(3/4), 279–292.

Federal Register, National Oil and Hazardous Substances Pollution Contingency Plan — Final Rule 40, CFR Parts 9 and 300, 1994.

General Oceanics, Product and Services Catalog. 1295 N.W. 163rd Street, Miami, FL, 33169, USA, 2006.

Harris, D.C., *Quantitative Chemical Analysis*, 4th ed. New York, NY: W.H. Freemantle & Company, 1995.

Henry, C.B., P.O. Roberts, and E.B. Overton, Advancing forensic chemistry of spilled oil: Self-normalizing fingerprint indexes. *In Proc. 1997 Intl. Oil Spill Conf.* 1997, Fort Lauderdale, FL, March 1997, pp. 936–937.

Hwang, R.J., D.K. Baskin, and S. Teerman, Allocation of commingled pipeline oils to field production. *Org. Geochem.*, 2000, **31**, 1463–1474.

ISO, Standard 8217, *Specifications of Marine Fuels*, in Petroleum products — Fuels (class F). International Organization for Standardization (ISO), Geneva, Switzerland, 1996.

Kauffman, R.L., A.S. Ahmed, and R.J. Elsinger, Gas chromatography as a development and production tool for fingerprinting oils from individual reservoirs: Applications in the Gulf of Mexico. *In Proc. 9th Annual Research Confe. Society of Econ. Paleontologists and Mineralogists* (D. Schumaker and B.F. Perkins, eds.), New Orleans, 1990, pp. 263–282.

Kvenvolden, K.A., F.D. Hostettler, P.R. Carlson, J.B. Rapp, C.N. Threlkeld, and A. Warden, Ubiquitous tar balls with a California-source signature on the shorelines of Prince William Sound, Alaska. *Environ. Sci. Technol.*, 1995, **29**, 2684–2694.

Lehr, W., C. Barker, and D. Simecek-Beatty, New developments in the use of uncertainty in oil spill forecasts. *In Proc. 22nd Artic and Marine Oilspill Program* (AMOP), Calgary, Canada, June 1999, pp. 271–284.

Lavine, B.K., D. Brzozowski, A.J. Moores, C.E. Davidson, and H.T. Mayfield, Genetic algorithm for fuel spill identification. *Analytica Chimica Acta.*, 2001, **43**, 233–246.

Marine Oil Spill Response Corporation (MSRC), Hydrocarbon chemistry analytical methods for oil spill assessments. Marine Spill Response Corporation, Research and Development Program. MSRC Technical Report Series. 95-032, 1995, pp. 77–114.

McCaffrey, M.A., H.A. Lagarre, and S.J. Johnson, Using biomarkers to improve heavy oil reservoir management: An example from the Cymric Field, Kern County. *Am. Assoc. Petrol. Geol. Bull.*, 1996, **80**(6), 904–919.

Mudge, S.M., Reassessment of the hydrocarbons in Prince William Sound and the Gulf of Alaska: Identifying the source using partial least squares. *Environ. Sci. Technol.*, 2002, **36**, 2354–2360.

NRC (National Research Council), *Oil in the Sea: Inputs, Fates, and Effects*. Washington DC: National Academic Press, 1985.

NRC (National Research Council), *Oil in the Sea III*. Washington DC: National Academic Press, 2003.

Nicolle, G.C., C. Boibien, H.L. ten Haven, E. Tegelaar, and P. Chavagnac, Geochemistry: A powerful tool for reservoir monitoring. Soc. Petrol. Engineers, Special Paper No. 37804, 1997, pp. 395–401.

Nordtest, Nordtest Method for Oil Spill Identification. NT Chem 001. Edition 2. NORDTEST. Esbo, Finland, 1991.

Overton, E.B., J.A. McFall, S.W. Mascarella, C.F. Steele, S.A. Antoine, L.R. Politzer, and J.L. laseter, Identification of petroleum residue sources after a fire and oil spill. In *Proce. 1981 Intl. Oil Spill Conf.* American Petroleum Institute, Washington, DC, 1981, pp. 541–546.

Page, D.S., P.D. Boehm, G.S. Douglas, and A.E. Bence, Identification of hydrocarbon sources in the benthic sediments of Prince William Sound and the Gulf of Alaska following the *Exxon Valdez* oil spill. In P.G. Wells, J.N. Butler, and J.S. Hughes (eds.), Exxon Valdez *Oil Spill: Fate and Effects in Alaskan Waters.* ASTM STP 1219. Philadelphia, PA: ASTM, 1995, pp. 41–83.

Page, D.S., P.D. Boehm, G.S. Douglas, A.E. Bence, W.A. Burns, and P.J. Mankiewicz, Pyrogenic polycyclic aromatic hydrocarbons in sediments record past human activity: A case study in Prince William Sound, Alaska. *Mar. Pollut. Bull.*, 1999, **38**, 247–260.

PERF, *Guidelines for the Scientific Study of Oil Spill Effects* — Draft Preface. Petroleum Environmental Research Forum, 1995.

Peters, K.E. and M.G. Fowler, Applications of petroleum geochemistry to exploration and reservoir management. *Organic Geochemistry*, 1992, **33**, 5–36.

Peters, K.E., G.L. Scheuerman, C.Y. Lee, J.M. Moldowan, R.N. Reynolds, and M.M. Pena, Effects of refinery processes on biological markers. *Energy Fuels*, 1992, **6**(5), 560–577.

Peters, K.E. and J.M. Moldowan, *The Biomarker Guide*. Englewood Cliffs, NJ: Prentice Hall, 1993.

Peters, K.E., C.C. Walters, and J.M. Moldowan, *The Biomarker Guide, Volume 2. Biomarker and Isotopes in Petroleum Exploration and Earth History.* Cambridge, UK: Cambridge University Press, 2005a.

Peters, K.E., C.C. Walters, and J.M. Moldowan, *The Biomarker Guide, Volume 1. Biomarker and Isotopes in the Environment and Human History.* Cambridge, UK: Cambridge University Press, 2005b.

Philp, R.P., *Fossil Fuel Biomarkers: Applications and Spectra.* Methods in Geochemistry and Geophysics, Vol. 23. Elsevier: Amsterdam, The Netherlands, 1985.

Prince, R.C., Biodegradation of petroleum and other hydrocarbons. In G. Bitton (ed.), *Encyclopedia of Environmental Microbiology.* New York: John Wiley, 2002, pp. 2402–2416.

Prince, R.C., D.L. Elmendorf, J.R. Lute, C.S. Hsu, C.E. Haith, J.D. Senius, G.J. Dechert, G.S. Douglas, and E.L. Butler, 17a (H) 21 β (H)-hopane as a conserved internal marker for estimating the biodegradation of crude oil. *Environ. Sci. Technol.*, 1994, **28**(1), 142–145.

Radke, M., D.H. Welte, and H. Willsch, Maturity parameters based on aromatic hydrocarbons: Influence of the organic matter type. *Org. Geochem.*, 1986, **10**, 51–63.

Stout, S.A., G.S. Douglas, A.D. Uhler, K.J. McCarthy, and S.D. Emsbo-Mattingly, Identifying the source of mystery waterborne oil spills — a case for quantitative chemical fingerprinting. *Environmental Claims Jour.*, 2005, **17**(1), 71–88.

Stout, S.A., W.P. Naples, A.D. Uhler, K.J. McCarthy, L.G. Roberts, and R.M. Uhler, Use of quantitative biomarker analysis in the differentiation and characterization of spilled oil. Society of Petroleum Engineers, Inc., Richardson, TX. SPE Publ. No. 61460. Stavanger, Norway, June 2000.

Stout, S.A., A.D. Uhler, and K.J. McCarthy, A strategy and methodology for defensibly correlating spilled oil to source candidates. *Environ. Forensics*, 2001, **2**, 87–98.

Uhler, A.D. and S.D. Emsbo-Mattingly, Environmental stability of PAH source indices in pyrogenic tars. *Bull. Environ. Contam. and Tox*, in press, 2006.

Tissot, B.P. and D.H. Welte, *Petroleum Formation and Occurrence*, 2nd ed. Berlin: Springer-Verlag, 1984.

Urdal, K., N.B. Vogt, S. Pl. Sporstol, R.G. Lichtenthaler, H. Mostad, K. Kolset, S. Nordenson, and K. Esbensen, Classification of weathered crude oils using multimethod chemical analysis, statistical methods and SIMCA pattern recognition. *Mar. Pollut. Bull.*, 1986, **17**(8), 366–373.

Wang, Z. and M. Fingas, Identification of the source(s) of unknown spilled oils. *In 1999 Intl. Oil Spill Conf.* Seattle, WA, Paper #162, 1999.

Wang, Z., M. Fingas, and D.S. Page, Oil spill identification. *J. Chromatogr. A.*, 1999, **842**, 369–411.

Wang, Z., M. Fingas, and L. Sigouin, Characterization and identification of a "mystery" oil spill from Quebec (1999). *J. Chromatogr. A.*, 2001, **909**, 155–169.

9 A Multivariate Approach to Oil Hydrocarbon Fingerprinting and Spill Source Identification

Jan H. Christensen[#] and Giorgio Tomasi[*]

[#]Department of Natural Sciences, Royal Veterinary and Agricultural University
[*]Department of Food Science, Royal Veterinary and Agricultural University

9.1 Introduction

Oil hydrocarbon fingerprinting was originally developed by geochemists in the petroleum industry to understand and track the source of crude oils and natural gases. In environmental forensics, methods basically similar to those of petroleum geochemistry have been applied to defensibly determine oil source(s) (Wang et al., 1999, 2002; Stout et al., 2001; Daling et al., 2002; Christensen et al., 2004, 2005d), distinguish spilled oil from background hydrocarbons of biogenic and pyrogenic origin (Boehm et al., 1997; Mudge, 2002), determine the weathering processes (Wang et al., 1998, 2001; Christensen et al., 2005a), and assess the ecosystem impact (risk assessment) (Porte et al., 2000; Barron and Holder, 2003).

A variety of analytical techniques have been used for oil hydrocarbon fingerprinting including gas chromatography-flame ionization detection (GC-FID), gas chromatography-mass spectrometry (GC-MS), and fluorescence spectroscopy. Oil hydrocarbon fingerprinting and spill source identification are, however, not limited to the chemical characterization using different analytical techniques, but consist of a combination of analytical techniques and methods for data preprocessing, analysis, and evaluation of the results. It is of fundamental importance for the defensibility of a combined fingerprinting approach that the process of extracting the desired information from a set of chemical data is objective and standardized. At the same time, the need to process an ever-increasing amount of samples and data implies that the data analysis should also be fast, comprehensive, and unsupervised.

Numerous methods for oil hydrocarbon fingerprinting and spill source identification have been described in the scientific literature since the 1980s (Munoz et al., 1997; Boehm et al., 1997; Burns et al., 1997; Wang et al., 1999, 2002; Stout et al., 2001; Page et al., 2002; Mudge, 2002; Daling et al., 2002; Christensen et al., 2004, 2005b, 2005d); most of them are elaborated on in this book. Standard methods are based on comparison of bulk oil properties such as total petroleum hydrocarbon concentration (TPH) (Reddy and Quinn, 1999; Wang et al., 2002), visual comparison of fluorescence spectra (Siegel et al., 1985; Siegel and Cheng, 1989) and of chromatograms obtained with GC-FID (Wang et al., 2000, 2002; Daling et al., 2002) or GC-MS (Jovancicevic et al., 1996; Ezra et al., 2000; Daling et al., 2002; Wang et al., 2002), concentrations of source-specific markers (Jovancicevic et al., 1996; Wang et al., 1999, 2002; Daling et al., 2002), bar plots of the concentrations of oil-characteristic polycyclic aromatic hydrocarbons (PAHs) (Wang et al., 1999, 2002), and lists and double plots of diagnostic ratios of PAHs and petroleum biomarkers (Wang et al., 1999, 2002).

Most of these methods, no matter how powerful and refined, rely on skill and expertise of the analyst. Crude oils and petroleum products are complex mixtures of chemical compounds; thus, it is not feasible to identify and quantify all individual compounds in the mixture. Without using appropriate methods for data preprocessing, analyzing, and evaluating, the sheer amount of data often infers so much that few variables are chosen based on *a priori* knowledge, which in turn can largely hinder the discovery of new or more informative patterns. Likewise, some differences may be so subtle that they become invisible even to the trained eye. This can increase the level of uncertainty in conclusions of spill source identity, which then become less defensible in a court of law.

Thus, one of the most important advances in oil hydrocarbon fingerprinting is the systematic use of multivariate statistical methods for comprehensive and objective comparison and classification of oil from single and multiple sources (Christensen et al., 2004, 2005b, 2005d). Although multivariate methods have been in use for several decades in the scientific community (Jolliffe, 1986; Smilde et al., 2004), their application for oil hydrocarbon fingerprinting is relatively new and has been quite sparse. In particular, multivariate methods have been used for data analysis in organic geochemistry since the 1980s (Øygard et al., 1984; Telnaes and Dahl, 1986), but their application for oil hydrocarbon fingerprinting and spill source identification is more recent and has mostly occurred since the middle of the 1990s (Aboul-Kassim and Simoneit, 1995a, 1995b; Burns et al., 1997; Stout et al., 2001; Lavine et al., 2001; Mudge, 2002; Li et al., 2004; Christensen et al., 2004, 2005b, 2005d).

9.1.1 Multivariate Methods and Oil Fingerprinting

The first evidence of multivariate statistical methods applied to oil hydrocarbon fingerprinting dates to the beginning of the 1980s, where, for example, Øygaard et al. (1984) and Telnaes and Dahl (1986) used PCA on the normalized distribution of hopanes for oil source differentiation in geochemistry. Multivariate methods were used later for oil hydrocarbon fingerprinting in environmental forensics, and in 1995 Aboul-Kassim and Simoneit used factor analysis to study particulate fallout samples in Alexandria (Aboul-Kassim and Simoneit, 1995a) and sediment samples from the Eastern Harbor of Alexandria (Aboul-Kassim and Simoneit, 1995b). In both works, multivariate statistical tools were employed to reduce the hydrocarbon datasets into their pollution sources based on the concentrations of aliphatic and aromatic hydrocarbons. In particular, the first analysis showed that two significant factors could explain 90% of the total variation among particulate fallout samples and confirmed petrochemical (79.6%) and thermogenic/pyrolytic (10.4%) sources. Analogously, Aboul-Kassim and Simoneit (1995b) determined that untreated sewage, rather than direct inputs from boating activities or urban runoff, was the main source of petroleum hydrocarbons in the Eastern Harbor.

In 1997, Burns et al. used PCA and a least-squares iterative matching procedure to allocate PAHs in intertidal and subtidal sediment samples from the Prince William Sound of Alaska to 30 potential sources (Burns et al., 1997). PCA was used to identify 18 possible sources, including diesel oil, diesel soot, spilled crude oil in various weathering states, natural background, creosote, and combustion products from human activities and forest fires. The least-squares model was subsequently used to estimate the source mix, with the best least-squares fit of 36 PAH analytes.

In 2001, Lavine et al. employed pattern recognition and PCA to study spilled jet fuel that had undergone weathering in a subsurface environment and classified them into five types (Lavine et al., 2001). Stout et al. (2001) analyzed a suite of diagnostic PAH and biomarker ratios with PCA. The ratios were selected on the basis of high analytical precision and low susceptibility to weathering. The analysis helped to identify the prime suspects for a heavy fuel oil (HFO) spill of unknown origin from 66 candidate sources.

In a relatively recent attempt to resolve the origin of background hydrocarbons in the sediments of Prince William Sound and the Gulf of Alaska, Mudge (2002) used partial least-squares regression. The percentage distribution of five possible sources — coal, seep oil, shales and input from two rivers — to the hydrocarbon loading in the Gulf of Alaska was estimated, with the individual contributions varying significantly across the sampling area.

Li et al. (2004) used PCA and point-to-point matching for screening and differentiation of nine oil samples based on their fluorescence emission spectra. The PCA model was able to distinguish all nine oil samples as well as the weathering extent of different oil samples, whereas only five of them were discriminated by point-to-point matching algorithms. A number of specific methods for oil hydrocarbon fingerprinting has been developed in our research group. In Christensen et al. (2005c), a semi-automatic method for processing complex first-order chromatographic data (namely GC-MS/SIM) is described that allows one to resolve convoluted peaks using mathematical functions with few parameters (i.e., Gaussian and exponential-Gaussian hybrid). The procedure was tested on chromatographic data from 20 replicate oil samples, and we found it to be less time-consuming and more objective compared to commercial software, while retaining comparable data quality. The same method was then applied in a forensic spill source identification study for the rapid and automated calculation of a large number of diagnostic ratios (Christensen et al., 2004). Four groups of diagnostic ratios derived from petroleum biomarkers (terpanes and steranes) and within homologue series of PAHs were thus evaluated simultaneously by weighted least-squares-principal component analysis (WLS-PCA), which was preferable to standard PCA as it can account for largely different variable uncertainties. The subsequent statistical testing of scores ensured an objective matching of oil spill samples with suspected source oils, and classification into positive match, probable match, and nonmatch. The sources of two spill samples (Norwegian crude oils from Oseberg East and Oseberg Field Centre) were identified and distinguished from closely related oil (crude oil from Oseberg South East).

However, some individual peaks in complex chromatograms such as the steranes at m/z 217 cannot be sufficiently resolved in the chromatograms to provide precise and accurate determinations of all peak areas. Furthermore, the number of possible diagnostic ratios is huge; in Christensen et al. (2004), ratios were limited to standard ratios used in geochemistry and a limited number of mostly binary ratios of alkylated naphthalenes and phenanthrenes based on pure combinatorics by calculating all possible binary isomer ratios within a compound group (e.g., $3 + 2 + 1 = 6$ ratios for C1-phenanthrenes consisting of four peaks). Thus, an alternative was sought by which most of the chemical information in one or several chromatograms could be analyzed without any prior peak identification and quantification. In order to eliminate variation unrelated to chemical composition, PCA was applied to the normalized first derivatives of aligned GC-MS chromatograms of m/z 217 (tricyclic and tetracyclic steranes) from a selection of crude oils, petroleum products, and oil spill samples collected from the coastal environment in the weeks after the *Baltic Carrier* oil spill, March 29, 2001, Grønsund, Denmark (Christensen et al., 2005d). Four reliable components could be observed in the data, and the spill samples were correctly assigned to the corresponding sources. Due to the large number of variables, the same WLS-PCA approach previously used on diagnostic ratios was employed here. At the same time a variable selection scheme was devised to obtain the most reliable results. This method has also been tested on oil samples from an *in vitro* biodegradation experiment in which a North Sea crude oil was exposed to three mixtures of bacterial strains over a 1-year period with five sampling times (Christensen et al., 2005a). The variation in degradability within groups of isomers of methylfluorene (*m/z* 180), methylphenanthrene (*m/z* 192), and methyldibenzothiophene (*m/z* 198) was used to evaluate the effects of microbial degradation on the oil composition. It was demonstrated that a mixture of strains of alkane degraders (and surfactant producers) and of PAH degraders affected the PAH isomer patterns differently than the two strain mixtures did independently.

Finally, the efficacy of an extension of PCA to higher-order data tensors for oil hydrocarbon fingerprinting was tested on fluorescence excitation-emission data (EEM) (Christensen et al., 2005b). One hundred twelve fluorescence landscapes of HFOs, light fuel oils (LFO), crude oils, lubricating oils, spill samples, and various mixed sources were analyzed simultaneously with parallel factor analysis (PARAFAC). With the exception of HFOs and crude oils, the method could discriminate between the four oil types and assign the spill samples to the corresponding sources.

9.1.2 Integrated Multivariate Oil Fingerprinting (IMOF)

A novel concept based on rapid, objective, and comprehensive multivariate statistical analysis of crude oils and refined petroleum products was developed in our research group between 2002 to 2005. The methodology has been used both for spill source identification (Christensen et al., 2004, 2005b, 2005d) and for the study of weathering of complex mixtures of oil hydrocarbons (Christensen, 2002; Christensen et al., 2005a). A number of methods are presented in this chapter as parts of the "Integrated Multivariate Oil Fingerprinting" (IMOF) methodology, which is proposed as a general framework for oil hydrocarbon fingerprinting. The IMOF methodology (Figure 9-1) is based on a combination of semiquantitative chemical analysis (cf. Section 9.2), automated and comprehensive data preprocessing (cf. Section 9.3), multivariate statistical analysis (cf. Section 9.4), and objective data evaluation (cf. Section 9.5). The data analysis step is more specifically based on multilinear decomposition methods (PCA being the most notorious, but also the parallel factor [PARAFAC] model when the data allow it) and statistical analyses, preceded by appropriate preprocessing and followed by rigorous data evaluation.

It is important to remark that IMOF is not intended as a closed set of techniques, but rather as a methodology that encompasses all aspects of oil hydrocarbon fingerprinting. Different analytical methods can provide different levels of detail of the analyzed chemical system. Hence, while in this chapter fluorescence spectroscopy is used for initial screening of oil samples and GC-MS for comprehensive compound-specific analyses, other analytical methods such as GC-FID and liquid chromatography-mass spectrometry (LC-MS) may well be incorporated into the IMOF methodology. Analogously, the goal of the chapter is not to exhaust the subject of applying multivariate statistical methods to oil fingerprinting. Nonetheless, we hope that this chapter manages to convey the great potential that these mathematical procedures have for rapid, reliable, and objective analysis of oils.

Procedures for data import, preprocessing, analysis, and evaluation in the specific methods for oil hydrocarbon fingerprinting developed in our research group and based on the IMOF methodology has been implemented in Delphi 4.0 (Borland) and Matlab 6.5 (The Mathworks). The procedure for chromatographic preprocessing of GC-MS data for analysis of complex mixtures of petroleum hydrocarbons (Christensen et al., 2005c) has been implemented in Borland Delphi 4.0 object-oriented programming, except for the extraction and sorting of data, which has been performed in Matlab 6.5 using the NetCDF software (http://my.unidata.ucar.edu). The procedures for data import, preprocessing, analysis, and evaluation in Christensen et al. (2004, 2005a, 2005b, 2005d) have been implemented in Matlab 6.5 (m-files). The relevant Matlab and Delphi files can be obtained by contacting the authors. Standard algorithms for data preprocessing and two-way data analysis can be downloaded from www.models.kvl.dk and

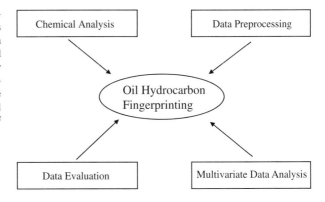

Figure 9-1 Illustration of the research strategy behind the IMOF methodology, which is based on four steps: chemical analysis, data preprocessing, multivariate data analysis and data evaluation. The aim of the work in our research groups has been to develop rapid, reliable and objective tools for comprehensive analysis and characterization of complex oil hydrocarbon mixtures using the IMOF methodology.

www-its.chem.uva.nl/research/pac. The *N*-way toolbox (Andersson and Bro, 2000) used for multiway data analyses (e.g., PARAFAC) can be downloaded from www.models.kvl.dk.

9.2 Sample Preparation and Chemical Analysis

Standard sample preparation and peak quantification procedures involve a series of time-consuming steps, such as extraction, evaporation, fractionation, addition of surrogate standards, peak detection, and integration and quantification based on response factors. The aim of this initial step of the IMOF methodology is to apply analytical procedures with limited time consumption and with consistently high-quality data. The latter can be ensured by comprehensive quality assurance and quality-control measures (QA/QC) and allows for a more adequate, comprehensive, and objective analysis of complex mixtures of oil hydrocarbons.

9.2.1 Sample Preparation

Sample preparation varies depending on the analyzed sample substrate. Thus, biota, soil, sediments, and pure oils all require different extraction, cleanup, and fractionation procedures. In our research group, the sampling steps have been limited to sampling of pure oil collected on-board ships and from vegetation and stones after oil spills (Christensen et al., 2002, 2004, 2005b, 2005d) and to sacrificing experimental units (Erlenmeyer flasks) during *in vitro* experiments (Christensen et al., 2005a). Hence, sample preparation has been limited to dilution of pure oil and extraction from stones and vegetation and to liquid/liquid extraction using dichloromethane. Subsequently, water and particles have been removed by cleanup through funnels with glass wool and sodium sulphate.

When analyzing more complex environmental matrices such as sediments, soils, and biota, semi-automated instrumental extraction techniques [e.g., microwave extraction (Jassie, 1995; Shu et al., 2000) and pressurized liquid extraction, PLE (Bandh et al., 2000)] should be preferred to less automated and more time-consuming extraction procedures such as Soxhlet extraction. PLE is especially attractive since extraction of a relatively large number of samples can be performed overnight with limited use of solvent (e.g., 10–30 ml per sample) (Bandh et al., 2000; Richter, 2000), and sample cleanup and fractionation may be integrated with the extraction procedure (Sporring et al., 2005; Nording et al., 2005). Fractionation into aliphatic, aromatic, and polar fractions, which is frequently used as a sample preparation step in oil hydrocarbon analysis (Wang et al., 1994a, 1994b), has been avoided throughout the present work to reduce the analysis time. However, chemical analysis of complex sample matrices may require this step, which can again be built into PLE procedures (Sporring et al., 2005).

9.2.2 Analytical Methods

A variety of analytical methods have been used for oil hydrocarbon analysis, including thin-layer chromatography (TLC), high-performance liquid chromatography (HPLC), gas chromatography-photo ionization detection (GC-PID), two-dimensional gas chromatography (GC-GC), gas chromatography-isotope ratio mass spectrometry (GC-IRMS), GC-FID and GC-MS; gravimetric measurements; and infrared, ultraviolet and fluorescence spectroscopy. Although all these methods have been used for oil hydrocarbon analysis, GC-FID and GC-MS are the preferred methods in most oil hydrocarbon fingerprinting studies (Aboul-Kassim and Simoneit, 1995a, 1995b; Jovancicevic et al., 1996; Munoz et al., 1997; Boehm et al., 1997; Burns et al., 1997; Wang et al., 1999, 2002; Ezra et al., 2000; Stout et al., 2001; Page et al., 2002; Mudge, 2002; Christensen et al., 2004, 2005d). Whereas GC-FID is the standard method for initial screening of oil samples, GC-MS is used for more comprehensive chemical characterization (Wang et al., 1999) since it can resolve a broad range of oil hydrocarbons including petroleum biomarkers and PAHs, and because of the low cost of quadrupole instruments. A method for initial screening of oil samples using fluorescence spectroscopy is suggested in this chapter as a complementary method to standard GC-FID screening.

9.2.3 Fluorescence Spectroscopy

Fluorescence excitation-emission matrices (EEMs) for initial screening in oil hydrocarbon fingerprinting can be obtained using a fluorescence spectrophotometer in scan mode. Fluorescence of oil is mainly caused by PAHs, which are highly fluorescent due to the presence of delocalized electrons within the aromatic rings, and because their rigid structure does not allow for efficient vibrational relaxation. Fluorescence is affected by quenching and energy-transfer processes and consequently is a complex process in multicomponent mixtures such as oil, which contains hundreds of individual PAHs.

The experimental procedure used in Christensen et al. (2005b) for fluorescence spectroscopy is based on sample dilution to avoid pronounced light absorption (i.e., the absorbances in the wavelength range between 240 and 600 nm were required to be below 0.05 absorbance units measured by UV-VIS) and thus reduces inner filter effects and effects of quenching and energy-transfer processes (Christensen et al., 2005b). The combination of high detector voltage (850 V), necessary to obtain a sufficient dilution, and a high scan speed (4800 nm/min) to reduce the analysis time, led to low signal-to-noise data. The PARAFAC model was, however, able to handle this by modeling the systematic variations and leaving the noise in the residuals.

In Christensen et al. (2005b), the EEMs were measured on a Varian Eclipse fluorescence spectrophotometer. A collection of emission scans from 250–600 nm with 2-nm increments was obtained at varying excitation wavelengths ranging from 240–475 nm with 5-nm increments. The bandwidths were 5 nm for both excitation and emission, and the scan rate was 4800 nm/min, the latter leading to a scan time of less than 10 min per sample. Each scan was comprised of 176 emission and 48 excitation wavelengths. Below an excitation wavelength of 240 nm, the spectra were noisy, and above 600 nm, the signals were negligible for the four oil types (crude oils, LFOs, HFOs, and lubricants). The fluorescence EEM measurements are thus consistent with the general goal of the IMOF methodology in being rapid and with only limited sample preparation (dilution and UV-VIS measurements). An excitation-emission scan of an HFO sample from the cargo tank of the *Baltic Carrier* is shown in Figure 9-2.

Rayleigh and Raman scatter show up in three-way fluorescence data as diagonal lines across EEMs (Andersen and Bro, 2003). The ridge due to the Rayleigh scatter is clearly visible in Figure 9-2. Rayleigh scatter is elastic (i.e., there is no energy loss); hence, the scattered emission wavelength is equal to that of excitation. Conversely, Raman scatter is

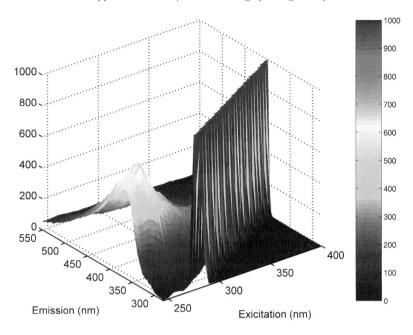

Figure 9-2 Fluorescence excitation-emission scans of an HFO. The vertical axis and scale bar show the fluorescence intensity. Modified from Christensen et al. (2005b). See color plate.

inelastic, and emission is shifted to longer wavelengths compared to excitation due to energy loss. In any case, scatter is unrelated to the chemical sample composition and cannot be modeled adequately by low-rank multilinear models (i.e., they require the fitting of numerous additional factors that have no chemical meaning) (Bro, 1997; Stedmon et al., 2003; Andersen and Bro, 2003; Christensen et al., 2005b). Several methods have been used to reduce detrimental effects due to scatter (Christensen et al., 2005b). The most immediate is to filter the corresponding signal. This can be done quite effectively by means of appropriate preprocessing before data analysis (cf. Section 9.3.3).

9.2.4 GC-MS

Mass spectra produced by GC-MS are one of the most valuable tools for identification of unknown compounds. Oil hydrocarbon fingerprinting by GC-MS is based on analysis of selected m/z ions with high sensitivity (base peaks in mass spectra) and selectivity for specific compound groups with common structure.

The selection of appropriate m/z ions for oil hydrocarbon fingerprinting in the IMOF methodology is based on prior work mainly by geochemists (Peters and Moldowan, 1993). We recommend that a large number of PAHs and petroleum biomarkers be included in the GC-MS analysis since these compounds are, generally, resistant to weathering processes and their fingerprints vary between oil sources due to the depositional environment, in-reservoir degradation, thermal maturity, and the refining process. Petroleum biomarkers and PAHs with common structures such as tri-pentacyclic terpanes, steranes, and the homologue series of C_0-C_4-phenanthrene isomers can be measured by detecting characteristic mass fragments or the molecular masses using GC-MS analysis with selected ion monitoring (SIM) (GC-MS/SIM) (Table 9-1).

A Finnigan Trace DSQ™ single quadrupole GC-MS operating in electron impact (EI) mode was used by Christensen et al. (2005d) to obtain chromatographic data for oil hydro-

Table 9-1 List of Mass Fragments (m/z) and Corresponding Compound Groups Analyzed with High Relevance for Oil Hydrocarbon Fingerprinting

Polycyclic Aromatic Hydrocarbons (PAHs)	Mass Fragments (m/z)
C_0-C_4-naphthalenes	128, 142, 156, 170, 184
C_0-C_4-phenanthrenes	178, 192, 206, 220, 234
C_0-C_3-fluorenes	166, 180, 194, 208
C_0-C_4-chrysenes	228, 242, 256, 270, 284
C_0-C_2-pyrenes and fluoranthrenes	202, 216, 230
Other PACs (5- and 6-ring)	252, 276, 278
Heterocyclic Aromatic Compounds	
C_0-C_4-benzothiophenes	134, 148, 162, 176, 190
C_0-C_4-dibenzothiophenes	184, 198, 212, 226, 240
C_0-C_1-naphthobenzothiophenes	234, 248
C_0-C_2-dibenzofuranes	168, 182, 196
Petroleum Biomarkers	
Sesquiterpanes	123
Terpanes	177, 191, 205
Steranes and diasteranes	217, 218, 259
Triaromatic steranes	231
Other Compounds	
n-alkanes and isoprenoids	85
Alkyltoluenes	105
C_0-C_3-biphenyls	154, 168, 182, 196

carbon fingerprinting. Capillary columns of 30 or 60 meters with nonpolar stationary phases (e.g., HP-5 ms, DB5, and Zebron ZB-5) are recommended for separation of oil hydrocarbons. A 60 m HP-5 ms (0.25-mm inner diameter × 0.25-μm film thickness) was used in Christensen et al. (2005d) injecting 1-μl aliquots in PTV splitless mode. The inlet temperature was 35°C during injection (1 min) and increased subsequently by 14.5°C/sec to 315°C during transfer (hold for 1 min during transfer). The column temperature program was 35°C (2 min), 60°C/min to 100°C, 5°C/min to 315°C (20 min), and transfer line and ion source temperatures: 300°C and 250°C, respectively. In Figure 9-3, a partial chromatographic profile of m/z 217 is shown using the injection and column temperature program just described.

Christensen et al. (2005d) suggest that increasing the sampling rate (e.g., focusing on fewer masses in each segment in the GC-MS/SIM and decreasing the dwell time) would improve the ability of the preprocessing to reduce variation unrelated to the chemical composition. For example, the initial parameterization and peak-fitting procedure applied for semiquantitative analysis would be affected positively since the number of data points in each peak would increase correspondingly (Christensen et al., 2005c). More specifically, the authors selected 48 mass fragments and analyzed them in six groups of 14–15 m/z ions GC-MS/SIM with a sampling rate of 1.27 scans/sec, and observed that early eluting chromatographic peaks such as naphthalene (m/z 128) consisted of relatively few data points (less than 10), which may be insufficient for determining the initial parameters for Gaussian peak fits (Christensen et al., 2005c). Likewise, Christensen et al. (2005d) concluded that an increased sampling rate

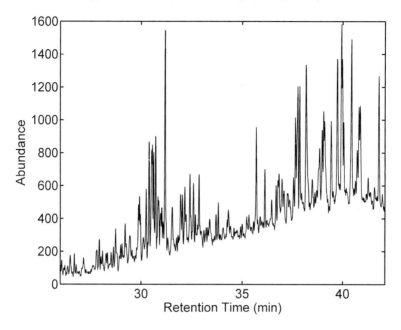

Figure 9-3 Partial GC-MS chromatogram of *m/z* 217, which contains tricyclic steranes (eluting between 26 and 34 min) and tetracyclic steranes (eluting between 35 and 42 min).

would allow for more refined retention time alignment. Thus, the analytical procedure was changed slightly for Christensen et al. (2005a, 2005c) such that 44 mass fragments were analyzed in 8 groups of 12 ions with a sampling rate of 2.34 scans/sec with 20 msec dwell times for each ion, which almost doubled the number of scans/sec.

9.2.5 Quality Assurance and Quality Control (QA/QC)

Semiquantitative analytical approaches are preferred to standard quantitative methods in IMOF due to the high reproducibility and low time consumption obtained with these methods. Fully quantitative approaches rely on quantification by the internal, external, or standard addition method. Conversely, a semiquantitative approach relies on frequent analyses of a laboratory reference sample, with comparable sample characteristics and chemical composition as the analytical samples. In Christensen et al. (2004, 2005a, 2005c, 2005d), the reference is a 1:1 mixture of a North Sea crude oil (Brent crude oil) and an HFO from the *Baltic Carrier* oil spill. In Christensen et al. (2005b), the reference is a mixture of Brent crude oil, light fuel oil (LFO), HFO, and a lubricant oil prepared in such a way that the four oils contribute approximately equally to the combined fluorescence signal. Hence, the sample characteristics of the references are comparable with those of the analytical samples (i.e., the GC-MS chromatograms contain most of the relevant peaks and the fluorescence excitation-emission spectra contain most of the fluorescence excitation-emission characteristics) of the four different oil types. The replicate analyses of laboratory reference samples have also been used for a variety of other purposes in the oil hydrocarbon fingerprinting methods elaborated in Christensen et al. (2004, 2005a, 2005c, 2005d), including calculation of the analytical uncertainty (Christensen et al., 2004, 2005a, 2005c, 2005d), optimization of data preprocessing parameters (Christensen et al., 2004, 2005a, 2005d), and normalization of diagnostic ratios (Christensen et al., 2004, 2005a, 2005c), as well as in the peak matching procedure described in Christensen et al. (2005c).

High-quality data are a prerequisite for meaningful and reliable data analysis. This is valid for any analytical method and is particularly true for the IMOF method. Multilinear models like PCA work under a number of assumptions and requirements (for example, peaks relative to the same compound are not shifted in different measurements), and it is important that all these requirements are met to a certain extent (Tauler et al., 1995; de Juan and Tauler, 2001; Smilde et al., 2004; Christensen et al., 2004, 2005a, 2005b, 2005d). Hence, while standard quantitative methods based on peak identification and quantification are relatively unaffected by variations in peak shape (e.g., from symmetrical to tailing) and retention time shifts, PCA on sections of GC-MS/SIM chromatograms can be heavily affected by such factors due to changes in the intensity distribution of peaks (e.g., fronting, symmetrical or tailing peaks) (Christensen et al., 2005a, 2005d). Limiting these factors by comprehensive QA/QC measures are, thus, of special importance when using the latter approach. Replicated measurements of a reference oil are used, among others, for general QA/QC measures in the laboratory and thus, to some extent, to verify the compliance of the data to the requirements of the subsequent data analysis. For example, the following QA/QC measures based on the replicated references are applied in Christensen et al. (2004, 2005a, 2005c, 2005d) for QA/QC of GC-MS data:

- The chromatographic peak shapes are checked regularly at selected ion masses [e.g., m/z 85 (n-alkanes and isoprenoids), 128 (naphthalene), 180 (methylfluorenes), 191 (terpanes), 192 (methylphenanthrenes), 198 (methyldibenzothiophenes), 217 (steranes), and 252 (five-ring PACs)]. Deterioration of the chromatographic column or worsening of the conditions in the inlet (e.g., dirty liner) often causes increased tailing.
- Changes in the sensitivity of the mass spectrometer (also checked by tuning the mass spectrometer each day of analysis).
- Mass discrimination due to changes in the inlet or a dirty ion source (also checked by tuning the mass spectrometer frequently).

Changes in the chromatographic peak shapes, mass discrimination, and a significant decrease in sensitivity (e.g., more than a factor of 10) led immediately to cleaning of the ion source, change of liner and septum, or trimming of the capillary column. The septum and liner of the injector system were changed, and the ion source of the mass spectrometer cleaned regularly throughout the analytical work.

In Christensen et al. (2004, 2005a, 2005c, 2005d), the QA/QC procedure is performed once for every 50 injections (i.e., including oil samples, blanks, and references), which results in chromatographic data of consistently high quality. Thus, relative analytical standard deviations (RSD_A) based on the references were kept consistently below 3% for diagnostic ratios of well-resolved peaks (Christensen et al., 2004, 2005a, 2005c), and the variability of replicated reference samples in score plots was significantly lower than the total variability within datasets (Christensen et al., 2005d). Moreover, more than 500 injections of oils, references, blanks, and quantification standards were performed as part of the *in vitro* biodegradation study in Christensen et al. (2005a), and less than 20% of samples were reanalyzed due to insufficient data quality (i.e., poor reproducibility or sensitivity, mass discrimination, and chromatographic resolution).

9.3 Data Preprocessing

The purpose of data preprocessing in IMOF is to reduce variation in the data unrelated to the relative chemical composition such as instrumental noise, retention time shifts, analytical variability, and absolute compound concentrations. The signal-to-noise ratios are improved by reducing these factors and, consequently, facilitate the extraction of meaningful chemical information. In turn, this leads to an improved ability for the integrated oil fingerprinting method to distinguish dissimilar samples (i.e., increased resolution power) in spill source identification cases. Furthermore, in order to be of any practical use, tools for data preprocessing have to be sufficiently rapid and should require limited human intervention. At the same time, preprocessing depends on the type of data. Thus, preprocessing of the three types of data that have been thus far analyzed in IMOF (namely, sections of GC-MS/SIM chromatograms, diagnostic ratios, and fluorescence EEM spectra) is presented separately in the ensuing sections.

9.3.1 Partial GC-MS/SIM Chromatograms

Three steps for preprocessing partial GC-MS/SIM chromatograms prior to multivariate statistical analysis were applied: baseline removal; time alignment; and normalization (Christensen et al., 2005a, 2005d). They are described in the ensuing paragraphs and have been implemented by the authors in MATLAB (The MathWorks) as one integrated methodology. The combined effects of the three preprocessing steps — baseline removal, retention time alignment, and normalization — in the oil hydrocarbon fingerprinting study (Christensen et al., 2005d) are illustrated in Figure 9-4.

9.3.1.1 Baseline Removal

Chromatographic baselines in GC-MS/SIM chromatograms are caused both by features unrelated to the chemical composition (e.g., sensitivity of the mass spectrometer) and by coelution of compounds in the complex oil mixture. Besides producing an increase in the number of factors necessary to describe the data using multilinear models, such baselines represent a problem for other preprocessing steps and should be removed prior to retention time alignment and normalization.

A number of methods can be used for this purpose. The most common are polynomial and piecewise-linear baseline fits and calculation of first or second derivatives with or without smoothing (Åberg et al., 2004). However, in complex mixtures such as crude oil and petroleum products, many peaks are not baseline-separated. Consequently, to manually select points for the baseline fit would be prone to subjective bias and is generally not viable. Although relatively refined methods exist to automate baseline removal in complex samples (e.g., using splines and convex hulls), numerical first derivatives have proven sufficient when directly applying multilinear models to oil hydrocarbon fingerprinting (Christensen et al., 2005a, 2005d). Therefore, this is the only method currently implemented in IMOF. The first derivatives (calculated through simple differences) of the partial GC-MS/SIM chromatogram (35.5–42 min) of tetracyclic steranes (m/z 217) are shown in Figure 9-4a for five reference oils and five source oils.

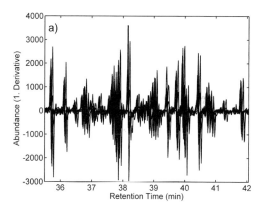
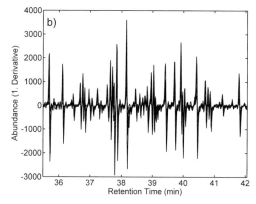

Figure 9-4 First derivative of a section of m/z 217 for five reference oils and five source oils: (a) before warping and (b) after warping using COW with segment length of 175 data points and a slack of three points. Modified from Christensen et al. (2005d).

With respect to smoothing, whether it is necessary or not when calculating the derivatives [e.g., using Savitsky-Golay (Åberg et al., 2004) or kernels] depends on the application and on the signal-to-noise ratio. Notwithstanding the fact that calculating derivatives through simple differences tends to increase noise in the data (Christensen et al., 2005d), if the signal-to-noise ratio is sufficiently good, it is still possible to extract meaningful information with limited data manipulation (Fraga et al., 2000; Johnson et al., 2003).

9.3.1.2 Retention Time Alignment

One of the most severe impediments to multivariate statistical analysis of partial GC-MS/SIM chromatograms is retention time shift, which is caused mainly by deterioration of the capillary column (Johnson et al., 2003). While recent technological advancements have helped to reduce the magnitude of this problem (Malmquist and Danielsson, 1994c; Witjes et al., 2001; Vogt and Booksh, 2004), even slight differences of one time point can affect the fitting of multilinear models (Wang and Isenhour, 1987; Malmquist and Danielsson, 1994b; Nielsen et al., 1998; Fraga et al., 2000; Rønn, 2001; Johnson et al., 2003; Eilers, 2004; Willse et al., 2005; Wong et al., 2005). Numerous methods have been proposed to correct for retention time shifts in chromatographic data in the scientific literature (van Nederkassel et al., 2005a, 2005b). While not necessarily the fastest, dynamic programming-based algorithms like dynamic time warping (DTW) and correlation optimized warping (COW) have been used with a certain degree of success for a broad range of chromatograms (Nielsen et al., 1998; Pravdova et al., 2002; Tomasi et al., 2004; van Nederkassel et al., 2005a, 2005b; Christensen et al., 2005d). While under some constraints the two algorithms may yield analogous results (Tomasi et al., 2004), only the COW algorithm is currently included in IMOF. The main reasons are that its MATLAB implementation is considerably faster than DTW and, while limited to piecewise linear corrections, COW is still capable of correcting the unwanted retention time shift for GC (Nielsen et al., 1998; Tomasi et al., 2004).

COW is a piecewise or segmented data preprocessing method, which operates on one sample at a time. It works under the assumption that corresponding subsections of two chromatograms should have the highest similarity when correctly aligned. Hence, correcting the retention time shift is reduced to identifying the boundaries of such sections and to moving them so that they occupy the same position (i.e., retention time) in all the samples as in the target.

In essence, COW aligns a sample (here a partial GC-MS/SIM chromatogram) to a target data vector (another GC-MS chromatogram) by splitting the two signals in an equal number of segments (whose length l is imposed by the analyst and depends on the data) and by finding the segments' boundaries in the sample according to a simple optimality criterion (i.e., the sum of the Pearson's correlation coefficients between corresponding segments). If the corresponding segments in the sample and in the target have different lengths, they are linearly interpolated to the same number of points. Interpolation is necessary both to be able to compute the correlation coefficient and to obtain the final alignment. Segment lengths are allowed limited changes and are restricted to integer values both to avoid extreme corrections and to keep the computational time manageable. The maximum length increase or decrease in a sample segment in terms of scan points is controlled by the slack parameter t (Eilers, 2004).

The correcting power of the algorithm is inversely proportional to the segment length and directly proportional to the slack. In order to better describe the function of COW and to illustrate when it may fail, the concept of a warping path is introduced. Alignment methods seek to find a relationship that associates the scan number (or the retention time) in the sample with a scan number in the target. This relationship, which can be an explicit function or simply represented as a set of indexes, is referred to as the *warping path* and can be visualized in a system of axes j_{sample} versus j_{target}, where j denotes the scan number. For example, quadratic models of the type $j_{sample} = c_2 j_{target}^2 + c_1 j_{target} + c_0$, where j denotes the scan number, are sometimes enough to correct for retention time shift (Tomasi et al., 2004). On the system of axes j_{sample} versus j_{target}, this appears as a parabola. In COW the warping path is formed by segments spanning I points on the reference axis and having the slope comprised of the interval between $(I - t)I^{-1}$ and $(I + t)I^{-1}$. Thus, if the correct warping path could be approximated by a parabola, the COW warping path (given t and I) would be the optimal piecewise linear approximation of such a parabola. The shorter the segments or the larger the allowed interval for the slope, the better the approximation becomes. However, there is a lower limit for I, which must be larger than the peak width at the base to avoid artifacts and peak deformations and vice versa, large values of I can correct the retention time shift only if the curvature of the "true" warping path (i.e., the one that would yield the correct alignment) is sufficiently small and can be approximated by long segments (Christensen et al., 2005d). This has important implications when aligning GC data, because the long segments commonly used with COW (with small values of t) are likely to yield optimal results only for compounds with similar physicochemical properties that are affected in the same way by changes in the column properties (e.g., chemical changes in the stationary phase) and for m/z ratios that refer to homogeneous classes of compounds [e.g., tri- and tetracyclic steranes in m/z 217 (Christensen et al., 2005d) and methylated PAC homologues in m/z 180, 192, and 198 (Christensen et al., 2005a)]. The partial GC-MS/SIM chromatograms (m/z 217) of five reference oils and five source oils are shown in Figure 9-4, before (4a) and after (4b) alignment using COW.

Conversely, chromatographic methods with less selective detectors such as GC-FID and liquid chromatography-diode array detection (LC-DAD) produce chromatograms with more irregular retention time shifts since the chromatograms contain compounds with very different physicochemical properties. In situations with large changes in the stationary phase of the column, peaks may even change elution order, and COW cannot adequately correct for such changes. In this case, optimal alignments may infer the selection of smaller Is, mostly of the same order as the width at the base of the smallest peak one wants to align (Malmquist and Danielsson, 1994a; Andersson et al., 2004; Tomasi G. et al., 2004).

Since only integer values of the segment lengths are checked in COW, the correction is typically less fine than other methods that allow fractional corrections or express the warping path parametrically (as in the quadratic function earlier). This is visible in the residual chromatographic shift of at most one scan point after time alignment is performed (Christensen et al., 2005a, 2005d). The main consequence of these two aspects is that, while the results are deterministic, it may be necessary to test different values of I and t in order to identify the best choice. Although this is not necessarily trivial and may require supervision, a simple automated procedure based on a grid search algorithm can be used if the same sample is frequently analyzed as part of the overall procedure (cf. Section 9.2.5) (Christensen et al., 2005a, 2005d). In particular, if one disregards noise and without mean centering, the rank of a matrix consisting of perfectly aligned chromatograms obtained from the same sample is 1. In this case, the optimal values of I and t would be those that reduce to 1 the number of nonzero singular values of such a matrix. In practice, since noise is always present in experimental data, the optimal choice of warping parameters is the one that maximizes the first singular value (Christensen et al., 2005a, 2005d).

Note that the residual shift mentioned earlier has an important secondary effect. Namely, as the rank of the fitted model (i.e., the number of principal components for a matrix) increases, the additional factors explain less and less of the chemical variation and more of the relatively small variation caused by residual shifts (Johansson et al., 1984).

9.3.1.3 Normalization

The chromatographic abundances are related to the sensitivity of the instrument and concentrations of the single constituents. It is affected by several parameters such as sampling, extraction, cleanup, fractionation, and the sensitivity of the mass spectrometer. Variations in data due to these factors are likely to mask the compositional information in the subsequent data analyses. Hence, normalization is a prerequisite for objective and automated comparison of sections of preprocessed GC-MS/SIM chromatograms by multivariate statistical analyses. Normalization to a constant Euclidean norm is a common procedure used to compensate for concentration effects and sensitivity changes [Eq. (9-1)]:

$$x_{nj}^N = \frac{x_{nj}}{\sqrt{\sum_{j=1}^{J} x_{nj}^2}} \quad (9\text{-}1)$$

where x_{nj} is the first derivative of the nth chromatogram at the jth retention time and J is the total number of retention times. This method is affected by the so-called closure of the dataset (i.e., if one peak increases, the size of the other peaks decrease) (Christensen et al., 2005d). In situations where the amount of information is limited (i.e., few peaks), this may lead to correlations in data that are only present due to the closure, which makes chemical interpretation of the results more difficult. In such cases, more complex normalization schemes using only a limited set of retention times referring to specific peaks could be adopted. To reduce the uncertainty and closure effects, these peaks should preferably be large and relatively constant in oil samples of different origin. Under such circumstances, the denominator in Eq. (9-1) is modified to the sum of the squared first derivatives of selected retention times instead of all retention times.

9.3.2 Diagnostic Ratios

As opposed to sections of GC-MS/SIM chromatograms, diagnostic ratios condense the compositional information in a smaller number of variables. Three types of diagnostic ratios are suggested within the IMOF methodology. They are listed in Eqs. (9-2)–(9-4).

$$DR = a_n^S / a_{n*}^S \quad (9\text{-}2)$$

$$DR = a_n^S / (a_n^S + a_{n*}^S) \quad (9\text{-}3)$$

$$DR = \frac{DR^S}{(DR^S + DR^R)}$$

$$DR^S = a_n^S / a_{n*}^S \quad \text{or} \quad DR^S = a_n^S / (a_n^S + a_{n*}^S)$$

$$DR^R = a_n^R / a_{n*}^R \quad \text{or} \quad DR^R = a_n^R / (a_n^R + a_{n*}^R)$$

$$(9\text{-}4)$$

where a_n^S and a_{n*}^S are the peak areas, peak heights, or concentrations of compound n and $n*$, respectively, or the sums of several compounds in an oil sample, and a_n^R and a_{n*}^R are the peak areas, peak heights, or concentrations of the corresponding compound(s) in a reference oil sample. If the peak values denote compound concentrations, the method of determining the diagnostic ratios is named "quantitative"; conversely, if peak areas or heights are employed, the procedure is termed "semiquantitative." The use of semiquantitative analyses as opposed to quantitative ones has been a matter of debate in the scientific community. The quantitative approach is often preferred to the semiquantitative one for oil hydrocarbon fingerprinting purposes (Wang et al., 1999). The main reasons are the inherent normalization based on the internal quantification procedure, which is generally thought to be more precise than a semiquantitative approach, and the potential for unraveling mixed oil signatures. In our experience, this is not necessarily the case, especially when employing multivariate statistical tools to comprehensive collections of diagnostic ratios or partial GC-MS/SIM chromatograms as described in Section 9.3.1. Furthermore, the semiquantitative method has proven to be sufficiently precise, less time-consuming, and simpler to implement (Christensen et al., 2004). In a recent oil hydrocarbon fingerprinting study, 137 diagnostic ratios based on petroleum biomarkers and PAHs were calculated from semiquantitative data (peak areas) (Christensen et al., 2004). RSD_A varied between 0.09% and 5.1% using Eq. (9-3) and decreased to 0.05% to 3.2% with external normalization to the oil reference analyzed closest in time [Eq. (9-4)]. In Christensen et al. (2005c), 72 diagnostic ratios were calculated from semiquantitative peak data (peak areas and heights). The RSD_A were below 5% for all diagnostic ratios based on well-separated and baseline distorted peaks using a new method based on peak modeling using mathematical formulas as well as commercial quantification software for peak quantification. For 10 diagnostic ratios based on incompletely resolved peaks, the RSD_A were between 2% and 8%. Finally, in an *in vitro* biodegradation study (Christensen et al., 2005a), the RSD_A for 19 diagnostic PAH ratios calculated from Eq. (9-3) varied between 0.5% and 4.8%. Direct comparisons between quantitative and semiquantitative approaches for calculating diagnostic ratios are difficult, since the RSD_As are rarely listed in oil hydrocarbon studies. However, in a study by Stout et al. (2001), the RSD_As of petroleum biomarker ratios were generally comparable or higher than those obtained in our studies. Note, however, that no matter which approach is used to calculate diagnostic ratios, the RSD_A should always be calculated as part of the QA/QC procedure and for justifying the results on sample identity.

9.3.3 Preprocessing of Fluorescence Spectra

Fluorescence data require an entirely different pretreatment than chromatographic ones. However, the purpose is still to reduce information that is not related to the chemical composition (i.e., what is relevant to oil hydrocarbon fingerprinting). The main problem with EEM signals is the presence of Rayleigh and Raman scatter in the data (e.g., Figure 9-2). These two physical phenomena are particularly problematic when modeling the data using low-rank multilinear models (i.e., PARAFAC), and how to remove their influence has been the subject of recent research (Andersen and Bro, 2003; Rinnan, 2004).

There are several ways of dealing with Rayleigh scatter; the most common is to insert missing values in a more or less wide diagonal band in the EEM landscape centered on the wavelengths at which such scatter is observed (i.e., at emission wavelengths close to the wavelength of the exciting light — see Figures 9-2 and 9-5) (Rinnan, 2004; Christensen et al., 2005b). Most algorithms for fitting multilinear models are capable of handling missing values using an expectation maximization approach (Bro, 1998; Tomasi and Bro, 2006). Conversely, Raman scatter effects are reduced, but not eliminated, by subtracting blanks (e.g., recorded in dichloromethane) (Stedmon et al., 2003; Tomasi and Bro, 2005).

As shown in Section 9.4, the PARAFAC decomposition essentially requires that the underlying model that gives rise to the data be inherently multilinear. Among others, this means that, in order for the mathematical model to be appropriate, the emission spectrum for one compound (or class of substances) must remain unaltered at all excitation wavelengths and in all samples. Likewise, that excitation spectrum must be the same at all emission wavelengths and in all samples (Andersen and Bro, 2003). While indeed an approximation, severe deviations from these assumptions may have important consequences on the usefulness of the fitted models (Bro, 1998). For this reason, instrument biases in EEM data are reduced by applying an excitation/emission correction spectrum derived from a combination of a Rhodamine spectrum and the spectrum from a ground quartz diffuser (Christensen et al., 2005b). The excitation/emission correction removes small artifacts in EEMs due to variations in detector efficiency as a function of wavelength. A preprocessed EEM is shown in Figure 9-5.

The fact that the signal recorded at emission wavelengths lower than the excitation one is physically zero may also negatively interact with the way EEM signals are modeled by PARAFAC (Andersen and Bro, 2003). Thus,

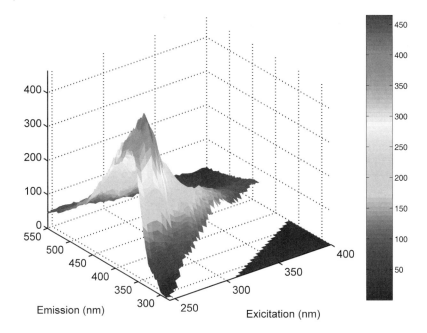

Figure 9-5 Preprocessed fluorescence EEM of the same oil sample as shown in Figure 9-2 after blank subtraction, insertion of missing values, and a small triangle of zeros and excitation/emission correction. The vertical axis and scale bar show the fluorescence intensity. Modified from Christensen et al. (2005b). See color plate.

the fact that the observed value is zero implies that for all compounds in a sample, the emission factor or the extinction coefficient at the corresponding wavelengths must be zero (assuming that negative contributions would make no sense from a physical point of view). It is obvious that this assumption is untenable for most fluorescent compounds, and some preprocessing is required. As for scatter, weighting or setting these values to missing are the most obvious choices. While it may be tempting to set all values "below" the Rayleigh scatter ridge to missing, this has proved to be problematic as the different factors are somehow allowed to interact in the missing values area, giving rise to artifacts and lack of convergence (Rinnan, 2004; Thygesen et al., 2004; Tomasi and Bro, 2005). The zeros can be considered to have a stabilizing effect, and the width of the missing values band should be the subject of optimization (Thygesen et al., 2004).

Finally, note that constraining the model parameters to be nonnegative or with certain choices of the loss function (i.e., instead of setting specific array elements to missing, assign a nonzero small weight to them) may also be beneficial (Bro et al., 2002; Andersen and Bro, 2003; Rinnan, 2004). Whether these choices are preferable to preprocessing depends on the data. We refer the reader to the original literature for further readings on the subject.

9.4 Multivariate Statistical Data Analysis

Technological advancement has made it possible to produce and treat datasets of increasing size and complexity. However, the number of underlying factors that give rise to a certain dataset is typically limited (within a relatively homogeneous dataset) compared to the number of observed variables. For example, a univariate chromatogram (GC-MS/SIM) may be considered as a vector of data in which each element corresponds to a certain scan time and may well contain thousands of data points, but the number of factors actually present in such data is at most equal to the number of peaks in the chromatogram. This is even more evident for a multivariate signal (as is the case for GC-MS/Full Scan data), where several tens or hundreds of measurements (m/z values) are taken at each time point in the chromatogram or when several samples are stacked into a matrix (or for EEM fluorescence data). To some extent, the peak quantification and calculation of diagnostic ratios, which often precedes the data analysis in oil hydrocarbon fingerprinting, serve the purpose of reducing the number of variables. The concept can be further extended if one takes into account that even the relative concentrations of the compounds may be related to one another in a way that may result from some common physicochemical phenomenon (e.g., depositional environment, thermal maturity, and in-reservoir degradation) affecting the composition. For example, the heptadecane/pristane and octadecane/phytane diagnostic ratios involve compounds whose physicochemical properties are almost identical and both describe biodegradation (because the branched alkanes at the denominator are less susceptible to microbial degradation than the *n*-alkanes that are at the numerator).

Multivariate statistical methods such as PCA and PARAFAC can be used to determine the most salient sources of variation and condense redundant information related to collinear variables. In this fashion, more complex patterns that might have been unknown or somewhat hidden prior to the data analysis may emerge and lead to a better understanding of the dataset and of the underlying factors. Moreover, these models have the capability of filtering some of the noise always present in experimental data, which may confound the data interpretation.

An important aspect of multilinear modeling is the determination of the correct number of factors (the so-called rank of the model). A number of tools exist for this purpose; some of them are generally applicable independently of the model (like cross-validation or the presence of relevant systematic variation in the residuals) while others may rely on the specific properties of a given model (like core consistency or the split-half analysis for PARAFAC) (Jolliffe, 1986; Bro, 1998; Rinnan, 2004).

It is also worth mentioning that the sources of systematic variation unrelated to the chemical composition should be reduced to a minimum prior to the multivariate analyses as suggested in the IMOF methodology. Otherwise, an increased number of factors are obtained whose interpretation will be difficult from a chemical point of view. Likewise, it may be necessary to select a subset of the variables or to downscale those with the highest level of noise in order to let certain phenomena become more pronounced. The main problem could be that the amount of information in the chosen dataset is insufficient to clearly establish important patterns in data. Additional information, such as the analytical and sampling uncertainties, can be used for automatic and objective selection or scaling of variables (cf. Section 9.4.2). This point is particularly relevant for complex mixtures like crude oil and petroleum products that have been exposed to weathering processes.

A necessary condition that is often overlooked is that, in order to yield interpretable results, the model that one decides to use has to be appropriate for the data at hand. For example, nonlinear behaviors cannot be modeled adequately using few principal components, and PARAFAC does not yield physically interpretable results if there are significant interactions between the factors. If these conditions are not met, it may be necessary to use a different model (Smilde et al., 2004). The appropriateness of the used model can be diagnosed by looking at the magnitude and systematicity of residuals which contain the variation not described by the model or at the scores and loadings. In general, if a multivariate model does not describe data sufficiently well or fails to identify systematic chemical information, the interpretation can lead to inadequate and even misleading conclusions in, for example, environmental forensic investigations. In this sense, in order to soundly establish the correctness of one's intuition, it may be necessary to validate the results based on *a priori* knowledge as well as testing the results from a statistical point of view (cf. Section 9.5.2). In the next two sections, PCA and the PARAFAC decomposition are outlined, as they have been used in IMOF.

9.4.1 Multilinear Models

9.4.1.1 Two-Way Case

A broad range of bilinear models exists for matrices (i.e., two-way datasets comprised of samples × variables) for both decomposition and calibration methods. The most common in chemometrics are PCA, principal component regression (PCR), partial least-squares regression (PLSR), and multivariate curve resolution (MCR) (Jolliffe, 1986; Wold et al., 1987; Martens and Næs, 1996; Smilde et al., 2004). In bilinear models, the basic assumption is that the systematic variation in data matrix $\mathbf{X}$ of dimensions $n \times p$ can be described by the product of two matrices of size $n \times F$ and $p \times F$ (Martens and Næs, 1996; Smilde et al., 2004). The different models then vary in how these two matrices are determined. In PCA, $\mathbf{X}$ is decomposed in the product of a score matrix $\mathbf{T}$ with a loading matrix $\mathbf{P}$ (with elements t_{if} and p_{jf}) that are column-wise orthogonal plus a matrix of residuals, $\mathbf{E}$ with elements e_{ij} (Bro, 1997) [see Eq. (9-5)]:

$$x_{ij} = \sum_{f=1}^{F} t_{if} p_{jf} + e_{ij} \quad (i = 1\ldots I; j = 1\ldots J)$$
$$\mathbf{X} = \mathbf{TP}^T + \mathbf{E}$$
(9-5)

where x_{ij} is the data point for the i-th sample and j-th variable, F is the number of principal components (PC's) and T identifies transposition. F is also referred to as the mathematical rank of the model, and the minimum value of F for which the residuals are zero is the mathematical rank of the $\mathbf{X}$. Moreover, the PCs are ordered in $\mathbf{T}$ and $\mathbf{P}$ in decreasing order of variance, and the columns of $\mathbf{P}$ have a norm equal to 1. Hence, the first PC is the most relevant source of variation, the second PC is the second-most relevant source of variation, and so forth, and for any given value of F the sum of the square of the residuals is minimized. All these conditions on the scores and loading matrices imply that the solution is uniquely determined. However, for F larger than 1, there is an infinite number of solutions that describe the data equally well (i.e., with the same value of the sum of squared residuals) that can be obtained from appropriate linear combinations of the columns of $\mathbf{T}$ and $\mathbf{P}$ (Bro, 1997; Riu and Bro, 2003). This is commonly referred to as "rotational freedom" and implies that there is no guarantee that the PCs also resemble the original factors that underlie the data. A clarifying example is provided by spectral data. If there are F spectrally active chemical species in a sufficiently diluted solution and the absorption at p wavelengths is measured for n samples, the physical model for the observed data is of the type described by Eq. (9-5), where t_{if} would be proportional to the concentration of the fth species in the ith sample, and p_{jf} would be proportional to the extinction coefficient at the jth recorded wavelength. Under these conditions, each column of the matrix $\mathbf{P}$ contains the absorption spectrum of one of the chemical species present in the solution and the columns of $\mathbf{T}$ of the corresponding concentrations. However, these spectra cannot be orthogonal, because orthogonality also implies that a varying number of elements in $\mathbf{P}$ are negative. In fact, MCR methods seek a rotation of the PCs such that all the columns of $\mathbf{P}$ are nonnegative.[1]

The PCs are themselves linear combinations of the original variables and can be obtained through a variety of methods. The most reliable from a numerical point of view is the singular value decomposition (SVD). The main problem with this method is that in its standard form it cannot handle missing values, which are a rather common occurrence. When these are present, other algorithms based on, for example, expectation maximization or weighted least squares, should be used (Grung

---

[1] Note that even this condition is generally not sufficient for uniqueness (Smilde et al., 2004).

and Manne, 1998; Walczak and Massart, 2001).

A final note regards the application of PCA (and other bilinear models) to higher-order tensors, which requires that the elements of the data array be rearranged in a matrix. However, while this so-called matricization is always possible, it also leads to a dramatic increase of the number of parameters compared to models such as PARAFAC that exploit the additional structural relationships that may be present in the dataset. In particular, for an $n \times p \times q$ array matriced to an $n \times pq$ matrix, each PC entails the estimation of $n + pq$ parameters as opposed to, for example, PARAFAC, which requires only $n + p + q$.

9.4.1.2 Higher-Order Arrays

As mentioned in the previous paragraph, bilinear models may require the estimation of an exceeding number of parameters (and likely incur overfitting) when applied to higher-order arrays (tensors) whose elements have been rearranged in a matrix. There are several extensions to the multiway case of PCA and of the other bilinear models (e.g., PARAFAC, Tucker, and nPLS). However, not all the properties of bilinear models are retained when they are extended to higher orders, and new ones may emerge. Hence, a bit of care is required to determine which is the best choice for a given problem. For example, Tucker and PARAFAC models can both be considered extensions of PCA, but have considerably different properties (Smilde et al., 2004). Due to space restrictions, only PARAFAC will be described in more detail. We refer the reader to the cited literature for additional information about Tucker and nPLS models and their relation with PARAFAC.

PARAFAC is perhaps the most straightforward extension of bilinear models and the PARAFAC equation for each element of the data array is obtained by simply adding a term for each additional order after the second. Thus, for a three-way array $\underline{\mathbf{X}}$ of dimensions $I \times J \times K$ (e.g., obtained by stacking on top of each other $J \times K$ EEM fluorescence measurements relative to I samples), the equation for PARAFAC is:

$$x_{ijk} = \sum_{f=1}^{F} a_{if} b_{jf} c_{kf} + e_{ijk}$$
$$(i = 1 \ldots I; j = 1 \ldots J; k = 1 \ldots K)$$
$$\underline{\mathbf{X}} = \mathbf{A} \circ \mathbf{B} \circ \mathbf{C} + \underline{\mathbf{E}}$$
(9-6)

where x_{ijk} and e_{ijk}, respectively, identify array elements and the corresponding residuals, and ∘ symbolizes the tensor product (Smilde et al., 2004). In order to highlight the symmetry in the role of these three matrices, they are oftentimes referred to solely as loading matrices (Bro, 1997; Thygesen et al., 2004). The appellation of scores is sometimes reserved to the mode relative to the sample being analyzed.

Unlike PCA, $\mathbf{A}$, $\mathbf{B}$, and $\mathbf{C}$ need not be columnwise orthogonal for the model to be identifiable under rather mild conditions (e.g., that no two columns of any of the loading matrices be proportional) up to trivial factor permutations and scaling (Smilde et al., 2004; Thygesen et al., 2004). The fact that orthogonality is not necessary to uniquely determine the model parameters has important consequences. The most important is that, provided that a minimum in the loss function is found and the rank of the model is correct, the model parameters can have, for several chemical data, a straightforward physical interpretation (Leurgans and Ross, 1992). For EEM fluorescence measurements of sufficiently diluted solutions, and if I is number of samples, J the number of emission wavelengths, and K the number of excitation wavelengths, the elements of the fth column of the first loading matrix are proportional to the concentration of one of the fluorescent species in the solution and the fth columns of $\mathbf{B}$ and $\mathbf{C}$ contain the emission and excitation spectrum, respectively, of such fluorophores (Leurgans and Ross, 1992; Andersen and Bro, 2003).

This identification between the PARAFAC model and the physical model, however, is subject to a number of restrictions, the first of which is that there are no observable interactions between the chemical species that are modeled by separate factors, and that the profile relative to such species in one mode does not change as the variable in the other modes changes. Thus, for EEM data, it is necessary that little or no quenching occurs and that emission and excitation spectra for the same fluorophore are identical in all samples. However, experimental data often deviate from these assumptions, which leads to inaccurate estimates for the physical model parameters. In some cases, it may be useful to enforce some constraints when they are known to apply (e.g., nonnegativity and unimodality on one or more of the loading vectors) (Bro, 1997; Stedmon et al., 2003; Andersen and Bro, 2003).

It is worth mentioning that, while constraints may improve the robustness of the modeling and are in accordance with chemical *a priori* knowledge (i.e., negative concentrations and fluorescence intensities have no physical meaning), they are by no means necessary to yield an interpretable solution provided that the deviations from the basic multilinearity assumptions are not too large. On the contrary, such artifacts may be quite informative on the presence of outliers and on the effect of the pattern of missing values. A detailed introduction to PARAFAC and examples of its applications to analysis of fluorescence EEMs have recently been published in relation to oil hydrocarbon fingerprinting (Christensen et al., 2005b).

Finally, as of this publication, there is no algorithm of fixed complexity that can be used to fit PARAFAC to three-way data that contain noise and, for particularly difficult problems, it may not be possible to retrieve a meaningful solution (Tomasi and Bro, 2006). One of the problems that adds to the difficulty of fitting a PARAFAC model is that PARAFAC models of increasing rank are not nested. That is, the first component in a rank-2 PARAFAC model is not equal to the only component of a rank-1 PARAFAC model. Hence, it is not possible, as it is in PCA, to fit a model with a large number of factors and then select the correct rank. Determination of the appropriate number of factors has mainly been based on split-half analysis, analysis of residuals (Bro, 1997, 1998; Andersen and Bro, 2003; Christensen et al., 2005b), and comparison of PARAFAC factors with EEMs of individual PAHs (Christensen et al., 2005b).

9.4.2 Variable Selection and Scaling

Although data are preprocessed prior to the multivariate statistical data analysis (Figures 9-4 and 9-5), some variables (e.g., chromatographic intensities, diagnostic ratios, or excitation–emission pairs) are more informative than others. There are at least two possible solutions to this problem. One is to select the variables that should be retained in the data array and exclude the noisiest, less informative ones; a second one is to downscale these variables or single measurements using a weighted least-squares fitting criterion (WLS). Both approaches have been tested as parts of IMOF (Christensen et al., 2004, 2005d).

The resolution power (r), intended as the ability of distinguishing dissimilar samples while keeping identical ones as close as possible to one another in the space defined by the principal components (score plots), has been used in IMOF as a criterion for optimizing variable selection and scaling (Christensen et al., 2005a, 2005d). Ideally, an optimal resolution power is obtained for the combination of variables with lowest variation over replicate samples compared to the total variance explained by the model (i.e., minimal r).

One option used in IMOF is to define r as the ratio between the variance of the PCA scores of replicate samples (or of samples that belong to a certain class) and the overall variance explained by the same model and to progressively include additional variables according to their analytical uncertainty (Christensen et al., 2005d). This method was applied to a section of m/z 217 chromatograms (i.e., associated to tri- and tetracyclic sterane biomarkers) and allowed to reduce the number of variables from 1251 to 351 with a twofold improvement on the resolution power for samples that had not been included in the calibration set. A somewhat more statistical treatment of the subject can be found in Pierce et al. (2005), in which ANOVA is used for feature selection.

Another option for selecting diagnostic ratios for oil hydrocarbon fingerprinting was suggested in Christensen et al. (2004) using the concept of diagnostic power (DP). The selection of diagnostic PAH and biomarker ratios for oil hydrocarbon fingerprinting was based on the analytical precision of individual ratios as well as their resistances to weathering (Christensen et al., 2004). The DP can be calculated from Eq. (9-7), and ratios with lowest DP can be excluded from the data analysis to reduce the analytical noise and sampling variability that can have negative effects on the spill source identification.

$$DP = \frac{RSD_V}{RSD_A} \quad \text{or} \quad DP = \frac{RSD_V}{RSD_S} \quad (9\text{-}7)$$

where the relative sampling standard deviation (RSD_S) is the combined random errors from the chemical analysis as well as the sample variation (relative sampling standard deviation). The latter includes the sample heterogeneity and the effect of weathering. DP is defined as the relative standard deviation of a diagnostic ratio in oils with different origin (RSD_V) divided by RSD_A or RSD_S. The variable selection method is based on the diagnostic powers used in Christensen et al. (2004) to select the most descriptive biomarker ratios for oil hydrocarbon fingerprinting. RSD_S, RSD_V, and DP were calculated for 17 diagnostic ratios and sorted with respect to their DP (Table 9-2). A comparable approach was suggested by Stout et al. (2001), who excluded ratios with RSD_A larger than 5% from the PCA.

Variable selection can be a rather time-consuming process that may introduce some degree of subjectivity in the data analysis. Therefore, a second method of improving the performance of PCA has been investigated in which all the variables are retained but are inversely scaled according to their analytical

Table 9-2 Application of a Variable-Outlier Detection Method to 17 Biomarker Ratios

Diagnostic Ratios	RSD_S	RSD_V	DP
C24TT/ (C24TT + 30ab)	1.5	37.7	25.7
25nor30ab / (25nor30ab + 29ab + 30ab)	2.2	43.3	20.0
25nor30ab / (25nor30ab + Ts+Tm)	1.9	33.5	17.8
C24TT / (C24TT + 29ab + 30ab)	2.2	34.6	15.9
(29ab+30ab)/(29ab+30ab + 27bbSt(S+R) + 28bbSt(S+R))	0.4	4.8	11.8
30G / (30G+30ab)	1.7	19.3	11.6
Ts / (Ts + Tm)	1.2	9.7	10.5
RC27TA / (RC27TA+ RC28TA)	1.0	8.3	8.5
32abS / (32abS + 32abR)	0.3	2.7	8.0
29ab / (29ab + 30ab)	1.5	10.0	6.9
29bb(R+S) / (29bb(R+S) + 29aa(R+S))	1.0	6.9	6.9
27db(R+S) / (27db(R+S) + 27bb(R+S))	2.7	15.3	5.7
29aaS / (29aaS + 29aaR)	1.7	8.4	5.0
28ab / (28ab+30ab)	6.7	29.0	4.3
29bbSt(S+R) / (27bbSt(S+R) + 28bbSt(S+R) + 29bbSt(S+R))	1.6	6.3	3.9
27bbSt(S+R) / (27bbSt(S+R) + 28bbSt(S+R) + 29bbSt(S+R))	1.5	3.5	2.4
28bbSt(S+R) / (27bbSt(S+R) + 28bbSt(S+R) + 29bbSt(S+R))	1.8	4.0	2.2

The values have been estimated from 24 oil spill samples from the *Baltic Carrier* oil spill (Christensen et al., 2004). The DPs were calculated from RSD_S and RSD_V using Eq. (9-7), whereas the diagnostic ratios were double normalized to the reference oil analyzed closest in time using Eq. (9-4). The normalization factor in Eq. (9-4) is omitted in the description for brevity. The individual compound abbreviations have been defined in the supporting information of Christensen et al. (2004).

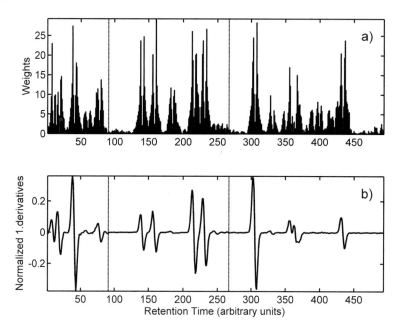

Figure 9-6 (a) Weights (RSD_A^{-1}) used for WLS-PCA and (b) the aligned and normalized mean combined chromatogram of unweathered Brent crude oil. The mean was calculated from four replicate samples. Modified from Christensen et al. (2005a).

(or sampling) uncertainty (Christensen et al., 2005a, 2005d). For example, three partial GC-MS/SIM chromatograms, C_1-fluorenes (m/z 180), C_1-phenanthrenes (m/z 192), and C_1-dibenzothiophenes (m/z 198), were combined and analyzed in Christensen et al. (2005a) to evaluate the effects of microbial degradation of oil in an *in vitro* experiment. The inverses of RSD_{AS} (weights) were calculated from the aligned and normalized reference set consisting of 33 replicate references, using the optimal warping parameters and the complete chromatographic sections for normalization (90, 175, and 225 data points). The weights are shown in Figure 9-6a and were used in the WLS-PCA to scale the importance of chromatographic intensities with respect to their analytical uncertainty. The first derivative of the aligned and normalized mean combined chromatogram of the unweathered Brent crude oil, used in Christensen et al. (2005a), are, for comparison, shown in Figure 9-6b. The peak and noise regions have, respectively, high and low weights. Thus, the importances of peak regions are high (but not the peak maxima, which are characterized by a higher uncertainty), compared to those of noise regions during fitting of the PCA model.

In Christensen et al. (2004, 2005a, 2005d), the WLS-PCA model gave comparable resolution power as the optimal variable selection followed by PCA. Hence, WLS-PCA was found to be a more attractive method, since it is highly objective and no data are excluded from the analysis. Yet it is important to acknowledge the fact that the weights should describe intrinsic properties of the dataset (e.g., analytical uncertainty). If this is not the case, WLS-PCA will result in a model with lower resolution power than PCA without variable selection and even to bias in the data. Algorithms for weighted PCA can be downloaded from www.models.kvl.dk and http://www-its.chem.uva.nl/research/.

9.5 Data Evaluation

A large number of methods can be used to evaluate the outputs from multivariate statistical analyses and to classify and match suspected source oils and spill samples in oil hydrocarbon fingerprinting. The methods described in the following sections are all based on the scores and loadings from multilinear models (again PCA and PARAFAC).

9.5.1 Visual Inspection of Score and Loading Plots

The most straightforward and intuitive method for evaluating a multilinear model is through score and loading plots, which may reveal relations between multiple samples and variables, respectively. The importance of visual inspection and comparison of score and loading plots can be illustrated from the correlations observed between scores and loadings in a recent oil hydrocarbon fingerprinting study, where preprocessed sections of GC-MS/SIM chromatograms of m/z 217 (i.e., tri- and tetracyclic steranes) for 101 oil samples were analyzed by WLS-PCA (Christensen et al., 2005d). The sample set used in this study was divided into a calibration set of 61 chromatograms from 51 source oils and 10 replicate samples (61 × 1231), a reference set containing 18 replicate reference samples (18 × 1231), and a test set comprised of 16 weathered oil samples collected in the spill area after the *Baltic Carrier* oil spill (Christensen et al., 2005d) and two spilled oils (analyzed in triplicate) from the study of Faksness et al. (2002). WLS-PCA was applied to the mean-centered calibration set, and the number of significant principal components was found by visual inspection of the chromatographic loadings, and confirmed by evaluating the variability of replicate samples (Christensen et al., 2005d). The visual inspection of loadings showed that some peaks in PC5 and subsequent principal components described the residual misalignment rather than systematic changes in chemical composition. Residual shifts show up in the cumulative sum of loadings as first derivative peaks. As an increasing number of principal components is extracted from data, the ratio between the systematic information and the variations caused by insufficient alignment increases,

until the latter becomes the most pronounced and the subsequent components describe residual shifts and additional instrumental noise.

It can be concluded from visual inspection and evaluation of the PC1 to PC4 scores in this study (e.g., Figure 9-7) that the *Baltic Carrier* oil spill samples and the corresponding source oil were clustered in PC1 through 4, despite weathering for up to 14 days. Likewise, two Round-Robin spill samples (Faksness et al., 2002), Spill I and Spill II, were grouped along the principal components with the corresponding sources, Oseberg East (E), and Oseberg Field Centre (FC). It is important to remark that, while principal components are ordered according to their explained variance and the differences along the first components are in general more significant than along subsequent ones, later components are not necessarily of lesser importance for the separation of dissimilar oil samples. Especially for large datasets that may require a larger number of components to be properly described, an important separation between similar samples is often found in later components. Note, for example, that although oil samples from Oseberg South East (SE) and Oseberg FC are closely related, and the samples could not be separated along the three most significant components (i.e., PC1 to PC3) (Christensen et al., 2005d), they were easily separated along PC4 (see Figure 9-7), the latter describing only 6.27% of the total variation in the dataset.

Interpretation of the correlations between samples in score plots, by visual inspection of the loading plots, further facilitates comparisons of source oils and spill samples. Loading plots show which original variables (e.g., retention times and diagnostic ratios) are responsible for the directions, changes, and groupings observed in the corresponding score plots, and their importance for oil hydrocarbon fingerprinting can be illustrated from Figure 9-8, which shows the cumulative sum of the fourth principal component, found to distinguish two closely related North Sea crude oils.

The PC4 loadings in Figure 9-8 are negative for ββ-isomers of C_{27} to C_{29}-rearranged steranes and positive for αα-isomers. The ratio of

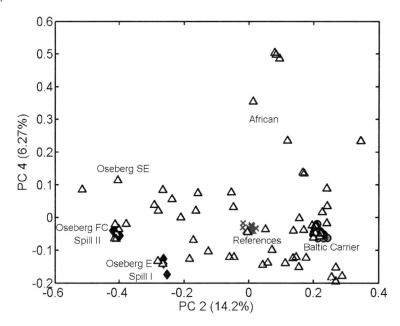

Figure 9-7 Score plot (PV2 versus PC4) using WLS-PCA on sections of preprocessed partial GC-MS/SIM chromatograms of m/z 217. The PCA model was calculated from the calibration set (61 × 1231), whereas a reference set (18 × 1231) of replicate references and test set (22 × 1231) were calculated by projecting the data onto the loadings. The test set was comprised of 16 *Baltic Carrier* oil spill samples and two spill samples from a Round-Robin exercise analyzed in triplicate (Spill I and Spill II) (Faksness et al., 2002).

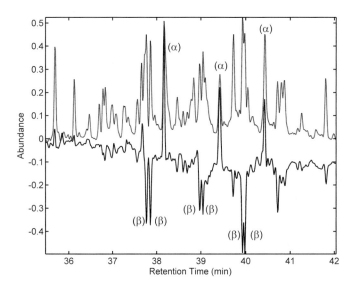

Figure 9-8 Integrated mean-chromatogram (dotted line) and integrated PC4 loadings (solid line) for WLS-PCA in the oil hydrocarbon fingerprinting study in Christensen et al. (2005d). The αα-steranes (αα) and ββ-steranes (ββ) are marked in the plot.

C_{29}-rearranged steranes (ββ / (ββ + αα)) is a highly specific parameter for maturity and appears to be independent of source organic matter input (Peters and Moldowan, 1993). The ββ isomers have a higher thermal stability than αα isomers; thus, the above ratio increases with thermal maturity. Since, the PC4 loadings are negative for ββ isomers and positive for αα isomers, the ββ / (ββ + αα) ratios are highest in oils located at low PC4 in the score plot (Figure 9-7). Thus, the main separation of oils from Oseberg SE and Oseberg FC can be explained by a higher thermal maturity of the oil from Oseberg FC. Likewise, oils from Oseberg E are slightly more mature than those from Oseberg FC, and the African crude oils studied have positive PC4 scores and thus have low relative maturity with respect to the oils in the calibration set.

The loadings of PC1, PC2, and PC3 (not shown) can be used to facilitate the correlation and differentiation of source oils and spill samples along PC1 to 3 in the same way as PC4 was used. The three components describe boiling point range, clay content of source rock, and carbon number distribution of sterols in the organic matter of the source rock, respectively (Christensen et al., 2005d). Thus, all four significant principal compounds in this study could be interpreted directly from *a priori* knowledge of the effects of source rock depositional environment and thermal maturity and the refining process on the sterane composition seen as variations in the chromatographic profile of m/z 217 (corresponding to the composition of tri- and tetracyclic steranes) (Christensen et al., 2005d).

When closely related source oils are present in a large dataset, it is likely that major trends, which are represented by the first PCs, mask the differences between these oils. The components that describe these minor differences may not be included in the optimal PCA model, as they represent a minimal variation as compared to the total (Jolliffe, 1986). To ensure that important but small variations are not masked by major trends, PCA can be applied to a subset of source oils that lie close to the spilled oil along the components retained in the original PCA model. In Christensen et al. (2004), local PCA models (i.e., modeling a subset of closely located samples) have been used to focus the data analysis on separating related samples with similar biomarker and PAH composition. Hence, the first few components describe variations relevant for the specific oil spill case by separating related samples in the first few PCs instead of in higher PCs. The latter are more affected by noise since they describe only a small percentage of the total variation in the dataset.

In Christensen et al. (2005b), normalized PARAFAC scores are used to characterize and match oil samples based on their relative PAH composition. The four oil types (crude oils, HFOs, LFOs, and lubricants) could, except for some overlap of HFOs and crude oils, be distinguished from two score plots where Factor 3 versus Factor 5 is shown in Figure 9-9. The evaluation showed that crude oils have the most regular distribution of factors and a large variability in the content of high-molecular-weight PAHs. In contrast, LFOs and lubricants have a high relative content of low-molecular-weight PAHs, whereas HFOs have a high relative content of high-molecular-weight compounds.

Since PARAFAC factors are essentially unique up to trivial permutation and scaling, if, upon validation [e.g., through split-half analysis and cross-validation (Smilde et al., 2004)], they appear stable and reliable, they can be expected to reflect some underlying phenomena that give rise to the data. In the case of EEM fluorescence, these phenomena can most often be assimilated to emission and excitation spectra of homogeneous classes of compounds (or even single fluorophores). In Christensen et al. (2005b), the PARAFAC factors were interpreted by comparing excitation and emission loadings with EEMs of selected PAHs and the fluorescence characteristics for a broad range of PAHs (Figure 9-10). The graphical comparison of PARAFAC loadings with the spectra of specific fluorescent compounds was sufficient to support the original hypothesis on the meaning of the factors and allowed to associate them to mixtures of PAHs with similar fluorescence characteristics: a mixture of naphthalenes and dibenzothiophenes, fluorenes, phenanthrenes, chrysenes, and five-ring PAHs.

9.5.2 Numerical Comparisons and Statistical Tests

The objectivity of the matching process of spill and source oil samples can be improved by

318 Oil Spill Environmental Forensics

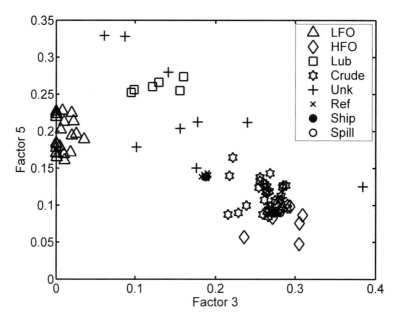

Figure 9-9 PARAFAC score plot. Factor 3 versus Factor 5. Symbols for LFOs, HFOs, lubes, crude oils, unknown oil samples, replicate reference oils, and the triplicate spill and ship samples are explained in the legend. The 17 replicate references are circled, and arrows mark the position of spill and *Baltic Carrier* ship samples.

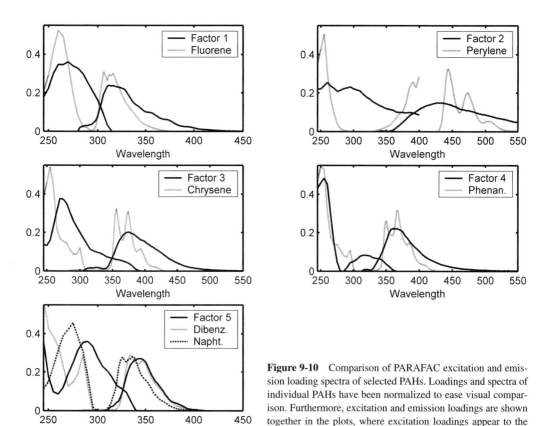

Figure 9-10 Comparison of PARAFAC excitation and emission loading spectra of selected PAHs. Loadings and spectra of individual PAHs have been normalized to ease visual comparison. Furthermore, excitation and emission loadings are shown together in the plots, where excitation loadings appear to the left of the corresponding emission loadings.

numerical comparisons (e.g., correlation coefficient) or statistical tests. Using the former approach, the similarity of oil samples can be calculated from the scores (Christensen et al., 2005b) or by point-to-point matching (Li et al., 2004).

In Christensen et al. (2005b), the similarity of oil samples is calculated using the correlation coefficient based on the similarity of normalized scores. More specifically, an oil sample collected in the spill area two weeks after the *Baltic Carrier* spill accident was compared to oil samples in the database. It was found that the triplicate samples from the cargo tank of the *Baltic Carrier* and three additional HFOs gave the highest match to spill samples ($r = 0.998 - 0.999$). Comparisons based on PCA scores could also have been used in Christensen et al. (2005) for objective spill/source matching of preprocessed chromatographic sections of biomarker hydrocarbons.

An even more objective method for matching oil samples is applied in Christensen et al. (2004) and is as such a highly objective alternative to visual inspection of score and loading plots as well as the use of similarity indices. The method consists of statistical evaluation based on the overall null hypothesis (H_0) that the spilled oil and the tested source oil are identical. The optimal number of principal components in a multivariate model (i.e., the retained components or factors) can be tested independently accepting a certain error level (often 5%, $\alpha = 0.05$). The method is used in Christensen et al. (2004) by independently testing the significant principal components using the inequality in Eq. (9-8) and accepting an error level of 5%, $\alpha = 0.05$. If the inequality is false in at least one of these tests, the overall H_0 is rejected and the tested source oil is "beyond reasonable doubt" not the source of the spill.

$$\frac{\left|\bar{t}_k^{(\text{spill})} - \bar{t}_k^{(\text{source})}\right|}{s_k^{(\text{pooled})}\sqrt{\frac{1}{n_{\text{spill}}} + \frac{1}{n_{\text{source}}}}} \leq q_{\alpha,\text{d.f.}} \quad (9\text{-}8)$$

where $s_k^{(\text{pooled})}$ is the pooled standard deviation, which can be calculated from either the analytical or the sampling standard deviation. In addition, n_{spill} and n_{source} are the number of replicates used to calculate the mean scores along the k-th principal component of the spilled oil ($\bar{t}_k^{(\text{spill})}$) and a source oil ($\bar{t}_k^{(\text{source})}$). $q_{\alpha,\text{d.f.}}$ is the α-quantile from t-student's distribution with d.f. degrees of freedom and $s_k^{(\text{pooled})}$ the pooled standard deviation for the k-th principal component.

There can be several possible outcomes of the classification of source oil with respect to the spilled oil in the multiple tests. Here are the criteria used in Christensen et al. (2004). However, other criteria can be used, such as those suggested in the modified Nordtest methodology (Daling et al., 2002).

Positive match: H_0 is acceptable (5% error level) for the tested source oil and the spill sample, and H_0 is rejected for all other source oils in the dataset.

Probable match: H_0 are acceptable for the tested source oil and spill sample, but the same holds for other source oils.

Nonmatch: H_0 is rejected.

The match criteria just described are based entirely on statistics and require a consistently high data quality. It is important to emphasize that caution against an overreliance on such unsupervised classification based on purely statistical criteria needs to be taken, and any matches need to be evaluated in light of other available data (e.g., additional compound groups).

9.6 Conclusions and Perspectives

Rapid, reliable, and objective tools are a requirement for the characterization of complex chemical mixtures such as oil. This chapter describes the development of such tools for oil hydrocarbon fingerprinting and spill source identification. An integrated multivariate oil hydrocarbon fingerprinting (IMOF) methodology comprised of four steps is described throughout the chapter: sample preparation and chemical analysis, data preprocessing, multivariate statistical analysis, and data evaluation. Figure 9-11 gives a schematic presentation of the specific oil hydrocarbon fingerprinting methods devel-

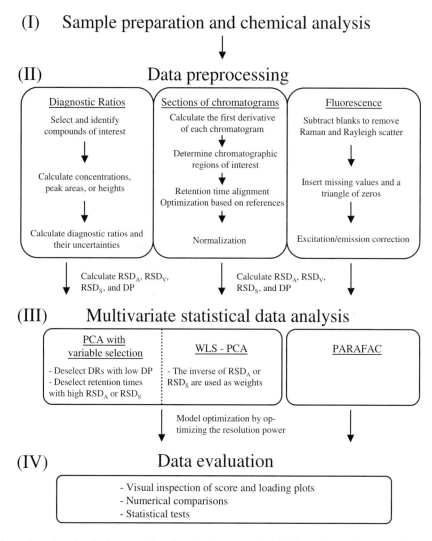

Figure 9-11 Flowchart for the IMOF methodology including the individual oil hydrocarbon fingerprinting methods developed by Christensen et al. (2004, 2005b, 2005c, 2005d).

oped in our research group and based on the IMOF framework.

The chemical analyses were based on fluorescence spectroscopy and GC-MS/SIM. The fluorescence spectroscopy method facilitated a rapid (less than 10 min per sample) screening and characterization of oil samples based on their main composition of PAHs. The technique provides a rapid and alternative, yet complementary (revealing PAH groups), technique to the more traditional GC-FID screening (revealing mainly paraffins). GC-MS/SIM has been used for more comprehensive compound-specific analysis. The GC-MS fingerprinting technique was based on a semiquantitative approach including frequent analysis of a reference sample. The semiquantitative approach, used to calculate diagnostic ratios, was rapid and with similar precision as a fully quantitative approach based on extensive use of internal and quantification standards. The replicate analysis of reference samples could furthermore be used for QA/QC, uncertainty estimations, automating the preprocessing, and external normalization. The semiquantitative approach based on GC-

MS/SIM constituted an important part of the IMOF methodology.

The preprocessing tools comprised semi-automated peak identification and quantification (Christensen et al., 2004, 2005c), analysis of sections of chromatograms by baseline removal, time warping and normalization (Christensen et al., 2005a, 2005d), and automated preprocessing of fluorescence EEMs (Christensen et al., 2005b). Time warping combined with PCA is a rapid and objective approach for oil hydrocarbon analysis compared to peak identification and quantification. The results of the research in our laboratory show, however, that the time warping approach is more affected by changes in the data quality than is peak quantification. Hence, the use of time warping in routine investigations requires extensive QA/QC measures to be taken. Furthermore, the ratio between the inherent variability in the dataset and the variability due to insufficient alignment are important criteria to determine whether or not the method provides an appropriate preprocessing tool for oil analysis.

The use of multivariate statistical methods such as PCA and PARAFAC are the cornerstone of the IMOF methodology. The multivariate methods enable the analysis and assessment of large datasets by extracting a number of principal components or factors that describe the prominent trends in data. A refined and more objective data analysis was obtained by WLS-PCA compared to PCA with variable selection. Yet, the weights should describe intrinsic properties of the dataset such as the analytical uncertainty. If this is not the case, the subjectivity of the data analysis increases, and the model becomes increasingly biased.

Rapid and objective oil hydrocarbon fingerprinting was attained using several evaluation techniques. Although excellent for monitoring and assessing the fate of complex oil hydrocarbon mixtures, visual interpretation of score and loading plots are often insufficient for proper analysis of chemical fingerprinting data. More objective methods for defensibly linking spilled oil with possible sources in an oil database are also presented in this chapter.

The analytical and sampling uncertainties have been used in Christensen et al. (2004) to test the null hypothesis and determine the source of spill samples. Thus, the conclusions are less dependent on data variations and subjective decisions. In summary, each of the four steps in the IMOF methodology contributes to rapid, objective, and comprehensive analyses of complex oil hydrocarbon mixtures.

The methods developed in our research group and described in this chapter can be employed in routine investigations, and they appear as significant improvements compared to standard qualitative and quantitative methods for oil hydrocarbon fingerprinting. The limited human intervention required — and the extended amounts of chemical information that can be generated, analyzed, and evaluated — are the major and obvious strengths of the IMOF methodology. More specifically, the methods described in Christensen et al. (2004, 2005b, 2005d) enable a more comprehensive and objective matching of oil samples than the standard methods, especially if the spilled oils have chemical characteristics related to several suspected source candidates. Furthermore, a sample database can be built over time that allows each new spill to be compared to the ever-growing database without the need for reanalysis or reprocessing of existing data. These methods can easily be implemented and used for routine investigations in forensic oil spill laboratories. Fluorescence spectroscopy combined with PARAFAC can be used for prescreening oil samples (Christensen et al., 2005b), while GC-MS/SIM combined with fast and objective preprocessing, data analysis, and data evaluation (Christensen et al., 2004, 2005d) can be used for compound-specific fingerprinting.

Acknowledgments

The authors acknowledge the co-authors of a recent oil hydrocarbon fingerprinting article as well as Lotte Frederiksen, Jørgen Avnskjold, and Peter Christensen for technical assistance. The work with developing the IMOF technology was financed by Roskilde University, the

National Environmental Research Institute, the Natural Sciences Research Foundation, all from Denmark, and the European Commission (contracts "BIOSTIMUL," QLRT-1999-00326, and "ALARM," GOCE-CT-2003–506675).

References

Åberg, K.M., R.J.O. Torgrip, and S.P. Jacobsson, Extensions to peak alignment using reduced set mapping. Classification of LC/UV data from peptide mapping. *J. Chemometrics*, 2004, **18**, 465–473.

Aboul-Kassim, T.A.T. and B.R.T. Simoneit, Aliphatic and aromatic-hydrocarbons in particulate fallout of Alexandria, Egypt — sources and implications. *Environmental Science & Tech.*, 1995a, **29**(10), 2473–2483.

Aboul-Kassim, T.A.T. and B.R.T. Simoneit, Petroleum hydrocarbon fingerprinting and sediment transport assessed by molecular biomarker and multivariate statistical-analyses in the Eastern harbor of Alexandria, Egypt. *Marine Pollution Bull.*, 1995b, **30**(1), 63–73.

Andersen, C.M. and R. Bro, Practical aspects of PARAFAC modeling of fluorescence excitation-emission data. *J. Chemometrics*, 2003, **17**(4), 200–215.

Andersson, C.A. and R. Bro, The N-way toolbox for MATLAB. *Chemometrics and Intelligent Laboratory Systems*, 2000, **52**(1), 1–4.

Andersson, F.O., R. Kaiser, and S.P. Jacobsson, Data preprocessing by wavelets and genetic algorithms for enhanced multivariate analysis of LC peptide mapping. *J. Pharmaceutical and Biomedical Analysis*, 2004, **34**(3), 531–541.

Bandh, C., E. Bjorklund, L. Mathiasson, C. Naf, and Y. Zebuhr, Comparison of accelerated solvent extraction and soxhlet extraction for the determination of PCBs in Baltic Sea sediments. *Environmental Science & Tech.*, 2000, **34**(23), 4995–5000.

Barron, M.G. and E. Holder, Are exposure and ecological risks of PAHs underestimated at petroleum contaminated sites? *Human and Ecological Risk Assessment*, 2003, **9**(6), 1533–1545.

Boehm, P.D., G.S. Douglas, W.A. Burns, P.J. Mankiewicz, D.S. Page, and A.E. Bence, Application of petroleum hydrocarbon chemical fingerprinting and allocation techniques after the *Exxon Valdez* oil spill. *Marine Pollution Bull.*, 1997, **34**(8), 599–613.

Bro, R., PARAFAC. Tutorial and applications. *Chemometrics and Intelligent Laboratory Systems*, 1997, **38**(2), 149–171.

Bro, R., Multi-way Analysis in the Food Industry. Models, Algorithms, and Applications, Ph.D., University of Amsterdam, 1998.

Bro, R., N.D. Sidiropoulos, and A.K. Smilde, Maximum likelihood fitting using ordinary least squares algorithms. *J. Chemometrics*, 2002, **16**(8–10), 387–400.

Burns, W.A., P.J. Mankiewicz, A.E. Bence, D.S. Page, and K.R. Parker, A principal-component and least-squares method for allocating polycyclic aromatic hydrocarbons in sediment to multiple sources. *Environ. Toxicology and Chem.*, 1997, **16**(6), 1119–1131.

Christensen, J.H., Application of multivariate data analysis for assessing the early fate of petrogenic compounds in the marine environment following the *Baltic Carrier* oil spill. *Polycyclic Aromatic Compounds*, 2002, **22**(3–4), 703–714.

Christensen, J.H., A.B. Hansen, G. Tomasi, J. Mortensen, and O. Andersen, Integrated methodology for forensic oil spill identification. *Environmental Science & Tech.*, 2004, **38**(10), 2912–2918.

Christensen, J.H., A.B. Hansen, U. Karlson, J. Mortensen, and O. Andersen, Multivariate statistical methods for evaluating biodegradation of mineral oil. *J. Chromatography A*, 2005a, **1090**(1–2), 133–145.

Christensen, J.H., A.B. Hansen, J. Mortensen, and O. Andersen, Characterization and matching of oil samples using fluorescence spectroscopy and parallel factor analysis. *Analytical Chem.*, 2005b, **77**(7), 2210–2217.

Christensen, J.H., J. Mortensen, A.B. Hansen, and O. Andersen, Chromatographic preprocessing of GC-MS data for analysis of complex chemical mixtures. *J. Chromatography A*, 2005c, **1062**(1), 113–123.

Christensen, J.H., G. Tomasi, and A.B. Hansen, Chemical fingerprinting of petroleum biomarkers using time warping and PCA. *Environ. Sci. Tech.*, 2005d, **39**(1), 255–260.

Daling, P.S., L.G. Faksness, A.B. Hansen, and S.A. Stout, Improved and standardized methodology for oil spill fingerprinting. *Environ. Forensics*, 2002, **3**(3–4), 263–278.

de Juan, A. and R. Tauler, Comparison of three-way resolution methods for non-trilinear chemical data sets. *J. Chemometrics*, 2001, **15**(10), 749–772.

Eilers, P.H.C., Parametric time warping. *Analytical Chem.*, 2004, **76**, 404–411.

Ezra, S., S. Feinstein, I. Pelly, D. Bauman, and I. Miloslavsky, Weathering of fuel oil spill on the east Mediterranean coast, Ashdod, Israel. *Organic Geochem.*, 2000, **31**(12), 1733–1741.

Faksness, L.G., P.S. Daling, and A.B. Hansen, Round Robin study — Oil spill identification. *Environ. Forensics*, 2002, **3**(3–4), 279–291.

Fraga, C.G., B.J. Prazen, and R.E. Synovec, Comprehensive two-dimensional gas chromatography and chemometrics for the high-speed quantitative analysis of aromatic isomers in a jet fuel using the standard addition method and an objective retention time alignment algorithm. *Anal. Chem.*, 2000, **72**(17), 4154–4162.

Grung, B. and R. Manne, Missing values in principal component analysis. *Chemometrics and Intelligent Laboratory Systems*, 1998, **42**, 125–139.

Jassie, L., Microwave technology in the analysis of contamination by petroleum. *Intl. Laboratory News*, 1995, 18.

Johansson, E., S. Wold, and K. Sjodin, Minimizing effects of closure on analytical data. *Anal. Chem.*, 1984, **56**(9), 1685–1688.

Johnson, K.J., B.W. Wright, K.H. Jarman, and R.E. Synovec, High-speed peak matching algorithm for retention time alignment of gas chromatographic data for chemometric analysis. *J. Chromatography A*, 2003, **996**(1–2), 141–155.

Jolliffe, I.T., *Principal Component Analysis*, New York: Springer-Verlag, 1986.

Jovancicevic, B.S., L.Z. Tasic, P.S. Polic, J.M. Nedeljkovic, A.K. Golovko, and D.K. Vitorovic, GC-MS in crude oil correlation studies — effects of biodegradation on sterane and terpane maturation parameters. *J. Serbian Chem. Soc.*, 1996, **61**(9), 817–821.

Lavine, B.K., D. Brzozowski, A.J. Moores, C.E. Davidson, and H.T. Mayfield, Genetic algorithm for fuel spill identification. *Analytica Chimica Acta*, 2001, **437**, 233–246.

Leurgans, S. and R.T. Ross, Multilinear models: Applications in spectroscopy. *Statistical Sci.*, 1992, **7**(3), 289–319.

Li, J.F., S. Fuller, J. Cattle, C.P. Way, and D.B. Hibbert, Matching fluorescence spectra of oil spills with spectra from suspect sources. *Analytica Chimica Acta*, 2004, **514**(1), 51–56.

Malmquist, G. and R. Danielsson, Alignment of chromatographic profiles for principal component analysis — a prerequisite for fingerprinting methods. *J. Chromatography A*, 1994, **687**(1), 71–88.

Martens, H. and T. Næs, *Multivariate Calibration*. Chichester, UK: John Wiley & Sons, 1996.

Mudge, S.M., Reassessment of the hydrocarbons in Prince William Sound and the Gulf of Alaska: Identifying the source using partial least-squares. *Environ. Sci. & Tech.*, 2002, **36**(11), 2354–2360.

Munoz, D., P. Doumenq, M. Guiliano, F. Jacquot, P. Scherrer, and G. Mille, New approach to study of spilled crude oils using high resolution GC-MS (SIM) and metastable reaction monitoring GC-MS-MS. *Talanta*, 1997, **45**(1), 1–12.

Nielsen, N.P.V., J.M. Carstensen, and J. Smedsgaard, Aligning of single and multiple wavelength chromatographic profiles for chemometric data analysis using correlation optimised warping. *J. Chromatography A*, 1998, **805**(1–2), 17–35.

Nording, M., S. Sporring, K. Wiberg, E. Bjorklund, and P. Haglund, Monitoring dioxins in food and feedstuffs using accelerated solvent extraction with a novel integrated carbon fractionation cell in combination with a CAFLUX bioassay. *Anal. Bioanal. Chem.*, 2005, **381**(7), 1472–1475.

Øygard, K., O. Grahl-Nielsen, and S. Ulvøen, Oil/oil correlation by aid of chemometrics. *Organic Geochem.*, 1984, **6**, 561–567.

Page, D.S., A.E. Bence, W.A. Burns, P.D. Boehm, J.S. Brown, and G.S. Douglas, A holistic approach to hydrocarbon source allocation in the subtidal sediments of Prince William Sound, Alaska, embayments. *Environ. Forensics*, 2002, **3**(3–4), 331–340.

Peters, K.E. and J.M. Moldowan, *The Biomarker Guide: Interpreting Molecular Fossils in Petroleum and Ancient Sediments*. Englewood Cliffs, NJ: Prentice Hall, 1993.

Pierce, K.M., J.L. Hope, K.J. Johnson, B.W. Wright, and R.E. Synovec, Classification of gasoline data obtained by gas chromatography using a piecewise alignment algorithm combined with feature selection and principal component analysis. *J. Chromatography A*, 2005, **1096**(1–2), 101–110.

Porte, C., X. Biosca, M. Sole, and J. Albaiges, The Aegean Sea oil spill on the Galician Coast (NW Spain). III: The assessment of long-term sublethal effects on mussels. *Biomarkers*, 2000, **5**(6), 436–446.

Pravdova, V., B. Walczak, and D.L. Massart, A comparison of two algorithms for warping of analytical signals. *Analytica Chimica Acta*, 2002, **456**(1), 77–92.

Reddy, C.M. and J.G. Quinn, GC-MS analysis of total petroleum hydrocarbons and polycyclic aromatic hydrocarbons in seawater samples after the North Cape oil spill. *Marine Pollution Bull.*, 1999, **38**(2), 126–135.

Richter, B.E., Extraction of hydrocarbon contamination from soils using accelerated solvent extraction. *J. Chromatography A*, 2000, **874**(2), 217–224.

Rinnan, Å., Application of PARAFAC on spectral data, Ph.D., The Royal Veterinary and Agricultural University, 2004.

Riu, J. and R. Bro, Jack-knife technique for outlier detection and estimation of standard errors in PARAFAC models. *Chemometrics and Intelligent Laboratory Systems*, 2003, **65**(1), 35–49.

Rønn, B.B., Nonparametric maximum likelihood estimation for shifted curves. *J. Royal Stat. Soc., Series B (Statistical Methodology)*, 2001, **63**(2), 243–259.

Shu, Y.Y., R.C. Lao, C.H. Chiu, and R. Turle, Analysis of polycyclic aromatic hydrocarbons in sediment reference materials by microwave-assisted extraction. *Chemosphere*, 2000, **41**(11), 1709–1716.

Siegel, J.A. and N.Z. Cheng, Fluorescence of petroleum-products 4. Three-dimensional fluorescence plots and capillary gas-chromatography of midrange petroleum-products. *J. Forensic Sciences*, 1989, **34**(5), 1128–1155.

Siegel, J.A., J. Fisher, C. Gilna, A. Spadafora, and D. Krupp, Fluorescence of petroleum products 1. Three-dimensional fluorescence plots of motor oils and lubricants. *J. Forensic Sciences*, 1985, **30**(3), 741–759.

Smilde, A.K., R. Bro, and P. Geladi, *Multi-Way Analysis. Applications in the Chemical Sciences*. Chichester, England: John Wiley & Sons Ltd, 2004.

Sporring, S., S. Bowadt, B. Svensmark, and E. Bjorklund, Comprehensive comparison of classic Soxhlet extraction with Soxtec extraction, ultrasonication extraction, supercritical fluid extraction, microwave assisted extraction and accelerated solvent extraction for the determination of polychlorinated biphenyls in soil. *J. Chromatography A*, 2005, **1090**(1–2), 1–9.

Stedmon, C.A., S. Markager, and R. Bro, Tracing dissolved organic matter in aquatic environments using a new approach to fluorescence spectroscopy. *Marine Chem.*, 2003, **82**(3–4), 239–254.

Stout, S.A., A.D. Uhler, and K.J. McCarthy, A strategy and methodology for defensibly correlating spilled oil to source candidates. *Environ. Forensics*, 2001, **2**(1), 87–98.

Tauler, R., A. Smilde, and B. Kowalski, Selectivity, local rank, 3-way data analysis and ambiguity in multivariate curve resolution. *J. Chemometrics*, 1995, **9**(1), 31–58.

Telnaes, N. and B. Dahl, Oil-oil correlation using multivariate techniques. *Organic Geochem.*, 1986, **10**(1–3), 425–432.

Thygesen, L.G., A. Rinnan, S. Barsberg, and J.K.S. Moller, Stabilizing the PARAFAC decomposition of fluorescence spectra by insertion of zeros outside the data area. *Chemometrics and Intelligent Laboratory Systems*, 2004, **71**(2), 97–106.

Tomasi, G. and R. Bro, A comparison of algorithms for fitting the PARAFAC model. *Computational Stat. Data Anal.*, 2006, **50**(7), 1700–1734.

Tomasi, G., F. van den Berg, and C. Andersson, Correlation optimized warping and dynamic time warping as preprocessing methods for chromatographic data. *J. Chemometr.*, 2004, **18**, 231–241.

Tomasi, G. and R. Bro, PARAFAC and missing values. *Chemometrics and Intelligent Laboratory Systems*, 2005, **75**(2), 163–180.

van Nederkassel, A.M., M. Daszykowski, D.L. Massart, and Y. Vander Heyden, Prediction of total green tea antioxidant capacity from chromatograms by multivariate modeling. *J. Chromatography A*, 2005a, **1096**(1–2), 176–186.

van Nederkassel, A.M., V. Vijverman, D.L. Massart, and Y. Vander Heyden, Development of a Ginkgo biloba fingerprint chromatogram with UV and evaporative light scattering detection and optimization of the evaporative light scattering detector operating conditions. *J. Chromatography A*, 2005b, **1085**(2), 230–239.

Vogt, F. and K. Booksh, Influence of wavelength-shifted calibration spectra on multivariate calibration models. *Appl. Spectroscopy*, 2004, **58**(5), 624–635.

Walczak, B. and D.L. Massart, Dealing with missing data, Part I. *Chemometrics and Intelligent Laboratory Systems*, 2001, **58**(1), 15–27.

Wang, C.P. and T.L. Isenhour, Time-warping algorithm applied to chromatographic peak matching gas-chromatography Fourier-transform infrared mass-spectrometry. *Anal. Chem.*, 1987, **59**(4), 649–654.

Wang, Z.D., M. Fingas, S. Blenkinsopp, G. Sergy, M. Landriault, L. Sigouin, J. Foght, K. Semple, and D.W.S. Westlake, Comparison of oil composition changes due to biodegradation and physical weathering in different oils. *J. Chromatography A*, 1998, **809**(1–2), 89–107.

Wang, Z.D., M. Fingas, and K. Li, Fractionation of a light crude-oil and identification and quantitation of aliphatic, aromatic, and biomarker compounds by Gc-Fid and Gc-Ms, 1. *J. Chromatographic Sci.*, 1994a, **32**(9), 361–366.

Wang, Z.D., M. Fingas, and K. Li, Fractionation of a light crude-oil and identification and quantitation of aliphatic, aromatic, and biomarker compounds by Gc-Fid and Gc-Ms, 2. *J. Chromatographic Sci.*, 1994b, **32**(9), 367–382.

Wang, Z.D., M. Fingas, E.H. Owens, L. Sigouin, and C.E. Brown, Long-term fate and persistence of the spilled Metula oil in a marine salt marsh environment — degradation of petroleum biomarkers. *J. Chromatography A*, 2001, **926**(2), 275–290.

Wang, Z.D., M. Fingas, and D.S. Page, Oil spill identification. *J. Chromatography A*, 1999, **843**(1–2), 369–411.

Wang, Z.D., M. Fingas, and L. Sigouin, Using multiple criteria for fingerprinting unknown oil samples having very similar chemical composition. *Environ. Forensics*, 2002, **3**(3–4), 251–262.

Wang, Z.D., M. Fingas, and L. Sigouin, Characterization and source identification of an unknown spilled oil using fingerprinting techniques by GC-MS and GC-FID. *Lc Gc North America*, 2000, **18**(10), 1058.

Willse, A., A.M. Belcher, G. Preti, J.H. Wahl, M. Thresher, P. Yang, K. Yamazaki, and G.K. Beauchamp, Identification of major histocompatibility complex-regulated body odorants by statistical analysis of a comparative gas chromatography/mass spectrometry experiment. *Anal. Chem.*, 2005, **77**, 2348–2361.

Witjes, H., M. Pepers, W.J. Melssen, and L.M.C. Buydens, Modelling phase shifts, peak shifts and peak width variations in spectral data sets: Its value in multivariate data analysis. *Analytica Chimica Acta*, 2001, **432**(1), 113–124.

Wold, S., K. Esbensen, and P. Geladi, Principal component analysis. *Chemometrics and Intelligent Laboratory Systems*, 1987, **2**(1–3), 37–52.

Wong, J.W.H., C. Durante, and H.M. Cartwright, Application of fast Fourier transform cross-correlation for the alignment of chromatographic and spectral datasets. *Anal. Chem.*, 2005, **77**, 5655–5661.

10 Chemical Heterogeneity in Modern Marine Residual Fuel Oils

Allen D. Uhler, Scott A. Stout, and Gregory S. Douglas

NewFields Environmental Forensics LLC, 100 Ledgewood Place Suite 302, Rockland MA 02370

10.1 Introduction

The vast majority of the vessels that make up the more than 700 million dry weight tonnes of the world's merchant marine fleet are powered by massive marine diesel engines that operate on heavy marine fuel oils derived from residual oil (Hiedeloff and Stockmann, 2005). In the maritime industry, ship fuels are commonly referred to as "bunker fuels" — derived from the notion than these fuels are stored in ships' holds that, during the coal-propulsion shipping era, were referred to as bunkers. Both routine commercial shipping operations (e.g., fueling and bilging) and maritime accidents (e.g., groundings, collisions, or sinking attributable to foul weather or mechanical failures) can lead to the operational discharge or accidental spillage of marine fuel oil, being carried as both fuel and cargo, into the sea. Annual estimates of petroleum discharged or spilled into the sea by worldwide commercial vessel operations approach 1 million tonnes (NRC, 2003). In the case of catastrophic maritime accidents involving commercial cargo vessels, many tens to hundreds of tons of fuel oil can be lost to the sea virtually at once, or chronically over days or months after an accident.

Oil spill investigators are often called upon to determine the source and fate of oil spilled in the sea. In some cases, investigators are asked to determine the likely source of "mystery" spills of oil found in shipping lanes, harbors, or the open ocean. In cases involving catastrophic spills of oil, investigators need to track the spilled oil in the environment and, ultimately, to differentiate that oil from other sources of fugitive petroleum in the environment (Stout et al., 2005; Hendrick and Reilly, 1993).

Because of their preponderant use as fuel in marine vessels, marine residual fuels are often the focus of maritime oil spill investigations. Residual fuels, often referred to generically as heavy fuel oil or HFO, pose a variety of challenges to oil spill investigators: from a fate and transport perspective, heavy fuel oils are typically low API gravity, with densities approaching, and sometimes exceeding, that of water (Neff et al., 2003; National Academy of Sciences, 1999). As such, heavy fuels may float on water, sink, or alternatively sink and then resurface, depending on meteorologic conditions (NOAA, 1997; Michele and Galt, 1995).

Heavy fuel oils are also distinctive in terms of their refining and production history. In fact, variability in the composition of modern heavy marine fuels provides unique opportunities for chemical "fingerprinting" of HFOs in the environment. In this chapter, we focus on the forensic chemistry of HFO — the most widely used of the commercial marine fuel oils — and chemical features of these fuels pertinent to oil spill investigations. A brief history of HFO production, use, and classification is presented. Distinctive chemical features of heavy fuels are described. The results of detailed chemical analyses for 71 modern worldwide IFO 380 HFOs are offered to demonstrate the variability in the composition of marine residual fuel oils.

10.1.1 Historical Perspective

The use of petroleum for ship propulsion began in the late 19th century, shortly after the discovery and production of commercially available petroleum. By the early 20th century, most commercial ships had converted from coal to oil-fired steam boiler propulsion plants. During the first half of the 20th century, the maritime propulsion fuel of choice was distillate-based; however, as the demand for distillate fuel and automotive gasoline grew, there was a strong economic incentive for maritime ship operators to develop boilers and diesel engines capable of burning cheaper fuels derived from less valuable refining residuals (Newbery, 1996). By the late 1950s, a majority of commercial ships were propelled by diesel engines fired by residual fuels, or so-called heavy fuel oils (HFOs). Until the Middle Eastern oil embargos of the early and mid-1970s, the consistency and quality of these vintage heavy fuels were, by all accounts, reasonable and predictable. These early HFOs were derived from the relatively simple residuals of atmospheric and vacuum distillation of crude oil.

In the wake of the 1970s oil embargos, worldwide petroleum refiners were forced to better utilize residual stocks to produce more distillate and automotive gasoline through more aggressive, complex refining — typically involving additional thermal and catalytic cracking processes. The quality of the cracked residuums resulting from these advanced refining processes varied, but in general were poorer quality than straight-run atmospheric or vacuum-distilled residuals. The viscosity of the cracked residuums increased, and the concentrations of undesirable constituents like sulfur, metals, and inorganic residues were greater than in simpler straight-run residuals. To achieve the performance specifications demanded by the marine heavy fuel market, refiners began blending and reformulating poor-quality cracked residuums with products such as cracked gas oils and other available refinery intermediates, so the resulting heavy fuel oils could be burned in modern marine diesel engines (Newbery, 1996; Winkler, 2003). Coincidently, maritime engineers improved marine diesel engines to reliably burn more chemically and physically diverse heavy fuel oils formulated from cracked, heavy residual blends.

10.1.2 Production of Heavy Fuel Oils

Heavy fuel oils are produced from residuals — the "leftovers" from the crude oil refining process (Leffler, 2000). As mentioned earlier, until the early 1970s, this primarily included the nondistillable residuum from the atmospheric distillation process (i.e., the fraction of crude oil that did not boil above 1050°F to 1100°F). Today, however, most refiners also utilize vacuum distillation techniques that "squeeze" even more desirable, lighter products from atmospheric distillation residuum. The residuum from vacuum distillation, often termed "flasher bottoms," is the primary feedstock in the current production of modern heavy fuel oils (HFOs). As the sophistication of the refining process increased to include additional steps aimed at squeezing more profit from residuums (e.g., thermal and catalytic cracking, visbreaking, coking), smaller volumes and lower-quality cracked HFO feedstocks resulted. Lower quality is typically expressed in terms of higher concentrations of sulfur, ash (sometimes associated with catalyst fines), and metals, and higher viscosity, pour point, and water content.

The reduction in the quality of residual fuel feedstocks has led to the need to blend other lower boiling residuals (e.g., cat-cracked gas oil), with the heavy distillation residuals in order to achieve HFO specifications. Though considered controversial, it has not been uncommon for producers of HFOs to use small amounts of used automotive and crankcase lubricating oil as a blending stock for residual fuels as a means of disposal (Mazur et al., 2004). Interestingly, only the most recent revisions of the ISO specifications for marine fuels expressly prohibit the use of lubricating oil (ULO) as blending components to marine fuels (ISO, 2005).

Table 10-1 ISO Specifications for Marine Residual Fuel Oils[1]

Parameter	Unit	Limit	RMA 30	RMB 30	RMD 80	RME 180	RMF 180	RMG 380	RMH 380	RMK 380	RMH 700	RMK 700
Density at 15°C	kg/m³	Max	960	975	980	991		991		1010	991	1010
Viscosity at 50°C	mm²/s	Max	30		80	180			380			700
Water	% V/V	Max	0.5		0.5	0.5			0.5			0.5
Microcarbon Residue	% m/m	Max	10		14	15	20	18		22		22
Sulfur[2]	% m/m	Max	3.5		4	4.5			4.5			4.5
Ash	% m/m	Max	0.1		0.1	0.1	0.15	0.15		0.15		0.15
Vanadium	mg/kg	Max	150		350	200	500	300		600		600
Flash point	°C	Min	60		60	60			60			60
Pour point, Summer	°C	Max	6	24	30	30			30			30
Pour point, Winter	°C	Max	0	24	30	30			30			30
Aluminum + Silicon	mg/kg	Max	80		80	80			80			80
Total Sediment, Potential	% m/m	Max	0.1		0.1	0.1			0.1			0.1
Zinc[3]	mg/kg	Max						15				
Phosphorus[3]	mg/kg	Max						15				
Calcium[3]	mg/kg	Max						30				

[1] Source: ISO 8217 Third Edition, Nov. 1, 2005. Petroleum products — Fuels (class F) — Specifications of marine fuels.
[2] A sulfur limit of 1.5% m/m will apply in SO× Emission Control Areas designated by the International Maritime Organization, when its relevant protocol comes into force. There may be local variations.
[3] The fuel shall be free of ULO. A fuel is considered to be free of ULO if one or more of the elements are below the limits. All three elements shall exceed the limits before being deemed to contain ULO.

While the specifications of marine residual fuels listed in Table 10-1 may appear stringent, they actually provide refiners with considerable latitude in the manufacturing of on-specification heavy fuel oils. Because no two refineries operate identically, and because HFO blending depends upon the current operating and economic considerations at a given refinery at a given time, the specific nature of HFOs produced within a single refinery, and certainly between refineries, is expected to be highly variable, while still meeting product specifications. While this may cause some consternation for manufacturers of HFOs, it is precisely this variability that oil spill investigators can use to their advantage in identifying and tracking fugitive marine residual fuels that have been spilled on water.

10.1.3 Marine Fuel Nomenclature and Classification

A complication in the discussion of marine bunker fuels is the naming convention used for the products. Both common and standardized nomenclature systems are regularly used to describe the different grades of marine fuels. Common usage of the catch-all term "bunker fuel" has been used for decades. Traditionally, three types of bunker fuels — A, B, and C — were used to describe these fuels. Bunker A was generally synonymous with No. 2 fuel oil, bunker B was generally synonymous with No. 4 or No. 5 fuel oils, and bunker C, the most commonly used bunker fuel, was generally synonymous with No. 6 fuel oil.

As the demand for marine fuels grew more diverse and intensive in the latter part of the 20th century, various national and international organizations such as the American Society for Testing and Materials (ASTM), the British Standards Institute (BS), and the International Organization for Standardization (ISO) recognized the need to classify and develop standardized specifications for marine fuels (Newbery, 2003; Thomas, 1981). Although a number of standardized marine fuel classification systems were developed, the most widely acknowledged is the structure developed by the ISO. The first ISO specifications for marine fuels were released in 1987, and revised in 1996, at which time the specifications included 19 grades of marine fuel (ISO, 1996). In 2005, the ISO released revised specifications for marine fuels, that now recognizes 14 grades of marine fuels, including 4 distillate and 10 residual grades — the latter of which are inventoried in Table 10-1 (ISO, 2005).

The chemical composition of the four types of distillate marine fuels are not the focus of this chapter. In practice, only two of these are commonly used, namely, marine diesel oil (MDO) and marine gas oil (MGO). MDO is a blend of gas oil and residual oil, and MGO is a high-quality distillate diesel fuel that contains no residual oil blending components.

The principal features that distinguish the 10 grades of residual marine fuel oils are viscosity, pour point, and sulfur, carbon, and metals residue content (Table 10-1); the latter of which is intended to prohibit the use of ULO (as described above). Of the 10 residual grades of marine fuels (Table 10-1), in practice, only five of these are commonly used for marine transportation. These five commonly used fuels occur within two "groups," the names of which are more commonly used within the maritime industry. These two groups are (1) intermediate fuel oil 180 (IFO 180s) and (2) intermediate fuel oil 380 (IFO 380s). These groups' names refer to the maximum permissible viscosities of these fuels at 50°C, 180 and 380 mm^2/s (cSt), respectively (Table 10-1). IFO 180s include ISO grades RME 180 and RMF 180, and IFO 380s include ISO grades RMG 380, RMH 380, and RMK 380 (Table 10-1). Within the IFO 180 and IFO 380 groups, the five individual grades are distinguished by differences in carbon residue, ash, and vanadium content (Table 10-1). Additional nuances have been introduced as recently as 2005 by the ISO to differentiate like-grade bunker fuels with differences in inorganic residuals.

In regard to the behavior of residual fuel oils, it is notable that the current ISO specifications for marine residual fuels includes provisions for two products (RMK 380 and RMK 700; Table 10-1) that closely approach the density of sea water (1030 kg/m^3; Stumm and Morgan, 1996). These very heavy fuels' behavior when spilled poses unique problems for oil spill investigators and responders because of their potential to sink, rather than float, once released into the aquatic environment (Castle et al., 1995; National Academy of Sciences, 1999).

10.2 Forensic Chemistry Considerations

10.2.1 General Chemical Fingerprinting

In the realm of marine heavy fuels, the two most popular groups of heavy fuel oils, IFO 180 and IFO 380, differ largely in their blending formulas. While on average, it is reported that IFO 180 is composed of residual oil blended with about 6% to 7% gas oil-range petroleum, while IFO 380, a more viscous fuel, is composed of residual oil blended with approximately 3% gas oil-range petroleum (Bunker World, 2006), our library of IFO 380 fuels indicates that a much greater percentages gas-oil range hydrocarbons are also possible in heavy fuel oils (see Table 10-2 and discussion that follows). From a forensic chemistry standpoint, it is the combination of the refining and blending processes that impose unique chemical "fingerprints" on IFO 380 HFOs, which oil spill investigators can use to identify and track spilled fuel in the environment.

One of the basic goals in an oil spill fingerprinting investigation is to determine the unique chemical characteristics of the spilled oil, so it may be tracked, linked to a source, and differentiated from other sources, including background petroleum. Collectively, the unique chemical features that describe a particular oil have been termed the products' "fingerprint" (Alimi et al., 2003; Stout et al., 2002; Wang and Fingas, 2003; Morrison, 2000). These "fingerprints" are the basis for qualitative oil spill assessment protocols such as ASTM D-3328, *Standard Test Methods for Comparison of Waterborne Petroleum Oils by Gas Chromatography* (ASTM, 1990); ASTM D-5739, *Standard Practice for Oil Spill Source*

Table 10-2 Weight Percentages of Total Hydrocarbons (THC) and Common Isoalkane Ratios for 71 Worldwide IFO 380 Heavy Fuel Oils

	Min	Max	95th percentile	Mean	Std Dev	% CV*
Gas chromatographable (THC; wt%)	23	100	74	58	13	22
Wt% C_{10}–C_{25}	6	64	49	35	9	26
Wt% C_{25}–C_{45}	6	70	41	23	10	45
Wt% C_{45+}	0	77	57	42	12	30
Isoalkane Ratios						
nC_{17}/Pr	0.6	3.0	2.3	1.3	0.5	36
nC_{18}/Ph	1.2	2.9	2.1	1.7	0.3	16

*Percent coefficient of variation.

Identification by Gas Chromatography and Positive Ion Electron Impact Low Resolution Mass Spectrometry (ASTM, 2000); and quantitative oil spill correlation methods, such as the revised *Nordtest Method for Oil Spill Identification* (Nordtest, 1991; Daling et al., 2002). The challenge for the oil spill investigator is to ascertain what constitutes the most useful chemical features of a "fingerprint" for any given oil.

Oil, once spilled onto water, undergoes a variety of physical and chemical changes due to the forces of environmental weathering (Stout et al., 2002; Rodgers et al., 2000; Fingas, 1988; Atlas, 1981). Evaporation, dissolution, and biodegradation alter or remove many chemical constituents from spilled oil, thereby changing its chemical makeup relative to the original, fresh product (Prince et al., 2003; Durell et al., 1995; Michel and Hayes, 1993; Bobra, 1992; Ostazeski et al., 1995a, 1995b; Shiu et al., 1990; Mackay and McAuliffe, 1988; Anderson et al., 1974; Atlas, 1981; Wang and Fingas, 1994). Because the properties of spilled oil may change in the environment, measures of spilled oils' basic physical characteristics (such as those used to classify the neat fuel; Table 10-1) are virtually meaningless for oil spill identification purposes. More advanced techniques, focused on molecular characterization of oil, have become the methods of choice for oil spill characterization (ASTM, 2000; Stout et al., 2001; Wang and Fingas, 2003; Daling et al., 2002; Nordtest, 1991; ASTM, 1990).

Most sophisticated analytical oil spill identification techniques rely upon gas chromatography (GC) methods. Gas chromatography provides several levels of compositional information. High-quality GC analysis of an oil or environmental sample that contains oil reveals the petroleum residue's overall compositional profile (Stout et al., 2005, 2002; Alimi et al., 2003; Wang and Fingas, 2003). The so-called gas chromatogram is a depiction of the boiling point distribution of the hydrocarbons that compose the oil. The gas chromatogram can allow the investigator to grossly classify the oil (e.g., as crude oil, residual fuel, diesel fuel, etc.). Often, distinctive features such as obvious blending characteristics, or the presence or absence of key hydrocarbon compounds, can be recognized in the gas chromatogram. Next, individual hydrocarbon compounds in the oil can be identified and quantified, usually using gas chromatography with mass spectrometry (GC/MS: Alimi et al., 2003; Stout et al., 2002; Wang and Fingas, 2003). Such compound-specific analyses are focused on measuring chemicals that have been determined to be representative of important components of oil in general and often diagnostic, or source-specific, to a particular oil or fuel blend. The target compounds typically include n-alkanes, acyclic isoprenoids, homo- and heteroatomic 2- through 6-ring parent and alkylated polycyclic aromatic hydrocarbons (PAH, Douglas et al., 1996), and numerous recalcitrant petroleum biomarkers, including hopanes, steranes, and triaromatic steroids (Peters et al., 2005;

Alimi et al., 2003; Stout et al., 2002; Wang and Fingas, 2003; Daling et al., 2002). The applications of biomarker analysis in oil spill assessment are covered thoroughly in Chapter 3 of this book.

Certain of these measured chemicals such as lower-molecular-weight n-alkanes, monoaromatics (e.g., BTEX), and lower-molecular-weight PAHs are labile because of weathering. When analyzed, they provide a qualitative assessment of the degree of alteration spilled petroleum has undergone (Stout et al., 2002; Ostazeski et al., 1995a, 1995b; Mackay and McAuliff, 1988; Fingas, 1988; Payne et al., 1987). Other chemicals, such as higher-molecular-weight PAH and biomarkers, are considered to be largely recalcitrant to weathering under most environmental conditions and timescales. These compounds provide diagnostic and reliable information for chemical fingerprinting of spilled oil because the individual compounds or associated compound source ratios remain largely unchanged, even in the face of environmental weathering (Stout et al., 2005, 2002, 2001; Peters et al., 2005; Douglas et al., 1996). Protocols such as those by Nordtest (Nordtest, 1991; Daling et al., 2002) and CEN (see Chapter 7) have been developed to guide the investigator toward use of those diagnostic chemicals most sensitive and reliable in the identification of a particular spilled oil.

10.2.2 Samples and Analytical Methods

In the remainder of this chapter, we will discuss the varied chemical features we have observed among 71 modern residual fuel oils — IFO 380s — in terms of the metrics often used in oil spill fingerprinting investigations. These 71 fuels were obtained directly from marine vessels, or from archived fuels that had been delivered to vessels in ports in the United States, Europe, Asia, and Australia over the past few years. All of the residual fuels studied were "fresh" in that they had not experienced any environmental weathering.

Each of the samples was analyzed for total chromatographable hydrocarbons, n-alkanes, and isoprenoid hydrocarbons, 2- through 6-ring PAH compounds, and petroleum biomarkers. The details of these analyses have been described elsewhere (Stout et al., 2002) and are only briefly described herein.

The samples were prepared for analysis by weighing approximately 50 mg of the oil into a tared 10-ml volumetric flask, bringing it to volume in dichloromethane (DCM). A 1-ml aliquot was spiked with surrogate internal standard (SIS; o-terphenyl, naphthalene-d_8, phenanthrene-d_{10}, chrysene-d_{10}) and recovery internal standard (RIS; 5α-androstane, acenaphthene-d_{10}, fluorene-d_{10}, benzo[a]pyrene-d_{12}) and split for quantitative GC/FID and GC/MS analysis as described next.

Gas chromatography-flame ionization detection (GC/FID) analysis was carried out as described by Stout et al. (2001). The target compounds consisted of selected n-alkanes and isoprenoids and chromatographic carbon ranges (i.e., C_{10}–C_{25}, C_{25}–C_{45}). Carbon range analysis was performed by employing a standard baseline integration technique. The "window" for each range was determined using the calibration standards. All calibration solution compounds that fell within the window were used to generate the average RF for that range. Areas for SIS and RIS compounds were subtracted. Gas chromatography/mass spectrometry (GC/MS) was conducted using a Hewlett-Packard 6890 Plus GC interfaced to a Hewlett-Packard 5973 mass selective detector (MSD) as described by Stout et al. (2001). Data acquisition was performed in the select ion monitoring (SIM) mode for greatest sensitivity and selectivity. Quantification of target compounds was performed by the method of internal standards using average response factor (RF) of the parent PAH compounds and those of a representative triterpane and sterane from within the 5-point initial calibration.

10.3 General Features of Modern Residual Marine Fuel Oils

Gas chromatographic analysis of petroleum fuels reveals the distinctive boiling point distribution of the chromatographable hydrocarbons that compose the fuels. Common distillates such as diesel or gas oils have predictable and largely similar gas chromatographic fingerprints, by virtue of the fact that they are distilled and blended from a molecular weight-constrained range of hydrocarbons (Speight, 1991; Gary and Handwerk, 1984).

Heavy fuel oils are not compositionally constrained like distillate fuels. Rather, as described earlier, heavy marine fuel oils are blended from residuums, gas oils, and other economically advantageous refinery stocks to achieve basic physical and chemical specifications (Table 10-1; see also Chapter 1 herein). Thus, it is not surprising to find that the gas chromatographic analysis of various residual fuel oils reveals substantial differences in their GC "fingerprints."

Figure 10-1 shows the gas chromatograms for six fresh IFO 380 fuel oils. The specific ISO type (as per Table 10-1) of each of these fuels is unknown to us, but the fuels were all collected from vessels using IFO 380. The variety in the chemical composition expressed in these chromatograms is remarkable and underscores the notion that there is no "typical" residual fuel oil. All six of these example fuels contain the chromatographic features of residuum, for example, a measurable unresolved complex mixture (UCM) of heavy hydrocarbons in the approximately C_{20} to C_{40+} range. However, it is clear from the relative masses of the UCMs in this family of chromatograms that some of the fuels (e.g., Figure 10-1D to F) contain substantially greater amounts of heavier residuum than others (e.g., Figure 10-1A to C). Figure 10-1A shows GC/FID chromatograms for a highly aromatic IFO 380 that is dominated by C_3–C_5 alkyl-benzenes and C_0–C_4 alkyl-naphthalenes within the gas oil range. Figure 10-1B shows the GC/FID chromatogram of another aromatic-rich IFO 380 enriched in C_0–C_4 alkyl-naphthalenes in the gas oil range and a greater proportion of residual range hydrocarbons, including a broad suite of n-alkanes. Figure 10-1C shows the GC/FID chromatogram for an IFO 380 predominantly comprised of residual range UCM with resolved compounds including alkyl-naphtalenes, phenanthrenes, and chyrsenes, while little gas oil range material is present. The IFO 380 chromatogram in Figure 10-1D is similar to that of Figure 10-1C but contains a greater proportion of gas oil compounds (e.g., C_3–C_5 alkyl-benzenes and alkyl-naphthalenes) and a greater proportion of unresolved residual range compounds. Figure 10-1E shows the GC/FID chromatograms for an IFO 380 that contains a significant residual range UCM, mostly boiling above C_{25}, the boiling range/viscosity of which is "offset" by a significant gas oil range component dominated by C_3–C_5 alkyl-benzenes and C_0–C_4 alkyl-naphthalenes. Finally, the GC/FID chromatogram in Figure 10-1F shows an IFO 380 dominated by a residual range UCM (with a distinctive profile) and a waxy component, as evidenced by the prominence of C_{30+} n-alkanes. Collectively, the GC/FID chromatograms of these IFO 380s demonstrate that they contain varying amounts of a highly aromatic, broad-cut gas oil range ($\sim C_{10}$–C_{25}) blending component — as is typical of modern residual fuel production practices (see above and Chapter 1 herein).

The dropoff in several of the UCMs at around C_{40} in Figure 10-1 indicates that these fuels contain a significant mass of high boiling material that is not chromatographable. This mass would include varying amounts of compounds boiling above C_{45}, including resins and asphaltenes, that are not detectable by conventional capillary gas chromatography. The amount of these non-gas chromatographable constituents can, if necessary, be determined by difference (see Figure 10-2) or by liquid chromatography.

Table 10-2 presents the summary statistics for the measurable total hydrocarbons and selected isoprenoid ratios for the 71 worldwide IFO 380 heavy fuel oils. These data were determined from the GC/FID analysis and integrations of individual peak areas and specific carbon ranges. The ratios of n-alkanes to isoprenoids (nC_{17}/Pr and nC_{18}/Ph) are commonly used to assess degrees of biodegradation in petroleums. These ratios are shown to

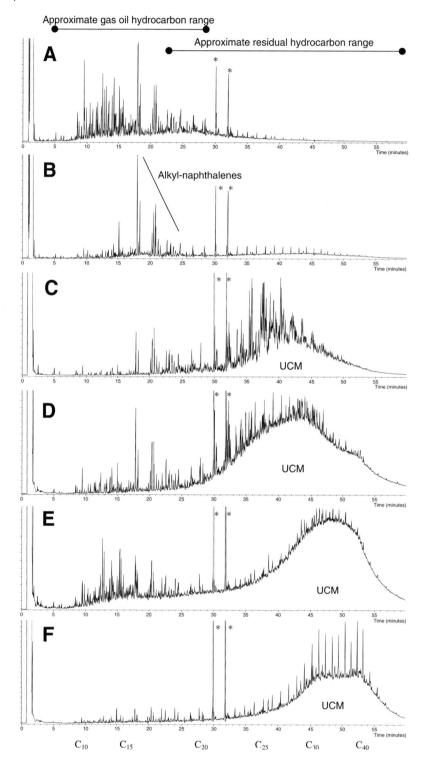

Figure 10-1 Gas chromatographic "fingerprints" of six IFO-380 HFOs, demonstrating the significant variability in chemical compositions in the sample fuel class. * — internal standards.

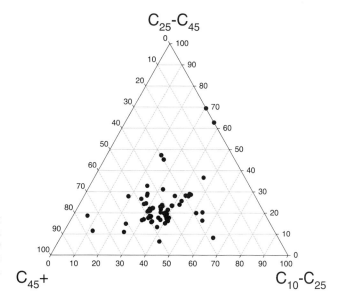

Figure 10-2 Ternary diagram depicting the variability in the bulk hydrocarbon composition among 71 worldwide IFO 380 heavy fuel oil samples. Weight percentages based upon quantitative GC/FID analysis and gravimetric analysis.

vary widely among these "fresh" (unweathered) IFO 380 fuel oils, which indicates that the use of these ratios to assess weathering in a spill investigation needs to consider the spilled oil's starting ratios before describing the degree of weathering among fugitive oils.

The mass of compounds eluting within the gas oil range diluent (C_{10}–C_{25}) versus the residual range residuum (C_{25}–C_{45}) were readily measured by integration of the GC/FID chromatograms. The sum of these two weight percentages provided the total weight percent mass of chromatographable hydrocarbons (THC). The weight percent C_{45+} was then determined by difference (100% minus the weight percent mass of THC, i.e., C_{10} through C_{45}). The statistical results for weight percentages in the 71 IFO 380 fuels studied are given in Table 10-2. These data clearly show what is evident in Figure 10-1, namely, that the proportions of gas oil (or light) diluent (C_{10}–C_{25}), residual range residuum (C_{25}–C_{45}), and nonchromatographable compounds (C_{45+}) vary widely among the IFO 380 fuels studied. The variability in the carbon mass range composition among the 71 worldwide IFO 380 heavy fuel oil samples can be seen in Figure 10-2. Here, the diversity in the carbon-range composition of the samples is clearly evident.

The quantitative measurements presented in Table 10-2 and shown in Figure 10-2 parallel the observations noted in the more qualitative gas chromatographic fingerprints shown in Figure 10-1. Inspection of Table 10-2 and Figure 10-2 reveals that two IFO 380 fuels contain very high (>75 wt%) mass percentages of chromatographable hydrocarbons (C_{45+}; Figure 10-2; Table 10-2). Such a high percent mass of C_{45+} compounds indicates that these fuels likely include a high proportion of asphaltenes or nonvolatile inorganic matter. On the other hand, two of the 71 IFO 380 fuels contained 100 weight percent of gas chromatographable material (Figure 10-2). Residual fuels containing no measurable non-chromatographable material clearly are blended using relatively high-quality petroleum stocks, and *de minimis* heavy residuum and <C_{45+} range compounds. This demonstrates that some fuels sold as IFO 380 fuels may actually be higher-quality fuels that are sold (due to local market pressure) as lower-quality IFO 380 fuels. The survey of 71 fuels, however, revealed that both extremes that are evident in Table 10-2 (i.e., excess of 75 wt% nonchromatographable or 0 wt% nonchromatographable mass), would seem atypical for most IFO 380 fuels. On average, IFO 380s studied contained 35 wt% gas oil range, 23 wt% residual range, and 42 wt% nonchromatographable range material (Table 10-2).

10.4 Molecular Variability among Modern Residual Fuel Oils

The variability in the bulk chromatographic features of the 71 IFO 380 heavy fuel oils discussed in Section 10.3 leads to the hypothesis that that there should be significant variability at the molecular level among disparate HFOs. In this section, characteristics of the petroleum biomarkers and polycyclic aromatic hydrocarbons (PAH) in the fuels are discussed, since these compounds often provide the chemical fingerprinting metrics most useful in oil spill investigations — and, in the case of PAH, studies of oil toxicity.

10.4.1 Petroleum Biomarkers

The concentrations and distributions of petroleum biomarkers were analyzed in most of the 71 IFO 380s studied and, also as expected, exhibited widely varying distributions. This variability reflects the inherent "primary" (genetic) variability among crude oil feedstocks and any "secondary" variability imparted during the refining process, for example, blending with cracked intermediate products (see Chapter 1 herein).

The absolute concentration of $17\alpha(H)$, $21\beta(H)$-hopane was measured in the 71 IFO 380s studied and was shown to vary from 28 mg/kg to 478 mg/kg (median: 134 mg/kg; average: 143 mg/kg; st. dev: 57 mg/kg). This range in concentration is likely the result of different degrees of blending of hopane-bearing residuum and (mostly) hopane-free gas oil range materials. A specific example of the biomarker distribution variability is shown in Figure 10-3. This diagram shows the relative abundances of C_{27}, C_{28}, and C_{29} $14\beta(H),17\beta(H)$ regular steranes measured from the *m/z* 218 mass chromatograms. The HFOs studied exhibit variation in the proportion of C_{29} steranes, a feature that likely reflects the "primary" characteristic related to the abundance of terrestrial organic matter in the parent crude oil feedstock's source rock (Moldowan et al., 1985). Alternatively, it is possible that the increase in C_{29} steranes in some IFO 380s results from an increase in their abundance brought about by distillation, resulting in a residuum enriched in the higher boiling C_{29} steranes — as was observed by Peters et al. (1992). Regardless, the variability among these fuel oils demonstrates the potential utility of biomarkers in distinguishing different residual fuel oils in oil spill investigations (e.g., Stout et al., 2001).

The issue of refining effects on biomarker distributions is reviewed in Chapter 1 herein. Briefly, studies on the effects of crude oil refining on biomarkers have shown that in addition to distillation effects, the thermal stresses experienced during refining, particularly vacuum distillation, can alter various biomarker-based thermal parameters (Peters et al., 1992, Pieri et al., 1996). In one study (Peters et al., 1992), a parent crude oil's biomarkers, with comparable boiling points, were variously depleted or enriched compared to daughter refinery streams (e.g., vacuum gas oil or vacuum residuum). Such changes were attributed to either (1) a preferential preservation (i.e., a greater thermal stability) of certain biomarkers or (2) preferential formation (cracking from bound precursors in the oil) of certain biomarkers during heating. For example, in that study the ratio of C_{29} sterane isomerization $(14\alpha(H),17\alpha(H)\ 20S/(20S + 20R)$ and $T_s/(T_s + T_m)$ were lower in the vacuum residuum than was observed in the crude oil feedstock, possibly due to the release of the less mature 20R and T_m epimers from high boiling precursors (e.g., asphaltenes) during vacuum distillation.

We evaluated some common thermally driven biomarker parameters for the 71 IFO 380 fuels studied. Cross-plots of two triterpane- and two sterane-based maturity ratios are shown in Figure 10-4. These data show that the C_{31} hopane isomerization ratio (22S/22S + 22R) in all of the HFOs are near equilibrium (~0.55) values found in crude oils (Peters et al., 2005), indicating that the refining

Chemical Heterogeneity in Modern Marine Residual Fuel Oils 337

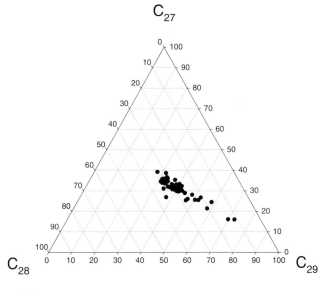

Figure 10-3 Ternary diagrams showing the distribution of C_{27}, C_{28}, and C_{29} regular steranes (5α, 14β, 17β(H) 20S+20R) calculated from absolute concentrations determined from the m/z 218 mass chromatogram for 71 IFO 380 residual fuel oils.

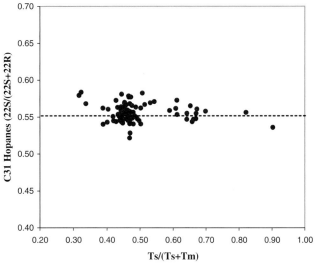

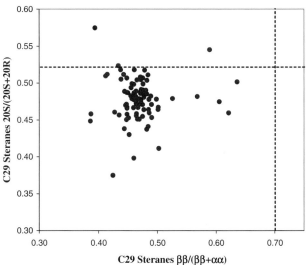

Figure 10-4 Cross-plots of (A) triterpane-based and (B) sterane-based thermal maturity parameters for 71 IFO 380 fuels. Dashed lines show typical maximum (equilibrium) values for crude oils (from Peters et al., 2005). Triterpane ratios both measured on the m/z 191 mass chromatograms. 14α(H),17α(H) C_{29} steranes measured on the m/z 217 mass chromatogram and 14β(H),17β(H) C_{29} steranes measured on the m/z 218 mass chromatogram.

process has not obviously affected this ratio. Similarly, any effect of refining on the other three biomarker ratios plotted is not obvious since most HFOs exhibit values within the ranges anticipated for crude oils. However, two HFOs exhibit a C_{29} sterane isomerization ratio (20S/20S + 20R) above the equilibrium (~0.57) values found in crude oils (Peters et al., 2005). The elevated 20S/(20S + 20R) ratios in these HFOs are likely due to distillation and/or heating effects during the refining process. Regardless, the heterogeneity in the thermal maturity ratios among the 71 IFO 380 fuels studied demonstrates that these (and other) biomarker ratios have value in chemical fingerprinting studies involving HFOs. Biomarker variability among the 71 fuels studied is not discussed further herein, but the reader is directed to Chapter 3 for more information on biomarkers.

10.4.2 Polycyclic Aromatic Hydrocarbons

Homo- and heteroatomic PAH compounds are important chemicals for the characterization and identification of petroleum in the environment, including spilled HFOs. The concentrations and distributions of these PAHs reflect the composition of the original crude oil feedstock from which the residuum is derived, alteration to the petroleum intermediates or residuum that occurs during refinery conversion processes (e.g., cracking, reforming, etc.), and blending of PAHs contained in any viscosity-lowering blending agents (e.g., cracked gas oil and other cutter stocks) into the final fuel mixes.

Table 10-3 presents the summary statistics for the concentrations of PAH and related compounds measured in the 71 worldwide IFO 380 heavy fuel oils studied. Figure 10-5 depicts the mean ±1 standard deviation for the concentrations of individual PAH measured in all of these samples. Significant variations in the PAH concentration among the fuels is evident from this compilation. Most of the PAH compounds are found at concentrations spanning more than an order of magnitude (Table 10-3).

Not surprisingly, the PAH distributions in the fresh fuels are, on average, dominated by lower-molecular-weight 2- and 3-ring PAHs such as the naphthalenes and phenanthrenes. (This was evident in several GC/FID chromatograms shown in Figure 10-1.) On average, the IFO 380s exhibit a decreasing abundance of PAH with increasing molecular weight. Each homologous series of alkylated PAH exhibits the anticipated "bell-shaped" profile that is characteristic of petrogenic materials. There are only very low concentrations of 5- and 6-ring PAHs, although these were detected in most samples (Table 10-3).

The prominence of alkyl-naphthalenes in most of the IFO 380 fuels arises largely from the common practice of blending lighter gas oil or other cutter stocks with distillation residuum to control the HFO's viscosity (see Section 10.1.2). In fact, for the majority of samples (~75%), the PAH content of the fuels is comprised of 60–80% C_0–C_4 alkyl-naphthalene, regardless of the total PAH content of the fuel (Figure 10-6). A significant number of the samples, however, diverge from this trend. As illustrated in Figure 10-6, approximately 25% of the samples contain less than 60% total naphthalenes, indicating these fuels were blended with less of a gas oil-range component — or with higher boiling gas oils or cutter stocks whose aromatic content contains proportionally greater amounts of higher molecular weight (3 or more rings) PAH.

The chemical impressions of the PAHs arising from a cracked gas oil cutter stock can easily be seen in the gas chromatographic analysis of blended HFOs (Figure 10-1) and in the PAH distributions. Figure 10-7 shows the PAH histograms for a low- (Figure 10-7A) and high-aromatic (Figure 10-7B) HFO. The relatively higher concentrations of N_0 to N_4 naphthalenes in the high-aromatic HFO is the direct result of blending residuum with an alkyl-naphthalene-rich cracked gas oil (Figure 10-7C). (This cracked gas oil was generated from the coking process at a refinery operating a petroleum coker.) The chemical impressions of the N_0 to N_4 alkyl-naphthalenes are readily

Table 10-3 Homo- and Heteroatomic PAH Composition for 71 Worldwide IFO 380 Heavy Fuel Oils (Concentrations in μg/kg)

	Abvr	Min	Max	95th percentile	Mean	Std Dev	% CV
Naphthalene	N0	108	9,502	3,263	2,129	1,255	59%
C1-Naphthalenes	N1	635	11,296	8,295	5,334	2,210	41%
C2-Naphthalenes	N2	671	10,839	10,085	5,881	2,220	38%
C3-Naphthalenes	N3	504	6,309	5,265	2,927	1,168	40%
C4-Naphthalenes	N4	291	2,823	2,215	1,141	533	47%
Biphenyl	Bph	19	1,731	481	295	215	73%
Acenaphthylene	Acl	ND	ND	NA	ND	ND	NA
Acenaphthene	Ace	20	1,334	325	195	154	79%
Dibenzofuran	DbF	7	215	161	85	40	47%
Fluorene	F0	31	665	311	150	92	61%
C1-Fluorenes	F1	77	767	560	270	145	54%
C2-Fluorenes	F2	146	1,252	874	441	222	50%
C3-Fluorenes	F3	182	1,442	789	438	228	52%
Anthracene	AN	9	120	89	47	21	46%
Phenanthrene	P0	7	1,025	735	358	196	55%
C1-Phenanthrenes/Anthracenes	P1	305	3,515	1,874	1,022	586	57%
C2-Phenanthrenes/Anthracenes	P2	504	5,718	2,542	1,329	831	62%
C3-Phenanthrenes/Anthracenes	P3	398	4,401	2,065	1,013	644	64%
C4-Phenanthrenes/Anthracenes	P4	169	1,870	872	451	265	59%
Dibenzothiophene	D0	19	324	190	83	55	66%
C1-Dibenzothiophenes	D1	86	1,049	536	250	153	61%
C2-Dibenzothiophenes	D2	144	2,002	877	374	275	73%
C3-Dibenzothiophenes	D3	196	1,654	795	348	231	66%
C4-Dibenzothiophenes	D4	104	744	408	190	118	62%
Fluoranthene	FL	5	65.5	40	20	10	50%
Pyrene	PY	25	469	243	125	75	60%
C1-Fluoranthenes/Pyrenes	FP1	100	1,807	630	350	257	73%
C2-Fluoranthenes/Pyrenes	FP2	174	2,718	780	481	370	77%
C3-Fluoranthenes/Pyrenes	FP3	101	2,187	639	373	280	75%
Benz[a]anthracene	BaA	8	275	91	47	33	69%
Chrysene	C0	20	479	156	84	56	67%
C1-Chrysenes	C1	69	1,820	557	303	218	72%
C2-Chrysenes	C2	94	1,893	732	344	242	70%
C3-Chrysenes	C3	27	692	287	126	93	73%
C4-Chrysenes	C4	9	231	137	43	37	87%
Benzo[b]fluoranthene	BbF	3	68.1	23	14	8	59%
Benzo[j/k]fluoranthene	BjkF	2	27.7	15	7	4	60%
Benzo[a]fluoranthene	BaF	ND	ND	NA	ND	ND	NA
Benzo[e]pyrene	BeP	9	159	60	31	20	63%
Benzo[a]pyrene	BaP	4	159	76	30	22	71%
Perylene	Per	2	83.8	52	20	14	69%
Indeno[1,2,3-c,d]pyrene	IND	1	15.1	7	3	3	100%
Dibenz[a,h]anthracene	DBA	1	23.3	15	6	4	66%
Benzo[g,h,i]perylene	Bghi	2	48.9	35	13	9	70%

apparent in the gas chromatographic "fingerprint" of the blended HFO (Figure 10-7D).

From a forensic chemistry standpoint, it is noteworthy that the IFO 380 HFOs contain measurable concentrations of 4-, 5-, and 6-ring PAHs such as fluoranthene, pyrene, the benzofluoranthenes, benzopyrenes, indeno(1,2,3-c,d) pyrene, dibenz(a,h)anthracene, and benzo (g,h,i)perylene (Table 10-3 and Figure 10-5 inset). These chemicals are usually not

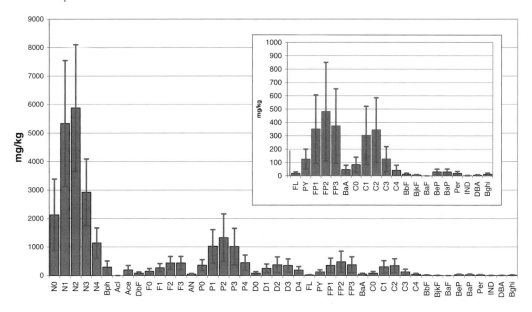

Figure 10-5 Average (±1 standard deviation) concentration of select PAH compounds in 71 worldwide IFO 380 HFOs. See Table 10-3 for compound abbreviations. Inset shows expanded view of 4- to 6-ring PAH.

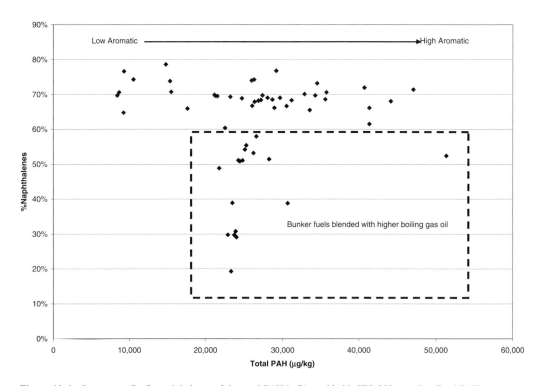

Figure 10-6 Percentage C_0–C_4 naphthalenes of the total PAH in 71 worldwide IFO 380 samples. Total PAH represents sum of 54 analytes from Table 10-3.

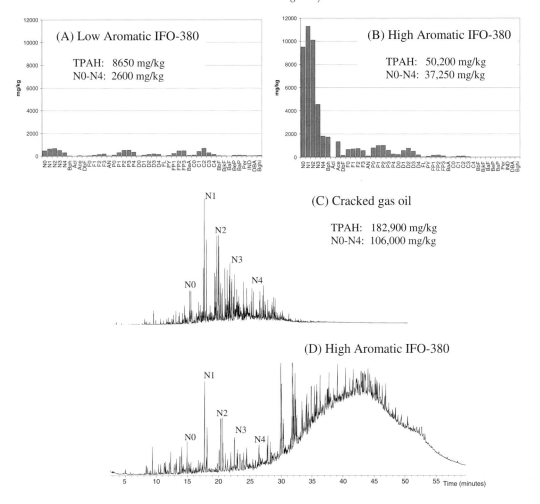

Figure 10-7 Histograms comparing distribution and concentration of PAH in (A) low aromatic IFO-380 and (B) high aromatic IFO-380. Gas chromatogram for (C) high aromatic, cracked gas oil blending stock, and (D) a high aromatic IFO-380 containing a prominent cracked gas oil blending stock.

detectable in most crude oils and distillate fuels because they either are not present or only exist at very low concentrations in such petroleum. These compounds are found in HFOs either because they have become concentrated within the distilled residuum or because they are produced during the distillation of the residuum and/or production of any cracked gas oil cutter stock. Regardless of their genesis, these 4-, 5-, and 6-ring PAH compounds are among the most environmentally stable of the PAHs, and their relative ratios are potentially useful for source identification (Emsbo-Mattingly et al., 2006; Uhler and Emsbo-Mattingly, 2006).

Oil spill investigators have utilized a variety of PAH-based indices and diagnostic ratios to characterize spilled oil in the environment (Stout et al., 2005, 2001; Wang and Fingas, 2003; Burns et al., 1997; Douglas et al., 1996; Henry et al., 1997; Page et al., 1995; ASTM, 1990). Table 10-4 contains a number of potentially diagnostic PAH parameters computed for the 71 worldwide IFO 380 fuel oils studied that provide additional insight into the compositional diversity among the samples. Bulk PAH compositional parameters presented in Table 10-4 offer additional evidence that the aromaticity among IFO 380 fuel oils varies dramatically. Interestingly, not only is there

Table 10-4 Summary Statistics for Certain PAH Diagnostic Metrics for 71 Worldwide IFO 380 Heavy Fuel Oils (See Table 10-3 for Compound Abbreviations.)

	Min	Max	95th percentile	Mean	Std Dev	% CV
PAH Metrics						
Total PAH (mg/kg, oil)[1]	8,414	51,345	42,757	27,171	8,974	33%
Priority Pollutant PAH (mg/kg, oil)[2]	616	12,575	4,768	3,227	1,539	48%
%TPAH as PP PAH	5.84	25.1	13.9	11.7	2.7	24%
Total Naphthalenes (mg/kg, oil)	0.26	3.73	2.90	1.74	0.67	39%
Total Dibenzothiophenes	0.06	0.58	0.28	0.12	0.08	65%
LPAH (mg/kg, oil)	5,465	49,289	39,480	24,751	8,591	35%
HPAH (mg/kg, oil)	791	13,148	4,076	2,419	1,666	69%
%LPAH	63.1	98.2	96.2	90.2	6.8	7%
%HPAH	1.77	36.9	25.4	9.8	6.8	69%
Selected Diagnostic Ratios						
MPI[3]	1.46	3.43	2.7	2.5	0.22	9%
MPR[3]	3.27	5.05	4.8	4.4	0.30	7%
%2MA/P1	4.21	8.86	8.1	6.9	0.97	14%
D2/P2	0.06	0.77	0.5	0.3	0.12	41%
D3/P3	0.08	0.86	0.7	0.4	0.15	40%
FL/PY	0.11	0.25	0.2	0.2	0.02	13%
BaA/CO	0.22	0.65	0.6	0.6	0.07	12%
AN/P0	0.06	0.18	0.2	0.1	0.02	18%

[1] Σ54 PAH and alkylated PAH per Table 3.
[2] Σ16 US EPA Priority Pollutant PAH.
[3] Methylphenanthrene index and ratio, respectively. See Radke et al. (1982).
[4] % 2-methyl-anthracene/Σmethyl-phenanthrene isomers.

significant spread in the total PAH content among HFOs (largely indicative of the proportions of high-aromatic, cracked gas oil blending stock used in the formulation of the fuel) but there is significant variation in the percentage of low-molecular-weight PAH (LPAH; i.e., sum of 2- through 3-ring PAHs) versus high-molecular-weight PAH (HPAH; i.e., sum of 4- through 6-ring PAHs). This almost certainly is consistent with the concept that the distribution of PAHs in HFOs is largely a function of the type and boiling range of cutter stock used to blend IFO 380 and the concentrations and types of PAHs that exist in the base residuum.

Inspection of Table 10-4 reveals that there are wide variations in the sulfur-containing dibenzothiophene concentrations among the samples that likely reflect the total sulfur content of the crude oil feedstocks used in the production of the residuum and cutter stocks (Song, 2000). Ratios of the C_2- and C_3-dibenzothiophenes relative to the respective C_2- and C_3-phenanthrenes are commonly used source ratios in oil spill studies (Stout et al., 2005, 2001; Burns et al., 1997; Douglas et al., 1996; Page et al., 1995). A plot of the source-specific D2/P2 versus D3/P3 ratios shows a wide spread in values among the 71 IFO 380 fuels studied, suggesting that these (and other) ratios hold the potential for differentiation and correlation among HFOs in oil spill studies (Figure 10-8). Notably, higher D2/P2 and D3/P3 values up to 1.6 and 2.2, respectively, have been observed in a high-sulfur residual fuel that was not among the 71 studied herein (Douglas, unpublished). It is interesting to note that the spread in the dibenzothiophene ratios among the 71 worldwide heavy fuel oils studied is not a strong function of total PAH concentration (Figure 10-8B). This is a favorable finding from a chemical fingerprinting standpoint, because it means that the diagnostic C2- and C3-dibenzothiophene to phenanthrene ratios are useful diagnostics for both lower aromatic and higher aromatic HFOs.

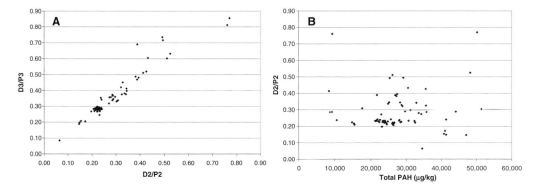

Figure 10-8 Cross-plots of (A) the range in of the diagnostic source ratios D2/P2 and D3/P3 and (B) the relationship between D2/P2 and total PAH in 71 worldwide IFO 380 heavy fuel oil samples.

10.5 Distinguishing Heavy Fuel Oils from Crude Oil

Many heavy fuel oils are readily recognized as manufactured products by the characteristic chromatographic (e.g., distillation) and molecular chemical features (e.g., presence of cracked petroleum products), some of which have been illustrated earlier in this chapter. For example, the clear presence of mixtures of residuum and broad (or narrow) boiling gas oil range cutter stocks, or the obvious presence of significant amounts of aromatic compounds such as alkyl-benzenes, alkyl-naphthalenes, and alkyl-phenanthrenes that arise from blending with high-aromatic, cracked gas oils (e.g., Figure 10-7C). However, among the diverse types of HFOs is a small subset that, at first inspection, shares many chromatographic and chemical similarities with evaporated or otherwise slightly weathered crude oil. Debate among oil spill fingerprinting experts can arise in instances where spilled oil — be it HFO or weathered crude oil — can be confounded by the possible presence of the other.

In cases where basic gas chromatographic analysis of a spilled oil residue is equivocal about the nature of the petroleum (i.e., weathered crude oil or manufactured HFO), we have relied upon in-depth analysis of PAH isomers and other compounds such as petroleum biomarkers, in an effort to identify subtle chemical markers that may serve to distinguish

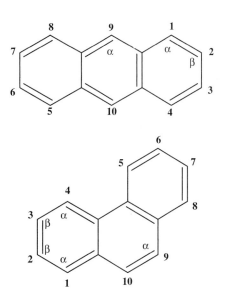

Figure 10-9 Structures of (A) linear 3-ring anthracene molecule and (B) nonlinear 3-ring phenanthrene molecule.

weathered crude oil from manufactured heavy fuel oil. A useful approach in such instances is careful inspection of PAH isomers whose presence is most likely due to refining — in particular, thermal and catalytic cracking. Consider the distribution of 3-ring methyl-PAHs (i.e., linear methyl-anthracenes and nonlinear methyl-phenanthrenes) (Figure 10-9). There are three possible isomers of methyl-anthracene and five isomers of methyl-phenan-

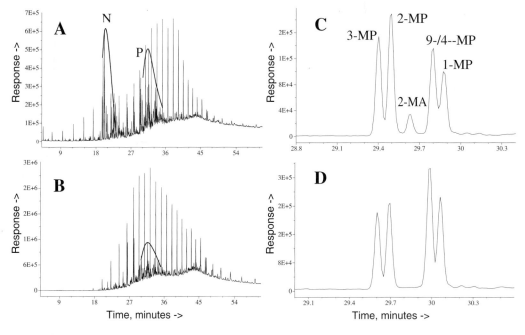

Figure 10-10 GC/FID chromatograms for (A) an unweathered IFO 380 fuel and (B) moderately evaporated crude oil and partial m/z 192 mass chromatograms for (C) the unweathered IFO 380 shown in (A) and (D) the moderately evaporated crude oil shown in (B). N — alkyl-naphthalenes, P — alkyl-phenanthrenes; MP — methylphenanthrene, MA — methylanthracene.

threnes that can be seen in gas chromatography/mass spectrometry m/z 192 mass chromatograms.

The proportions of the five methyl-phenanthrene isomers (which appear as four prominent peaks in the m/z 192 mass chromatogram of most oils; see Figure 10-10) have been used extensively to monitor the thermal stresses imposed under geologic heating of coals and oils, with the 3- and 2-methylphenanthrene (β-type) isomers proving more stable than the 9-, 4-, and 1-methylphenanthrene (α-type) isomers (e.g., Radke et al., 1982, 1990). These studies have shown that although methyl-anthracenes are present in low-rank coal extracts, they quickly are decreased (relative to methyl-phenanthrenes) or lost in the extracts of higher-rank coals (Radke et al., 1982). As such, it is reasonable that the linear methyl-anthracenes are considerably less thermodynamically stable than the nonlinear methyl-phenanthrenes. Therefore, methyl-anthracenes generally are absent or present in only low concentrations (relative to phenanthrenes) in most crude oils, which are produced slowly over long periods of time under geologic heating conditions. On the other hand, methyl-anthracenes, particularly 2-methyl-anthrancene (the only β-type isomer), are prominent components in pyrolytic organic materials that are produced under rapid heating conditions (e.g., coal tar). By analogy, 2-methyl-anthracene may be produced under some pyrolytic conditions experienced in the course of refining (e.g., thermal cracking).

2-Methyl-anthracene is easily resolved in the m/z 192 mass chromatograms for most polar GC capillary columns (Figure 10-10). As such, the presence of 2-methyl-anthracene (in conjunction with other particular aromatic compounds) might serve as an indicator of the presence of a cracked gas oil blending stock in the HFO — and thereby allow distinction from crude oil. In the 71 IFO 380 fuels studied herein, 2-methyl-anthracene was detected in all of them and comprised between 4.2%

and 8.9% of the total C2-phenanthrenes/anthracenes measured on the m/z 192 mass chromatogram (Table 10-4).

An example of this potential is shown in Figure 10-10. The GC/FID chromatograms for an unweathered IFO 380 and a moderately evaporated crude oil are shown in Figure 10-10A and 10-10B, respectively. These two oils share many common features, which is not surprising since the IFO 380 was produced from the same type of crude oil feedstock as is depicted in Figure 10-10B. Qualitative comparison between these would reveal a relative prominence of alkyl-naphthalenes and alkyl-phenanthrenes in the IFO 380 (relative to the n-alkanes), which are much lower in abundance in the crude oil. This difference is attributable to the presence of a cracked gas oil component in the IFO 380 — which, of course, is absent in the crude oil. Inspection of the corresponding m/z 192 mass chromatograms (Figure 10-10C and D) reveals a distinct difference in the distribution of methyl-phenanthrene isomers and in the prominence of 2-methyl-anthracene. The IFO 380 fuel oil contains a greater proportion of 2- and 3-methyl-phenanthrene (i.e., the more thermally stable β-type isomers) and 2-methyl-anthracene than the crude oil. This difference would also be consistent with the presence of a cracked gas oil component in the IFO 380, in which thermal stability of the 2- and 3-methyl-phenanthrene (β-type) isomers and the formation of 2-methyl-anthracene produce a fingerprint (Figure 10-10C) that is distinct from the crude oil (Figure 10-10D).

In an oil spill scenario in which the IFO 380 shown in Figure 10-10A were to experience evaporative weathering, the presence of the cracked gas oil component in the GC/FID would not be nearly as obvious. Under this circumstance, the weathered IFO 380 might be confused with a weathered crude oil (such as in Figure 10-10B). However, despite weathering, the presence of 2-methyl-anthracene among the methyl-phenanthrenes would provide evidence that the spilled oil was actually HFO and not crude oil.

As described in Section 10.4.1, biomarker thermal maturity indicators may also provide a means to distinguish HFO from crude oils, where the samples in question otherwise share many chromatographic and chemical similarities. The basis for this is that HFO contains distillation residuums that are subject to alteration under the pressure and temperature conditions, particularly during vacuum distillation, that can alter thermal maturity parameters (Peters et al., 1992). Sometimes this alteration may be beyond reasonable for naturally heated crude oils (e.g., see outliers in Figure 10-4B).

An example of this potential is demonstrated in the biomarker-based thermal maturity ratios shown in Table 10-5. These data were collected from a crude oil and from five asphalt samples made from the vacuum distillation residuums of that crude oil. (Admittedly, petroleum asphalts are not HFOs, but both

Table 10-5 Biomarker-Based Thermal Maturity Parameters Measured in Five Asphalts (S_1–S_5) Manufactured with Vacuum Distillation Residuum Derived from the Parent Crude Oil Source

Thermal Parameters	Asphalt Samples					Parent Crude Oil	
	S1	S2	S3	S4	S5	Original	Duplicate
Ts[1]/(Ts + Tm[2])	0.52	0.50	0.48	0.56	0.44	0.29	0.33
X[3]/(X + Ts)	0.24	0.27	0.29	0.35	0.39	0.00	0.00
C29 Steranes (ββ)/(ββ + αα)	0.49	0.49	0.49	0.51	0.49	0.27	0.26

[1] Ts = 18α(H)-22,29,30-Trisnorneohopane.
[2] Tm = 17α(H)-22,29,30-Trisnorhopane.
[3] X = 17α(H)-Diahopane.

products contain the vacuum distillation residuums of crude oil that has been subjected to extensive heating and pressures.) Each of the thermal parameters in the asphalts is elevated relative to the parent crude oil. The increase in each of these parameters argues for the formation or release of the more thermally stable Ts, 17α(H)-diahopane, or 14β(H), 17β(H) C_{29} steranes during heating. In a spill scenario, the increased values of these parameters might allow for the distinction between a refined product containing vacuum residuums (in this case, asphalt) and crude oil.

10.6 Conclusion

Modern residual marine fuels are economically important fuels used throughout the world in the commercial maritime shipping industry. Once a simpler fuel composed largely of atmospheric residuums, modern (post-1970s) HFOs are mixtures of highly refined, atmospheric and vacuum residuum, blended with thermally and catalytically cracked gas oils or other (typically low-value) refinery intermediate stocks. The complexities and economics of producing modern HFOs have resulted in heavy fuel products of remarkable chemical diversity, which can prove useful in oil spill investigations relying upon chemical fingerprinting.

The compositional variety of 71 modern IFO 380 fuel oils obtained from vessels and ports around the world has been illustrated in this chapter. Often, distinctive gas chromatographic fingerprints of these fuels include a prominence of aromatic-rich gas oil range (C_{10}–C_{25}) components, admixed with a broad-boiling residual range (C_{25}–C_{45}) component consisting mostly of an unresolved complex mixture (UCM) and varying amounts of resolved paraffins and PAHs. As such, the unique gas chromatographic fingerprints of an HFO can potentially be used as a reliable indicator in an oil spill investigation — particularly in the early phase of a spill before significant weathering has occurred. However, the gas chromatographic fingerprint of an HFO will alter due to postspill weathering, thereby ultimately confounding the use of simple qualitative gas chromatographic comparisons in oil spill identification and tracking. Fortunately, the detailed molecular chemistry of HFOs — including petroleum biomarkers and PAHs — have been shown in this chapter to be very diverse, meaning that unique and reliable chemical metrics — source ratios — can be developed for use in sophisticated oil spill tracking and source correlation analyses. The use of appropriate forensic chemistry techniques, coupled with knowledge of the refining chemistry of HFOs and their behavior on water, affords oil spill investigators the best opportunity to confidently identify and track the fate of spilled heavy fuel oils in the environment.

References

Alimi, H., T. Ertel, and B. Schug, Fingerprinting of hydrocarbon fuel contaminants: Literature review. *Environ. Foren*, 2003, **4**, 25–38.

ASTM, D-3328, *Standard Test Methods for Comparison of Waterborne Petroleum Oils by Gas Chromatography*. ASTM International, 100 Bar Harbor Drive, PO Box C700, West Conshohocken, PA, USA, 1990.

ASTM, D-5739, *Standard Practice for Oil Spill Source Identification by Gas Chromatography and Positive Ion Electron Impact Low Resolution Mass Spectrometry*. ASTM International, 100 Bar Harbor Drive, PO Box C700, West Conshohocken, PA, USA, 2000.

Anderson, J.W., J.M. Neff, B.A. Cox, H.E. Tatem, and G.M. Hightower, Characteristics of dispersions of water soluble extracts of crude and refined oils and their toxicity to estuarine crustaceans and fish. *Mar. Biol.*, 1974, **27**, 75–88.

Atlas, R.M., Microbial degradation of petroleum hydrocarbons: An environmental perspective. *Microbio. Rev.*, 1981, 180–209.

Bobra, M., Solubility Behavior of Petroleum Oils in Water. Environmental Emergency Manuscript Report Number EE-130, Environment Canada, Ottawa, Ontario, 1992.

BunkerWorld, Bunker Fuel Grades — How do the four main fuel grades differ from each other? www.bunkerworld.com. London, UK, 2006.

Burns, W.A., P.J. Mankiewicz, A.E. Bence, D.S. Page, and K.R. Parker, A principal component and least squares method for allocating polycyclic aromatic hydrocarbons in a sediments to multiple sources. *Environ. Toxicol. Chem.*, 1997, **16**, 1119–1131.

Castle, R.W., F. Wehrenberg, J. Bartlett, and J. Nuckols, Heavy oil spills: Out of site, out of mind. *Proc. 1995 Oil Spill Conf*, U.S. Coast Guard, Amer. Petrol. Inst., U.S. EPA, Washington, DC, 1995, 565–571.

Daling, P.S., L.F. Faksness, A.B. Hansen, and S.A. Stout, Improved and standardized methodology for oil spill fingerprinting. *J. Env. Foren.*, 2002, **3**, 263–278–4.

Douglas, G.S., A.E. Bence, R.C. Prince, S.J. McMillen, and E.L. Butler, Environmental stability of selected petroleum hydrocarbon source and weathering ratios. *Environ. Sci. Technol.*, 1996, **30**, 2332–2339.

Durell, G.S., A.D. Uhler, S.A. Ostazeski, and A.B. Nordvik, An integrated approach to determining physico-chemical and molecular chemical characteristics of petroleum as a function of weathering. *Proc. Eighteenth Arctic and Marine Oil Spill Confe*. Environment Canada, Edmonton, Alberta, 1995.

Emsbo-Mattingly, S.M., A.D. Uhler, S.A. Stout, G.S. Douglas, K.J. McCarthy, and A. Coleman, Determining the source of PAHs in sediment. Second International Symposium and Exhibition on the Redevelopment of Manufactured Gas Plant Sites (MGP 2006), 2006.

Fingas, M., Heavy oil behavior in the ocean. *Proc. Tech. Assessment Research Prog. Offshore Minerals Operations Workshop*, Minerals Management Service, Herndon, VA, 1988, pp. 144–147.

Gary, J.H. and G.E. Handwerk, *Petroleum Refining*, 2nd ed. New York: Marcel Dekker, 1984.

Heildoff, C. and D. Stockman, ISL market analysis 2005. Major shipping countries. *Shipping Statistics Market Research*, October issue, 2005, pp. 3–8.

Hendrick, M.S. and T.R. Reilly, Evolution of the U.S. Coast Guard's oil identification system. *Proc. 1993 Intl. Oil Spill Conf.*, 1993, p. 873.

Henry, C.B., P.O. Roberts, and E.B. Overton, Advanced forensic chemistry of spilled oil: Self-normalizing fingerprint indices. In *Proc. 1997 Intl. Oil Spill Conf.*, Ft. Lauderdale, FL, 1997, pp. 936–937.

ISO, Standard 8217 *Specifications of marine fuels*, in Petroleum products — Fuels (class F) 3rd ed. 2005-11-01. International Organization for Standardization (ISO), Geneva, Switzerland, 2005.

ISO, Standard 8217 *Specifications of marine fuels*, in Petroleum products — Fuels (class F). International Organization for Standardization (ISO), Geneva, Switzerland, 1996.

Leffler, W.L., In *Petroleum Refining in Non-Technical Language*. Tulsa, OK: Penwell Corp., 2000.

Mackay, D. and C.D. McAuliffe, Fate of hydrocarbons discharged at sea. *Oil Chem. Pollut.*, 1988, **5**, 1–20.

Mazur, L., C. Milanes, K. Randles, and C. Salocks, Used oil in bunker fuel: A review of potential human health implications. Prepared for California EPA Office of Environmental Health Hazard Assessment, Sacramento, CA, 2004.

Michel, J. and J. Galt, Conditions under which floating oil slicks can sink in marine settings. *Proc. 1995 Intl. Oil Spill Conf.*, Washington, DC, 1995, pp. 573–576.

Michel, J. and M.O. Hayes, Persistence and weathering of *Exxon Valdez* oil in the intertidal zone 3.5 years later. *Proc. 1993 Intl. Oil Spill Conf.*, American Petroleum Institute, Washington DC, 1993, pp. 279–286.

Moldowan, J.M., W.K. Seifert, and E.J. Gallegos, Relationship between petroleum composition and depositional environment of petroleum source rocks. *The Amer. Assoc. Petroleum Geologists Bull.*, 1985, **69**(3), 1255–1268.

Morrison, R.D., *Environmental Forensics*. Boca Raton, FL: CRC Press, 2000.

National Academy of Sciences, *Spills of Nonfloating Oils: Risk and Response*. Committee on Marine Transportation of Heavy Oils. Washington, DC: National Academy Press, 1999.

Neff, J.M., S.A. Stout, P. Beall, and G.S. Durell, Review of environmental challenges of heavy oils. Report prepared by Battelle Memorial Institute for Statoil, Trondheim, Norway, 2003.

Newbery, P.J., Ongoing development of the international marine fuel standard. *Trans. IMarE.*, 1996, **108**, 241–257.

NOAA, Oil beneath the water surface and review of currently available literature on Group V oils: An annotated bibliography. Report HMRAD 95-8. January 1997 update. Seattle, WA: Hazardous Materials Response and Assessment Division, NOAA, 1997.

National Research Council (NRC), *Oil in the Sea: Inputs, Fates, and Effects*. Washington, DC: National Academy Press, 2003.

Nordtest, Nordtest Method for oil spill identification. NORDTEST, Esbo, Finland, 1991.

Ostazeski, S.A., G.S. Durell, K.J. McCarthy, S.C. Macomber, A.D. Uhler, and A. Nordvik, Oil weathering study of Maya crude oil. Tech. Report Series 95-026. Washington DC: Marine Spill Response Corporation, 1995a.

Ostazeski, S.A., G.S. Durell, K.J. McCarthy, S.C. Macomber, A.D. Uhler, and A. Nordvik, Oil weathering study *Morris J. Berman* cargo oil. Tech. Report Series 95-024. Washington DC: Marine Spill Response Corporation, 1995b.

Page, D.S., P.D. Boehm, G.S. Douglas, and A.E. Bence, Identification of hydrocarbon sources in the benthic sediments of Prince William Sound and the Gulf of Alaska following the *Exxon Valdez* oil spill. In P.G. Wells, J.N. Butler, and J.S. Hughes (eds.), Exxon Valdez *Oil Spill: Fate and Effects in Alaskan Waters.* ASTM STP 1219, Philadelphia, PA: American Society for Testing and Materials, 1995.

Payne, J.F., J.R. Phillips, and W. Hom, Transport and transformations: Water column processes. In *Long-Term Environmental Effects of Offshore Oil and Gas Development*, D.F. Boesch and N.N. Rabalais (eds.), London, UK: Elsevier Applied Science Publishers, 1987, pp. 175–231.

Peters, K.E., G.L. Scheuerman, C.Y. Lee, J.M. Moldowan, R.N. Reynolds, and M.M. Pena, Effects of refinery processes on biological markers. *Energy Fuels*, 1992, **6**, 560–577.

Peters, K.E., C.C. Walters, and J.M. Moldowan, *The Biomarker Guide.* Vol. 1. *Biomakers and Isotopes in the Environment and Human History.* Englewood Cliffs, NJ: Prentice Hall Inc., 2005.

Pieri, N., F. Jocquot, G. Mille, J.P. Planche, and J. Kister, GC-MS identification of biomarkers in road asphalts and in their parent crude oils. Relationship between crude oil maturity and asphalt reactivity towards weathering. *Org. Geochem.*, 1996, **25**(17), 51–68.

Prince, R.C., Petroleum spill biodegradation in the marine environment. *Crit. Rev. Microbiol.*, 1993, **19**, 217–242.

Prince, R.C., R.M. Garrett, R.E. Bare, M.J. Grossman, G.T. Townsend, J.M. Suflita, K. Lee, E.H. Owens, G.A. Sergy, J.F. Braddock, J.E. Lindstrom, and R.R. Lessard, The roles of photooxidation and biodegradation in long-term weathering of crude and heavy fuel oils. *Spill Sci. Technol. Bull.*, 2003, **8**, 145–156.

Radke, M., P. Garrigues, and H. Willsch, Methylated dicyclic and tricyclic aromatic hydrocarbons in crude oil from the Handil Field, Indonesia. *Org. Geochem.*, 1990, **15**(1), 17–34.

Radke, M., H. Willsch, and D. Leythaeuser, Aromatic components of coal: Relation of distribution pattern to rank. *Geochim. Cosmochim. Acta*, 1982, **46**, 1831–1848.

Rodgers, R., E.N. Blumer, M.A. Freitas, and A.G. Marshall, Complete compositional monitoring of the weathering of transportation fuels based on elemental compositions from Fourier transform ion cyclotron resonance mass spectrometry. *Environ. Sci. Technol.*, 2000, **34**, 1671–1678.

Shiu, W.Y., A.M. Bobra, L. Marjanen, L. Suntio, and D. Mackay, The water solubility of crude oils and petroleum products. *Oil Chem. Pollut.*, 1990, **7**, 57–84.

Stumm, W.M. and J.J. Morgan, *Aquatic Chemistry*, 3rd ed. New York: John Wiley & Sons, 1996.

Song, C., Introduction to chemistry of diesel fuels. In C. Song, C.S. Hsu, and I. Mochida (eds.), *Chemistry of Diesel Fuels.* Boca Raton, FL: Taylor & Francis, Inc., 2000, pp. 1–60.

Speight, J.G., *The Chemistry and Technology of Petroleum*, 2nd ed., New York: Marcel Dekker, Inc., 1991.

Stout, S.A., G.S. Douglas, A.D. Uhler, K.J. McCarthy, and S.D. Emsbo-Mattingly, Identifying the source of mystery waterborne oil spills — a case for quantitative chemical fingerprinting. *Env. Claims J.*, 2005, **17**(2), 71–88.

Stout, S.A., A.D. Uhler, K.J. McCarthy, and S.D. Emsbo-Mattingly, Chemical fingerprinting of hydrocarbons. In *Introduction to Environmental Forensics*, B. Murphy and R. Morrison (eds.), New York: Academic Press, 2002, pp. 135–260.

Stout, S.A., A.D. Uhler, and K.J. McCarthy, A strategy and methodology for defensibly correlating spilled oil to source candidates. *Environ. Forensics*, 2001, **2**, 87–98.

Thomas, R.F., Development of marine fuel standards. *Trans IMareE*, 1981, **93**, 1–10.

Uhler, A.D. and S.M. Emsbo-Mattingly, Environmental stability of PAH source indices in pyrogenic tars. *Bull. Environ. Contam. Tox.* 2006, **76**, 689–696.

Wang, Z. and M. Fingas, Development of oil hydrocarbon fingerprinting and identification techniques. *Mar Pollut Bull.*, 2003, **47**(9–12), 423–452.

Wang, Z. and M.F. Fingas, Study of the 22-year-old Arrow oil samples using biomarker compounds by GC/MS. *Environ. Sci. Technol.*, 1994, **28**, 1733–1746.

Winkler, M.F., Introduction to marine petroleum fuels. In S.R. Westbrook and R. Shah (eds.), *Fuels and Lubricants Handbook: Technology, Properties, Performance and Testing.* ASTM Manual Series: MNL37WCD. West Conshohocken, PA: ASTM International, 2003.

11 Biodegradation of Oil Hydrocarbons and Its Implications for Source Identification

Roger C. Prince[1] and Clifford C. Walters[2]

[1] ExxonMobil Biomedical Sciences Inc. and [2] ExxonMobil Research and Engineering Co., 1545 Route 22 East, Annandale, NJ 08801-0998.

11.1 Introduction

Hydrocarbons have been part of the biosphere from its inception, produced initially by prebiotic processes, and subsequently both by living organisms and during the generation of fossil fuels. As highly reduced forms of carbon, hydrocarbons provide a rich source of energy and carbon to those organisms, typically microorganisms, which are able to consume them. Indeed, almost all hydrocarbons are readily degraded under appropriate conditions. There is, nevertheless, a clear preference for the catabolism of some molecules before others; hence, the composition of a fuel or crude oil changes as biodegradation proceeds.

This chapter addresses the fundamentals of hydrocarbon biodegradation, especially of liquid fossil fuels, and attempts to bring together the conclusions from two rather disparate areas of research. One is from the community of microbiologists and environmental scientists studying biodegradation in the laboratory and the field. The other is from the community of geochemists studying petroleum in reservoirs. The potential timescales for these processes may be quite different. Laboratory and environmental studies of hydrocarbon biodegradation occur over days to years whereas deep subsurface reservoirs may be filled with oil for millions of years.

Nevertheless, both groups agree that biodegradation leaves distinctive molecular and isotopic fingerprints in the residual material that reflect the initial nondegraded composition and the nature and extent of the microbial alteration. For environmental scientists, these molecular signatures can provide information on the type and source of contamination at spill sites, allow extrapolation of how much future biodegradation may be expected, and perhaps suggest ways of speeding up this process. For geochemists, the fingerprints can provide important information on the source rocks that matured to form the petroleum and the processes that may be altering oil quality along migration pathways or while contained in reservoir rocks.

11.2 Biochemistry of Petroleum Biodegradation

Hydrocarbons pervade the biosphere, albeit usually at low levels. They are made by many plants, animals, and microorganisms, and are a significant part of the great carbon cycle of our planet (Berner, 2003). They also are present in fossil fuels, especially petroleums (literally rock oils), which result from the combined effects of temperature and pressure on buried biomass (kerogen) over prolonged time (Tissot and Welte, 1984). Petroleum has been leaking to the surface for a very long time. There is good evidence for active petroleum generation, and likely leakage, during the early Precambrian, 2.63 to 3.2 billion years ago (Rasmussen, 2005), even before earth's atmosphere had any appreciable oxygen (Kasting, 2004).

Biogenic and petrogenic hydrocarbons provide a rich source of energy and carbon for those organisms able to degrade them. Species in some 90 genera of bacteria grow on hydrocarbons (Prince, 2005), and many more degrade hydrocarbons while growing on other substrates. Species in more than 100 genera of fungi also have been shown to degrade hydrocarbons (Prince, 2005). Just which of these organisms, or indeed others yet to be characterized, are likely to be present during active petroleum biodegradation is probably determined by the availability of suitable terminal electron acceptors, the actual hydrocarbons present, and perhaps the history of the local environment. Some organisms have the ability to grow on a broad range of hydrocarbons, others on only a few, and the microbial ecology likely changes as the most readily degradable hydrocarbons are consumed from a petroleum mixture, and/or the availability of oxidants changes.

Before beginning a discussion of the biodegradation of petroleum hydrocarbons, it is necessary to have an overview of the composition of the various hydrocarbon fuels in commerce. Crude oils are produced all over the world, and are typically transported long distances through pipelines, by rail, and by sea before they are refined. These oils range in quality from clear, volatile fluids to near-solid, highly viscous asphalts. The density of a crude oil is the principal determinant of its value, and this is described by the API (American Petroleum Institute) gravity.

$$\text{API Gravity} = \frac{141.5}{\text{specific gravity}} - 131.5$$

expressed as degrees (°). Thus, water has an API gravity of 10°. Oils with API gravities greater than 40° are usually said to be light oils, while those with API gravities of less than about 17° are said to be heavy. Note that almost all transported oils float on water; only those with API gravities <10 will sink in fresh water.

Light oils and condensates range from orange-yellow to clear in color and are composed mostly of volatile ($<C_{10}$) saturated and aromatic hydrocarbons. These light fluids may contain minor to trace amounts of large polynuclear aromatic hydrocarbons (PAHs) and sulfur-containing compounds such as mercaptans. Crude oils are typically brown to black, principally due to large polycyclic aromatic hydrocarbons and light-absorbing heterocyclic molecules such as aliphatic and aromatic sulfides, nitrogen-containing pyrroles and pyridines, and oxygen-containing phenols, acids, and furans. These heteroatomic molecules, known variously as polars, NSO-compounds, resins, or asphaltenes, are still principally composed of carbon and hydrogen, but contain one or more sulfur, nitrogen, and/or oxygen atoms. Most are not amenable to gas chromatography and remain relatively uncharacterized (Sirota, 2005). The average composition of crude oils is ~57% saturated hydrocarbons, 29% aromatic hydrocarbon, and ~14% polars (Tissot and Welte, 1984). Aromatic hydrocarbons are defined as those containing one or more aromatic ring and include compounds with substantial alkyl substitution and/or the addition of saturated (alicyclic) rings. The saturated hydrocarbons are themselves approximately half acyclic (known as paraffins in the oil industry) and half alicyclic (known as naphthenes in the oil industry), so there is typically a rough parity between the concentration of paraffins, naphthenes, and aromatics in most crude oils, albeit with wide variation in unusual or altered examples.

Refineries convert crude oils into a range of valuable products, especially fuels. Refining starts with distillation, and the simplest distinction of the various refined products can be related to boiling-point distributions. The most volatile liquid fuel is aviation gasoline, followed by automobile gasoline, jet fuels, diesel and heating oils, and then the heavy oils used for fueling ships and some electrical generation. All sizeable ships contain quite large volumes of heavy fuel oil, often known as Bunker C, which is barely liquid at ambient temperatures and must be kept warm to be pumped into the engines. Distillation cuts correspond roughly with carbon number distribution. Most of the molecules in gasoline have between 4 and 10 carbons, most in diesel fuel have between 9 and 20, and heavy fuel oils typically have very few molecules with less than 15 carbon atoms except for those added as diluent to achieve the appropriate viscosity. The lightest products, those with the lowest boiling points such as gasolines and diesels, are almost entirely hydrocarbons, while the heavy fuel oils are enriched in the polar constituents such as asphaltenes; this is reflected in the color of the products.

It is important to recognize that fuels are graded and sold based on their properties, such as octane- or cetane-rating or viscosity, and not on their molecular composition. There are many petroleum mixtures that meet specific product specifications. Refineries manipulate the chemical composition of the initial distillates to satisfy these requirements, and fuels with the same name can have very different chemical compositions even though all meet their specifications.

Despite the best efforts of producers, refiners, and consumers, petroleum and petroleum products invariably get spilled into the environment. Fortunately, the vast majority of the hydrocarbons in petroleum and refined products is biodegradable — the focus of this chapter. Understanding this process is important for effective cleanup of accidental oil spills, explaining the alteration that occurs in natural oil seeps, and predicting the quality of oil in potential new discoveries.

Since aerobes activate hydrocarbons by the introduction of free oxygen, whereas anaerobes must use alternative pathways, we will address aerobes separately. Environmental conditions, such as moisture, temperature, salinity, availability of trace nutrients, and composition of the fuel, further constrain and define the microbial ecology.

11.2.1 Aerobic Biodegradation of Hydrocarbons

Oxygen is both an essential reactant in the initial activation of hydrocarbons under aerobic conditions and the terminal electron acceptor for microbial growth. Hydrocarbons are initially activated through the addition of either one or both atoms of diatomic oxygen by enzymes known, respectively, as monooxygenases and dioxygenases (Figures 11-1 and 11-2) (Fritsche and Hofrichter, 2000; Kanaly

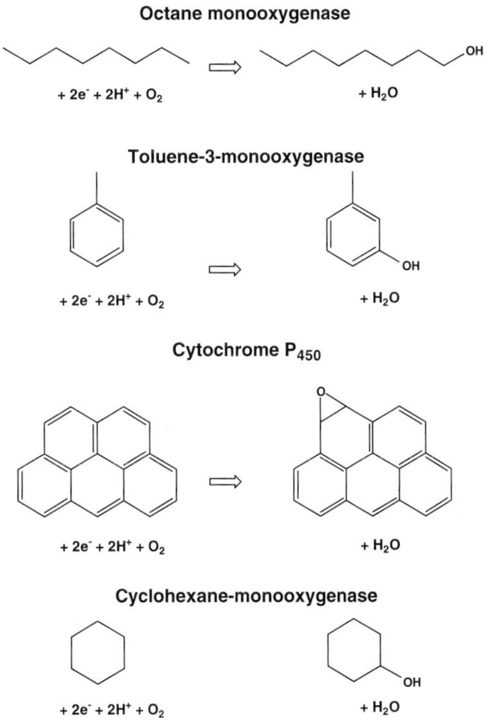

Figure 11-1 Typical reactions catalyzed by hydrocarbon monooxygenases. Octane monooxygenase of *Pseudomonas putida* catalyzes the oxidation of octane to octanol (van Beilen et al., 1994), toluene-3-monooxygenase of *Ralstonia pickettii* catalyzes the oxidation of toluene to 3-cresol (Tao et al., 2004), the cytochrome P450 of *Mycobacterium vanbaalenii* apparently catalyzes the oxidation of benzo[*a*]pyrene to benzo[*a*]pyrene-11,12-epoxide (Moody et al., 2004), and the cyclohexane monooxygenase of *Brachymonas petroleovorans* oxidizes cyclohexane to cyclohexanol (Brzostowicz et al., 2005). Other enzymes catalyze the subterminal oxidation of alkanes (Ludwig et al., 1995), and the oxidation of the 2- or 3-position of toluene (Yeager et al., 1999; Tao et al., 2004).

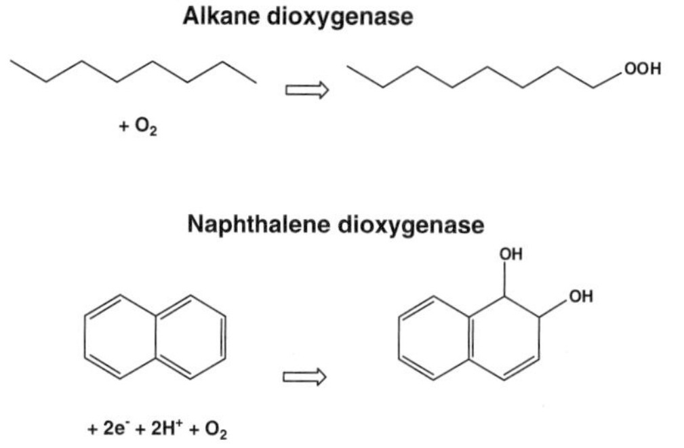

Figure 11-2 Typical reactions catalyzed by hydrocarbon dioxygenases. The alkane dioxygenase of an *Acinetobacter* apparently converts alkanes to alkanehydroperoxides (Maeng et al., 1996). Naphthalene dioxygenase of *Pseudomonas putida* oxidizes naphthalene to *cis*-naphthalene-1,2-dihydrodiol (Karlsson et al., 2003).

and Harayama, 2000; Arp et al., 2001; Parales et al., 2002; Prince, 2002; Karlsson et al., 2003; Leahy et al., 2003; Van Hamme et al., 2003; Hlavica, 2004; van Beilen and Witholt, 2005). The alkane dioxygenase of Figure 11-2 is enigmatic; it was initially proposed by Finnerty (1977) and received some support from the work of Maeng et al. (1996), but is otherwise unexplored. All other enzymes shown in Figures 11-1 and 11-2 require a source of reductant, in the form of NADH (reduced nicotinamide adenine dinucleotide), but subsequent oxidation of the oxygenated products returns this investment.

Once at least one oxygen atom has been added to a hydrocarbon, the molecule is generally amenable to manipulation by the central metabolism of the cell. Alkane alcohols (Figure 11-1) are oxidized to acids, attached to Coenzyme A, and directed to the lipid catabolic pathways. Saturated oxygenated rings, such as cyclohexanol (Figure 11-1) or cyclododecanol, are oxidized to the ketones and then by a monooxygenase to the lactone, which is hydrolyzed to the dicarboxylic acid (Cheng et al., 2002; Brzostowicz et al., 2003). These, too, are directed into the lipid catabolic pathways.

Complex branching hinders both the initial oxidation and the subsequent lipid catabolism, apparently because tertiary and quaternary carbon atoms interfere by steric hindrance with the oxidation enzymes. Thus, branched hydrocarbons are generally more resistant to degradation than linear ones, although they are eventually consumed. One way an organism can bypass tertiary and quaternary carbon atom blockage of β-oxidation is by degrading the chain from both ends, ω-oxidation. Another way is to produce different enzymes with the ability to degrade branched molecules, as elegantly demonstrated by Pirnik et al. (1974) with a *Brevibacterium*. When provided with both *n*-hexadecane and pristane (2,6,10,14-tetramethylpentadecane), the organism did not degrade pristane until the *n*-hexadecane had been degraded to less than 5% of the total hydrocarbon; afterwards, pristane was completely consumed.

This phenomenon has been used to identify the onset of biodegradation in the field. Crude oils, diesels, and heavy fuel oils usually contain both *n*-heptadecane and *n*-octadecane, together with pristane and phytane (2,6,10,14-tetramethylhexadecane), which elute just after the respective *n*-alkanes in typical gas chromatography. Miget et al. (1969) seem to have been the first to realize that a decrease in the ratio of *n*-heptadecane to pristane or *n*-octadecane to phytane presaged the onset of biodegradation.

Aromatic compounds pose the additional hurdle of opening the ring. Many pathways have been described that convert the products of initial monooxygenase activation (Figure 11-1) to dihydroxybenzoates or catechols (Ellis et al., 2006), and dioxygenases produce catechols directly (Figure 11-2). Catechols are then cleaved by extradiol or intradiol dioxygenases to linear acids that can enter classical catabolic pathways (Figure 11-3).

Most natural petroleums contain only trace amounts of alkenes. A few crude oils contain minor amounts from radiolysis (Frolov et al., 1998), but alkenes can be quite abundant in refined products such as gasoline. These are readily degraded (Solano-Serena et al., 1999, 2001). Small alkenes such as propylene and ethene are activated by specific monooxygenases to generate epoxides, which are then carboxylated with the aid of Coenzyme M and

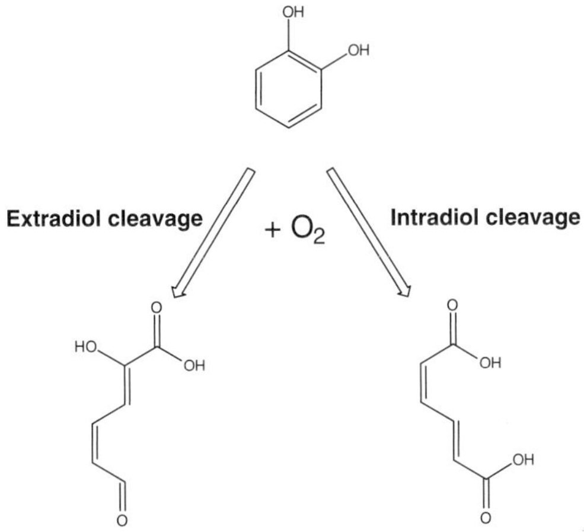

Figure 11-3 The opening of aromatic rings by extradiol and intradiol dioxygenases. Examples include the extradiol protocatechuate 4,5-dioxygenase of *Sphingomonas paucimobilis* (Sugimoto et al., 1999) and the intradiol protocatechuate 3, 4-dioxygenase of *Acinetobacter* (Vetting et al., 2000).

delivered to central metabolism (Hartmans et al., 1989; Allen et al., 1999; Ensign, 2001). Larger alkenes are dioxygenated by toluene dioxygenase (Lange and Wackett, 1997).

Further diversity in hydrocarbon catabolism is seen in the degradation of molecules that combine aromatic, alicyclic, and/or alkyl parts. For example, an *Alcanivorax* initially isolated by its ability to degrade *n*-alkanes starts degrading long-chain alkyl substituted cyclohexanes and benzenes with the oxidation of the terminal alkyl carbon (Dutta and Harayama, 2001). Whether such compounds also can be degraded by an initial oxidation of the ring is unresolved, but seems likely (Dutta, 2005). The initial oxidation of tetralin (1,2,3,4-tetrahydronaphthalene), which consists of an aromatic ring joined to a saturated cyclohexane, is by a monooxygenase on the alicyclic ring in a *Pseudomonas stutzeri*, but by a dioxygenase on the aromatic ring in *Sphingomonas macrogolitabida* (Martínez-Pérez et al., 2004).

An interesting caveat of the foregoing discussion is that many (although not all) of the enzymes capable of hydrocarbon oxidation have quite broad specificities. For example, the well-studied alkane monooxygenase (also known as alkane hydroxylase) of *Pseudomonas putida* (formerly *Ps. oleovorans*) of Figure 11-1 is capable of oxidizing *n*-alkanes with 5 to 12 carbons (*n*-pentane to *n*-dodecane), along with methyl- and, in many cases, dimethyl-forms (van Beilen et al., 1994). It is also capable of oxidizing cyclopentane, cyclohexane, methyl- and ethyl-cyclohexane, ethylbenzene, and several other aromatics. In fact, the enzyme oxidizes 1,3-diethylbenzene faster than it oxidizes *n*-nonane, its preferred alkane substrate. It is also a very good oxidizer of 1,4-diethylbenzene, but has very low activity with the 1,2-isomer (van Beilen et al., 1994).

Similarly, aromatic dioxygenases, such as those exemplified in Figure 11-2, often have a very broad specificity that typically extends to heterocyclic aromatics, saturated rings, and linear alkenes (Gülensoy and Alvarez, 1999; Gibson and Parales, 2000; Boyd et al., 2001; Kasai et al., 2003) although not, as far as is known, to alkanes. The potential evolutionary relationship of these enzymes to the alkane dioxygenase of Figure 11-2 is unknown. Many aromatic monooxygenases also have quite broad specificities (Sazinsky et al., 2004). For example, the appetites of four different organisms, all grown on toluene, for the C3-benzenes in gasoline (trimethyl, methyl-ethyl, propyl, and isopropyl benzenes) are distinctly different (Figure 11-4).

Among the most resistant resolvable saturated hydrocarbons under aerobic conditions are the hopanes and steranes, which are molecular fossils of the biomass that gave rise to the original kerogen (Ourisson and Albrecht, 1992; Peters et al., 2005). Some, such as the cholestanes, degrade at about the same rate as the methylated polycyclic aromatics (Prince et al., 2002). Others, such as C_{30}-hopane, are so sufficiently resistant to biodegradation that they have proved useful conserved internal markers for following biodegradation in the field. Nevertheless, several reports of hopane biodegradation have now appeared (Tritz et al., 1999; Bost et al., 2001; Frontera-Suau et al., 2002; Watson et al., 2002; Huesemann

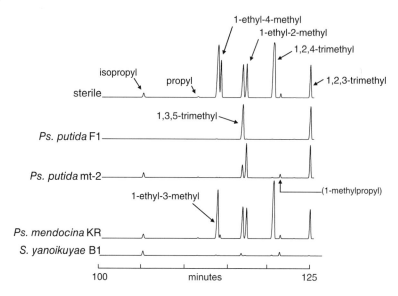

Figure 11-4 The biodegradation of trimethyl, methyl-ethyl, propyl, and isopropyl benzenes by four organisms. *Pseudomonas putida* F1 degrades toluene with a dioxygenase (Cho et al., 2000), while *Ps. putida* mt-2 (Bühler et al., 2000), and *Ps. mendocina* KR (Tao et al., 2004) use monooxygenases, directed at the methyl group and the *para*-position, respectively. *Sphingomonas yanoikuyae* B1 possesses a biphenyl dioxygenase and a xylene monooxygenase (Kim and Zylstra, 1999), and it is certainly possible that all strains contain other hydrocarbon-oxidizing enzymes. These experiments (Prince, V.L., Zylstra, G.J., and Prince, R.C., unpublished) used cells initially grown with vapor phase toluene that were washed and resuspended in minimal medium with 1 μl of gasoline in 10 ml of culture in a 40-ml vial and incubated for 90–113 hr. The traces are for the $m/z = 105$ ion.

et al., 2003). Interestingly, several of the papers report biodegradation products such as carboxylic acids (Watson et al., 2002) or olefins (Tritz et al., 1999), which are quite different from the putative biodegradation products seen in biodegraded crude oils from reservoirs that we discuss ahead.

Thus, most petroleum hydrocarbons are biodegradable under aerobic conditions. Refined products such as gasoline, diesel, and jet fuel are essentially completely degradable (Eriksson et al., 1998; Solano-Serena et al., 1999, 2001; Marchal et al., 2003, Penet et al., 2004). Crude oils and heavy fuel oils, however, contain molecules that are very resistant to biodegradation, at least in the short term. For example, McMillen et al. (1995) compared the short-term biodegradability of 17 crude oils in soil microcosms and found that the API gravity was the most useful predictor of biodegradability, at least for the most degradable fraction of the oils. With 0.5 wt.% oil in a loam soil, with appropriate moisture, nutrients, and aeration, more than 61% of the most degradable oil (API = 46°) was lost in four weeks, while only 10% of the least degradable oil (API = 15°) was consumed under the same conditions. Further degradation occurs on a longer timescale; for example, some samples collected from the Baffin Island Oil Spill (BIOS) site (Sergy and Blackall, 1987) 20 years after the deliberate experimental spill of a medium crude oil (Lago Medio, API = 32°), had lost more than 87% of their initial hydrocarbons (Prince et al., 2002).

Aerobic biodegradation of crude oils and refined products seems to follow a clear progression. The most readily degraded compounds are the normal alkanes larger than hexane as well as aromatics, especially simply substituted benzenes and naphthalenes (up to at least four- or five-pendant carbons). The smaller normal and the branched alkanes, such as pristane and phytane, and the monocycloalkanes are biodegraded at a slightly slower rate, as are the phenanthrenes and dibenzothiophenes. Slower still are the larger polycyclic aromatics, such as the chrysenes and pyrenes, and the larger heterocyclics. A clear pattern of biodegradation is seen within the alkylated three-ring polycyclic aromatics; the parent compound is degraded more rapidly than the various methyl forms, which in turn are degraded more rapidly than the dimethyl forms, in turn more rapidly than the trimethyl forms, etc. (Elmendorf et al., 1994; Douglas et al., 1996; Prince et al., 2003). While individual organisms show clear preferences for degrading particular benzene isomers (Figure 11-4), samples collected from the field or laboratory samples from experiments with enrichment cultures do not usually show such obvious preferences.

11.2.2 Anaerobic Biodegradation of Hydrocarbons

Because no compounds can substitute for molecular oxygen in hydrocarbon-activating oxygenase enzymes, it was long thought that hydrocarbons could not be biodegraded under anoxic conditions. We now know that many simple hydrocarbons are consumed under a variety of oxygen-free conditions, including sulfate-reducing, nitrate-reducing, perchlorate-reducing, ferric ion-reducing, humic acid-reducing, and methanogenic conditions (Cervantes et al., 2000; Meckenstock et al., 2000; Spormann and Widdel, 2000; Zwolinski et al., 2000; Widdel and Rabus, 2001; Chakraborty and Coates, 2004; Meckenstock et al., 2004; Gieg and Suflita, 2005; Rabus, 2005). Despite this diversity of anaerobic metabolic conditions, the activation of hydrocarbons appears to follow certain themes. Most common seems to be the addition of fumarate, yielding a substituted succinate (Figure 11-5), which has been demonstrated for alkanes, aromatics, and naphthenes (Widdel and Rabus, 2001; Rios-Hernandez et al., 2003; Meckenstock et al., 2004; Cravo-Laureau et al., 2005). Subsequent metabolism and addition of Coenzyme A deliver the hydrocarbon to the lipid catabolic pathways of the cell. Apparently less common, but implicated for both alkanes and polycyclic aromatics, is the addition of inorganic bicarbonate/carbonate/CO_2 (Figure 11-6), perhaps after a transient initial hydroxylation of the substrate, as has been shown for benzene (Chakraborty and Coates, 2005; Ulrich et al., 2005). There are suggestions that methylation also can occur (Rabus, 2005; Ulrich et al., 2005; Safinowski and Meckenstock, 2006). *Azoarcus* and related organisms oxidize ethylbenzene, propylbenzene, and 3-methyl-2-pentene (but not 3-methyl-1-pentene) to ethanols, with the oxygen coming from water (Figure 11-6). So far, this activity seems restricted to these substrates, and the activation of these substrates by fumarate addition is more common (Figure 11-5).

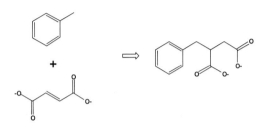

Figure 11-5 Anaerobic hydrocarbon activation by fumarate addition. The denitrifying bacterium HxN1 catalyzes the production of 1-methylpentylsuccinate from hexane (Rabus et al., 2001) and Biegert et al. (1996) were the first to demonstrate the production of benzylsuccinate under anaerobic conditions with *Thauera aromatica*. Similar reactions have been seen under a variety of conditions by a diverse array of organisms (Rabus, 2005).

Although anaerobic hydrocarbon microbiology is still in its infancy, a broad range of compounds have been found to be degraded,

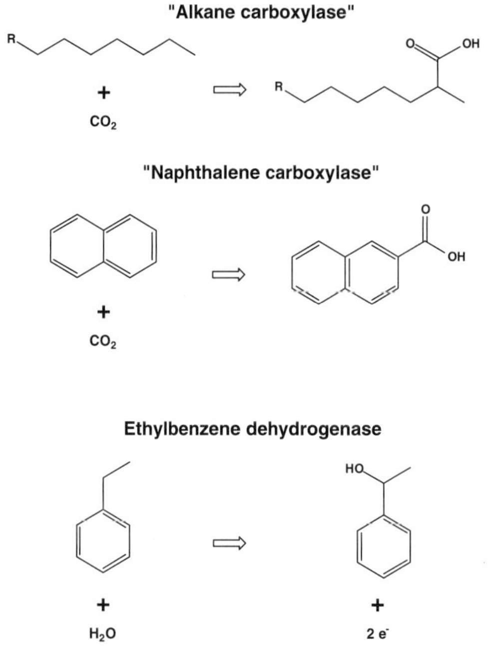

Figure 11-6 Anaerobic hydrocarbon activation by carboxylation and dehydration. Carboxylation at the 2-carbon of hexadecane by the sulfate-reducing strain AK-01 was reported by So and Young (1999). Carboxylation at the 3-position was subsequently reported for strain Hxd3 (So et al., 2003). Carboxylation of naphthalene by a sulfate-reducing enrichment culture was reported by Zhang and Young (1997). Recent work on benzene degradation (Chakraborty and Coates, 2005; Ulrich et al., 2005) suggests that a transient hydroxylation precedes carboxylation. While it seems clear that labeled bicarbonate is incorporated into the hydrocarbons, the enzymes responsible have not been characterized. In contrast, ethylbenzene dehydrogenase from *Azoarcus* has been well-characterized (Johnson et al., 2001; Kniemeyer and Heider, 2001); it contains the molybdopterin cofactor.

including alkanes up to n-C_{34} (Caldwell et al., 1998), phytane (Grossi et al., 2000), tetralin (Annweiler et al., 2002), polycyclic aromatic compounds with three and more rings (Chang and Shiung, 2002; Rothermich et al., 2002; Chang et al., 2003; Meckenstock et al., 2004), thiophenes and dibenzothiophenes (Annweiler et al., 2001; Onodera-Yamada et al., 2001; Marcelis et al., 2003), and cycloalkanes (Townsend et al., 2004). Interestingly, different strains seem to show clear steric preferences in their biodegradation; for example, various microorganisms have quite distinct preferences for the different xylene isomers (Suflita et al., 2004; Rabus, 2005).

One clear distinction from aerobic degradation is the apparent preference for degrading alicyclic compounds. These were among the most rapidly degraded compounds under sulfate-reducing and methanogenic conditions in the experiments of Townsend et al. (2004), yet they are among the last to go under aerobic conditions. Furthermore, Townsend et al. (2004) saw significant biodegradation of only *trans*-1,2-dimethylcyclopentane and *cis*-1,3-dimethylcyclohexane of all the dimethyl isomers of these alicyclic compounds in the gasoline they were using, and only under sulfate-reducing, not methanogenic conditions. The parent, methyl-, and ethyl-substituted compounds also were degraded. We know of no reports of such remarkable specificity in aerobic experiments with mixed microbial populations, but note that this may reflect a limited microbial population in the contaminated subsurface aquifer used as the source of inoculum in the anaerobic experiments compared to the potentially much broader diversity in well-mixed aerobic environments. We will return to this topic later.

The number of specific organisms known to degrade hydrocarbons under anaerobic conditions (Rabus, 2005) is so far much smaller than the number known to degrade hydrocarbons by aerobic respiration (Prince, 2005). All are Proteobacteria, a super-phylum of gram-negative bacteria whose members grow with a broad diversity of catabolic pathways and terminal electron acceptors, not to mention by anaerobic photosynthesis (Gupta, 2000). No isolated archaeal culture, grown under oxic or anoxic conditions, has been shown to degrade hydrocarbons (see Prince, 2005), but methanogenic archaea must be members of hydrocarbon-degrading consortia that operate under methanogenic conditions (Caldwell et al., 1998; Zengler et al., 1999; Townsend et al., 2003). Three groups of syntrophic microorganisms may be required: hydrocarbon-degrading bacteria; acetogenic bacteria; and methanogens (Figure 11-7). Although it has not been clearly demonstrated to date, it

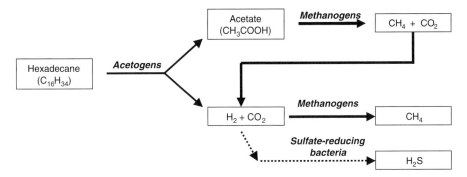

Figure 11-7 Anaerobic degradation of hexadecane under methanogenic conditions. A consortium of syntrophic eubacteria and archaea can degrade hexadecane to CH_4, CO_2, and H_2S. [Figure adapted from Parkes (1999) and Zengler et al. (1999)].

is likely that methanogenesis, including CO_2 reduction, only occurs in the absence of any other terminal electron acceptor, where the methanogen is acting as the "electron acceptor" of the otherwise anaerobically respiring hydrocarbon degrader.

The apparent paucity of anaerobic hydrocarbon degraders may reflect the bias of laboratory culture experiments. The identified Proteobacteria that grow in pure cultures are capable of utilizing specific hydrocarbons as their sole carbon source, but prefer other organic substrates, and growth on hydrocarbons is slow and considered stressed. Other bacteria are capable of degrading hydrocarbons under anaerobic conditions, but not as their sole carbon source. For example, Grishchenkov et al. (2002) reported degradation of biphenyl by *Citrobacter freundii* (a nitrate-reducing facultative anaerobe of the Proteobacteria) when grown along with culture media. Several recent studies have shown that anaerobic enrichment cultures are capable of growing on hydrocarbon mixtures (e.g., Rothermich et al., 2002; Eriksson et al., 2003; Noh et al., 2003; Rios-Hernandez et al., 2003) or whole oils (Townsend et al., 2003), and it may turn out that these cultures rely on interactions between organisms, and, thus, not be amenable to growth as pure cultures. The other potential explanation, that we will discuss in more detail ahead, is that almost by definition, anaerobic environments are not readily accessible to pelagic and air-borne invasion, so each site may have only a limited subset of the total biodiversity for this ecotype; perhaps there are far more types of anaerobes than we currently expect (Fraser, 2004), but they are geographically isolated.

11.3 Subsurface Biodegradation of Petroleum

The amount of biodegraded oil, worldwide, exceeds that of nondegraded, conventional oil. For example, the Orinoco Heavy Oil Belt in eastern Venezuela and the collective tar sands of Western Canada (e.g., Athabasca, Cold Lake, Wabasca, and Peace River) are estimated to contain ~1.2×10^{12} barrels of degraded oil each, while the supergiant oil field of Ghawar in Saudi Arabia contains ~1.9×10^{11} barrels of nondegraded oil (Roadifer, 1987). How is this biodegradation related to our understanding of biodegradation in the laboratory and the field that was discussed above?

The first thing to note is that the scales of the processes are vastly different, both in time and in volume. The vast majority of even large oil spills disappears from marine spill sites in a few years even without human help (Kingston, 2002), and although some of that oil may still be in the biosphere, its concentration is so low that it is rarely found. In contrast, biodegradation in reservoirs may occur over many thousands if not millions of years.

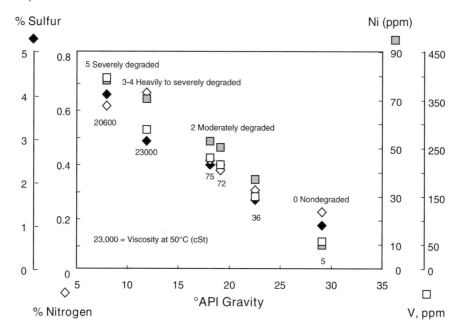

Figure 11-8 Bulk oil properties for a suite of oils (La Luna source) from Eastern Venezuela. Sulfur, nitrogen, nickel, and vanadium increase proportionally with decreasing API gravity. The degree of biodegradation is indicated by the numerical ranking of the Biomarker Biodegradation Scale (Peters et al., 2005) and by descriptive terms used by Wenger et al. (2002).

Economic reservoirs contain millions to billions of barrels of oil that even if severely biodegraded are worthwhile resources. Only the largest oil releases, such as occurred during the first Persian Gulf War or during the Ixtoc-1 blowout, are comparable in volume to minor subsurface reserves.

The second difference between laboratory, field, and reservoir biodegradation is that our understanding of the types of biodegradation becomes less clear as we move from laboratory biodegradation experiments, through monitoring biodegradation of subsurface spills in the field, to assessing biodegradation in oil reservoirs. We usually know the initial chemical composition of most surface spills and can be fairly confident in assigning biodegradative losses (Prince et al., 2002). In contrast, oil in reservoirs is usually fairly homogeneous, and while there are sometimes differences in composition from different levels or wells within a given play, it is uncommon to find equivalent, undegraded oil to establish the initial composition (although see Figures 11-8 and 11-9). Obviously, the redox conditions, whether aerobic or anaerobic, and available terminal electron acceptors are well-controlled in laboratory studies and can often be inferred from subsurface spills into near-surface aquifers. Redox conditions and the availability of electron acceptors are generally not known in deep subsurface reservoirs and, furthermore, may have varied over time. Thus, there is not yet a clear view of the relative contributions of aerobic and anaerobic biodegradation to the total biodegradative process in reservoirs, although forensic geochemistry is beginning to be applied to unravel these effects.

In any case, the oil exploration industry has considerable interest in biodegradation, as it can greatly impact the economic value and producibility of petroleum. Just as in laboratory experiments, saturated and aromatic hydrocarbons are consumed preferentially, and heterotomic compounds increase in relative abundance. Hence, petroleum quality

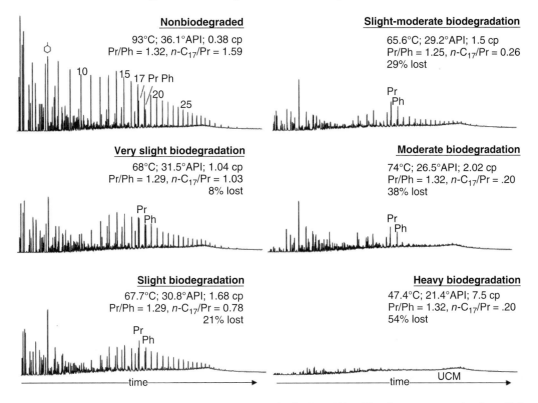

Figure 11-9 Gas chromatograms for a suite of selected crude oils from Africa. The oils were generated and expelled from the same source rock under comparable thermal conditions, but indicate quite different amounts of *in situ* biodegradation. Shown are reservoir temperature, API gravity, viscosity, pristane/phytane (Pr/Ph) and nC_{17}/ pristane (nC_{17}/Pr) ratios, and mass percentage lost calculated using hopane as a conserved internal standard. Number labels indicate *n*-alkane carbon number [modified from Wenger et al. (2002)].

decreases (i.e., the residuum has higher specific gravity, viscosity, and concentrations of heteroatoms and trace metals), and the volume of producible liquids decreases as gases and nonproducible solid bitumens increase.

The first indications of oil biodegradation typically involve the selective removal of C_6–C_{12} normal alkanes and small aromatics (Figure 11-9), although the latter may also be lost by water washing. As biodegradation proceeds, saturated hydrocarbons outside the initial range are selectively removed, with normal alkanes removed more rapidly than mono- and multimethylated alkanes. As the major resolved compounds diminish, the chromatographic baseline hump becomes more prominent. This hump is called the "unresolved complex mixture" (UCM) and consists of bioresistant compounds, including highly branched and cyclic saturated, aromatic, naphthenoaromatic, and polar compounds (Sutton et al., 2005). Depending on the amount of high-molecular-weight polar compounds and asphaltenes that are initially present in nondegraded petroleum, the UCM may account for nearly all to less than half of the total mass of a highly degraded oil. We reiterate that most of the polar and asphaltene compounds do not volatilize and are not detected by whole-oil gas chromatographic analysis.

Since normal and branched paraffins typically constitute ~35–50% of the hydrocarbons of a nonbiodegraded oil, their removal greatly alters the physical properties and economic value of a crude oil. Wenger et al. (2002) illustrated the impact of biodegradation on oil quality by analyzing a suite of petroleum samples that were generated from the same marine shale source facies and expelled at about the same level of maturity (Figure 11-9). Correlation of the oils was made using bioresistant cyclic saturated and aromatic hydrocarbons, including many biomarkers (Peters et al., 2005), that are unaltered at this level of biodegradation. The differences in physical properties between the nondegraded and heavily degraded oils are considerable; API gravity changes from 36.1–21.4° and viscosity increases from 0.38–7.5 centiPoise, respectively. Even mild to moderate levels of microbial alteration can have a strong influence on whether reservoir accumulations are deemed suitable for economic development.

Saturated and aromatic biomarkers are biodegraded only after consumption of *n*-alkanes, most simple branched alkanes, and some of the alkylated aromatics (Seifert and Moldowan, 1979; Seifert et al., 1984; Moldowan et al., 1992; Peters et al., 2005). Laboratory and empirical observations indicate that these compounds also are consumed in a preferential order. Regular steranes and alkylated aromatics are the most susceptible to biodegradation, followed by hopanes, aromatic steroidal hydrocarbons, diasteranes, and tricyclic terpanes. At advanced stages of alteration, certain biomarkers, such as 25-norhopanes and secohopanes, appear to be created.

Because the polar (NSO) fraction is enriched by biodegradation, these molecules were long considered to be very resistant to microbial alteration. In fact, while some species, such as porphyrins, are highly conserved, others such as phenols are removed at rates comparable to aromatic hydrocarbons, while pyrroles may persist with altered alkyl sidechains. Total acid abundance increases with increasing biodegradation, but some acyclic and monocyclic acids are consumed (Kim et al., 2005). Because of their differential resistance to biodegradation, comparisons of the relative amounts of compound classes can be used to rank extent of biodegradation of petroleum on a scale of 1–10 (Peters et al., 2005; Figure 11-10). Different degrees of bioresistance noted within chemical classes (Figure 11-11) are attributed to conformational selectivity of specific enzymes involved in the degradation process.

11.3.1 The Biodegradation of Hopanes and the Formation of 25-Norhopanes

The biodegradation of hopanes, a homologous series of C_{27}–C_{35} pentacyclic triterpanes, remains something of a conundrum. We previously discussed the evidence from laboratory experiments, where degradation to hopanoic acids is sometimes reported. Here we address the evidence from reservoir oils. Some severely biodegraded oils contain 25-norhopanes, and hopanes were apparently removed prior to steranes, while other severely biodegraded oils contain no 25-norhopanes, and steranes were apparently removed prior to hopanes. From such studies, Brooks et al. (1988) concluded that two pathways for hopane biodegradation occur in the subsurface: one where 25-norhopanes are generated from microbial alteration of hopanes and another where hopanes are more completely catabolized.

The presence of 25-norhopanes is enigmatic as they appear to be generated from the selective removal of the methyl group at the C-10 position (Rullkötter and Wendisch, 1982; Trendel et al., 1990) (Figure 11-12). This produces a homologous series of 25-norhopanes with a distribution similar to the parent hopanes (Figure 11-13), and their presence is thought to be indicative of severe biodegradation even though no laboratory culture has yet shown such activity, nor have they been reported to be produced at seeps or at oil spill sites. Somewhat puzzling is that some oils contain these compounds yet appear to be relatively nonbiodegraded as indicated by the predominance of *n*-alkanes and acyclic isoprenoids; Volkman et al. (1983) proposed that such oils were mixtures of biodegraded oil residues that were subsequently dissolved by nondegraded oil during accumulation in the reservoir. Such mixed oils are common in

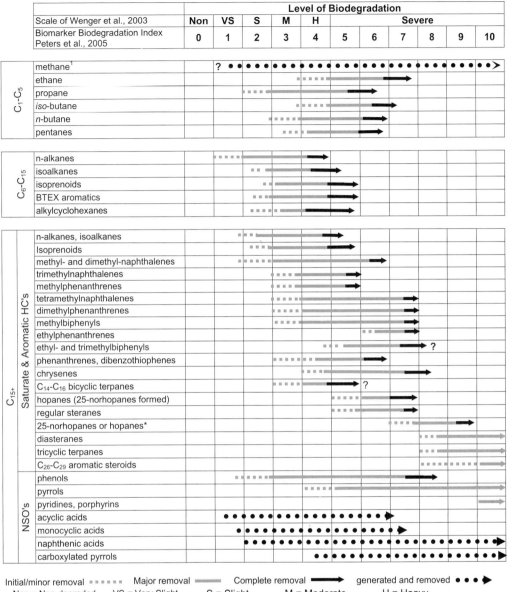

Figure 11-10 The extent of biodegradation of mature crude oil can be ranked on a scale of 1–10 based on differing resistance of compound classes to microbial attack. Biodegradation is quasi-sequential because some of the more labile compounds in the more resistant compound classes can be attacked prior to complete destruction of less resistant classes [from Peters et al. (2005)]. Arrows indicate where compound classes are first altered (dashed lines), substantially depleted (solid gray), and completely eliminated (black). Sequence of alteration of alkylated PAHs is based on work by Fisher et al. (1998) and Trolio et al. (1999). Qualitative descriptions of the degree of biodegradation from Wenger et al. (2002) reflect changes in oil quality.

	Class	Biodegradation Susceptibility
Most ↑ Common order of susceptibility to Biodegradation ↓ **Least**	n-Alkanes	$C_3 \sim C_8\text{-}C_{12} > C_6\text{-}C_8 \sim C_{12}\text{-}C_{15} > C_{6\text{-}} \sim C_{15+}$
	Branched alkanes	Monomethyl > polymethyl > highly branched (e.g., pristane>>2,3,4-trimethylpentane)
	Acyclic Isoprenoids	Lower molecular weight (e.g., C_{10}) > higher molecular weight (e.g., C_{20}). Acyclic isoprenoids degraded before major alteration of polycyclic biomarkers.
	Alkylated benzenes and PAHs	1-Ring > 2-Ring > 3-Ring > 4-Ring. Methyl and dimethyl > trimethyl or extended alkylated species
	Alkylbiphenyls and Alkyldiphenylmethanes	Alkylation at C-4 > C-2 or C-3
	Hopanes (25-norhopanes present)	17α-hopanes: $C_{27}\text{-}C_{32} > C_{33} > C_{34} > C_{35}$, $C_{31}\text{-}C_{35}$ 22R > 22S (Peters and Moldowan, 1991). Exceptions do occur (e.g., Rullkötter and Wendisch, 1982).
	Steranes (25-norhopanes present)	ααα 20R and αββ 20R > ααα 20S and αββ 20S C_{27} Cholestane > C_{28} > C_{29} > C_{30} Alkylcholestanes
	Steranes (25-norhopanes absent)	$ααα20R(C_{27}\text{-}C_{29}) > ααα20S(C_{27}) > ααα20S (C_{28}) > ααα20S(C_{29}) \geq αββ(20S+20R)(C_{27}\text{-}C_{29})$
	Hopanes (25-norhopanes absent)	17α-hopanes: $C_{35} > C_{34} > C_{33} > C_{32} > C_{31} > C_{30} > C_{29} > C_{27}$ and 22R > 22S.
	Diasteranes	$C_{27} > C_{28} > C_{29}$
	Non-hopanoid triterpanes	Gammacerance and oleanane are more resistant to biodegradation than the 17α-hopanes.
	Aromatic steroids	$C_{20}\text{-}C_{21}$ TA > $C_{27}\text{-}C_{29}$ 20R MA ~ $C_{26}\text{-}C_{28}$ 20R TA > C_{21}, C_{22} MA. TA = triaromatic MA = monoaromatic
	Porphyrins	No evidence for significant biodegradation of porphyrins (e.g., Sundararaman and Hwang, 1993)

Figure 11-11 Differential biodegradation within compound classes. Modified from Peters et al. (2005).

basins with shallow (<80°C) reservoirs and may develop either as a continuous process, where the rates of biodegradation are comparable to the rates of reservoir charging, or as episodic events, where a reservoir is charged and biodegraded, and then recharged with nondegraded oil.

11.4 Factors Limiting Biodegradation

Many different factors control the extent and nature of petroleum biodegradation. Furthermore, biodegradation is likely a dynamic process with changing conditions, microbiota, and petroleum components as the process proceeds. When conditions are appropriate, large volumes of oil can be degraded in a relatively short time compared to geologic and geochemical processes. In order for petroleum biodegradation to occur

1. Terminal electron acceptors (e.g., molecular oxygen, nitrates, sulfates, ferric iron, or carbon dioxide) must be present. As we shall discuss, contaminated near-surface aquifers exhibit clear progressions as the more favorable oxidants are exhausted (Bekins et al., 1999, 2001; Haack and Bekins, 2000; Lovley and Anderson, 2000; Chapelle et al., 2002a; Kleikemper et al.,

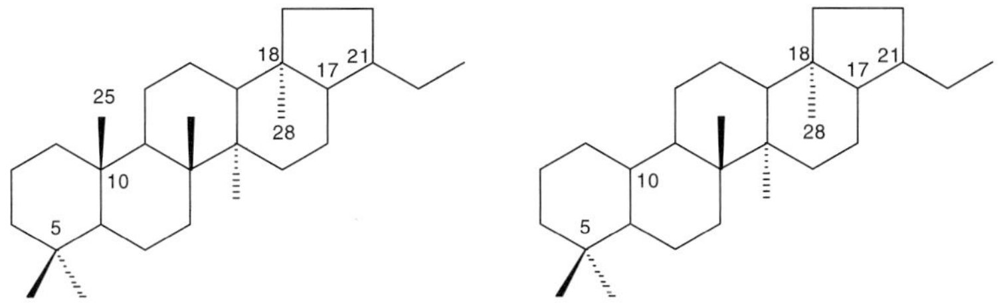

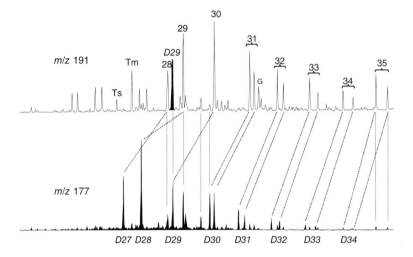

Figure 11-12 25-norhopanes have the same optical configuration as hopanes. They differ only by the removal of the methyl group attached to the C-10 carbon.

Figure 11-13 Mass chromatograms of hopanes (m/z 191) and norhopanes (m/z 177) from Eastern Venezuela. Ion traces are scaled proportional to their relative response. The 25-norhopanes (indicated by D-carbon number) are believed to originate by loss of a methyl group from C-10 in hopanes. Thus, the single epimer of C_{30} 17α,21β (H)-hopane (top) has been partially altered to C_{29} 25-nor-17α-hopane (bottom), while each of the C_{31}–C_{35} 17α-hopane (22S + 22R) epimers correspond to two C_{30} to C_{34} 25-norhopane epimers. Vertical lines indicate some peaks that yield both m/z 191 and 177 ions. Modified from Peters et al. (2005).

2002). Adding terminal acceptors such as sulfate, nitrate, or chelated iron can stimulate anaerobic biodegradation (Da Silva et al., 2005; Tang et al., 2005).

2. Essential nutrients must be available. Hydrocarbons provide a rich source of carbon and energy, but in general they do not provide the other essential elements of life, such as useful nitrogen, phosphorus, and trace metals, and these can limit biodegradation. Bioremediation by at least partially overcoming these limitations is a useful tool in responding to oil spills at sea and on land (Prince and Clark, 2004).

3. Microbes must be able to access the oil. Soluble components such as benzene, toluene, ethylbenzene, and the xylenes (BTEX) are washed out of petroleum and fuels, but the very low aqueous solubility of most hydrocarbons mandates that most biodegradation takes place at the oil–water interface. This implies that *in situ* biodegradation is at least partially dependent on the geometry of the reservoir (Larter et al.,

1999), and the rate at which the most biodegradable hydrocarbons can diffuse to the oil–water interface (Huang et al., 2004). Increasing a spilled oil's surface area with dispersants is an important tool in responding to oil spills at sea (Prince and Clark, 2004). When the oil is in soil or a reservoir, the matrix must have sufficient porosity and permeability to allow the diffusion of nutrients, electron acceptors, and bacteria. For example, Brooks et al. (1988) found greater biodegradation in coarse- than in fine-grained reservoir lithologies in heavy oil accumulations in western Canada.

4. Water must be present (Holden et al., 2001), and it must not be too saline for the indigenous microbes. Biodegradation in the soil experiments of McMillen et al. (1995) was severely inhibited at a soil conductivity of 41.7 mSiemens/cm, about 35 ppt ionic solids, although this was not a salt-acclimated sample. Halophilic bacteria that can degrade hydrocarbons in the presence of several molar salts have been described (Gauthier et al., 1992), but it seems that salinity should generally be less than ~100–150 parts per thousand for optimal biodegradation.

5. Temperatures must remain within limits that support life. Although no hyperthermophilic oil-degraders have yet been found, extreme thermophiles such as *Thermus* and *Bacillus* species degrade polycyclic aromatic hydrocarbons and long-chain alkanes at up to 83°C (Kato et al., 2001; Hao et al., 2004), and empirical evidence suggests that this is an upper limit for biodegradation in reservoirs (Shi Ji-Yang et al., 1982; Connan, 1984). Under typical geothermal gradients, such temperatures correspond to depths <2000 m. Significant biodegradation occurs below 0°C (Rike et al., 2003).

6. Microorganisms capable of degrading hydrocarbon must be present. Exposure to temperatures >80°C may have "sterilized" subsurface reservoirs that have subsequently cooled to more moderate temperatures (Wilhelms et al., 2001).

7. The environment must lack bacteriocides. For example, H_2S is highly toxic to aerobic microbes, and concentrations above ~5% H_2S inhibit anaerobic sulfate reducers. Indeed, sulfate-reducers are known to poison themselves with H_2S if there is no sink for this toxic waste, such as iron that can be precipitated as pyrite, or zinc that can be precipitated as sphalerite (Labrenz et al., 2000).

Microbiologists also recognize two distinct ways in which the chemical composition of a hydrocarbon mixture can affect its biodegradation. One is related to gene regulation, where the synthesis of a biodegradation enzyme or pathway is controlled by the presence of one or more of the substrates. Toluene is a prime example of a molecule that upregulates systems for its biodegradation (Leuthner and Heider, 1998; Diaz and Prieto, 2000; Tropel and van der Meer, 2004), and as we saw in Figure 11-4 and discussed above, some toluene-degrading systems have a broad substrate range. Thus, it is reasonable to expect that the biodegradation of a broad range of compounds will be enhanced in the presence of toluene, and indeed this is seen. Gülensoy and Alvarez (1999) reported enhanced aerobic biodegradation of benzene, *p*-xylene, and naphthalene by several pseudomonads when toluene was present, and a similar phenomenon occurred under anaerobic conditions (Prince and Suflita, 2006).

The other microbiological complication is the phenomenon of co-metabolism, a process usually defined as "any oxidation of substances without utilization of the energy derived from the oxidation to support microbial growth" (Horvath, 1972). Many fungi degrade hydrocarbons in this way, oxidizing large polycyclic aromatic hydrocarbons with their lignin-degrading systems (Gramss et al., 1999), but the phenomenon is well known for smaller substrates as well, such as xylenes (Tsao et al., 1998; Prenafeta-Boldu et al., 2002), thiophene, and benzothiophene (Dyreborg et al., 1996; Rivas et al., 2003). In pure cultures, such co-metabolism tends to lead to oxidized byproducts rather than complete mineralization to CO_2, but these products are usually biodegradable by other organisms and are rarely found in mixed cultures or in the field. Unraveling the complexities of cometabolism will require much work, but it seems likely that it plays an important role in hydrocarbon degradation in the biosphere, implying that the extent of biodegradation of some hydrocarbons will depend on the presence of others.

11.5 Microbial Ecology of Petroleum Biodegradation

Microbial ecology may be defined as the study of interrelationships between microorganisms and their environment, both living and inert. For about a century, indeed since the days of the great 19th-century microbial ecologist Beijerinck, it has been widely accepted that "Microbes are everywhere, the environment selects" (Papke et al., 2003). In other words, the environment is an open system and microbes possessing traits making them suited for a particular environment will inhabit those niches in abundance while populations of unsuited organisms will be minimal. The apparent ubiquity of bacteria and archaea in many environments can be attributed to their immense population size and to their ability to disperse through the atmosphere, hydrosphere, and possibly even the lithosphere. Griffin et al. (2002) estimated that ~10^{18} bacterial cells are transported annually from one continent to the next through the air. Transport in the ocean is slower, but still open to the entire microbial biota. Because of this high dispersion, the probability of local extinction of any microbial species is low and even if the environment is totally unsuitable, some cells will be present. For example, thermophilic bacteria are found in cold sea water (Isaksen et al., 1994) and newly emerging, deep-sea hydrothermal vents are quickly colonized by thermophilic organisms (Tunnicliffe et al., 1997).

While the ubiquitous dispersion of microbial species appears true for environments on or near the surface of the earth, this may not be true for deep subsurface environments that are essentially closed systems, or for particularly unusual environments. Hyperthermophilic organisms, in particular, may not be widely dispersed. Papke et al. (2003) and Whitaker et al. (2003) studied, respectively, the genetic diversity of thermophilic cyanobacteria (*Synechococcus* and *Oscillatoria*) and archaea (*Sulfolobus solfataricus*) in hot springs from around the world and discovered genetic differences that increased with geographic distance. The genetic drift suggests that cellular exchange is not rapid between these isolated habitats. The biotic diversity and transport within the lithosphere is much lower than on the surface. Thus, microbes inhabiting contaminated aquifers and oil reservoirs are likely to be a limited subset of the surface microbes able to survive burial during deposition or transport by basinal fluid flow. Unlike the surface, the microbial population is probably not constantly renewed, and ecological principles suggest that the microbial community in a reservoir will change over time by a succession process. Similarly, the microbial ecology of a reservoir rock will likely undergo a succession of population changes as the reservoir fills with hydrocarbons.

11.5.1 The Succession of Microbial Communities

Typically, microorganisms are limited by available carbon sources or other nutrients needed for growth, such as available nitrogen or terminal electron acceptors. In the former case, incipient starvation is immediately relieved by the introduction of spilled or migrating oil or other organic pollutants. Under such conditions, microbial growth would likely be limited mostly by the availability of an oxidant unless the organisms were in a very porous environment, such as an oiled gravel shore. Nitrate, sulfate, perchlorate, carbon dioxide, organic substrates such as quinones, and oxidized metal ions, such as Mn^{IV} and Fe^{III} can all serve as oxidants in the absence of molecular oxygen. Common oxi-

Table 11-1 Comparison of Aerobic and Anaerobic Respiration Reactions

e-Acceptor	Reaction (Toluene)	$\Delta G°$, kJ/mol Toluene	Molar Ratio	Mass Ratio
O_2	*Aerobic respiration* $C_7H_8 + 9O_2 \rightarrow 7CO_2 + 4H_2O$	−3913	9	3.1
NO_3^-	*Denitrification* $C_7H_8 + 7.2NO_3^- + 0.2H^+ \rightarrow 3.6N_2 + 7HCO_3^- + 0.6H_2O$	−3554	7.2	4.8
Mn(IV)	*Manganese reduction* $C_7H_8 + 21MnO_2 + 14H^+ \rightarrow 7MnCO_3 + 14MnO + 7H_2O$	−3502	21	27
Fe(III)	*Iron reduction* $C_7H_8 + 94Fe(OH)_3 \rightarrow 7FeCO_3 + 29Fe_3O_4 + 145H_2O$	−3398	94	109
$SO_4^=$	*Sulfate reduction* $C_7H_8 + 4.5SO_4^{2-} + 3H_2O \rightarrow 7HCO_3^- + 2.5H^+ + 4.5HS^-$	−205	4.3	4.5
$H_2O\ (CO_2)$	*Methanogenesis* $C_7H + 7.5H_2O \rightarrow 2.5HCO_3^- + 2.5H^+ + 4.5CH_4$	−131	7.5	1.5

Source: Modified from Heider et al. (1998) and Zwolinski et al. (2000). Note that these values reflect the overall free energy available, which may be harnessed by consortia of organisms, each capable of catalyzing only part of the process. For example, some reducing nitrate to nitrite, others reducing nitrite to nitrogen. Methanogenesis involves intermediate production and reduction of HCO_3^-.

dants, also referred to as terminal electron acceptors, used in microbial metabolisms are listed in Table 11-1 by order of decreasing energy yield. The amount of energy obtained varies appreciably. Microorganisms able to use terminal electron acceptors that yield the greatest amount of energy are likely to be most competitive and will persist until that terminal electron acceptor is exhausted. The microbial community is then replaced with organisms that can use a different terminal electron acceptor, presumably the one yielding the next-greatest energy yield. This process continues as oxidants are depleted, and the microbial community will develop as a succession of populations dictated by the availability of terminal electron acceptors.

The best example of the succession of different microbial populations driven by the availability of terminal electron acceptors is in aquifers contaminated with oil, especially the water-soluble BTEX (benzene, toluene, ethylbenzene, and the xylenes) (Bekins et al., 1999, 2001; Haack and Bekins, 2000; Lovley and Anderson, 2000; Chapelle et al., 2002a; Kleikemper et al., 2002, 2005). Distinct zones of microbial processes are formed as hydrocarbons move through the aquifer and various electron acceptors are consumed, usually in order of decreasing redox potential. At the leading edge of the flow path, oxygen is quickly consumed by aerobic organisms. Oxygen solubility in water limits its concentration to a range of 300 to 400 μM depending on temperature. After the oxygen is exhausted, nitrate-reducing bacteria become dominant if there is any nitrate. Some nitrate-reducers are facultative aerobes and switch their metabolism as oxygen is depleted; other nitrate-reducers are obligate anaerobes. With the exhaustion of nitrate, manganese- and iron-reducing bacteria become predominant and, when these metal ions are depleted, sulfate-reducing bacteria thrive if sulfate is available. Murphy and Schramke (1998) showed that sulfate may be produced through the oxidation of reduced sulfur species in the aquifer and need not be derived from gypsum/anhydrite dissolution. Sulfate-reducing bacteria outcompete methanogens in environments with even trace amounts of available sulfate, such as freshwater lakes (Lovley and Klug, 1983). The last surviving microbial population, called the "climax community," is methanogenic with carbon dioxide serving as the terminal electron acceptor being reduced to methane.

Beyond the fundamental energetics (Table 11-1), physiological controls enforce the segregation. Obligate and facultative aerobes regulate their metabolism to utilize oxygen when available. The segregation of iron reduction, sulfate reduction, and methane production into discrete zones seems to be controlled by the levels of acetate and hydrogen, which are used by sulfate-reducers and methanogens. Iron-reducers keep these chemical species so low in concentration that sulfate-reducers cannot use them, but when iron-reducers start to fail for lack of ferric ion, the levels of these substrates rise and the sulfate-reducers thrive. Sulfate-reducers, in turn, keep the levels too low for the methanogens, and only when the available sulfate runs out can the methanogens emerge. Measurements of the concentrations of sulfate, iron, hydrogen, and acetate in subsurface environments have been used to predict which anaerobic microbial process is taking place (Chapelle et al., 1996; Cozzarelli et al., 2000; Kleikemper et al., 2002, 2005).

11.5.2 Deep Subsurface Ecology

The microbial ecology of a contaminated aquifer suggests that an analogous scenario occurs as oil fills a reservoir formation, creating a succession of different groups of bacteria utilizing the available terminal electron acceptors in preferential order. The aquifer model also suggests that if microbial activity persists in a reservoir with no external inputs of terminal electron acceptors (such as sulfate), it will eventually become methanogenic.

11.5.2.1 Aerobic Respiration

Aerobic biodegradation certainly takes place in seeps and spills where oil is in direct contact with the atmosphere or oxic waters. Near-subsurface reservoirs that are in contact with abundant quantities of oxygen-bearing meteoric waters also are likely to be degraded by aerobic bacteria. Transporting large amounts of dissolved oxygen into deep subsurface reservoirs is more problematic. Typical near-surface groundwaters contain less than 5 ppm dissolved oxygen and most (all?) of the oxygen is likely to be consumed before reaching a deep reservoir, either biotically or abiotically, as the water passes through soils and aquifers with reactive organic matter and minerals such as pyrite. The stripping process is very efficient as studies of shallow contaminated aquifers show that even small plumes of organic contaminants are sufficient to remove free oxygen from near-surface groundwater (e.g., Baedecker et al., 1993; Chapelle et al., 2002a). On the other hand, if meteoric water is constrained to flowing into deep reservoirs along fractures, unconformities, or highly permeable rocks, oxygen-bearing water could be transported to deep strata once all reducing agents are consumed along the pathway.

Many petroleum systems with active aquifers have flushed completely, or partially — all connate brines. Hanor et al. (2004), for example, showed that the entire sedimentary section of the National Petroleum Reserve, Alaska (NPRA), east of the Meade Ridge has been flushed with meteoric water to a depth of 2 km or more. Few studies, however, have attempted to determine the volume of water encountered per volume of oil. Horstad et al. (1992) estimated that several thousand reservoir volumes of meteoric water would be required to supply the oxygen demand required for aerobic biodegradation of oil in the Gullfaks field. Such volumes, while staggering, are not inconceivable. Mauk and Burruss (2002) estimated water:oil ratios of 300:1 to over 9000:1 in the Midcontinent rift system based on the hydrocarbon compositions of fluid inclusions. The water:oil ratio is believed to be typically <300 in sedimentary basins with commercial accumulations (Ballentine et al., 1996), and the current consensus is that nearly all deep-reservoir strata can only support anaerobic communities.

Less certain is whether large near-surface accumulations of tar sands have been biodegraded by aerobic or anaerobic processes. Connan et al. (1997) concluded that aerobic biodegradation is a dominant process in shallow petroleum reservoirs and tar sands as the degree of alteration exceeds that observed in deeper, anaerobic reservoirs. Nonetheless, some have questioned how accumulations the size of the Orinoco or Alberta tar sands could be degraded solely by aerobic microbes, when small plumes of organic contaminants are effective at removing all dissolved oxygen. Trace metabolites consistent with an anaerobic biodegradation pathway for naphthalene (Aitken et al., 2004) and biogenic methane associated with the tar sands provide indirect evidence that these large accumulations were at least partially biodegraded by anaerobic microbial processes (Head et al., 2003).

Also unresolved is the effect that low to negligible oxygen levels may have on the subsurface microbial ecology. Facultative aerobes and obligate anaerobes would use only anaerobic pathways. Less certain is the role of microaerophilic bacteria that are capable of maintaining aerobic pathways under a low O_2 condition via biochemical pathways termed *oxidative metabolic gearing* (Ludwig, 2004). This adaptation allows microaerophilic bacteria to quickly respond to changing O_2 levels with minimal energy costs and provides a selective advantage that allows them to exploit an ecological niche not favorable for either normal aerobic or anaerobic respirers. However, little work has yet been done on hydrocarbon-degrading microaerophiles (Holden et al., 2001; Yerushalmi et al., 2001) or on the capabilities of facultative aerobes to function under suboxic conditions (Berthe-Corti and Fetzner, 2002).

11.5.2.2 Anaerobic Respiration

Although almost any oxidized pair of a redox couple, inorganic or organic, may serve as a terminal electron acceptor for microbial growth, few are readily available in reservoir rocks to support large-scale biodegradation. Of the inorganic oxidizers, only sulfate may occur in significant amounts in reservoir fluids. Although it is one of the least energetic of the inorganic terminal electron acceptors, sulfate-reducing bacteria capable of catabolizing hydrocarbons are common in oil-field formation waters (Magot, 2005). It is reasonable to conclude that sulfate-reducing bacteria are responsible for petroleum biodegradation, where sulfate is available from connate waters or gypsum/anhydrite dissolution.

Of the metals, only iron is likely to be present in any appreciable abundance in the deep subsurface. The significance of iron-reducing bacteria is just beginning to be appreciated (e.g., Coleman et al., 1993; Schmitt et al., 1996; Prommer et al., 1999; Lovley et al., 2000), and these microorganisms may play a significant role in degrading oil. Iron-reducing bacteria seem to be particularly adapted to life in the deep subsurface. All hyperthermophilic microorganisms appear capable of iron reduction, suggesting that this may have been the earliest form of anaerobic respiration (Vargas et al., 1998). Iron-reducing bacteria have been identified in enrichment cultures of oil field brines (Nazina et al., 1995; Slobodkin et al., 1999; Magot, 2005), but the extent of oil biodegradation in the deep subsurface may be severely limited by the availability of iron oxide surfaces. The abundance of Fe^{III} in sandstones is in the order of 0.4 to 4.0 oxide wt.%, found on grain coatings or pore filling material (Pettijohn, 1963). While the energy derived from reduction of Fe^{III} is comparable to that obtained from aerobic respiration or the reduction of NO_3^-, the amount of oxidant needed is appreciably more (Table 11-1). On a volume bases, ~10 L of hematite would be needed to biodegrade ~1 L of hydrocarbon. Given the very low bioavailability of ferric iron in hematite, such a process would likely be very slow and eventually hampered further by the fact that hematite tends to be removed during diagenesis and much of the iron in sedimentary reservoir rocks is bound in even less reactive silicates and sulfides.

Geobacter, the most common iron-reducer in the subsurface, must adhere to Fe^{III} oxides to reduce them (Nevin and Lovley, 2000a). Interestingly, the electron transfer may occur *via* pili that act as biological nanowires (Reguera et al., 2005). However, other bacteria have evolved several adaptions to utilize iron minerals while not in intimate contact (Nevin and Lovley, 2000b; Rosso et al., 2003). Lovley et al. (1994) showed that anaerobic degradation of benzene in sediments was greatly stimulated by the addition of humic substances, which they reasoned shuttled electrons from the bacteria to Fe^{III} oxides (Lovley et al., 1996, 1998). Quinones (Scott et al., 1994) and humic matter are widespread (Coates et al., 1998), although not all iron-reducing bacteria can use them (Nevin and Lovley, 2000b). Some methanogenic archaea have been shown to use electron shuttles to oxidize Fe^{III} (Bond and Lovley, 2002).

Another intriguing possibility is that various oil components may serve as electron shuttles, electron donors, or terminal electron acceptors. For example, species with humic-like properties, such as asphaltenes and resins, could serve as the electron acceptors when hydrocarbons are oxidized. Phenazines are present in oil and have been proposed to act in this way (Hernandez and Newman, 2001).

Methanogens utilize CO_2 as a terminal electron acceptor. While the archaea do not appear to be capable of degrading hydrocarbons directly (Prince, 2005), methanogens do participate in bacterial consortia that collectively degrade hydrocarbons. Known consortia consist of at least three interacting groups: (1) various fermentative bacteria that consume complex compounds and excrete volatile fatty acids, H_2, and CO_2; (2) acetogenic and other bacteria that oxidize the higher acids to acetate or formate and H_2; and (3) methanogens that use several enzymatic pathways to form microbial methane, one of which is carbon dioxide reduction, where CO_2 or CO_3^{2-} is converted to CH_4 using electrons from H_2 or formate.

Since CO_2 is present as a buffered species in all reservoir brines, CO_2-reducing methanogens would not be limited by the availability of a terminal electron acceptor. Methanogenic archaea are certainly present in the deep subsurface (Obraztsova et al., 1987; Orphan et al., 2000), and biogenic methane is a component of reservoir gases that are in association with biodegraded oils (e.g., Troll field; Horstad and Larter, 1997). Such methane has been termed *secondary biogenic methane* (Scott et al., 1994) and has been proposed as a way of harnessing productive hydrocarbons in cases involving advanced biodegradation or in reservoirs that have come to the end of their productive life (Larter et al., 1999; Suflita et al., 2004).

The presence of biogenic methane with biodegraded oils, however, is not necessarily coupled as there are alternative sources of hydrogen for reducing CO_2. Nonbiotic hydrogen may form through mineral reactions, radiolysis of hydrocarbons or water, or thermal conversion of organic compounds (Kotelnikova, 2002). In fact, the deep subsurface may be dominated by hydrogen-based ecosystems where the hydrogen is derived from inorganic processes. For example, Chapelle et al. (2002b) described such an ecosystem in deep hydrothermal waters (200 m below ground) from Lidy Hot Springs in Idaho, where methanogens dominate the ecosystem and appear to thrive exclusively on geothermal hydrogen and carbon dioxide. No reduced form of carbon is available. The existence of pure subsurface lithoautotrophic microbial ecosystems, or SLiMEs, is likely but not proven (Nealson et al., 2005).

11.6 Conclusions; Implications of Biodegradation on Identification

We have seen that biodegradation can have a major impact on the composition of petroleum products and crude oils. Indeed, very few hydrocarbons are totally resistant to the process in the long term. The only other significant routes for hydrocarbons to leave the biosphere are combustion and photochemical oxidation. Photooxidation is a very important process in the atmosphere (Griffin et al., 1999; Spaulding et al., 2003), and while it does convert aromatic compounds in slicks to oxygenated species (Garrett et al., 1998), it accounts for relatively little loss of nonvolatile hydrocarbons. Burning seeps, such as the Fire Mountains on the Absheron Peninsula, have been known for millennia, and deliberate ignition is sometimes an appropriate oil spill response tool (Buist, 2003), Nevertheless, combustion pales in comparison to microbial degradation as the principal fate of hydrocarbons in the biosphere.

Biodegradation seems to follow clear patterns that can be exploited in identifying spills and reservoir oils from similar sources, determining how far biodegradation has proceeded in a particular sample and predicting how much further biodegradation can be expected at a spill site. Linear alkanes and simple one- and two-ring aromatic compounds are degraded first under all conditions studied to date. Branched alkanes are degraded somewhat less rapidly, so ratios of linear to simply branched alkanes (e.g., octadecane to phytane) change rapidly during the early stages of biodegradation, making such ratios good indicators of the initiation of biodegradation, and poor ones for identifying sources. Multiple branched alkanes are degraded more slowly, especially those with tertiary carbons, so ratios of these molecules to each other (e.g., 2,2-dimethylbutane to 2,2-dimethylpentane) have the potential to serve as hydrocarbon fingerprints of different gasolines. 2,2,4-trimethylpentane has proven a good conserved internal marker for following the biodegradation of gasoline, but even this is eventually biodegraded completely.

Alicyclic molecules are degraded rapidly under some anaerobic conditions (Townsend et al., 2004), but apparently rather more slowly in most aerobic systems. Thus, changes in the ratios of simple alicyclic compounds to more stable ones (e.g., 2,2,4-trimethylpentane), especially if they are more rapid than the biodegradation of, for example, multiple substituted benzenes that are degraded rapidly under aerobic conditions (Solano-Serena et al., 1999), may imply that anaerobic biodegradation is the dominant process.

Three-ring aromatic hydrocarbons (phenanthrene, acenaphthene, anthracene, dibenzothiophene, etc.) are next on the seriatim, and within these compounds there is a clear pattern that increasing alkylation, at least of multiple methyl groups, slows degradation. Douglas et al. (1996) showed that the degradation of alkyl phenanthrenes and dibenzothiophenes in field samples occurs at similar rates such that the ratios of these compounds (e.g., C_3-phenanthrenes to C_3-dibenzothiophenes, where C_3 implies trimethyl-, methylethyl-, and propyl-substituents) stayed constant during substantial biodegradation, allowing these compounds to act as reliable fingerprints for discriminating different oils in the environment. Chrysenes and larger polycyclic aromatic hydrocarbons are degraded more slowly than the three-ringed aromatics.

Some steranes are degraded contemporaneously with the alkylated three-ring aromatics but are still among the least degradable components of most diesel fuels and can act as conserved markers for following biodegradation. Hopanes are more resistant than steranes in surface conditions and may be equal or more resistant under reservoir conditions. Since hopanes and oleananes are among the last saturated molecules to be biodegraded in oil spills, the distributions of these compounds can act as fingerprints for identifying oils from the same source, and they can serve as conserved internal markers for following biodegradation, and indeed photooxidation, evaporation, and partial combustion (Prince and Douglas, 2005).

Further useful forensic indicators will likely emerge as our understanding of the various different pathways of biodegradation of hydrocarbons in the environment continues to grow.

References

Aitken, C.M., D.M. Jones, and S.R. Larter, Anaerobic hydrocarbon biodegradation in deep subsurface oil reservoirs. *Nature*, 2004, **431**, 291–294.

Allen, J.R., D.D. Clark, J.G. Krum, and S.A. Ensign, A role for coenzyme M (2-mercaptoethanesulfonic acid) in a bacterial pathway of aliphatic epoxide carboxylation. *Proc. Nat. Acad. Sci. USA*, 1999, **96**, 8432–8437.

Annweiler, E., W. Michaelis, and R.U. Meckenstock, Anaerobic cometabolic conversion of benzothiophene by a sulfate-reducing enrichment culture and in a tar oil contaminated aquifer. *Appl. Environ. Microbiol.*, 2001, **67**, 5077–5083.

Annweiler, E., W. Michaelis, and R.U. Meckenstock, Identical ring cleavage products during anaerobic degradation of naphthalene, 2-methylnaphthalene, and tetralin indicate a new metabolic pathway. *Appl. Environ. Microbiol.*, 2002, **68**, 852–858.

Arp, D.J., C.M. Yeager, and M.R. Hyman, Molecular and cellular fundamentals of aerobic cometabolism of trichloroethylene. *Biodegradation*, 2001, **12**, 81–103.

Baedecker, M.J., I.M. Cozzarelli, R.P. Eganhouse, D.I. Siegel, and P.C. Bennett, Crude oil in a shallow sand and gravel aquifer-III. Biogeochemical reactions and mass balance modeling in anoxic groundwater. *Appl. Geochem.*, 1993, **8**, 569–586.

Ballentine, C.J., R.K. O'Nions, and M.L. Coleman, A magnus opus: Helium, neon, and argon isotopes in a North Sea oilfield. *Geochimica et Cosmochimica Acta*, 1996, **60**, 831–849.

Bekins, B.A., I.M. Cozzarelli, E.M. Godsy, E. Warren, H.I. Essaid, and M.E. Tuccillo, Progression of natural attenuation processes at a crude oil spill site: II. Controls on spatial distribution of microbial populations. *J. Contaminant Hydrol.*, 2001, **53**, 387–406.

Bekins, B.A., E.M. Godsy, and E. Warren, Distribution of microbial physiologic types in an aquifer contaminated by crude oil. *Microbial Ecol.*, 1999, **37**, 263–275.

Berner, R.A., The long-term carbon cycle, fossil fuels and atmospheric composition. *Nature*, 2003, **426**, 323–326.

Berthe-Corti, L. and S. Fetzner, Bacterial metabolism of *n*-alkanes and ammonia under oxic, suboxic and anoxic conditions. *Acta Biotechnologica*, 2002, **22**, 299–336.

Biegert, T., G. Fuchs, and J. Heider, Evidence that anaerobic oxidation of toluene in the denitrifying bacterium *Thauera aromatica* is initiated by formation of benzylsuccinate from toluene and fumarate. *Eur. J. Biochem.*, 1996, **238**, 661–668.

Bond, D.R. and D.R. Lovley, Reduction of Fe(III) oxide by methanogens in the presence and absence of extracellular quinones. *Environ. Microbiol.*, 2002, **4**, 115–124.

Bost, F.D., R. Frontera-Suau, T.J. McDonald, K.E. Peters, and P.J. Morris, Aerobic biodegradation of hopanes and norhopanes in Venezuelan crude oils. *Organic Geochem.*, 2001, **32**, 105–114.

Boyd, D.R., N.D. Sharma, and C.C.R. Allen, Aromatic dioxygenases: Molecular biocatalysis and applications. *Current Opinion in Biotech.*, 2001, **12**, 564–573.

Brooks, P.W., M.G. Fowler, and R.W. Macqueen, Biological marker and conventional organic geochemistry of oil sands/heavy oils, Western Canada Basin. *Organic Geochem.*, 1988, **12**, 519–538.

Brzostowicz, P.C., D.M. Walters, R.E. Jackson, K.H. Halsey, H. Ni, and P.E. Rouvière, Proposed involvement of a soluble methane monooxygenase homologue in the cyclohexane-dependent growth of a new *Brachymonas* species. *Environ. Microbiol.*, 2005, **7**, 179–190.

Brzostowicz, P.C., D.M. Walters, S.M. Thomas, V. Nagarajan, and P.E. Rouviere, mRNA differential display in a microbial enrichment culture: Simultaneous identification of three cyclohexanone monooxygenases from three species. *Appl. Environ. Microbiol.*, 2003, **69**, 334–342.

Bühler, B., A. Schmid, B. Hauer, and B. Witholt, Xylene monooxygenase catalyzes the multistep oxygenation of toluene and pseudocumene to corresponding alcohols, aldehydes, and acids in *Escherichia coli* JM101. *J. Biol. Chem.*, 2000, **275**, 10085–10092.

Buist, I., Window-of-opportunity for *in situ* burning. *Spill Sci. & Tech. Bull.*, 2003, **8**, 341–346.

Caldwell, M.E., R.M. Garrett, R.C. Prince, and J.M. Suflita, Anaerobic biodegradation of long-chain *n*-alkanes under sulfate-reducing conditions. *Environ. Sci. & Tech.*, 1998, **32**, 2191–2195.

Cervantes, F.J., S. van der Velde, G. Lettinga, and J.A. Field, Quinones as terminal electron acceptors for anaerobic microbial oxidation of phenolic compounds. *Biodegradation*, 2000, **11**, 313–321.

Chakraborty, R. and J.D. Coates, Anaerobic degradation of monoaromatic hydrocarbons. *Appl. Microbiol. Biotech.*, 2004, **64**, 437–446.

Chakraborty, R. and J.D. Coates, Hydroxylation and carboxylation — two crucial steps of anaerobic benzene degradation by *Dechloromonas* strain RCB. *Appl. Environ. Microbiol.*, 2005, **71**, 5427–5432.

Chang, B.V. and L.C. Shiung, Anaerobic biodegradation of polycyclic aromatic hydrocarbon in soil. *Chemosphere*, 2002, **48**, 717–724.

Chang, B.V., S.W. Chang, and S.Y. Yuan, Anaerobic degradation of polycyclic aromatic hydrocarbons in sludge. *Adv. Environ. Res.*, 2003, **7**, 623–628.

Chapelle, F.H., P.M. Bradley, D.R. Lovley, and D.A. Vroblesky, Measuring rates of biodegradation in a contaminated aquifer using field and laboratory methods. *Ground Water*, 1996, **34**, 691–698.

Chapelle, F.H., P.M. Bradley, D.R. Lovley, K. O'Neill, and J.E. Landmeyer, Rapid evolution of redox processes in a petroleum hydrocarbon-contaminated aquifer. *Ground Water*, 2002a, **40**, 353–360.

Chapelle, F.H., K. O'Neill, P.M. Bradley, B.A. Methe, S.A. Ciufo, L.L. Knobel, and D.R. Lovley, A hydrogen-based subsurface microbial community dominated by methanogens. *Nature*, 2002b, **415**, 312–315.

Cheng, Q., S.M. Thomas, and P. Rouvière, Biological conversion of cyclic alkanes and cyclic alcohols into dicarboxylic acids: Biochemical and molecular basis. *Appl. Microbiol. Biotech.*, 2002, **58**, 704–711.

Cho, M.C., D.-O. Kang, B.D. Yoon, and K. Lee, Toluene degradation pathway from *Pseudomonas putida* F1: Substrate specificity and gene induction by 1-substituted benzenes. *J. Industrial Microbiol. Biotech.*, 2000, **25**, 163–170.

Coates, J.D., D.J. Ellis, E.L. Blunt-Harris, C.V. Gaw, E.E. Roden, and D.R. Lovley, Recovery of humic-reducing bacteria from a diversity of environments. *Appl. & Environ. Microbiol.*, 1998, **64**, 1504–1509.

Coleman, M.L., D.B. Hedrick, D.R. Lovley, D.C. White, and K. Pye, Reduction of Fe(III) in sediments by sulphate-reducing bacteria. *Nature*, 1993, **361**, 436–438.

Connan, J., Biodegradation of crude oils in reservoirs. In *Advances in Petroleum Geochemistry*, Vol. 1, J. Brooks and D.H. Welte (eds.), London: Academic Press, 1984, 299–335.

Connan, J., G. Lacrampe-Couloume, and M. Magot, Anaerobic biodegradation of petroleum in reservoirs: A widespread phenomenon in nature. *8th International Meeting on Organic Geochemistry, September 22–26, 1997, Maastricht, The Netherlands*, pp. 5–6.

Cozzarelli, I.M., J.M. Suflita, G.A. Ulrich, S.H. Harris, M.A. Scholl, J.L. Schlottmann, and S. Christenson, Geochemical and microbiological methods for evaluating anaerobic processes in an aquifer contaminated by landfill leachate. *Environ. Sci. & Tech.*, 2000, **34**, 4025–4033.

Cravo-Laureau C., V. Grossi, D. Raphel, R. Matheron, and A. Hirschler-Réa, Anaerobic *n*-alkane metabolism by a sulfate-reducing bacterium, *Desulfatibacillum aliphaticivorans* strain CV2803^T. *Appl. Environ. Microbiol.*, 2005, **71**, 3458–3467.

Da Silva, M.L.B., G.M.L. Ruiz-Aguilar, and P.J.J. Alvarez, Enhanced anaerobic biodegradation of BTEX-ethanol mixtures in aquifer columns amended with sulfate, chelated ferric iron or nitrate. *Biodegradation*, 2005, **16**, 105–114.

Diaz, E. and M.A. Prieto, Bacterial promoters triggering biodegradation of aromatic pollutants. *Curr. Opin. Biotech.*, 2000, **11**, 467–475.

Douglas, A.G., A.E. Bence, S.J. McMillen, R.C. Prince, and E.L. Butler, Environmental stability of selected petroleum hydrocarbon source and weathering ratios. *Environ. Sci. & Tech.*, 1996, **30**, 2332–2339.

Dutta, T., Origin, occurrence, and biodegradation of long-side-chain alkyl compounds in the environment: A review. *Environ. Geochem. Health*, 2005, **27**, 271–284.

Dutta, T.K. and S. Harayama, Analysis of long side chain alkylaromatics in crude oil for evaluation of their fate in the environment. *Environ. Sci. & Tech.*, 2001, **35**, 102–107.

Dyreborg, S., E. Arvin, K. Broholm, and J. Christensen, Biodegradation of thiophene, benzothiophene and benzofuran with eight different primary substrates. *Environ. Toxicol. Chem.*, 1996, **15**, 2290–2292.

Ellis L.B.M., D. Roe, and L.P. Wackett, The University of Minnesota, Biocatalysis/biodegradation database: The first decade. *Nucleic Acids Res.*, 2006, **34**, D517–D521.

Elmendorf, D.L., C.E. Haith, G.S. Douglas, and R.C. Prince, Relative rates of biodegradation of substituted polycyclic aromatic hydrocarbons. In *Bioremediation of Chlorinated and Polycyclic Aromatic Hydrocarbon Compounds*, R.E. Hinchee, A. Leeson, L. Semprini, and S.K. Ong (eds.), Boca Raton, FL: Lewis Publishers, 1994, 188–202.

Ensign, S.A., Microbial metabolism of aliphatic alkenes. *Biochem.*, 2001, **40**, 5845–5853.

Eriksson, M., E. Sodersten, Z. Yu, G. Dalhammar, and W.W. Mohn, Degradation of polycyclic aromatic hydrocarbons at low temperature under aerobic and nitrate-reducing conditions in enrichment cultures from northern soils. *Appl. Environ. Microbiol.*, 2003, **69**, 275–284.

Eriksson, M., A. Swartling, and G. Dalhammar, Biological degradation of diesel fuel in water and soil monitored with solid-phase micro-extraction and GC-MS. *Appl. Microbiol. Biotech.*, 1998, **50**, 129–134.

Finnerty, W.R., The biochemistry of microbial alkane oxidation: New insights and perspectives. *Trends in Biochem. Sci.*, 1977, **2**, 73–75.

Fisher, S.J., R. Alexander, R.I. Kagi, and G.A. Oliver, Aromatic hydrocarbons as indicators of biodegradation in north Western Australian reservoirs. *Sedimentary Basins of Western Australia: West Australian Basins Symp.*, 1998, 185–194.

Fraser, C.M., All things great and small. *Trends in Microbiol.*, 2004, **12**, 7–8.

Fritsche, W. and M. Hofrichter, Aerobic degradation by microorganisms. In *Biotechnology,* Vol. 11b, *Environmental Processes II — Soil Decontamination*, J. Klein (ed.), 2000, 146–164.

Frolov, E.B., M.B. Smirnov, V.A. Melikhov, and N.A. Vanyukova, Olefins of radiogenic origin in crude oils. *Org. Geochem.*, 1998, **29**, 409–420.

Frontera-Suau, R., F.D. Bost, T.J. McDonald, and P.J. Morris, Aerobic biodegradation of hopanes and other biomarkers by crude oil-degrading enrichment cultures. *Environ. Sci. & Tech.*, 2002, **36**, 4585–4592.

Garrett, R.M., I.J. Pickering, C.E. Haith, and R.C. Prince, Photooxidation of crude oils. *Environ. Sci. & Tech.*, 1998, **32**, 3719–3723.

Gauthier, M.J., B. Lafay, R. Christen, L. Fernandez, M. Acquaviva, P. Bonin, and J.C. Bertrand, *Marinobacter hydrocarbonoclasticus* gen. nov., sp. nov., a new, extremely halotolerant, hydrocarbon-degrading marine bacterium. *Int. J. Systematic Bacteriol.*, 1992, **42**, 568–579.

Gibson, D.T. and R.E. Parales, Aromatic hydrocarbon dioxygenases in environmental biotechnology. *Curr. Opin. Biotech.*, 2000, **11**, 236–243.

Gieg, L.M. and J.M. Suflita, Metabolic indicators of anaerobic hydrocarbon biodegradation in petroleum-laden environments. In *Petroleum Microbiology*, B. Oliver and M. Magot (eds.), Washington, DC: ASM Press, 2005, 337–356.

Gramss, G., K.-D. Voigt, and B. Kirsche, Degradation of polycyclic aromatic hydrocarbons with three to seven aromatic rings by higher fungi in sterile and unsterile soils. *Biodegrad.*, 1999, **10**, 51–62.

Griffin, D.W., C.A. Kellogg, V.H. Garrison, and E.A. Shinn, The global transport of dust — an intercontinental river of dust, microorganisms and toxic chemicals flows through the Earth's atmosphere. *Amer. Scientist*, 2002, **90**, 228–235.

Griffin, R.J., D.R. Cocker III, R.C. Flagan, and J.H. Seinfeld, Organic aerosol formation from the oxidation of biogenic hydrocarbons. *J. Geophysi. Res.*, 1999, **104**, 3555–3568.

Grishchenkov, V.G., A.V. Slepen'kin, and A.M. Boronin, Anaerobic degradation of biphenyl by the facultative anaerobic strain *Citrobacter freundii* BS2211. *Appl. Biochem. Microbiol.*, 2002, **38**, 125–128.

Grossi, V., D. Raphel, A. Hirschler-Réa, M. Gilewicz, A. Mouzdahir, J.-C. Bertrand, and J.-F. Rontani, Anaerobic biodegradation of pristane by a marine sedimentary bacterial and/or archaeal community. *Org. Geochem.*, 2000, **31**, 769–772.

Gülensoy, N. and P.J.J. Alvarez, Diversity and correlation of specific aromatic hydrocarbon biodegradation capabilities. *Biodegradation*, 1999, **10**, 331–340.

Gupta, R.S., The phylogeny of proteobacteria: Relationships to other eubacterial phyla and eukaryotes. *FEMS Microbiol. Rev.*, 2000, **24**, 367–402.

Haack, S.K. and B.A. Bekins, Microbial populations in contaminant plumes. *Hydrogeology J.*, 2000, **8**, 63–76.

Hanor, J.S., J.A. Nunn, and Y. Lee, Salinity structure of the central North Slope foreland basin, Alaska, USA: Implications for pathways of past and present topographically driven regional fluid flow. *Geofluids*, 2004, **4**, 152–168.

Hao, R., A. Lu, and G. Wang, Crude-oil-degrading thermophilic bacterium isolated from an oil field. *Can. J. Microbiol.*, 2004, **50**, 175–182.

Hartmans, S., J.A. de Bont, and W. Harder, Microbial metabolism of short-chain unsaturated hydrocarbons. *FEMS Microbiol. Rev.*, 1989, **63**, 235–264.

Head, I.M., D.M. Jones, and S.R. Larter, Biological activity in the deep subsurface and the origin of heavy oil. *Nature*, 2003, **426**, 344–352.

Heider, J., A.M. Spormann, H.R. Beller, and F. Widdel, Anaerobic bacterial metabolism of hydrocarbons. *FEMS Microbiol. Rev.*, 1998, **22**, 459–473.

Hernandez, M.E. and D.K. Newman, Extracellular electron transfer. *Cell. Mol. Life Sci.*, 2001, **58**, 1562–1571.

Hlavica, P., Models and mechanisms of O–O bond activation by cytochrome P450: A critical assessment of the potential role of multiple active intermediates in oxidative catalysis. *Eur. J. Biochem.*, 2004, **271**, 4335–4360.

Holden, P.A., L.E. Hersman, and M.K. Firestone, Water content mediated microaerophilic toluene biodegradation in arid vadose zone materials. *Microbial Ecol.*, 2001, **42**, 256–266.

Horstad, I. and S.R. Larter, Petroleum migration, alteration, and remigration within Troll Field, Norwegian North Sea. *Amer. Assoc. Petrol. Geol. Bull.*, 1997, **81**, 222–238.

Horstad, I., S.R. Larter, and N. Mills, A quantitative model of biological petroleum degradation within the Brent Group reservoir in the Gullfaks Field,

Norwegian North Sea. *Org. Geochem.*, 1992, **19**, 107–117.

Horvath, R.S., Microbial co-metabolism and the degradation of organic compounds in nature. *Bacteriol. Rev.*, 1972, **36**, 146–155.

Huang, H., S.R. Larter, B.F.J. Bowler, and T.B.P. Oldenburg, A dynamic biodegradation model suggested by petroleum compositional gradients within reservoir columns from the Liaohe basin, NE China. *Org. Geochem.*, 2004, **35**, 299–316.

Huesemann, M.H., T.S. Hausmann, and T.J. Fortman, Biodegradation of hopane prevents use as conservative biomarker during bioremediation of PAHs in petroleum contaminated soils. *Bioremediation J.*, 2003, **7**, 111–117.

Isaksen, M.F., F. Bak, and B.B. Jørgensen, Thermophilic sulfate-reducing bacteria in cold marine sediment. *FEMS Microbiol. Ecol.*, 1994, **14**, 1–8.

Johnson, H.A., D.A. Pelletier, and A.M. Spormann, Isolation and characterization of anaerobic ethylbenzene dehydrogenase, a novel Mo-Fe-S enzyme. *J. Bacteriol.*, 2001, **183**, 4536–4542.

Kanaly, R.A. and S. Harayama, Biodegradation of high-molecular-weight polycyclic aromatic hydrocarbons by bacteria. *J. Bacteriol.*, 2000, **182**, 2059–2067.

Karlsson, A., J.V. Parales, R.E. Parales, D.T. Gibson, H. Eklund, and S. Ramaswamy, Crystal structure of naphthalene dioxygenase: Side-on binding of dioxygen to iron. *Science*, 2003, **299**, 1039–1042.

Kasai, Y., K. Shindo, S. Harayama, and N. Misawa, Molecular characterization and substrate preference of a polycyclic aromatic hydrocarbon dioxygenase from *Cycloclasticus sp.* strain A5. *Appl. Environ. Microbiol.*, 2003, **69**, 6688–6697.

Kasting, J.F., Palaeoclimatology: Archaean atmosphere and climate. *Nature*, 2004, **432**, 1.

Kato, T., M. Haruki, T. Imanaka, M. Morikawa, and S. Kanaya, Isolation and characterization of long-chain-alkane degrading *Bacillus thermoleovorans* from deep subterranean petroleum reservoirs. *J. Bioscience Bioeng.*, 2001, **91**, 64–70.

Kim, E. and G.J. Zylstra, Functional analysis of genes involved in biphenyl, naphthalene, phenanthrene, and *m*-xylene degradation by *Sphingomonas yanoikuyae* B1. *J. Indus. Microbiol. Biotech.*, 1999, **23**, 294–302.

Kim, S., L.A. Stanford, R.P. Rodgers, A.G. Marshall, C.C. Walters, K. Qian, L.M. Wenger, and P. Mankiewicz, Microbial alteration of the acidic and neutral polar NSO compounds revealed by Fourier transform ion cyclotron resonance mass spectrometry. *Org. Geochem.*, 2005, **36**, 1117–1134.

Kingston, P.F., Long-term environmental impact of oil spills. *Spill Sci. & Tech. Bull.*, 2002, **7**, 53–61.

Kleikemper, J., S.A. Pombo, M.H. Schroth, W.V. Sigler, M. Pesaro, and J. Zeyer, Activity and diversity of methanogens in a petroleum hydrocarbon-contaminated aquifer. *Appl. Environ. Microbiol.*, 2005, **71**, 149–158.

Kleikemper, J., M.H. Schroth, W.V. Sigler, M. Schmucki, S.M. Bernasconi, and J. Zeyer, Activity and diversity of sulfate-reducing bacteria in a petroleum hydrocarbon-contaminated aquifer. *Appl. Environ. Microbiol.*, 2002, **68**, 1516–1523.

Kniemeyer, O. and J. Heider, Ethylbenzene dehydrogenase, a novel hydrocarbon-oxidizing molybdenum/iron-sulfur/heme enzyme. *J. Biol. Chem.*, 2001, **276**, 21381–21386.

Kotelnikova, S., Microbial production and oxidation of methane in deep subsurface. *Earth-Sci. Rev.*, 2002, **58**, 367–395.

Labrenz, M., G.K. Druschel, T. Thomsen-Ebert, B. Gilbert, S.A. Welch, K.M. Kemner, G.A. Logan, R.E. Summons, G. De Stasio, P.L. Bond, B. Lai, S.D. Kelly, and J.F. Banfield, Formation of sphalerite (ZnS) deposits in natural biofilms of sulfate-reducing bacteria. *Science*, 2000, **290**, 1744–1747.

Lange, C.C. and L.P. Wackett, Oxidation of aliphatic olefins by toluene dioxygenase: Enzyme rates and product identification. *J. Bacteriol.*, 1997, **179**, 3858–3865.

Larter, S., A. Hockey, A. Aplin, N. Telnaes, A. Wilhelms, I. Horstad, R. di Primio, and O. Sylta, When biodegradation preserves petroleum! Petroleum geochemistry of N. Sea oil rimmed gas accumulations (ORGAs), AAPG Hedberg Research Conference on "Natural Gas Formation and Occurrence," Durango, CO, 1999.

Leahy, J.G., P.J. Batchelor, and S.M. Morcomb, Evolution of the soluble diiron monooxygenases. *FEMS Microbiol. Rev.*, 2003, **27**, 449–479.

Leuthner, B. and J. Heider, A two-component system involved in regulation of anaerobic toluene metabolism in *Thauera aromatica*. *FEMS Microbiol. Lett.*, 1998, **166**, 35–41.

Lovley, D.R. and R.T. Anderson, Influence of dissimilatory metal reduction on fate of organic and metal contaminants in the subsurface. *Hydrogeol. J.*, 2000, **8**, 77–88.

Lovley, D.R., J.D. Coates, E.L. Blunt-Harris, E.J.P. Phillips, and J.C. Woodward, Humic substances as electron acceptors for microbial respiration. *Nature*, 1996, **382**, 445–448.

Lovley, D.R., J.L. Fraga, E.L. Blunt-Harris, L.A. Hayes, E.J.P. Phillips, and J.D. Coates, Humic substances as a mediator for microbially catalyzed metal reduction. *Acta Hydrochimica et Hydrobiologica*, 1998, **26**, 152–157.

Lovley, D.R., K. Kashefi, M. Vargas, J.M. Tor, and E.L. Blunt-Harris, Reduction of humic substances and Fe(III) by hyperthermophilic microorganisms. *Chem. Geol.*, 2000, **169**, 289–298.

Lovley, D.R. and M.J. Klug, Sulfate reducers can outcompete methanogens at freshwater sulfate concentrations. *Appl. Environ. Microbiol.*, 1983, **45**, 187–192.

Lovley, D.R., J.C. Woodward, and F.H. Chapelle, Stimulated anoxic biodegradation of aromatic hydrocarbons using Fe(III) ligands. *Nature*, 1994, **370**, 128–131.

Ludwig, B., A. Akundi, and K. Kendall, A long-chain secondary alcohol dehydrogenase from *Rhodococcus erythropolis* ATCC 4277. *Appl. & Environ. Microbiol.*, 1995, **61**, 3729–3733.

Ludwig, R.A., Microaerophilic bacteria transduce energy via oxidative metabolic gearing. *Res. Microbiol.*, 2004, **155**, 61–70.

Maeng, J., Y. Sakai, Y. Tani, and N. Kato, Isolation and characterization of a novel oxygenase that catalyzes the first step of *n*-alkane oxidation in *Acinetobacter* sp. strain M-1. *J. Bacteriol.*, 1996, **178**, 3695–3700.

Magot, M., Indigenous microbial communities in oil fields. In *Petroleum Microbiology*, B. Oliver and M. Magot (eds.), Washington, DC: ASM Press, 2005, 21–34.

Marcelis, C.L.M., A.E. Ivanova, A.J.H. Janssen, and A.J.M. Stams, Anaerobic desulphurization of thiophenes by mixed microbial communities from oilfields. *Biodegradation*, 2003, **14**, 173–182.

Marchal, R., S. Penet, F. Solana-Serena, and J.P. Vandecasteele, Gasoline and diesel oil biodegradation. *Oil & Gas Sci. Tech. — Rev. IFP.*, 2003, **58**, 441–448.

Martínez-Pérez, O., E. Moreno-Ruiz, B. Floriano, and E. Santero, Regulation of tetralin biodegradation and identification of genes essential for expression of *thn* operons. *J. Bacteriol.*, 2004, **186**, 6101–6109.

Mauk, J.L. and R.C. Burruss, Water washing of Proterozoic oil in the Midcontinent rift system. *Amer. Assoc. Petrol. Geol. Bull.*, 2002, **86**, 1113–1127.

McMillen, S.J., A.G. Requejo, G.N. Young, P.M. Davis, P.D. Cook, J.M. Kerr, and N.R. Gray, Bioremediation potential of crude oil spilled on soil. In *Microbial Processes for Bioremediation, Proc. Third Intl. in Situ and On-Site Bioreclamation Symp., San Diego, CA, April 24–27*, R.E. Hinchee, C.M. Vogel, and F.J. Brockman (eds.), Columbus, OH: Battelle Press, 1995, 91–99.

Meckenstock, R.U., E. Annweiler, W. Michaelis, H.H. Richnow, and B. Schink, Anaerobic naphthalene degradation by a sulfate-reducing enrichment culture. *Appl. Environ. Microbiol.*, 2000, **66**, 2743–2747.

Meckenstock, R.U., M. Safinowski, and C. Griebler, Anaerobic degradation of polycyclic aromatic hydrocarbons. *FEMS Microbiol. Ecol.*, 2004, **49**, 27–36.

Miget, J.R., C.H. Oppenheimer, and H.I. Kator, Microbial degradation of normal paraffin hydrocarbons in crude oil. In *Proc. API/FWPCA Joint Conf. on Prevention and Control of Oil Spills*, New York: American Petroleum Institute, December 15–17, 1969, 327–331.

Moldowan, J.M., C.Y. Lee, P. Sundararaman, T. Salvatori, A. Alajbeg, B. Gjukic, G.J. Demaison, N.E. Slougui, and D.S. Watt, Source correlation and maturity assessment of select oils and rocks from the central Adriatic Basin (Italy and Yugoslavia). In *Biological Markers in Sediments and Petroleum; A Tribute to Wolfgang K. Seifert*, J.M. Moldowan, P. Albrecht, and R.P. Philp (eds.), Englewood Cliffs, NJ: Prentice Hall, 1992, 370–401.

Moody, J.D., J.P. Freeman, P.P. Fu, and C.E. Cerniglia, Degradation of benzo[*a*]pyrene by *Mycobacterium vanbaalenii* PYR-1. *Appl. Environ. Microbiol.*, 2004, **70**, 340–345.

Murphy, E.M. and J.A. Schramke, Estimation of microbial respiration rates in groundwater by geochemical modeling constrained with stable isotopes. *Geochemica et Cosmochimica Acta*, 1998, **62**, 3395–3406.

Nazina, T.N., A.E. Ivanova, O.V. Golubeva, R.R. Ibutulin, S.S. Belyaev, and M.V. Ivanova, Occurrence of sulfate-reducing and iron-reducing bacteria in stratal waters of the Romashkinskoe oil-field. *Microbiol. (Mikrobiologiya)*, 1995, **64**, 203–208.

Nealson, K.H., F. Inagaki, and K. Takai, Hydrogen-driven subsurface lithoautotrophic microbial ecosystems (SLiMEs): Do they exist and why should we care? *Trends in Microbiol.*, 2005, **13**, 405–410.

Nevin, K.P. and D.R. Lovley, Lack of production of electron-shuttling compounds or solubilization of Fe(III) during reduction of insoluble Fe(III) oxide by *Geobacter metallireducens*. *Appl. Environ. Microbiol.*, 2002a, **66**, 2248–2251.

Nevin, K.P. and D.R. Lovley, Mechanisms for Fe(III) oxide reduction in sedimentary environments. *Geomicrobiol. J.*, 2002b, **19**, 141–159.

Noh, S.-L., J.-M. Choi, Y.-J. An, S.-S. Park, and K.-S. Cho, Anaerobic biodegradation of toluene coupled to sulfate reduction in oil-contaminated soils: Optimum environmental conditions for field applications. *J. Environ. Sci. and Health Part A, Toxic Hazardous Substances and Environ. Eng.*, 2003, **38**, 1087–1098.

Obraztsova, A.Y., O.V. Shipin, L.V. Bezrukova, and S.S. Belyaev, Properties of the coccoid methylotrophic methanogen, *Methanococcoides euhalobius* sp.-nov. *Microbiol. (Mikrobiologiya)*, 1987, **56**, 523–527.

Onodera-Yamada, K., M. Morimoto, and Y. Tani, Degradation of dibenzothiophene by sulfate-reducing bacteria cultured in the presence of only nitrogen gas. *J. Biosci. Bioeng.*, 2001, **91**, 91–93.

Orphan, V.J., L.T. Taylor, D. Hafenbradl, and E.F. de Long, Culture-dependent and culture-independent characterization of microbial assemblages associated with high-temperature petroleum reservoirs. *Appl. Environ. Microbiol.*, 2000, **66**, 700–711.

Ourisson, G. and P. Albrecht, Hopanoids. 1. Geohopanoids: The most abundant natural products on Earth? *Accounts of Chem. Res.*, 1992, **25**, 398–402.

Papke, R.T., N.B. Ramsing, M.M. Bateson, and D.M. Ward, Geographical isolation in hot spring cyanobacteria. *Environ. Microbiol.*, 2003, **5**, 650–659.

Parales, R.E., N.C. Bruce, A. Schmid, and L.P. Wackett, Biodegradation, biotransformation, and biocatalysis (B3). *Appl. Environ. Microbiol.*, 2002, **68**, 4699–4709.

Parkes, R.J., The deep hot biosphere. *Nature*, 1999, **401**, 644–644.

Penet, S., R. Marchal, A. Sghir, and F. Monot, Biodegradation of hydrocarbon cuts used for diesel oil formulation. *Appl. Microbiol. Biotechnol.*, 2004, **66**, 40–47.

Peters, K.E. and J.M. Moldowan, Effects of source, thermal maturity, and biodegradation on the distribution and isomerization of homohopanes in petroleum. *Org. Geochem.*, 1991, **17**, 47–61.

Peters, K.E., C.C. Walters, and J.M. Moldowan, *The Biomarker Guide, Biomarkers and Isotopes in Petroleum Exploration and Earth History*, Vol. 1 & 2, New York: Cambridge Univ. Press, 2005, 1155.

Pettijohn, F.J., Chemical composition of sandstones; excluding carbonate and volcanic sands. *U.S. Geol. Survey Prof. Paper 440-S*, 1963, 21.

Pirnik, M.P., R.M. Atlas, and R. Bartha, Hydrocarbon metabolism by *Brevibacterium erythrogenes*: Normal and branched alkanes. *J. Bacteriol.*, 1974, **119**, 868–878.

Prenafeta-Boldu, F.X., J. Vervoort, J.T.C. Grotenhuis, and J.W. van Groenestijn, Substrate interactions during the biodegradation of benzene, toluene, ethylbenzene, and xylene (BTEX) hydrocarbons by the fungus *Cladophialophora* sp. strain T1. *Appl. & Environ. Microbiol.*, 2002, **68**, 2660–2665.

Prince, R.C., Petroleum and other hydrocarbons, biodegradation of. In *Encyclopedia of Environmental Microbiology*, G. Bitton (ed.), New York: John Wiley, 2002, 2402–2416.

Prince, R.C., The microbiology of marine oil spill bioremediation. In *Petroleum Microbiology*, B. Oliver and M. Magot (eds.), Washington, DC: ASM Press, 2005, 317–335.

Prince, R.C. and J.R. Clark, Bioremediation of marine oil spills. In *Petroleum Biotechnology — Studies in Surface Science and Catalysis*, Vol. 151, R. Vazquez-Duhalt and R. Quintero-Ramirez (eds.), Elsevier, 2004, 495–512.

Prince, R.C. and G.S. Douglas, Quantification of hydrocarbon biodegradation using internal markers. In *Manual of Soil Analysis — Monitoring and Assessing Soil Bioremediation*, R. Margesin and F. Schinner (eds.), Berlin: Springer-Verlag, 2005, 179–188.

Prince, R.C., R.M. Garrett, R.E. Bare, M.J. Grossman, T. Townsend, J.M. Suflita, K. Lee, E.H. Owens, G.A. Sergy, and J.F. Braddock, The roles of photooxidation and biodegradation in long-term weathering of crude and heavy fuel oils. *Spill Sci. Tech. Bull.*, 2003, **8**, 145–156.

Prince, R.C., E.H. Owens, and G.A. Sergy, Weathering of an Arctic oil spill over 20 years: The BIOS experiment revisited. *Mar. Poll. Bull.*, 2002, **44**, 1236–1242.

Prince, R.C. and J.M. Suflita, Submitted for publication, 2006.

Prommer, H., G.B. Davis, and D.A. Barry, Geochemical changes during biodegradation of petroleum hydrocarbons: Field investigations and biogeochemical modeling. *Org. Geochem.*, 1999, **30**, 423–435.

Rabus, R., Biodegradation of hydrocarbons under anoxic conditions. In *Petroleum Microbiology*, B. Oliver and M. Magot (eds.), Washington, DC: ASM Press, 2005, 277–300.

Rabus, R., H. Wilkes, A. Behrends, A. Armstroff, T. Fischer, A.J. Pierik, and F. Widdel, Anaerobic initial reaction of *n*-alkanes in a denitrifying bacterium: Evidence for (1-methylpentyl)succinate as initial product and for involvement of an organic radical in *n*-hexane metabolism. *J. Bacteriol.*, 2001, **183**, 1707–1715.

Rasmussen, B., Evidence for pervasive petroleum generation and migration in 3.2 and 2.63 Ga shales. *Geol.*, 2005, **33**, 497–500.

Reguera, G., K.D. McCarthy, T. Mehta, J.S. Nicoll, M.T. Tuominen, and D.R. Lovley, Extracellular electron transfer via microbial nanowires. *Nature*, 2005, **435**, 1098–1101.

Rike, A.G., K.B. Haugen, M. Børresen, B. Engene, and P. Kolstad, In situ biodegradation of petroleum hydrocarbons in frozen arctic soils. *Cold Regions Sci. Tech.*, 2003, **37**, 97–120.

Rios-Hernandez, L.A., L.M. Gieg, and J.M. Suflita, Biodegradation of an alicyclic hydrocarbon by a sulfate-reducing enrichment from a gas condensate-contaminated aquifer. *Appl. Environ. Microbiol.*, 2003, **69**, 434–443.

Rivas, I.M., H. Mosbaek, and E. Arvin, Product formation from thiophene by a mixed bacterial culture. Influence of benzene as growth substrate. *Water Res.*, 2003, **37**, 3047–3053.

Roadifer, R.E., Size distributions of the world's largest known oil and tar accumulations: Section I. Regional resources. In *American Association of Petroleum Geologists Studies in Geology 25*, R.F. Meyer (ed.), 1987, 3–23.

Rosso, K.M., J.M. Zachara, J.K. Fredrickson, Y.A. Gorby, and S.C. Smith, Non-local bacterial electron transfer to hematite surfaces. *Geochimica et Cosmochimica Acta*, 2003, **67**, 1081–1087.

Rothermich, M.M., L.A. Hayes, and D.R. Lovley, Anaerobic, sulfate-dependent degradation of polycyclic aromatic hydrocarbons in petroleum-contaminated harbor sediment. *Environ. Sci. Tech.*, 2002, **74**, 4811–4817.

Rullkötter, J. and D. Wendisch, Microbial alteration of 17α(H)-hopanes in Madagascar asphalts; removal of C-10 methyl group and ring opening. *Geochimica et Cosmochimica Acta*, 1982, **46**, 1545–1553.

Rullkötter, J. and D. Wendisch, Microbial alteration of 17alpha (H)-hopanes in Madagascar asphalts; removal of C-10 methyl group and ring opening. *Geochimica et Cosmochimica Acta*, 1982, **46**, 1545–1553.

Safinowski, M. and R.U. Meckenstock, Methylation is the initial reaction in anaerobic naphthalene degradation by a sulfate-reducing enrichment culture. *Environ. Microbiol.*, 2006, **8**, 347–352.

Sazinsky, M.H., J. Bard, A. Di Donato, and S.J. Lippard, Crystal structure of the toluene/*o*-xylene monooxygenase hydroxylase from *Pseudomonas stutzeri* OX1 — Insight into the substrate specificity, substrate channeling, and active site tuning of multicomponent monooxygenases. *J. Biol. Chem.*, 2004, **279**, 30600–30610.

Schmitt, R., H.-R. Langguth, W. Püttmann, H.P. Rohns, P. Eckert, and J. Schubert, Biodegradation of aromatic hydrocarbons under anoxic conditions in a shallow sand and gravel aquifer of the Lower Rhine Valley, Germany. *Org. Geochem.*, 1996, **25**, 41–50.

Scott, A.R., W.R. Kaiser, and W.B.J. Ayers, Thermogenic and secondary biogenic gases, San-Juan Basin, Colorado and New Mexico — implications for coalbed gas producibility. *Ame. Assoc. Petroleum Geologists Bull.*, 1994, **78**, 1186–1209.

Seifert, W.K. and J.M. Moldowan, The effect of biodegradation on steranes and terpanes in crude oils. *Geochimica et Cosmochimica Acta*, 1979, **43**, 111–126.

Seifert, W.K., J.M. Moldowan, and G.J. Demaison, Source correlation of biodegraded oils. In *Advances in Organic Geochemistry 1983*, P.A. Schenck, J.W. de Leeuw, and G.W.M. Lijmbach (eds.), New York: Pergamon Oxford, International, 1984, **6**, 633–643.

Sergy, G.A. and P.J. Blackall, Design and conclusion of the Baffin Island Oil Spill Project. *Arctic*, 1987, **40** (Suppl. 1), 1–9.

Shi, Ji-Y., A.S. MacKenzie, R. Alexander, G. Eglinton, A.P. Gowar, G.A. Wolff, and J.R. Maxwell, A biological marker investigation of petroleums and shales from the Shengli oilfield, the People's Republic of China. *Chem. Geol.*, 1982, **35**, 1–31.

Sirota, E.B., Physical structure of asphaltenes. *Energy & Fuels*, 2005, **19**, 1290–1296.

Slobodkin, A.I., C. Jeanthon, S. L'Haridon, T. Nazina, M. Miroshnichenko, and E. Bonch-Osmolovskaya, Dissimilatory reduction of Fe(III) by thermophilic bacteria and archaea in deep subsurface petroleum reservoirs of Western Siberia. *Current Microbiol.*, 1999, **39**, 99–102.

So, C.M., C.D. Phelps, and L.Y. Young, Anaerobic transformation of alkanes to fatty acids by the sulfate-reducing bacterium, strain Hxd3. *Appl. Environ. Microbiol.*, 2003, **69**, 3892–3900.

So, C.M. and L.Y. Young, Initial reactions in anaerobic alkane degradation by a sulfate reducer, strain AK-01. *Appl. Environ. Microbiol.*, 1999, **65**, 5532–5540.

Solano-Serena, F., M.R. Marchal, R.J.-M. Lebeault, and J.P. Vandecasteele, Biodegradation of gasoline: Kinetics, mass balance and fate of individual hydrocarbons. *J. Appl. Microbiology*, 1999, **86**, 1008–1016.

Solano-Serena, F., R. Marchal, and J.P. Vandecasteele, Biodegradation of gasoline in the environment: From overall assessment to the case of recalcitrant hydrocarbons. *Revue de l'Institut Francais du Petrole*, 2001, **56**, 479–498.

Spaulding, R.S., G.W. Schade, A.H. Goldstein, and M.J. Charles, Characterization of secondary atmospheric photooxidation products: Evidence for biogenic and anthropogenic sources. *J. Geophys. Res.*, 2003, **108**, 4247–4264.

Spormann, A.M. and F. Widdel, Metabolism of alkylbenzenes, alkanes, and other hydrocarbons in anaerobic bacteria. *Biodegradation*, 2000, **11**, 85–105.

Suflita, J.M., I.A. Davidova, L.M. Gieg, M. Nanny, and R.C. Prince, Anaerobic hydrocarbon biodegradation and the prospects for microbial enhanced energy production. In *Petroleum Biotechnology — Studies in Surface Science and Catalysis*, R. Vazquez-Duhalt and R. Quintero-Ramirez (eds.), Elsevier, 2004, **151**, 283–306.

Sugimoto, K., T. Senda, H. Aoshima, E. Masai, M. Fukuda, and Y. Mitsui, Crystal structure of an aromatic ring opening dioxygenase LigAB, a protocatechuate 4,5-dioxygenase, under aerobic conditions. *Structure with Folding and Design*, 1999, **7**, 953–965.

Sundararaman, P. and R.J. Hwang, Effect of biodegradation on vanadylporphyrin distribution. *Geochimica et Cosmochimica Acta*, 1993, **57**, 2283–2290.

Sutton, P.A., C.A. Lewis, and S.J. Rowland, Isolation of individual hydrocarbons from the unresolved complex hydrocarbon mixture of a biodegraded crude oil using preparative capillary gas chromatography. *Org. Geochem.*, 2005, **36**, 963–970.

Tang, Y.J., S. Carpenter, J. Deming, and B. Krieger-Brockett, Controlled release of nitrate and sulfate to enhance anaerobic bioremediation of phenanthrene in marine sediments. *Environ. Sci. Tech.*, 2005, **39**, 3368–3373.

Tao, Y., A. Fishman, W.E. Bentley, and T.K. Wood, Oxidation of benzene to phenol, catechol, and 1,2,3-trihydroxybenzene by toluene 4-monooxygenase of *Pseudomonas mendocina* KR1 and toluene 3-monooxygenase of *Ralstonia pickettii* PKO1. *Appl. Environ. Microbiol.*, 2004, **70**, 3814–3820.

Tissot, B. and D.H. Welte, *Petroleum Formation and Occurrence*, New York: Springer-Verlag, 1984, 699.

Townsend, G.T., R.C. Prince, and J.M. Suflita, Anaerobic oxidation of crude oil hydrocarbons by the resident microorganisms of a contaminated anoxic aquifer. *Environ. Sci. Tech.*, 2003, **37**, 5213–5218.

Townsend, G.T., R.C. Prince, and J.M. Suflita, Anaerobic biodegradation of alicyclic constituents of gasoline and natural gas condensate by bacteria from an anoxic aquifer. *FEMS Microbiol. Ecol.*, 2004, **49**, 129–135.

Trendel, J.M., J. Guilhem, P. Crisp, D.J. Repeta, J. Connan, and P. Albrecht, Identification of two C-10 demethylated C_{28} hopanes in biodegraded petroleum. *J. Chem. Soci. Chem. Commun.*, 1990, 424–425.

Tritz, J.-P., D. Herrmann, P. Bisseret, J. Connan, and M. Rohmer, Abiotic and biological hopanoid transformation: Towards the formation of molecular fossils of the hopane series. *Org. Geochem.*, 1999, **30**, 499–514.

Trolio, R., K. Grice, R. Alexander, R.I. Kagi, and S.J. Fisher, Alkylbiphenyls and alkyldiphenylmethanes as indicators of petroleum biodegradation. *Org. Geochem.*, 1999, **30**, 1241–1253.

Tropel, D. and J.R. van der Meer, Bacterial transcriptional regulators for degradation pathways of aromatic compounds. *Microbiol. Mol. Biol. Rev.*, 2004, **68**, 474–500.

Tsao, C.-W., H.-G. Song, and R. Bartha, Metabolism of benzene, toluene, and xylene hydrocarbons in soil. *Appl. Environ. Microbiol.*, 1998, **64**, 4924–4929.

Tunnicliffe, V., R.W. Embley, J.F. Holden, D.A. Butterfield, G.J. Massoth, and S.K. Juniper, Biological colonization of new hydrothermal vents following an eruption on Juan de Fuca Ridge. *Deep Sea Res. Part I: Oceanographic Res. Papers*, 1997, **44**, 1627–1644.

Ulrich, A.C., H.R. Beller, and E.A. Edwards, Metabolites detected during biodegradation of $^{13}C_6$-benzene in nitrate-reducing and methanogenic enrichment cultures. *Environ. Sci. Tech.*, 2005, **39**, 6681–6691.

van Beilen, J.B. and B. Witholt, Diversity, function, and biocatalytic applications of alkane oxygenases. In *Petroleum Microbiology*, B. Oliver and M. Magot (eds.), Washington, DC: ASM Press, 2005, 259–276.

van Beilen, J.B., M.G. Wubbolts, and B. Witholt, Genetics of alkane oxidation by *Pseudomonas oleovorans*. *Biodegradation*, 1994, **5**, 161–174.

Van Hamme, J.D., A. Singh, and O.P. Ward, Recent advances in petroleum microbiology. *Microbiol. Mol. Biol. Rev.*, 2003, **67**, 503–549.

Vargas, M., K. Kashefi, E.L. Blunt-Harris, and D.R. Lovley, Microbiological evidence for Fe(III) reduction on early Earth. *Nature*, 1998, **395**, 65–67.

Vetting, M.W., D.A. D'Argenio, L.N. Ornston, and D.H. Ohlendorf, Structure of *Acinetobacter* strain ADP1 protocatechuate 3, 4-dioxygenase at 2.2 A resolution: Implications for the mechanism of an intradiol dioxygenase. *Biochem.*, 2000, **39**, 7943–7955.

Volkman, J.K., R. Alexander, R.I. Kagi, R.A. Noble, and G.W. Woodhouse, A geochemical reconstruction of oil generation in the Barrow Sub-basin of Western Australia. *Geochimica et Cosmochimica Acta*, 1983, **47**, 2091–2105.

Watson, J.S., D.M. Jones, R.P.J. Swannell, and A.C.T. van Duin, Formation of carboxylic acids during aerobic biodegradation of crude oil and evidence of microbial oxidation of hopanes. *Org. Geochem.*, 2002, **33**, 1153–1169.

Wenger, L.M., C.L. Davis, and G.H. Isaksen, Multiple controls on petroleum biodegradation and impact on oil quality. *SPE Reservoir Eval. Eng.*, 2002, **5**, 375–383.

Whitaker, R.J., D.W. Grogan, and J.W. Taylor, Geographic barriers isolate endemic populations of hyperthermophilic archaea. *Science*, 2003, **301**, 976–978.

Widdel, F. and R. Rabus, Anaerobic biodegradation of saturated and aromatic hydrocarbons. *Curr. Opin. Biotech.*, 2001, **12**, 259–276.

Wilhelms, A., S.R. Larter, I. Head, P. Farrimond, R. Di-Primio, and C. Zwach, Biodegradation of oil in uplifted basins prevented by deep-burial sterilization. *Nature*, 2001, **411**, 1034–1037.

Yeager, C.M., P.J. Bottomley, D.J. Arp, and M.R. Hyman, Inactivation of toluene 2-monooxygenase in *Burkholderia cepacia* G4 by alkynes. *Appl. Environ. Microbiol.*, 1999, **65**, 632–639.

Yerushalmi, L., J.-F. Lascourreges, C. Rhofir, and S. Guiot, Detection of intermediate metabolites of benzene biodegradation under microaerophilic conditions. *Biodegradation*, 2001, **12**, 379–391.

Zengler, K., H.H. Richnow, R. Roselló-Mora, W. Michaelis, and F. Widdel, Methane formation from long-chain alkanes by anaerobic microorganisms. *Nature*, 1999, **401**, 266–269.

Zhang, X. and L.Y. Young, Carboxylation as an initial reaction in the metabolism of naphthalene and phenanthrene by sulfidogenic consortia. *Appl. Environ. Microbiol.*, 1997, **63**, 4759–4764.

Zwolinski, M.D., R.F. Harris, and W.J. Hickey, Microbial consortia involved in the anaerobic degradation of hydrocarbons. *Biodegradation*, 2000, **11**, 141–158.

12 Identification of Hydrocarbons in Biological Samples for Source Determination

Jeffrey W. Short[1] and Kathrine R. Springman[2]

[1] Auke Bay Laboratory, Alaska Fisheries Science Center, National Marine Fisheries Service, NOAA, 11305 Glacier Highway, Juneau, Alaska USA 99801-8626.
[2] University of California, Davis, Civil & Environmental Engineering, One Shields Avenue, Davis, California, USA 95617.

12.1 Introduction

Aquatic fauna are often examined for evidence of hydrocarbon exposure, as it may pose a direct threat to their health. Whether petroleum-derived hydrocarbons are released from chronic sources or from isolated, intermittent spills, biological tissues are routinely sampled and analyzed to evaluate the spatial extent and persistence of pollution, as well as the potential for adverse effects. Evidence obtained from these analyses can be paramount in litigation, where estimates of lost resource value associated with the extent of biological injury can form the basis of damage claims. Such estimates are rarely straightforward because multiple hydrocarbon sources must nearly always be distinguished, making it difficult to establish a clear link between the evidence of biological impact and the sources responsible for it.

In addition to the challenges associated with fingerprinting neat product samples following environmental release, analysis of hydrocarbons in biological tissues must also account for the route of hydrocarbon uptake and biotransformation. The four most important uptake pathways are (1) physical coating on the external surfaces of biota, possibly followed by passive partitioning through the skin, (2) ingestion of whole particulate oil or oil-contaminated prey, (3) passive partitioning of dissolved hydrocarbon components into biological tissues of aquatic biota, and (4) inhalation of hydrocarbon vapors. Absorbed hydrocarbons may then undergo biochemical transformation, depending on the metabolic capacity of the organism involved. Invertebrates are often less well-equipped to metabolize hydrocarbons compared with vertebrates, but there is considerable phylogenetic variability in the distribution of this competence, even within phyla (Livingstone, 1998). When it can be shown that hydrocarbon composition changes mediated by partitioning among phases or by biochemical transformations are not important, then biota often provide a convenient sampling matrix for evaluating the extent, persistence, and source of hydrocarbon contamination.

Clearly, the utility of tissue hydrocarbon analyses depends on the questions asked. When biota are sampled and analyzed to evaluate biological exposure or injury, the same data may provide evidence that constrains prospective hydrocarbon sources. As physical and biochemical processes can modify an already complex hydrocarbon profile, it is important to establish the limitations of data derived from the chemical analysis of biota with respect to the issue at hand. In some cases, analyses of biota may furnish evidence for source identification that is comparable with sediments or neat product for hydrocarbon source identification. In others, only rough estimates of hydrocarbon concentrations in the aqueous environment may be inferred from concentrations found within organisms. Even when the hydrocarbon fingerprint of the source is extensively modified by partitioning and biochemical transformation, biochemical evidence of hydrocarbon exposure may provide spatial and temporal patterns that implicate particular sources or eliminate others. In every case, all processes having the potential to alter the original hydrocarbon source signature must be considered carefully if the resulting evidence is to be applied defensibly.

Our objective here is to provide a succinct overview of the main processes that affect the interpretation of hydrocarbons in biota, with respect to inferences regarding hydrocarbon source. The primary issue in this endeavor is determining when definitive indicators of the hydrocarbon source are preserved in the biological matrix sampled. When they are, then inferences regarding sources may proceed along the lines that have been developed for fingerprinting environmental samples affected only by physical weathering processes, the subject of extensive literature discussed elsewhere in this volume (e.g., Chapters 3 and 13). When they are not, the associated sample collection information may still furnish evidence regarding prospective sources, albeit less conclusively. We therefore begin with a summary of the physical processes altering hydrocarbon composition that are associated with each of the main routes of hydrocarbon exposure for biota, which determines the degree of source-specific information retained by accumulated hydrocarbons. This is followed by a discussion of the biochemical pathways that may transform the hydrocarbons accumulated, which may stimulate biochemical responses such as cytochrome P450 induction that can be very sensitive indicators of exposure. These pathways typically reflect biochemical adaptations to mitigate the toxicity of accumulated hydrocarbons, so their relation to hydrocarbon mechanisms of toxicity is also presented. Finally, many of these principles are illustrated by the comprehensive studies of the *Exxon Valdez* oil spill (see ahead and Chapter 15). While our coverage of the relevant literature is far from exhaustive, we have tried to include references to studies and reviews that have been helpful in gaining a broader understanding of the processes involved while providing ready identification of more focused studies.

12.2 Determination of the Primary Route of Hydrocarbon Accumulation by Biota

As soon as a petroleum-derived product is released into the environment, its composition changes in response to physical and biological processes, including evaporation, dissolution, microbial oxidation, and photooxidation, collectively referred to as "weathering" (Garrett et al., 1998; Ezra et al., 2000). Initially, evaporation is the key process, followed by dissolution if the spilled material is in contact with water. While microbial processes require more time, they can lead to profound compositional changes that are more difficult to predict because isomer-specific degradation rates vary among microbial species (e.g., see Chapter 11 herein), so composition changes caused by microbial degradation depend on the composition of the ambient microbial flora. Photooxidation is inherently slower, affecting mainly the polycyclic aromatic hydrocarbons (PAHs, herein used to include heterocyclic aromatics) on oil surfaces exposed to strong sunlight and high oxygen concentrations (Hansen, 1977).

Substantial changes in the hydrocarbon profile, such as the loss of the most volatile components, can result from evaporation from thin oil films on time scales of hours to days (Payne et al., 1984). For crude oils, evaporation may account for losses of ~25% or more, with considerably greater losses for lighter distillates, but less for heavy residual fuels such as bunker oils (Fingas, 2004). Due to their lower vapor pressures, evaporative losses of less volatile components are sometimes dismissed as negligible but can be substantial. For example, when the ratio of surface area to volume of an oil $(A/V)_{oil}$ is large and the oil is exposed to the atmosphere for long periods (years), considerable proportions of semivolatile components may ultimately be lost through evaporation (Short and Heintz, 1997).

Dissolution is a more gradual process than evaporation because the aqueous boundary layer adjacent to the oil parcel surface provides more diffusive resistance to the flux of dissolved components away from the oil. Dissolution also results in different phase-partitioning compared with evaporation. Evaporative component losses are determined primarily by vapor pressure. For example, naphthalene, with a molecular mass of 128, has a vapor pressure of 3.95×10^{-4} atm[1] at 25°C (Hinckley et al., 1990), equivalent to a gas-phase concentration of 1.62×10^{-5} mole/L, whereas nonane, an alkane of the same molecular weight, has a much higher vapor pressure[2] (6.07×10^{-3} atm) and gas-phase concentration (2.49×10^{-4} mole/L). The lower vapor pressure of naphthalene is due to the stronger attractive Van der Waals force associated with its less tightly bound aromatic π-electrons. Hence, the aliphatic hydrocarbons in petroleum products have a much greater tendency to be lost through evaporation compared with aromatic hydrocarbons of similar molecular masses. Just the opposite situation occurs with dissolution. The solubility of naphthalene[1] in water at 25°C is 9.44×10^{-4} mole/L (Shiu et al., 1988), or ~58 times greater than its "solubility" in air, while the solubility of nonane is 1.72×10^{-6} mole/L (McAuliffe, 1969), which is *lower* than its solubility in air by a factor of ~150. This comparison demonstrates that water acts like a sponge for aromatic hydrocarbons, but as a lid for aliphatics, when compared with the atmosphere. As a result, the water-soluble fraction (WSF) of oil is enriched with the more soluble aromatic hydrocarbons but deficient in aliphatics.

Microbial oxidation may cause dramatic changes in oil composition on time scales of days to years following release to the environment (e.g., see Chapter 11 herein), depending on the favorability of growth conditions. Warm temperatures, a steady supply of essential nutrients, and high initial populations of competent microbes are the most important factors promoting rapid and sustained oxidation rates. The *n*-alkanes are the most readily oxidized substrate in petroleum products, followed by branched alkanes, unsubstituted PAH, and alkyl-substituted PAH (Pirnik, 1977; Schaeffer et al., 1979; Payne et al., 1984; Wang et al., 1996). Other components of oil, even including some of the most persistent compounds such as the complex alicyclic hopanes, other triterpanes, and steranes, are susceptible to eventual degradation by microbial oxidation although at much slower rates (see, e.g., Wang et al., 2000).

The weathering processes summarized above lead to hydrocarbon degradation patterns that are nearly always observed following oil spills. First, evaporation rapidly purges the most volatile hydrocarbons, including aliphatics with molecular weights below *n*-dodecane, and the monocyclic aromatics (especially benzene, toluene, ethyl-benzene, and the xylenes, or BTEX compounds). These compounds are consequently lost to the atmosphere within the first week or two of a spill (e.g., Payne et al., 1984). Second, if the spilled oil is dispersed in an oxic environment, the *n*-alkanes are consumed by microbes relatively rapidly. The more resistant isoprenoids such as pristane and phytane remain somewhat longer as resolvable peaks on aliphatic chromatograms, usually associated with a substantial unresolved complex mixture (UCM), which appears as a hump (see, e.g., Payne et al., 1984; Boehm et al., 1987). Indeed, the ratio of *n*-octadecane/phytane has often been used as a measure of the extent of aliphatic

---

[1] Values are for the subcooled liquid, i.e., the value that would be observed if the compound were in the liquid state at 25°C (see Shiu et al., 1988).
[2] Estimated from application of the Clausius–Clapeyron equation to vapor pressure data tabulated in Weast (1977).

biodegradation since phytane is rarely found apart from petroleum-derived products (Dean and Whithehead, 1961; Blumer and Snyder, 1965).

Finally, if oil is in an aquatic environment, the water will extract PAH at rates that decrease with increasing number of rings and degree of alkyl substitution. This pattern is the result of the effects of molecular volume on solubility (McAuliffe, 1966). At equilibrium, the extent of PAH loss from oil is determined by the distribution coefficient K of the PAH and the relative volumes of oil and water involved. Shiu et al. (1988) show that the equilibrium concentration C_w of a PAH in water is approximately $C_{so}/(K + Q)$, where C_{so} is the initial concentration of the PAH in the oil, K is ratio of the equilibrium concentration of the PAH in the oil (C_o) and C_w, and Q is the water-to-oil volume ratio V_w/V_o. The distribution coefficient K is rarely measured, but may be approximated by the octanol-water distribution coefficient K_{ow}. For a 0.1-mm-thick oil slick in equilibrium with a 10-m mixed layer of water beneath it, $Q = 100,000$, implying that more than 50% depletion will occur for PAH with K_{ow} smaller than this value. This would include all the naphthalenes bearing as many as three alkyl carbon atoms, and phenanthrenes with as many as one, whereas dissolution losses of four-ring PAH would be considerably smaller. At smaller values of Q, the extent of PAH loss from oil is determined principally by the distribution coefficient K. The value of K increases rapidly with increasing molecular volume, leading to aqueous PAH distribution patterns that are strongly dominated by the most soluble components. Conversely, the aqueous distribution pattern will approach that initially observed in the oil when Q becomes large relative to K of the least soluble PAH (which may be $\sim 10^6$). Under these conditions, there is sufficient water available to extract nearly all the PAH initially present in the oil. However, despite the large water-to-oil volumes that may be applicable during oil spills, such extreme depletion is rarely observed. This is because oil releases usually occur in open systems that are far from equilibrium, so dissolution kinetics are also important. Hence, in addition to being less soluble, larger PAH also dissolve more slowly, further enhancing relative concentrations of the lower-molecular-weight PAH in the aqueous phase.

Selective PAH loss from dissolution, often called "water-washing," is the dominant cause of changes in the PAH profile of spilled oil and follows a course that may be predicted from the initial PAH composition of the oil and the molecular surface area of the PAH. A simple model of this process (Short, 2002) accurately reproduces the dominant PAH weathering patterns observed following oil spills. For example, application of this model to the PAH composition of *Exxon Valdez* cargo oil matched the analytical results of nearly all of the hundreds of samples of oiled sediments and mussels within analytical precision (Short and Heintz, 1997), implying that other PAH weathering processes were minor compared with evaporation and dissolution after that spill. The PAH weathering sequence predicted by the model is depicted in Figure 12-1. This characteristic pattern of PAH losses produces a complementary pattern of dissolved PAH in the receiving water, illustrated in Figure 12-2, calculated from the difference in PAH concentrations represented in the two uppermost panels of Figure 12-1. The enrichment of the more readily soluble PAH is apparent and reflects the preferential partitioning of PAH into water discussed above. Collectively, these weathering processes provide crucial guidance for deducing the exposure route based on hydrocarbon concentration patterns observed in biota.

An uptake pathway involving accumulation of whole oil by an organism is indicated if the hydrocarbon distribution pattern matches the source oil, allowing for weathering and provided the organism's metabolism does not alter the pattern further. Oil adhering to the external surfaces of animals such as marine mammals and birds is an obvious example. Less obvious is oil accumulated by invertebrate suspension feeders such as mussels (*Mytilus sp.*) or in the intestinal tracts of fish. Chemical hallmarks of whole-oil accumulation include a complete set

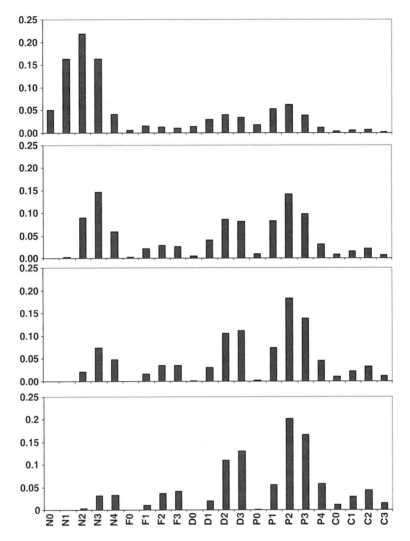

Figure 12-1 Typical weathering sequence of PAH in Alaska North Slope crude oil. N = naphthalene, F = fluorene, D = dibenzothiophene, P = phenanthrene/anthracene, C = chrysene; numbers indicate substituent alkyl carbon atoms. Weathering increases from top panel (unweathered oil) to bottom panel (moderately weathered oil). Abscissa: proportion of total PAH.

of petroleum biomarkers[3] (e.g., hopanes, other triterpanes, and steranes) that have ratios

---

[3] The term "biomarker" is used in two different ways in oil pollution studies. Petroleum biomarkers, sometimes called "molecular fossils," are alicyclic hydrocarbons that persist on geologic time scales and are usually associated with specific evolutionary events such as the appearance of flowering plants. Biological biomarkers refer to biochemical responses to pollutant exposure that are readily identifiable, such as the detoxification pathways involving cytochrome P450 enzymes.

matching the corresponding ratios of the source oil, a water-washed PAH distribution pattern comparable with those shown in Figure 12-1 (reflecting the weathering state of the oil when ingested), a prominent aliphatic and aromatic UCM, and often a variable profile of n-alkanes, perhaps with pristane and phytane prominent. When these criteria are met, the source identification fingerprinting methods applicable to whole-oil samples (see Wang et al., 1999, for a review) may be confidently

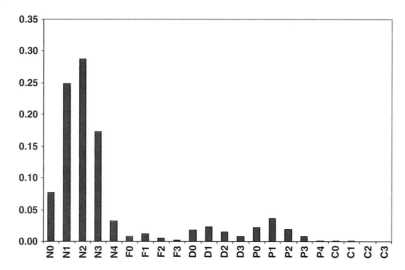

Figure 12-2 Relative composition of aqueous PAH dissolved from initially unweathered Alaska North Slope oil. Abbreviations and axes as in Figure 12-1.

applied to biological samples. Indeed, such samples are often invaluable in source identification.

Hydrocarbon results for mussels collected two months following the *Exxon Valdez* oil spill provide an example of hydrocarbon distribution patterns indicative of whole-oil accumulation (Figure 12-3). Comparison of PAH patterns of surface oil collected 11 days following the spill with those of mussels collected about six weeks later clearly shows the water-washed PAH pattern depicted in Figure 12-1 in the mussels, but phytane as a proportion of total PAH is little changed. Because the aqueous solubility of phytane is negligible, it serves as a petroleum biomarker for the weathered whole oil.

Much less source information is available from organisms that accumulate hydrocarbons from the dissolved phase. Uptake of dissolved hydrocarbons involves passive partitioning between the aqueous phase of the exposure medium and the lipid compartment(s) of the organism. An organism exposed to the dissolved-phase PAH distribution illustrated in Figure 12-2 will acquire a PAH burden resembling it. The higher-molecular-weight PAHs that are less readily extracted from the oil by the water are less available for subsequent partitioning into biotic lipids. Hence, mussels exposed to water-soluble fractions of crude oil typically contain scant four-ring PAH, for example, because so little of these partition into the water from the oil. After partitioning from crude oil via an intermediate aqueous phase into biota, PAH distributions resemble patterns that result from diesel oil, since the initial dissolution process truncates the higher boiling and less soluble PAH in the crude oil (which are removed during the refining process in diesel oil). Consequently, the resulting hydrocarbon signature in biota that accumulates only dissolved hydrocarbons will usually retain little information regarding the source.

Of course, aquatic biota may be simultaneously exposed to both dispersed oil droplets and their dissolved components, but this is most likely soon after (and close to) a spill event. Once an oil–water dispersion of droplets is created, it will promote dissolution of the soluble components because of the increase of $(A/V)_{oil}$, and the droplets will respond to buoyancy forces. The dissolved components, though, will not, which usually leads to separation of oil droplets from the most soluble oil components in the affected water masses.

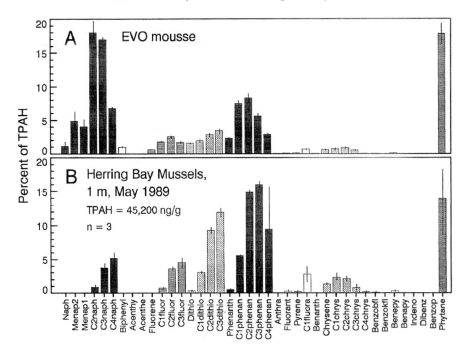

Figure 12-3 Relative abundance of PAH and phytane in (A) *Exxon Valdez* oil mousse 11 days following the spill, and (B) in mussels collected six weeks later. Note the PAH weathering evident in the mussels compared with the oil (compare with Figure 12-1), and the prominent phytane in both, indicating ingestion of whole weathered oil droplets by the mussels. Abbreviations: Naph = naphthalene, Menaph = methylnaphthalene, Acenthy = acenaphthylene, Acenthe = acenaphthene, Fluor = fluorene, Dithio = dibenzothiophene, Phenan = phenanthrene, Anthra = anthracene, Fluorant = fluoranthene, Benanth = benzo[*a*]anthracene, Chrys = chrysene, Benzobfl = benzo[*b*]fluoranthene, Benzokfl = benzo[*k*]fluoranthene, Benepyr = benzo[*e*]pyrene, Beneapyr = benzo[*a*]pyrene, Indeno = indeno[1,2,3-*c,d*]pyrene, Dibenz = dibenzo[*a,h*]anthracene, Benzop = benzo[*g,h,i*] perylene (from Short and Harris, 1996b).

Considerably less is known about tissue hydrocarbon patterns that characterize exposure pathways involving vapor inhalation and absorption across the skin. These pathways are thought to be important for marine mammals (Engelhardt, 1987; Geraci and St. Aubin, 1987; Spraker et al., 1994), but are difficult to study. Inhalation of hydrocarbon vapors may lead to reversible narcosis when not fatal, so evidence of exposure is usually either ephemeral or associated with a carcass on the sea floor. PAHs that are absorbed through the skin (or through the intestinal wall) are subject to the efficient biochemical detoxification pathways found in mammals, which may obscure evidence of dermal absorption. Although the biochemical detoxification responses may themselves become the primary indication of exposure, in these cases almost all information associated with hydrocarbon distribution patterns that reflect those of the original hydrocarbon sources is lost. A summary of these pathways is presented in the next section.

12.3 Catabolic Degradation of Hydrocarbons Accumulated by Biota

12.3.1 Catabolic Degradation of PAH

Catabolism is the breakdown of complex substances into simpler molecules. At least some capacity to catabolically degrade PAH is widely distributed among the animal kingdom (Livingstone, 1998), resulting from the need to detoxify poisons produced by prey. The evolutionary competition between predators and prey has produced chemical defenses of generally increasing sophistication among prey, especially plants, which has in turn selected for detoxification pathways in predators that are commensurately sophisticated. These detoxification pathways are more complex and efficient in more evolutionarily advanced organisms, which leads to large differences in detoxification rates among organisms. Before considering these differences, it will be helpful to summarize the basic molecular biology of the most important PAH detoxification pathways, which involve the cytochrome P450-dependent monooxygenases.

Cytochrome P450 enzymes are a superfamily of heme-containing proteins that catalyze the oxidation of their substrates and are so named because their complex with carbon monoxide has an absorption maximum at 450 nm (Ortiz de Montellano, 1995). Some of these enzymes are inducible by their substrates, facilitating a biochemical adaptation to changing environmental conditions (Denison and Whitlock, 1995). One of these, cytochrome P450 1A (CYP 1A hereafter), catalyzes the oxidation of planar aromatic molecules including PAH, dioxins, and some polychlorinated biphenyls (PCB) (Nelson et al., 1996). In most species, this protein is found at low constitutive levels in the cytosol due to its suppression by a repressor protein. In the presence of an inducer, however, both the derepression and activation of transcription by the aryl hydrocarbon receptor (AhR) are involved. The AhR is normally complexed in a 1:2 ratio with heat shock protein (hsp 90), which dissociates on binding of a planar aromatic ligand to the AhR. The activated AhR may cross the nuclear membrane, where it can bind with the aryl hydrocarbon nuclear translocator protein (ARNT). The resulting heterodimer then binds to DNA regulatory sequences located upstream of the CYP 1A gene, thus upregulating its expression (Whitlock, 1993). This cascade of events stimulates the production of messenger RNA, followed by protein translation on the endoplasmic reticulum to produce CYP 1A in the cytosol (Figure 12-4). The CYP 1A catalyzes several types of oxidative reactions making its substrates, such as PAH or other planar aromatics in the cytosol, more water-soluble and readily excreted. These reactions can introduce an epoxy group to the aromatic ring, which may then undergo further metabolism to form a diol or diol epoxide (Parkinson, 2001). Depending on the parent compound, the metabolism of PAHs can increase their toxicity by activating the toxin, leading to protein and DNA adducts, and possibly cancer (Varanasi et al., 1986) or other adverse effects (Guengerich and Liebler, 1985). The substrate specificity of CYP 1A is less restrictive than is the ligand-AhR binding specificity, so that not all PAHs are effective inducers of CYP 1A (Barron et al., 2004), but once induced, CYP 1A may oxidize a broader spectrum of PAH (Hawkins et al., 2002).

This initial oxidation by CYP 1A is often referred to as phase I metabolism to distinguish it from phase II metabolism, which involves condensation of the hydroxylated aromatic with a sugar or other highly soluble compound to increase the water solubility while decreasing the toxicity of the phase I metabolite. In vertebrates and in some invertebrates, the same ligand-AhR binding that triggers CYP 1A production also activates other genes that code for phase II catabolic pathways (Foureman, 1989). These include glutathione S-transferase, uridine diphosphate-glucuronyl and -glucosyl transferase, sulphotransferase, and amino acid conjugases (Whitlock, 1993).

These phase I and II detoxification pathways are fully developed in the vertebrates, but are much less so in the more primitive phyla. For example, an inducible CYP 1A system is absent in cnidarians (Kaji et al., 1983; Newman et al., 1990) and probably in sipunculid worms (Lee, 1981) and is generally less active in annelids, mollusks, and echinoderms when compared to crustaceans, fish, and vertebrates (Lee, 1981; Livingstone, 1998; Chaty et al., 2004). When present in mollusks, echinoderms, and species of other less advanced phyla, CYP 1A activity is localized in the digestive tract (Lee, 1981; Livingstone, 1998) but is more concentrated in the hepatopancreas of crustaceans. In vertebrates, CYP 1A

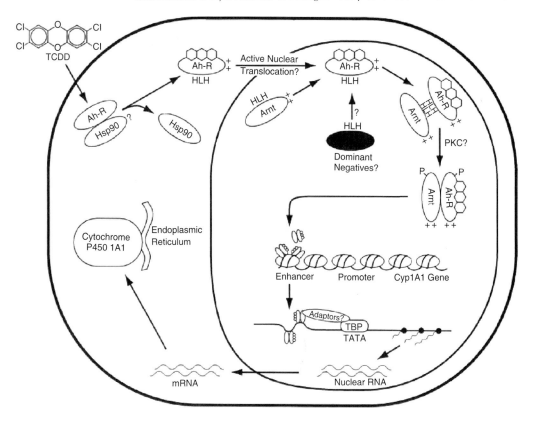

Figure 12-4 Cytochrome P450 1A induction pathway (from Whitlock, 1993; see text in section 12.3.1 for abbreviation).

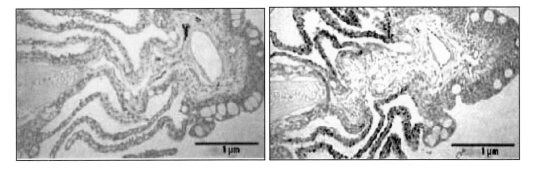

Figure 12-5 Gill epithelial cells under normal conditions (left), and under CYP 1A induction (right) (from Malins et al., 2004).

activity can be found in a broad array of vertebrate cell types, including gill epithelial cells (Figure 12-5), with the liver as the focus of CYP 1A activity (Livingstone, 1998).

Almost every aspect of the CYP 1A induction sequence has been used as a basis for evaluating induction, including antibody reactivity with the CYP 1A protein, detection of the mRNA sequence, and direct assay of catalytic activity (Stegeman and Hahn, 1994; Bucheli and Fent, 1995; Chaty et al., 2004). Of these, direct assay of catalytic activity is most widespread, due to its specificity, relative ease, and low cost. The most popular assay is based on

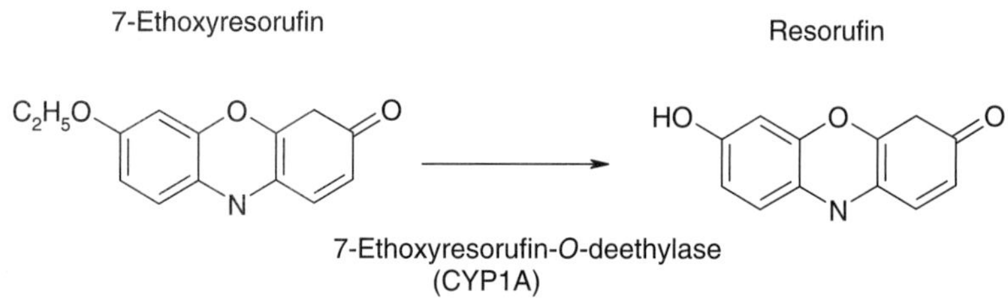

Figure 12-6 Catalysis of 7-ethoxyresorufin to resorufin by cytochrome P450 1A (CYP 1A). Resorufin is fluorescent, and its production rate is the basis for the EROD assay for CYP 1A activity.

the de-ethylation of 7-ethoxyresorufin to form resorufin (Figure 12-6), usually denoted as the EROD assay (Whyte et al., 2000). The substrate is mixed with a microsomal suspension isolated from homogenized liver tissue, along with NADPH, Mg^{2+}, and a buffer. The resorufin produced may be analyzed by absorption spectrophotometry or by fluorescence, with excitation at 530 nm and emission at 580 nm.

Various aspects of the CYP 1A induction sequence provide the basis for a suite of biological biomarkers indicative of exposure to planar aromatics, including PAH. Measurement of high specific CYP 1A activity (i.e., picomoles of resorufin/mg protein/min) is one of these, but several others are in current use in pollution studies [see reviews by Stegeman and Hahn (1994), Bucheli and Fent (1995), and van der Oost et al. (2003)]. The amount of CYP 1A enzyme induced may also be measured immunochemically by binding with functionalized antibodies (e.g., Smolowitz et al., 1991), by enzyme-linked immunosorbent assay (ELISA), or by other related immunological techniques (see reviews by Stegeman and Hahn, 1994, and by Bucheli and Fent, 1995). Detection of the mRNA encoding the CYP 1A protein may be separated by electrophoresis and detected with a complementary oligonucleotide sequence (northern blot analysis; Bucheli and Fent, 1995). Products of both phase I and phase II metabolism accumulate in the bile of vertebrates, which is a rather convenient compartment to sample in fish. Because these metabolites retain the intact PAH ring structure, they are fluorescent,

making them readily detectable. Hence, high-performance liquid chromatography analysis of fluorescent aromatic compounds in biliary fluid can distinguish 2–3-ring, 3–4-ring, and 4–5-ring PAH (Krahn et al., 1992), which may provide information regarding the source of the PAH exposure. Electrophilic PAH metabolites form covalent adducts with DNA, which may be detected using ^{32}P radiolabeling methods (Varanasi et al., 1989), and which provide direct evidence of DNA damage that may lead to carcinogenesis and other symptoms. Another biological biomarker of chromosome damage is the electrophoretic separation of DNA, termed the "comet assay." Broken strands of DNA produce skewed migration patterns that resemble comets in appearance (Fairbairn et al., 1995). However, while very sensitive, the comet assay is less specific because agents other than polycyclic aromatics can cause chromosome breakage.

12.3.2 Effects of Catabolism on PAH Accumulation, Persistence, and Depuration

Establishing a link between evidence of exposure to hydrocarbons and their potential source requires careful consideration of numerous factors. These include the age, life stage, gender, reproductive status, and nutritional status of the target species, the hydrocarbon exposure route, and the characteristics, as well as idiosyncrasies, of the biological measurements involved. Hydrocarbons accumulated as whole-oil droplets (usually as microdroplets) by suspension feeders with little, if any, capacity to biotransform these hydrocarbons retain considerable information regarding sources. In vertebrates such as fish, which may rapidly metabolize hydrocarbons, the main value biological biomarkers of PAH exposure will usually be as evidence that the organism has been exposed to compounds that can elicit the anticipated effects. Spatial patterns of the intensity of biological biomarker responses in relation to the suspected source may bolster a case, as may temporal patterns, especially if data are available prior to the suspected incident at both control and impacted sites, permitting use of statistically powerful "before/after-control/impact" (BACI) sampling designs (Green, 1979). But in all these applications, it is necessary to have an appreciation for the characteristic timescales of hydrocarbon accumulation, biological biomarker induction, depuration, and persistence of biomarker responses, which are also dependent on target species and exposure route.

Mussels of the genus *Mytilus* are by far the most widely used organisms for assessing marine pollution and oil spills, in part because they are sessile suspension feeders with a limited capacity for altering hydrocarbon distribution patterns when exposed to concentrated pollution sources (i.e., oil spills). These bivalves filter on the order of 1 L of seawater per hour (Foster-Smith, 1975) and capture particles as small as 2 µm (Vahl, 1972). Consequently, their uptake of dispersed oil droplets is very efficient. Mussels may accumulate hydrocarbons associated with dispersed whole oil in excess of 1000 times exposure levels (Fossato and Canzonier, 1976), compared with increases on the order of 20- to 80-fold for 2- to 3-ring PAH when dissolved (Hansen et al., 1978). An important consequence of this effectiveness is that ingestion of oil droplets may rapidly overwhelm the limited ability of mussels to depurate or metabolize these hydrocarbons, thereby preserving source information contained in hydrocarbon distribution patterns. Other advantages of mussels include their large size, making them easy to collect and process for analysis; their relatively high tolerance to oil pollution (Smith, 1968), so they do not usually function abnormally when exposed; and a cosmopolitan distribution. Also, mussels provide a time-integrated sample of their exposure, which captures transient pulses of hydrocarbons that might easily be missed by discrete water sampling. As a result of these advantages, there have been a wealth of studies that have employed them, which facilitates comparisons with other pollution incidents (Bayne, 1976; Donkin et al., 2003; Aarab et al., 2004).

Mussels do have their disadvantages. Their availability may be limited or nonexistent in the desired locations; the variability of accumulated hydrocarbons among identically exposed individuals may be substantial, due to differences in size, lipid content, and reproductive state (and the reproductive state is neither obvious without sacrifice of the animal nor necessarily seasonally predictable); and their pumping rate depends strongly on temperature and food supply (Bayne, 1976). But these disadvantages are all manageable and are usually far outweighed by the convenience of sampling, straightforward and well-characterized chemical analysis methods, and their propensity to preserve hydrocarbon source information that derives from their preferential accumulation of particles (which may be oiled) and scant ability to metabolically transform hydrocarbons.

Compelling evidence of mussels' ability to preserve information regarding hydrocarbon source was demonstrated by a study of the *Exxon Valdez* oil spill, where over a thousand samples collected during a 3-year period following the incident showed PAH distribution patterns that were not distinguishable from those in oiled sediments (Short and Heintz, 1997). The many advantages of using mussels have led to their selection as the marine monitoring species of choice when practical. Mussels have also been of fundamental importance in several large-scale monitoring programs such as the U.S. National Status and Trends program (Wade et al., 1998). By analogy, other suspension feeders with a similar capacity to metabolically transform ingested hydrocarbons would be alternate candidates for hydrocarbon source identification studies, for example, oysters and clams (Bender et al., 1988) or freshwater mussels.

The time scale of hydrocarbon accumulation and depuration by mussels is generally similar when exposed to oil dispersions or to dissolved hydrocarbons. Typically, 10 to 20 days are required for hydrocarbon concentrations in mussels to approach equilibrium with exposure concentrations (Fossato and Canzonier, 1976; Pruell et al., 1986). Depuration usually follows a two-compartment model, with rapid depuration from one compartment followed by much slower from the other (see review by Meador et al., 1995). Depuration is slower when the exposure route is ingestion of oil droplets compared with dissolved hydrocarbons, suggesting that hydrocarbons absorbed through the gills are less effectively incorporated into somatic tissues than when absorbed through the digestive system. Also, the proportion of hydrocarbons incorporated in the slowly depurating compartment tends to increase with exposure time, probably because of slow exchange between depot lipids and the circulatory system. The depuration half-life of hydrocarbons ingested by mussels is on the order of 3 days for losses from the rapid-depuration compartment (Fossato and Canzonier, 1976). For PAH accumulated from aqueous solutions, depuration half-lives increase with the K_{ow}, are faster at higher temperature, and vary with season and physiological state (Meador et al., 1995). These half-lives typically range from 2 d for naphthalene to as long as 2 weeks for 4-ring PAH.

Vertebrates with substantial capacity to metabolize hydrocarbons can still exhibit hydrocarbon distribution patterns that are little altered from source distributions if the hydrocarbon exposure overwhelms the metabolic capacity of the organism. One of the clearest examples of this occurred following the *Exxon Valdez* oil spill, when contamination of prey across hundreds of square km presented juvenile salmon with a choice between ingesting oiled prey and starvation. Surviving juveniles obviously chose the former, and the PAH distribution of the ingested oil was identical with that of the weathered oil for weeks following the incident, especially in gut samples (Carls et al., 1996). But apart from catastrophic oil spills, the more usual exposure route is through branchial and skin absorption of dissolved hydrocarbons. Even aliphatic hydrocarbons may be absorbed across the intestine when ingested, but fish and other vertebrates readily metabolize this class of hydrocarbons as well, so that only traces of ingested aliphatics persist in the lipid compartment of fish (Cravedi and Tulliez, 1982).

Once incorporated somatically, fish and other vertebrates may transform accumulated PAH into metabolites within hours. Fish metabolize up to 99% of PAH within 24 h of uptake, with most of the metabolites being excreted into the bile (Varanasi et al., 1989). The distribution of PAH among tissues of oiled marine mammals implies metabolic transformation rates that are similarly rapid, with most of the PAH in the liver or in depot lipids of animals that continued to be exposed (e.g., Frost et al., 1994). While it is possible to detect a small proportion of accumulated hydrocarbons in the depot lipids of these animals for considerable periods following exposure, most of the PAH will be present as metabolites, if at all. In fish, biliary PAH metabolites may persist for one to several weeks following cessation of exposure (Krahn et al., 1986). The PAH adducts in DNA are more persistent and may remain detectable for months (Sikka et al., 1991).

The time course of CYP 1A induction and persistence may provide important clues regarding the source of the induction agent. Intraperitoneal injection studies with fish demonstrate that induction is rapid once an inducer has been incorporated into tissues, with elevated CYP 1A levels detectable within hours of injection (e.g., Sleiderink and Boon, 1996) and are paralleled by similar mammalian responses (Renwick et al., 2000). Longer response times of one to a few days are required for field-exposed organisms due to the time required for sufficient contaminant accumulation to stimulate induction. For readily metabolized inducers such as PAH, the time course of CYP 1A induction typically increases for a few days following the onset of PAH exposure, followed by a decline as the accumulated PAHs are more rapidly metabolized. If the PAH exposure level is steady and chronic, the CYP 1A level declines to a steady-state value determined by a balance between sufficient PAH to maintain induction and the rate of its catabolic transformation (Munkittrick et al., 1995). Once an inducer is no longer available, CYP 1A levels may decline rapidly, with a half-life on the order of a day or less (Sadar and Andersson, 2000). In field studies, CYP 1A induction may be evident for weeks or months because of repeated exposure, or because of release from lipid depots as they are being drawn down (Kennish et al., 1992). But in contrast with PAH, less readily metabolizable inducers such as polychlorinated biphenyls or dioxin may continue to stimulate induction for weeks following a discrete exposure event (van der Weiden et al., 1994; Beyer et al., 1997), so the persistence of CYP 1A following the end of exposure to the inducing medium provides an indication of the chemical class of the inducing agent responsible.

Environmental and other factors can modify the time course of CYP 1A induction by PAH, which should be considered when interpreting these induction patterns (reviewed by Whyte et al., 2000). Even among vertebrates, the strength of the induction may vary considerably among species, within species of differing sizes and especially with life stage and reproductive condition, with lower induction occurring in early life stages and in reproductively active females (Lindström-Seppä and Stegeman, 1995). Lower temperatures retard induction, as does poor nutritional status of the organism. Nonetheless, the sensitivity of the response to PAH exposure is sufficiently robust that most vertebrates will exhibit detectable induction following a sufficiently large exposure, with the advantage that the CYP 1A response may well remain evident when PAHs are not, due to metabolic transformation. For example, rockfish collected after the *Braer* oil spill had CYP 1A levels nearly an order of magnitude above constitutive levels for months after the incident when PAHs were undetectable, and the CYP 1A response increased with proximity to the accident site (George et al., 1995). These results clearly demonstrate protracted exposure of these fish to bioavailable PAH in the subtidal zone of the spill-affected area.

In summary, the spatial and temporal patterns of hydrocarbon accumulation by biota, or of biological biomarkers of exposure to hydrocarbons, may provide direct or supporting evidence regarding the sources responsible. When it can be shown that the biota sampled have a negligible effect on the composition of the hydrocarbons accumulated, these data may be sufficient to identify the source unambiguously. Biological biomarker responses may provide supporting evidence that may be used to reduce the number of plausible sources, as may biomarkers that are based on higher levels of biochemical and physiological integration, through that of the individual organism. The following section summarizes these higher-level biological responses of organisms to hydrocarbon exposure, which may be used in a similar manner with the biological biomarkers as evidence for determining the source(s) responsible.

12.4 Modes of Toxic Action of Accumulated Hydrocarbons

A number of organism-level responses to hydrocarbon exposure have been described, and some of these have a specificity approaching that of the biological biomarkers discussed in Section 12.3. Exposure to hydrocarbons, and especially to PAH, causes a variety of organism-level responses, depending on the mode of exposure, the species, the developmental status of the target, and the time scale of observation. Perhaps the most widely applied, yet least sensitive, of these exposure indicators is cumulative mortality induced through narcosis. Oil pollution and PAH had once been regarded as minimally toxic based on the relatively high exposure concentrations required to kill test organisms within a few days. But within the last decade, the ecological consequences of reduced fitness have become more fully appreciated, with the recognition that even a small reduction in the ability to acquire prey or to avoid predators in the wild may lead to mortality prior to the first reproductive opportunity, and that these effects have the potential to harm a population as certainly as mortality from acute toxicity (Rose et al., 2003). Such eventual adverse effects may result from impaired development, growth inhibition by ingested hydrocarbons, immune suppression, or carcinogenesis. As with the biological biomarkers, the spatial and temporal patterns of these responses may indicate particular hydrocarbon sources of exposure.

The aqueous concentrations of hydrocarbons that are necessary to kill aquatic biota through narcosis are so high that such mortality is not usually observed in conjunction with exposure to oil pollution in the field. Narcosis-induced mortality arises when nonpolar contaminants such as hydrocarbons accumulate in the neural membrane lipids, inhibiting nerve transmission, and causing death through asphyxiation or heart failure. The PAH concentrations required to cause lethal narcosis are on the order of 5×10^{-5} mole/g lipid, corresponding to about 2.5×10^{-6} mole/g wet weight of the whole organism if it is 5% lipid (DiToro et al., 2000). Assuming again that the K_{ow} is a reasonable approximation of the equilibrium distribution of a hydrocarbon in the lipid and the aqueous exposure water, a lethal exposure concentration may be expressed as $(5 \times 10^{-2}/K_{ow})$ moles/L.

Comparison of the ratio $(5 \times 10^{-2}/K_{ow})$ moles/L with the expected concentration of a hydrocarbon in equilibrium with the oil phase ($\approx C_{so}/(K_{ow} + Q)$; see Section 12.2) shows that lethal narcosis would only occur when the oil phase contains a high proportion of the most soluble hydrocarbons (i.e., BTEX) in contact with a limited amount of the aqueous phase. For example, an oil phase containing 10% benzene (about 0.13 mole/L, K_{ow} of 135) would produce a lethal aqueous concentration only at water-to-oil volume ratios smaller than about $Q = 200$. Most oil-derived products do not contain such high concentrations of BTEX and are not confined with such limited quantities of water following environmental release. Consequently, narcosis-induced mortality is usually associated with products with high BTEX contents such as gasoline and some diesel oils, when energetically mixed in confined waterways. Because of the high vapor pressures of the BTEX compounds, the potential for narcosis-induced mortalities is also ephemeral, as these hydrocarbons readily evaporate on time scales of hours to a day or so.

Chronic exposure to PAH-contaminated sediments can elicit a complex of cancerous tumors in fish. A particularly clear linkage was established between liver neoplasms of brown bullhead catfish (*Ameiurus nebulosus*) and PAH associated with a coking facility on the Black River of Ohio (Baumann and Harshbarger, 1995). Sediment PAH and liver cancers in the catfish declined simultaneously following closure of the facility in 1982. Disturbance of the sediments from dredging in 1990 led to an increase of tumors during the following 2 to 3 years, and the population age structure of the affected fish was consistent with exposure during 1990 (Baumann and Harshbarger, 1998). These results corroborated extensive studies of Puget Sound, Washington, that implicate exposure to PAH in sediments as contributing agents to carcinogenesis in fish there (Myers et al., 1990).

More subtle but insidious biological manifestations of exposure to hydrocarbons are the effects of PAH on the early developmental stages of fish. Research motivated by field observations following the *Exxon Valdez* oil spill, and by concerns associated with pulp mill effluents, has led to a number of recent studies implicating the role of the more environmentally persistent 3- and 4-ring PAHs, which elicit a suite of symptoms similar to those produced by dioxin exposure (Incardona et al., 2004, 2005). This syndrome, sometimes denoted as "blue-sac disease," can include pericardial and yolk sac edema, hemorrhages,

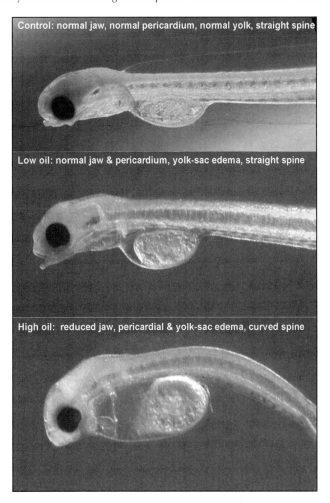

Figure 12-7 Some effects of oil exposure in herring larvae (photos courtesy of NOAA).

craniofacial and spinal malformations, and mortality (Billiard et al., 1999; Carls et al., 1999; Heintz et al., 1999; Couillard, 2002) and may occur following embryonic exposures to PAH concentrations in the range of 1–30 μg/L. Its severity is dose-response-related, as shown in Fig. 12-7. Unlike narcosis-mediated mortality, these concentrations may be readily attained following environmental releases of oil products. For example, the combined concentration of parent and alkyl-substituted phenanthrenes and dibenzothiophenes, both of which are implicated in blue-sac disease syndrome, in Alaska North Slope (ANS) crude oil is about 3×10^6 μg/L (Wang et al., 2003), and their K_{ow} is on the order of 10^5. The expression $C_w = C_{so}/(K_{ow} + Q)$ implies that water equilibrated with ANS oil could attain combined concentrations of these PAH of ~10 μg/L at water-to-oil volume ratios (Q) on the order of 2×10^5. Hence, 1 L of ANS oil could contaminate 200 m³ of water with this concentration. This suggests that the appearance of this syndrome may be a sensitive indicator of PAH exposure, either in natural populations sampled from their natal habitat or in captive organisms deliberately exposed to assess field conditions (Couillard, 2002), provided other potential inducers such as dioxins can be discounted.

Exposure to hydrocarbons may cause other less observable adverse effects on biota. Ingestion of aliphatic hydrocarbons may cause marked inhibition of growth in fish (Luquet et al., 1984), and although the metabolic mechanism underlying this is not clear, such growth inhibition has been documented in fish that have ingested oil following the *Exxon Valdez* oil spill (Carls et al., 1996). Some of the PAHs in oil products are also phototoxic, meaning that they may catalyze production of reactive oxygen species inside cells that accumulate them when irradiated by the ultraviolet component of sunlight (Pelletier et al., 1997).

These macroscopic responses of biota may provide essential clues of exposure to hydrocarbons when their involvement is not already confirmed by other evidence, such as proximity to an accidental oil spill. For example, the appearance of tumors and deformities in aquatic biota might serve as the first indication that the ambient water may be receiving a cryptic input of hydrocarbons, prompting more definitive sampling. Examination of biological biomarkers in exposed biota may then provide indications of the route of exposure and perhaps its spatial location. From this perspective, these macroindicators may serve as the initial indicators of potential hydrocarbon exposure, prompting use of more involved but reliable environmental sampling methods.

Many of the advances in our understanding of how hydrocarbons may permeate the biotic components of natural ecosystems stemmed from studies of the *Exxon Valdez* oil spill, easily the most intensively studied oil pollution event in history. The following section summarizes the most noteworthy findings of that event in regard to the biological responses and indicators of exposure to the oil, as an example of the processes and responses described above.

12.5 Case Study: The *Exxon Valdez* Oil Spill

The opportunity provided by the *Exxon Valdez* oil spill to study the environmental effects of oil pollution was unique, and not likely ever to be repeated. As a large perturbation in an otherwise mostly pristine region, its effects were rarely confounded by other sources of pollution. Although remote, the impacted area within Prince William Sound, Alaska, is at the edge of accessibility by road, three ports, and commercial air service, and is within 100 km of a large population center, facilitating long-term scientific studies. Finally, unprecedented funds were made available for such studies, permitting examination of questions that otherwise would have been neglected. These unique circumstances have led to a particularly clear illustration of the biological responses to the spilled oil as it made its way through the ecosystem. In retrospect, these responses show how hydrocarbons accumulated by biota — and the biotic responses to those accumulations — are related to the environmental dispersion of the oil.

To appreciate the biotic responses to the oil, it will be helpful to summarize the dispersion of the oil in the environment and the physical factors underlying the dispersive processes. The overall fate of the oil was largely determined by weather events during the first few weeks. Oil discharge began shortly after midnight on Friday, March 24, 1989, during calm conditions at seawater temperatures of about 5°C. Wind speeds remained variable and below about 5 m/sec for the next three days, allowing the oil slick to remain a compact and roughly circular pool that spread rapidly at rates approaching $2000\,m^2$/sec to an area of nearly $400\,km^2$, with thicknesses that decreased from generally less than 1 mm after the first few hours of discharge to perhaps 0.1 mm by Sunday. The rapid increase in the surface area of the oil accelerated evaporation of the most volatile components, leading to losses of ~15% by weight, including nearly all of the BTEX, and of the saturated hydrocarbons with vapor pressures greater than that of dodecane (Payne et al., 1991).

An intense storm with winds at speeds of up to 35 m/sec disrupted the slick and drove it toward beaches during the next 3 days. The first landfall of the slick occurred with breaking waves of 1–3 m. Initially, the main effects of the high seas associated with the storm were to disperse oil as small droplets into the water column and to promote water incorporation into the oil. Oil droplet dispersion occurred at least to depths of 25 m (Short and Harris, 1996a), and probably considerably deeper because the density of the water column varies little with depth near the end of winter in Prince William Sound (Vaughan et al., 2001). The large surface-to-volume ratio associated with these small oil droplets promoted dissolution of the more readily soluble oil components into the seawater column, leading to concentrations of PAH (mainly dissolved) that totaled a few parts per billion weeks after the storm had passed (Neff and Stubblefield, 1995; Short and Harris, 1996b).

The high storm winds caused slick thicknesses to increase adjacent to the initially impacted beaches, and the breaking waves caused substantial accumulations of oil to be stranded on beaches during falling tides, often blanketing the ~4-m vertical tidal excursion with oil. On porous beaches, this process allowed pools of stranded oil to percolate into subsurface sediments as the water table lowered during the outgoing tide. Even the fine-grained sediments of these beaches have high permeabilities (Bragg and Yang, 1995), so the intertidal water table follows the tide level closely, permitting the more viscous oil hours to seep into the underlying sediments, to depths of 1 m in some locations (Neff et al., 1995). Subsequent rising tides failed to completely remove this subsurface oil because the increased oil viscosity inhibited displacement from finer-grained sediments. This initial percolation of oil into subsurface sediments set the stage for long-term persistence there.

The dispersion and environmental persistence of the spilled oil were clearly reflected in the associated biota. Mussels were the most extensively sampled sentinel organisms, and when elevated, the PAH distributions found in them were nearly always consistent with the weathering patterns that characterized the whole oil (Short and Heintz, 1997), including association with phytane (e.g., Short and Harris, 1996a). Corroborating patterns were also evident in other suspension-feeders, such as clams (Short and Heintz, 1997), and in the guts of juvenile salmonids and other animals that ingested oiled prey (Bence and Burns, 1995; Carls et al., 1996). Both the spatial and temporal patterns of hydrocarbon accumulation by biota showed clear relationships with proximity to beaches that were heavily oiled (e.g., Short and Harris, 1996a).

Exposure of biota to dissolved hydrocarbons from the spilled oil was less clear, because of the large capacity for dilution and because of difficulties distinguishing PAH concentration patterns indicative of dissolved species from those characteristic of diesel oil. Enormous effort was expended during attempts to clean oiled beaches and the sea surface, which involved a host of support vessels that were potential ancillary sources of hydrocarbon contamination. Diesel oil from these vessels, as well as from scientific support vessels, almost certainly contaminated some of the samples collected (Bence and Burns, 1995). When available, data for associated aliphatic hydrocarbons such as phytane may be used to differentiate between exposure to diesel oil vs. the water-soluble components of crude oil (as in the decision-tree approach of Bence and Burns, 1995).

Longer term, oil in a semi-liquid state remained sequestered within the most heavily oiled beaches throughout the 1990s, providing a reservoir that could contaminate organisms that forage by disturbing these sediments (Short et al., 2004). Studies of sea ducks and otters consistently found biochemical evidence of exposure to hydrocarbons in the form of elevated CYP 1A levels in animals that increased with proximity to the lingering oil and that steadily declined with time as the years progressed (Trust et al., 2000; Esler et al., 2002; Bodkin et al., 2002), reflecting either the gradual dissipation of the lingering oil on the beaches, or possibly learned avoidance behavior.

A more subtle PAH-exposure pathway was revealed by CYP 1A measurements on wild pink salmon embryos. Salmon spawning streams usually escaped direct oiling, even on otherwise heavily oiled beaches, because the freshwater stream flow diverted oil away from them. Hence, while PAH concentrations in sediments of these streams were usually well below 1000 ng/g (Brannon et al., 1995), they were higher by factors of 10–1000 in sediments adjacent to the stream channels (Murphy et al., 1999). Induction of CYP 1A was found in 13 of the 16 samples of pink salmon eggs incubating within gravels of four streams where the adjacent beaches had been heavily oiled. In contrast, no induction was noted in any of the seven samples from five reference streams on unoiled beaches. A field study subsequently confirmed that groundwater movement following hydraulic gradients within these beaches as the tides fall would readily transport PAH dissolved from adjacent oil to the subsurface stream flow where the eggs were incubating (Carls et al., 2003). This unexpected exposure pathway would likely not have been detected without the indications provided by the CYP 1A measurements of the developing embryos. As with the sea duck and otter studies, these spatial and temporal patterns of biochemical responses to oil exposure demonstrate the power of these responses to implicate pollution sources. In fact, it was the unexpected appearance of these responses in biota during the 1990s that motivated the study to evaluate the long-term persistence of the oil on the beaches in 2001 (Short et al., 2004).

12.6 Summary

Whether direct or indirect, assessments of hydrocarbons in biota may provide evidence that is crucial to identifying pollution sources. Biological samples are often collected and analyzed for hydrocarbons, or for evidence of exposure to hydrocarbons, as part of damage assessments associated with pollution-monitoring efforts, and useful information regarding pollution sources may often be gained from relatively little additional chemical or data analysis. When the hydrocarbon distribution pattern is not substantially altered along the exposure pathway or by the biota through biochemical transformation, then these patterns may provide direct evidence of the source, comparable with environmental samples of the released hydrocarbon source material. But even when considerable alteration occurs, spatial and temporal patterns of biological biomarkers indicative of hydrocarbon exposure may still furnish evidence regarding potential sources. The most widely used biological biomarker has been associated with the phase I metabolic transformation pathway involving induction of the CYP 1A detoxification enzyme (e.g., the EROD assay), but other biomarkers associated with phase II condensation reactions (e.g., bile analytes of anthropogenic PAH), as well as toxic responses at the organismal level of affected biota, have also proved useful. The source information available from such approaches and analyses has the potential to be pivotal when evaluating potential sources of environmental hydrocarbon pollution.

References

Aarab, N., C. Minier, S. Lemaire, E. Unruh, P.D. Hansen, B.K. Larsen, O.K. Andersen, and J.K. Narbonne, Biochemical and histological responses in mussel (*Mytilus edulis*) exposed to North Sea oil and to a mixture of North Sea oil and alkylphenols. *Mar. Environ. Res.*, 2004, **58**, 437–441.

Barron, M.G., R. Heintz, and S.D. Rice, Relative potency of PAHs and heterocycles as aryl hydrocarbon receptor agonists in fish. *Mar. Environ. Res.*, 2004, **58**, 95–100.

Baumann, P.C. and J.C. Harshbarger, Decline in liver neoplasms in wild brown bullhead catfish after coking plant closes and environmental PAHs plummet. *Environ. Health Perspect.*, 1995, **103**, 168–170.

Baumann, P.C. and J.C. Harshbarger, Long term trends in liver neoplasm epizootics of brown bullhead in the Black River, Ohio. *Environ. Monitor. Assess.*, 1998, **53**, 213–223.

Bayne, B.L., *Marine Mussels — Their Ecology and Physiology*. London: Cambridge Univ. Press, 1976.

Bence, A.E. and W.A. Burns, Fingerprinting hydrocarbons in the biological resources of the *Exxon Valdez* spill area. In *Exxon Valdez Oil Spill: Fate and Effects in Alaskan Waters*, P.G. Wells, J.N. Butler, and J.S. Hughes (eds.), Philadelphia: American Society for Testing and Materials, 1995, 84–140.

Bender, M.E., W.J. Hargis, R.J. Huggett, and M.H. Roberts, Effects of polynuclear aromatic hydrocarbons of fishes and shellfish: An overview of research in Virginia. *Mar. Environ. Res.*, 1988, **24**, 237–241.

Beyer, J., M. Sandvik, J.U. Skåre, E. Egaas, K. Hylland, R. Waagbo, and A. Goksøyr, Time- and dose-dependent biomarker responses in flounder (*Platichthys flesus* L.) exposed to benzo[*a*] pyrene, 2,3,3′,4,4′,5-hexachlorobiphenyl (PCB-156) and cadmium. *Biomarkers*, 1997, **2**; 35–44.

Billiard, S.M., K. Querbach, and P.V. Hodson, Toxicity of retene to early life stages of two freshwater fish species. *Environ. Toxicol. Chem.*, 1999, **9**, 2070–2077.

Bodkin, J.L., B.E. Ballachey, T.A. Dean, A.K. Fukuyama, S.C. Jewett, L. MacDonald, D.H. Monson, C.E. O'Clair, and G.R. VanBlaricom, Sea otter population status and the process of recovery from the 1989 *Exxon Valdez* oil spill. *Mar. Ecol. Prog. Ser.*, 2002, **241**, 237–253.

Boehm, P.D., M.S. Steinhauer, D.R. Green, B. Fowler, B. Humphrey, D.L. Fiest, and W.J. Cretney, Comparative fate of chemically dispersed and beached crude oil in subtidal sediments of the arctic nearshore. *Arctic*, 1987, Suppl. 1, **40**, 133–148.

Blumer, M. and D.W. Snyder, Isoprenoid hydrocarbons in recent sediments: Presence of pristane and probable absence of phytane. *Science*, 1965, **150**, 1588–1589.

Bragg, J.R. and S.H. Yang, Clay-oil flocculation and its role in natural cleansing in Prince William Sound following the *Exxon Valdez* oil spill. *Exxon Valdez Oil Spill: Fate and Effects in Alaskan Waters*, P.G. Wells, J.N. Butler, and J.S. Hughes (eds.), Philadelphia: American Society for Testing and Materials, 1995, 178–214.

Brannon, E.L., L.L. Moulton, L.G. Gilbertson, A.W. Maki, and J.R. Skalski, An assessment of oil-spill effects on pink salmon populations following the *Exxon Valdez* oil spill — Part I. Early life history. *Exxon Valdez* Oil Spill: Fate and Effects in Alaskan Waters, P.G. Wells, J.N. Butler, J.S. Hughes (eds.), Philadelphia: American Society for Testing and Materials, 1995, 548–584.

Bucheli, T.D. and K. Fent, Induction of cytochrome P450 as a biomarker for environmental contamination in aquatic ecosystems. *Crit. Rev. Environ. Sci. Tech.*, 1995, **25**, 201–268.

Carls, M.G., S.D. Rice, and J.E. Hose, Sensitivity of fish embryos to weathered crude oil: Part I. Low-level exposure during incubation causes malformations, genetic damage, and mortality in larval Pacific herring *Clupea pallasi*. *Environ. Toxicol. Chem.*, 1999, **18**, 481–493.

Carls, M.G., R.E. Thomas, M.R. Lilly, and S.D. Rice, Mechanism for transport of oil-contaminated groundwater into pink salmon redds. *Mar. Ecol. Prog. Ser.*, 2003, **248**, 245–255.

Carls, M.G., A.C. Wertheimer, J.W. Short, R.M, Smolowitz, and J.J. Stegeman, Contamination of juvenile pink and chum salmon by hydrocarbons in Prince William Sound after the *Exxon Valdez* oil spill. *Proc. Exxon Valdez Oil Spill Symp.* S.D. Rice, R.B. Spies, D.A. Wolfe, and B.A. Wright (eds.), Bethesda, MD: American Fisheries Society, 1966, 593–607.

Chaty, S., F. Rodius, and P. Vasseur, A comparative study of the expression of CYP 1A and CYP 4 genes in aquatic invertebrate (freshwater mussel, *Unio tumidus*) and vertebrate rainbow trout, *Oncorhynchus mykiss*. *Aquatic Toxicolo.*, 2004, **69**, 81–93.

Couillard, C.M., A microscale test to measure petroleum oil toxicity to mummichog embryos. *Environ. Toxicol.*, 2002, **17**, 195–202.

Cravedi, J.P. and J. Tulliez, Accumulation, distribution and depuration in trout of naphthenic and isoprenoid hydrocarbons (dodecylcyclohexane and pristane). *Bull. Environ. Contam. Toxicol.*, 1982, **28**, 154–161.

Dean, R.A. and E.V. Whitehead, The occurrence of phytane in petroleum. *Tetrahedron Lett.*, 1961, **21**, 768–770.

Denison, M.S. and J.P. Whitlock, Jr., Xenobiotic-inducible transcription of cytochrome P450 genes, *J. Biol. Chem.*, **270**, 18175–18178.

DiToro, D.M., J.A. McGrath, D.J. Hansen, Technical basis for narcotic chemicals and polycyclic aromatic hydrocarbon criteria. I. Water and tissue, *Environ. Toxicol. Chem.*, 2000, **19**, 1951–1970.

Donkin, P., E.L. Smith, and S.J. Rowland, Toxic effects of unresolved complex mixtures of aromatic hydrocarbons accumulated by mussels, *Mytilus edulis*, from contaminated field sites. *Environ. Sci. Tech.*, 2003, **37**, 4825–4830.

Engelhardt, F.R., Assessment of the vulnerability of marine mammals to oil pollution; fate and effects of oil in marine ecosystems. J. Kuiper and W.J. Van Den Brink (eds.), Boston: Martinius Nijhoff, 1987, 101–115.

Esler, D., T.D. Bowman, K.A. Trust, B.E. Ballachey, T.A. Dean, S.C. Jewett, and C.E.

O'Clair, Harlequin duck population revocery following the *Exxon Valdez* oil spill: Progress process and contraints. *Mar. Ecol. Prog. Ser.*, 2002, **241**, 271–286.

Ezra, S., S. Feinstein, I. Pelly, D. Bauman, and I. Miloslavsky, Weathering of fuel oil spill on the east Mediterranean coast, Ashdod, Israel. *Org. Geochem.*, 2000, **31**, 1733–1741.

Fairbairn, D.W., P.L. Olive, and K.L. O'Neill, The comet assay — a comprehensive review. *Mutat. Res.*, 1995, **339**, 37–59.

Fingas, M.F. Modeling evaporation using models that are not boundary-layer regulated. *J. Haz. Mat.*, 2004, **107**, 27–36.

Fossato, V.U. and W.J. Canzonier, Hydrocarbon uptake and loss by the mussel *Mytilus edulis*. *Mar. Biol.*, 1976, **36**, 243–250.

Foster-Smith, R.L. The effect of concentration of suspension on the filtration rates and pseudofaecal production for *Mytilus edulis* L., *Cerastoderma edule* (L.), and *Venerupis pullastra* (Montagu), *J. Exp. Mar. Biol. Ecol.*, 1975, **17**, 1–22.

Foureman, G.L., Enzymes involved in metabolism of PAH by fishes and other aquatic animals: Hydrolysis and conjugation enzymes (or phase II enzymes). *Metabolism of PAH in the Aquatic Environment*, U. Varanasi (ed.), Boca Raton, Fl: CRC, 1989, 185–202.

Frost, K.J., C.A. Manen, and T.L. Wade, Petroleum hydrocarbons in tissues of harbor seals from Prince William Sound and the Gulf of Alaska. *Marine Mammals and the Exxon Valdez*, T.R. Loughlin (ed.), San Diego: Academic Press 1994, 331–358.

Garrett, R.M., I.J. Pickering, C.E. Haith, and R.C. Prince, Photooxidation of crude oils. *Environ. Sci. Tech.*, **32**, 3719–3723.

George, S.G., J. Wright, and J. Conroy, Temporal studies of the impact of the *Braer* oilspill on inshore feral fish from Shetland, Scotland. *Arch. Environ. Contam. Toxicol.*, 1995, **29**, 530–534.

Geraci, J.R. and D.J. St. Aubin, Effects of offshore oil and gas development on marine mammals and turtles. In *Long-Term Environmental Effects of Offshore Oil and Gas Development*, D.F. Boesch and N.N. Rabalais (eds.), New York: Elsevier Applied Science 1987, 587–617.

Green, R.H., *Sampling Design and Statistical Methods for Environmental Biologists*. New York: Wiley-Interscience, 1979.

Guengerich, F.P. and D.C. Liebler, Enzymatic activation of chemicals to toxic metabolites. *Crit. Rev. Toxicol.*, 1985, **14**, 259–307.

Hansen, H.P. Photodegradation of hydrocarbon surface films. In *Reun. Cons. Int. Explor. Mer.* P.-V. Rapp (ed.), 1977, **171**, 101–106.

Hansen, N., V.B. Jensen, H. Appelquist, and E. Mørch, The uptake and release of petroleum hydrocarbons by the marine mussel *Mytilus edulis*. *Prog. Wat. Tech.*, 1978, **10**, 351–359.

Hawkins, S.A., S.M. Billiard, S.P. Tabash, R.S. Brown, and P.V. Hodson, Altering cytochrome P4501A activity affects polycyclic aromatic hydrocarbon metabolism and toxicity in rainbow trout (*Oncorhynchus mykiss*). *Environ. Toxicol. Chem.*, 2002, **9**, 1845–1853.

Heintz, R.A., J.W. Short, and S.D. Rice, Sensitivity of fish embryos to weathered crude oil: Part II. Increased mortality of pink salmon (*Oncorhynchus gorbuscha*) embryos incubating downstream from weathered *Exxon Valdez* crude oil. *Environ. Toxicol. Chem.*, 1999, **18**, 494–503.

Hinckley, D.A., T.F. Bidleman, and W. T. Foreman, Determination of vapor pressures for nonpolar and semipolar organic compounds from gas chromatographic retention data. *J. Chem. Eng. Data*, 1990, **35**, 232–237.

Incardona, J.P., T.K. Collier, and N.L. Scholz, Defects in cardiac function precede morphological abnormalities in fish embryos exposed to polycyclic aromatic hydrocarbons. *Toxicol. Appl. Pharmacol.*, 2004, **196**, 191–205.

Incardona, J.P., M.G. Carls, H. Teraoka, C.A. Sloan, T.K. Collier, and N.L. Scholz, Aryl hydrocarbon receptor-independent toxicity of weathered crude oil during fish development. *Environ. Health Perspect.*, 2005, **113**, 1755–1762.

Kaji, H., S.A. Weiss, E.M. Johnson, and B.E.G. Gabel, Protein distribution and endogenous P-450 in intact and dissociated hydra. *Teratology*, 1983, **27**, A54.

Kennish, J.M., R.A. Bolinger, K.A. Chanbers, and M.L. Russell, Xenobiotic metabolizing enzyme activity in sockeye salmon (*Oncorhynchus nerka*) during spawning migration. *Mar. Environ. Res.*, 1992, **34**, 293–298.

Krahn, M.M., L.J. Kittle, Jr., and W.D. MacLeod, Jr., Evidence for exposure of fish to oil spilled into the Columbia River. *Mar. Environ. Res.*, 1986, **20**, 291–298.

Krahn, M.M., D.G. Burrows, G.M. Ylital, D.W. Brown, C.A. Wigren, T.K. Collier, S.L. Chn, and U. Varanasi, Mass spectrometric determination of metabolites of aromatic compounds in the bile of fish captured from Prince William Sound, Alaska, after the *Exxon Valdez* oil spill. *Environ. Sci. Tech.*, 1992, **26**, 116–126.

Lindström-Seppä, P. and J.J. Stegeman, Sex differences in cytochrome P4501A induction by environmental exposure and b-naphthoflavone in liver and extrahepatic organs of recrudescent winter flounder. *Mar. Environ. Res.*, **39**, 219–223.

Livingstone, D.R., The fate of organic xenobiotics in aquatic ecosystems: Quantitative and qualitative differences in biotransformation by invertebrates and fish. *Comp. Biochem. Physiol. A*, 1998, **120**, 43–49.

Luquet, P., J.P. Cravedi, J. Tulliez, and G. Bories, Growth reduction in trout induced by naphthenic and isoprenoid hydrocarbons (dodecylcyclohexane and pristane). *Ecotoxicol. Environ. Safety*, 1984, **8**, 219–226.

Malins, D.C., J.J. Stegeman, J.W. Anderson, P.M. Johnson, J. Gold, and K.M. Anderson, Structural changes in gill DNA reveal the effects of contaminants on Puget Sound fish. *Environ. Health Perspect.*, 2004, **112**, 511–515.

McAuliffe, C. Solubility in water of paraffin, cycloparaffin, olefin, acetylene, cycloolefin, and aromatic hydrocarbons. *J. Phys. Chem.*, 1966, **70**, 1267–1275.

McAuliffe, C. Solubility of normal C9 and C10 hydrocarbons. *Science*, 1969, **163**, 478–479.

Meador, J.P., J.E. Stein, W.L. Reichert, and U. Varanasi, Bioaccumulation of polycyclic aromatic hydrocarbons by marine organisms. *Revi. Environ. Contam. Toxicol.*, 1995, **143**, 79–165.

Munkittrick, K.R., B.R. Blunt, M. Leggett, S. Huestis, and L.H. McCarthy, Development of a sediment bioassay to determine bioavailability of PAHs to fish. *J. Aquat. Ecosyst. Health*, 1995, **4**, 169–181.

Myers, M.S., J.T. Landahl, M.M. Krahn, L.L. Johnson, and B.B. McCain. Overview of studies on liver carcinogenesis in English sole from Puget Sound: Evidence for a xenobiotic chemical etiology I: Pathology and epizootiology. *Sci. Total Environ.*, 1990, **94**, 33–50.

Neff, J.M., E.H. Owens, S.W. Stoker, and D.M. McCormick, Shoreline oiling conditions in Prince William Sound following the *Exxon Valdez* oil spill. In Exxon Valdez *Oil Spill: Fate and Effects in Alaskan Waters*, P.G. Wells, J.N. Butler, and J.S. Hughes (eds.), Philadelphia: American Society for Testing and Materials, 1995, 312–346.

Neff, J.M. and W.A. Stubblefield, Chemical and toxicological evaluation of water quality following the *Exxon Valdez* oil spill. In Exxon Valdez *Oil Spill: Fate and Effects in Alaskan Waters*, P.G. Wells, J.N. Butler and J.S. Hughes (eds.), Philadelphia: American Society for Testing and Materials, 1995, 141–177.

Nelson, D.R., L. Koymans, T. Kamataki, J.J. Stegeman, R. Feyereisen, D.J. Waxman, M.R. Waterman, O. Gotoh, M.J. Coon, R.W. Estabrook, I.C. Gunsalus, and D.W. Nebert, P450 superfamily: Update on new sequences, gene mapping, accession numbers and nomenclature. *Pharmacogenetics*, 1996, **6**, 1–42.

Newman, I.M., E.M. Johnson, R.L. Giacobbe, and L. Fu, The in vitro activation of cyclophosphamide in the Hydra developmental toxicology assay. *Fundamental Appl. Toxicol.*, 1990, **15**, 488–499.

Ortiz de Montellano, P.R., *Cytochrome P450: Structure, Mechanism and Biochemistry*, New York: Plenum, 2nd ed., 1995.

Parkinson, A. Biotransformation of xenobiotics. In *Toxicology, the Basic Science of Poisons*, 6th ed., C.D. Klaassen (ed.), New York: McGraw Hill Medical Publishing Division, 2001, Chapter 6, pp. 141–144.

Payne, J.R., B.E. Kirstein, G.D. McNabb, J.L. Lambach, R. Redding, R.E. Jordan, W. Hom, C. de Oliveira, G.S. Smith, D.M. Baxter, and R. Geagel, Multivariate analysis of petroleum weathering in the marine environment — subarctic. *Vol. I, Technical Results; Final Reports of Principal Investigators*, U.S. Department of Commerce, National Oceanic and Atmospheric Administration, Ocean Assessment Division, Juneau; Alaska, NTIS Accession Number PB85-215796, 1984, Vol. 21.

Payne, J.R., J.R. Clayton, G.D. McNabb, Jr., and B.E. Kirstein, *Exxon Valdez* oil weathering fate and behavior: Model predictions and field observations. *Proc. 1991 Int. Oil Spill Conf.*, Washington, DC: American Petroleum Institute Publication No. 4529, American Petroleum Institute, 641–654.

Pelletier, M.C., R.M. Burgess, K.T. Ho, A. Juhn, R.A. McKinney, and S.A. Ryba, Phototoxicity of individual polycyclic aromatic hydrocarbons and petroleum to marine invertebrate larvae and juveniles. *Environ. Toxicol. Chem.*, 1997, **16**, 2190–2199.

Pirnik, M.P., Microbial oxidation of methyl branched alkanes. *Crit. Rev. Microbiol.*, 1977, **5**, 413–422.

Pruell, R.J., J.L. Lake, W.R. Davis, and J.G. Quinn, Uptake and depuration of organic contaminants by blue mussels (*Mytilus edulis*) exposed to environmentally contaminated sediment. *Mar. Biol.*, 1986, **91**, 497–507.

Renwick, A.B., P.S. Watts, R.J. Edwards, I.G. Barton, R.J. Price, M. Tredger, O. Pelkonen, A.R. Boobis, and B.G. Lake, Differential maintenance of cytochrome P450 enzymes in cultured precision-cut human liver slices. *Amer. Soc. Pharmacol. Exp. Therapeutics*, 2000, **28**, 1202–1209.

Rose, K.A., C.A. Murphy, S.L. Diamond, L. Fuiman, and P. Thomas, Using nested models and laboratory data for predicting population effects of contaminants on fish: A step toward a bottom-up approach for establishing causality in field studies. *Human Ecol. Risk Assess*, 2003, **9**, 231–257.

Schaeffer, T.L., S.G. Cantwell, J.L. Brown, D.S. Watt, and R.R. Fall, Microbial growth on hydrocarbons: Terminal branching inhibits biodegradation. *Appl. Env. Microbiol.*, 1979, **38**, 742–746.

Shiu, W.Y., A. Maijanen, A.L.Y. Ng, and D. Mackay, Preparation of aqueous solutions of sparingly soluble organic substances: II. Multicomponent systems — hydrocarbon mixtures and petroleum products. *Environ. Toxicol. Chem.*, 1988, **7**, 125–137.

Short, J.W. Oil identification based on a goodness-of-fit metric applied to hydrocarbon analysis results. *Environ. Forensics*, 2002, **3**, 349–355.

Short, J.W. and P.M. Harris, Petroleum hydrocarbons in caged mussels deployed in Prince William Sound after the *Exxon Valdez* oil spill. In *Proc. Exxon Valdez Oil Spill Symp.* S.D. Rice, R.B. Spies, D.A. Wolfe, and B.A. Wright (eds.), Bethesda, MD: American Fisheries Society, 1996a, 29–39.

Short, J.W. and P.M. Harris, Chemical sampling and analysis of petroleum hydrocarbons in near-surface seawater of Prince William Sound after the *Exxon Valdez* oil spill. In *Proc. Exxon Valdez Oil Spill Symp.*, S.D. Rice, R.B. Spies, D.A. Wolfe, and B.A. Wright (eds.), Bethesda, MD: American Fisheries Society, 1996b, 17–28.

Short, J.W. and R.A. Heintz, Identification of *Exxon Valdez* oil in sediments and tissues from Prince William Sound and the northwestern Gulf of Alaska based on a PAH weathering model. *Environ. Sci. Tech.*, 1997, **31**, 2375–2384.

Short, J.W., M.R. Lindeberg, P.M. Harris, J.M. Maselko, J.J. Pella, and S.D. Rice, Estimate of oil persisting on the beaches of Prince William Sound 12 years after the *Exxon Valdez* oil spill. *Environ. Sci. Tech.*, 2004, **38**, 19–25.

Sikka, H.C., A.R. Steward, C. Kandaswami, J.P. Rutkowski, J. Zaleski, K. Earley, and R.C. Gupta, Metabolism of benzo[a]pyrene and persistence of DNA adducts in the brown bullhead (*Ictalurus nebulosus*). *Comp. Biochem. Physiol.*, 1991, **100C**, 25–28.

Sleiderink, H.M. and J.P. Boon, Temporal induction pattern of hepatic cytochrome P450 1A in thermally acclimated dab (*Limanda limanda*) treated with 3,3′,4,4′-tetrachlorobiphenyl (CB77). *Chemosphere*, 1996, **32**, 2335–2344.

Smith, J.E. (ed.), *"Torrey Canyon" Pollution and Marine Life*. London: Cambridge Univ. Press, 1968.

Smolowitz, R.M., M.E. Hahn, and J.J. Stegeman, Immunohistochemical localization of cytochrome P-450IA1 induced by 3,3′,4,4′-tetrachlorobiphenyl and by 2,3,7,8-tetrachlorodibenzofuran in liver and extrahepatic tissues of the teleost *Stenotomus chrysops* (Scup). *Drug Metabol. Dispos.*, 1991, **19**, 113–123.

Spraker, T.R., L.F. Lowry, and K.J. Frost, Gross necropsy and histopathological lesions found in harbor seals. In *Marine Mammals and the Exxon Valdez*, T.R. Loughlin (ed.), San Diego, CA: Academic Press, 1994, 281–311.

Stegeman, J.J. and M.E. Hahn, Biochemistry and molecular biology of monooxygenases: Current perspectives on forms, functions, and regulation of cytochrome P450 in aquatic species. In *Aquatic Toxicology: Molecular, Biochemical and Cellular Perspectives*, D.C. Malins and G.K. Ostrander (eds.), Boca Raton, FL: Lewis Publishers, 87–206.

Trust, K.A., D. Esler, B.R. Woodin, and J.J. Stegeman, Cytochrome P450 1A induction in sea ducks inhabiting nearshore areas of Prince William Sound, Alaska. *Mar. Poll. Bull.*, 2000, **40**, 397–403.

Vahl, O., Pumping and oxygen consumption rates of *Mytilus edulis* L. of different sizes. *Ophelia*, 1973, **12**, 45–52.

van der Oost, R., J. Beyer, and N.P.E. Vermeulen, Fish bioaccumulation and biomarkers in environmental risk assessment: A review. *Environ. Toxicol. Pharmacol.*, 2003, **13**, 57–149.

van der Weiden, M.E.J., R. Bleumink, W. Seinen, and M. van den Berg, Relative potencies of polychlorinated dibenzo-*p*-dioxins (PCDDs) dibenzofurans (PCDFs), and biphenyls (PBCs), for cytochrome P450 1A induction in the mirror carp

(*Cyprinus carpio*). *Aquat. Toxicol.*, 1994, **29**, 163–182.

Varanasi, U., M. Nishimoto, W.L. Reichert, and B-T. Le Eberhart, Comparative metabolism of benzo[*a*]pyrene and covalent binding to hepatic DNA in English sole, starry flounder, and rat. *Cancer Res.*, 1986, **46**, 3817–3824.

Varanasi, U., W.L. Reichert, and J.E. Stein, ^{32}P-postlabelling analysis of DNA adducts in liver of wild English sole (*Parophrys vetulus*) and winter flounder (*Pseudopleuronectes americanus*). *Cancer Res.*, 1989, **49**, 1171–1177.

Vaughan, S.L., C.N.K. Mooers, and S.M. Gay III, Physical variability in Prince William Sound during the SEA study (1994–98). *Fish. Oceanogr.*, 2001, **10**(Suppl. 1), 58–80.

Wade, T.L., J.L. Sericano, P.R. Gardinali, G. Wolff, and L. Chambers, NOAA's "mussel watch" project: Current use organic compounds in bivalves. *Mar. Pollut. Bull.*, 1998, **37**, 20–26.

Wang, Z., M.F. Fingas, S.A. Blenkinsopp, G.A. Sergy, M. Landriault, L. Sigouin, J. Foght, and D.W.S. Westlake, Oil composition changes due to biodegradation and differentiation between these changes and those due to weathering. In *Nineteenth Arctic and Marine Oilspill Program Technical Seminar*, Environment Canada, Ottawa, 1996, 163–183.

Wang, Z., M.F. Fingas, E.H. Owens, and L. Sigouin, Study of long-term spilled *Metula* oil: Degradation and persistence of petroleum biomarkers. In *Twentythird Arctic and Marine Oilspill Program Technical Seminar*, Environment Canada, Ottawa, 2000, 99–122.

Wang, Z., M.F. Fingas, and D.S. Page, Oil spill identification. *J. Chromatog. A*, 1999, **843**, 369–411.

Wang, Z., B.P. Hollebone, M. Fingas, B. Fieldhouse, and L. Sigouin, Characteristics of spilled oils, fuels, and petroleum products 1. Composition and properties of selected oils. Report Number EPA/600/R-03/072, Environmental Protection Agency, Research Triangle Park, NC; 2003.

Weast, R.C. (ed.), *CRC Handbook of Chemistry and Physics*, Cleveland, OH: CRC Press; 1977.

Whitlock, J.P. Jr., Mechanistic aspects of dioxin action. *Chem. Res. Toxicol.*, 1993, **6**, 754–763.

Whyte, J.J., R.E. Jung, C.J. Schmitt, and D.E. Tillitt, Ethoxyresorufin-O-deethylase (EROD) activity in fish as a biomarker of chemical exposure. *Crit. Rev. Toxicol.*, 2000, **30**, 347–570.

Wiedmer, M., J.J. Fink, J.J. Stegeman, R. Smolowitz, G.D. Marty, and D.E. Hinton, Cytochrome P-450 induction and histopathology in preemergent pink salmon from oiled sites in Prince William Sound. In *Proc. Exxon Valdez Oil Spill Symp.*, S.D. Rice, R.B. Spies, D.A. Wolfe, and B.A. Wright (eds.), Bethesda, MD: American Fisheries Society, 509–517.

13 Trajectory Modeling of Marine Oil Spills

Debra Simecek-Beatty and William J. Lehr

Hazardous Materials Response Division, National Oceanic and Atmospheric Administration, 7600 Sand Point Way NE, Seattle, Washington.

13.1 Introduction

While fingerprinting of oil is a common aid in identifying the source of an oil spill, techniques that estimate the past history of a "mystery" spill can also assist the forensic investigation. These techniques can include such procedures as estimating the age of the slick, based upon the degree of environmental weathering, and tracking the oil back to its possible source by time-reversing standard oil trajectory forecast models. In order to understand the usefulness and limitations of these techniques, it is necessary to understand current oil spill trajectory and fate forecasting methods.

Even a novice observer will notice that, when spilled on water, most oils quickly spread into a thin film. The oil will drift with certain persistence, mostly likely downwind or, in the case of a strong current, with the surface flow. Eventually, waves rupture and tear the film into smaller and smaller patches. Because the sea is in constant motion, random swirls and eddies disperse the oil patches further apart. While water moves horizontally and vertically leading to divergences and convergences at the surface, patches of floating oil primarily move horizontally at the sea surface and drift closer to areas where the water converges. Oil patches are often fluid and sticky. If they float into an area of convergence, they may adhere to each other and coalesce into a larger patch. The spreading, dispersion, and coalescence of oil fluctuate constantly. Describing such chaotic phenomena is not easy, but if we think in terms of averages, the movement and behavior of the spill are predictable. Today, computerized models can estimate the "most likely" behavior of an oil spill, in spite of the chaotic nature of the wind and sea conditions. This has not always been the case.

Before the 1960s, finding a mathematical model that predicted the movement of oil spilled on water was difficult. However, several events led to the rapid progress in oil spill modeling and research. In the United States, oil spill damage from the 1967 well blowout of Union Oil Platform A in the Santa Barbara Channel resulted in more oil spill regulation and increased research funding. There were also a series of "super" spills from tankers in 1967 and 1968 with the *Torrey Canyon* casualty being the most notable (Biglane, 1969). With the influx of oil spill research funding, a multitude of theoretical models was developed for predicting oil spill fate and behavior. The review by Fallah and Stark (1976) of the work prior to the mid-1970s is admirable, but for the most part, Stolzenbach et al. (1977) is considered the classic work of its time due to its comprehensive discussions on wind fields, advection, and oil weathering.

A rash of spills in the 1970s led to an influx of new ideas and methods for oil spill modeling. The *Argo Merchant* grounding in 1976 was one of the most studied oil spills in history, with over 200 scientists participating in the response effort. Five independent research teams provided operational forecasting of the oil distribution (Grouse and Mattson, 1977; Pollack and Stolzenbach, 1978). During the 1979 IXTOC-1 well blowout, then considered the world's largest spill, operational forecasting was fully integrated into the response effort and was used to direct cleanup activities (Hooper, 1981). In each incident, approaches to the forecasting problem varied from deterministic, via simple vector addition of the wind forecast and tidal current predictions, to statistical, using probability matrices of the historical wind and current data. Observations of the oil distribution and, if available, satellite imagery were used to initialize and update the model forecasts. These early forecasting efforts tried to predict oil spill trajectories operationally in a manner similar to the way that meteorological offices use atmospheric models to routinely update weather forecasts. Other, and just as significant, modeling efforts occurred (Haung, 1983; Spaulding, 1988), but the *Argo Merchant* and IXTOC responses best illustrate the state-of-the-art for the times.

If a spill were to occur today, the "best guess" would probably be a compilation of outputs from different models (Daniel et al., 2002) or even from the same model if using different boundary conditions and data choices. More than one model may be used because a particular model may perform better in certain situations. Performance varies because the mathematical equations describing oil movement are complex, making an analytical solution impossible. To circumvent this problem, all spill models simplify the underlying equations to a certain extent by making assumptions. Therefore, one model's simulation of a particular aspect of an oil spill's fate and behavior may be rigorous, but it is likely to be weaker in other aspects. Discussions of the strengths and weaknesses of the current state-of-the-art oil spill models can be found in Yapa and Shen (1994), Anon (1996), Cekirge and Palmer (2001), and French-McCay (2004).

To some, oil spill forecasting is still considered a seat-of-the pants operation that relies on experience to manipulate both model input and parameters. Part of this perception is due to the use of nonrigorous mathematics, as simple physical laws cannot describe many aspects of oil spill behavior. Another complication is the general lack of real-time environmental data for model initialization. Even the initial release details about the location and amount of oil spilled are often sketchy. Therefore, any uncertainty in the release is directly translated into uncertainties in the forecast concentrations. At best, most spill modelers rely on empirically derived models to some extent and expect large uncertainties in the model input data. Despite these limitations, it is clear there has been some degree of success in oil spill forecasting.

13.2 Forecasting and Hindcasting Oil Spill Movement

Most spill models are flexible enough to allow for a wide variety of modeling techniques. In general, the models run in either a tactical mode, used for short-term forecasting from a known source or in a statistical mode, using historical environmental conditions to generate a long-term forecast. Occasionally, the modeler uses a forensic approach to investigate the past trajectory of an oil spill. Modelers call this approach a "hindcast" of the spill. After the fact, the trajectory of the oil is analyzed to determine a possible oil source and the most probable path of the spill. In hindcasting, the idea with either a forward or backward model run is to "re-forecast" the oil movement for a particular spill event and provide investigators an indication when and where the spill occurred. Selecting either a forward or backward option depends upon the question being asked.

The first option, running the model forward in time, is used if the location of the oil source is known or needs verification. Depending on the spill conditions, small errors in the location and time can cause significant changes in the trajectory. For example, a 30-minute error in the release time can be important in a tidal dominated area due to changes in the tidal phase. By comparing the spill trajectory to the reported release site with the timing and location of, for example, shoreline oiling, the modeler can further refine the likely timing of the release. Besides the location and time of the release, model inputs will include on-scene wind and current observations. Astronomical tides and historical current data are used if nothing else is available. Using this technique, the modeler can investigate where the spill occurred.

The second option involves running the model backwards in time from the location where the oil is found. This particular modeling approach attempts to answer the question, "Where could the oil have come from to contact this resource?" Perhaps an oiled shoreline or, even, oiled birds are discovered, but the source of the spill is unknown. Here, the model is run backwards using historical environmental data and statistical techniques to determine possible spill locations and times. Searches of the potential spill locations may expose a source, such as a submerged vessel or a vessel transiting the area and dumping bilge oil.

Even when using models in a hindcast mode, the modeler will be faced with incomplete environmental information. Uncertainty in the input data translates as uncertainty in the trajectory calculation, either in the forward or backward time sense. The traditional approach to hindcasting trajectories uses the most probable set of environmental data in a deterministic answer that provides a "best-guess" oil path and potential release point. However, doing so neglects valuable statistical information contained in environmental history for the spill area, a point recognized by meteorologists (Roulston and Smith, 2002) when making weather predictions.

There are two nonexclusive ways of incorporating statistical uncertainty into hindcasts. One way is to assume that the underlying major environmental parameters (e.g., general wind and current patterns) are exact and that internal parameters in the model vary according to known statistical distributions (Lehr et al., 1995, 1999). These "internal" parameters can be components of the spill model itself, such as drift factor, or wind and current fluctuations that occur at a smaller spatial or temporal scale than that used to define the major wind and current patterns. Jones (1999) has shown that the spill trajectory is particularly sensitive to temporal resolution of the winds. However, the time series of wind velocity at a data-recording site may be linked only in a statistical sense to the wind time series experienced by floating oil at a different location. Winds from a measurement location may not accurately represent winds over the entire area covered by the spill.

Another common method to incorporate uncertainty is to include the statistical variation directly within the large-scale wind and current data (Paluszkiewicz and Marshall, 1989). Usually, this is done in a forecasting mode but can also apply to hindcasts, particularly when one recognizes that the exact location and time of the spilled oil are uncertain. Paluszkiewicz and Marshall recommend using prior wind histories in any sample run rather than generating artificial wind histories from climatological data. When using either or both methods for including uncertainty in the output display, the modeler is faced with a significant challenge in presenting the resulting probability information in a manner that is easily understood by the model user. This is an area of active research.

13.3 Oil Spill Transport

Winds and currents play an important role in oil spill transport and, occasionally, oil moves in a direction that results in unexpected outcomes. For instance, during the 1967 grounding of the *Argo Merchant*, 27,000 tonnes of heavy fuel oil were spilled. There was a certain expectation of extensive shoreline oiling because the vessel was near shore. However, the spill essentially drifted out to sea due to offshore winds and currents. In contrast, the offshore sinking of the *Erika* in 1999 released approximately 14,000 tonnes of heavy oil and resulted in more than 100 km of oiled coastline.

There are two accepted methods for modeling oil movement on water: Eulerian and Lagrangian. The Eulerian representation of oil movement records the concentration of the oil patches or particles flowing past a fixed point. The Eulerian approach solves a simplified form of the classic advection-diffusion equation. Assuming two-dimensional flow, the Eulerian form of the equation follows:

$$\frac{\partial c}{\partial t} = -\nabla(cV) + \nabla(D\nabla c) + S$$

where:

c = the concentration of oil
t = time
V = advection velocity
D = horizontal diffusion coefficient
S = sources and sinks of oil
∇ = gradient operator

There are significant challenges to solving the equation. For example, most spills begin as small releases or point sources, such as a series of damaged tanks from a vessel aground. Over a period of days to weeks, the oil spreads over hundreds of square kilometers. The advection due to winds, currents, and turbulence has a different time and length scale a few weeks into the spill as compared to the initial release. There is an additional problem of numerical diffusion associated with hydrodynamic modeling, but this is a minor issue compared to the overall errors associated with oil spill model uncertainty (Barker, 2005). Solving the advection-diffusion problems often requires various statistical methods, which increases the computational complexity of the model.

The Lagrangian method represents a patch of oil as a set of particles and follows the path taken by each particle of oil as it moves relative to the earth. The velocity and direction are calculated as the particle changes position with time. In addition to tracking the overall movement of the spill, the particles also track the weathering of the oil (e.g., evaporation and dissolution).

Traditionally, oil spill modelers use a combination of Eulerian and Lagrangian methods. The velocity field for the currents and winds are derived using Eulerian techniques and are represented as individual velocity vectors at fixed points in the model domain. The method is useful for areas with historical data, such as tidal records or salinity and temperature, to predict the flow passed a fixed point. Oil patches are represented by individual particles that are sometimes called Lagrangian elements (LEs), spillets (French-McCay, 2003), or splots (Beegle-Krause, 2003). A major problem is translating numerous oil particles into a continuous concentration function. Usually, this is done by setting up a grid and counting the number of particles in each grid box. This may make concentration dependent on grid resolution. To adequately represent the concentration, a large number of particles and a fine grid are needed (James, 2002).

The modeler must take care to ensure the number of particles selected is sufficient to assure accurate and stable statistics. The number will vary with each spill situation. For a very large spill, 1% of the oil spilled would be significant and, consequently, a large number of particles should be used to model the spill movement. As an example, 1% of the 1989 *Exxon Valdez* spill represented about 410 m^2, a significant amount of oil. Therefore, 10,000 particles were used to represent the spill (Galt, 1991). In the event of a minor release, 1% may not be significant. Typically, 100 particles are used to represent smaller spills. Further analysis, with multiple model runs, can help determine the appropriate number of particles. To demonstrate this concept, an oil spill model was run twice with the same environmental conditions, but the number of particles chosen to represent the spill was different (Figure 13-1). Both outputs have similar movement, but the trajectory with 100 particles, Figure 13-1A, does not show a shoreline threat, whereas the model run with 1000 particles, Figure 13-1B, indicates shoreline contact.

In some instances, the modeler contours the particles and presents the output as the "best

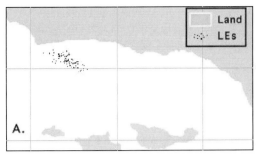 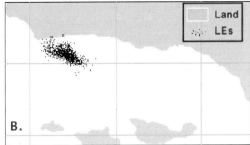

Figure 13-1 Comparison of a trajectory forecast with oil representing 100 (a) and 1000 (b) particles or Lagrangian elements (LEs).

guess" of the oil movement (Galt, 1997a). Since the particle separation is a measure of concentration, the analysis is very sensitive to the particle location. Determining concentrations for particles that are very close together or overlay each other becomes problematic. As Barker (2005) points out, it is not clear if two particles overlaying each other represent the same patch of oil. Modelers may contour the particles and present the best guess with a confidence level that, in general, is the bounding polygon of 99% of the particles. Selecting the appropriate number of particles is critical to the analysis, as one particle cannot give any uncertainty bound and, given a lengthy processing time, one million particles are not practical.

The oil spill trajectory should represent more than one model run (Galt, 1997b). The first set of model runs represents the standard trajectory forecast or "best guess" and uses the best-available data. The second set of model runs, the "uncertainty analysis," uses the expected errors in the input data. Each model run will have slightly different initial conditions that represent various "what-if" scenarios. This is particularly important because model input data are generally imprecise and limited during an emergency response. Real-time meteorological and oceanographic measurements are notoriously sparse. In these situations, the investigator will explore alternative spill scenarios, such as the effects on the slick movement if the surface current was 50% less than initial reports.

Other possible errors in the model input data are the location of the spill site and the distribution and amount of the floating oil. Initial estimates of the amount of oil spilled are often unreliable due to calculating release rates under adverse conditions (e.g., bad weather, grounding, or collision). Visually observing the resulting slick to estimate the amount of oil floating on the surface is inaccurate (Research Institute, 1983). Other potential errors result from reports of oil that are actually naturally occurring phenomena, such as kelp beds, silt plumes, algae, and jellyfish (NOAA, 1996). It is not unusual for emergency responders to collect oil spill observations using nonstandardized techniques and, as a result, observational reports vary between observers. Standard practice for reporting visual observations of oil on water is described in the American Standards for Testing Materials (ASTM, 1997).

Unfortunately, techniques for estimating uncertainty in observational errors have yet to be developed. To account for these types of uncertainty and others, multiple model runs with varying inputs are commonly used. In general, the model runs are averaged and the response community is presented with a "best guess" of where the oil is likely to go. The model runs that provide answers outside the average are interpreted as model "uncertainty."

13.3.1 Wind

When the wind blows across the sea surface, the oil slick drifts with the waves and currents.

Traditionally, oil spill modelers couple oil transport by wind-driven waves and the shear stress on the oil slick into one term called the "wind drift factor." Perhaps the best-known rule of thumb in oil spill modeling is the "3% rule" (Smith, 1976; Huang, 1983). This rule actually has some theoretical basis (Wu, 1983) and has been verified in observations of accidental oil spills and field and laboratory experiments (Fallah and Stark, 1976). The 3% rule represents average conditions, but in the experience of the authors and others, the actual factor ranges from 1% to 6% (Lehr and Simecek-Beatty, 2000). For example, heavier oils are often subject to overwash. While submerged, they will only drift at the speed of the water current and, hence, will have a net lower drift speed than that given by the 3% rule. On the other hand, oil caught in the "valleys" or convergences in the windrows will move faster than normal (Leibovich, 1997). This uncertainty is modeled by randomly selecting a slick drift between 1% and 6% of the wind speed for each particle at each time step.

Oil drifts at an angle of about 10° (Madsen, 1977) to the wind (to the right in the northern hemisphere and left in the southern hemisphere). Trajectory modelers may not include the deflection angle, as the marine forecast for the wind direction is generally not accurate to within 10°.

In general, the largest input error for the oil spill model is the wind forecast. The accuracy of the forecast depends on, among other things, special weather features, length of the forecast period, and ability of the forecaster to localize his/her prediction to the spill site. Optimum forecast periods are usually between 6 to 24 hours. Beyond 5 days, a wind forecast based on numerical predictions is generally no better than climatology. Lehr et al. (1999) discuss typical ranges of forecast errors in the wind direction and speed. They report two different approaches for entering wind uncertainty into oil spill models: (1) ask the marine forecaster about wind uncertainty; or (2) use an algorithm to simulate forecast uncertainty. For the forensic investigator, this is an important consideration if the on-scene weather observations are not available near the spill site and the only other available information is archived weather forecasts.

The first approach, and by far the easiest, requires a good verbal briefing by the meteorologist, who can provide information about wind shift timing, the strength of the pressure gradient, location of high/low fronts, and local land effects (i.e., sea breeze, topography). With this approach, a constant, time-dependent wind vector is assumed to represent the entire spill area. The wind file containing the meteorologist's best estimate and error estimate can then be fed directly into the model. The challenge with this approach is converting a verbal or written forecast into a file form to enter into an oil spill trajectory model. Most forecasters use standard words and phrases that can be easily translated to a digital format. For example, the wind forecast may indicate winds from the south at 10 m/s for 12 hours, becoming southwest at 5 m/s. If the forecaster indicates the forecast wind shift could be off by 3 hours, the wind direction off by 20°, and the speeds by 3 m/s, the original wind file is modified or an additional file is created with this data. Further details on this approach can be found in Lehr et al. (2003). The second approach adopts a more automated slant to estimating wind data error. This approach uses algorithms based on past weather forecast accuracy records (Lehr et al., 1995, 1999, 2003). While this method may be easier to implement, there is no confirmation yet that it produces superior results than that achieved by a skilled weather forecaster and an experienced modeler.

For the investigator, using a single wind vector based on a verbal briefing of the marine forecast may be sufficient in offshore areas over short length scales. As the oil moves closer to shore, a single constant wind vector could misrepresent effects due to topographical steering of the winds and other localized phenomena. Discussions with the forecaster about these effects can reduce the overall error in the analysis.

An approach that is becoming more accepted in the oil spill modeling community involves importing a time-dependent and spatially varying wind field from an atmospheric model. Careful consideration is needed before importing winds from an atmospheric model. Discussions with a local meteorologist can provide insight to the investigator about the availability of models for a specific area and the model limitations for the time frame of a particular spill event. Atmospheric model resolution ranges from a kilometer to several hundred kilometers (Table 13-1). For most instances, the regional meso-scale models are suited for oil spill trajectory modeling.

13.3.2 Currents

The ocean has three types of current systems that can transport oil: density-driven; wind-driven; and tidal. Offshore, the density-driven currents are often regarded as slow and not varying very much in speed or direction. Therefore, their variance is of less importance for most oil spills since small spills have time scales of a few hours to several days. Density-driven flow can be important in estuaries and in the nearshore coastal areas due to freshwater outflow. Generally, the investigator's focus will be on spills in estuarine environments and areas with coastal flow over the continental shelf with well-mixed conditions. In these locations, the wind-driven and tidally driven currents are of the most interest and have length and time scales that range from tens to hundreds of kilometers and a few hours to weeks (Table 13-2).

One of the more important advances in the development of oil spill modeling capabilities in the last 20 years has been in ocean circulation forecasting. In a few regions, oil spill modelers have the ability to import time and spatially varying current forecasts that are automatically updated every few hours (Beegle-Krause, 2003). Numerical models are initialized with real-time observation data (e.g., water level, satellite altimetry, and sea-surface temperature), and a circulation forecast is made using meteorological and, if needed, river flow forecasts. Other technical advances are nested grid systems that use a low-resolution, global ocean model to provide boundary conditions for high-resolution, regional models. In general, the modeler ensures quality-assurance standards are in place before importing the current predictions. Using standardized data exchange protocols (Beegle-Krause et al., 2003), these types of current predictions are now routinely imported into oil spill trajectory models.

In areas without a real-time regional circulation model, simulating the current becomes complicated. Ideally, a three-dimensional hydrodynamic model (e.g., Blumberg and Mellor, 1987; Blumberg and Herring, 1987) would be modified for the spill site. The models are sophisticated and require extensive oceanographic data for input. In a spill response situation, acquiring relevant real-time data, such as salinity and temperature at depth, is highly unlikely. To work around this problem, modelers may use a combination of real-time measurements, such as the wind

Table 13-1 Grid Resolution of Atmospheric Models Modified from Kalnay (2003)

Atmospheric models	Grid resolution
Climate	Several hundred kilometers
Global weather	50–100 km
Regional meso-scale	10–50 km
Storm scale	1–10 km

Table 13-2 Estimated Length and Time Scale of Oil Movement Due to Surface Current Transport

Surface current	Length scale	Time scale
Ocean circulation	1000's of kilometers	Months to years
Coastal flow	100's of kilometers	Weeks
Estuarine circulation	10's of kilometers	1 to 2 days
River	10's of kilometers	Hours to days

direction and velocity, astronomical tidal predictions, and historical data for the ocean measurements at depth [e.g., *NODC* (Levitus) *World Ocean Atlas*, 1994]. The difficulty is that the historical records are often short and collected under environmental conditions very different from that for the spill. For example, the data may be collected for a few weeks during a research cruise years before the spill incident. Even if the data were available to run the model, it is very difficult to adjust the current distribution or patterns from on-scene observations. If such adjustments are made, they must be done in such a way that ensures continuity is preserved. Failure to consider conservation of mass when modifying the currents can result in unintended convergences or divergences in the modeled circulation.

Two-dimensional hydrodynamic models provide a relatively simple and quick method for creating a circulation pattern based on bathymetry and fluid conservation laws. Key advantages for using a 2D model are the minimal data requirements while still conserving mass. The models are usually depth-averaged and assume the circulation is slowly varying and, therefore, steady-state. Updating the model from on-scene observations of the currents is generally straightforward. Because the current patterns conserve mass, the modeler can add multiple patterns to describe specific characteristics of the spill area. Examples of 2D hydrodynamic models used for spill response can be found in Galt (1975, 1980), Galt and Payton (1981), Proctor et al. (1994), and Sankaranarayanan and French McCay (2003).

It is important for the investigator to carefully select the best available tool that is based on fluid conservation laws. Convergences and divergences in the flow, which are important for spreading and concentrating oil, need to be accurately represented to forecast or hindcast oil movement. Two especially problematic approaches used by the operational response community are "smoothing" of the circulation pattern and high-frequency radar (HF). The first approach involves the linear interpolation of one or two on-scene current observations to develop a spatially varying current. This may or may not involve smoothing "errant" vectors along the shoreline. The main issue with the approach is the blatant disregard to fluid conservation and, therefore, the modeling community routinely rejects it.

The second approach, the remote sensing tool HF radar, is used to measure and present the spatial distribution of the upper layers of the surface current direction and velocity by averaging small areas of the sea surface. In order to derive the current circulation, a complex process is used to remove the wind from the signal. The investigator should recognize that HF systems are not predictive tools. Oil spill modelers need to know the currents at least 24 to 36 hours into the future. To turn the system into a predictive tool, modelers need to know something about the forcing mechanisms responsible for transporting the water. At this time, modelers do not have the tools to analyze the HF data to understand the forcing mechanisms. Other problems include that placement of the HF antennas often results in blind spots in the nearshore areas. In the circulation pattern, this appears as areas without current. In addition, the output may show on-shore current vectors. In reality, there is likely a long-shore current transporting oil near the shoreline. This could lead to errors in predicting oil contact on the shoreline. Finally, the system cannot provide 3D current observation: the system measures the water movement at the surface.

Subsurface circulation models developed for response tend to be very simple due to the constraints involved with 3D hydrodynamic models. Pseudo-3D models, which are essentially modified 2D models, are the accepted practice. Since near-bottom drift data and higher resolution bathymetry (e.g., depressions or deep holes in the bottom) may not be readily available, oil moving along the bottom is often parameterized with a simple resuspension term. Unfortunately, issues with modeling and tracking submerged oil as described in Conomos (1975) are not much different than present-day capabilities (NOAA, 2004).

13.3.3 Turbulent Diffusion

Modeling turbulence in the ocean is a complex process that requires complicated computer simulations. Most oil spill models use simplified formulas to simulate mixing. In general, a particle-tracking scheme is used with turbulent diffusion represented by random movement of the particles. A common approach is to represent turbulence using a constant diffusion coefficient, but there are other options. James (2002) offers an alternative to uniform turbulence, but the formulas may not be practical. Elliot (1992) has developed empirical formulas that relate horizontal diffusion to the tides and wind. Thibodeaux (1979) suggests the horizontal dispersion coefficient can be estimated by comparing successive aerial observations of the surface slicks. In practice, the method is not used because the dispersion coefficient estimated on the first day of a spill event will be different seven days into the spill due to advection of the oil by larger-sized eddies. Some researchers suggest that diffusion is proportional to the depth of the region. Most spill models assume constant and uniform dispersion coefficients, which are based on historical measurements such as those presented in Okubo (1971). The appropriate coefficient is dependent on the scale of the spill and model prediction and the location of the release. The oil spill modeler may adjust the turbulent mixing term between model runs if the model output does not completely match the observations of the oil distribution.

13.4 Evolution of an Oil Spill

Due to oil spreading, evaporation, dispersion, emulsification, dissolution, oxidation, oil particle interaction, and biodegradation, oil changes its chemical and physical properties almost immediately when spilled into the water. Eventually, the amount of oil in the surface slick decreases over time.

13.4.1 Spreading

The major mechanism that rapidly changes the surface slick is spreading of the oil onto the water surface. The rate at which oil spreads is interactive with the other major weathering processes, both affecting and being affected by them. Oil begins to spread as soon at it is spilled, but it does not spread uniformly. Any shear in the surface current will cause stretching, and even a slight wind will cause a thickening of the slick in the downwind direction. Most spills quickly form a comet shape with a relatively smaller black- or brown-colored region trailed by a much larger sheen of colors varying from dull-colored to a rainbow or silver sheen. While the sheen covers a much larger area than the black or brown region, most of the oil is found in the latter since the sheen is orders of magnitude thinner (ASTM, 1997). This is unfortunate as formulas exist to estimate the thickness, and hence volume, of the sheen. However, no such formulas exist for the thick part. Experimental oil spills have confirmed a commonly used rule-of-thumb that approximately 90% of the oil volume is located in about 10% of the slick area (Research Institute, 1983). The personal experiences of the authors indicate that this is not unusual during the early stages of the spill for 90% of the oil to be found in the leading edge of the slick rather than in the trailing, thin, silver and rainbow sheens.

Oil spreading is a complicating factor when attempting to learn where the oil could have initially been spilled. The comet-shaped appearance and elongation of the slick are due to winds, sea state, currents, and random eddies. As a diffusive process, spreading is not amenable to accurate hindcast predictions. Small variations in the environmental parameters can cause large variations in the time-reversed outcome.

Spill researchers have attempted to model spreading by two separate approaches. One deterministic approach balances spreading and retarding forces in a fluidic application of Newton's law. The most widely used such deterministic model was developed by James Fay of M.I.T. (Fay, 1969, 1971). Fay divided the spreading process into three separate phases, depending upon the major driving and retarding forces. When the slick is relatively thick, gravity causes the oil to spread laterally; later, interfacial tension at the periphery will be the dominant spreading force. The main retarding force is primarily inertia followed later by the viscous drag of the water. Fay therefore labeled the three phases: gravity-inertial, gravity-viscous, and surface tension-viscous.

As an alternative approach, researchers have attempted to model slick spreading as strictly a water turbulence phenomenon with the oil acting as a neutral tracer (Murray, 1972). Elliot (2004) showed that, for oil, such a process is non-Fickian, requiring a time-dependent diffusion parameter. While oil spreading is a diffusive process, it occurs in two dimensions and therefore, unlike three-dimensional spreading such as smoke from a fire, it is possible under certain circumstances to undergo an increase in oil concentration as the oil moves farther away from the spill site.

There are several different mechanisms that may cause these "collection" zones to occur. Wind and surface waves frequently induce windrow formation. This interaction results in vortices in the surface mixed layer of the ocean that are aligned in the general direction of the wind. Between the vortices, the surface water either diverges or converges. These vortices are called "Langmuir cells" and the flow associated with the cells, "Langmuir circulation" (LC). A conceptual diagram of LC is shown in Figure 13-2.

Operational oil spill models that incorporate Langmuir circulation have shown only limited success (Simecek-Beatty et al., 2001). Therefore, most spill models have generally neglected LC, even though such models may, mathematically, be quite complex (Payne et al., 1984; Lehr et al., 2000). Thorpe (1995), Leibovich (1997), and Li (2000) are among the few researchers who attempted to model LC effects on oil spills.

If LC does have a measurable influence on oil slick behavior, it is likely that three major weathering processes would be altered. These are surface spreading of the oil, dispersion of oil droplets in the water column, and transport of the surface slick. It is worthwhile for the investigator to consider this process when hindcasting the oil movement and behavior.

Those attempting to track mystery spills back to their origin face two obstacles from spill spreading. In the first case, spreading is a dispersive phenomenon that obscures the spill source. Secondly, concentrations of floating oil may have been caused by natural mechanisms such as LC that are independent of the spill event.

13.4.2 Oil Weathering

For forensics, the chemical properties of spilled oil cause a greater challenge than a spill of a pure substance. However, the behavior of the mixture of hundreds of different organic compounds that make up the typical oil or refined oil product provides a useful tool for the investigator to both identify and estimate the age of the spill. The methods used to identify, or fingerprint, oil are described elsewhere in this book. This section will describe the significant changes in the chemical nature of oil exposed to the environment that are referred to as *"oil spill weathering."*

Major short-term weathering processes include evaporation, dispersion, and emulsification. These are the important drivers in both the mass balance of the spill and the overall physical behavior of the surface slick. Nevertheless, there are minor processes that, over a longer term, can cause significant alterations to the spilled oil. These minor processes comprise such changes as dissolution,

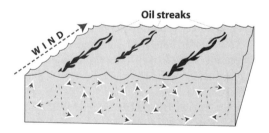

Figure 13-2 A cross-sectional view of an idealized oil spill in the sea. The arrows below the oil streaks represent the motion of these vortices. The oil is shown collecting in areas with converging surface currents.

sedimentation, biodegradation, and photooxidation. Discussions of these processes are found in Lehr (2001) and Fingas (1995).

For most spills, evaporation is the major mechanism for mass removal from the surface slick (National Research Council, 2003). This includes both natural processes and cleanup attempts. It is quite possible to lose half of a light crude spill just due to evaporation. Small, light, refined product spills will typically disappear in less than a day due to evaporation unless high sea states drive the oil into the water column. Also, evaporation changes the chemical mixture of the slick as the lighter components evaporate more quickly than the heavier hydrocarbons. The structure of the molecule is of some importance. However, the major factor is molecular weight. The vapor pressures of hydrocarbons with a carbon number of 10 or above are orders of magnitude smaller than the vapor pressures of hydrocarbons with a carbon number of less than 10 and, hence, older spilled oil will contain disproportionate amounts of the larger hydrocarbon molecules. By comparing spilled oil sample characteristics to fresh oil, forensic investigators can estimate the age of the sample. For example, for an oil of known evaporation characteristics, the evaporations formulas in Wang's Section 7.4 can be inverted to give a guess as to the age of any surface slick samples.

Other major weathering processes, such as dispersion into the water column or the formation of a stable water-in-oil emulsion, do not affect the oil chemistry but are important for determining the amount of surface oil and the oil's rheological properties. Most weathering software programs use some version of the dispersion model developed by Delvigne and Sweeney (1988). They assume dispersion is affected by sea state, slick viscosity, and surface tension of the oil. While oil may not dissolve in water to any great extent, Delvigne and Sweeney recognized that it could certainly disperse as a cloud of droplets when subject to turbulent wave energy. These droplets will be in various sizes and will be subject to the conflicting forces of buoyancy and turbulence.

For the smallest oil droplets (~50 to 70 μm), turbulence will win the battle and the droplet will not refloat to rejoin the slick. Dispersion becomes a dominant mechanism for low-viscosity oils under high sea state conditions (Lehr, 2001).

If the receiving water contains large quantities of particulate matter, dispersed oil may attach to these particulates in a process called oil-particle interaction, or sedimentation. The combined oil-sediment particles will typically have a different buoyancy than either alone. Usually, the buoyancy will be negative, and turbulence will be required to keep the oil-sediment particle from settling to the bottom. Very little research is available regarding the tracking and weathering of oil in the water column or on the bottom.

In one way, emulsification can be thought of as the reverse of dispersion. Rather than oil droplets dispersing into the water column, water is entrained in the oil. This causes significant changes in the volume, density, and, especially, viscosity of the slick. It is not uncommon for the viscosity of an emulsified oil to be two or three orders of magnitude larger than the viscosity of the fresh oil (Lehr, 2001). Emulsification will also change the way slicks drift with the wind when the oil is no longer a thin film. At this point, the slick has been torn apart into smaller pieces, some of which maybe overwashed by waves. This complicates hindcasting the movement and weathering of the oil. Not all oils will emulsify, and some oils will emulsify only after they have undergone some weathering. It appears that resins (Fingas et al., 2003) and, most importantly, asphaltene content play the dominant role in determining whether emulsification will occur. A common, but not necessarily reliable, rule-of-thumb is that crude oil will emulsify when these component contents reach 5% of the mass of the oil.

The two long-term mechanisms for the breakdown of hydrocarbons in the environment are photooxidation and biodegradation. The combination of hydrocarbons with oxygen is called "oxidation" with, in this case, photons providing the energy source. The newly formed oxidized compounds may affect the oil slick by increasing dissolution, dispersion, the formation of tar balls, or emulsification. Hydrocarbons, including those found in oil slicks, are a food source for many microorganisms. Examining the stages of the biodegradation can indicate the period of toxic exposure for the ecosystem due to the spilled oil.

13.5 Conclusions and Challenges

We have tried to provide the forensic investigator with a brief overview of present oil spill modeling capabilities. To understand the current state of knowledge, a historical overview of oil spill modeling development was presented. This was important as much of the original work is still used in today's models.

Understanding the difference between "forecasting" and "hindcasting" is key to identifying the best approach for investigating the movement and behavior of spilled oil. We attempted to clarify some of the problems with stochastic modeling and pointed out the challenges with presenting probability information to the end user.

We attempted to demystify the spill modeling process by disclosing some of the more practical modeling techniques. Oil spill modelers are notorious for using simple parameterizations to simulate oil movement based on the significant errors that exist in forecast fields. The challenge is to incorporate more rigorous approaches to emergency response.

Since oil changes its physical and chemical characteristics over time, we briefly described the evolution of an oil spill and oil weathering. To aid the investigator, we offered "rules-of-thumb" that are common in the oil spill response community but may not necessarily appear in peer-reviewed literature. The challenge is presenting useful knowledge gained by experienced responders to newcomers.

Acknowledgments

The findings and conclusions in this chapter are those of the author(s) and do not necessarily represent the views of the National Oceanic and Atmospheric Administration. The authors wish to thank C. J. Beegle-Krause and Glen Watabayashi for their helpful comments.

References

ASTM METHOD F1779-97, *Standard Practice for Reporting Visual Observations of Oil on Water*, Philadelphia: American Standards for Testing Materials, 1997.

Anon., State-of-the-art review of modeling transport and fate of oil spills. *J. Hydraulic Eng.*, 1996, **122**(11), 594–609.

Barker, C., *Personal communication*, 2005.

Beegle-Krause, C.J., Advantages of separating the circulation model and trajectory model: GNOME trajectory model used with outside circulation models. In *Proc. Arctic and Marine Oil Spill Program (AMOP) Technical Seminar*, Environment Canada, British Columbia, 2003, 825–840.

Beegle-Krause, C.J., J. Callahan, and C. O'Connor, NOAA model extended to use nowcast/forecast currents. In *Proc. 2003 Inte. Oil Spill Conf.*, API Publication No. 14730, British Columbia, Canada, IOSC 2003 Proceedings, Vancouver, BC, 2003, 991–994.

Blumberg, A.F. and G.L. Mellor, Description of a three-dimensional coastal ocean circulation model. In *Three-Dimensional Coastal Model*, N.S. Heaps (ed.), Coastal and Estuarine Sciences 4, AGU, Washington, DC, 1987, 1–16.

Blumberg, A.F. and H. Herring, Circulation modeling using orthogonal curvilinear coordinates. In *Three-Dimensional Models of Marine and Estuarine Dynamics*, J. Nihoul and B. Jamart (eds.), Elsevier Oceanography Series, 1987, **45**, 55–88.

Biglane, K.E., A history of major oil spill incidents. In *Proc. Joint Conf. Prevention and Control of Oil Spills*, New York, 1969, 5–6.

Cekirge, H.M. and S.L. Palmer, Mathematical modeling of oil spilled into marine waters. In *Oil Spill Modeling and Process*, C.A. Brebbia (ed.), Southampton, UK: WIT Press, 2001, 1–15.

Conomos, T.J., Movement of spilled oil as predicted by estuarine nontidal drift. *Limnology and Oceanography*, March 1975, **20**(2), 159–173.

Daniel, P., P. Dandin, P. Josse, C. Skandrani, R. Benshila, C. Tiercelin, and F. Cabioch, Towards better forecasting of oil slick movement at sea based on information from the *Erika*. In *Proc. Third R&D Forum on High-Density Oil Spill*

Response, Brest, France: International Maritime Organization, 2002.

Delvigne, G.A.L. and C.E. Sweeney, Natural dispersion of oil. *Oil & Chemical Pollution*, 1988, **4**, 281–310.

Elliot, A.J., A probabilistic description of the wind over Liverpool Bay with application to oil spill simulations. *J. Estuarine Coast and Shelf Sci.*, 2004, **61**(4), 569–581.

Elliot, A.J., A.C. Dale, and R. Proctor, Modelling the movement of pollutants in the UK shelf seas. *Mar. Poll. Bull.*, 1992, **24**(12), 614–619.

Fallah, M.H. and R.M. Stark, Random drift of an idealized oil patch. *Ocean Eng.*, 1976, **3**, 83–97.

Fay, J.A., *The Spread of Oil Slicks on a Calm Sea*, Fluid Mechanics Laboratory, Dept. of Mech. Eng., MIT, Cambridge, MA, 1969.

Fay, J.A., Physical processes in the spread of oil on a water surface. In *Proc. Joint Conf. on Prevention and Control of Oil Spill*, Washington, DC, 1971, 463–467.

Fingas, M., A literature review of physics and predictive modeling of oil spill evaporation. *J. Hazardous Materials*, 1995, **42**, 157–175.

Fingas, M.F. and B. Fieldhouse, *Studies of the formation of water-in-oil emulsions. Mar. Poll. Bull.*, 2003, **47**, 369–396.

French-McCay, D.P., Development and application of an oil toxicity and exposure model, OilToxEx. *Environ. Toxicol. Chem.*, 2002, **21**(10), 2080–2094.

French-McCay, D., Development and application of damage assessment modeling: Example assessment for the *North Cape* oil spill. *Mar. Poll. Bull.*, 2003, **47**, 341–359.

French-McCay, D.P., Oil spill impact modeling: Development and validation. *Environ. Toxicol. Chem.*, 2004, **23**(10), 2441–2456.

Galt, J.A., The integration of trajectory models and analysis into spill response information systems. *Spill Sci. Tech.*, 1997a, **4**(2), 123–129.

Galt, J.A., Uncertainty analysis related to oil spill modeling. *Spill Sci. Tech.*, 1997b, **4**(4), 231–238.

Galt, J.A., *Development of a simplified diagnostic model for the interpretation of oceanic data*, NOAA Technical Report ERL 339-PMEL 25, National Oceanic and Atmospheric Administration, Seattle, WA, 1975.

Galt, J.A., A finite solution procedure for the interpolation of current data in complex regions. *J. Phys. Oceanogr.*, 1980, **10**(12), 1984–1997.

Galt, J.A. and D.L. Payton, Finite element routines for the analysis and simulation of near shore circulation. In *Proc. 1981 Oil Spill Conf.*, Atlanta, GA, 1981.

Galt, J.A., W.J. Lehr, and D.L. Payton, Fate and transport of the *Exxon Valdez* oil spill. *Environ. Sci. Tech.*, 1991, **25**(2), 202–209.

Grouse, P.L. and J.S. Mattson, *The* Argo Merchant *Oil Spill*, National Oceanic and Atmospheric Administration, Environmental Research Laboratory, Boulder, CO, 1977.

Hooper, C.H., *The IXTOC 1 Oil Spill: The Federal Scientific Response*, National Oceanic and Atmospheric Administration, 1981.

Huang, J., A review of the state-of-the-art of the oil spill fate/behavior models. In *Proc. 1983 Oil Spill Conf.*, San Antonio, TX, 1983, 313–323.

James, I.D., Modelling pollution dispersion, the ecosystem and water quality in coastal waters: A review. *Environ. Modelling & Software*, 2002, **17**, 365–385.

Jones, B., The use of numerical weather prediction model output in spill modeling. *Spill Sci. Tech.*, 1999, **5**(2), 153–159.

Kalnay, E., *Atmospheric Modeling, Data Assimilation and Predictability*, London: Cambridge Univ. Press, 2003, 127–129.

Lehr, W.J., Review of modeling procedures for oil. In *Oil Spill Modeling and Processes*, C.A. Brebbia (ed.), WIT Press, 2001.

Lehr, W., C. Barker, and D. Simecek-Beatty, *New developments in the use of uncertainty*. In *Proc. Twenty-Second Artic Marine Oil Spill (AMOP) Technical Seminar*, Alberta, Canada, 1999, 271–284.

Lehr, W.J., J.A. Galt, and R. Overstreet, Handling uncertainty in oil spill modeling. In *Proc. Fifteenth Arctic and Marine Oil Spill Program (AMOP) Technical Seminar*, Environment Canada, Ottawa, Ontario, 1995, 759–767.

Lehr, W.J. and D. Simecek-Beatty, The relation of Langmuir circulation processes to the standard oil Spill spreading, dispersion, and transport algorithms. *Spill Sci. Tech. Bull.*, 2000, **6**, 247–253.

Lehr, W.J., D. Simecek-Beatty, and M. Hodges, Wind uncertainty in long range trajectory forecasts. In *Proc. 2003 Int. Oil Spill Conf.*, British Columbia, Canada, 2003, 435–439.

Leibovich, S., *Surface and near-surface motion of oil in the sea*, Contract 14-35-0001-30612, Minerals Management Service, U.S. Department of the Interior, 1997.

Li, M., Estimating horizontal dispersion of floating particles in wind-driven upper-ocean. *Spill Sci. Tech.*, 2000, **6**(3–4), 255–261.

Madsen, O.S., A realistic model of the wind-induced Ekman boundary, *J. Phys. Oceanogr.*, 1977, **7**, 248–255.

Murray, S.P., Turbulent diffusion of oil in the ocean. *Limnology and Oceanogr.*, 1972, **27**, 651–660.

National Research Council, *Oil in the Sea III: Inputs, Fates and Effects*, The National Academies Press, 2003.

NOAA Report, *Submerged Oil Assessment — Athos I Oil Spill*, Report from Submerged Oil Assessment Unit to the *Athos 1* Oil Spill Unified Command, 11 December 2004.

NODC, *NODC (Levitus) World Ocean Atlas*, NOAA-CIRES ESRL/PSD Climate Diagnostics Branch, Boulder, CO, http://www.cdc.noaa.gov/, 1994.

Okubo, A., *Diffusion and Ecological Problems: Mathematical Models*, New York: Springer-Verlag, 1980.

Paluszkiewicz, T. and C. Marshall, Comparison of techniques for forcing an oil spill trajectory model. In *Proc. 1989 Oil Spill Conf.*, Washington, DC, 1989, 547–553.

Payne, J.R., B.E. Kirstein, G.D. McNabb, J.L. Lambach, R. Redding, R.R. Jordan, W. Hom, C. Oliveira, G.S. Smith, D.M. Baxter, and R. Gaege, *Multivariate Analysis of Petroleum Weathering in the Marine Environment — Sub Arctic*, Outer Continental Shelf Environmental Assessment Program of the National Oceanic and Atmospheric Administration, 1984.

Pollack, A.M. and K.D. Stolzenbach, *Crisis Science: Investigations in Response to the* Argo Merchant *Oil Spill*, Sea Grant Program, Report No. MITSG 78-8, Cambridge, MA: M.I.T., 1978.

Proctor, R., R.A.F. Flather, and A.J. Elliot, Modelling tides and surface drift in the Arabian Gulf — application to the Gulf oil spill. *Continental Shelf Res.*, 1994, **14**(5), 531–545.

Research Institute, *Final Report on Estimating Spill Size by Visual Observation*, Project No. 24028, Dhahran, Saudi Arabia: University of Petroleum and Minerals, 1983.

Roulston, M. and L. Smith, Evaluating probabilistic forecasts using information theory. *Monthly Weather Rev.*, 2002, **130**, 1653–1660.

Sankaranarayanan, S. and D. French-McCay, Applications of a two-dimension depth averaged hydrodynamic tidal model. *J. Ocean Eng.*, 2003, **30**(14), 1807–1832.

Schwartzberg, H.G., The movement of oil spills. In *Proc. Int. Conf. Prevention and Control of Oil Spills*, Washington, DC, 1971, 484–494.

Simecek-Beatty, D., W.J. Lehr, R. Lai, and R. Overstreet (eds.), Special issue: Langmuir circulation and oil spill modeling. *Spill Sci. Tech. Bull.*, 2001, **6**(3/4).

Smith, C.L., Determination of the leeway of oil slicks. In *Fate and Effects of Petroleum Hydrocarbons in Marine Ecosystems and Organisms*, D.A. Wolfe, J.W. Anderson, D.K. Button, D.C. Malins, T. Roubal, and U. Varanasi (eds.), New York: Pergamon Press, 1976, 351–362.

Stolzenbach, K.D., O.S. Madsen, E.E. Adams, and C.K. Cooper, *A review and evaluation of basics techniques for predicting the behavior of surface oil slicks*, Report No. 22, Cambridge, MA: M.I.T., 1977.

Spaulding, M., A state-of-the-art review of oil spill trajectory and fate modeling. *Oil and Chem. Poll.*, 1988, **4**, 39–55.

Thibodeaux, L., *Chemodynamics: Environmental Movement of Chemicals in Air, Water, and Soil*, New York: John Wiley & Sons, 1979.

Thorpe, S.A., On the meandering and dispersion of a plume of floating particles caused by Langmuir circulation and a mean current. *J. Phys. Oceanogr.*, 1995, **25**, 685–690.

Yapa, P.D. and H.T. Shen, Modeling river oil-spills — a review. *J. Hydr. Res.*, 1994, **32**(5), 765–782.

Wang, Z., B. Hollebone, M. Fingas, B. Fieldhouse, M. Landriault, and P. Smith, Development of a physical and chemical property database for ten EPA-selected oils. In *Proc. Twenty-Sixth Arctic and Marine Oil Spill Program (AMOP) Technical Seminar*, 2003, 117–142.

Wu, J., Sea-surface drift currents induced by wind and waves. *J. Phys. Oceanogr.*, 1983, **13**, 1441–1450.

14 Oil Spill Remote Sensing: A Forensic Approach

Merv Fingas and Carl E. Brown

Emergencies Science and Technology Division, Environment Canada, Environmental Technology Centre, Ottawa, Ontario, Canada K1A 0H3.

14.1 Introduction

Spills of oil and related petroleum products in the marine environment can have serious biological and economic impacts (Wiese, 2002; Wiese and Ryan, 2003; Wiese et al., 2004). Public and media scrutiny is usually intense after a spill, with demands that the source of the oil spill be determined and the responsible party be brought to justice. Remote sensing is playing an increasingly important role in oil spill response efforts and will no doubt play an important role in forensics related to identifying "mystery spills" or discharges at sea. Through the use of modern remote sensing instrumentation, oil spillage can be monitored on the open ocean around the clock. Certain developments, particularly that of the laser fluorosensor, can provide enough forensic evidence to confirm that oil has indeed been spilled and that it is of a given type.

Even though the design and electronics of sensors are becoming increasingly sophisticated and sensors are becoming much less expensive, the operational use of remote sensing equipment lags behind the development of the technology. The most common forms of oil spill surveillance and mapping are done with simple still or video photography, which provide little, if any, forensic data. Remote sensing from an aircraft is still the most common form of oil spill tracking. Attempts to use satellite remote sensing for oil spills, although successful, are not necessarily as claimed and are generally limited to identifying features at sites of known oil spills. Forensic capability is not as good from satellite systems as from aircraft.

Several general reviews of oil spill remote sensing have been prepared (Fingas and Brown, 2000a, b, 2002a, 2005). These reviews show that progress is being made in oil spill remote sensing, although the progress is slow. The reviews also show that off-the-shelf sensors have very limited forensic application to oil spills, whereas specialized sensors offer advantages to oil spill remote sensing and an evolving forensic capability. Remote sensors for forensic application to oil spills are reviewed in this chapter. Many common sensors are available; however, they have limited forensic application. An infrared camera or an infrared/ultraviolet (IR/UV) system can detect oil under a variety of conditions and distinguish oil from some backgrounds but has limited forensic application. The inherent weaknesses of these systems include the inability to distinguish oil on beaches and among weeds or debris and to detect oil under certain lighting conditions. Lack of positive discrimination between oil and some backgrounds implies a lesser forensics applicability.

The laser fluorosensor is a most useful instrument to forensics because of its unique capability to positively identify oil against most backgrounds, including water, soil, weeds, ice, and snow. The availability of a fluorescent spectrum further enhances the forensic capability, particularly for identifying light oils that have different features.

Radar offers the only potential for searching in large areas and carrying out remote sensing during foul weather conditions, but offers very poor positive detection characteristics and thus low forensic capability. In addition, radar is prone to numerous interferences and false targets can be as high as 95%.

Equipment that measures relative slick thickness is still under development. Passive microwave has been studied for several years, but many commercial instruments lack sufficient spatial resolution to be practical, operational instruments. A prototype laser-acoustic instrument has been developed that provides the only technology to measure absolute oil thickness, but it may be too early to assess its potential for use in forensics. However, estimating spill volumes from slick thickness can have forensic implications.

Equipment operating in the visible spectrum, such as cameras and scanners, is useful for documentation or to provide a basis for the overlay of other data. This is the traditional forensic approach material; however, it is largely reliant on an expert observer rather than the discriminating power of the sensor and its output. It must be noted that oil shows no distinguishing spectral characteristics in the visible region; thus, documentation is by visual characteristics only.

Satellite-borne sensors are useful, particularly radar; however, their frequency of overpass and low spatial resolution make them of marginal use for spills. Radar satellites have similar forensic limitations as their airborne counterparts.

Sensors for detecting oil in ice are new and there are some promising concepts that require further research. Their potential for use in forensics, however, remains unknown at this time.

14.2 Visible Indications of Oil

Traditionally, much of the forensic work has been carried out by an "expert" observer trained to judge oil from its appearance and relate this to a vessel either discharging this oil or very close to where the oil was discharged. Often, however, oil on the water's surface is not visible to the eye (Fingas et al., 1999). Other than the obvious conditions of nighttime and fog, there are many other circumstances when oil cannot be seen. For example, oil is difficult to see if it is a thin slick such as from ship discharges or if it is masked by other materials on the water surface such as seaweed, ice, and floating debris. Often there are conditions or substances on the sea that may appear to be oil, but actually are not. These include wind shadows from land forms, surface wind patterns on the sea, surface dampening by submerged objects or weed beds, natural oils or biogenic material, and oceanic fronts. With large oil spills, the area covered by oil may be too great to be mapped visually. Due to all these factors, more sophisticated remote sensing systems must be used to assist in locating and tracking oil for forensic purposes.

14.3 Optical Sensors

14.3.1 Visible

Human vision alone is still the most common technique for oil spill surveillance and delivery of forensic evidence. In the past, major campaigns using only human vision were carried out with varying degrees of success (Taft et al., 1995). Optical techniques using the same range of the visible spectrum are the most common means of remote sensing. Cameras, both still and video, are common because of their low price and commercial availability. Documentation by camera images still forms the pillar of forensic evidence in court cases on oil discharges. Typically, these are presented by an expert observer who interprets the evidence to the court.

In recent years, visual and camera observation has been enhanced by the use of GPS (Global Positioning Systems) (Lehr, 1994). Systems are now available to directly map remote sensing data onto base maps.

In the visible region of the electromagnetic spectrum (approximately 400 to 700 nm), oil has a higher surface reflectance than water, but shows limited nonspecific absorption tendencies. Oil generally manifests throughout the entire visible spectrum. Sheen shows up silvery and reflects light over a wide spectral region down to the blue. As there is no strong information in the 500- to 600-nm region, this region is often filtered out to improve contrast (O'Neil et al., 1983). Overall, however, oil has no specific characteristics that distinguish it from the background (Brown et al., 1996).

Taylor studied oil spectra in the laboratory and the field and observed flat spectra with no usable features distinguishing it from the background (Taylor, 1992). Therefore, techniques that separate specific spectral regions within the visible spectrum do not increase detection capability. It has been found that high contrast in visible imagery can be achieved by setting the camera at the Brewster angle (53° from vertical) and using a horizontally aligned polarizing filter that passes only that light reflected from the water surface. This is the component that contains the information on surface oil (O'Neil et al., 1983). It has been reported that this technique increases contrast by up to 100%. Filters with bandpass below 450 nm can be used to improve contrast. These techniques, however, do not result in imagery that proves that oil is present nor can they define any physical or chemical characteristics of the oil.

On land, hyperspectral data (use of multiple spectral bands, typically 10 to 100) have been used to delineate the extent of an oil well blowout (Bianchi et al., 1995). The technique used was spectral reflectance in the various channels as well as the usual black coloration. This would provide some benefit in discriminating oil from land, but would not provide additional discrimination at sea.

Video cameras are often used in conjunction with filters to improve the contrast in a manner similar to that noted for still cameras. This technique has had limited success for oil spill remote sensing because of poor contrast and lack of positive discrimination. Despite this, video systems have been proposed as remote sensing systems for monitoring oil spills (Bagheri et al., 1995).

Scanners have been used as sensors in the visible region of the spectrum. A rotating mirror or prism sweeps the field-of-view (FOV) and directs the light toward a detector. Before the advent of CCD (charge-coupled device) detectors, this sensor provided much more sensitivity and selectivity than a video camera. Another advantage of scanners is that signals are digitized and processed before display. Newer technology now allows similar digitization to be achieved without scanning by using a CCD imager and continually recording all elements, each of which is directed to a different field-of-view on the ground. This type of sensor, known as a pushbroom scanner, has many advantages over the older types. It can overcome several types of aberrations and errors, the units are more reliable than mechanical ones, and all data are collected simultaneously for a given line perpendicular to the direction of the aircraft's flight.

Several types of scanners were developed. In Canada, the MEIS (Multi-Detector Electrooptical Imaging Scanner) (O'Neil et al., 1983) and the CASI (Compact Airborne Spectrographic Imager) (Palmer et al., 1994) have been developed, and the Caesar system was developed in the Netherlands (Wadsworth et al., 1992). The CASI is now a commercial unit.

The use of visible techniques in oil spill remote sensing is largely restricted to documentation of the spill because there is no mechanism for positive oil detection and there are many interferences or false alarms. Sun glint and wind sheens can be mistaken for oil sheens. Biogenic material such as surface seaweeds or sunken kelp beds can be mistaken for oil. Oil on shorelines is difficult to identify positively because seaweeds look similar to oil and oil cannot be detected on darker shorelines.

In summary, the usefulness of the visible spectrum for oil detection is limited. It is, however, an economical way to document oil spills and provide baseline data on shorelines or relative positions. As there is no spectrum associated with oil, there is less forensic application with photography or visible cameras than with other sensors that provide a spectral signature.

14.3.2 Infrared

Oil, which is optically thick, absorbs solar radiation and re-emits a portion of this radiation as thermal energy, primarily in the 8- to 14-μm region. In infrared (IR) images, thick oil appears hot, intermediate thicknesses of oil appear cool, and thin oil or sheens are not detected. The thickness at which these transitions occur is not well understood, but evidence indicates that the transition between the hot and cold layer is between 50 and 150 μm, and the minimum detectable layer is between 10 and 70 μm (Hurford, 1989; Goodman, 1989; Belore, 1982; Neville et al., 1979).

The reason for the appearance of the "cool" slick is not fully understood. A plausible theory is that a moderately thin layer of oil on the water surface causes destructive interference of the thermal radiation waves emitted by the water, thereby reducing the amount of thermal radiation emitted by the water (Fingas et al., 1999). This may be analogous to the appearance of the rainbow sheen, which is explained in Section 14.6.2. The cool slick would correspond to the thicknesses as observed above because the minimum destructive thickness would be about two times the wavelength, which is between 8 to 10 μm. This would yield a destructive onset of about 16 to 20 μm to about four wavelengths, or about 32 to 40 μm. The destructive area is usually only seen with test slicks, which is explained by the fact that the more rapidly spreading oil is more transparent than the remaining oil. In theory, the onset of the hot thermal layer would then be at thicknesses greater than this or at about 50 μm. Infrared sensors do not provide enough information to establish either the thickness or presence of oil on a forensic basis.

Infrared devices cannot detect emulsions (water-in-oil emulsions) under most circumstances (Bolus, 1996). This is probably due to the high thermal conductivity of emulsions as they typically contain 70% water and thus do not show a temperature difference.

Infrared cameras are now very common and commercial units are available from several manufacturers. Scanners with infrared detectors have been used recently. The older type of infrared detectors, however, required cooling to avoid thermal noise, which would overwhelm any useful signal. Liquid nitrogen, which provides about 4 hours of service, was traditionally used to cool the detector. New, smaller sensors use closed-cycle or Sterling coolers, which operate on the cooling effect created by an expanding gas. While a gas cylinder or compressor must be transported with this type of cooler, refills or servicing may not be required for days at a time (Goodman, 1988). In the past few years, uncooled detectors have entirely replaced the older, cooled detectors.

Most infrared sensing of oil spills takes place in the thermal infrared at wavelengths of 8 to 14 μm. One sensor, which is designed as a fixed-mounted unit, uses the differential reflectance of oil and water at 2.5 and 3.1 μm (Seakem Oceanography, 1988). Tests of a midband infrared system (3.4 to 5.4 μm) over the *Tenyo Maru* oil spill showed no detection in this range; however, ship scars were visible (Rogne and Smith, 1992; Rogne et al., 1992; Kennicutt et al., 1992). Specific studies in the thermal infrared (8 to 14 μm) show that there is no useful spectral structure in this region (Salisbury et al., 1993). Tests of a number of infrared systems show that spatial resolution is extremely important when the oil is distributed in windrows and patches, emulsions are not always visible in the IR, and cameras operating in the 3- to 5-μm range are only marginally useful (Hover, 1994).

The relative thickness information in the thermal infrared can be used to direct countermeasures equipment to thicker portions of the oil slick, but is not useful forensically. Oil detection in the infrared is not positive, however, as several false targets can interfere, including seaweed, shoreline, and oceanic fronts (Brown et al., 1998). Thus, the presence or absence of infrared spectral information about spilled oil has limited application in a forensic sense.

14.3.3 Ultraviolet

Ultraviolet sensors can be used to map sheens of oil, as oil slicks display high reflectivity of ultraviolet (UV) radiation even at thin layers (<0.1 μm). Overlaid UV and IR images are often used to produce a relative thickness map of oil spills. Although inexpensive, ultraviolet cameras are not often used in this process, however, as it is difficult to overlay camera images (Goodman, 1988). Data from infrared scanners and that derived from push-broom scanners can be easily superimposed to produce these IR/UV overlay maps. Ultraviolet data are also subject to many interferences or false images such as wind slicks, sun glints, and biogenic material. Since these interferences are often different than those for infrared sensing, combining IR and UV can provide a more positive indication of oil than using either technique alone, but does not offer high forensic proof.

14.3.4 Night Vision Cameras

The night vision camera has the ability to capture the image of the water's surface under low light levels such as would be expected on a clear, starlit night. Oil is not positively detected using this technology, but it might be possible to discern a contrast between oil and the surrounding water.

With new light-enhancement technology (low lux), video cameras can be operated even in darkness. Tests of a Generation III night vision camera show that this technology is capable of providing imagery in very dark night conditions (Brown et al., 2004a, 2005a). Earlier, low-light level cameras were used for this purpose, but they required at least moonlight to produce an image (O'Neil et al., 1983). Under very low-light level conditions, the Generation III night vision camera now in use has enough sensitivity to observe the ultraviolet laser pulses from the Scanning Laser Environmental Airborne Fluorosensor (SLEAF) on the water's surface. Furthermore, it is postulated that it should be possible to observe visible fluorescence induced by the SLEAF ultraviolet laser if oil is present on the surface. This would be observable as brighter spots on the surface.

In 1993, a less advanced unit (Generation II) was used to image oil in shallow manmade pools onto which oil was deposited (Hover and Plourde, 1994). Although a contrast was observed between the oil and water, it was suspected that some of the contrast was due to the shallow depth of the pools and might not be evident in deeper waters.

A night vision camera may provide substantiating imagery similar to cameras and other photographic devices. The application of this technology to oil spill detection has not been studied extensively.

14.4 Laser Fluorosensors

Laser fluorosensors are active sensors that take advantage of the fact that certain compounds in petroleum oils absorb UV light and become electronically excited. This excitation is rapidly removed through the process of fluorescence emission, primarily in the visible region of the spectrum. Since very few other compounds show this tendency, fluorescence is a strong indication of the presence of oil. Natural fluorescing substances, such as chlorophyll, fluoresce at sufficiently different wavelengths than oil to avoid confusion.

As different types of oil yield slightly different fluorescent intensities and spectral signatures, it is possible to differentiate between classes of oil under ideal conditions (Balick et al., 1997; Brown and Fingas, 2003; Brown et al., 1994a, 2001a, 2002a, b, 2003a, b, 2004a, b, 2005a; Hengstermann and Reuter, 1990; Pantani et al., 1995). Laser fluorosensors are the most important sensor in terms of the forensics of oil spill detection as they provide positive detection capability, a classification, and, most importantly, a spectrum, which is somewhat akin to a gas chromatogram (i.e., a chemical fingerprint).

Another phenomenon, known as Raman scattering, involves energy transfer between the incident light and the water molecules. When the incident UV light interacts with the water molecules, Raman scattering occurs. The water molecules absorb some of the energy as rotational-vibrational energy and scatter light at wavelengths that are the difference between the incident radiation and the vibration-rotational energy of the molecule. The Raman signal for water occurs at 344 nm when the incident wavelength is 308 nm (XeCl laser). The water Raman signal is useful for maintaining wavelength calibration of the fluorosensor in operation, but has also been used in a limited way to estimate oil thickness (thicknesses less than 10 μm depending on the oil type) because the strong absorption by oil on the surface will suppress the water Raman signal in proportion to thickness (Hoge and Swift, 1980; Piskozub et al., 1997). This is shown in Eq. (14-1).

$$\text{transmittance} = \text{EXP (thickness} \times \text{absorption coefficient)} \quad (14\text{-}1)$$

The point at which the Raman signal is entirely suppressed depends on the type of oil, since each oil has a different absorption coefficient. Details of the use of Raman scattering to measure oil slick thickness can be found in the early work of Hoge and Swift (1980) and the recent studies by Patsayeva et al. (2000).

The principle of fluorescence can also be used on a smaller scale. A handheld UV light has been developed to detect oil spills at night at short range (Fingas, 1982). The "Fraunhofer Line Discriminator" is another related instrument, which is essentially a passive fluorosensor using solar irradiance instead of laser light (O'Neil et al., 1983). This instrument was not very successful because of the limited discrimination and the low signal-to-noise ratio.

Laser fluorosensors have been developed over the past 30 years to allow for the exploration of petroleum resources and the airborne surveillance and monitoring of oil spills. The first airborne laser fluorosensors were flown in the early 1970s, including those in Ottawa reported in Measures and Bristow (1971). Another pioneering development was NASA's Airborne Oceanographic Lidar, details of which can be found on its website (NASA, 2001).

Laser fluorosensors are active sensors that provide their own source of excitation and can therefore operate equally well during full daylight conditions and at night. The high sensitivity of many of these instruments allows even small amounts of the fluorescent substances, e.g., petroleum oils, being investigated to be detected. Recently, the ability to deal with the large quantities of data collected by laser fluorosensors has improved dramatically. True real-time operating systems and the increased processing power provided by modern computer processors can produce usable data in real time on-board the aircraft. In the case of a response to an oil spill situation, fluorescence information can be rapidly transferred to personnel on the ground or at sea to aid in the effective interdiction of any polluter.

Most laser fluorosensors used for oil spill detection employ an ultraviolet laser emitting between 300 and 355 nm (Anderson, 1994; Barbini et al., 1991; Brown et al., 1996; Calleri and Bernardi, 1993; Castagnoli et al., 1986; Diebel et al., 1989; Geraci et al., 1993; Gruner et al., 1991; Koechler et al., 1992). These excitation wavelengths are a compromise in that they can excite all three classes of oil with reasonable efficiency. Shorter-wavelength lasers would excite lighter oils efficiently but are less efficient at exciting crude and heavy refined oils. Figure 14-1 shows the discrimination in spectra obtained using a fluorosensor targeting three fuels with nearly identical physical properties. Such discrimination is not the case with heavier oils.

There are several reasonably priced, commercially available ultraviolet lasers in the 300- to 355-nm region, including the XeCl excimer laser (308 nm), the nitrogen laser (337 nm), the XeF excimer laser (351 nm), and the frequency-tripled Nd:YAG (355 nm). With excitation in this wavelength region, there exists a spectrally broad fluorescent return due

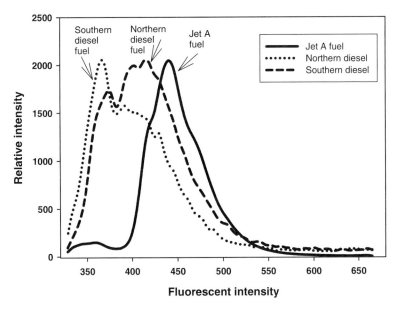

Figure 14-1 Spectra of three fuels with similar physical properties showing the spectral differences in them using a fluorosensor.

to organic matter, centered at 420 nm. This is known as Gelbstoff or yellow matter and must be accounted for. The signal due to Gelbstoff disappears when the oil layer is optically thick (10 to 20 μm). It can, however, be an interfering signal when attempting to detect thin films of light oils on water. Chlorophyll yields a sharp peak at 685 nm. Typically, crude oil fluorescence return is in the region of 400 to 550 nm, with the maximum centered in the 480-nm region.

Laser fluorosensors have significant potential for the remote sensing of petroleum oils because they can discriminate between oiled and unoiled weeds and detect oil in a variety of marine and terrestrial environments including on water, snow, ice, and beaches. Tests on shorelines show that this technique has been very successful (Dick et al., 1992). This coupled with their forensic capability extends the capability to shoreline and land as well as at sea.

Algorithms have been developed for the detection of oil on shorelines (James and Dick, 1996). Work has been conducted on detecting oil in the water column such as occurs with the product Orimulsion (Brown et al., 2002b, 2003a, b). The fluorosensor is also the only reliable means of detecting oil in certain ice and snow situations. Recent usage shows that the laser fluorosensor is a powerful tool for oil spill remote sensing for forensic purposes (Brown et al., 1997a).

Environment Canada's current system, the Scanning Laser Environmental Airborne Fluorosensor (SLEAF), was designed to detect, characterize, and map oil contamination in marine coastal and shoreline environments (Brown et al., 2000). Excitation is provided by a 100-mJ/pulse, 308-nm XeCl excimer laser (Lambda Physik, LPX140i) operating at rates of up to 400 Hz. Laser-induced fluorescence is detected with a spectrometric receiver consisting of a 20-cm-diameter, $f/3$ Newtonian telescope, with a 1×3-mrad field-of-view, a concave holographic grating, and a gated intensified diode-array detector (Princeton Instruments, 64 spectral channels, 330 to 610 nm). The detector is range-gated with digital lidar circuitry in order to collect only the laser-induced fluorescence, while rejecting most of the background solar radiation.

Two scanner heads (Optech Inc.) are used to provide a choice of narrow or wide swath coverage (1/6 or 1/3 of the operating altitude of 300 to 600m). Full spectral resolution georeferenced, fluorescence data are collected for each laser pulse and recorded directly to computer hard disk. Individual fluorescence spectra are analyzed in real time to determine the presence or lack of oil in the sensor field-of-view. Oil is classified using a principal component analysis (James and Dick, 1996) as light, medium, or heavy, and the extent of oil coverage in the field-of-view is estimated as clean, light, moderate, or heavy.

Oil contamination is displayed on the operator's display monitor and on hard-copy maps. Oil class is illustrated on the operator's display by color and with a text message on the hard-copy map. Oil coverage is represented by the length of a line perpendicular to the flight path of the aircraft. Since the amount of oil classification and coverage information is considerable, results are averaged over an area approximately 50m long by one-half of the swath width on either side of the aircraft. This averaging allows for a more concise and readable summary of oil contamination on the map display.

14.5 Microwave Sensors

14.5.1 Radiometers

Microwave radiometers detect the presence of an oil film on water by measuring an interference pattern excited by the radiation from free space. The apparent emissivity factor of water is 0.4 compared to 0.8 for oil (O'Neil et al., 1983; Ulaby et al., 1989). A passive device can detect this difference in emissivity and could therefore be used to detect oil. In addition, as the signal changes with thickness, in theory, the device could be used to measure thickness.

This detection method has not been very successful in the field, however, as several environmental and oil-specific parameters must be known. In addition, the signal return is dependent on oil thickness but in a cyclical fashion. A given signal strength can imply any one of two or three signal film thicknesses within a given slick. Microwave energy emission is greatest when the effective thickness of the oil equals an odd multiple of one-quarter of the wavelength of the observed energy. Biogenic materials also interfere, and the signal-to-noise ratio is low. In addition, it is difficult to achieve high spatial resolution (Goodman, 1994a).

The Swedish space agency has done some work with different systems, including a dual-band, 22.4- and 31-GHz device and a single-band 37-GHz device (Fäst, 1986). A two-channel device operating at 37.5 and 10.7GHz is described in Skou et al. (1994). Mussetto and coworkers described the tests of 44- to 94-GHz and 94- to 154-GHz, two-channel devices over oil slicks (Mussetto et al., 1994). They showed that correlation with slick thickness is poor and suggest that factors other than thickness also change surface brightness. They suggest that a single-channel device might be useful as an all-weather, relative-thickness instrument.

Tests of single-channel devices over oil slicks have also been described in the literature, specifically a 36-GHz (Zhifu and Wiesbeck, 1988) and a 90-GHz device (Süss et al., 1989). A new method of microwave radiometry has recently been developed in which the polarization contrasts at two orthogonal polarizations are measured in an attempt to measure oil slick thickness (Pelyushenko, 1995, 1997). A series of frequency-scanning radiometers have been built and appear to have overcome the difficulties with the cyclical behavior (McMahon et al., 1995, 1997).

In summary, passive microwave radiometers may have potential as all-weather oil sensors. Their potential as a reliable device for measuring slick thickness, however, is uncertain at this time, and thus their forensic application is limited.

14.5.2 Radar

Capillary waves on the ocean reflect radar energy, producing a "bright" image known as sea clutter. Since oil on the sea surface dampens some of these capillary waves, the presence of an oil slick can be detected as a "dark" sea or one with an absence of this sea clutter. Unfortunately, oil slicks are not the only phenomena that are detected in this way. There are many interferences or false targets, including fresh water slicks, wind slicks (calms), wave shadows behind land or structures, seaweed beds that calm the water just above them, glacial flour, biogenic oils, and whale and fish sperm (Frysinger et al., 1992; Hühnerfuss et al., 1989; Poitevin and Khaif, 1992). As a result, radar can be ineffective in locations such as Prince William Sound, Alaska, where dozens of islands, freshwater inflows, ice, and other features produce hundreds of such false targets. Its forensic capability is therefore low. Despite these limitations, radar is an important tool for oil spill remote sensing because it is the only sensor that can be used to search large areas and it is one of the few sensors that can "see" at night and through clouds or fog.

The two basic types of radar that can be used to detect oil spills and for environmental remote sensing in general are synthetic aperture radar (SAR) and side-looking airborne radar (SLAR). The latter is an older, but less expensive, technology, which uses a long antenna to achieve spatial resolution. Synthetic aperture radar uses the forward motion of the aircraft to synthesize a very long antenna, thereby achieving very good spatial resolution, which is independent of range, but with the disadvantage of requiring sophisticated electronic processing.

While inherently more expensive, the SAR has greater range and resolution than the SLAR. In fact, comparative tests show that SAR is vastly superior (Bartsch et al., 1987; Brown and Fingas, 2003b; Mastin et al., 1994). Search radar systems, such as those frequently used by the military, cannot be used for oil spills as they usually remove the clutter signal, which is the primary signal of interest. Furthermore, the signal processing of this type of radar is optimized to pinpoint small, hard objects, such as periscopes. This signal processing is very detrimental to oil spill detection.

SLAR has predominated oil spill remote sensing, primarily because of its lower price (Dyring and Fäst, 2004; Zielinski and Robbe, 2004). Operators recognize that SLAR is very susceptible to false hits, but solutions are not offered. SLAR is not typically used for forensic purposes, but rather is used to provide wide swath-width detection.

Experimental work on oil spills has shown that X-band radar yields better data than L- or C-band radar (C-CORE, 1981; Intera Technologies, 1984). It has also been shown that vertical antenna polarizations for both transmission and reception (V,V) yield better results than other configurations (Bartsh et al., 1987; Kozu et al., 1987; Macklin, 1992; Madsen et al., 1994). The ability of radar to detect oil is also limited by sea state. Sea states that are too low will not produce enough sea clutter in the surrounding sea to contrast to the oil, and very high seas will scatter radar sufficiently to block detection inside the troughs. Indications are that minimum wind speeds of 1.5 m/s (~3 knots) are required to allow detectability, and a maximum wind speed of 6 m/s (~12 knots) will again remove the effect (Hühnerfuss et al., 1996; Hielm, 1989). This limits the environmental window of application for using radar to detect oil slicks.

Gade et al. (1996) studied the difference between extensive systems from a space-borne mission and a helicopter-borne system. They found that at high winds, it was not possible to discriminate biogenic slicks from oil. At low wind speeds, it was found that images in the L-band showed discrimination. Under these conditions, the biogenic material showed greater damping behavior in the L-band. Okamoto et al. (1996) studied the use of ERS-1 using artificial oil (oleyl alcohol) and found that an image was detected at a wind speed of 11 m/s but not at 13.7 m/s.

Radar has also been used to measure currents and predict oil spill movements by observing frontal movements (Forget and Brochu, 1996). Work has shown that frontal currents and other features can be detected by SAR (Marmorino et al., 1997).

Shipborne radar has similar limitations and the additional handicap of low altitude, which restricts its range to 8 to 30 km, depending on the height of the antenna. Ship radars can be adjusted to reduce the effect of sea clutter de-enhancement. Shipborne radar successfully detected a surface slick in the Baltic Sea from 8 km away and during a trial off the coast of Canada at a maximum range of 17 km (Tennyson, 1985). During the *Prestige* spill, a Netherlands vessel successfully used this technique to guide a recovery vessel into slicks. The technique is, however, very limited by sea state and, in all cases where it was used, the presence and location of the slick were already known or suspected. Ship radars will provide little forensic evidence.

Gangeskar (2004) has proposed an automatic system that could be mounted on oil-drilling platforms. This system would use standard X-band ship navigation units and provide an alert if an oil spill is present. The system includes an extensive post-processing system to provide both a user-friendly GUI (graphical user interface) and an automatic detection and alert system. The system has not been fully tested to date.

In summary, radar optimized for oil spills is useful in oil spill remote sensing, particularly for searches of large areas and for nighttime or foul weather work. The technique is highly prone to false targets, however, and is limited to a narrow range of wind speeds. Radar therefore offers little in the way of forensic data.

14.5.3 Microwave Scatterometers

A microwave scatterometer measures the scattering of microwave or radar energy by a target. The presence of oil reduces the scattering of the microwave signals just as it does for radar sensors, however, and this device is adversely affected by the same large number of false targets. One radar scatterometer was flown over several oil slicks and used a low-power transmitter operating in the Ku band (13.3 GHz) (O'Neil et al., 1983). The scatterometer detected the oil, but discrimination was poor. The "Heliscat," a device with five frequencies, has been used to investigate capillary wave damping (Hühnerfuss et al., 1996).

The advantage of a microwave scatterometer is that it has an aerial coverage similar to optical sensors and it operates in a nadir geometry, i.e., it looks straight down. The main disadvantages include the lack of discrimination for oil and the lack of imaging capability. As with radar, there is little of relevance for forensic applications.

14.6 Determination of Slick Thickness

There are presently no reliable methods, either in the laboratory or in the field, for accurately measuring the thickness of an oil slick on water. There has long been a need to measure the thickness of oil slicks, both within the oil spill response community and among academics in the field. Knowledge of slick thickness would make it possible to determine the effectiveness of certain oil spill countermeasures including dispersant application and *in situ* burning. Indeed, the effectiveness of individual dispersants could be determined quantitatively if the oil remaining on the water surface after dispersant application could be accurately measured (Goodman and Fingas, 1988). The knowledge of oil spill thickness may also be useful in forensic applications, as the quantity of oil at sea is often a relevant issue in court. The quantity of oil can only be estimated if a reliable estimate or measurement of oil thickness is available.

14.6.1 Visual Thickness Indications

A very important tool for working with oil spills has been the relationship between appearance and thickness. A series of experiments were conducted in the 1930s, and charts produced then are still used today (Congress, 1930). It had already been recognized that slicks on water were consistent or nearly consistent in appearance. Only a few experiments have been done in recent years.

14.6.2 Theoretical Approaches

Horstein (1972, 1973) reviewed theoretical approaches and used interference phenomenon to correlate the threshold of rainbow colors to slick thickness. The appearance of the rainbow colors is the result of constructive and destructive interference of the lightwaves reflected from the air–oil interface with those reflected from the oil–water interface. The difference in optical path lengths for these two waves depends on the refractive index of the oil. The refractive index of a given wavelength results in a difference in optical path length. This difference can be given as

$$\Delta L = 2t \, (\mu^2 = \sin^2 i)^{1/2} \quad (14\text{-}2)$$

where

ΔL is the difference in optical path length
t is the film thickness
μ is the refractive index of the film
i is the angle of light incidence

Horstein points out that if ΔL contains a whole number of wavelengths, then maximum destructive interference will occur. If ΔL contains an odd number of half-wavelengths, then maximum constructive interference will occur.

Then the maximum destructive interferences occur at

$$\lambda = \Delta L / x \quad (14\text{-}3)$$

where λ is the wavelength under consideration, and x is a whole even number such as 2, 4, 6, etc. The maximum constructive interferences occur at

$$\lambda = 2\Delta L / x \quad (14\text{-}4)$$

where x is a whole odd number such as 1, 3, 5, 7, etc.

Tables of constructive and destructive wavelengths can be written. These then result in the following color chart for visible oil:

thickness less that 0.15 μm — no color apparent
thickness of 0.15 μm — warm tone apparent
thickness of 0.2 to 0.9 μm — variety of colors (e.g., rainbow)
thickness greater than 0.9 μm — colors of less purity, heading toward gray

Horstein calculated the differential reflectivity of oil and water. He calculated that, at an incidence angle of 30°, the reflectivity of oil is 0.041 and that of water is 0.021. At 60° oil shows a reflectivity of 0.09 and water of 0.06 and at 75° oil has a reflectivity of 0.25, and water that of 0.21. These angles are calculated as the angle of light incidence from the normal and thus show that reflectivity decreases as the angle of viewing becomes less vertical. The reflectivity may explain the visibility of very thin films of oil (less than shown by coloration) on the water surface. This calculation demonstrates that viewing angle is important and that the greatest contrast is seen from near-vertical angles.

14.6.3 Literature Review of Visual Indications of Oil Slick Thickness

Literature results are presented in chronological order and numerical values are summarized in Table 14-1. In 1930, scientists from the U.S. Navy and the National Bureau of Standards conducted both laboratory and field studies to examine the visibility and fate of oil slicks (Congress, 1930). Oil was spilled near Hawaii and off New York, and then followed until it was no longer visible. Weather, area, etc. were recorded. It was concluded that the minimum visible thickness of an oil slick is 0.1 μm.

Allen and Schlueter (1969, 1970) developed a slick thickness algorithm for the purpose of estimating the discharges from the Santa Barbara oil seeps. In 1969, the American Petroleum Institute compiled a visibility chart using much of the previous literature (API, 1969). This chart has been used in many subsequent documents. Hollinger and Mennella (1973) conducted a series of eight controlled oil spills off the coast of Virginia to investigate the use of microwave radiometry to delineate oil spills. It was noted that the sheens were typically 2- to 4-μm thick. It was found that 90% of the oil was in 10% of the slick area. Horstein (1972) studied the relationship of slick

Table 14-1 Relationships between Appearance and Oil-Slick Thickness

Author	Year	Oil	Type	Number	Height (m)	Viewing Angle	Visibility Thresholds (μm)						Other Information
							Minimum	Silvery	Rainbow	Darkening Colors	Dull Colors	Brown	
Congress	1930	Various incl. Bunker, fuel oil	Experiments	>15	Ship board	Oblique	0.1						
Allen and Schlueter	1969	Crude—Santa Barbara	Experiments	Multiple	ns	ns		0.05 to 0.18	0.23 to 0.75	1 to 2.5	2.5 to 5.5		Probably done at close proximity
API	1969	General	Literature		ns	ns	0.04	0.08	0.15 to 0.3	1	2		
Horstein	1972	Arabian and Louisiana crudes	Experiments	>20	1 to 2	Various	<0.15	up to 0.15	0.15 to 0.9	0.9 to 1.5	1.5 to 3		
Horstein, Parker, and Cormack	1973	Various	Lit & experiments		Ship & aerial	Various	0.038	0.076	0.15 to 0.31	1	2		
	1979	North Sea and Arabian crudes	Experiments	2	Ship & aerial	Various	0.1						
ITOPF	1981	General	Literature		Aerial	ns	0.1	0.1	0.3	0.1			
Schriel	1987	General	Lit & experiments		Aerial	Various	0.05	0.1	0.15	0.3	1	2	Combination of experiments and literature
Schriel	1987	General	Lit & experiments		Aerial	Various							Modified table
Duckworth	1993	Various crudes	Experiments	Several	ns	ns	0.1	0.1	0.3	1	5	15	Wave dampening threshold at 0.1 μm
Brown et al.	1995	Crude—Norman Wells	Experiments	32	30 m	Nadir	0.094						
		Diesel	Experiments	25	30 m	Nadir	0.165						
		Lubricating oil	Experiments	16	30 m	Nadir	0.077						
		Hydraulic oil	Experiments	13	30 m	Nadir	0.159						
Coast Guard	1996	General—tar codes	Literature			Nadir	0.04	0.075	0.15	0.3	1	3	
						Average	*0.09*	*0.1*	*0.59*	*0.91*	*2.7*	*8.5*	

ns = not specified

thickness and appearance theoretically and in the laboratory as well as from spills in the field. Horstein calculated that oil would show rainbow colors beginning at 200 nm (0.2 μm) on the basis of wave interference calculations. Onset of any color would be at 150 nm for the same reason.

In 1981, the International Tanker Owners Pollution Federation published a visibility guide based on existing literature (ITOPF, 1981). Schriel (1987) presented a table of thickness estimation based both on literature and on experiments conducted in the North Sea. MacDonald et al. (1993) used photography to define slicks in an area of the Gulf of Mexico, offshore from Louisiana. Work performed from the space shuttle, surface ships, aircraft, and a submarine confirmed oil in some of the slicks. In the case of the shuttle photography, pictures of the surface were best using the sun glint, i.e., when the sun was shining onto the surface creating the most glint. Unpublished thickness data, based on small-scale experiments and literature, were used to calculate the oil on the surface.

Brown et al. (1995, 1996) conducted 120 experiments in a large wave basin to measure the visibility of oil slicks. It was found that the detection ability decreased by over 50% for most oils and for the cameras when the angle was changed from 90° to 55° from the horizontal. Detectability degraded to 70% and sometimes to nil as the viewing angle was decreased past 55° through 35°. It is concluded that under optimal conditions, a sheen of about 0.15 μm can be detected visually. Sun angle is important, and it was found that the best angle is with the sun at right angles to the viewing plane. This degrades as viewing angle decreases and depends on oil type. The Canadian Coast Guard, as well as many other organizations around the world, has issued color charts for their operational personnel (Canadian Coast Guard, 1996). These data are based on many of the historical works described here.

On the basis of the literature results discussed in this section, it can be seen that visual or photographic evidence for slick thickness cannot be used outside the range of about 0.5 to 2 μm in the color regime.

14.6.4 Oil Slick-Thickness Relationships in Remote Sensors

Hollinger and Mennella (1973) conducted a series of eight controlled oil spills off the coast of Virginia to investigate the use of microwave radiometry to delineate oil spills. They used 19.4- and 69.8-GHz radiometers on the spills. Measurements using sorbents were used to calibrate the radiometer. It was noted that the sheens were typically 2- to 4-μm thick. It was found that 90% of the oil was in 10% of the slick area and that the microwave threshold was about 0.1 mm (100 μm).

A series of experiments were carried out in 1979 to evaluate infrared (IR) and side-looking airborne radar (SLAR) for oil spill detection (Parker and Cormack, 1979). The imagery was correlated against visual and sorbent measurements, which were used to derive a thickness estimate. It was concluded that the infrared threshold was between 25 and 50 μm, and 100 nm for SLAR. Manipulation of data showed that a mass balance could be achieved if the thickness at which the infrared showed oil to be colder at the sea occurred at 100 μm and at 1000 μm for the heated portion of the oil.

The United Kingdom conducted Isowake Experiments in 1982 (Hurford and Martinelli, 1982, 1984). On the basis of estimations and calculations, it was concluded that the lowest detectable slick thickness for IR was between 10 and 50 μm, whereas hot spots in the IR image could be as much as 1000 μm.

MacDonald et al. (1993) used photography from the space shuttle to define up to 124 slicks in an area of the Gulf of Mexico, offshore from Louisiana. Similarly, a thematic image from Landsat showed at least 66 slicks in one large area. Some of the thickness relationships were based on unpublished experimental data from Duckworth (1993).

Brown et al. (1995, 1996) conducted experiments to measure the visibility of oil slicks. The observers and an ultraviolet and visible camera were mounted in a crane basket 30 m over the slick. It was found that the detection ability decreased by over 50% for most oils and for the cameras when the angle was changed from 90° to 55° from the horizontal (equivalent incidence angle of 0° to 35°). Detectability degraded to 70% and sometimes to nil as the viewing angle was decreased past 55° through 35°. Brown et al. (1998) conducted several experiments to ascertain the relationship between thickness of slicks and the density (or intensity) of the infrared image. The thicknesses varied between 1 to 10 mm, and thicknesses were measured using an acoustic system. No relationship was found between slick thickness and infrared brightness.

The results of some of the experiments reviewed are summarized in Table 14-2.

It is readily apparent from the discussion in this section that most available oil spill remote sensors do not provide forensic-quality oil-thickness information. Furthermore, the thresholds for many of the sensors vary over two orders-of-magnitude.

14.6.5 Specific Oil-Thickness Sensors

The suppression of the water Raman peak in laser fluorosensor data discussed in Section 14.4 has not been fully exploited or tested. This technique may work for thin slicks, but not necessarily for thick ones, at least not with a single excitation frequency. Attempts have been made to calibrate the thickness appearance of infrared imagery, but also without success. It is suspected that the temperatures of the slick as seen in the IR are highly dependent on oil type, sun angle, and weather conditions. If so, it may not be possible to use IR as a calibrated tool for measuring thickness. As accurate ground-truth methods do not exist, it is very difficult to calibrate existing equipment. The use of sorbent techniques to measure surface thickness yields highly variable results (Goodman and Fingas, 1988). As noted in Section 14.5.1, the signal strength measured by microwave radiometers can imply one of several thicknesses. This methodology does not appear to have potential, other than for measuring relative oil thickness.

A variety of electrical, optical, and acoustic techniques for measuring oil thickness have been investigated (Goodman et al., 1997; Reimer and Rossiter, 1987). Two promising techniques were pursued in a series of laboratory measurements. In the first technique, known as "thermal mapping," a laser is used to heat a region of oil and the resultant temperature profiles created over a small region

Table 14-2 Relationship between Oil Thickness and Detection Limits in Remote Sensing Instruments

Author	Year	Oil	Number	Visible Minimum	In UV	Invisible Camera	In IR	In SLAR	In Microwave
Congress	1930	Various fuels & bunker	>15	0.1					
Horstein	1972	Arab & Louisiana crudes	>20	<0.15					
Hollinger	1973	Diesel, fuel oil, bunker C	8				Microwave threshold = 100 µm		
Parker and Cormack	1979	North Sea & Arab crudes	2	0.1			25–30	0.1	
						0.1			
Brown et al.	1995	Crude—Norman Wells	32	0.094	0.11	0.15			
		Diesel	25	0.165	0.11	0.15			
		Lubricating oil	16	0.077	0.11	0.15			
		Hydraulic oil	13	0.159	0.11	0.15			
Brown et al.	1998	Crude—Federated and Hondo	5–40				No relationship between IR brightness and thickness		
			Average	0.09	0.11	0.14	25–30	0.1	

near this heating are examined using an infrared camera (Aussel and Monchalin, 1989). The temperature profiles created are dependent on the oil thickness.

Thickness sensors are still in the developmental stages, and it would be premature to use their information for forensic purposes at this time.

A more promising technique involves laser acoustics (Brown et al., 1994; Choquet et al., 1993; Krapez and Cielo, 1992). The laser ultrasonic remote sensing of oil thickness (LURSOT) has been developed under contract to the Industrial Materials Institute (IMI) of the National Research Council of Canada in Boucherville, Quebec. Although complete details of the LURSOT system are beyond the scope of this chapter, a brief overview of the system is presented here.

The LURSOT sensor is a three-laser system with one laser coupled to an optical interferometer to accurately measure oil thickness (Brown et al., 1997b, 2001b, 2005b; Choquet et al., 1993). The measurement process is started by the absorption of a powerful infrared carbon-dioxide (CO_2, ~10 µm) laser pulse that creates a thermal pulse in the oil layer. Rapid thermal expansion of the oil occurs near the surface where the laser beam was absorbed. This leads to a step-like rise of the sample surface and creates an acoustic pulse of high frequency and large bandwidth (<15 MHz for oil). The acoustic pulse travels down through the oil layer to the oil–water interface, where it is largely transmitted (~85%) and partially reflected back (up to 15% depending on the oil) to the oil–air interface, where it causes another small displacement of the oil surface.

The amount of time required for the acoustic pulse to travel through the oil and back to the surface again is a function of the thickness of the oil layer and the acoustic velocity (speed of sound) of the oil. The displacement of the oil layer is measured by a second laser probe beam (Nd:YAG, 1064 µm) aimed at the surface. Motion of the surface causes a phase or frequency shift (Doppler shift) in the reflected probe beam. The modulation of the probe beam is subsequently demodulated with an optical interferometer, either a confocal Fabry–Perot interferometer or a photorefractive interferometer (Monchalin, 1986). The absolute oil slick thickness can be determined from the time of propagation of the acoustic wave between the upper and lower surfaces of the oil layer and the knowledge of the speed of sound in oil. The system uses a third laser (cw HeNe laser, 632.8 nm) to monitor the water surface and generate a trigger pulse when the correct surface geometry for measurement exists.

The initial attempt to test the LURSOT system in an airborne environment was not successful for a number of reasons (Brown et al., 1994a, b). In brief, the extreme operating environment provided by a moving platform was radically different from the laboratory setting in which the prototype was first developed. The broad range of temperatures encountered in the aircraft was found to induce optical beam-path differences in the probe laser (Nd:YAG) that were outside acceptable limits. To correct this, a novel laser system was developed and mounted on a zero thermal expansion carbon-epoxy optical breadboard. In addition, system alignment is conducted at the same temperature expected during flight conditions.

A second alignment problem with the probe laser was corrected by replacing certain mirror mounts with stiffer mounts. The intense vibrations experienced in the airborne environment initially led to the removal of the Fabry–Perot interferometer in favor of a photorefractive interferometer for probe-laser frequency demodulation. The photorefractive interferometer was developed by IMI to provide a demodulation device that is insensitive to vibrations. The photorefractive interferometer was, however, found to be sensitive to frequency changes caused by motion of the target relative to the airborne laser source. To compensate for these frequency (Doppler) shifts, an optical frequency compensation device (OFCD) was designed and tested. To use the OFCD on-board an aircraft, a measurement of vertical velocity was needed. This was fulfilled through the development of a vertical velocity sensor (VVS) by the Institute for Aerospace Research at the National Research Council of Canada in Ottawa. The VVS couples a differential global positioning system (DGPS) receiver with an accelerometer in order to provide real-time vertical velocity measurement. The VVS was tested on-board the aircraft and found to operate satisfactorily when isolated mechanically from the floor of the aircraft and filtered appropriately (Brown et al., 1997a, b). Although this system of compensation was encouraging, the inability to make this system operational in real time prevents its use at this time. It was decided to construct a device to measure the instantaneous optical frequency of the returning target laser beam. This device provides a diagnostic as to how well the optical frequency compensation device is working. Finally, a decision was made to revert to the Fabry–Perot system with enhanced vibrational and acoustic shielding.

There were also concerns about maintaining the colinearity of the three laser beams used in the LURSOT system during flight. A thorough theoretical analysis of the support structure that houses the optical and laser components was undertaken at the University of Toronto's Aerospace Institute in order to understand the effect of vibrations on the colinearity of the optical system. The investigation uncovered several shortcomings with the existing structure and provided the information required to design a structure that provides the stability required to ensure colinearity of the laser beams at the operating altitude of 300 ft.

A technique was developed to measure the in-flight colinearity of the three laser beams used in the LURSOT system. A mirror is used to image the three laser beams onto a thin plate of blackened aluminum. The laser beams in turn heat a spot on the aluminum plate and the infrared camera (sensitive in the thermal infrared) captures an image of these "hot" spots. A suitable delay is added so that each beam can be observed sequentially. During testing, a mirror is moved into position to divert the laser beams onto the aluminum plate. The positions of the three beams are subsequently determined through data analysis of the infrared camera images, and beam colinearity/overlap is confirmed. The redesigned LURSOT system was re-assembled at IMI and tested extensively in a laboratory environment to confirm functionality of individual components and the successful measurement of oil-slick thickness.

The LURSOT was mounted, tested, and certified for airborne operation in Environment Canada's DC-3 aircraft. The LURSOT then participated in a series of flights designed to produce the first airborne measurement of oil-slick thickness. Instead of constructing an expensive large-scale outdoor test tank operation for the LURSOT oil-slick measurement flights, a more economical approach was taken. Five industrial waste dumpsters, 8 ft wide by 20 ft long, were placed on the tarmac at the MacDonald-Cartier International Airport in Ottawa, Canada. The dumpsters were lined with two one-piece polyethylene bag liners (8-mil thick) and filled with water to a depth of approximately 6 in. The water in the test tanks was warmed with heaters (Sinking Tank Heaters, Allied Precision Industries Inc.) to prevent the water from freezing during cold spring evenings. The test tanks were covered with polyethylene covers to help retain heat while not being overflown. Waves were made in the test tanks with pieces of lumber 2 in. $\times$ 4 in. $\times$ 3 ft) suspended by nylon ropes into the water column. The pieces of lumber were weighted down with steel flat-bars screwed to the underside of the 2 $\times$ 4s. Using the nylon ropes, the lumber was lifted manually at regular intervals (~1 Hz) to create waves of the approximate period required in the tanks.

The oil used in the airborne slick measurement experiments was Alberta Sweet Mixed Blend (ASMB), a typical western Canadian crude oil. Oil was added to four of the five tanks to provide nominal slick thicknesses of 1, 3, 6, and 12 mm (based on area and volume of oil added). Physical and chemical properties of ASMB can be found on Environment Canada's Oil Properties Database on the Internet (Environment Canada, 2005).

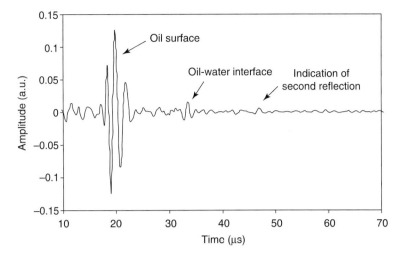

Figure 14-2 The return signals from the LURSOT thickness sensor showing a measurement of 9 mm from a light crude oil. Measurement was taken 61 m above ground with aircraft at 200 km/hr.

These flights were designed to confirm the operation of the triggering mechanism of the LURSOT system. Following the verification of acceptable airborne operation of the individual components of the LURSOT system, a final set of test flights were undertaken to acquire an airborne measurement of oil-slick thickness in the manmade test tanks. Numerous flight lines were flown over the oil-covered test tanks, and the measurement of oil-slick thickness from an airborne platform was successful. A resulting thickness measurement is shown in Figure 14-2.

14.7 Acoustic Systems

Pogorzelski (1995) has shown that acoustic means can be used to measure oil viscosities on the surface. A directional acoustic system employing high-frequency forward specular scattering was used in the laboratory and at sea. Signals scattered are related to the rheological film properties. It is not known at this time if the system is scalable or exactly what the limitations are.

14.8 Satellite Remote Sensing

Optical satellite remote sensing has been used several times to detect oil spills. The slick from the IXTOC I well blowout in Mexico was detected using GOES (Geostationary Operational Environmental Satellite) and by the AVHRR (Advanced Very High Resolution Radiometer) on the LANDSAT satellite (O'Neil et al., 1983). A blowout in the Persian Gulf was subsequently detected. The massive *Exxon Valdez* slick was detected on SPOT (Satellite pour l'Observation de la Terre) satellite data (Dean et al., 1990). Oiled ice in Gabarus Bay resulting from the *Kurdistan* spill was detected using LANDSAT data (Alfoldi and Prout, 1982; Dawe et al., 1981). Several workers were able to detect the Arabian Gulf War Spill in 1991 (Al-Ghunaim et al., 1992; Al-Hinai et al., 1993; Cross, 1992; Rand et al., 1992). The *Haven* spill near Italy was also monitored by satellite (Cecamore et al., 1992). A spill in the Barents Sea was tracked using an IR band on NOAA 10 (Voloshina and Sochnev, 1992). It is significant to note that, in all these cases, the position of the oil was known and data had to be processed to actually see the oil, which usually took several weeks.

Several problems are associated with relying on satellites for oil spill remote sensing. The first is the timing and frequency of overpasses (Clark, 1989) and the absolute need for clear skies to perform optical work.

The chances of the overpass and the clear skies occurring at the same time give a very low probability of seeing a spill on a satellite image. This point is well illustrated in the case of the *Exxon Valdez* spill (Noerager and Goodman, 1991). Although the spill covered vast amounts of ocean for over a month, there was only one clear day that coincided with a satellite overpass, and that was on April 7, 1989. Another disadvantage of satellite remote sensing is the difficulty in developing algorithms to highlight the oil slicks and the long time required to do so. With the *Exxon Valdez* spill, it took more than 2 months before the first group managed to "see" the oil slick in the satellite imagery, although its location was precisely known.

In its present state, optical satellite imagery does not offer much potential for oil spill remote sensing. Radar satellites, including ERS-1 and -2, Radarsat, and ENVISAT, are useful for detecting large offshore spills and for spotting anomalies (Brown et al., 2002c). Radarsat has been used for detecting oil seeps (Biegert et al., 1997) and smaller spills resulting from an oil barge (Werle et al., 1997), although the relative location of these smaller slicks was known before the detection. A novel application of Radarsat has been the study of oil lakes in the deserts of Kuwait (Kwarteng et al., 1997). Radar satellites are now used routinely by a number of nations to provide imagery for larger spills and to give indications of ship discharges. ERS-1 and -2 have been used for mapping of oil spills in the Caspian Sea (Ivanov and Ermoshkin, 2004). Fortuny et al. (2004) describe the use of ERS-2 and ENVISAT to provide imagery during the *Prestige* incident off the coast of Spain.

Several "automatic" systems have been designed for slick detection (Solberg and Theophilopoulos, 1997). Limited testing with ERS-1 has shown that many false signals are present in most locations (Bern et al., 1993; Wahl et al., 1993). Extensive efforts on data processing appear to improve the chances of oil detection (Yan and Clemente-Colon, 1997).

There are problems with resolution and timeliness with all satellite data. A comparison of the use of satellite- versus airborne-derived data showed that satellite data lack resolution and timeliness for many oil spill applications (Fingas and Brown, 1997). It is not clear at this time whether satellite systems can provide any information that is useful for forensic purposes.

14.9 Detection of Oil under Ice

The difficulties in detecting oil in or under ice are numerous (Fingas and Brown, 2002b; Bryce, 2000; Dickins, 2000). Ice is never a homogeneous material but rather it incorporates air, sediment, salt, and water, many of which may present false oil-in-ice signals to the detection mechanisms. In addition, snow on top of the ice or even incorporated into the ice adds further complications. During freeze-up and thaw in the spring, there may not be distinct layers of water and ice. There are many different types of ice and different ice crystalline orientations, which complicate the situation for any potential use in forensics.

The feasibility of various technologies for detecting oil in ice was extensively reviewed by Gill (1979). Based on this feasibility study, some of the technologies were tried on oil under ice in a test tank (Remotec Applications Ltd., 1981; Stapleton et al., 1981). This led to the pursuit of acoustic technologies, which were taken as far as the field testing of a prototype. Many of the other technologies have not been tried since. Much of the literature on the topic is now two decades old.

The acoustical properties of ice are variable (Fingas and Brown, 2002b). Experimenters found that standard acoustic units designed for metal and concrete inspection could be used for oil-in-ice detection (Remotec Applications Ltd., 1981; Stapleton et al., 1981). Initially, this was a surprise because the attenuation of ice and the source of the reflected signal for oil were not readily apparent from the data. Subsequent studies have shown that the physics of sound–oil interaction is relatively simple. There are two sources of signal from oil in or under ice. First, oil reflects the standard compressive (*p*) wave, and this signal is received by standard acoustic units just like the interfaces in metal or other building materials. But oil behaves acoustically like a non-Newtonian fluid and will also reflect the shear or (*s*) wave (also called the transverse wave). The *s*-waves travel at about half the velocity of the compressive waves and could be distinguished by their time delay. One could develop a more discriminating oil-in-ice detector by developing a unit that selectively detected shear waves. In theory, only sediment would propagate similar shear waves.

Jones and Kwan (1982, 1983) and Jones et al. (1984, 1986) developed a detection device consisting of a phased array detector that was capable of detecting shear waves directly and thus determining whether oil was present, with a high factor of reliability. Knudsen Engineering Ltd. (1984) studied the acoustic and electric properties of transducers and the coupling to ice. Goodman et al. (1985a) reviewed the technology and noted the differences in using low- and high-acoustic frequency, the latter yielding better spatial resolution, but with lower penetration capability.

Ice has variable transparency to radiowaves (Fingas and Brown, 2000c). Freshwater ice is relatively transparent, whereas saline first-year ice is highly attenuating. Low frequencies (less than 1 MHz) are best suited to the task of penetrating ice. An important facet of radiofrequency is the dielectric constants of oil, ice, and water. Oil has a dielectric constant of 2 to 3, snow of 1 to 2, and seawater of about 80 (Gill, 1979). Multiyear ice has a dielectric constant of about 3, and first-year ice of 3 to about 5. This differential in dielectric constants has led many theorists to predict that oil should be detectable in ice because of the phase reversals that should be apparent when a wave passes through a dielectric constant of 2 (oil) and immediately hits the seawater with a dielectric constant of 80. If the oil was not there, the dielectric constant would slowly change from that of ice (2 to 5) to that of seawater. This should produce a return due to the strong reflection caused by the dielectric change.

Four types of signal return might be used to detect the presence of oil under ice: (1) out-of-phase returns due to the low conductivity of oil; (2) large amplitude returns due to constructive interference effects; (3) spatial dependence of amplitude-of-return signals due to interference effects; and (4) conductivity differences (Goodman and Fingas, 1983).

Resonance scattering theory was proposed as a means of explaining the signals that might be achieved from plane dielectric layers of oil and ice (Jackins et al., 1982; Dean, 1983; Goodman et al., 1985b). Subsequent analysis by Tunaley (1985), Tunaley and Moorcroft (1986), and MacDougall and Tunaley (1986) showed this to be an inappropriate model. Moorcroft and Tunaley (1985) summarize this in calculations that show there would be essentially no electromagnetic resonance effect in sea ice at frequencies above about 0.2 GHz. This is because of the combined effects of absorption in the conducting sea ice and variations in its thickness. Additional effects are present that serve to eliminate resonances, including scattering of the electromagnetic wave by small-scale surface structure. As a result, there is no possibility of using resonances to detect the presence of oil under sea ice, confirming the findings from the tank experiments.

Several early workers proposed that the oil–ice boundary should be seen in impulse radar outputs (Gill, 1979). Tests during a field test in the Beaufort Sea showed anomalies in the output when the oil and gas were located under the ice (Butt et al., 1981). In another field test, an anomaly was observed in a field test where oil was present (Goobie et al., 1981). The interface was not seen, however, in subsequent tank tests (Remotec Applications Ltd., 1981).

The technology for detecting oil in or under ice is still evolving. Of the many potential technologies reviewed, only acoustic techniques have potential and have been successfully tested in the field. The potential radiofrequency techniques are still awaiting testing in the field or in test tanks. Their

14.10 Real-Time Displays and Printers

The production of data that can be quickly and directly used by operations people is a very important aspect of remote sensing. For forensic applications, the data must also be secure and time-location-stamped in order to be presentable. Real-time displays are important so that remote sensor operators can adjust instruments directly in flight and provide information quickly on the location or state of the spill.

A major concern of the client is that data be rapidly available (Goodman, 1994b). An additional concern is that the data from various sensors be available in a combined or fused form (Zielinski and Robbe, 2004). There is also a need to correct this data for aircraft motion and to annotate the data with time and position. At this time, existing hardware and software must be adapted, as commercial off-the-shelf equipment for directly outputting and printing sensor data is not yet available.

14.11 Future Trends

Advances in sensor technology will continue to influence the use of remote sensors as operational oil spill forensic tools in the future. In the next decade, advances in solid-state laser technology, in particular diode-pumped solid-state lasers, will greatly reduce the size and energy consumption of laser-based remote sensors. This will promote the use of these sensors in smaller, more economical aircraft that are within the budget of many more regulatory agencies and maritime countries. Rapidly improving computer capabilities will allow for true real-time processing. At the present time and for the foreseeable future, there is no single "magic bullet" sensor that will provide all the information required to detect, classify, and quantify oil in the marine and coastal environment for forensic purposes. The laser fluorosensor is currently the only sensor that provides significant forensic benefit.

Recommendations are based on the above considerations and include economy as a major factor. Table 14-3 shows the considerations related to the state of development, cost, and use of the sensor. Table 14-4 shows the suitability of the sensor to various types of missions. The laser fluorosensor offers the only potential for discriminating between oiled and unoiled weeds or shoreline and for positively identifying oil pollution on ice, among ice, and in a variety of other situations. This sensor provides the best potential for forensic use at this time. Most other sensors are experimental or do not offer good potential for forensic applications.

Table 14-3 Attributes for Sensor Selection

Sensor	State of Development	Amount of Experience in Use	Specific to Oil	Immunity to False Targets	Forensic Application	Acquisition Cost Range (k$)	Aircraft Physical Requirements
Still camera—Film	High	High	Poor	Poor	Documentation	0.25 to 5	No
Still camera—CCD	High	High	Poor	Poor	Documentation	1 to 20	No
Video	High	High	Poor	Poor	Documentation	1 to 10	No
IR camera (3 to 5 µm)	High	Medium	Poor	Poor	Documentation	4 to 20	No
IR camera (8 to 14 µm)	Medium	Medium	Medium	Medium	Documentation	20 to 200	No
UV camera	Medium	Medium	Poor	Poor	Documentation	4 to 20	No
Multi-spectral scanner	Medium	Medium	Poor	Poor	Documentation	100 to 300	Some
Radar	High	High	Medium	Poor	Little	1200 to 8000	Yes—Dedicated
Microwave radiometer	Medium	Medium	Medium	Medium	Very little	400 to 2000	Yes—Dedicated
Laser fluorosensor	Medium	Limited	Good	Good	Good	300 to 2000	Yes—Dedicated

Table 14-4 Sensor Suitability for Various Missions

Sensor	Support for Cleanup	Night and Fog Operation	Detection of Oil with Debris	Oiled Shoreline Survey	Spill Mapping	Ship Discharge Surveillance	Enforcement and Prosecution
Still camera—film	n/a	n/a	n/a	1	1	3	3
Still camera—CCD	2	n/a	1	2	2	2	2
Video	2	n/a	1	2	2	2	2
IR camera (3 to 5 µm)	3	2	1	n/a	3	2	2
IR camera (8 to 14 µm)	4	2	1	n/a	3	3	3
UV camera	2	n/a	n/a	n/a	3	2	1
UV/IR scanner	4	2	1	n/a	4	3	3
Multi-spectral scanner	1	n/a	n/a	1	2	1	1
Radar	n/a	4	n/a	n/a	4	3	2
Microwave radiometer	1	3	n/a	n/a	2	2	1
Laser fluorosensor	4	3	5	5	1	5	5

Key: n/a = not applicable; numerical values represent a scale from 1 = poorly suited to 5 = ideally suited.

References

Al-Ghunaim, I., M. Abuzar, and F.S. Al-Qurnas, Delineation and monitoring of oil spill in the Arabian Gulf using Landsat thematic mapper (TM) data. In *Proc. First Thematic Conf. Remote Sensing for Marine and Coastal Environ.* ERIM Conferences, Ann Arbor, MI, 1992, 1151–1160.

Al-Hinai, K.G., M.A. Khan, A.E. Dabbagh, and T.A. Bader, Analysis of Landsat thematic mapper data for mapping oil slick concentrations — Arabian Gulf oil spill 1991. *The Arabian J. Sci. Eng.*, 1993, **18**(2), 85–93.

Alfoldi, T.T. and N.A. Prout, The use of satellite data for monitoring oil spills in Canada Environment Canada Report EPS 3-EC-82-5, Ottawa, ON, 1982.

Allen, A.A. and R.S. Schlueter, Estimates of surface pollution resulting from submarine oil seeps at platform A and coal oil point. General Research Corp., prepared for Santa Barbara County, Santa Barbara, CA, 1969.

Allen, A.A. and R.S. Schlueter, Natural oil seepage at Coal Oil Point, Santa Barbara, California. *Science*, 1970, **170**, 974–977.

Anderson, S.G., Dual laser system provides real-time fluorescence images. *Laser Focus World*, 1994, **August**, 15.

API, *Manual on Disposal of Refinery Wastes, Volume on Liquid Wastes*, American Petroleum Institute, 1969.

Aussel, J.D. and J.-P. Monchalin, Laser-ultrasonic measurement of oil thickness on water from aircraft, feasibility study. Industrial Materials Research Institute Report, Boucherville, Quebec, Canada, 1989.

Bagheri, S., M. Stein, and C. Zetlin, Utility of airborne videography as an oil spill-response monitoring system. In *Encyclopedia of Environ. Control Tech.*, Houston, TX: Gulf Publishing Company, 1995, 367–376.

Balick, L., J.A. DiBenedetto, and S.S. Lutz, Fluorescence emission spectral measurements for the detection of oil on shore. In *Proc. Fourth Thematic Conf. Remote Sensing for Marine and Coastal Environ.* Ann Arbor, MI: Environmental Research Institute of Michigan, 1997, **I**, 13–20.

Barbini, R., R. Fantoni, A. Palucci, S. Ribezzo, and H.J.L. Van der Steen, Fluorosensor lidar for environmental diagnostics. In *Proc. Conf. — Quantum Electronics and Plasma Physics*, Bologna: SIF, 1991, **29**, 383–387.

Bartsch, N., K. Grüner, W. Keydel, and F. Witte, Contribution to oil spill detection and analysis with radar and microwave radiometer: Results of the Archimedes II campaign. *IEEE Trans. Geosci. Remote Sensing*, 1987, **GE.25**(6), 677–690.

Belore, R.C., A device for measuring oil slick thickness. *Spill Tech. Newsletter*, 1982, **7**(2), 44–47.

Bern, T-I., T. Wahl, T. Anderssen, and R. Olsen, Oil spill detection using satellite-based SAR: Experience from a field experiment. *Photogrammetric Eng. Remote Sensing*, 1993, **59**(3), 423–428.

Bianchi, R., R.M. Cavalli, C.M. Marino, S. Pignatti, and M. Poscolieri, Use of airborne hyperspectral images to assess the spatial distribution of oil spilled during the Trecate blow-out (Northern Italy). *SPIE*, 1995, **2585**, 352–362.

Biegert, E.K., R.N. Baker, J.L. Berry, S. Mott, and S. Scantland, Gulf offshore satellite applications project detects oil slicks using Radarsat. In *Proc. Intl. Symp. Geomatics in the Era of Radarsat*, Ottawa, Ontario, Canada, 1997.

Bolus, R.L., Airborne testing of a suite of remote sensors for oil spill detecting on water. In *Proc. Second Thematic Intl. Airborne Remote Sensing Conf. Exhibition*, Ann Arbor, MI: Environmental Research Institute of Michigan, 1996, III-743–III-752.

Brown, C.E. and M.F. Fingas, Review of the development of laser fluorosensors for oil spill application. *Mar. Poll. Bull.*, 2003, **47**, 477–484.

Brown, C.E., M. Fruhwirth, M.F. Fingas, R.H. Goodman, M. Choquet, R. Héon, G. Vaudreuil, J.-P. Monchalin, and C. Padioleau, Laser ultrasonic remote sensing of oil thickness: Absolute measurement of oil slick thickness. In *Proc. First Intl. Airborne Remote Sensing Conf. Exhibition*, Ann Arbor, MI: Environmental Research Institute of Michigan, 1994, **1**, I-567–I-578.

Brown, C.E., M. Fruhwirth, Z. Wang, P. Lambert, and M. Fingas, Airborne oil spill sensor test program. In *Proc. Second Thematic Conf. Remote Sensing for Mar. Coastal Environ.: Needs, Solutions and Applic.*, Ann Arbor, MI: ERIM Conferences, 1994a, I-19–33.

Brown, C.E., Z. Wang, M. Fruhwirth, and M. Fingas, May 1993 oil-spill sensor test program: Correlation of laser fluorosensor data with chemical analysis. In *Proc. Seventeenth Arctic and Marine Oilspill Tech. Sem.*, Ottawa, Ontario: Environment Canada, 1994b, 1239–1261.

Brown, C.E., M. Fruhwirth, R. Nelson, and M.F. Fingas, Real-time response with a laser fluorosensor: The integration of fluorescence data with down-looking video images and the prompt delivery of hard-copy maps. In *Proc. Second Intl. Airborne Remote Sensing Conf. Exhibition*, Ann Arbor, MI: ERIM Conferences, 1996, III-673–681.

Brown, C.E., R. Nelson, M.F. Fingas, and J.V. Mullin, Airborne laser fluorosensing: Overflights during lift operations of a sunken oil barge. In *Proc. Fourth Thematic Conf. Remote Sensing for Mar. Coastal Environ.*, Ann Arbor, MI: Environmental Research Institute of Michigan, 1997a, **I**, 23–30.

Brown, C.E., M.F. Fingas, R.H. Goodman, M. Choquet, A. Blouin, D. Drolet, J.-P. Monchalin, and C.D. Hardwick, The LURSOT sensor: Providing absolute measurements of oil slick thickness. In *Proc. Fourth Thematic Conf. Remote Sensing for Mar. Coastal Environ.*, Ann Arbor, MI: Environmental Research Institute of Michigan, 1997b, **I**, 393–397.

Brown, C.E., R. Marois, M.F. Fingas, and J.V. Mullin, Preliminary testing of the scanning laser environmental airborne fluorosensor. In *Proc. Twenty-Third Arctic and Mar. Oilspill Program Techn. Sem.*, Ottawa, ON: Environment Canada, 2000a, 519–523.

Brown, C.E., M.F. Fingas, and J. An, Laser fluorosensors: A survey of applications and developments of a versatile sensor. In *Proc. Twenty-Fourth Arctic and Mar. Oilspill Program Tech. Sem.*, Ottawa, ON: Environment Canada, 2001a, 485–493.

Brown, C.E., R. Marois, M.F. Fingas, M. Choquet, J-P. Monchalin, J. Mullin, and R. Goodman, Airborne oil spill sensor testing: Progress and recent developments. In *Proc. 2001 Intl. Oil Spill Conf.*, Washington, DC: American Petroleum Institute, 2001b, 917–921.

Brown, C.E., M.F. Fingas, R.M. Gamble, and A.E. Myslicki, The remote detection of submerged oil. In *Proc. Third R&D Forum on High-Density Oil Spill Response*. London: International Maritime Organization, 2002a, 46–54.

Brown, C.E., R. Marois, G. Myslicki, and M.F. Fingas, Initial studies on the remote detection of submerged orimulsion with a range-gated laser fluorosensor. In *Proc. Twenty-Fifth Arctic and Marine Oilspill Program Tech. Sem.*, Ottawa, ON: Environment Canada, 2002b, 773–783.

Brown, C.E., M.F. Fingas, and T.J. Lukowski, Airborne and space-borne synergies: The old dog teaches tricks to a new bird. In *Proc. Fifth Intl. Airborne Remote Sensing Conf. Exhibition*, Ann Arbor, MI: Veridien, 2002c, 8.

Brown, C.E., R. Marois, R.M. Gamble, and M.F. Fingas, Further studies on the remote detection of submerged orimulsion with a range-gated laser fluorosensor. In *Proc. Twenty-Sixth Arctic and Marine Oilspill Program Tech. Sem.*, Ottawa, ON: Environment Canada, 2003a, 279–286.

Brown, C.E., R. Marois, G. Myslicki, M.F. Fingas, and R. MacKay, Remote detection of submerged orimulsion with a range-gated laser fluorosensor. In *Proc. 2003 Intl. Oil Spill Conf.*, Washington, DC: American Petroleum Institute, 2003b, 779–784.

Brown, C.E., M.F. Fingas, and R. Marois, Oil spill remote sensing: Laser fluorosensor demonstration flights off the east coast of Canada. In *Proc. Twenty-Seventh Arctic and Marine Oilspill Program Tech. Sem.*, Ottawa, ON: Environment Canada, 2004a, 317–334.

Brown, C.E., M. Fingas, R. Marois, B. Fieldhouse, and R.L. Gamble, Remote sensing of water-in-oil emulsions: Initial laser fluorosensor studies. In *Proc. Twenty-Seventh Arctic and Marine Oilspill Program Tech. Sem.*, Ottawa, ON: Environment Canada, 2004b, 295–316.

Brown, C.E., M.F. Fingas, and R. Marois, Oil spill remote sensing flights in the coastal waters around newfoundland. In *Proc. Eighth Intl. Conf. Remote Sensing for Marine and Coastal Environ.*, Ann Arbor, MI, 2005a, 8.

Brown, C.E., M.F. Fingas, J-P. Monchalin, C. Neron, and C. Padioleau, Airborne oil slick thickness measurements: Realization of a dream. In *Proc. Eighth Intl. Conf. Remote Sensing for Marine and Coastal Environ.*, Ann Arbor, MI: Altarum, 2005b, 8.

Brown, H.M. and Goodman, R.H. In-situ burning of oil in ice leads. In *Proc. Ninth Annual Arctic and Marine Oilspill Program Tech. Sem.*, Ottawa, ON: Environment Canada, 1986, 245–256.

Brown, H.M., J.P. Bittner, and R.H. Goodman, Visibility limits of spilled oil sheens. Imperial Oil Internal Report, Calgary, AB, 1995.

Brown, H.M., J.P. Bittner, and R.H. Goodman, The limits of visibility of spilled oil sheens. In *Proc. Second Thematic Intl. Airborne Remote Sensing Conf. Exhibition*, Ann Arbor, MI: Environmental Research Institute of Michigan, 1996, III-327–III-334.

Brown, H.M., J.J. Baschuk, and R.H. Goodman, The limits of visibility of spilled oil sheens. In *Proc. Twenty-First Arctic and Marine Oilspill Program Tech. Sem.*, Ottawa, ON: Environment Canada, 1998, 805–810.

Bryce, P., Design considerations for Arctic subsea leak detection systems. In *Proc. Intl. Oil and Ice Workshop*, Anchorage, AK: ACS, 2000.

Butt, K., P. O'Reilly, and E. Reimer, A field evaluation of impulse radar for detecting oil in and under sea ice. *Oil and Gas Under Sea Ice Experiment*, Dome Petroleum, APOA Contract 169, 1981.

Calleri, F. and P.L. Bernardi, Airborne fluorescence lidar system. In *Proc. Intl. Symp. Operationalization of Remote Sensing*, Enschede: ITC, 1993, **17**, 87–97.

Canadian Coast Guard, Appearance and thickness of an oil slick. *Operations Manual*, Section 3, Annex C, Ottawa, ON, 1996.

Castagnoli, F., G. Cecchi, L. Pantani, I. Pippi, B. Radicati, and P. Mazzinghi, A fluorescence LIDAR for land and sea remote sensing. *SPIE*, **663**, *Laser Radar Techn. Applic.*, 1986, 212–216.

C-CORE (Centre for Cold Ocean Resources Engineering), Microwave systems for detecting oil slicks in ice-infested waters: Phase I — literature review and feasibility study, Environment Canada Report EPS 3-EC-81-3, Ottawa, ON, 1981, 353.

Cecamore, P., A. Ciappa, and V. Perusini, Monitoring the oil spill following the wreck of the tanker *HAVEN* in the Gulf of Genoa through satellite remote sensing techniques. In *Proc. First Thematic Conf. Remote Sensing for Marine and Coastal Environ.*, Ann Arbor, MI: ERIM Conferences, 1992, 183–189.

Choquet, M., R. Héon, G. Vaudreuil, J.-P. Monchalin, C. Padioleau, and R.H. Goodman, Remote thickness measurement of oil slicks on water by laser ultrasonics. In *Proc. 1993 Intl. Oil Spill Conf.*, Washington, DC: American Petroleum Institute, 1993, 531–536.

Clark, C.D., Satellite remote sensing for marine pollution investigations. *Mar. Poll. Bull.*, 1989, **26**(7), 92–96.

Congress, Report on oil-pollution experiments — behaviour of fuel oil on the surface of the sea. Hearings before the Committee on River and Harbors, 71st Congress, 2nd Session, H.R. 10625, Part I, 41–9, Washington, DC, May 2, 3, and 26, 1930.

Cross, A., Monitoring marine oil pollution using AVHRR data: Observations off the Coast of Kuwait and Saudi Arabia during January 1991. *Intl. J. Remote Sensing*, 1992, **13**, 781–788.

Dawe, B.R., S.K. Parashar, T.P. Ryan, and R.O. Worsfold, The use of satellite imagery for tracking the *Kurdistan* oil spill. Environment Canada Report EPS 4-EC-81-6, Ottawa, ON, 1981, 31.

Dean, A.M., Investigating the practical applications of the resonant scattering theory for the detection of oil under sea ice. In *Proc. Sixth Annual Arctic Marine Oilspill Program Tech. Sem.*, Ottawa, ON: Environment Canada, 1983, 235–260.

Dean, K.G., W.J. Stringer, J.E. Groves, K. Ahlinas, and T.C. Royer, The *Exxon Valdez* oil spill: Satellite analyses. In *Oil Spills: Management and Legislative Implications*, M.L. Spaulding and M.

Reed (eds.), New York: American Society of Civil Engineers, 1990, 492–502.

Dick, R., M. Fruhwirth, M.F. Fingas, and C.E. Brown, Laser fluorosensor work in Canada. In *Proc. First Thematic Conf.: Remote Sensing for Marine and Coastal Environ.*, Ann Arbor, MI: Environmental Research Institute of Michigan, 1992, 223–236.

Dickins, D.F., Detection and tracking of oil under ice. In *U.S. Minerals Mgt. Report*, Herndon, VA, 2000.

Diebel, D., T. Hengstermann, R. Reuter, and R. Willkomm, Laser fluorosensing of mineral oil spirits. In *The Remote Sensing of Oil Slicks*, A.E. Lodge (ed.), Chichester, UK: John Wiley and Sons, 1989, 127–142.

Duckworth, R., unpublished data report in MacDonald et al., 1993.

Dyring, A. and O. Fäst, MSS puts the aircraft in the oil spill tracking network. In *Proc. Interspill*, Trondheim, Norway, 2004.

Environment Canada, Oil properties database, http://www.etc-cte.ec.gc.ca/databases/OilProperties/Default.aspx, 2005.

Fäst, O., Remote sensing of oil on water — air and space-borne systems. In *Proc. DOOS Sem.*, SINTEF, Trondheim, Norway, 1986.

Fingas, M.F., A simple night time oil slick detector. *Spill Tech. Newsletter*, 1982, 7(1), 137–141.

Fingas, M.F. and C.E. Brown, Airborne oil spill remote sensors — do they have a future? In *Proc. Third Intl. Airborne Remote Sensing Conf. Exhibition*, Ann Arbor, MI: Environmental Research Institute of Michigan, 1997, 715–722.

Fingas, M.F. and C.E. Brown, Review of oil spill remote sensing. In *Proc. Fifth Intl. Conf. Remote Sensing for Marine and Coastal Environ.*, Ann Arbor, MI: Environmental Research Institute of Michigan (ERIM), 2000a, I211–218.

Fingas, M.F. and C.E. Brown, Review of oil spill remote sensing. In *Proc. SPILLCON 2000*, Australian Marine Safety Authority, Sydney, Australia, www.meetingplanners.com.au/spillcon/, 2000b.

Fingas, M.F. and C.E. Brown, A review of the status of advanced technologies for the detection of oil in and with ice. *Spill Sci. Tech. Bull.*, 2000c, 6(5/6), 295–302.

Fingas, M.F. and C.E. Brown, Review of oil spill remote sensors. In *Proc. Seventh Intl. Conf. Remote Sensing for Marine and Coastal Environ.*, Ann Arbor, MI: Veridien, 2002a, 9.

Fingas, M.F. and C.E. Brown, Detection of oil in and under ice. In *Proc. Twenty-Fifth Arctic and Marine Oilspill Program Tech. Sem.*, Ottawa, ON: Environment Canada, 2002b, 199–214.

Fingas, M.F. and C.E. Brown, An update on oil spill remote sensors. In *Proc. Twenty-Eighth Arctic and Marine Oilspill Program Tech. Sem.*, Ottawa, ON: Environment Canada, 2005, 825–860.

Fingas, M.F., C.E. Brown, and L. Gamble, The visibility and detectability of oil slicks and oil discharges on water. In *Proc. Twenty-Second Arctic and Marine Oilspill Program Tech. Sem.*, Ottawa, ON: Environment Canada, 1999, 865–886.

Forget, P. and P. Brochu, Slicks, waves and fronts observed in sea coastal area by an X-band airborne synthetic aperture radar. *Remote Sensing of the Environ.*, 1996, 57, 1–12.

Fortuny, J., D. Tarchi, G. Ferraro, and A. Sieber, The use of satellite radar imagery in the prestige accident. In *Proc. Interspill*, Trondheim, Norway, 2004.

Frysinger, G.S., W.E. Asher, G.M. Korenowski, W.R. Barger, M.A. Klusty, N.M. Frew, and R.K. Nelson, Study of ocean slicks by nonlinear laser processes in second-harmonic generation. *J. Geophys. Res.*, 1992, 97(C4), 5253–5269.

Gade, M., W. Alpers, H. Hüehnerfuss, and V. Wismann, Radar signatures of different oceanic surface films measured during the SIR-C-X-SAR missions. *Remote Sensing (96)*, 16th Symp. Eur. Associ. of Remote Sensing Lab. (EARsel), Rotterdam, Holland, 1996, 233–240.

Gangeskar, R., Automatic oil-spill detection by marine X-band radars. *Sea Tech.*, 2004, **Aug**, 40–45.

Geraci, A.L., F. Landolina, L. Pantani, and G. Cecchi, Laser and infrared techniques for water pollution control. In *Proc. 1993 Oil Spill Conf.*, Washington, DC: American Petroleum Institute, 1993, 525–529.

Gill, R., Feasibility of surface detection of oil under ice. Environ. Protection Service Report EPS 3-EC-79-11, Environment Canada, Ottawa, ON, 1979.

Goobie, G.I., T.W. Laidley, and E.M. Reimer, C-CORE oil spill research activities. In *Proc. Fourth Annual Arctic Marine Oilspill Program Tech. Sem.*, Ottawa, ON: Environment Canada, 1981, 623–643.

Goodman, R.H., Simple remote sensing system for the detection of oil on water. Environmental Studies Research Fund Report Number 98, Ottawa, ON, 1988, 31.

Goodman, R.H., Application of the technology in North America. In *The Remote Sensing of Oil Slicks*, A.E. Lodge (ed.), Chichester, UK: John Wiley and Sons, 1989, 39–65.

Goodman, R.H., Remote sensing resolution and oil slick inhomogeneities. In *Proc. Second Thematic Conf. Remote Sensing for Marine and Coastal Environ.: Needs, Solutions and Applica.*, Ann Arbor, MI: ERIM Conferences, 1994a, I-1–17.

Goodman, R.H., Overview and future trends in oil spill remote sensing. *Spill Sci. Tech.*, 1994b, **1**(1), 11–21.

Goodman, R.H. and M.F. Fingas, Detection of oil-under-ice — a joint Esso/EPS project. In *Proc. Sixth Annual Arctic Marine Oilspill Program Tech. Sem.*, Ottawa, ON: Environment Canada, 1983, 207–214.

Goodman, R.H. and M.F. Fingas, The use of remote sensing for the determination of dispersant effectiveness. In *Proc. Eleventh Arctic and Marine Oilspill Program Tech. Sem.*, Ottawa, ON: Environment Canada, 1988, 377–384.

Goodman, R.H., H. Jones, and M.F. Fingas, The detection of oil under ice using acoustics. In *Proc. Conf. Port and Ocean Eng. under Arctic Conditions*, 1985a, 903–916.

Goodman, R.H., A. Dean, and M.F. Fingas, The detection of oil under ice using electromagnetic radiation. In *Proc. Conf. Port and Ocean Eng. under Arctic Conditions*, 1985b, 895–902.

Goodman, R., H. Brown, and J. Bittner, The measurement of thickness of oil on water. In *Proc. Fourth Thematic Conf. Remote Sensing for Marine and Coastal Environ.*, Ann Arbor, MI: Environmental Research Institute of Michigan, 1997, **I**, 31–41.

Gruner, K., R. Reuter, and H. Smid, A new sensor system for airborne measurements of maritime pollution and of hydrographic parameters. *GeoJournal*, 1991, **24**(1), 103–117.

Hengstermann, T. and R. Reuter, Lidar fluorosensing of mineral oil spills on the sea surface. *Appl. Opt.*, 1990, **29**, 3218–3227.

Hielm, J.H., NIFO comparative trials. In *The Remote Sensing of Oil Slicks*, A.E. Lodge (ed.), Chichester, UK: John Wiley and Sons, 1989, 67–75.

Hoge, F.E. and R.N. Swift, Oil film thickness measurement using airborne laser-induced water Raman backscatter. *Appl. Opt.*, 1980, **19**(19), 3269–3281.

Hollinger, J.P. and R.A. Mennella, Oil spills: Measurements of their distributions and volumes by multifrequency microwave radiometry. *Science*, 1973, **181**, 54–56.

Horstein, B., The appearance and visibility of thin oil films on water. Environmental Protection Agency Report, EPA-R2-72-039, Cincinnati, OH, 1972.

Horstein, B., The visibility of oil-water discharges. In *Proc. 1973 Intl. Oil Spill Conf.*, Washington, DC: American Petroleum Institute, 1973, 91–99.

Hover, G.L., Testing of infrared sensors for U.S. Coast Guard oil spill response applications. In *Proc. Second Thematic Conf. Remote Sensing for Marine and Coastal Environ.: Needs, Solutions and Applications*, ERIM Conferences, Ann Arbor, MI, 1994, I-47–58.

Hover, G.L. and J.V. Plourde, Evaluation of night capable sensors for the detection of oil on water. USCG Report Number CG-D-009-94, available as NTIS Report Number ADA281728, 1994.

Hühnerfuss, H., W. Alpers, and F. Witte, Layers of different thicknesses in mineral oil spills detected by grey level textures of real aperture radar images. *Intol. J. Remote Sensing*, 1989, **10**, 1093–1099.

Hühnerfuss, H., W. Alpers, H. Dannhauer, M. Gade, P.A. Lange, V. Neumann, and V. Wismann, Natural and man-made sea slicks in the North Sea investigated by a helicopter-borne 5-frequency radar scatterometer. *Int. J. Remote Sensing*, 1996, **17**(8), 1567–1582.

Hurford, N., Review of remote sensing technology. In *The Remote Sensing of Oil Slicks*, A.E. Lodge (ed.), Chichester, UK: John Wiley and Sons, 1989, 7–16.

Hurford, N. and F.N. Martinelli, Use of an infrared line scanner and a side-looking airborne radar to detect oil discharges from ships. Warren Spring Laboratory Report, Stevenage, UK, 1982.

Hurford, N. and F.N. Martinelli, Use of an infrared line scanner and a side-looking airborne radar to detect oil discharges from ships. In *Remote Sensing for the Control of Marine Poll.*, J. M. Massin (ed.), New York: Plenum Press, 1984, 405–421.

Intera Technologies, Radar surveillance in support of the 1983 COATTF oil spill trials. Environment Canada Manuscript Report EE-51, Ottawa, ON, 1984, 48.

ITOPF, Aerial observation of oil at sea. Intl. Tanker Owners Poll. Fed., London, 1981.

Ivanov, A.Y. and I.S. Ermoshkin, Mapping of oil spills in the Caspian Sea using the ERS-1.ERS-2

SAR image quick-looks and GIS. In *Proc. Interspill*, Trondheim, Norway, 2004.

Jackins, P.D., G.C. Gaunaurd, and C.D. McKindra, Radar resonance reflection from sets of plane dielectric layers. In *Proc. Fifth Annual Arctic Marine Oilspill Program Tech. Sem.*, Ottawa, ON: Environment Canada, 1982, 365–390.

James, R.T.B. and R. Dick, Design of algorithms for the real-time airborne detection of littoral oilspills by laser-induced fluorescence. In *Proc. Nineteenth Arctic and Marine Oilspill Tech. Sem.*, Ottawa, ON: Environment Canada, 1996, 1599–1608.

Jones, H.W. and H.W. Kwan, The detection of oil spills under arctic ice by ultrasound. In *Proc. Fifth Annual Arctic Marine Oilspill Program Tech. Sem.*, Ottawa, ON: Environment Canada, 1982, 391–411.

Jones, H.W. and H.W. Kwan, The detection of oil spills under seawater in the Arctic Ocean. In *Proc. Sixth Annual Arctic Marine Oilspill Program Tech. Sem.*, Ottawa, ON: Environment Canada, 1983, 241–252.

Jones, H.W., H.W. Kwan, and E.M. Yeatman, On the design of an apparatus to detect oil trapped under sea ice. In *Proc. Seventh Annual Arctic Marine Oilspill Program Tech. Sem.*, Ottawa, ON: Environment Canada, 1984, 295–305.

Jones, H.W., H.W. Kwan, T. Hayman, and E.M. Yeatman, The detection of oil under ice by ultrasound using multiple element phased arrays. In *Proc. Ninth Annual Arctic Marine Oilspill Program Tech. Sem.*, Ottawa, ON: Environment Canada, 1986, 475–484.

Kennicutt, M.C., I.R. MacDonald, T. Rogne, C. Giammona, and R. Englehardt, The Tenyo Maru oil spill: A multi-spectral and sea truth experiment. In *Proc. 1992 Arctic and Marine Oilspill Tech. Sem.*, Ottawa, ON: Environment Canada, 1992, 349–356.

Knudsen Engineering Ltd., Experiments in the detection of oil under ice. Environment Canada Manuscript Report, Ottawa, ON, 1984.

Koechler, C., J. Verdebout, G. Bertolini, A. Gallotti, E. Zanzottera, G. Cavalcabo, and L. Fiorina, Determination of aquatic parameters by a time-resolved laser fluorosensor operating from a helicopter. *SPIE*, **1714**, *Lidar for Remote Sensing*, 1992, 93–107.

Kozu, T., T. Umehara, T. Ojima, T. Suitsu, H. Masuyko, and H. Inomata, Observation of oil slicks on the ocean by X-Band SLAR. In *Proc. IGARSS '87 Sym.*, Ann Arbor, MI, 1987, 735–740.

Krapez, J.C. and P. Cielo, Optothermal evaluation of oil film thickness. *J. Appl. Phys.*, 1992, **72**(4), 1255–1261.

Kwarteng, A.Y., V. Singhroy, R. Saint-Jean, and D. Al-Ajmi, Radarsat SAR data assessment of the oil lakes in the Greater Burgan Oil Field, Kuwait. In *Proc. Intl. Symp.: Geomatics in the Era of Radarsat*, Ottawa, ON, 1997.

Lehr, W.J., Oil spill monitoring using a field microcomputer-GPS receiver combination. In *Proc. Second Thematic Conf. Remote Sensing for Marine and Coastal Environ.: Needs, Solutions and Applications*, Ann Arbor, MI: ERIM Conferences, 1994, I-435–439.

MacDonald, I.R., N.L. Guinasso, Jr., S.G. Ackleson, J.F. Amos, R. Duckworth, R. Sassen, and J.M. Brooks, Natural oil slicks in the Gulf of Mexico visible from space. *J. Geophys. Res.*, 1993, **98**(C9), 16351–16364.

MacDougall, J.W. and J.K.E. Tunaley, The complex permittivity of crude oil. In *Proc. Ninth Annual Arctic Marine Oilspill Program Tech. Sem.*, Ottawa, ON: Environment Canada, 1986, 413–420.

Macklin, J.T., The imaging of oil slicks by synthetic aperture radar. *GEC J. Res.*, 1992, **10**(1), 19–28.

Madsen, S., N. Skou, and B.M. Sorensen, Comparison of VV and HH polarized SLAR for detection of oil on the sea surface. In *Proc. Second Thematic Conf. Remote Sensing for Marine and Coastal Environ.: Needs, Solutions and Applications*, Ann Arbor, MI: ERIM Conferences, 1994, I-498–503.

Marmorino, G.O., D.R. Thompson, H.C. Graber, and C.L. Trump, Correlation of oceanographic signatures appearing in synthetic aperture radar and interferometric synthetic aperture radar imagery with in-situ measurements. *J. Geophys. Res.*, 1997, **102**(C8), 18723–18736.

Mastin, G.A., J.J. Mason, J.D. Bradley, R.M. Axline, and G.L. Hover, A comparative evaluation of SAR and SLAR. In *Proc. Second Thematic Conf. Remote Sensing for Marine and Coastal Environ.: Needs, Solutions and Applications*, Ann Arbor, MI: ERIM Conferences, 1994, I-7–17.

McMahon, O.B., E.R. Brown, G.D. Daniels, T.J. Murphy, and G.L. Hover, Oil thickness detection using wideband radiometry. In *Proc. Intl. Oil Spill Conf.*, Washington, DC: American Petroleum Institute, 1995, 15–20.

McMahon, O.B., T.J. Murphy, and E.R. Brown, Remote measurement of oil spill thickness. In *Proc. Fourth Thematic Conf. Remote Sensing for Marine and Coastal Environ.*, Ann Arbor, MI: Environmental Research Institute of Michigan, 1997, **1**, 353–360.

Measures, R.M. and M. Bristow, The development of a laser fluorosensor for remote environmental probing. *Can. Aeron. Space J.*, 1971, **17**, 421–422.

Monchalin, J.P., Optical detection of ultrasound. *IEEE Transa. Ultrasonics, Ferroelectrics and Frequency Control*, 1986, **UFFC-33**(5), 485–499.

Moorcroft, D.R. and J.K.E. Tunaley, Electromagnetic resonance in layers of sea ice over sea water. In *Proc. Eighth Annual Arctic Marine Oilspill Program Tech. Sem.*, Ottawa, ON: Environment Canada, 1985, 269–277.

Mussetto, M.S., L. Yujiri, D.P. Dixon, B.I. Hauss, and C.D. Eberhard, Passive millimeter wave radiometric sensing of oil spills. In *Proc. Second Thematic Conf. Remote Sensing for Marine and Coastal Environ.: Needs, Solutions and Applications*, Ann Arbor, MI: ERIM Conferences, 1994, I-35–46.

NASA, Airborne oceanographic lidar. http://aol.wff.nasa.gov/html/aoldes.html, 2001.

Neville, R.A., V. Thompson, K. Dagg, and R.A. O'Neil, An analysis of multispectral line scanner imagery from two test spills. In *Proc. First Workshop Sponsored by Working Group I of the Pilot Study on the Use of Remote Sensing for the Control of Marine Poll.*, NATO Challenges of Modern Society, 1979, **6**, 201–215.

Noerager, J.A. and R.H. Goodman, Oil tracking, containment and recovery during the *Exxon Valdez* response. In *Proc. 1991 Oil Spill Conf.*, Washington, DC: American Petroleum Institute, 1991, 193–203.

O'Neil, R.A., R.A. Neville, and V. Thompson, The Arctic marine oilspill program (AMOP) remote sensing study. Environment Canada Report EPS 4-EC-83-3, Ottawa, ON, 1983, 257.

Okamoto, K., T. Kobayashi, H. Masuko, S. Ochiai, H. Horie, H. Kumagai, K. Nakamua, and M. Shimada, Results of experiments using synthetic aperture radar onboard the European remote sensing satellite 1–4. Artificial oil pollution detection. *J. Comm. Res. Lab.*, 1996, **43**(3), 327–344.

Palmer, D., G.A. Borstad, and S.R. Boxall, Airborne multi spectral remote sensing of the January 1993 Shetlands oil spill. In *Proc. Second Thematic Conf. Remote Sensing for Marine and Coastal Environ.: Needs, Solutions and Applications*, Ann Arbor, MI: ERIM Conferences, 1994, II-546–558.

Pantani, L., G. Cecchi, and M. Bazzani, Remote sensing of marine environments with the high spectral resolution fluorosensor, FLIDAR 3. *SPIE*, 1995, **2586**, 56–64.

Parker, H.D. and D. Cormack, Evaluation of infrared line scan (IRLS) and side-looking airborne radar (SLAR) over controlled oil spills in the North Sea. Warren Spring Laboratory Report, 1979.

Patsayeva, S., V. Yuhakov, V. Varlamov, R. Barbini, R. Fantoni, C. Frassanito, and A. Palucci, Laser spectroscopy of mineral oils on the water surface. In *Proc. Fourth EARSeL Workshop Lidar Remote Sensing of Land and Sea*, http://las.physik.uni-oldenburg.de/projekte/earsel/4th_workshop.html#proceedings, 2000.

Pelyushenko, S.A., Microwave radiometer system for the detection of oil slicks. *Spill Sci. Tech. Bull.*, 1995, **2**(4), 249–254.

Pelyushenko, S.A., The use of microwave radiometer scanning system for detecting and identification of oil spills. In *Proc. Fourth Thematic Conf. Remote Sensing for Marine and Coastal Environ.*, Ann Arbor, MI: Environmental Research Institute of Michigan, 1997, **I**, 381–385.

Piskozub, J., V. Drozdowska, and V. Varlamov, A lidar system for remote measurement of oil film thickness on sea surface. In *Proc. Fourth Thematic Conf. Remote Sensing for Marine and Coastal Environ.*, Ann Arbor, MI: Environmental Research Institute of Michigan, 1997, **1**, 386–391.

Pogorzelski, S.J., Ultrasound scattering for characterization of marine crude oil spills. In *Encyclopedia of Environmental Control Technology*, Houston, TX: Gulf Publishing Company, 1995, 485–537.

Poitevin, J. and C. Khaif, A numerical study of the backscattered radar power in presence of oil slicks on the sea surface. In *Proc. First Thematic Conf. Remote Sensing for Marine and Coastal Environ.*, Ann Arbor, MI: ERIM Conferences, 1992, 171–182.

Rand, R.S., D.A. Davis, M.B. Satterwhite, and J.E. Anderson, Methods of monitoring the Persian Gulf oil spill using digital and hardcopy multiband data. U.S. Army Corps of Engineers Report, TEC-0014, 1992, 33.

Reimer, E.R. and J.R. Rossiter, Measurement of oil thickness on water from aircraft; A: Active microwave spectroscopy; B: Electromagnetic thermoelastic emission. Environmental Studies Revolving Fund Report Number 078, Ottawa, ON, 1987.

Remotec Applications Ltd., Laboratory experiments in the detection of oil under ice. Environment Canada Manuscript Report EE-26, Ottawa, ON, 1981.

Rogne, T.J. and A.M. Smith, *Tenyo Maru* oil spill remote sensing data analysis. Washington, DC: Marine Spill Response Corporation, MSRC Technical Report Series 92-003, 1992, 97.

Rogne, T., I. Macdonald, A. Smith, M.C. Kennicutt, and C. Giammona, Multi-spectral remote sensing and truth data from the *Tenyo Maru* oil spill. In *Proc. First Thematic Conf. Remote Sensing for Marine and Coastal Environ.*, Ann Arbor, MI: ERIM Conferences, 1992, 37–48. (Also published in *Photogrammetric Engineering and Remote Sensing*, 1993, **59**(3), 391–397.)

Salisbury, J.W., D.M. D'Aria, and F.F. Sabins, Thermal infrared remote sensing of crude oil slicks. *Remote Sensing in the Environ.*, 1993, **45**, 225–231.

Schriel, R.C., Operational air surveillance and experiences in the Netherlands. In *Proc. 1987 Intl. Oil Spill Conf.*, Washington, DC: American Petroleum Institute, 1987, 129–136.

Seakem Oceanography, Remote sensing chronic oil discharges. Environment Canada Manuscript Report EE-108, Ottawa, ON, 1988, 46.

Skou, N., B.M. Sorensen, and A. Poulson, A new airborne dual frequency microwave radiometer for mapping and quantifying mineral oil on the sea surface. In *Proc. Second Thematic Conf. Remote Sensing for Marine and Coastal Environ.*, Ann Arbor, MI: ERIM Conferences, 1994, II-559–565.

Solberg, R. and N. Theophilopoulos, ENVISYS — a solution for automatic oil spill detection in the Mediterranean. In *Proc. Fourth Thematic Conf. Remote Sensing for Marine and Coastal Environ.*, Ann Arbor, MI: Environmental Research Institute of Michigan, 1997, **I**, 3–12.

Stapleton, G.F., S.K. Parashar, J.B. Snellen, and R.D. Worsfold, Detection of oil under ice — a laboratory program. In *Proc. Fourth Annual Arctic Marine Oilspill Program Tech. Sem.*, Ottawa, ON: Environment Canada, 1981, 587–605.

Süss, H., K. Grüner, and W.J. Wilson, Passive millimeter wave imaging: A tool for remote sensing. *Alta Frequenza*, 1989, **LVIII**(5–6), 457–465.

Taft, D.G., D.E. Egging, and H.A. Kuhn, Sheen surveillance: An environmental monitoring program subsequent to the 1989 *Exxon Valdez* shoreline cleanup. In *Exxon Valdez Oil Spill: Fate and Effects in Alaskan Waters*, ASTM STP 1219, Philadelphia, PA: American Society for Testing and Materials, 1995, 215–238.

Taylor, S., 0.45 to 1.1 µm Spectra of Prudhoe crude oil and of beach materials in Prince William Sound, Alaska. CRREL Special Report No. 92–5, Cold Regions Research and Engineering Laboratory, Hanover, NH, 1992, 14.

Tennyson, E.J., Shipborne radar as an oil spill tracking tool. In *Proc. Eleventh Arctic and Marine Oilspill Program Tech. Sem.*, Ottawa, ON: Environment Canada, 1985, 385–390.

Tunaley, J.K.E., The scattering of electromagnetic waves from the sea-ice oil and sea-water interfaces. In *Proc. Eighth Annual Arctic Marine Oilspill Program Tech. Sem.*, Ottawa, ON: Environment Canada, 1985, 292–302.

Tunaley, J.K.E. and D.R. Moorcroft, Aspects of the detection of oil under sea ice using radar methods. In *Proc. Ninth Annual Arctic Marine Oilspill Program Tech. Sem.*, Ottawa, ON: Environment Canada, 1986, 463–474.

Ulaby, F.T., R.K. Moore, and A.K. Fung, *Microwave Remote Sensing: Active and Passive*, Norwood, MA: Artech House Inc., 1989, 1466–1479.

Voloshina, I.P. and O.Y. Sochnev, Observations of surface contamination of the region of the Kol'shii Gulf from IR measurements. *Soviet J. Remote Sensing*, 1992, **9**(6), 996–1000.

Wadsworth, A., W.J. Looyen, R. Reuter, and M. Petit, Aircraft experiments with visible and infrared sensors. *Intl. J. Remote Sensing*, 1992, **13**(6–7), 1175–1199.

Wahl, T., K. Eldhuset, and Å. Skøelv, Ship traffic monitoring and oil spill detection using ERS-1. In *Proc. Intl. Symp. Operationalization of Remote Sensing*, ITC Enschede, The Netherlands, 1993, 97–105.

Werle, D., B. Tittley, E. Theriault, and B. Whitehouse, Using Radarsat-1 SAR imagery to monitor the recovery of the *Irving Whale* oil barge. In *Proc. Intl. Symp: Geomatics in the Era of Radarsat*, Ottawa, ON, 1997.

Wiese, F.K., Seabirds and Atlantic Canada's ship-source oil pollution: Impacts, trends, and solutions. World Wildlife Fund Canada Contract Report, 2002, 85.

Wiese, F.K. and P.C. Ryan, The extent of chronic marine oil pollution in Southeastern Newfoundland waters assessed through beached bird surveys 1984–1999. *Mar. Poll. Bull.*, 2003, **46**, 1090–1101.

Wiese, F.K., G.J. Robertson, and A.J. Gaston, Impacts of chronic marine oil pollution and the Murre Hunt in Newfoundland on thick-billed Murre *Uria lomvia* populations in the Eastern Canadian Arctic. *Biol. Conserv.*, 2004, **116**, 205–216.

Yan, X-H. and P. Clemente-Colon, The maximum similarity share matching (MSSM) method applied to oil spill feature tracking observed in SAR imagery. In *Proc. Fourth Thematic Conf. Remote Sensing for Marine and Coastal Environ.*, Ann Arbor, MI: Environmental Research Institute of Michigan, 1997, **I**, 43–55.

Zhifu, S. and W. Wiesbeck, A study of passive microwave remote sensing. In *Proc. 1988 Intl. Geoscience and Remote Sensing Symp.*, European Space Agency, Paris, France, 1988, 1091–1094.

Zielinski, O. and N. Robbe, Past and future of airborne pollution control. In *Proc. Interspill*, Trondheim, Norway, 2004.

15 Advances in Forensic Techniques for Petroleum Hydrocarbons: The *Exxon Valdez* Experience

A. Edward Bence,[1,2] David S. Page,[3] and Paul D. Boehm[4]

[1] ExxonMobil Upstream Research Co., URC.URC.S169, PO Box 2189, Houston, TX 77252.
[2] Currently, AEB Services, LLC, 703 Tanglewood Drive, Friendswood, TX 77546.
[3] Department of Chemistry, Bowdoin College, 6600 College Station, Brunswick, ME 04011.
[4] Exponent, 3 Clock Tower Place, Suite 205, Maynard, MA 01754.

15.1 Introduction

The March 25, 1989, *Exxon Valdez* oil spill (EVOS) in Prince William Sound (PWS), Alaska, is the most extensively studied spill in history. In the years immediately following the spill and continuing today, PWS and the adjacent Gulf of Alaska (GOA) have proven to be natural laboratories for the development and testing of new chemical forensic techniques for oil spill impact studies.

EVOS released approximately 258,000 barrels of a total cargo of 1.26 million barrels of Alaska North Slope (ANS) crude oil. It is estimated that ~40% of that oil was stranded along 783 km of shorelines in southwestern PWS (Jahns et al., 1991; Owens, 1991; Wolfe et al., 1993). By the summer of 1990, ~55% of the oil was removed due to cleanup efforts and to the effects of winter storms (Michel et al., 1991) and other natural weathering processes. By 1992, the length of oiled shoreline had been reduced by ~90%. A 2001 NOAA study (Short et al., 2004) estimated that 55.6 mT (~336 barrels) of *Exxon Valdez* crude (EVC) residues remained, with the majority of that oil classified as light and buried in the middle- and upper-intertidal zones on boulder-cobble beaches. Furthermore, that study, using a first-order kinetics model, estimated that the residues were being removed at a rate of 20–26% per year. If the assumptions in the model are correct, by the end of 2005, less than 100 barrels of oil, or <0.05% of the volume originally spilled, remain. By comparison, it is estimated that 0.016% of the oil spilled from the *Florida* (1969), and ~0.2% of the *Arrow* spill (1970) remain after more than 35 years (Peacock et al., 2005; Ed Owens, personal communication, 2006).

As oiling levels decreased, the significance of prespill hydrocarbon sources, both natural and anthropogenic, became increasingly important. This early finding that multiple hydrocarbon sources contributed to the PWS marine environment underscored the need for tools that could quantitatively resolve them in order to properly assess the potential long-term environmental impact from the spill. Of particular interest was the discovery that the prespill benthic sediments in the sound contained a substantial background of petrogenic hydrocarbons derived from Tertiary hydrocarbon systems consisting of petroliferous shales, oil seeps, and coals, located in the eastern GOA. It has been suggested that 10–12 years after the spill, polycyclic aromatic hydrocarbons (PAH) from those residues were bioavailable at levels sufficient to impact populations of fish and wildlife (e.g., Bodkin et al., 2002; Trust et al., 2000). Consequently, identification and quantification of degraded spill remnants in the potential exposure pathways, i.e., water column, food chain, and benthic sediments, were required. New forensic environmental geochemistry tools were developed to address hydrocarbon source resolution in sediments and tissues and concentrations of spill hydrocarbons in the potential exposure pathways.

An oil spill in U.S. waters initiates a sequence of scientific events under the injury assessment phase of the U.S. federal natural resource damage assessment (NRDA) process. The Federal NRDA regulations under The Comprehensive Environmental Response, Compensation, and Liability Act of 1980 (CERCLA) defines recovery period as the "length of time required to return the services of the injured resource to their baseline condition," where baseline services "should reflect conditions that would have been expected at the assessment area had the discharge of oil not occurred, taking into account both natural processes and those that are the result of human activities" (U.S. Code of Federal Regulations, 2001). Therefore, the application of chemical forensics to the identification and quantification of all hydrocarbon sources in a spill zone is an essential part of the injury assessment process. Injury assessment has to be based upon well-defined dose-response relationships linking measured biological effects to clearly defined PAH sources.

15.2 Identification of Hydrocarbon Sources in PWS

15.2.1 Multiple Sources of Hydrocarbons

Multiple hydrocarbon inputs at oil spill sites are well-documented, and an objective forensic approach should consider the possibility other hydrocarbon sources, both anthropogenic and natural (e.g., Boehm et al., 1997; Neff et al., 2005), may exist. Examples of spills in which there are hydrocarbon inputs from sources other than the primary spill include the *Metula* spill site in the Straits of Magellan (Baker et al., 1975), the Antarctic *Bahia Paraiso* spill (Kennicutt et al., 1992), and the 1978 *Amoco Cadiz* spill on the Brittany coast (Gundlach et al., 1981; Page et al., 1988). This earlier experience underscored the importance of considering multiple sources of petroleum in designing postspill injury assessment studies of the *Exxon Valdez* oil spill in PWS and the Gulf of Alaska (GOA).

Numerous studies of the fate and effects of EVOS show that the impacted area has many sources of hydrocarbons other than EVC (e.g., Kvenvolden et al., 1995; Bence et al., 1996; Page et al., 1995, 1996, 1999, 2002a; Boehm et al., 1997, 2001; Van Kooten et al., 2002). PWS has a long history of human activity, and there are numerous sites of both present and past human activity that provide localized inputs to the marine environment of hydrocarbons derived from fossil fuel use (Page et al., 1999, 2002a; Wooley, 2002). These localized hydrocarbon sources are superimposed on a regional background consisting of natural petroleum hydrocarbons derived from petroleum hydrocarbon systems in the eastern GOA (Page et al., 1996; Bence et al., 1996; Boehm et al., 2001) and globally sourced combustion products. The application of advanced fingerprinting and geochemical forensic methods to chemistry data for intertidal and subtidal sediment samples identified and quantitatively resolved a number of hydrocarbon sources that contribute to the PWS marine environment:

1. Petrogenic hydrocarbons
   a. Weathered EVC petroleum residues
   b. Petroleum and refined products from the Miocene Monterey Formation (CA)
   c. Hydrocarbons derived from eroding organic sediments and oil seep sources east of PWS
   d. Refined petroleum products, such as marine diesel and heavy fuel oils

released during commercial and recreational boating activities
2. Pyrogenic hydrocarbons derived from the burning of petroleum products and wood
   a. Global and regional atmospheric sources
   b. Point sources at active and historical sites of human activity
3. Biogenic hydrocarbons generated by natural biologic or diagenetic processes

Each of these hydrocarbon sources was identified through the application of chemical fingerprinting methods that focused on the distributions of the polycyclic aromatic hydrocarbons (PAH) and the saturate and aromatic biomarkers (steranes, triterpanes, and triaromatic steroids).

Because comparable PAH data have been available through sampling and analytical studies beginning in 1989, and because the PAH distributions are highly diagnostic of many of the hydrocarbon sources in the region, the use of PAH analyte distributions to distinguish hydrocarbon sources has been the most important fingerprinting tool. PAH profiles for examples of the major hydrocarbon sources in PWS are shown in Figure 15-1. Analysis of the distributions of PAH analytes is a useful means for source discrimination because some PAH are more resistant to weathering than the aliphatic fraction (NRC, 1985) and their distributions vary among sources. However, in situations where the samples are extremely degraded, loss of diagnostic PAH may preclude source identification. In those cases, the more refractory saturate and aromatic biomarkers are used. The chemical criteria used to distinguish petrogenic, biogenic, and pyrogenic sources are discussed in detail elsewhere (Page et al., 1995; Bence et al., 1996; Burns et al., 1997, 2006) and are briefly reviewed here.

15.2.2 Petrogenic Hydrocarbons

These include crude oil and its refined products and are characterized by homologous families of related 2- to 4-ring PAH (naphthalenes, fluorenes, phenanthrenes, dibenzothiophenes, and chrysenes), where the unsubstituted parent PAH for each family is less abundant than the alkylated homologues. In many crude oils and middle distillates, the PAH composition is dominated by the light, 2- to 3-ring compounds. With the exception of the alkyl chrysenes, 4- to 6-ring PAH are usually at levels below their limits of detection in oil. Sources of petrogenic PAH to the PWS marine environment include weathered and unweathered EVOS crude, an Alaska North Slope (ANS) pipeline crude oil released in the 1989 spill (Figures 15-1A and B); diesel fuel refined from ANS crude oil (Figure 15-1C); natural oil and gas seeps in the eastern Gulf of Alaska (GOA); heavy fuel oils and asphalts brought in from other parts of the world, including Cook Inlet, southern California (Figure 15-1E), the Far East, etc.; and kerogens derived from eroding Tertiary shales and coals in the eastern GOA and brought into the sound as suspended sediments by the counterclockwise circulation of the Alaskan Coastal Current (Royer et al., 1990). The regional petrogenic background in the benthic sediments of PWS (Figure 15-1D) is formed through the deposition of these suspended particulates.

Diesel refined from ANS and other crude oils is a periodic contaminant to PWS due to ongoing marine incidents. The diesel PAH fingerprint is similar to that of ANS crude oil but lacks the 4–6 ring PAH (i.e., alkyl chrysenes) that are removed in the refining process. Refining also removes the majority of the saturate and aromatic biomarkers.

Prior to the onset of oil production from the Cook Inlet (1958) and North Slope fields (1977), petroleum and petroleum products were imported from the lower 48 states. Major components of those shipments were crude oil and refined products, including asphalt for highway and airport runway construction, produced from the Miocene Monterey Formation in California (Lethcoe and Lethcoe, 2001). Remnants from those shipments are found at numerous locations throughout Prince William Sound. In the early 1990s, Kvenvolden et al. (1993, 1995) reported widely distributed tar deposits on shorelines in northern and western

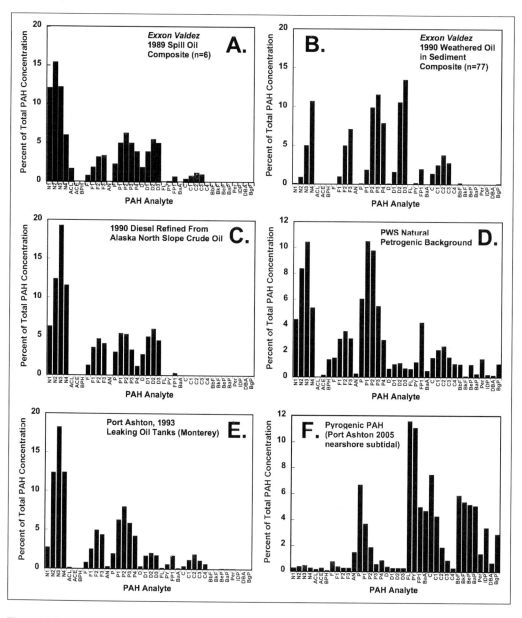

Figure 15-1 PAH compositions of major hydrocarbon source inputs to the marine environment of PWS. A PAH analyte may be a single compound, such as C_0-phenanthrene, or the sum of a group of isomers, such as C_1-phenanthrene, which has four isomers. N1, N2, N3, N4 = C_1, C_2, C_3, C_4-naphthalenes; ACL = acenaphthylene; ACE = acenaphthene; BPH = biphenyl; F, F2, F2, F3 = C_0, C_1, C_2, C_3-fluorenes; P, P1, P2, P3, P4 = C_0, C_1, C_2, C_3, C_4-phenanthrenes/anthracenes; D, D1, D2, D3 = C_0, C_1, C_2, C_3-dibenzothiophenes; FL = fluoranthene; PY = pyrene; FP1 = C_1-fluoranthenes/pyrenes; BaA = benzo(*a*)anthracene; C, C1, C2, C3, C4 = C_0, C_1, C_2, C_3, C_4-chrysenes; BbF = benzo(*b*)fluoranthene; BkF = benzo(*k*)fluoranthene; BeP = benzo(*e*)pyrene; BaP = benzo(*a*)pyrene; Per = perylene; IDP = indeno(1,2,3-cd)pyrene; DBA = dibenzo(*a,h*)anthracene; BgP = benzo(*g,h,i*) perylene.

parts of PWS having a ^{13}C-enriched carbon isotope composition (~−24 per mil as compared to the EVOS value of ~−29 per mil) that was consistent with Monterey oils. It is hypothesized that the 1964 Alaskan earthquake released some of this material from storage tanks that ruptured at Valdez and at other locations throughout PWS (Kvenvolden et al., 1995). In addition to their distinctive carbon isotopic composition, these products are readily identified on the basis of their saturate biomarker (primarily the triterpanes) and PAH (Figure 15-1E) compositions.

15.2.3 Biogenic Hydrocarbons

Biogenic hydrocarbons are generated by biologic processes or in the early stages of diagenesis in marine sediments (e.g., perylene, a 4-ring unsubstituted PAH). Perylene can be seen as an addition to the predominantly petrogenic PAH profile in Figure 15-1D and is present in most subtidal sediment samples. Retene, a conifer-derived alkyl-substituted phenanthrene, is of considerable interest because it is abundant in the PWS marine environment and it has been known to induce cytochrome P4501A (CYP1A) activity in rainbow trout (Fragaso et al., 1998; Hawkins et al., 2003). Retene is also produced through the thermal alteration of diterpenoids in conifer wood (Rogge et al., 1998). Thus, it is a common product in wood smoke, and regional atmospheric inputs from forest fires are a likely source of retene to PWS. The pentacyclic triterpane 17 beta(H),21 beta(H)-hop-22(29)-ene (diplotene) is a bacterially derived biomarker that is formed in soils.

15.2.4 Pyrogenic Hydrocarbons

These are generated by the combustion of fossil fuels, such as coal, oil, and diesel fuel, and of recent organic material such as wood. Pyrogenic sources have a PAH distribution dominated by the 4- to 6-ring PAH, e.g., fluoranthene, pyrene, and benzo(a)pyrene, and by the parent compounds of the 3-, 4-, and 5-ring PAH, with decreasing abundances of alkyl homologs in each group (Figure 15-1F). They are of particular concern in PWS because some, e.g., chrysene, benzo(a)pyrene, and benzo(a)anthracene, are strong inducers of CYP1A in fish (e.g., Lee and Anderson, 2005). Inputs of pyrogenic PAH to the PWS marine environment include atmospheric fallout of particulates from both global and regional sources, point source inputs at both active and historic population and industrial centers, and the ubiquitous inputs from fuel consumption by commercial and pleasure boats.

15.3 Composition of Exxon Valdez Crude and Its Weathering Products

15.3.1 Bulk Composition and Trace Chemistry

The *Exxon Valdez* crude (EVC) was a 1989 Alaskan pipeline crude consisting of a mixture of oils from the Alaskan North Slope fields. In 1989, that mixture was in the approximate proportions of 76% Prudhoe Bay, 17% Kuparuk, 5% Endicott, and 2% Lisburne (Bence and Burns, 1995). In the years since the spill, the composition of the pipeline crude has changed as production from some fields declined and production from newly discovered fields was brought on line.

EVC is a medium sulfur crude (~1.3–1.4% S) consisting of, in approximate weight proportions, C_{1-10} volatiles (20%), $>C_{10}$ saturated hydrocarbons (30%), aromatic and naphthenoaromatic hydrocarbons (30%), and asphaltenes (resins and nitrogen, sulfur, and oxygen compounds) (20%) (Figure 15-2A). In the first few days following the spill, most of the volatile fraction was lost due to evaporation (Wolfe et al., 1993). The PAHs, including the heterocyclic sulfur-containing dibenzothiophenes, comprise approximately 12,000–14,000 ppm (1.2–1.4 wt.%).

EVC and its weathering products are readily distinguishable from other natural and anthropogenic hydrocarbon sources occurring in the region (e.g., Cook Inlet, Yakataga, Katalla, and Monterey (CA) oils, refined products,

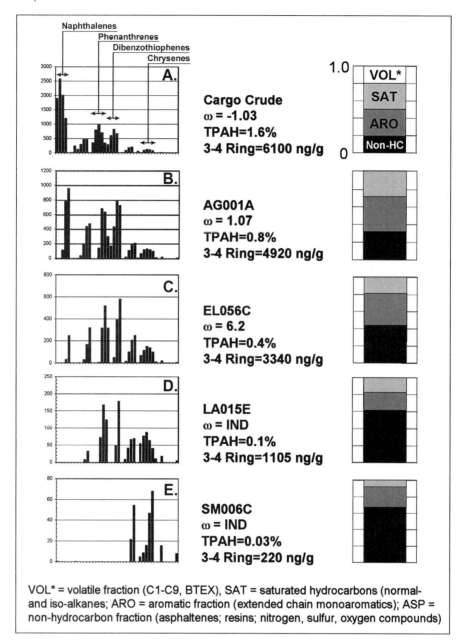

Figure 15-2 PAH distributions, TPAH and 3–4-ring PAH concentrations, major fraction compositions, and weathering parameters for EVC and four variably weathered PWS shoreline oil samples. The weathering parameter, ω, calculated from the distributions of 14 PAH analytes (Short and Heintz, 1997) becomes indeterminate (IND) at advanced weathering states when one or more analytes are not detected.

kerogens from eastern Gulf of Alaska Tertiary sediments, and combustion products) on the basis of their PAH and saturate and aromatic biomarker compositions (e.g., Kvenvolden et al., 1993; Page et al., 1995; Bence et al., 1996; Short et al., 1996; Van Kooten et al., 2002). The primary distinguishing feature of the PAH fingerprint of EVC is the approximately equal abundances of the phenanthrene and dibenzothiophene analytes—a consequence of its higher sulfur content compared to other petrogenic hydrocarbon sources in the region—which is reflected in the C_2-Db/C_2-Ph ratio (Table 15-1A). Moderate weathering of EVC in the shoreline regime does not significantly alter this relationship (Figure 15-1B). Consequently, this has been a useful parameter in the quantitative deconvolution of mixtures of hydrocarbon sources in shoreline and benthic sediment samples (Burns et al., 1997). However, because there are differences in the relative proportions of the alkylated PAH groups reported for EVC among laboratories, deconvolution of sources in mixed samples analyzed at different laboratories may not be possible.

Table 15-1A Major Hydrocarbon Fractions, TOC, and PAH Characteristics of EVC, Weathered EVC, PWS Prespill Benthic Sediments, Oil Residues at Historical Human Activity (HA) Sites, and Potential Background Sources

Hydrocarbon Source Type	Sat:Aro:NSO	TOC (%)	TPAH/TOC (μg/g OC)	TPAH (% of oil or extract)	LPAH/ TPAH	C_2-Db/ C_2-Ph	C_2-Ch/ C_2-Ph	Pyrogenic Index FP/(FP + C 24Ph)
EVC (N = 25) (ω = −1.0)	38:38:24*	85	15,300	1.3 ± 0.1	0.94 ± 0.01	0.86 ± 0.06	0.13 ± 0.01	0.01 ± 0.002
Moderately weathered EVC EL056C (ω = 6.2)	18:38:43	ND	ND	0.4	0.71	1.23	0.48	0.01
Heavily weathered EVC LA015E (ω = IND)	8:25:67	ND	ND	0.03	0.00	Undefined	∞	Undefined
HA site oil residues								
Unakwik	29:35:36	ND	—	0.30	0.72	0.48	0.35	0.02
Port Audrey	1:60:39	ND	—		0.67	0.06	0.42	0.79
Latouche Town site	27:36:37	ND	—	0.55	0.64	0.53	0.53	0.10
Pt. Ashton	ND	ND	—	1.12	0.89	0.27	0.20	0.06
Katalla wellhead oil		84.9	6,240	0.53	0.95	0.13	0.07	0.01
Johnson Crk seep		79.2	30,800	2.44	0.77	0.14	0.18	0.02
Munday Crk seep		76.6	8,000	0.61	0.50	0.08	1.45	0.03
Yakataga FM shale (N = 6)	ND	0.38 ± 0.1	13,170 ± 13,250	4.16 ± 3.4	0.85 ± 0.09	0.20 ± 0.12	0.15 ± 0.14	0.15 ± 0.08
Kulthieth FM coal (N = 7)	8:23:69	58.6 ± 9.8	600 ± 300	4.43 ± 2.89	0.78 ± 0.04	0.04 ± 0.04	0.29 ± 0.08	0.06 ± 0.03
Bering River coal (low volatile bituminous)	ND	85.5	242	0.42	0.78	0.34	0.40	0.11
PWS prespill benthic sediments (N = 4)	ND	0.64 ± 0.22	1780 ± 240	0.41 ± 0.14	0.74 ± 0.01	0.13 ± 0.02	0.23 ± 0.02	0.10 ± 0.01

*Excluding volatile fraction.

Table 15-1B Saturate and Aromatic Biomarker Indices for EVC, Weathered EVC, PWS Prespill Benthic Sediments, Oil Residues at Historical Human Activity (HA) Sites, and Potential Background Sources

Hydrocarbon Source Type	C_{30}-Hopane ($\mu g/g$)	Trisnorhopane Index TS/(TS + TM)	Oleanane Index Ol/(Ol + C_{30}-hopane)	$C_{24}Tt/(C_{24}Tet + C_{26}Tri(S + R))$	$C_{27}:C_{28}:C_{29}$ ($\alpha\alpha\alpha$) (20S + 20R)	$C_{29}\alpha\alpha\alpha$ (20S)/ (20S + 20R))	TAS Index $C_{20-21}/(C_{20-21} + C_{26-28})$
EVC (N = 25) ($\omega = -1.0$)	120 ± 13	0.42 ± 0.02	0.00	0.31 ± 0.02	0.47:0.24:0.29	0.51 ± 0.04	0.18 ± 0.01
Moderately weathered EVC EL056C ($\omega = 6.2$)	224.7	0.44	0.00	0.36	0.47:0.27:0.27	0.42	0.20
Heavily weathered EVC LA015E (ω = IND)	252.8	0.44	0.00	0.33	0.47:0.25:0.29	0.48	0.13
HA site oil residues							
Unakwik	216.7	0.35	0.19	0.44	0.42:0.35:0.23	0.31	0.09
Port Audrey	17.3	0.04	0.07	0.20	0.40:0.31:0.29	0.39	0.70
Latouche town site	201.3	0.33	0.19	0.45	0.42:0.35:0.24	0.32	0.09
Pt. Ashton	132.2	0.29	0.18	0.20	0.42:0.34:0.24	0.32	0.20
Katalla wellhead	514	0.48	0.10	0.28	0.40:0.25:0.35	0.57	0.12
Johnson Creek seep	463	0.20	0.25	0.98	0.04:0.19:0.77	0.55	0.33
Munday Ck seep	556	0.19	0.24	0.95	0.04:0.20:0.76	0.57	0.29
Yakataga FM (N = 6)	0.014 ± 0.008	0.16 ± 0.11	0.16 ± 0.08	0.61 ± 0.20	0.22:0.22:0.56	0.05 ± 0.03	0.11 ± 0.12
Kulthieth FM coal (N = 7)	4.9 ± 4.1	0.17 ± 0.05	0.10 ± 0.03	1.00	0.28:0.20:0.52	0.51 ± 0.02	0.40 ± 0.08
Bering River Coal (low-volatile bituminous)	0.0	Undefined	Undefined	Undefined	Undefined	Undefined	Undefined
PWS prespill benthic sediments (N = 4)	0.007 ± 0.003	0.26 ± 0.03	0.21 ± 0.01	0.60 ± 0.03	0.32:0.28:0.40	0.29 ± 0.04	0.13 ± 0.16

ND = not determined.

EVC also has saturate and aromatic biomarker (steranes, triterpanes, and triaromatic steroids) compositions that can be used to distinguish it and its weathered equivalents from other hydrocarbon sources in the region. In the petroleum industry, these biomarkers are used to identify depositional environments of petroleum source rocks, to constrain the thermal histories of those source rocks, and to evaluate the extent of biodegradation of oils. The greater stability of these compounds relative to many of the PAHs make them useful for quantitative source allocations (e.g., Burns et al., 1997; see also Chapter 8 herein). Specific distinguishing characteristics of the triterpanes in EVC (Table 15-1) include the absence of the source indicator $18\alpha(H)$-oleanane ($C_{30}H_{52}$), extended pentacyclics, TS/(TS + TM) = 0.4, C_{24} tetracyclics/(C_{24} tetracyclics + C_{26} tricyclics) = 0.3 (e.g., Bence et al., 1996; Kvenvolden et al., 1993). The absence of oleanane and $C_{24}:C_{26}$ serves to discriminate EVC and its weathering products from Monterey (CA) tars and residual fuel oils and from natural seep oils, the organic shales from which the seeps were sourced, and low-to-high thermal maturity coals associated with the Tertiary sedimentary province in the eastern Gulf of Alaska (e.g., Bence et al., 1996; Kvenvolden et al., 1995; Van Kooten et al., 2002). Specific features of the steranes include the relative proportions of the source indicator $C_{27}:C_{28}:C_{29}(\alpha\alpha\alpha)(20S + 20R)$ and the thermal maturity indicator $C_{29}\alpha\alpha\alpha(20S/(20S + 20R))$. These discriminate EVC from Monterey (CA) tars and oils and low-thermal maturity Kuskutz coals from the Kulthieth Formation (Table 15-1B). The triaromatic steroid (TAS) index $(C_{20+21})/((C_{20-21}) + (C_{26-28}))$ is useful in distinguishing EVC from various Monterey tars and oils, from the prespill benthic background, and from low-thermal maturity Koskutz coals (Table 15-1B). It further discriminates the Koskutz coals from the prespill benthic sediment background.

15.3.2 Weathering Trends

15.3.2.1 Data Sources

Oil stranded on the shorelines was subjected to variably intense physical and chemical weathering depending upon the extent of sequestration (i.e., buried beneath a boulder-cobble veneer, trapped in the wave shadow behind protective outcrops, etc.). Residual oil suites collected from both surface and subsurface regimes define the EVOS natural weathering trends. Laboratory studies, conducted both by NOAA (Heintz et al., 1999; Carls et al., 1999) and the University of Idaho (Brannon et al., 2006), involving water circulation through oiled gravel columns, reveal the compositional changes that occur under controlled conditions.

15.3.2.2 Major Fraction Trends

In the first few days following the spill, the volatile fraction, comprising approximately 20% of the oil by weight, was lost (Wolfe et al., 1993). Over the years, oil stranded on the shoreline underwent chemical, biological, and mechanical weathering at varying rates dependent upon the degree of exposure to environmental conditions. Changes in the major oil fraction include loss of the saturate and aromatic fractions accompanied by a concomitant increase in the nonhydrocarbon component (asphaltenes and resins) (Figure 15-2). At the most extreme stages of weathering, the nonhydrocarbon component comprises 70% or more of the residual oil.

15.3.2.3 PAH Trends

With natural weathering on PWS shorelines the EVC PAHs, as a group, are removed from the oil. In the most extreme cases, TPAH is reduced to less than 0.05 wt.% of the residual oil (Table 15-1A, Figure 15-2). Weathering produces predictable changes in the PAH distributions based upon the relative solubilities of the individual compounds. This is evident in the progression of the calculated values of the Short and Heintz (1997) weathering parameter, ω, which increases with increasing degree of weathering (Figure 15-2) and becomes indeterminate at higher weathering states when one or more of the PAH analytes used in the calculation are no longer detected. The 2- and 3-ring PAH (light PAH) are removed more quickly than the 4-ring PAH (heavy PAH or HPAH). Consequently, the relative concentration of the HPAH fraction (fluoranthenes/pyrenes and chrysenes) in the total PAH fraction increases and LPAH/TPAH decreases with degree of weathering. However, the concentration of HPAH as a fraction of the residual oil decreases (Figure 15-2; Table 15-1A). Consequently, weathered oil does not become increasingly toxic due to buildup of 3- and 4-ring PAH, as has been suggested (Bue et al., 1996; Heintz et al., 1999).

The relative rate of removal of the PAH groups as the oil weathers is naphthalenes > fluorenes > phenanthrenes $\cong$ dibenzothiophenes $\cong$ fluoranthenes/pyrenes > chrysenes. Within each PAH group, the preferential removal is parent > C1 > C2 > C3.

The PAH pyrogenic index (fluoranthene + pyrene)/(fluoranthene + pyrene + C_{2-4} phenanthrene) or (FP)/(FP + C24Ph) was used to discriminate combustion products (fluoranthene and pyrene) from EVC and its weathering products (Neff et al., 2006). This ratio in the cargo crude is 0.01 (Table 15-1A). With both natural weathering on the shoreline and artificial weathering in the laboratory, that ratio increases gradually to a maximum of ~0.05 (Figure 15-3). Fluoranthene, pyrene, and C_4 phenanthrene (the most stable of the alkylated phenanthrenes) are removed from the oil at essentially the same rate such that when mass loss approaches 60%, the ratio is undefined. A conservative upper limit of 0.20 was selected by Neff et al. (2006). Samples having TPAH >100 ng/g and (FP)/(FP + C24Ph) values

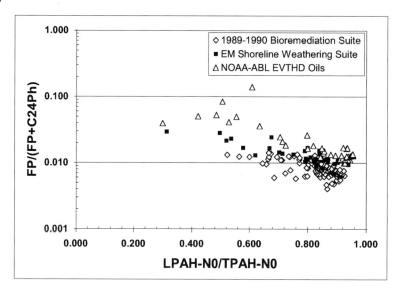

Figure 15-3 Pyrogenic index FP/(FP + C24Ph) vs. the ratio of low-molecular-weight (2–3-ring PAH) to total PAH, LPAH/TPAH, for naturally and artificially weathered shoreline oils. LPAH/TPAH decreases with increased weathering. 1989–1990 bioremediation suite oils are from Prince et al. (1994). EM shoreline oil weathering suite from samples collected in 2001–2002. NOAA EVTHD oils from the 2003 version of *Exxon Valdez* Trustees Oil Database on NOAA-NMFS, Auke Bay Laboratory webpage (www.afsc.noaa.gov/abl/oilspill). These include oils that were weathered in the laboratory.

>0.20 were interpreted to contain combustion products.

15.2.3.4 Mass Loss during Weathering

Utilizing the methodology of Douglas et al. (1996), which assumes conservation of C_{30}-hopane during weathering, a conservative estimate of the minimum mass loss can be made using the following equation: percent oil depletion = $[1 - Ho/Hs] \times 100\%$, where Ho is the concentration of C_{30}-hopane in EVC and Hs is the concentration of C_{30}-hopane in the weathered oil sample. When applied to the two weathered samples of EVC reported in Table 15-1B, minimum mass losses of 39% (sample EL056C) and 46% (sample LA015E) are calculated. Shoreline oil samples collected in 2001–2002 have minimum mass losses of up to 60% of the oil. Inert asphaltenes and resins comprise the largest component of the heavily weathered oils (Figure 15-2). This method of calculating mass loss cannot be used when weathering is so severe that C_{30}-hopane is not conserved. Significant loss of C_{30}-hopane is observed in subsurface oil residues at a number of locations. Weathered oils collected at Point Helen (shoreline segment KN-405A) in 2005 have only 20 ppm C_{30}-hopane, 0.01% PAH (consisting only of C_3- and C_4-chrysenes), 13% saturates, 29% aromatics, and 57% NSOs.

15.4 Resolution of Inputs to the Natural Background

Studies conducted between 1989 and 1991 in PWS and the GOA involved the collection and chemical analysis of over 10,000 seafloor sediment samples taken from water depths ranging from 0 m to over 750 m (e.g., Rapp et al., 1990; Manen et al., 1993; Boehm et al., 1995; Sale and Short 1995; Page et al., 1995, 1996; Short et al., 1996). One of the objectives of these investigations was to determine if the rapidly diminishing shoreline oil residues were being transferred to subtidal sediments where, potentially, they could enter the food chain.

The result is a large body of chemical data on benthic sediments from stations both inside and outside the spill zone.

Three components were essential to the resolution of the natural background: (1) an analytical program that could differentiate similar petrogenic sources with high precision; (2) an extensive sampling program to collect seafloor and subseafloor benthic sediments and samples of the numerous potential hydrocarbon sources to the PWS marine environment; and (3) application of a quantitative statistical allocation technique. PAH, triaromatic steroids (TAS), triterpane, and sterane distributions for the PWS subsurface (prespill) petrogenic background and hydrocarbons sources that contribute to that background are shown in Figures 15-4 and 15-5.

The identification of 18α(H)-oleanane ($C_{30}H_{52}$) in deep seafloor sediment samples taken after the spill in 1989 (Rapp et al., 1990) was one of the earliest indications of nonspill origins of seafloor petrogenic hydrocarbons in PWS. Oleanane is a saturate biomarker for petroleum, having a post-latest Cretaceous terrigenous source and is not present in ANS crude oils (Bence et al., 1996; Peters et al., 2005).

Page et al. (1995, 1996) reported the results of sampling and analytical programs carried out in 1989, 1990, and 1991 whose goals were to detect EVC contributions to seafloor sediments, to quantify the amount of EVC present relative to the pre-existing petrogenic background, and to characterize historical inputs of petrogenic and nonpetrogenic hydrocarbons. Seven separate surveys sampled 649 subtidal locations in PWS and the GOA and collected more than 2300 samples that were subsequently analyzed for PAH, and for the 1991 samples, for saturate biomarkers as well. These stations ranged from near-shore locations (<100 m from the nearest shoreline), stations in protected bays, offshore stations (~100 to 1000 m offshore), to deep subtidal stations (>1000 m offshore). Chemistry data were obtained for seafloor surface (0–2 cm) sediments and from ^{210}Pb age-dated sediment cores providing samples from depths of up to 32 cm below the seafloor. These lower sedimentary layers were deposited up to ~150 years prior to the 1989 spill (Page et al., 1995).

These studies showed that EVC and its residues were readily distinguishable from the regional background PAH by their very different C_2-Dbt/C_2-Ph ratios (Table 15-1A and Figure 15-6). However, identification of the individual components that made up the regional background was not possible. The 1991 study linked the regional background in PWS to hydrocarbon systems (petroleum source rocks and natural oil seeps) east of PWS along the GOA coast, an area of oil production in the early 1900s.

The observation that both benthic sediments and coals from the Katalla area of the GOA had low values for the refractive index (RI) (def., triaromatic steroids/1-methylchrysene) was used by Hostettler et al. (1999) to conclude that the source of the natural hydrocarbon background in the benthic sediments of the GOA and PWS was most likely due to coal (see, also, Short et al., 1999; Van Kooten et al., 2002). However, coal is only one of several possible explanations for the measured low RI values in PWS sediments (Bence et al., 2000), and other chemical constraints indicate that these coals cannot be important components of the PWS natural background (Tables 15-1A and 15-1B). For example, high maturity (Ro > 1.5) coals from the Bering River coal fields have high concentrations of total organic carbon (TOC) and low concentrations of saturate biomarkers. The very low TOC concentrations of PWS benthic sediments and relatively high concentrations of saturate biomarkers (e.g., hopane) preclude these high-maturity coals from being important components of the background (Table 15-1A; Figures 15-4 and 15-5). Inputs of low-maturity Koskutz coals from the Kulthieth formation (see Van Kooten et al., 2002) are constrained to relatively small amounts by low TPAH/TOC, C_2-Dbt/C_2-Ph, and oleanane indices and high TAS index (Table 15-1A, 15-1B; Figures 15-4K and 15-5K).

At those intertidal sites in PWS where residual subsurface oil remains, the spatial relationships of hydrocarbons introduced from

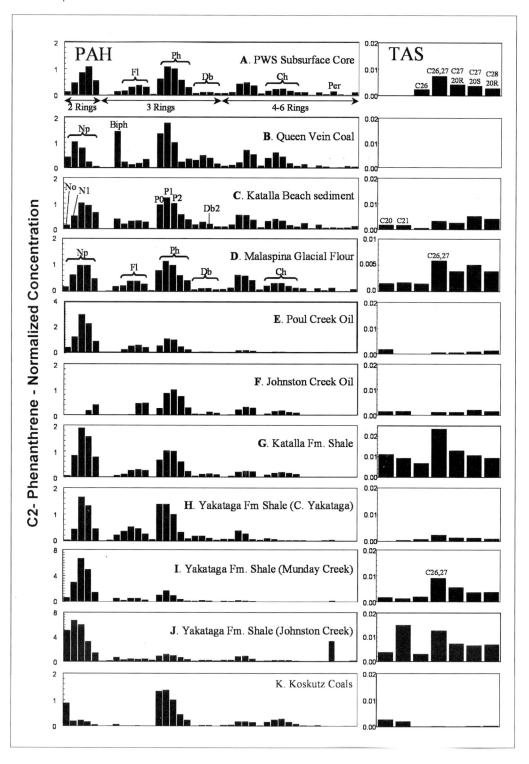

Figure 15-4 PAH and triaromatic steroid (TAS) compositions (C_2-phenanthrene-normalized) for (A) PWS prespill segment of a benthic core, (B–J) potential source inputs to the natural background. Abbreviations: Np = naphthalenes, Fl = fluorenes, Ph = phenanthrenes, Db = dibenzothiophenes, Ch = chrysenes, Per = perylene, N0 = C_0-naphthalene, N1 = C_1-naphthalene, P0 = C_0-phenanthrene, P1 = C_1-phenanthrene, P2 = C_2-phenanthrene, Db2 = C_2-dibenzothiophene. Adapted with permission from Boehm et al. (2001). Copyright 2001, American Chemical Society.

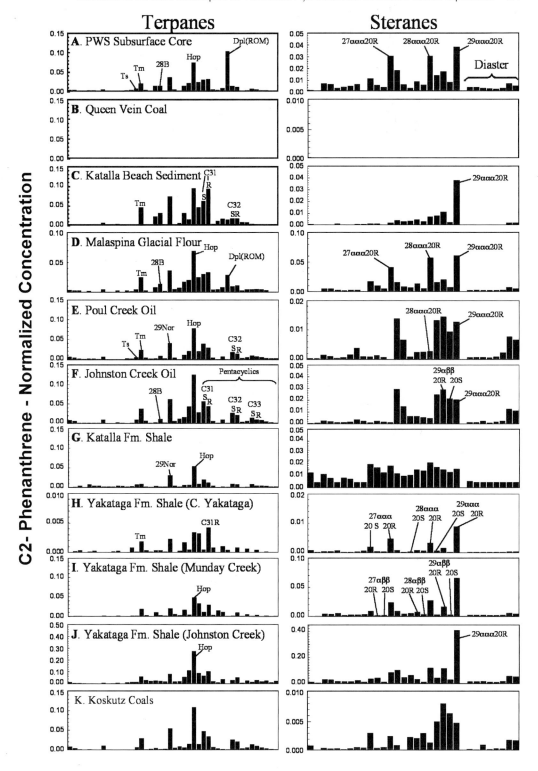

Figure 15-5 Triterpane and sterane compositions (C2-phenanthrene-normalized) for (A) PWS benthic core, (B–J) potential source inputs to the natural background. Abbreviations: Ts = 17α(H)-22,29,30-trisnorhopane, Tm = 18α(H)-22,29,30-trisnorhopane, 28B = $C_{28,30}$-bisnorhopane, Dpl(ROM) = diplotene, recent organic matter (the Copper River is a major source), 29Nor = norhopane, Hop = C_{30}-hopane, Diaster = diasteranes. Adapted with permission from Boehm et al. (2001). Copyright 2001, American Chemical Society.

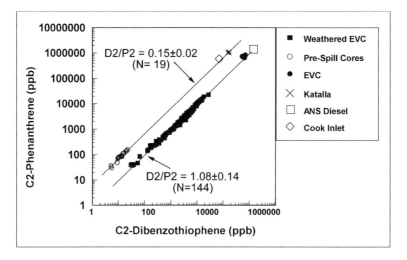

Figure 15-6 C_2-Dibenzothiophenes *versus* C_2-phenanthrenes (D2/P2) relationship of weathered EVC, diesel refined from ANS crude feed stock at a Kenai Refinery, PWS regional background hydrocarbons from deep subtidal cores, and Katalla and Cook Inlet crude oils.

different sources to the intertidal, zero tide, to the near-shore shallow subtidal, and to offshore sediments are well documented in data collected during Exxon-supported and NOAA surveys. The PAH compositions, reported by NOAA in the PWSOIL database, of the seafloor sediments at various water depths for one of those sites sampled in 2001, Northwest Bay, Eleanor Island, are shown in Figure 15-7. Intertidal subsurface sediments in Figure 15-7 have a PAH distribution dominated by weathered EVC. At the zero tide level TPAH is markedly lowered, the heavily weathered EVC component is reduced to ppb levels, and the 4- to 6-ring PAHs of a pyrogenic component have increased. From 3–6-m water depths, the petrogenic component in seafloor sediments is further reduced and now contains a recognizable contribution from the petrogenic background (high naphthalenes/fluorenes ratio) and the 4- to 6-ring PAH pyrogenic component is enhanced. Heavily weathered EVC, indicated by the comparable concentrations of C_3-Ph and C_3-Db, is a minor component at low ppb levels. At 10 m, the pyrogenic (4- to 6-ring) component dominates. The regional background (low C_2-Db/C_2-Ph) is a minor component and no EVC is detected. At 20 and 60 m, regional background, pyrogenic PAH, and perylene dominate the fingerprints. At a water depth of 100 m, the regional petrogenic background signal dominates. The increase of TPAH with water depth in the subtidal sediments is due to the close association of the PAH with the fine-grained fraction (clay/mud) in the sediments.

15.5 Hydrocarbon Source Allocations

15.5.1 Source Allocation Models

The techniques used for source identification and the quantitative allocation of hydrocarbon sources in mixtures have been developed and evolved in environmental studies and the petroleum industry largely over the last 20 years (Burns et al., 1997, 2006; Mudge, 2002). Methods to quantitatively allocate the multiple source components found in the PWS marine environment evolved as understanding of the complexity of hydrocarbon inputs improved. These methods included simple analyte-ratio mixing models, principal component analysis, and quantitative multivariate techniques such as partial least squares (PLS).

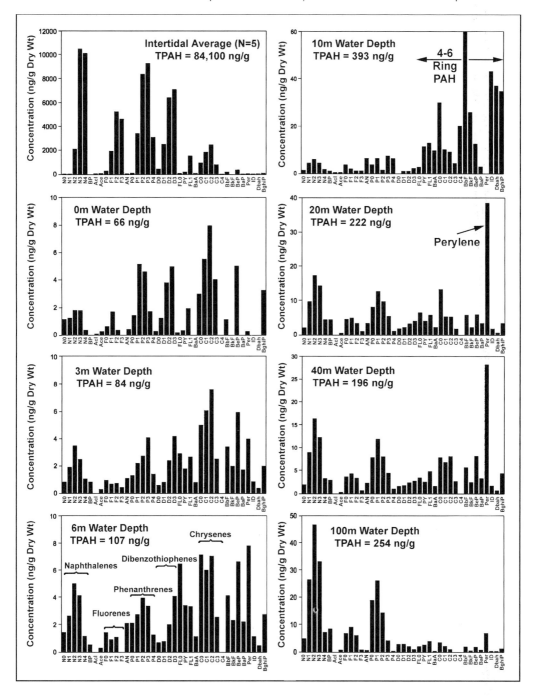

Figure 15-7 Sediment PAH compositions for a profile from the mid-tide zone to 100-m water depths at a location containing buried EVC residues on the beach, Northwest Bay, Eleanor Island. (Data from S. Rice, personal communication, 2003).

15.5.2 Qualitative Allocation Models

Short and Heintz (1997) developed a first-order loss-rate kinetics weathering model, calibrated with the results of experiments on gravels coated with oil and washed in the laboratory for six months, to describe the extent of weathering of EVOS in a sample and to discriminate EVOS and the natural petrogenic background. The TPAH-normalized concentrations of 14 PAH analytes (13 alkylated PAH and unsubstituted chrysene) in a sample are compared to their concentrations in the unaltered cargo crude. A suite of 26 weathered oil samples containing all 14 analytes was used to describe the PAH analyte variability with extent of weathering. Discrepancies between the measured and model-predicted PAH concentrations in EVOS samples were compared with a probability distribution of these discrepancies derived from the experimentally weathered samples. The weathering parameter, ω, summarizes the exposure history of the volume of oil.

The model can be used with relatively fresh oils when all 14 analytes are present. However, if any of the 14 analytes is absent, the model fails. This could happen at any weathering condition for sediment samples having low PAH concentrations or for samples that contain sufficiently weathered oil. For EVC shoreline oil that condition is reached when the mass loss from the original oil approaches 50% and all of the alkylated naphthalenes have been removed (Figure 15-2).

The model is also used to determine the probability that a specific sediment sample contains EVOS PAH or PAH from the natural regional background. The background is defined by the normalized PAH distributions for the median value of 14 benthic sediment samples from Constantine Harbor. The error distribution for this pattern is obtained by comparing the other 13 patterns with the median pattern. The model then uses the error distributions determined for the weathering model (from the 26 samples in the weathering suite) and for the natural background model (from the 14 samples in the background suite) to determine the probability that a given benthic sediment sample contains PAH from one source or the other. The model cannot apportion those sources in mixed samples.

15.5.3 Quantitative Models

15.5.3.1 PAH Ratios

The PAH-based fingerprinting and source-allocation method used by Page et al. (1995, 1996) was based on the observation that the regional background PAH had low ratios of alkyl dibenzothiophene to alkyl phenanthrene (e.g., C_2-Dbt/C_2-Phe) relative to the spill oil and its weathering products (Table 15-1A and Figure 15-6). Douglas et al. (1996) showed that the ratios for the respective isomers remain constant over a range of weathering states up to approximately 70% depletion of the total petroleum hydrocarbons present. Diesel fuel, refined from Alaska North Slope crude oil, has the same C_2-Dbt/C_2-Phe as EVC, but can readily be differentiated by the lack of PAH with 4 rings or greater (C_2-Ch/C_2-Phe = 0.0) and the lack of saturate and aromatic biomarkers, all having boiling points higher than the distillate fuel boiling-point range. The EVC TPAH contribution to a benthic sediment sample containing only two hydrocarbon components (petrogenic background and EVC) is determined from the following equation developed by Page et al. (1995). Subtracting this value from the total petrogenic PAH yields the total background-sourced petrogenic PAH.

$$\text{TPAH}_{\text{ANS}} = \frac{C_2 - \text{Phe}_{\text{sample}}}{0.093} \left[1 - \left[\frac{1.07 - \left(\frac{C_2 \text{Dbt}}{C_2 \text{Phe}} \right)_{\text{Sample}}}{0.92} \right] \right]$$

This source allocation model is based entirely on the 3-ring PAH. Consequently, it cannot distinguish ANS crude from ANS-derived diesel. Nor can it resolve the multiple petrogenic inputs that contribute to the natural hydrocarbon background. Due to variabilities in the distinguishing parameters, Page et al. (1996) concluded that low levels (less than

3 to 8%) of total petrogenic PAH having an ANS signature generally cannot be differentiated from the natural regional petrogenic background.

15.5.3.2 Statistical Models

The use of a combination of principal components analysis (PCA) and partial least-squares (PLS) analysis was developed for specific use in apportioning the natural petrogenic background to its component sources. While there are some differences in the PAH compositions of the various sources (Table 15-1A and Figure 15-3), they alone cannot be used to resolve those sources. However, there are small, but important, differences in the saturated and TAS biomarkers among these sources (Table 15-1B and Figure 15-4). Consequently, statistical programs that use both the PAH and biomarkers can resolve them. Because some of the source inputs are distinguished by relatively small compositional differences at low concentration levels, highly precise analyses with lower method detection limits (MDLs) were required. Refinements to standard analytical methods including gas chromatography-mass spectrometry (GCMS) analysis in the selected-ion-monitoring mode (SIM), increased sample size, column cleanup of the extract, and decreased pre-injection volume (volume of final extract prior to injection into instrument) lowered MDLs by factors of 10 to 1000 (Douglas et al., 2004). In sediment samples, MDLs < 0.5 ng/g for the PAH, including the alkylated homologs, and biomarkers were achieved.

Principal components analysis (PCA) is a multivariate statistical technique where sets of data are subjected to statistical analysis to reveal underlying relationships between groups of parameters and hence groups of samples in which these parameters are measured. (See Chapter 9 herein.) Based upon extensive field sampling programs in PWS and the eastern GOA, Burns et al. (1997) selected 18 potential sources contributing to PWS shoreline and benthic sediments in the path of the spill. They included EVOS in various weathering states, diesel oil and diesel soot, creosote and other combustion sources, and the natural PAH background in PWS sourced in the Gulf of Alaska. PCA analysis (Figure 15-8) was used to categorize the samples

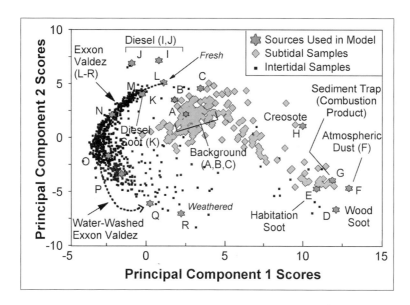

Figure 15-8 Factor score plot resulting from the PCA analysis of PAH in shoreline and benthic sediments in PWS. Modified after Burns et al. (1997).

according to similarity of PAH composition and, by inference, source inputs.

The PCA results subsequently were used to determine the combination of PAH and biomarker analytes that best resolved the potential sources of hydrocarbons. Both in the earlier PWS studies (Burns et al., 1997) and in the later GOA studies (Boehm et al., 2001; Burns et al., 2000, 2006), the choice of analytes and numbers of sources were selected by PCA for use in the constrained least-squares analysis. PCA cannot on its own be used to quantify the various source components in a given sample. Furthermore, PCA can be highly influenced by outliers, samples with radically different source compositions that affect the analysis and the clustering of the samples on the PCA plots. A more quantitatively rigorous apportionment methodology was needed and a partial least-squares model was employed (Burns et al., 1997, 2006).

15.5.3.3 Statistical Models

15.5.3.3.1 Multivariate Methods — Constrained Least Squares A constrained least-squares (CLS) model to allocate sources of PAH in PWS was used by Burns et al. (1997) in which iterative matching was used to determine the best fit of 36 PAH analytes in benthic sediment samples to 18 possible sources. For each sample, the model initially equalizes the contribution of total PAH from each source. Then the amount of total PAH from each source is varied using an iterative procedure that starts from the initial equalized contribution and systematically changes source contributions (the x_j below) by small steps until the following expression is minimized:

$$\sum_{i=1}^{36}\left[\sum_{j=1}^{18}(S_{i,j}x_j/\text{TPAH}_j)-d_i/\text{TPAH}_{\text{samp}}\right]^2$$

In this expression, i refers to the PAH analytes (1–36), j refers to the PAH sources (1–18), $s_{i,j}$ is the concentration of the ith analyte of source j, x_j is the fraction of the sample's total PAH that is due to source j, TPAHj is the total PAH of source j, d_i is the concentration of the ith analyte of the sample, and TPAHsamp is the total PAH of the sample. The $s_{i,j}$ and TPAHj are described in Burns et al. (1997) for 18 possible sources. The procedure fits only the analytes that are detected in the analysis; analytes reported as nondetects are skipped so it is imperative that low detection limits (Douglas et al., 2004) be achieved on all samples. The least-squares algorithm always converges to the same results from different starting values. For each sample, the least-squares model calculates the contributions of each of the 18 sources and the goodness of fit. The contributions to a sample from any major source type (such as spill oil, diesel oil and diesel soot, natural background, or combustion products and creosote) can be determined by summing the contributions of appropriate sources and multiplying by the total PAH (TPAH) of the sample.

Burns et al. (2002, 2006), recognizing the limitations of using just the PAH analytes and the important additional chemical constraints provided by the saturate and aromatic biomarkers, incorporated them into the CLS model. They noted, however, that care must be taken to ensure that accurate and meaningful solutions are obtained. Results are affected by many factors including the selection of analytes and sources, the use of data from different labs and/or instruments, the presence of analytical noise in the data, and the need to be consistent with other sediment chemical constraints such as total organic carbon (TOC). For example, including too many low-concentration analytes, i.e., those well below the method detection limit (MDL), from a dataset containing concentration-dependent noise risks driving the solution in an incorrect direction. On the other hand, including too few can cause the solution to miss important contributing sources entirely. The use of data from different labs, even labs ostensibly following the same protocols, can result in incorrect source attributions due to different instrument sensitivities and differences in quantification procedures.

The expanded CLS model was used to quantitatively allocate sources in the prespill benthic sediments of PWS and in the benthic sediments of the GOA east of PWS (Boehm et al., 2001, 2002). Based upon PCA results, 38 PAH analytes and biomarker compounds were selected and up to 18 GOA sources were identified. The constrained least-squares iterative method was then used to find the linear combination of analyte distributions from the 18 different sources that best matched the analyte distribution of the sample. The results of that analysis are shown in Figure 15-9. They show that the dominant contributors to both GOA and PWS sediments are the organic components (kerogens) of eroding shales and residues from the natural seeps. Seep contributions are particularly important in the GOA near the mouths of the streams, where seeps have been identified. Coal inputs are important

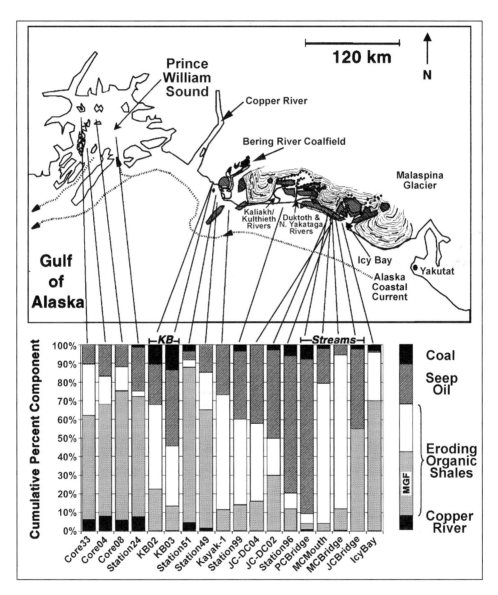

Figure 15-9 Source apportionments for GOA and prespill PWS benthic sediments. KB = Katalla Beach; MGF = Malaspina Glacier Flour. Reprinted with permission from Boehm et al. (2001). Copyright 2001, American Chemical Society.

only at the mouths of streams whose drainage areas include coal seams. The important contributors to prespill benthic sediments in PWS are the shales and seep residues.

The current version of the CLS model includes 136 PAH and biomarker analytes and up to 30 sources (Burns et al., 2006).

15.5.3.3.2 Multivariate Analysis — Partial Least Squares.

A partial least-squares (PLS) model was used by Mudge (2002) to assess the percentage contribution of coal, seep oil, shales, and rivers to the hydrocarbon loading of benthic sediments in the Gulf of Alaska. Previously published compositions of benthic sediment samples were analyzed using selected sites as sources in order to develop signatures. These signatures are based on the PAH and saturate biomarkers.

The principal components describing these sources are then fitted to the data for other sites around Prince William Sound (PWS) and Gulf of Alaska (GOA) to determine the proportion of the variability described by each source. A mixed source of coal, seep oil, eroding shales, and river sediments (two groups) described 13%, 18%, 24%, 26%, and 20%, respectively, of the variance in PWS and GOA benthic sediment data.

The procedure involves addition of a small constant to each analyte to permit log transformation of nondetected analytes. A principal components analysis (PCA) is conducted, followed by a PLS study. PCA identifies orthogonal data projections (principal components 1 and 2) that explain much of the variance in the data. The PCA results identified the five source types (oil, coal, shale, and river sediments groups 1 and 2). Source-type selection is based on the first two principal components, and some individual samples were excluded.

PLS is used to determine loadings factors for the source types and the amount of variance in target samples explained by each source-type loadings signature. If variance explained by the various source types totals more than 100%, variance explained is then normalized to 100%. To obtain percent contribution, CLS fits relative concentrations of all individual analytes rather than variance and is therefore more consistent with material balance constraints on individual analytes.

Different sources may explain the same amount of variance but have different degrees of fit to the relative concentrations. Consequently, PLS and CLS can yield different results for the same data set (Burns et al., 2006). The PLS model used by Mudge (2002) provides results for percent variance explained by the identified source types, not by individual source samples. By contrast, CLS determines the contributions of individual source samples, which can then be summed in various groupings to show contributions by source type. Source apportionment by both the CLS and PLS methods yield comparable results in that they both recognize that coal is not a major contributor to the background in benthic sediments of PWS. However, the 13% coal estimate produced by application of PLS to the data is certainly too high, given the high organic carbon content of coal (>50%) and the low total organic carbon in PWS benthic sediments (<1.0%), which is much lower than a 13% coal contribution would predict, as discussed in the next section.

15.5.3.4 Total Organic Carbon (TOC) Constraints on Source Allocations

The application of multivariate statistical models to source allocation always provides a solution. But how realistic is that solution? It must be geologically reasonable, i.e., the source inputs to the model must actually be contributors given the geological setting. Furthermore, all potentially contributing sources must be included in the model. For some types of samples, the reasonableness of a solution can be checked independently using TOC mass-balance constraints. Those are samples in which the PAH and chemical biomarkers are in the TOC-contributing components (usually fossil forms of carbon such as kerogen, asphalt, oil) and are strongly correlated. For example, the results of the source allocations obtained by Boehm et al. (2001, 2002) were checked using TOC as a constraint. The TOC content of all potential sources and the environmental samples are required for this check. The TOC contents and the relative amounts of the sources in the calculated best fit are used to calculate the TOC content of the mathematically reconstructed sample. This calculated value is then compared with the measured TOC content of the true sample. If recent organic matter (ROM) such as plant debris and animal remains is not an important component of the environmental sample, a close match between the calculated and measured values for TOC will indicate that the solution is permissible and may be correct. If the two do not agree within appropriate error limits, the source allocation cannot be correct. This mismatch could be due to the presence of ROM, which is a contributor of TOC but not of any of the analytes used in the source allocation, or it may indicate that all of the potential source inputs have not been considered. Mass-balance assessments of source input based on TOC, PAH, and chemical biomarkers are useful because sources having different thermal maturity levels and different organic facies, characteristics of the eastern GOA tertiary sedimentary section, can exhibit large differences in TOC-nomalized PAH and biomarker concentrations.

15.6 Allocation of Anthropogenic Sources of PAH

Prince William Sound has had an extensive history of human and industrial activity in PWS going back into the 1800s and earlier. Those activities, particularly ones involving fuel use, leave chemical signatures in near-shore and intertidal sediments (e.g., see Page et al., 1999, 2002a, 2004). Consequently, understanding the industrial and cultural history of the spill zone in an area like PWS is a key part of a comprehensive approach to environmental forensics. Pyrogenic and petrogenic hydrocarbons, unrelated to the 1989 spill, are present in the vast majority of the subtidal sediment samples analyzed (Page et al., 1995, 1996, 1999, 2002a, 2004; Bence et al., 1996). Localized shoreline sources of PAH that contribute to intertidal and near shore subtidal sediments are characterized by relatively high concentrations of 4- to 6-ring PAH (pyrogenic sources) or by their petroleum signatures (e.g., Monterey petroleum from unused fuel in derelict fuel tanks). These anthropogenic sources are located at villages, fish hatcheries, fish camps, recreational campsites as well as now-abandoned settlements, canneries, sawmills, and mines. Petroleum was the major energy source at these facilities.

The 1964 earthquake and the accompanying tsunami destroyed settlements and structures at sites throughout PWS (Lethcoe and Lethcoe, 2004). This destruction released significant quantities of petroleum products (Kvenvolden et al., 1995). Some entered the PWS marine environment from onshore locations and were dispersed into intertidal and near-shore subtidal areas. Because these chronically contaminated sites are located above the tide zone, hydrocarbon inputs have continued to the present time.

Field studies were conducted in 1999 and 2000 to characterize the near-shore subtidal sediments in embayments adjacent to sites of past industrial activity inside the *Exxon Valdez* spill zone in western PWS as well as bays that were heavily oiled in 1989 and an embayment outside the spill zone (Page et al., 2002a, 2004). Embayment locations are shown in Figure 15-10. Sediment samples were analyzed for saturate and aromatic biomarkers, PAH, total organic carbon (TOC), and grain size. Included were 36 2- to 6-ring PAH analytes, 44 C_{19}-C_{35} tri-tetra- and pentacyclic triterpanes, demethylated hopanes (*m/z* 191), 14 C_{21}–C_{29} regular steranes (*m/z* 217, 218), 8 C_{27}–C_{29} diasteranes (*m/z* 259), and C_{20}–C_{28} tri-aromatic steroids (*m/z* 231). Diagnostic patterns of PAH analytes and biomarker compounds were used to discriminate hydrocarbon sources in near-shore embayment sediments using an expanded version of the least-squares iterative methodology described by Burns et al. (1997). For this study, four categories of sources (weathered NSC, the prespill natural regional background,

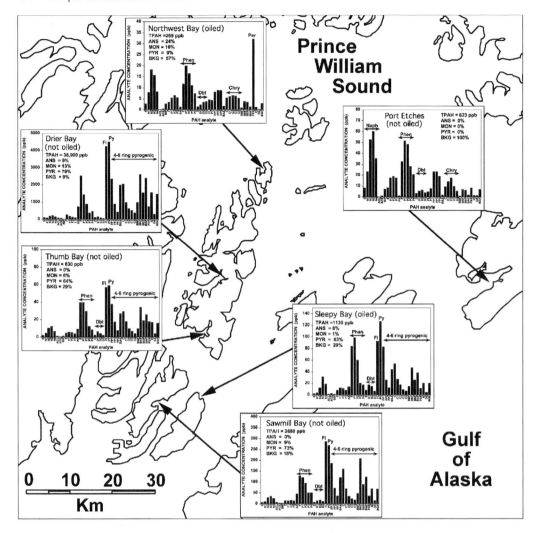

Figure 15-10 1999–2000 embayment benthic sediment sampling program with site locations, oiling category, representative sediment PAH profiles, and the results of the CLS source allocations, where ANS = Alaska North Slope oil, Mon = Monterey petroleum, PYR = pyrogenic PAH, and BKG = regional petrogenic background.

Monterey tars, and combustion products) were represented by 16 specific samples (see Page et al., 2002a) and 86 PAH and biomarker analytes were used to describe the hydrocarbon sources in the model.

The source allocation results of the CLS analysis are summarized in Table 15-2. They show that pyrogenic and regional petrogenic background hydrocarbons are the dominant sources, even in bays that were heavily oiled in 1989. In all cases, the relative amount of dibenzothiophenes is low, consistent with small or no inputs of ANS petroleum, either from the 1989 spill or other sources. Even at Sleepy Bay, which was heavily oiled in 1989, pyrogenic PAH, possibly from fuel use by fishing vessels known to anchor there, is the dominant source.

15.7 Identification of Hydrocarbons in Biological Samples

To assess the impact of the spill on the biological resources of the region, Bence and

Advances in Forensic Techniques for Petroleum Hydrocarbons: The *Exxon Valdez* Experience 471

Table 15-2 Summary of Source Allocation Results for Subtidal Sediments Sampled in Six PWS Embayments in 1999 and 2000

Source Type	Sleepy Bay	Port Etches	Sawmill Bay	NW Bay	Drier Bay	Thumb Bay
Spill path?	Yes	No	No	Yes	No	No
Industrial activity site?	No	No	Yes	No	Yes	Yes
Measured TOC (wt.%)	0.37	0.49	0.49	0.98	10.85	1.75
TPAH (ppb dry wt.)	1130	623	2650	269	38,900	630
ANS %	8	0	0	24	0	0
Monterey %	1	0	9	10	13	6
Pyrogenic %	63	0	73	9	79	64
Background %	29	100	18	57	9	29

Burns (1995) used forensic chemistry to analyze tissue PAH data for more than 1500 tissue samples reported by the Oil Spill Health Task Force (OSHTF) and nearly 4700 tissue samples in PWSOIL, the government's damage-assessment database. OSHTF was formed shortly after the spill to assess subsistence food safety reported results for samples collected in 1989, 1990, and 1991. The November 1993 version of PWSOIL studied by Bence and Burns (1995) also reported data only for samples collected during the 1989–1991 period. Samples reported in that database were analyzed either at the NOAA-NMFS Auke Bay Laboratory (NOAA/ABL) or at the Geological and Environmental Research Group (GERG) of Texas A&M University. The public version of the government database EVTHD (*Exxon Valdez* Trustees' Hydrocarbon Database) was first released in 1996. The PWSOIL version became the general repository database for all samples, including those collected but not analyzed. Both continue to be maintained through funding support from the *Exxon Valdez* Trustee Council. In more recent years, studies by both *Exxon Valdez* Trustee- and Exxon-supported scientists have focused on the bioavailability of weathered EVOS residues to marine organisms. These have included analyses of intertidal biota at beaches where EVOS residues are buried beneath a boulder-cobble surface layer (Boehm et al., 2004; Neff et al., 2006) and of fish eggs and embryos that have been exposed to varying doses of oil in the laboratory (e.g., Heintz et al., 1999; Carls et al., 1999). The 2004 version of EVTHD, downloaded from the NOAA-NMFS-Auke Bay Lab website, contains additional chemical data for samples collected in the period 1991–1996 that were not evaluated by Bence and Burns (1995).

For their analysis of the OSHTF and PWSOIL (November 1993 version) databases, Bence and Burns (1995) developed a rigorous procedure for classifying the PAH signatures, where eight categories of PAH fingerprints were identified in the tissue databases.

1. EVC. *Exxon Valdez* crude and its weathering products (Figures 15-11A, 15-12).
2. WSF. Water-soluble fraction of EVC (Figure 15-14A) or of diesel (Figure 15-1).
3. DIESEL. EVC signature absent the alkylated chrysenes and usually confirmed by the truncated normal alkane distribution (Figures 15-11B, 15-12).
4. PETRO. Diesel signature, but alkylated chrysenes are below detection limits.
5. Non-ANS. Contains alkylated PAH, but no C_2-dibenzothiophene when C_2-phenanthrene > 10 ppb.
6. UNRESOLVED. Contains alkylated PAH, but concentrations are too low to be definitive.
7. ALKYL NAPH (ALKNP). Contains only alkylated naphthalenes.
8. PROCEDURAL ARTIFACT (PA). Consists solely of 25 required reporting

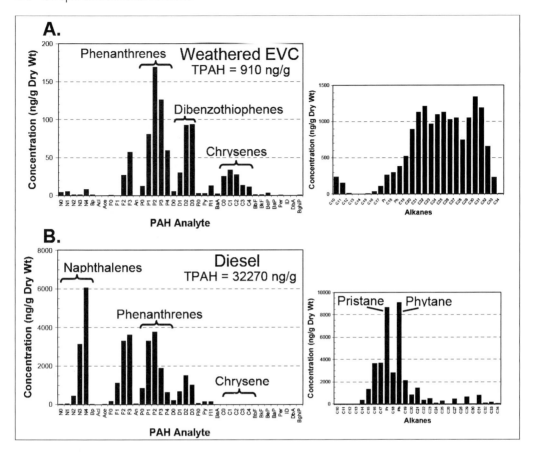

Figure 15-11 PAH and alkane compositions of bald eagle eggshells. (A) Weathered EVC on eggshell collected at Italian Bay on June 8, 1989, EVTHD ID #23363. (B) Relatively fresh diesel on eggshell collected from Hells Hole in eastern PWS outside the spill zone on July 18, 1990, EVTHD ID #20029.

analytes including the two C_1-, one C_2-, and one C_3-naphthalene isomers, and 1-methylphenanthrene (Figures 15-12C, 15-13A, B). This PAH signature reflects an analytical protocol that required the reporting of an integrated signal for each of the 25 analytes when the signal was below the analyte MDL. No baseline correction was made. Consequently, it is integrated electronic noise. It is present in all analyses but is recognizable only in samples having PAH signals that are at their limits of detection.

All assignments of PAH classifications were made conservatively. That is, if a fingerprint contained a mixed signature, e.g., EVC and PETRO, it was classified as EVC.

When applied to the nearly 6200 analyses of tissues (excluding shellfish) in the two databases, Bence and Burns (1995) found the procedure to be highly diagnostic of EVC for samples of external surfaces, where selective uptake and metabolism were not issues and applicable for samples of the gastrointestinal tract. They found that, excluding shellfish, only a small fraction of the samples contained recognizable traces of EVC. Most of those samples were collected in 1989. Samples collected in 1990 rarely contained identifiable EVC. Samples reported to have been collected in 1991 never contained identifiable traces of

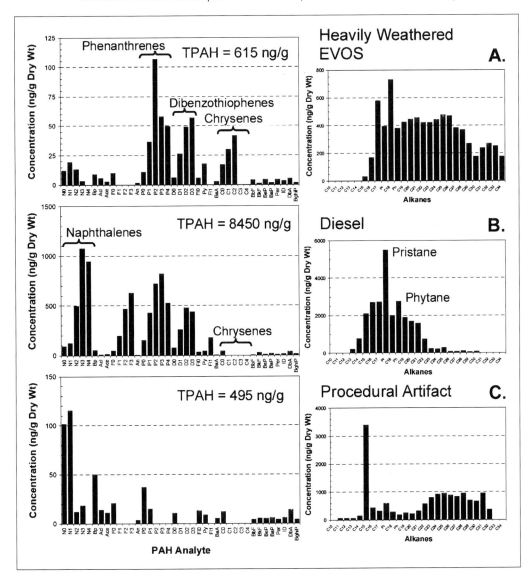

Figure 15-12 PAH and alkane compositions for sea otter tissues. (A) Heavily weathered EVOS, sea otter skin collected April 12, 1989, EVTHD ID #27757, site not documented. (B) Relatively fresh diesel, sea otter fur collected August 1, 1990, site not documented, EVTHD ID #22851. (C) Procedural artifact, sea otter blood, Ogden Passage, SE Alaska, EVTHD ID #20044.

EVC. These results showed that subsistence foods were safe for human consumption shortly after the spill.

Procedural artifacts formed the dominant fingerprint in the EVTHD database. Diesel, presumably from field contamination, constituted the most common petroleum-sourced fingerprint.

The methodology developed by Bence and Burns was criticized by Carls et al. (2002) for failure to identify EVC in tissues of herring, including eggs, which had been exposed to EVC and its water-soluble fraction in laboratory experiments. However, those experiments were conducted after 1991 (Carls et al., 1999) and were not included in the 1993 version of

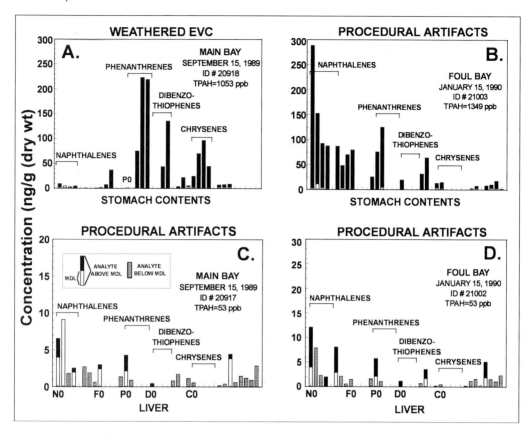

Figure 15-13 PAH compositions for Harlequin duck stomach contents (A–B) and corresponding liver tissue (C–D) from the same bird collected by State of Alaska investigators. Adapted, with permission, from *STP 1219 Exxon Valdez Oil Spill: Fate and Effects in Alaskan Water*, copyright ASTM International, 100 Barr Harbor Drive, West Conshohocken, PA 19428.

PWSOIL that was evaluated by Bence and Burns (1995). Our subsequent analysis of the additional tissue data reported for 1991 through 1996 in the 2004 version of EVTHD is reported in Table 15-3. Results for field and laboratory-exposed samples are reported separately.

Because EVOS raised questions about the solubility of Alaska North Slope crude in water and the mechanisms by which fish and wildlife could be exposed to toxic components in the oil, a method was developed to discriminate an analysis of water that contained dispersed oil droplets from an analysis of the water-soluble fraction (WSF) of that oil. That method invoked the relative solubilities of the petrogenic PAH and their equilibrium octanol-water partition coefficients (Miller et al., 1985; Mackay et al., 1992) and was tested on the products of laboratory oil-exposure experiments. Briefly, an EVC-WSF designation is assigned when C_2-dibenzothiophene is present, C_2-phenanthrene > 1 ng/g, C_0-Db/C_1-Db > 1.0, C_0-Ph/C_1-Ph > 1.0, and C_2–C_4 chrysenes and other 4- to 6-ring combustion products, e.g., anthracene, benzo(*a*)pyrene, etc., are not detected. The presence of C_2–C_4 chrysenes at concentrations above their MDLs indicates that dispersed oil droplets, in addition to dissolved components, are present. (C_0/C_1-chrysene can be used to confirm the WSF designation when the methyl chrysenes are above their MDL.)

Laboratory exposure studies conducted by NOAA/ABL in the pink salmon and herring egg and embryo investigations involved direct exposure to artificially weathered EVC and to

Table 15-3 Interpretations of Tissue (Excluding Shellfish) PAH Compositions Reported in the 2004 Version of EVTHD (See Text for Definition of EVTHD)

A. Laboratory-Exposure Samples

Year	Tissue (No. of Analyses)	EVC & EVC + WSF	WSF	DIESEL	PETRO	UNRES	ALK NP	PA
1991	P. Salm. viscera (54)	5				45	4	
	P. Salm. carcass (111)	13				85	13	
1992	P. Salm. alevin (11)	3	2			6		
	P. Salmon egg (4)		2			2		
1993	P. Salm. alevin (29)	9				20		
	P. Salm. egg (11)	7	1			3		
	P. Salmon fry (3)	2				1		
1994	P. Salmon fry (14)	3	11					
	Herring egg (15)		14			1		
	Herr gonad (45)		18			27		
	Herr muscle (84)	52				32		
1995	P. Salmon (6)	6						
	Herring egg (8)	3				5		
1996	P. Salmon (9)	7				2		

B. Field-Exposure Samples

Year	Tissue (No. of Analyses)	EVC & EVC + WSF	WSF	DIESEL	PETRO	UNRES	ALK NP	PA
1991	BLKI (1)				1			
1992	None							
1993	HADU Fat (10)				2	8		
	HADU Gut (15)			1	1	12		1
	HADU Liv (14)				3	9	2	
1994	None							

P. Salm. = pink salmon; BLKI = black kittiwake; HADU = harlequin duck.

naturally weathered EVC as well as to the water-soluble fractions (WSF) of the former. The chemistry data from those experiments are reported in EVTHD. Interpretations of the tissue analyses, following the decision-tree approach developed by Bence and Burns (1995), are reported in Table 15-3. A number of the samples designated as UNRES (UNRESOLVED) have some WSF characteristics, but the individual PAH analyte concentration levels are very close to their MDLs (ppb to sub-ppb in tissues) making their classification uncertain.

Many of the analyses classified as EVC also contain a WSF component as evidenced by C_0-Db/C_1-Db > 1.0 or C_0/C_1-chrysene > 1.0. Samples having these characteristics are placed in the EVC & (EVC+WSF) column of Table 15-3. A WSF fingerprint for pink salmon eggs exposed to weathered EVC in a 1994 NOAA/ABL experiment is shown in Figure 15-14A. It has the characteristic $C_0 > C_1 > C_2 > C_3$ pattern for the fluorenes, phenanthrenes, and dibenzothiophenes, and the chrysenes are at or below detection limits. A mixed EVC+WSF PAH composition for herring eggs exposed to artificially weathered EVC in NOAA/ABL 1993 laboratory experiments is shown in Figure 15-14B. These observations indicate that, for many of the laboratory exposure experiments conducted by NOAA/ABL in gravel column generators (Heintz et al., 1999; Carls et al., 1999), direct exposure to dispersed oil droplets in the water phase was the probable cause of observed harmful effects rather than exposure to low (≤1 ppb) total concentrations of water-soluble PAH. Brannon et al. (2006) used the same experimental design as

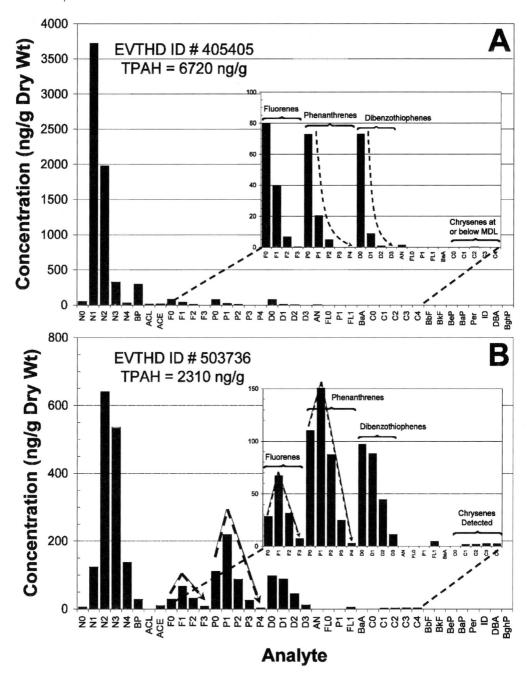

Figure 15-14 PAH compositions of tissue exposed to weathered oil in column generator experiments conducted by NOAA/ABL, Juneau, AK. A. Pink salmon eggs after 36 days of exposure to artificially weathered EVC at a gravel loading of 281 ppm oil showing characteristic WSF composition. B. Herring eggs showing mixed oil + WSF composition.

used in the NOAA experiments to assess the toxicity of weathered EVC to pink salmon embryos. They observed that the experimental design of the generators did not exclude dispersed oil droplets from the aqueous phase and concluded that toxicity was not limited to the water-soluble fraction.

The results of the laboratory gravel generator experiments in which pink salmon embryos were exposed to very weathered oil (Heintz et al., 1999) formed the foundation of a changing paradigm in ecotoxicology proposed by Peterson et al. (2003). Those workers concluded that wildlife exposed to as little as 1 ppb TPAH dissolved in the aqueous phase can experience chronic sublethal effects, including population reductions and cascading indirect effects resulting in delayed population recovery. Because the source of the toxicity measured in those experiments was not due solely to the dissolved fraction, the experimental foundation upon which the changing paradigm is based is suspect and the experiments should be repeated with more rigorous controls to ensure that dispersed oil droplets are eliminated.

15.8 Applications of Forensic Methods to Assessments of Oil Bioavailability

15.8.1 PAH Uptake in Biota

Measurements of petroleum fractions in sediments alone are insufficient to establish that there is a risk to wildlife because the spill remnants may be in forms and locations that are not available for uptake by biota. Therefore, the identification and quantification of tissue hydrocarbons, in conjunction with the analyses of the spill-contaminated sediments, is needed to establish the bioavailability of those spill remnants. Indicator species, such as bivalve molluscs, are often used to assess the bioavailability of petroleum residues present at a spill site because they concentrate bioavailable lipid-soluble material from the ambient water column. Consequently, when PAH levels in the water column are too low to measure using conventional methods, mussel PAH can be used to obtain an estimate of the PAH concentrations dissolved in the water. The method uses lipid-water partitioning coefficients to estimate the equilibrium water concentration of dissolved PAH (Neff and Burns, 1996). In addition, in those situations where PAH exposure to higher-life forms *via* the food chain is suspected, prey biota can be analyzed to assess exposure risk to foraging wildlife.

Because mussels are ubiquitous and integrate exposure over time, they are useful indicators of bioavailable hydrocarbon inputs and have served as the basis for many water-quality monitoring programs such as Mussel-watch (O'Connor, 2002). Many studies of the *Exxon Valdez* spill have used the hydrocarbon chemistry of mussels and other biota to assess inputs of bioavailable hydrocarbons from residues of the spill, from sites of past and current industrial activity, and from natural petroleum hydrocarbon sources (e.g., see Babcock et al., 1996, 1998; Boehm et al., 1997, 2004; Carls et al., 2001, 2004; Neff et al., 2006; Page et al., 2005, 2006).

To assess the bioavailability of buried oil residues remaining on some beaches, a comprehensive mussel sampling program was conducted in 1998, 2000, 2001, and 2002 (Boehm et al., 2004). Using the fingerprinting methods described above to discriminate spill residues from other PAH sources, Boehm et al., (2004) found that petrogenic PAHs, attributable to the spill, were in low concentrations in mussels at sites where the underlying sediments have much higher concentrations of spill-related TPAH (Figure 15-15A, B, C, D). This illustrates the relative lack of bioavailability of the buried oil when the subsurface oil deposits are not disturbed. When the oil is disturbed, such as in the digging of pits for the deployment of SPMDs, mussels in the immediately adjacent area have elevated TPAH levels (Boehm et al., 2005). At sites having lower EVC-attributable concentrations in underlying sediments, low levels of pyrogenic compounds dominate the mussel PAH composition (Figure 15-15E, F, G, H).

Comparisons of TPAH levels in mussels at oiled sites with unoiled reference sites led

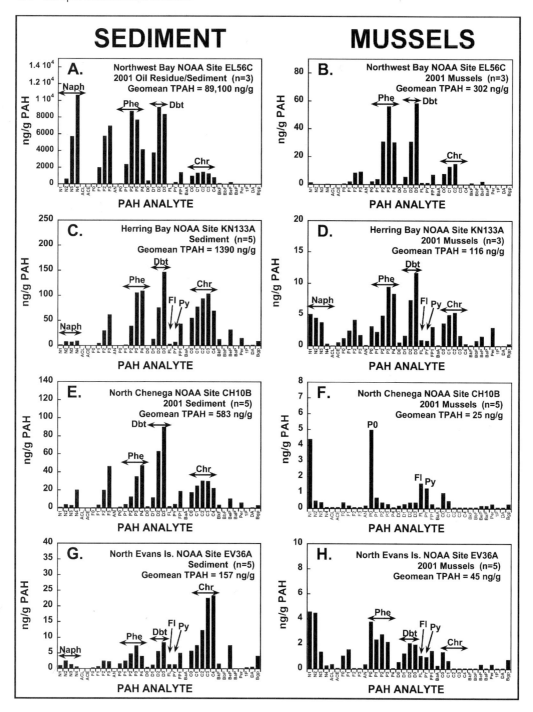

Figure 15-15 Composition and concentration of PAH in mussel tissues from NOAA oiled mussel bed sites compared with the PAH composition and concentration of underlying sediments for samples collected in 2001. The data are presented as the geometric means of the analyte distributions of replicate samples at each site. Note that the mussel TPAH concentrations are factors of 3 to 300 lower than the TPAH concentrations in the corresponding sediments.

Boehm et al. (2004) to further conclude that bioavailable hydrocarbons in the *Exxon Valdez* oil spill zone in PWS were at or near background levels by 2002. This conclusion is in agreement with Carls et al. (2004), who reported that mussel beds that were heavily oiled in 1989 were at background levels by 1999. Page et al. (2005) repeatedly sampled and analyzed mussels from 11 oiled sites over the period 1990–2002 to quantify the changes in bioavailable PAH with time. From those time-series data, they concluded that the annual mean loss of bioaccumulated PAH in mussels was 25%.

In 2002, Neff et al. (2006) tested 7 species of intertidal biota at 16 sites having buried EVOS residues to determine if PAHs from those residues were sufficiently bioavailable to intertidal prey organisms that they might pose a health risk to populations of birds and wildlife that forage on the shore. Forensic fingerprinting methods were used to discriminate EVC PAH from other PAH sources. Mean tissue TPAH from all sources in clams, worms, and intertidal fish from oiled sites range from 24 to 36 ng/g dry weight; mean TPAH concentrations in sea lettuce, whelks, and hermit crabs, and intertidal range from 5 to 11 ng/g. Mean TPAHs in biota at unoiled reference sites and sites having a record of historical human and industrial activity (HA sites) range from 6 to 15 ng/g and 7 to 2200 ng/g, respectively. Oil residues at oiled sites are concentrated above +1.8 m in the middle- (upper half) and upper-intertidal zones, whereas the biota are located largely in the lower-tidal zone (below +1.3 m). Tide-zone averages for the mean sediment TPAH at the 16 oiled sites are 2500 ng/g upper-tidal zone, 1600 ng/g middle-tidal zone, and 70 ng/g lower-tidal zone. Figure 15-16 shows the TPAH levels in clams and sediments from the three categories of sites (oiled, reference, and HA).

A 2005 study analyzed clams collected from shallow subtidal and lower intertidal locations at sites that had been oiled in 1989 and where sea otters have actively foraged since. Results for two locations are shown in Figure 15-17. Combustion products (parent compounds dominant relative to the alkylated homologs; presence of 4- to 6-ring PAHs) dominate the PAH compositions of both. The Bay of Isles

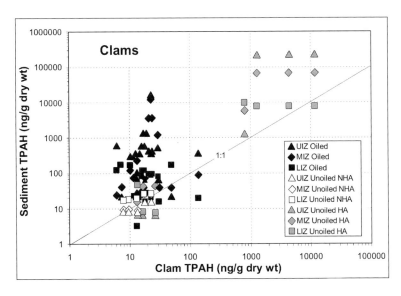

Figure 15-16 Mean sediment TPAH versus composite clam TPAH for 16 oiled, 3 reference, and 4 HA sites in western PWS. UIZ = upper intertidal zone (+2 m mllw), MIZ = middle intertidal zone (+1–+2 m mllw), LIZ = lower intertidal zone (0–+1 m mllw).

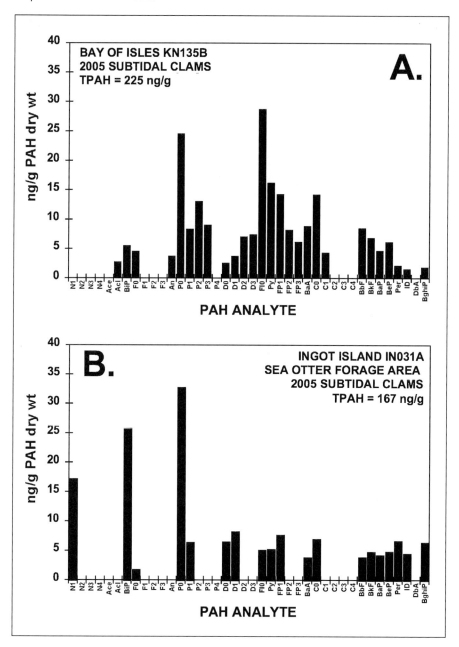

Figure 15-17 PAH composition of composite clam tissues from two sea otter foraging sites that had been oiled in 1989.

sample (Figure 15-17A) also has a minor, heavily weathered, petrogenic component. The absence of alkylated chrysenes suggests that it is diesel. These results indicate that sea otters that feed on these clams probably would experience low levels of CYP 1A induction. However, that induction would be due to the presence of the strongly CYP 1A-inducing pyrogenic 4- to 6-ring PAHs present as part of the background and not to EVC residues.

15.8.2 Passive Sampling of PAH in Water

Passive samplers such as semipermeable membrane devices (SPMDs) and low-density polyethylene strips (LDPEs) are used as surrogates for mussels. They are stable, easily deployed and retrieved, moderately easy to process and analyze, and can be used to obtain water-quality data at locations where mussels cannot live. However, because they partition the PAH differently than mussels, results from these devices should be interpreted with caution. In a direct comparison of mussels and SPMDs as monitoring tools at a range of oiled and unoiled sites in PWS, Boehm et al. (2005) found that TPAH concentrations in mussels and in surface-deployed SPMDs were correlated, but that lower-molecular-weight PAHs were relatively more abundant in the SPMDs than in the mussels. At sites with buried oil, disruption of the beach substrate during deployment, such as placing the SPMDs in hand-dug pits, artificially enhances the bioavailability of the oil residues and results in anomalously high TPAH concentrations. The study concluded that when mussels are available, their collection and analysis are the recommended monitoring tools. This is especially true when the assessment involves estimations of risk to foraging wildlife, because the sampling of biota provides a more realistic evaluation of the composition and concentrations of PAH to which foragers might be exposed.

15.8.3 Biological Markers

Elevated biological markers (CYP 1A activity and bile fluorescent aromatic compounds) in fish and near-shore vertebrate predators sampled at some locations within the EVOS spill path have been cited as circumstantial evidence that PAHs in the residues of the spill are bioavailable to wildlife (Trust et al., 2000; Bodkin et al., 20002; Jewett et al., 2002). Those studies further concluded that this exposure has had harmful effects at the population level for some species. However, it should be pointed out that the levels of CYP 1A observed in those studies are generally low and are pervasive throughout the PWS region. For the study reported by Trust et al. (2000), the maximum values of CYP 1A, as measured by ethoxyresorufin 0-deethylase (EROD) activity, in harlequin ducks from control sites (386 pmol/min/mg protein) were comparable to those from oiled sites (377 pmol/min/mg protein). Although certain PAH compounds such as benzo(a)pyrene and benzo(k)fluoranthene are strong inducers of CYP 1A (Lee and Anderson, 2005), they are either absent or at trace concentrations in EVC. Chrysene, present at ~50 ppm in EVC (Bence et al., 1996), is a weak to moderate inducer of CYP 1A activity. However, its solubility in water is so low ($\log K_{ow}$ ~6–8) (McKay et al., 1992; Neff et al., 1994; Neff and Burns, 1996) as to make exposure via the water column unlikely. Linking elevated biomarkers directly to a given source is problematic because there is no source specificity in the biomarker measurements. In PWS, this problem is compounded by the multiplicity of PAH sources that, potentially, could be the cause of elevated biomarkers in some wildlife species. For example, Huggett et al. (2003) conclusively demonstrated that sources other than EVOS spill residues are responsible for elevated CYP 1A and bile fluorescent aromatic contaminants (FAC) in five fish species collected in PWS and in the GOA at sites far-removed from the spill zone. The source of the biomarker induction is believed to be the regional background.

Elevated levels of CYP 1A and bile FAC were detected in some near-shore fish species caught at embayment locations in the spill zone in PWS that were also associated with past human and industrial activity (Huggett et al., 2003). Pyrogenic PAH comprise more than 50% of benthic sediment PAH in Drier Bay, Mummy/Thumb Bays, Sawmill Bay, and Sleepy Bay (Table 15-2), all sites of past or current industrial activity. Except for Sleepy Bay, these embayments were not oiled in 1989 (Page et al., 2002a, 2004). Subtidal sediments in Snug Harbor, an oiled bay, had ~12,000 ppb of pyrogenic PAH consistent with past fish

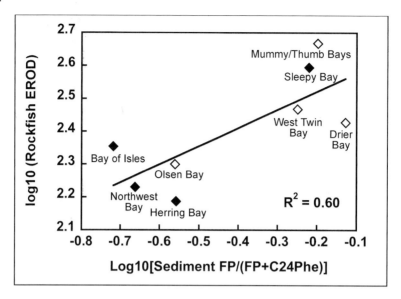

Figure 15-18 Correlation of rockfish liver ethoxyresorufin *O*-deethylase (EROD) activity, a measure of the level of CYP 1A induction, with benthic sediment pyrogenic index (FP)/(FP + C_{24}Ph). Solid symbols = sites oiled in 1989. Open symbols = unoiled reference sites.

processing activity (Page et al., 1999). Pyrogenic PAHs with four or more rings are much more potent CYP 1A inducers than the 2–3-ring PAH that dominate most of the PAH fraction of EVC and its residues (Neff, 2001; Lee and Anderson, 2005). A positive correlation of the pyrogenic PAH indicator ratio (FP)/(FP + $C_2 = C_4$) in embayment benthic sediments with CYP 1A activities in rockfish at the same sites suggests that pyrogenic PAHs are the source of the induction (Huggett et al., 2003; Page et al., 2004) (Figure 15-18). There are no correlations of biomarker activity with 1989 oiling or with concentrations of EVC residues (Jewett et al., 2002; Page et al., 2004). Pyrogenic PAH from past and current industrial activities are the likely sources of CYP 1A induction in fish from the spill zone locations studied.

15.9 Summary

In the more than 16 years of research conducted to assess the environmental impact of the *Exxon Valdez* oil spill, forensic techniques have evolved and new ones have been developed to address a host of issues, some of which appear unique to this incident and the Prince William Sound environment. These advances include the development of analytical, interpretive, and statistical tools to quantitatively resolve the multiplicity of hydrocarbon source inputs to benthic and intertidal sediments. Analytical protocols were refined to be more precise at lower concentrations, and saturate and aromatic chemical biomarkers were merged with the PAH, making it possible to resolve distinct hydrocarbon sources having relatively small differences in composition. Mass balance of total organic carbon (TOC), that compared calculated with measured chemical compositions, was invoked to provide independent assessments of the reasonableness of a given source allocation. Results showed that the prespill benthic sediments in the sound contained a substantial background of petrogenic hydrocarbons derived from Tertiary hydrocarbons systems consisting of petroliferous shales, oil seeps, and coals located in the eastern GOA.

Methods were developed for the quantification of both total mass loss and the chemical changes that occur in various fractions of shoreline oils as they weather. They confirmed that the inert asphaltenes and resin fractions make up an increasing fraction of the bulk oil as it weathers. Furthermore, they showed that all of the PAHs, including the 3- and 4-ring PAHs, are removed as weathering progresses, a feature of EVC not previously recognized. Although the alkylated chrysenes become an increasingly more important component of the PAH fraction, they decrease in the bulk oil as the total PAH fraction is removed during weathering. This is consistent with field observations that show oil toxicity decreases with weathering due to an overall decrease in toxic fractions with time.

A first-order kinetic model was developed to determine the rate of removal of subsurface oil residues. It showed that oil mass loss in 2001 was occurring at an annual rate of 20–26%, a rate similar to the annual mean loss of bioaccumulated PAH in mussels. If the assumptions made in that model are correct, at year-end 2005, less than 100 barrels of subsurface oil (<0.05% of the amount initially spilled) remain scattered along ~783 km of shoreline initially oiled. The proportion of these oil remnants to the amount initially released is comparable to those observed for the *Arrow* and *Florida* spills more than 35 years after those spills occurred.

Shortly after the spill, forensic fingerprinting tools were developed to identify sources of PAH in tissues of wildlife living in the spill zone and assess the suitability of those tissues for subsistence consumption. Those tools showed that most samples with identifiable EVC were samples of surfaces or the gastrointestinal tract collected in the first two years after the spill. Due to preferential uptake and differential metabolism, EVC was rarely recognized in internal tissues. PAH fingerprinting, when applied to tissues exposed to oil in laboratory experiments, revealed that in many cases the exposure was not solely to a water-soluble fraction of the oil as assumed but to dispersed oil microdroplets as well. Because the results of these experiments form the foundation of a proposed changing paradigm in ecotoxicology — that exposure to aqueous TPAH at ≤1 ppb concentrations can cause chronic sublethal effects in wildlife populations — they need to be repeated with more rigorous controls to ensure that a free oil phase in the form of microdroplets is not present.

Forensic fingerprinting tools facilitated correlations of biomarker (CYP 1A and FAC) levels in fish with the composition and amounts of PAH found in the region's natural hydrocarbon background showed that a component of the background was bioavailable. Slightly enhanced CYP 1A levels in sea otters and harlequin ducks that inhabit the spill zone of PWS are advanced as evidence for continuing exposure of wildlife to spill residues. However, low and variable levels of CYP 1A in those species are pervasive throughout the region that includes areas far beyond the area impacted by the spill. Natural and anthropogenic sources of PAH found throughout PWS and the GOA are likely responsible for this CYP 1A induction.

Concerns that PAH from the buried remnants of spill oil could be bioavailable to intertidal predators *via* the food chain at levels sufficient to have population-level impacts were addressed through the analysis of intertidal prey species at sites having buried oil residues. In this application of forensic techniques, it was shown that levels of PAH uptake in those prey species, including mussels, clams, worms, and intertidal fish, are far below levels known to cause harm to an individual intertidal forager, let alone entire populations. Furthermore, more than 99% of the shoreline in PWS now contains no oil residues. Consequently, claims of population-level impacts to wildlife in PWS seem implausible.

Acknowledgments

This work was funded by the ExxonMobil Corporation.

References

Babcock, M.M., P.M. Harris, M.G. Carls, C.C. Brodersen, and S.D. Rice, Persistence of oiling in mussel beds three to four years after the *Exxon*

Valdez oil spill. In *Proc. Exxon Valdez Oil Spill Symp.*, AFS Symposium 18, S.D. Rice, R.B. Spies, D.A. Wolfe, and B.A. Wright (eds.), Bethesda, MD: American Fisheries Society, 1996, 290–298.

Babcock, M.M., P.M. Harris, M.G. Carls, C.C. Brodersen, and S.D. Rice, Mussel bed restoration and monitoring. *Exxon Valdez* Oil Spill Restoration Project Final Report (Restoration Project 95090). National Oceanic & Atmospheric Administration, National Marine Fisheries Service, Auke Bay Laboratory, Juneau, AK, 1998.

Baker, J.M., I. Campodonico, L. Guzman, J.J. Texera, B. Texera, C. Venegas, and A. Sabhueza, An oil spill in the Straits of Magellan. In *Marine Ecology and Oil Pollution*, J.M. Baker (ed.), New York: J. Wiley and Sons, 1975, 441–471.

Bence, A.E. and W.A. Burns, Fingerprinting hydrocarbons in the biological resources of the *Exxon Valdez* spill area. In *Exxon Valdez Oil Spill: Fate and Effects in Alaskan Waters*. ASTM Special Technical Publication #1219, P.G. Wells, J.N. Butler, and J.S. Hughes (eds.), Philadelphia, PA: American Society for Testing and Materials, 1995, 84–140.

Bence, A.E., K.A. Kvenvolden, and M.C. Kennicutt, Organic geochemistry applied to environmental assessments of Prince William Sound, Alaska, after the *Exxon Valdez* oil spill — a review. *Org. Geochem.*, 1996, **24**, 7–42.

Bence, A.E., W.A. Burns, P.J. Mankiewicz, D.S. Page, and P.D. Boehm, Comment on "PAH refractory index as a source discriminant of hydrocarbon input from crude oil and coal in Prince William Sound Alaksa," by Hostettler et al., *Org. Geochem.*, 2000, **31**, 931–938.

Bodkin, J.L., B.E. Ballachey, T.A. Dean, A.K. Fukuyama, S.C. Jewett, L. McDonald, D.H. Monson, C.E. O'Clair, and G.R. VanBlaricom, Sea otter population and status and the process of recovery from the 1989 *Exxon Valdez* oil spill. *Maine. Ecology Progress Series*, 2002, **241**, 237–253.

Boehm, P.D., D.S. Page, E.S. Gilfillan, W.A. Stubblefield, and E.J. Harner, Shoreline ecology program for Prince William Sound, Alaska, following the *Exxon Valdez* oil spill: Part 2 — chemistry. In *Exxon Valdez Oil Spill: Fate and Effects in Alaskan Waters*, ASTM Special Technical Publication #1219, P.G. Wells, J.N. Butler, and J.S. Hughes (eds.), Philadelphia, PA: American Society for Testing and Materials, 1995, 347–397.

Boehm, P.D., G.S. Douglas, W.A. Burns, P.J. Mankiewicz, D.S. Page, and A.E. Bence, Application of petroleum hydrocarbon chemical fingerprinting and allocation techniques after the *Exxon Valdez* oil spill. *Mar. Poll. Bull.*, 1997, **34**, 599–613.

Boehm, P.D., D.S. Page, W.A. Burns, A.E. Bence, P.J. Mankiewicz, and J.S. Brown, Resolving the origin of the petrogenic hydrocarbon background in Prince William Sound, Alaska. *Environ. Sci. Tech.*, 2001, **35**, 471–479.

Boehm, P.D., W.A. Burns, D.S. Page, A.E. Bence, P.J. Mankiewicz, J.S. Brown, and G.S. Douglas, Total organic carbon, an important tool in a holistic approach to hydrocarbon fingerprinting. *J. Environ. Forensics*, 2002, **3**, 243–250.

Boehm, P.D., D.S. Page, J.S. Brown, J.M. Neff, and W.A. Burns, Polycyclic aromatic hydrocarbon levels in mussels from Prince William Sound, Alaska, USA, document the return to baseline conditions. *Environ. Toxicol. Chem.*, 2004, **23**, 2916–2929.

Boehm, P.D., D.S. Page, J.S. Brown, J.M. Neff, and A.E. Bence, Comparison of mussels and semipermeable membrane devices as intertidal monitors of polycyclic aromatic hydrocarbons at oil spill sites. *Mar. Poll. Bull.*, 2005, **50**, 740–750.

Brannon, E.L., K.M. Collins, J.S. Brown, J.M. Neff, K.R. Parker, and W.A. Stubblefield, Toxicity of weathered *Exxon Valdez* crude oil to pink salmon embryos. *Environ. Toxicol. Chem.*, 2006, **25**, 962–972.

Bue, B.G., S. Sharr, S.D. Moffitt, and D. Craig, Effects of *Exxon Valdez* oil spill on pink salmon embryos and preemergent fry. In *Proc. Exxon Valdez Oil Spill Symp.* AFS Symposium 18, S.D. Rice, R.B. Spies, D.A. Wolfe, and B.A. Wright (eds.), Bethesda, MD: American Fisheries Society, 1996, 619–627.

Burns, W.A., P.J. Mankiewicz, A.E. Bence, D.S. Page, and K. Parker, A principal-component and least-squares method for allocating sediment hydrocarbons to multiple sources. *Environ. Toxicol. Chem.*, 1997, **16**, 1119–1131.

Burns, W.A., A.E. Bence, P.D. Boehm, J.S. Brown, D.S. Page, G.S. Douglas, and K.R. Parker, Source allocation by least-squares hydrocarbon fingerprint matching. *SETAC*, 23rd Annual Meeting Abstract Book, Society of Environmental Toxicology and Chemistry, Orlando, FL, 2002, 16–20.

Burns, W.A., S.M. Mudge, A.E. Bence, P.D. Boehm, J.S. Brown, D.S. Page, and K.R. Parker,

Source allocation by least-squares fingerprint matching. *Environ. Sci. Tech.*, 2006, (in press).

Carls, M.G., S.D. Rice, and J.E. Hose, Sensitivity of fish embryos to weathered crude oil: Part I. Low level exposure during incubation causes malformations, genetic damage, and mortality in larval pacific herring (*Clupea pallasi*). *Environ. Toxicol. Chem.*, 1999, **18**, 481–493.

Carls, M.G., M.M. Babcock, P.M. Harris, G.N. Irvine, J.A. Cusick, and S.D. Rice, Persistence of oiling in mussel beds after the *Exxon Valdez* oil spill. *Mar. Environ. Res.*, 2001, **51**, 167–190.

Carls, M.G., G.D. Marty, and J.E. Hose, Synthesis of the toxicological impacts of the *Exxon Valdez* oil spill on Pacific herring (*Clupea pallasi*) in Prince William Sound, Alaska, U.S.A. *Can. J. Fisheries and Aquatic Sci.*, 2002, **59**, 153–172.

Carls, M.G., P.M. Harris, and S.D. Rice, Restoration of oiled mussel beds in Prince William Sound, Alaska. *Mar. Environ. Res.*, 2004, **57**, 359–376.

Douglas, G.S., A.E. Bence, R.C. Prince, S.J. McMillen, and E.L. Butler, Environmental stability of selected petroleum source and weathering ratios. *Environ. Sci. Tech.*, 1996, **30**, 2332–2339.

Douglas, G.S., W.A. Burns, A.E. Bence, D.S. Page, and P.D. Boehm, Optimizing detection limits for the analysis of petroleum hydrocarbons in complex environmental samples. *Environ. Sci. Tech.*, 2004, **38**, 3958–3964.

Fragaso, N.M., J.L. Parrott, M.E. Hahn, and P.V. Hodson, Chronic retene exposure causes sustained induction of CYP 1A activity and protein in rainbow trout (*Oncorhynchus mykiss*). *Environ. Toxicol. Chem.*, 1998, **17**, 2347–2353.

Gundlach, E.R., S. Berné, L. d'Ozouville, and J.A. Topinka, Shoreline oil two years after *Amoco Cadiz*: New complications from *Tanio*. In *Proc. 1981 Oil Spill Conf.*, Washington, DC: American Petroleum Institute, 1981, 525–534.

Hawkins, S.A., S.M. Billiard, S.P. Tabash, R.S. Brown, and P.V. Hodson, Altering cytochrome P450 1A activity affects polycyclic aromatic hydrocarbon metabolism and toxicity in rainbow trout (*Oncorhynchus mykiss*). *Environ. Toxicol. Chem.*, 2003, **21**, 1845–1853.

Heintz, R.A., J.W. Short, and S.D. Rice, Sensitivity of fish embryos to weathered crude oil: Part II. Increased mortality of pink salmon (*Oncorhynchus gorbuscha*) embryos incubating downstream from weathered *Exxon Valdez* crude oil. *Environ. Toxicol. Chem.*, 1999, **18**, 494–503.

Hostettler, F.D., R.J. Rosenbauer, and K.A. Kvenvolden, PAH refractory index as a source discriminant of hydrocarbon input from crude oil and coal in Prince William Sound Alaska. *Org. Geochem.*, 1999, **30**, 87–879.

Huggett, R.J., J.J. Stegeman, D.S. Page, K.R. Parker, B.R. Woodin, and J.S. Brown, Biomarkers in fish from Prince William Sound and the Gulf of Alaska: 1999–2000. *Environ. Sci. Tech.*, 2003, **37**, 4043–4051.

Jahns, H.O., J.R. Bragg, L.C. Dash, and E.H. Owens, Natural cleaning of shorelines following the *Exxon Valdez* oil spill. In *Proc. 1991 Intl. Oil Spill Conf.*, San Diego, CA, Washington, DC: American Petroleum Institute, 1991, 167–176.

Jewett, S.C., T.A. Dean, B.R. Woodin, M.K. Hoberg, and J.J. Stegeman, Exposure to hydrocarbons 10 years after the *Exxon Valdez* oil spill: Evidence from cytochrome P450 1A expression and biliary FACs in nearshore demersal fishes. *Mar. Environ. Res.*, 2002, **54**, 21–48.

Kennicutt, M.C., III, T.J. McDonald, G.J. Denoux, and S.J. McDonald, Hydrocarbon contamination the Antarctic Peninsula I. Arthur Harbor — subtidal sediments. *Mar. Poll. Bull.*, 1992, **24**, 499–506.

Kvenvolden, K.A. and P.R. Carlson, Possible connection between two Alaskan catastrophes occurring 25 yr apart (1964 and 1989). *Geology*, 1993, **21**, 813–816.

Kvenvolden, K.A., F.D. Hostettler, P.R. Carlson, J.B. Rapp, C.N. Threlkeld, and A. Warden, Ubiquitous tar balls with a California-source signature on the shorelines of Prince William Sound, Alaska. *Environ. Sci. Tech.*, 1995, **29**, 2684–2694.

Lee, R.F. and J.W. Anderson, Significance of cytochrome P450 system responses and levels of bile fluorescent aromatic compounds in marine wildlife following oil spills. *Mar. Poll. Bull.*, 2005, **50**, 705–723.

Lethcoe, J. and N. Lethcoe, *History of Prince William Sound, Alaska. Revised Second Edition*, Valdez, AK: Prince William Sound Books, 2001.

Mackay, D., W.Y. Shiu, and K.C. Ma, *Illustrated Handbook of Physical–Chemical Properties and Environmental Fate for Organic Chemicals*. Chelsea, MI: Lewis, 1992.

Manen, C.A., J.R. Price, S. Korn, and M.G. Carls, Natural Resource Damage Assessment: Database Design and Structure. National Oceanic and Atmospheric Administration, U.S. Department of

Commerce, Rockville, MD, NOAA Technical Memorandum NOS/ORCA, 1993.

Michel, J., M.O. Hayes, W.J. Sexton, J.C. Gibeaut, and C. Henry, Trends in natural removal of the *Exxon Valdez* oil spill in Prince William Sound from September 1989 to May 1990. In *Proc. 1991 Intl. Oil Spill Conf. Prevention, Behavior, Control, Cleanup.* Washington, DC: American Petroleum Institute, 1991, 181–187.

Miller, M.M., S.P. Wasik, G.L. Huang, W.Y. Shiu, and D. Mackay, Relationship between octanol-water partition coefficient and aqueous solubility. *Environ. Sci. Tech.*, 1985, **19**, 522–529.

Mudge, S.M. Reassessment of the hydrocarbons in Prince William Sound and the Gulf of Alaska: Identifying the source using partial least-squares. *Environ. Sci. Tech.*, 2002, **36**, 2354–2360.

Neff, J.M. and W.A. Burns, Estimation of polycyclic aromatic hydrocarbon concentrations in the water column based on tissue residues in mussels and salmon: An equilibrium partitioning approach. *Environ. Toxicol. Chem.*, 1996, **15**, 2240–2253.

Neff, J.M., D.E. Langseth, E.M. Graham, T.C. Sauer, Jr., and S.C. Gnewuch, *Transport and Fate of Non-BTEX Petroleum Chemicals in Soil and Groundwater.* Publication 4593. Washington, DC: American Petroleum Institute, 1994.

Neff, J.M., S.A. Stout, and D.G. Gunster, Ecological risk assessment of polycyclic aromatic hydrocarbons in sediments: Identifying sources and ecological hazard. *Integrated Environ. Assessment and Mgt.*, 2005, **1**, 22–33.

Neff, J.M., A.E. Bence, K.R. Parker, D.S. Page, J.S. Brown, and P.D. Boehm, Bioavailability of PAH from buried shoreline oil residues 13 years after the *Exxon Valdez* oil spill: A multispecies assessment. *Environ. Toxicol. Chem.*, 2006, **25**, 947–961.

NRC. *Oil in the Sea: Inputs, Fates, and Effects.* National Research Council. Washington, DC: National Academy Press, 1985.

O'Connor, T. National distribution of chemical concentrations in mussels and oysters in the USA. *Mar. Environ. Res.*, 2002, **53**, 117–143.

Owens, E.H. Shoreline conditions following the *Exxon Valdez* spill as of fall 1990. Woodward-Clyde Consultants, Seattle, WA.

Page, D.S., J.C. Foster, P.M. Fickett, and E.S. Gilfillan, Identification of petroleum sources in an area impacted by the *Amoco Cadiz* oil spill. *Mar. Poll. Bull.*, 1988, **19**, 107–115.

Page, D.S., E.S. Gilfillan, P.D. Boehm, and E.J. Harner, Shoreline ecology program for Prince William Sound, Alaska, following the *Exxon Valdez* oil spill: Part 1 — Study design and methods. *Exxon Valdez Oil Spill: Fate and Effects in Alaskan Waters*, ASTM STP #1219, P.G. Wells, J.N. Butler, and J.S. Hughes (eds.), Philadelphia, PA: American Society for Testing and Materials, 1995, 263–295.

Page, D.S., P.D. Boehm, G.S. Douglas, A.E. Bence, W.A. Burns, and P.J. Mankiewicz, The natural petroleum hydrocarbon background in subtidal sediments of Prince William Sound, Alaska. *Environ. Toxicol. Chem.*, 1996, **15**, 1266–1281.

Page, D.S., P.D. Boehm, G.S. Douglas, A.E. Bence, W.A. Burns, and P.J. Mankiewicz, Pyrogenic polycyclic aromatic hydrocarbons in sediments record past human activity: A case study in Prince William Sound Alaska. *Mar. Poll. Bull.*, 1999, **38**, 247–260.

Page, D.S., A.E. Bence, W.A. Burns, P.D. Boehm, J.S. Brown, and G.S. Douglas, A holistic approach to hydrocarbon source allocation in the subtidal sediments of Prince William Sound, Alaska, embayments. *Environ. Forensics*, 2002a, **3**, 331–340.

Page, D.S., R.J. Huggett, J.J. Stegeman, K.R. Parker, B.R. Woodin, J.S. Brown and A.E. Bence, Polycyclic aromatic hydrocarbon sources related to biomarker levels in fish from Prince William Sound and the Gulf of Alaska. *Environ. Sci. Tech.*, 2004, **38**, 4928–4936.

Page, D.S., P.D. Boehm, J.S. Brown, J.M. Neff, W.A. Burns, and A.E. Bence, Mussels document loss of bioavailable polycyclic aromatic hydrocarbons and the return to baseline conditions for oiled shorelines in Prince William Sound, Alaska. *Mar. Environ. Res.*, 2005, **60**, 422–436.

Page, D.S., J.S. Brown, P.D. Boehm, A.E. Bence, and J.M. Neff, A hierarchical approach measures the aerial extent and concentration levels of PAH-contaminated shoreline sediments at historic industrial sites in Prince William Sound, Alaska. *Mar. Poll. Bull.*, 2006, in press.

Peacock, E.E., R.K. Nelson, A.R. Solow, J.D. Warren, J.L. Baker, and C.M. Reddy, The West Falmouth oil spill: 100 kg of oil found to persist decades later. *Environ. Forensics*, 2005, **6**, 273–281.

Peters, K.E., C.C. Walters, and J.M. Moldowan, *The Biomarker Guide*, 2nd ed. Cambridge: Cambridge University Press, 2005.

Peterson, C.H., S.D. Rice, J.W. Short, D. Esler, J.L. Bodkin, B.E. Ballachey, and D.B. Irons, Long-term ecosystem response to the *Exxon Valdez* oil spill. *Science*, 2003, **302**, 2082–2086.

Prince, R.C., J.R. Clark, J.E. Lindstrom, E.L. Butler, E.J. Brown, G. Winter, M.J. Grossman, P.R. Parris, R.E. Bare, J.F. Braddock, W.G. Steinhauer, G.S. Douglas, J.M. Kennedy, P.J. Barter, J.R. Bragg, E.J. Harner, and R.M. Atlas, Bioremediation of the *Exxon Valdez* oil spill: Monitoring safety and efficacy. In *Hydrocarbon Bioremediation,* R.E. Hinchee, B.C. Alleman, R.R. Hoeppel, and R.N. Miller (eds.), Ann Arbor, MI: Lewis Publishers, 1994, 107–124.

Rapp, J.B., F.D. Hostettler, and K.A. Kvenvolden, Comparison of *Exxon Valdez* oil with extractable material from deep-water bottom sediments in Prince William Sound and the Gulf of Alaska. *Bottom Sediment Along Oil Spill Trajectory in Prince William Sound and Along Kenai Peninsula, Alaska.* P.R. Carlson and E. Reimnitz (eds.), U.S. Geological Survey Open File Report 90-39B, 1990.

Rogge, W.F., L.M. Hildemann, M.A. Mazurek, G.R. Cass, and B.R.T. Simoneit, Sources of fine organic aerosol. 9. Pine, oak and synthetic log combustion in residential fireplaces. *Environ. Sci. Tech.,* 1998, **32**, 13–22.

Royer, T.C., J.A. Vermersch, T.J. Weingartner, H.J. Niebauer, and R.D. Muench, Ocean circulation influencing the *Exxon Valdez* oil spill. *Oceanography,* 1990, **3**, 3–10.

Sale, D.M., J.C. Gibeaut, and J. Short, Nearshore subtidal transport of hydrocarbons and sediments following the *Exxon Valdez* oil spill. Final Report to Exxon Valdez Oil Spill State/Federal Natural Resources Damage Assessment Final; Report (Subtidal Study Number 3B) Alaska Department of Environmental Conservation, Juneau, AK, 1995.

Short, J.W. and M.M. Babcock, Prespill and postspill concentrations of hydrocarbons in mussels and sediments in Prince William Sound. In *Proc. Exxon Valdez Oil Spill Symp.,* AFS Symposium 18, S.D. Rice, R.B. Spies, D.A. Wolfe, and B.A. Wright (eds.), American Fisheries Society, Bethesda, MD. 149–168.

Short, J.W., D.M. Sale, and J.C. Gibeaut, Nearshore transport of hydrocarbons and sediments after the *Exxon Valdez* oil spill. In *Proc. Exxon Valdez Oil Spill Symp.,* AFS Symposium 18. S.D. Rice, R.B. Spies, D.A. Wolfe, and B.A. Wright (eds.), Bethesda, MD: American Fisheries Society, 1996, 40–60.

Short, J.W. and R.A. Heintz, Identification of *Exxon Valdez* oil in sediments and tissues from Prince William Sound and the Northwestern Gulf of Alaska based on a PAH weathering model. *Environ. Sci. Tech.,* 1997, **31**, 2375–2384.

Short, J.W., K.A. Kvenvolden, P.R. Carlson, F.D. Hostettler, R.J. Rosenbauer, and B.A. Wright, Natural hydrocarbon background in benthic sediments of Prince William Sound, Alaska: Oil versus coal. *Environ. Sci. Tech.,* 1999, **33**, 34–42.

Short, J.W., M.R. Lindeberg, P.M. Harris, J.M. Maselko, J.J. Pella, and S.D. Rice, Estimate of oil persisting on the beaches of Prince William Sound 12 years after the *Exxon Valdez* oil spill. *Environ. Sci. Tech.,* 2004, **38**, 19–25.

Trust, K.A., D. Esler, B.R. Woodin, and J.J. Stegeman, Cytochrome P450 1A induction in sea ducks inhabiting nearshore areas of Prince William Sound, Alaska. *Mar. Poll. Bull.,* 2000, **40**, 397–403.

U.S. Code of Federal Regulations. 43 C.F.R. Part 11, as amended by: 53 FR 5171, 5175, Feb. 22, 1988: 59 FR 14281, 14283, Mar. 25, 1994: 61 FR 20612, May 7, 1996 (current; CERCLA: 43CFR11. 14, 2001: 43CFR11.72, 2001), US Government Printing Office, Washington, DC, USA, 2001.

Van Kooten, G.K., J.W. Short, and J.J. Kolak, Low-maturity Kulthieth Formation coal: A possible source of polycyclic aromatic hydrocarbons in benthic sediment of the northern Gulf of Alaska. *Environ. Forensics,* 2002, **3**, 227–241.

Wolfe, D.A., M.J. Hameedi, J.A. Galt, G. Watabayashi, J. Short, C. O'Clair, S. Rice, J. Michel, J.R. Payne, J. Braddock, S. Hanna, and D. Sale, Fate of the oil spilled from the T/V *Exxon Valdez* in Prince William Sound Alaska. *Environ. Sci. Tech.,* 1994, **28**, 561A–568A.

Wooley, C. The myth of the "pristine environment": Past human impacts in Prince William Sound and the northern Gulf of Alaska. *Spill Sci. Tech. Bull.,* 2002, **4**, 89–104.

16 Case Study: Oil Spills in the Strait of Malacca, Malaysia

Mohamad Pauzi Zakaria[1]* and Hideshige Takada[2]

[1]Faculty of Environmental Studies, Universiti Putra Malaysia, 43400 UPM Serdang, Selangor, Malaysia; [2]Laboratory of Organic Geochemistry (LOG), Institute of Symbiotic Science and Technology, Tokyo University of Agriculture and Technology, Fuchu, Tokyo 183-8509, Japan.

16.1 Strait of Malacca, Malaysia: Introduction

The Strait of Malacca is located between the east coast of Sumatra and the west coast of Peninsular Malaysia (Figure 16-1). The Strait of Malacca is an important sea route for local and international trade and shipping traffic. It is also a major shipping route for oil tankers transporting crude oil from the Middle East countries to the northeast Asian countries.

Strategically located and rich in coastal resources, the Strait of Malacca through the centuries has been a center of international trade and civilization. More than 500 years ago, merchants from the State of Gujerat, India, and Arab Islamic missionaries flocked to the area and turned Malacca into an important trading center. Eventually, the Europeans established a spice trade in the whole of the Malay Archipelago, transforming the Strait of Malacca into a busy trade route. Today, the southeast Asian economy remains closely linked with maritime trade and activities such as shipping, fisheries, and eco-tourism.

The Strait of Malacca has valuable and vulnerable marine and coastal resources. The Strait has been constantly exposed to a variety of sea-based and land-based pollutants (Abdullah et al., 1999). With increased maritime transportation in the area, it is expected that the impact of oil pollution will increase. The importance and contributions of the coastal and marine environmental resources of the Strait are being increasingly recognized. With extensive mangrove and forest resources, coral reefs, and sea grass along the path of a major migratory bird route from Siberia to Australasia, the area contains a number of ecosystems values for their biodiversity and cultural significance by the surrounding littoral states.

16.1.1 Hydro-Oceanographic Condition of the Strait

The length of the Strait of Malacca is approximately 600 nautical miles, while its width ranges between 8 nautical miles near Riau Archipelago to 220 nautical miles at the northwest entrance. The northern part of the Strait is deeper but becomes shallower southwards. In the southern part are many sandbars that are elongated in shape, following the shape of the Strait itself, and hence they imperil navigation. The water in the middle of the Strait is deeper but shoals closer to shore.

The oceanographic characteristics of the Strait of Malacca are strongly affected by the adjacent oceans. Surface currents in the Strait are generated as a result of interactions

490 Oil Spill Environmental Forensics

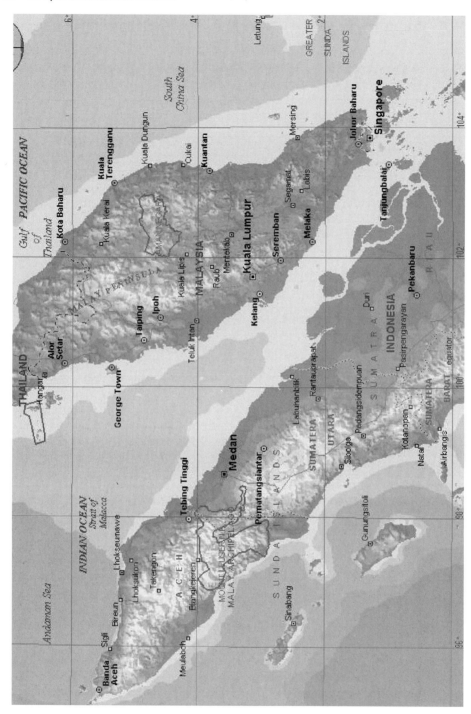

Figure 16-1 Map of Peninsular Malaysia showing the Strait of Malacca.

between oceanic currents and weather conditions in the Indian Ocean and the South China Sea. The current flows mainly in two directions for both monsoon seasons and depends on the sea-level differences between the southeast and northwest entrances (Wyrtki, 1961). The dominant direction is northwestward from the southeast entrance (Andaman Sea). Another surface current enters the Strait of Malacca from the northwest entrance (Andaman Sea) with a southeasterly flow, but reverses to the northwest direction near the northern tip of the Peninsular Malaysia.

16.1.2 Ship Traffic in the Strait of Malacca: Historical and Present

The Strait of Malacca are recognized as one of the busiest shipping lanes in the world. Historically, the importance of the Strait was heightened by the opening of the Suez Canal in 1869, which had the effect of diverting ships away from the southern route around the Cape of Good Hope and chanelling ships through Colombo and the Strait. Naidu (1997) noted that the estimated traffic through the Strait was 43,633 vessels in 1982 and 84,414 in 1991. In 1997, the vessel traffic increased to 131,003 vessels, which amounts to about 358 vessels per day. The total number of ships that passed through the Strait in 2003 was 62,334. The Marine Department of Malaysia estimated that about 34% of these vessels in 2003 were oil tankers.

16.2 Chronic and Acute Oil Spill Events in the Strait

The threat of oil spills as a result of accidents has been recognized over the years, as can be seen in Figure 16-2. Therefore, we can deduce that there was an increasing number of oil spill incidents in the Strait. The impacts of oil spills on marine habitats in the Strait, such as the mangroves, coral reefs, and seagrass beds, are major concerns. Coral reefs in the Straits have been largely affected by years of spills, and their recovery has been very slow (Dow, 1997).

There were several major oil spill incidents involving oil tankers in the Strait of Malacca from 1976 to 1997. Examples of such spills are *Diego Silang* (1976), *Asian* (1977), *M. T. Ocean Treasure* (1981), *Mv. Pantas* (1986), *Nagasaki Spirit* (1992), *An Tai* (1997), and *Sun Vista* (1999).

16.2.1 Contribution of Oil Pollution Sources in Malaysia

Although Malaysia has a relatively short history of industrialization and modernization as compared to the more established and developed countries, several factors may contribute to petroleum. First, the Malacca Strait is a major international tanker route transporting crude oil from the Middle East to northeast Asia. Oil spills and tanker accidents are frequent in the Strait. Second, a major

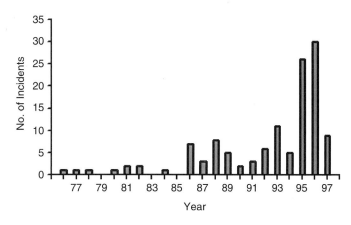

Figure 16-2 Oil spill incidents in the Strait of Malacca (Malaysian Marine Department, 2004).

oil field was discovered along the eastern seaboard of Peninsular Malaysia that produces and exports crude oil to other countries. Third, Malaysia is undergoing rapid industrialization, and petroleum is the most important source of energy for this process. The demand for petroleum has increased very rapidly in the past few decades with increasing population and urbanization. Motor-vehicle ownership has also quadrupled in recent years. Furthermore, current lax enforcement of environmental laws and regulations, especially in the dumping of used oil products, can lead to serious oil pollution problems in Malaysia. Industrial activities may also be important sources. Routine car maintenance, petrol stations, and motor-vehicle workshops are other important contributors; contributions from ports, harbors, marinas, and tanker accidents are also becoming more important in Malaysian coastal environments.

16.3 Methodology

16.3.1 Sample Collection

Figure 16-3 shows sampling locations for tar balls, while Figure 16-4 shows sampling locations for sediment on the west coast of Peninsular Malaysia.

16.3.2 Source Petroleum

Crude oil samples including three Middle East oils (Marban, Arabian Light, and Ummu Shaif) and four southeast Asian oils (Tapis, Labuan, Miri, and Sumatra) were kindly supplied by the Maritime Agency of Japan.

16.3.3 Tar-Ball Samples

Twenty tar-ball samples were collected by hand using clean plastic gloves in the tidal zones from 19 locations of the west and east coasts of Peninsular Malaysia (Figure 16-3). Two samples, MYKETR-T1 and MYKETR-T2, were taken from the same location but on two different dates. The samples were wrapped with aluminum foil, stored in clean plastic Zippered-lock bags, kept in a cooler box, transported to the laboratory, and stored at −18°C until analysis.

16.3.4 Sediment Samples

Twelve sediment samples were collected from 10 rivers on the west coast of Peninsular Malaysia in 1998 and 1999 (Figure 16-4). The locations were selected as such to cover the whole west coast including both rural (e.g., Teluk Intan, Nibong Tebal) and urban (e.g., Pinang Estuary, Port Klang, Malacca city, Johor Bahru) areas. Five estuarine sediments (St.A, St.C, St.E, St.G, and St.H) were collected along the Klang Estuary in both 2000 and 1998. Four inshore sediment samples (St.2, St.9, St.11, and St.15) were collected off the Klang Estuary. Eight offshore sediment samples (St.2, St.8, St.14, St.17, St.18, St.21, St.22, and St.23) were collected in the Strait of Malacca. The sampling locations cover a broad range of the narrow Strait. River and estuarine sediments were collected using an Eckman Dredge. The collected sample of the sediments was placed on a stainless steel pan, and the top 0–5 cm of sediments were taken using a precleaned stainless steel scoop. Because of both active input of terrestrial material to the rivers and estuaries and strong flushing of the bottom sediments caused by the frequent and strong rain in the tropical area, the top 5-cm layers of the sediments are thought to represent modern input. The coastal samples were collected using a Smith-McIntyre sampler. Immediately after the sampler was raised onto the boat, the top 0–2-cm layers were taken using a precleaned stainless steel scoop. Considering sediment accumulation rate in the coastal area, the coastal sediments are also expected to represent accumulated modern pollution input (e.g., during the last 10–20 years). For example, the sediment accumulation rate for the coastal area off the Klang Estuary was reported to be ~2 mm/year (Ibrahim, 1988), indicating that the 0–2-cm layers of the inshore sediment samples correspond to input of the last 10 years. However, the effects of bioturbation of the sediment samples were not accounted. The samples were placed in tight-sealed, solvent-

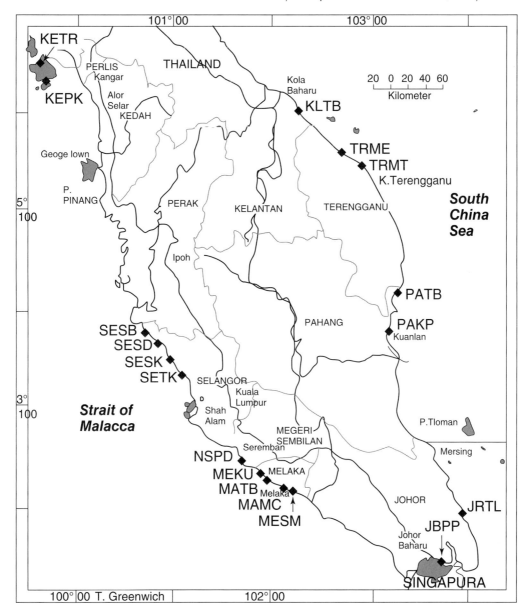

Figure 16-3 Sampling stations for tar ball samples (solid diamond) in Peninsular Malaysia.

rinsed stainless steel containers and transported on ice to the laboratory. The samples were then stored at −35°C until further analysis.

16.3.5 Street Dust Samples

Three street dust samples (KL-1, KL-2, KL-3) were collected from three heavy-trafficked streets in Kuala Lumpur in 1998. The samples were collected using a clean straw brush and stored in solvent-rinsed, tight-sealed vials with a Teflon-lined cap. The samples was transported to the laboratory and stored at −18°C until further analysis.

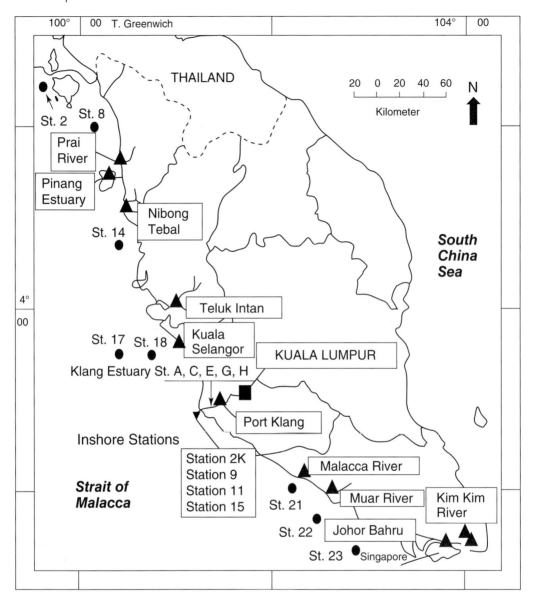

Figure 16-4 Sampling stations for sediment in Peninsular Malaysia. Solid triangles represent river stations and solid circles represents inshore stations.

16.3.6 Asphalt Samples

Fresh asphalt samples (AS-1 and AS-2) were obtained from two pieces of asphalt detached from the surface of an asphalt-paved road near the campus of the Universiti Putra Malaysia, a suburb of Kuala Lumpur. The samples were wrapped with aluminum foil, stored in a clean, plastic Zippered-locked bag, transported to the laboratory, and stored in the freezer until further analysis.

16.3.7 Fresh Crankcase Oil

Two fresh crankcase oil samples (Petronas® and BP®) were randomly purchased from the Malaysian market.

16.3.8 Used Crankcase Oil

Four used crankcase oil samples were taken from storage tanks in a gas station, an automotive workshop, garages, and a motorcycle workshop in Malaysia. The samples were collected using pre-cleaned 10-ml glass pipettes and placed in previously baked 30-ml amber vials. The sample was transported to the laboratory, and stored in the freezer until further analysis.

16.3.9 Automobile Tire Rubber

Tire rubber data and NCBA data were provided by Dr. Hidetoshi Kumata of Tokyo University of Pharmacy and Life Sciences, Hachioji, Tokyo, Japan. Tire particles from four types of used tires were obtained by abrasion of their surfaces with a stainless steel file.

16.3.10 Aerosol Samples

Aerosols data were kindly provided by Dr. Hidetoshi Kumata of Tokyo University of Pharmacy and Life Sciences, Hachioji, Tokyo, Japan. The aerosol samples were collected in Universiti Putra Malaysia (UPM), Petaling Jaya (PJ), Port Dickson (PD), and Langkawi. Aerosol samples from UPM were acquired in both 1997 and 1998, whereas aerosol samples from PJ were acquired in 1998. The aerosol samples were acquired using a high-volume sampler (HVS) (Ecotech, Australia) fitted with a glass fiber filter (Filter #6424596, UPM, exposed during haze in 1997; #6424527, UPM; exposed during haze in 1998; #39472, PJ, exposed during haze in 1997; #40060, PJ, exposed during haze in 1997) (20.3 cm × 25.4 cm, Whatman EPM 2000, England). The HVS was operated at a flow of ~$1.0 m^3 min^{-1}$, and ~$750 m^3$ of air passed through the filter during the 24-hr sampling. Before weighing, all fresh and exposed filters were conditioned for 24 hr in a dry cabinet where the temperature was maintained at 25°C, with relative humidity at 45%. Exposed filters were folded, stored in pre-cleaned aluminum foil, kept in a glass jar, and kept refrigerated until further analysis.

Aerosol samples from PD and Langkawi were acquired in 1998 using a high-volume sampler (Staplex TFIA-2) fitted with quartz fiber filters (Pallflex 2500QAT-UP) (filter #0001, Langkawi, and filter #0003, PD). Approximately $1500 m^3$ of air passed through the filter during the 24-hr sampling.

16.4 Analytical Procedure

16.4.1 Chemicals

Authentic standards of n-C_{16}, C_{18}, C_{20}, C_{22}, C_{24}, C_{26}, C_{28}, C_{32}, and C_{36} and deuterated PAHs (acenaphthene-d_8, chrysene-d_{12}, anthracene-d_{10}, p-terphenyl-d_{14}, naphthalene-d_8, benz(a)anthracene-d_{12}, perylene-d_{14}), and PAHs (phenanthrene, anthracene, methylphenanthrenes, fluoranthene, pyrene, benz(a)anthracene, chrysene, benzo(b)fluoranthene, benzo(j)fluoranthene, benzo(k)fluoranthene, benzo(e)pyrene, benzo(a)pyrene, perylene, indeno[1,2,3-cd]pyrene, benzo(ghi)perylene, and coronene) were purchased from Wako pure Chemical Ltd. 17β, 21(H)β-hopane, 17α(H)-22,29,30-trisnorhopane (Tm), 17β(H), 21α(H)-norhopane (C_{29} 17β), 17α(H), 21β(H)-hopane (C_{30} 17α), 17bβ(H), 21α(H)-hopane (C_{30} 17α), 18α(H)-oleanane, and 18β(H)-oleanane were purchased from Chiron, Norway. A 100–200-mesh silica gel (F.C.923) from Davison Chemical Corp. was baked at 400°C for ~4 hr, cooled, and then activated at 200°C overnight and deactivated with 5% (w/w) of distilled water. A 60–200-mesh silicic acid (SIL-A-200) from Sigma Chemical Company, USA, was rinsed with methanol and dichloromethane (DCM) using Soxhlet extraction, dried at 50°C, and fully activated at 200°C overnight, and stored in dry n-hexane. Organic solvents were distilled in glass before use. All glassware was washed with detergents and tap water and rinsed successively with methanol, DCM, and distilled hexane. Elemental copper (granular, 0.2–0.9 mm) was purchased from Kishida Chemicals Co. Ltd., Osaka, Japan.

16.4.2 Extraction and Fractionation

Crude oil and tar-ball samples (ca. 10 mg) were precisely weighted, dissolved in 1 ml of DCM/hexane (1:3, v/v), and transferred onto the top of 5% H_2O deactivated silica gel column (1 cm i.d. × 9 cm). Hydrocarbons ranging from n-alkanes to PAHs with 7 rings were eluted with 20 ml of DCM/hexane (1:3, v/v). The sample was then reduced in volume to ~5 ml, where approximately 5 g of activated copper were added and allowed to stand overnight to react with elemental sulfur. The solution was passed through a glass funnel plugged with quartz wool, which trapped copper and the copper sulfide on it. The filtrates were roto-evaporated just to dryness and were subsequently dissolved into 0.4 ml of n-hexane, and transferred onto the fully activated silica gel column (0.47 cm i.d. × 18 cm). Three fractions were eluted. The first fraction containing aliphatic and alicyclic hydrocarbons was eluted with 4 ml of hexane. Since tar balls do not contain LABs, the second fraction was combined with PAHs fraction and eluted with 18 ml of hexane and 7-ml hexane/DCM (3:1, v/v), respectively. This fraction contained the higher-molecular-weight PAHs with 3–7 benzene rings. The first hexane fraction was evaporated and transferred to a 1.5-ml glass ampule. The hexane was evaporated just to dryness under nitrogen and redissolved in 50 μl–1 ml of isooctane for subsequent instrumental analysis. Subsequent treatment of the PAHs fraction is dealt with in the following sections of this chapter.

Sediment samples were freeze-dried and Soxhlet-extracted using DCM over 8 hr. Activated copper was used to remove elemental sulfur. Wet mussel tissue (ca. 10 g) was macerated/extracted with DCM and pre-baked sodium sulfate anhydrous in a glass centrifuge tube by Polytron (RT2000; Kinebatica). The sediment and mussel extracts were purified and fractionated using the same procedure as the oil samples. For the mussel samples, some co-eluting materials such as esters, etc., in the alkane fractions were hydrolyzed with KOH/MeOH prior to alkane and hopane analyses.

16.5 Instrumental Analysis

16.5.1 Analysis of Alkanes and Hopanes

Normal and isoprenoid alkanes were analyzed on Shimadzu 14 B gas chromatograph with flame ionization detector. A J&W Scientific Durabond DB-5, 30-m fused silica capillary column, 0.25-mm i.d., and 0.25-μm film thickness, was used with helium as the carrier gas at 200 kPa. The injection port was maintained at 300°C, and the sample was injected with splitless mode followed by a 1-min purge after the injection. The column temperature was held at 70°C for 1 min, then programmed at 30°C/min to 150°C, 5°C/min to 310°C, and held for 10 min. The detector temperature was held at 310°C. The concentrations of C_{16} to C_{36} normal alkanes, pristane, and phytane were determined. Identification of sample peaks was based on comparison of retention time of the authentic standard run on the same day and confirmed by GC-MS. Quantification was based on peak height, and the response factor was calculated through the standard run. Response factors for compounds with no corresponding standards were estimated by proportional allotment using standards that bracket the target peak on the chromatogram.

After the alkane analysis, the aliphatic and alicyclic hydrocarbon fraction was evaporated to dryness and re-dissolved into an appropriate volume (50–200 μL) of isooctane containing 5 ppm of 17β, 21β(H)-hopane as an internal injection standard for triterpane analysis. Hopane analyses were made using a Hewlett Packard 5972A quadrupole mass spectrometer integrated with an HP5890 gas chromatograph equipped with a J&W Scientific Durabond HP-5MS, 30-m fused silica capillary column, 0.25-mm i.d., and 0.25-μm film thickness, using helium as the carrier gas on a constant-flow rate mode at 1 ml/min. GC-MS operating conditions were 70-eV ionization potential with the source at 200°C and electron multiplier voltage at ~2000 eV. The injection port was maintained at 300°C, and the sample was injected with splitless mode followed by a purge 1 min after the injection. The column temperature was held at 50°C for 2 min, then programmed at 6°C/min to 300°C, and held for 15 min. A selected ion monitoring method was employed after a delay of 4 min, and triterpanes were quantified at $m/z = 191$. Peaks were identified by comparison of their retention times with those for the standards and their mass spectra, which were obtained on a different GC-MS run on Scan mode, with those in the literature (Peters and Moldowan, 1993; Wang et al., 1994). Quantification was made by peak area. The response factor of 18α(H)-22,29,30-trisnorneohopane (Ts) was assumed to be the same as Tm, 17α(H), 21β(H)-30-norhopane (C_{29} 17α) as C_{29} 17β, and homohopanes ranging from C_{31}–C_{35} of carbon number as C_{30} 17β. Because 18α(H)-oleanane and 18β(H)-oleanane cannot be separated on the gas chromatogram, an average of response factors of both isomers obtained from independent runs was applied to the samples. Differences in response factors of both oleanane isomers were within 10% and, therefore, using the averaged factor has not caused significant errors. For another set of stereo isomers (C_{30} 17α and C_{30} 17β), the differences in their response factors were within 20%. These suggest that the expected differences in response factors for the other set of stereoisomers (e.g., Ts and Tm, C_{29} 17α and C_{29} 17β) are within 20%.

Reproducibility of the whole analytical procedure was checked through the triplicate analysis of an oil sample. Relative standard deviations of individual alkanes and triterpane concentrations were less than 11% and 13%, respectively. Recoveries were determined by spiking a mixture of the standards listed above to the oil samples followed by the entire analytical procedure. Recovery of the spiked standards was more than 82% for alkanes and over 87% for triterpanes.

16.5.2 N-Cyclohexyl-2-Benzothiozolamine (NCBA)

NCBA data were provided by Dr. Hidetoshi Kumata of Tokyo University of Pharmacy and Life Sciences, Hachioji, Tokyo, Japan. The organic extracts of NCBA were purified by liquid–liquid extraction with sulfuric acid and fractionated with 5% deactivated silica gel column chromatography. The NCBA was quantified by a Hewlett-Packard 5890 Series II gas chromatograph equipped with an HP-35 fused silica capillary column (30 m × 0.25-mm i.d., 0.25-μm film thickness) interphased with a flame photometric detector (FPD).

16.5.3 Analysis of PAHs

Polycyclic aromatic hydrocarbon fraction was evaporated to approximately 1 ml, transferred to a 1.5-ml amber ampule, and evaporated to dryness under a gentle stream of nitrogen and re-dissolved into an appropriate volume (50–200 μL) of isooctane containing acenaphthene-d_8 and chrysene-d_{12} as an internal injection standard (IISTD) for PAH analysis. PAHs were analyzed by GC-MS using a 30-m fused silica column (HP-5MS) installed in a gas chromatograph (HP5890) interfaced with a Hewlett-Packard 5972A quadrupole mass selective detector (SIM mode), using helium as the carrier gas on a constant pressure at 60 kg/cm². GC-MS operating conditions were 70-eV ionization potential with the source at 200°C and electron multiplier voltage at ~2000 eV. The injection port was maintained at 300°C, and the sample was injected with splitless mode followed by purge 1 min after the injection. The column temperature was held at 70°C for 2 min, then programmed at 30°C/min to 150°C, 4°C/min to 310°C, and held for 10 min. A selected ion monitoring method was employed after a delay of 4 min. PAHs were monitored at $m/z = 178$ (phenanthrene, anthracene), $m/z = 192$ (methylphenanthrenes), $m/z = 202$ (fluoranthene, pyrene), $m/z = 228$ (benz(a)anthracene, chrysene), $m/z = 252$ (benzo(b)fluoranthene, benzo(j)fluoranthene, benzo(k)fluoranthene, benzo(e)pyrene, benzo(a)pyrene, perylene), $m/z = 276$ (indeno[1,2,3-cd]pyrene, benzo(ghi)perylene), and $m/z = 300$ (coronene).

Individual PAHs were quantified by comparing the integrated peak area of the selected ion with the peak area of the IISTD. Acenaphthene-d_8 and chrysene-d_{12} were used as IISTD for the quantification of PAHs ranging from phenanthrene to 1-methylphenanthrene and for PAHs from fluoranthene to coronene, respectively. Corrections for relative response at the corresponding mass/charge ratio were made by analyzing a PAH standard (phenanthrene, anthracene, 1-methylphenanthrene, fluoranthene, pyrene, chrysene, benzo[b]fluoranthene, benzo[e]pyrene, benzo[a]pyrene, perylene, benzo[ghi]perylene, coronene) under the same instrumental conditions as the sample analyses. No 2-, 3-, or 9-methylphenanthrene standards were available; therefore, an estimated response factor for these compounds was based on a response for 1-methylphenanthrene. Similarly, benz[a]anthracene, benzo[j]fluoranthene + benzo[k]fluoranthene, and indeno[1,2,3-cd]pyrene concentrations were based on the chrysene/triphenylene, benzo[b]fluoranthene, and benzo[ghi]perylene response, respectively. Since naphthalenes were difficult to quantify due to their high volatility, the present study focused on PAHs with three or more benzene rings (i.e., phenanthrene-benzo[ghi]perylene). PAHs concentrations were recovery-corrected using the spiked surrogates, and recoveries were in the range of 64–83%. The precision of the method was determined through four replicated analyses of the heavy residual oil sample. The relative standard deviation (RSD; $n = 4$) of individual PAHs identified in sample extracts was <10%.

16.6 Establishment and Application of Biomarker Analysis for Source Identification of Oil Pollution Sources in the Strait of Malacca

Identification of oil pollution sources is an important task in order to assess environmental damage, and to understand the fate and behavior of spilled oils on the environment. Since spilled oils may cause extensive damage to marine and terrestrial organisms including human health, identification of the exact source of oil pollution can provide quite an accurate prediction of the long-term impact of spilled oil on the environment. Recently, Wang and Fingas (2003) reviewed development of chemical analysis methodologies that are most frequently used in oil spill characterization and identification studies and environmental forensic investigations.

In the Strait of Malacca, Zakaria et al. (2000) investigated oil pollution where various samples including Malaysian oil, Middle East crude oils (MECO), southeast Asia crude oils (SEACO), tar balls, sediments, and mussels were collected and analyzed. Case Study I is introduced explaining the development of the analytical method for oil pollution source identification using biomarkers in the Straits of Malacca. The major finding from Case Study I was that approximately 30% of the sea-based petroleum hydrocarbon pollution was derived from MECO, while the rest is of domestic source. Therefore, the remaining 70% of the domestic source of either sea- or land-based petroleum hydrocarbon pollution can be very significant and important (see Section 16.8).

The above scenario was examined in Case Study II, where a major finding was that very serious petroleum hydrocarbon pollution is occurring in Peninsular Malaysia. Polycyclic aromatic hydrocarbons (PAHs) of petrogenic origin are widely distributed in Malaysian riverine and coastal sediments, and used crankcase oil and street dust appear to be the major contributors of the pollution. It has been proposed that rapid transfer of land-based pollution into the aquatic environments was due to intense rainfall and runoff waters. If PAHs of terrestrial origin could be transferred rapidly to aquatic environments, this would bring a serious threat to Malaysia. A brief summary of Case Studies I and II follows. Case Study I is based on studies conducted by Zakaria et al. (1999), (2000), and (2001), while Case Study II is based on studies conducted by Zakaria et al. (2002).

16.7 Case Study 1: Development of the Analytical Method for Oil Pollution Source Identification Using Biomarkers in the Strait of Malacca

Crude oil samples typical of petroleum sources included four southeast Asian crode oils

(SEACO), three Middle East (MECO) crude oils, and two Malaysian fresh crankcase oils, which were analyzed for alkanes and hopanes. SEACO were characterized by lower C_{29}/C_{30}-hopane and ΣC_{31}–C_{35}/C_{30} hopane ratios, and the reverse was true for MECO. The C_{29}/C_{30} and ΣC_{31}–C_{35}/C_{30} hopane ratios were thus potential source identifiers to distinguish between SEACO and MECO. In addition, PAHs were also used as additional source identifiers to get information on weathering of tar balls.

16.7.1 Weathering of Tar Balls

PAHs and alkane compositions were utilized to evaluate the weathering of the tar ball samples. Table 16-1 lists concentrations of PAHs and alkanes and the compositional indices for the tar balls collected from the west and east coasts of Malaysia together with data on crude oils.

PAHs in the tar-ball samples were more abundant in higher-molecular-weight compounds (more depleted in lower-molecular-weight compounds) as compared to the crude oil. Evaporation, dissolution, and biodegradation enhance losses of low-molecular-weight PAHs (Heitkamp and Cerniglia, 1987). The difference in molecular-weight distributions is parameterized as the L/H-PAH ratio, as defined in Table 16-1. L/H-PAH ratios ranged from 7.5–44 in crude oil, compared with 0.23–8.57 in the tar-ball samples. The lower ratios indicate that the tar-ball samples had undergone variable weathering, from slightly to severely. Tar-ball samples SESB, SESK, MEKU, MESM, JBPP, MAMC-T2, JRTL, and SESD showed much lower L/H-PAH ratios (0.23–1.48) than those for crude oil samples (8–44). This is consistent with low PAH contents in the eight tar-ball samples (10–343 μg/g) compared with those in the crude oil samples (193–3069 μg/g). Alkane data are consistent with the PAH compositions. The above eight tar-ball samples were extremely depleted in lower-molecular-weight alkanes, and their L/H-alkane ratios (0.00–1.11) were lower than those for the crude oil samples (1.62–4.60). Furthermore, the unresolved complex mixture (UCM), an indicator of weathering of petroleum (Volkman et al., 1997), was observed for three of the eight samples (i.e., SESB, SESK, and MEKU). However, the other five tar-ball samples (i.e., MESM, SESD, MAMC-T2, JBPP, and JRTL) showed no UCM. This means that lower-molecular-weight components of PAHs and n-alkanes might have been depleted due to some other processes than weathering. One of the possible causes is tank washing and discharges from oil tankers. This possibility will be discussed in the following subsection. Other molecular ratios of PAHs such as fluoranthene/pyrene, benzo[a]pyrene/benzo[e]pyrene, and benz[a]anthracene/chrysene had also been examined as weathering indicators for the tar balls. However, those ratios did not show clear trends due to variation among crude oils.

Alkane and hopane data were very powerful molecular markers for source identification of oil pollution in the Strait of Malacca. However, integration of alkane and hopanes with PAHs could facilitate for more specific source identification and may provide further information on weathering.

The analytical results from the present study have established and successfully applied biomarkers as a molecular tool in tracing and determining the sources of oil pollution (MECO vs. SEACO) in the Strait of Malacca. This study has shown that C_{29}/C_{30} and ΣC_{31}–C_{35}/C_{30} ratios are not only powerful but reliable source identifiers of oil pollution. This study has also shown that high-oleanane concentration in tar balls indicates domestic petroleum origin (i.e., SEACO). However, low concentrations of oleanane in tar balls could not be identified as SEACO or MECO because both crude oils contain low concentrations of oleanane. This is probably due to the variability in oleanane contents in SEACOs. The results imply that low concentration of oleanane does not always indicate Middle East petroleum contribution although at high concentration oleanane can still be useful biomarker for Southeast Asian oil sources. Also,

Table 16-1 PAHs and Alkanes Compositions for Crude Oils and Malaysian tar-ball

	PAHs			Alkanes			
	Total PAHs $(ug/g)^a$	MP/P ratio[b]	L/H-PAH[c]	Total Alkanes[d] (mg/g)	Pr/Ph[e]	L/H-Alkane[f]	UCM[g]
Crude oils							
MECO (Middle East Crude Oil)							
Arabianligh	193	3.39	12.13	25	0.64	4.60	x
Umushyfu	700	4.34	20.35	38	0.97	3.35	x
Marban	341	3.88	15.68	49	0.94	3.92	x
SEACO (South East Asian Crude Oil)							
LABUAN	3069	2.43	44.46	99	4.01	2.43	x
MIRI	495	2.45	12.43	52	4.03	2.45	x
TAPIS	952	3.17	21.52	128	3.66	4.52	x
Sumatra	702	2.24	7.50	109	2.91	1.62	x
Tar-ball							
West Coast							
KETR-T2	469	6.83	6.95	101	2.36	1.24	x
KETR-T1	463	8.91	3.96	11	1.60	1.59	x
KEPK	60	3.05	2.48	51	1.38	0.32	x
SESB	18	1.40	0.36	0.4	n.c.[b]	0.09	o
SESD	31	0.16	1.48	106	n.c.[h]	0.46	x
SESK	73	2.68	0.23	1	n.c.[h]	0.07	o
SETK	112	1.70	2.34	4	0.42	1.72	x
NSPD	189	6.24	5.61	128	3.85	0.83	x
METB	557	3.63	8.57	175	4.14	1.07	x
MEKU	180	6.75	0.43	1	n.c.[h]	0.00	o
MESM	75	8.56	0.27	192	3.05	0.62	x
MAMC-T2	343	3.99	0.97	54	1.14	0.21	x
JBPP	41	12.09	0.75	8	0.83	1.11	x
East Coast							
JRTL	10	9.00	0.72	16	1.70	0.75	x
PAKP	248	8.47	2.79	33	0.69	1.29	x
PATB	195	8.03	3.13	37	1.46	1.19	x
TRMT	95	5.56	4.77	108	2.10	1.31	x
TRMT-T2	57	12.86	3.07	103	1.80	1.21	x
TRME	42	5.16	7.20	20	2.19	0.97	x
KLTB	476	8.86	4.57	62	0.84	0.80	x

[a] Total PAHs = sum of concentrations of phenanthrene + anthracene + 3-methylphenanthrene + 2-methylphenanthrene + 9-methylphenanthrene + 1-methylphenanthrene + fluoranthene + pyrene + benza(a)anthracene + chrysene + benzofluo thenes + benzo(e)pyrene + benzo(a)pyrene + perylene + indeno(1,2,3-cd)pyrene. [b] MP/P ratio = a ratio of sum of 3-methylphenanthrene + 2-methylphenanthrene + 9-methylphenanthrene + 1-methylphenanthrene relative to phenanthrene concentrations. [c] L/H = sum of concentrations of phenanthrene to pyrene relative to sum of concentrations of benz(a)anthracene to benzo(ghi)perylene in a. [d] Σ(C16–C36) = sum of n-C16 alkane to n-C36 alkane. ealkane/(pr + py) = sum of alkane relative to sum of pristane and phytane. [f] L/H = sum of n-C16 alkane relative to sum of n-C27–n-C36 alkane. [g] UCM = unresolved complex mixture. X = not present, o = present

Source: Zakaria, M. P., T. Okuda, and H, Takada. "Polycyclic Aromatic Hydrocarbon (PAHs) and Hopanes in Stranded Tarballs on the Coast of Peninsular Malaysia: Applications of Biomarkers for Identifying Sources of Oil Pollution. Marine Pollution Bulletin, 2001, **42**(12), 1357–1366.

the homohopane index was shown to be useful as a source identifier in this study since it clearly distinguishes MECO from SEACO. This study, however, did not find that the Tm/Ts ratio is a useful source identifier of the oil spill in the Strait of Malacca since the ratio is inconsistent between MECO tar balls and SEACO tar balls.

This study has also shown that PAHs provide useful additional information on source identification of petroleum pollution, and their molecular distributions are useful as an index of weathering. Thus, the combination of biomarkers of triterpanes, alkanes, and PAHs has become an extremely powerful tool for source identification of petroleum pollution.

16.8 The Application of Molecular Markers for Source Identification of Tar-Ball Pollution in Malaysia

Using the source identifiers, the C_{29}/C_{30} and $\Sigma C_{31}–C_{35}/C_{30}$ ratios, the origins of 20 tar-ball samples collected from 18 locations in the coastal beaches of Peninsular Malaysia were determined (Figure 16-5).

PAH profiles were also utilized to obtain information on weathering of tar balls, and the absence of PAHs was found to be a good indicator of weathering. Using the molecular marker approach, we demostrated that approximately 30% of tar-ball samples collected from the west coast of Peninsular Malaysia originated from the Middle East crude oil (MECO). This suggests that a tanker-derived source has significantly contributed to the petroleum pollution in the Strait of Malacca.

On the other hand, about 70% of the petroleum hydrocarbon input was actually domestically derived. Hence, determination of the distribution and sources of compound-specific petroleum hydrocarbons in Malaysian rivers, coastal sediments, and the Strait of Malacca sediments is necessary. Therefore, the next goal was to identify the distribution and sources of PAH pollution in Malaysia.

The present study clearly demonstrated a substantial contribution of MECO spilled from tankers passing through the Strait to the tar balls (e.g., SETK, SESB, SESK, KETR-T1). Some international action is therefore necessary to reduce the petroleum pollution from the tankers. In addition, domestic sources have been identified as major contributors, and identification of the specific sources and regulations of these sources are even more important. It was thought that oil spills from the petroleum platform off the east coast have largely contributed to the tar-ball pollution on the east coast (e.g., PAKP, PATB, TRMT, TRMT-T2, TRME, KLTB) and perhaps even to some extent on the west coast (e.g., MEKU). More stringent controls and regulations on the offshore platforms to prevent unnecessary spillages could reduce the tar-ball pollution on the Malaysian coasts. Contributions from tank washing and ballasting discharges from the tankers transporting the SEACO could also have contributed significantly to the overall tar-ball pollution along the Strait of Malacca (e.g., SESD, MESM, MAMC-T2). Appropriate treatment of the tank washings and ballast water from the tankers should be enforced. Oil pollution from other domestic sources such as leakage from oil refineries on the west coast and the Sumatran oil fields has been

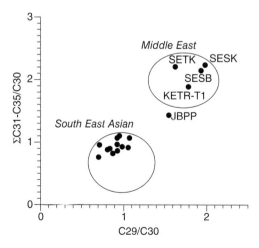

Figure 16-5 Cross-plots for source identification of oils. The circles indicating both categories were established through the analysis of crude oils in Zakaria et al. (2000).

suggested, although their contributions have not specifically been identified. To identify such domestic sources, more efforts are required, and the molecular marker approach proposed in the present study could be a powerful tool.

16.9 Case Study 2: Distribution and Sources of Polycyclic Aromatic Hydrocarbons (PAHs) in Rivers and Estuaries in Malaysia

In order to determine the distribution and sources of PAHs in Malaysia, sediment samples collected in rivers, estuaries, and the Strait of Malacca were analyzed for PAHs. For source discrimination of petroleum-related sources, triterpanes (petroleum biomarker) together with PAHs were measured. Prospective sources, including fresh and used crankcase oil, street dust, and asphalt were also analyzed. To examine the contribution from automobile-related sources, a molecular marker, n-cyclohexyl-2-benzothiozolamine (NCBA; Kumata et al., 2000), was utilized in polluted sediment samples and street dust.

Parent PAHs (3 to 7 rings) were analyzed by gas chromatography mass spectrometry (GC-MS). Total PAH concentrations in the sediment ranged from 4 to 924 ng/g, and alkylated homologues were abundant for most of the sediment samples. The ratio of the sum of methylphenanthrenes to phenanthrene (MP/P), an index of petrogenic PAH contributions, was more than unity for 26 of 29 sediment samples and more than 3 of 7 samples for urban rivers covering a broad range of locations.

The MP/P ratio showed a strong correlation with total PAH concentrations with an r^2 of 0.74. This ratio and all other compositional features indicated that Malaysian urban sediments are heavily impacted by petrogenic PAHs. The finding is unique in comparison to other studies reported in many industrialized countries where PAHs are mostly of pyrogenic origin. The MP/P ratio was also significantly correlated with higher-molecular-weight PAHs such as benzo[a]pyrene, suggesting a unique PAH source in Malaysia that contains both petrogenic PAHs and pyrogenic PAHs. PAH and hopane fingerprints indicated that used crankcase oil is one of the major contributors of the sedimentary PAHs. Two major routes of inputs were identified to aquatic environments: (1) spillage and dumping of waste crankcase oil; (2) leakage of crankcase oils from vehicles onto road surfaces with subsequent washout by street runoff. Input from street dust was confirmed by detection of N-cyclohexyl-2-benzothiazolamine (NCBA) in polluted sediment samples.

Oil pollution in coastal environments continues to be a major issue. The Strait of Malacca is no exception. The present study expands our limited understanding of possible sources of oil pollution to the fragile ecosystems in the Strait. The Strait have always been a source of valuable living and nonliving resources. Given the region's growing dependence upon Middle Eastern oil-producing nations and increasing trade and transport of fossil fuels, the Strait of Malacca has become a major conduit of world trade. Proper management such as preventive and remedial measures as well as conservation of the Strait will greatly contribute to the benefit of Malaysia and various other countries for years to come.

16.10 Conclusions and Future Scenario

This study has shown that C_{29}/C_{30} and ΣC_{31}–C_{35}/C_{30} ratios are not only powerful but reliable source identifiers of the oil pollution. This study has also shown that high oleanane concentration in tar balls indicates domestic petroleum origin (i.e., SEACO). However, low concentrations of oleanane in tar balls could not be identified as SEACO or MECO because both crude oils contain low concentrations of oleanane. This is probably due to the variability in oleanane contents in SEACOs. The results imply that low concentration of oleanane does not always indicate Middle East petroleum contribution although at high concentration oleanane can still be useful biomarker for Southeast Asian oil sources. Also, the homohopane index was shown to be useful as a source identifier in this study since it clearly distinguishes MECO from SEACO. This study, however, did not find that the Tm/Ts ratio is a useful source identifier of the oil spill in the Strait of Malacca since the ratio is inconsistent between MECO tar balls and SEACO tar balls.

The MP/P ratio showed a strong correlation with total PAH concentrations with an r^2 of 0.74. This ratio and all other compositional features indicated that Malaysian urban sediments are heavily impacted by petrogenic PAHs. The finding is unique in comparison to other studies reported in many industrialized countries where PAHs are mostly of pyrogenic origin. The MP/P ratio was also significantly correlated with higher-molecular-weight PAHs such as benzo[a]pyrene, suggesting a unique PAH source in Malaysia that contains both petrogenic PAHs and pyrogenic PAHs. PAH and hopane fingerprints indicated that used crankcase oil is one of the major contributors of the sedimentary PAHs. Two major routes of inputs were identified to aquatic environments: (1) spillage and dumping of waste crankcase oil; (2) leakage of crankcase oils from vehicles onto road surfaces with subsequent washout by street runoff. Input from street dust was confirmed by detection of N-cyclohexyl-2-benzothiazolamine (NCBA) in polluted sediment samples.

References

Abdullah, A.R., N.R. Tahir, T.S. Loong, T.M. Hoque, and A.H. Sulaiman. The GEF/UNP/IMO Malacca Straits demonstration project: Sources of pollution. *Mar. Pollut. Bull.*, 1999, **39**, 229–233.

Clemons, J.H., L.M. Allan, C.H. Marvin, Z. Wu, B.E. McCarry, and D.W. Bryant. Evidence of estrogen- and TCDD-like activities in crude and fractionated extracts of PM10 air particulate material using in vitro gene expression assays, *Environ. Sci. Tech.*, 1998, **32**(12), 1853–1860.

Dow, K. An overview of pollution issues in the Straits of Malacca. *In The Straits of Malacca: International Co-operation in Trade, Funding and Navigational Safety*, H. Ahamed (ed.), 1997, Selangor, Malaysia: Pelanduk Publications, pp. 61–102.

Heitkamp, M.A. and C.E. Cerniglia. "Effects of Chemical Structure and Exposure on the Microbial Degdradation of Polycyclic Aromatic Hydrocarbons in Freshwater and Estuarine Ecosystems." *Environmental Toxicology and Chemistry*, 1987, **6**, 535–546.

Kumata, H., Y. Sanada, H. Takada, and T. Ueno. Historical trends of N-cyclohexyl-2-benzothiazolamine, 2-(4-morpholinyl)-benzothiazole and other anthropogenic contaminants in the urban reservoir sediment core. *Environ. Sci. Techn.*, 2000, **34**(2), 246–253.

Malaysian Marine Department, *Malaysian Ministry of Transport Occasional Publication*, 2004, Total number of vessels passing through the Straits of Malacca (1987–2004) (observations made from one fanthom bank lighthouse).

Naidu, G. The Straits of Malacca in the Malaysian economy. *In The Straits of Malacca: International Cooperation in Trade, Funding and Navigational Safety*, Hamzah Ahmad (ed.), 1997, Kuala Lumpur: Maritime Institute of Malaysia (MIMA).

Neff, J.M. *Polycyclic Aromatic Hydrocarbons in the Aquatic Environment. Source, Fate and Biological Effects*. 1979, Applied Science Publishers, Ltd.: London.

Okuda, T., H. Kumata, H. Takada, M.P. Zakaria, H. Naraoka, and R. Ishiwatari. Source identification of polycyclic aromatic hydrocarbons in forest fire smoke using molecular and isotopic signature, *Atmos. Environ.*, 2002, **36**, 611–618.

Volkman, J.K., A.T. Revill, and A.P. Murray. Applications of biomarkers for identifying sources of natural and pollutant hydrocarbons in aquatic environments. *In Molecular Markers in Environmental Geochemistry*, R.P. Eganhouse (ed.), 1997, Washington, DC: American Chemical Society, pp. 279–313.

Wang, Z.D. and M. Fingas. Development of oil hydrocarbon fingerprinting and identification techniques, *Mar. Pollut. Bull.*, 2003, **47**, 423–452.

Wyrtki, K. (1961). *In Scientific Results of Marine Investigations of the South China Sea and the Gulf of Thailand 1959–1961*. Naga Report, La Jolla, CA: Scripps Institution of Oceanography, p. 195.

Zakaria, M.P., H. Takada, A. Horinouchi, S. Tanabe, and A. Ismail. The use of pentacyclic triterpanes as biomarkers for source identification of oil pollution in the Straits of Malacca. *American Chemical Society (ACS) Division of Environmental Chemistry*, Preprints, 1999, **39**(2), 6–9.

Zakaria, M.P., A. Horinouchi, S. Tsutsumi, H. Takada, S. Tanabe, and A. Ismail. Application of biomarkers for source identification of oil pollution in the Straits of Malacca, Malaysia, *Environ. Sci. Tech.*, 2000, **34**(7), 1189–1196.

Zakaria, M.P., T. Okuda, and H. Takada. Polycyclic aromatic hydrocarbon (PAHs) and hopanes in stranded tar-balls on the coasts of Peninsular Malaysia: Applications of biomarkers for identifying sources of oil pollution, *Mar. Poll. Bull.*, 2001, **42**(12), 1357–1366.

Zakaria, M.P., H. Takada, S. Tsutsumi, K. Ohno, J. Yamada, E. Kouno, and H. Kumata. The distribution of polycyclic aromatic hydrocarbons (PAHs) in rivers and estuaries in Malaysia: A widespread input of petrogenic PAHs, *Environ. Sci. Tech.*, 2002, **36**, 1907–1918.

17 Case Study: Evaluation of Hydrocarbon Sources in Guanabara Bay, Brazil

Maria de Fatima G. Meniconi and Silvana M. Barbanti

Petrobras S.A. Av. Jequitiba, 950, Cidade Universitaria, Ilha do Fundao, Rio de Janeiro, RJ, Brazil

17.1 Guanabara Bay and Hydrocarbon Apportioning

Hydrocarbons are present in worldwide environmental ecosystems, and their potential to cause adverse effects is usually associated with the concentration of polycyclic aromatic hydrocarbons (PAHs) (Neff, 1979). It must be emphasized, however, that these compounds may be introduced by many different mechanisms, including natural and anthropogenic processes (Philp, 1985; Kennicutt II et al., 1994; Kennicutt II, 1995; Simoneit, 1998; Lipatou and Albaigés, 1994).

Sources of naturally occurring PAH include natural fires, natural oil seepage, and recent biological or diagenetic processes — biogenic origin (Hites and Biemann, 1975; Youngblood and Blumer, 1975; Kennicutt II et al., 1994; Kennicutt II, 1995; Yunker et al., 2000; Yunker et al., 2002).

Anthropogenic sources of PAH are from direct runoff and discharges and indirect atmospheric deposition, i.e., from waste and releases/spills of petroleum and derivatives such as river runoff, sewage outfalls, maritime transport, pipelines, and combustion or pyrolysis of organic matter such as petroleum, coal, and wood (Lipatou and Albaigés, 1994; Budzinski et al., 1997; Elias et al., 2000; Wang et al., 1999a; Yunker et al., 2000; Yunker et al., 2002; Stout et al., 2001; Readman et al., 2002).

These compounds tend to interact with different types of environments and are subjected to many processes that lead to geochemical fates such as physical-chemical transformation, biodegradation, and photooxidation. Therefore, PAH characterization and the correlation of PAH to known or suspected sources become a challenge.

Another class of hydrocarbons present in worldwide environmental ecosystems is saturated hydrocarbons derived from fossil fuels (petroleum and its refined products). The trace organic compounds that provide information about source, maturation, and depositional environment are the so-called biological fossils or biomarkers (Eglinton et al., 1964; Eglinton and Calvin, 1967; Peters et al., 2005). Molecular parameters based on GC-MS analysis of these biomarkers, primarily terpanes and steranes, provide a method to relate the environment samples and crude oils. Therefore, the source of a certain sample containing fossil hydrocarbons can be determined by applying both biomarker ratios and profiles (see Chapter 3 herein).

Numerous studies for apportioning sources of PAH in the environment that apply diagnostic indexes based on chemical fingerprinting have been made mainly in temperate climate countries (Readman et al., 2002; Budzinski et al., 1997; Sicre et al., 1987; Gschwend and Hites, 1981). In the present study, the sources of PAH and other hydrocarbons have been investigated on tropical estuarine ecosystem in the Guanabara Bay, Brazil, in which the scenarios of chronic and acute anthropogenic events were taken into consideration. Our study sought to identify hydrocarbon sources using both PAH diagnostic indexes and biomarker ratios, in combination with multivariate statistics based on principal component analysis (PCA) for intertidal and subtidal area sediments investigated in two campaigns: the years 2000 and 2003.

17.1.1 Regional Setting

Guanabara Bay is located in Greater Rio de Janeiro, where there exists a highly urban and industrialized ecosystem that receives intense chronic anthropogenic pollution and was the scene of an oil spill of $1300\,m^3$ of marine heavy fuel oil in January 2000 (Gabardo et al., 2001; Meniconi et al., 2002).

The bay is the centerpoint of a complex river drainage basin receiving water from about 50 rivers and channels and is used to dispose extensive municipal sewage, usually with minimal or no treatment, and receives urban runoff and industrial waste from the country's second largest city, with almost 10 million people (JICA, 1995; FEEMA, 1998). The bay drainage basin encompasses 15 districts with about 14,000 industries, 14 oil terminals, 2 commercial ports, 32 dock yards, more than 1000 gas stations, and 2 refineries. All of these inputs are responsible for an annual inflow of approximately 6.6 tons of hydrocarbons into the bay (FEEMA, 2005; Ferreira, 1995) (Figure 17-1). About 50% of the hydrocarbon input in the bay is due to the municipal sewage ($17\,m^3/s$), while 27% corresponds to the urban runoff (Ferreira, 1995). Furthermore, 13,000 tonnes/day of solid waste are produced by the Great Rio de Janeiro area, of which 5,000 tonnes/day are ejected directly into the bay, forming a permanent trash line along the beaches and mangroves of the bay.

The bay has a water surface area of about $400\,km^2$ and a 156-km shoreline. It is a system with strong currents along the central channel due to tides and weaker currents in the north area of the bay. A shallow area of up to 5 m corresponds to about 46% of the total water surface of the bay. However, the average flushing time to renew 50% of the bay water is about 11 days (Kjerve et al., 1997). The central channel presents a faster renewal of the water due to the strong currents. On the other hand, in the north, northwest, and northeast areas of the bay, the renewal is low since the tidal currents are weaker than the central channel. Even so, the wind can increase the currents in the north, contributing to the renewal of water in this area. Spread along the shallow north region of the bay are approximately 350 fish traps, which form an important social economic aspect of the bay, the fishing communities.

The mangrove area, located on the northwest and northeast shorelines of the bay, is about $86\,km^2$, most of it significantly degraded (FEEMA, 2005; Amador, 1997). The largest part of the bay mangrove is situated inside an environmental protected area, established in the bay's northeast region.

17.1.2 January 2000 Heavy Fuel Oil Spill

The acute event of January 2000 was an accidental oil release of about $1300\,m^3$ of marine heavy fuel oil into the bay, due to a pipeline rupture at the Duque de Caxias Refinery of Petrobras, the Brazilian state oil company. An oil slick up to $56\,km^2$ formed on the bay surface (Bentz and Miranda, 2001), as can be seen in Figure 17-2. An extensive response for on-water oil recovery and shoreline cleanup was implemented following the accident. The main portion of the oil, transported by tidal currents and wind, reached the beaches and some islands in the north and northeast parts of the bay. Mangroves in the vicinity of the spill emission point also were affected.

Immediately following the oil spill response, 22 water and 57 sediment samples were collected from the Guanabara Bay area. These were characterized by measuring the concentrations of the n-alkanes, unresolved complex mixture (UCM), and polycyclic aromatic hydrocarbons (PAHs) biomarkers (terpanes and steranes) and by conducting toxicological assays for three species (*Artemia*

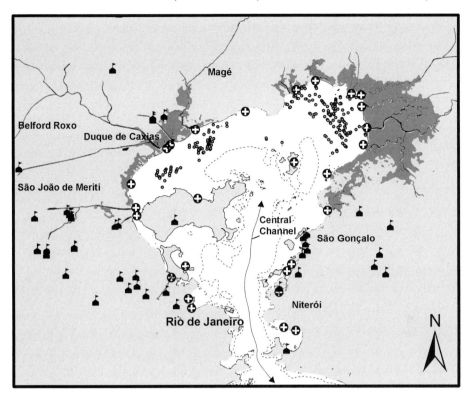

Legend
- ○ Fish Traps
- ⊕ Sewage
- ▲ Industries
- ------ Shallow Area (-5 m)
- ⌒ hydro
- ▨ Mangrove

Figure 17-1 Social, economical, and physical characteristics of Guanabara Bay.

sp., *Mysidium gracile*, and *Vibrio fisheri* for Microtox® system). Composite fish tissue samples from two species were also analyzed, being chosen based on their abundance, marketability, and feeding habits. The fish were collected in the affected area and then analyzed and compared to the same species' frozen-dry samples collected 1 year prior to the spill by a local university (reference samples). The spilled heavy fuel oil was also characterized, revealing a relatively high density (0.9817 g/mL), 12°API, and viscosity (20°C) of 5313 cP. Approximately 21% of the fuel oil had evaporated after 72 hours of exposure, as calculated from the gas chromatographic comparison of the solvent-rinsed sand samples collected on the affected beaches and the original fuel oil. The biomarker analysis of the spilled marine heavy fuel oil confirmed that it was refined from a crude oil from a Brazilian marginal basin, as expected (Meniconi et al., 2002). Based on the results of this diagnostic study, no significant impact on the sediments, water column, and fish tissue of Guanabara Bay was observed due to the oil spill.

17.2 Methodology for Hydrocarbon Determination and Source Evaluation

17.2.1 Sampling Design

The strategy for this work on hydrocarbon source apportionment in the tropical estuarine ecosystem of Guanabara Bay was based on analysis of hydrocarbons in the sediments.

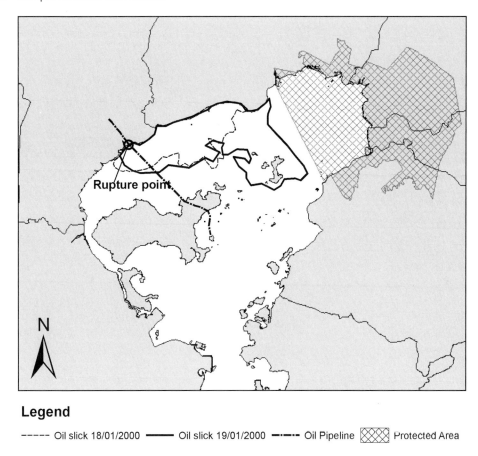

Legend

----- Oil slick 18/01/2000 ——— Oil slick 19/01/2000 —·—·— Oil Pipeline   ▩▩▩ Protected Area

Figure 17-2 Oil slick after January 2000 oil spill accident in Guanabara Bay.

These data included a large number of compounds, including 38 PAH and 61 petroleum biomarkers (42 terpanes and 17 steranes). Among the 57 sediment stations analyzed immediately after the oil spill, 21 stations were chosen to develop the present study, representing areas potentially affected and unaffected by the spilled oil in January 2000 (Meniconi et al., 2002). The samples were collected using cores and dredges from the intertidal (stations T1 to T28) and subtidal regions (stations T31 to T57) of the bay. A subsample of the top 3 cm of each sediment was transferred into wide-mouth glass jars with Teflon caps and then freeze-stored for subsequent PAH analysis. Figure 17-3 shows the geographical location of the 21 sample stations in Guanabara Bay.

In the first sampling campaign (January 2000), the samples were collected from 21 stations (Figure 17-3), just 10 days after the accident. In the second sampling campaign (2003), the samples were collected from the same 21 stations studied in 2000, which permitted a good temporal investigation in the region.

17.2.2 Chemical Analysis

17.2.2.1 Sediment Sample Extraction

The majority of the samples was extracted following the methodology EPA Method 3540 (Soxhlet) and in some cases EPA 3550B (sonication). The sediment samples were thawed and homogenized, and a subsample (5 g) was mixed with sodium sulphate (1:4; w/w). Extraction was performed in a Soxhlet appa-

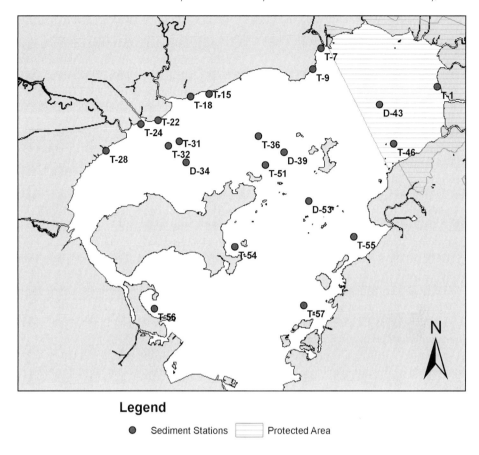

Figure 17-3 Geographical location of the 21 sediment stations in Guanabara Bay sampled in 2000 and 2003.

ratus for 4 hr with 200 mL of dichloromethane. Before the extraction, perdeuterated paraffins (n-eicosane-d_{42}, n-tetracosane-d_{50}, n-triacontane-d_{62}) and an aromatic hydrocarbon (p-terphenyl-d_{14}) were added to the sediment samples as surrogates. Sulfur was removed during extraction with granulated copper. A rotary evaporator was used to reduce the bulk extract volume to 5 mL, and then the solvent was changed to n-hexane in a Kuderna-Danish apparatus to a final volume of 1 mL.

17.2.2.2 Extract Cleanup

Sediment extracts were fractionated by adsorption chromatography, based on EPA Method 3630 modified. A column filled with 2.5 g of combusted sand (400°C), 20 g of 5% deactivated silicagel, 10 g of 1% deactivated basic alumina, 2.5 g of combusted sand, and 2.5 g of combusted sodium sulfate was used. The concentrated extract (1 mL) was carefully added to the top of the column, and the following fractions were collected: (1) 50 mL of n-hexane (aliphatic hydrocarbons, isoprenoids, and UCM); and (2) 200 mL of 1:1 dichloromethane:hexane (aromatic compounds). Both fraction volumes were reduced by rotary evaporator and a gentle stream of pure nitrogen to a final volume of 1 mL and stored under refrigeration until the analysis.

17.2.2.3 PAH Analysis of Sediment Samples

Gas chromatography-mass spectrometry (GC-MS) analysis for polycyclic aromatic hydrocarbons and their homologues followed the

EPA Method 8270-C, with modifications. This was carried out in an HP 5890-series II GC coupled to a Finnigan GCQ mass detector system. The mass spectrometer was operated in the electron impact (EI) mode (70 eV), and the data were acquired in full scan mode (15–400 amu). GC conditions were similar to those reported in aliphatic analyses, except for the oven temperature (40°C for 5 min and from 40 to 280°C at 8°C/min, with a final time hold of 30 min). Quantification was performed by internal standard method using a solution with acenaphthene-d_{10}, naphthalene-d_8, phenanthrene-d_{10}, chrysene-d_{12}, and perylene-d_{12}. The analytical method was certified by the extraction, in triplicate, of a marine sediment reference material and dry frozen mussels from NIST (SRM 1941a and SRM 2974), and blanks were periodically run to check contamination problems. The Canadian Association for Environmental Analytical Laboratories (CAEAL) certifies the laboratory twice a year. Detection limits were calculated as $1 \, ng \cdot g^{-1}$ for sediment, dry weight.

17.2.2.4 Biomarkers

Biomarker analyses were performed by using a gas chromatograph coupled to a mass selective detector (MSD 5973N, Agilent), electron impact as ionization method, and selective ion monitoring as data acquisition mode for some diagnostic ions of these compounds [m/z 191, 177, 398, 412 for terpanes; m/z 217, 218 for steranes; m/z 231 for methylsteranes; m/z 259 for diasteranes and tetracyclic terpane polyprenoids (TPP); (Peters et al., 2005; Holba et al., 2000, 2003)]. GC conditions were similar to those reported in aromatic analyses, except for the oven temperature (55°C for 2 min, from 55 to 150°C at 15°C/min, from 150 to 320°C at 1.5°C/min, with a final time hold of 20 min). Quantification was performed by internal standard method using cholestane-d_4 or 5β-cholane. The analytical method was certified by analyzing the saturated hydrocarbon fraction of a reference crude oil, and running blanks periodically in order to check contamination problems.

17.2.3 Source Identification Techniques

The hydrocarbon source identification of the sediment samples from Guanabara Bay has been made by using different approaches: PAH diagnostic indexes, PAH multivariate analysis, and biomarker ratios, described in the following sections.

17.2.3.1 PAH Diagnostic Ratios

As previously reported, many PAH molecular ratios have been used to help to identify the PAH sources in environmental samples, whether petrogenic, biogenic, or pyrolytic (Youngblood and Blumer, 1975; Gschwend and Hites, 1981; Sicre et al., 1987; Lipatou and Albaigés, 1994; Kennicutt II, 1995; Budzinski et al., 1997; Baumard et al., 1998, Wang et al., 1999a; Yunker et al., 2000, 2002; Readman et al., 2002). The difficulty remains with the complexity of the samples themselves and the weathering effects on the composition of the original source of the compounds.

The use of parent PAH diagnostic ratios to interpret PAH sources is based on the relative stability of the PAH isomers, which have been evaluated from the relative heat of formation of the compounds. The heat of formation is the energy difference for each PAH isomer relative to the most stable isomer within a given molecular mass. The calculation of the heat of formation for a restricted mass (m/z 178, m/z 202, m/z 252, m/z 276) would prevent or minimize interferences such as differences in volatility, water solubility, and adsorption characteristics of the compounds (Yunker et al., 2000, 2002). It would be expected that ratios based on isomers with the highest heat of formation differences (i.e., greatest range in stability) would provide the highest discrimination capacity. For instance, the heat of formation differences between phenanthrene (Phe) and anthracene (An) and for fluoranthene (Fl) and pyrene (Py) are 5.48 kcal/mol and 20.58 kcal/mol, respectively, and the ratio Fl/Py would then have higher discrimination ability (Yunker et al., 2000, 2002). Software such as AM1 (Hyperchem, V4, 5, Hypercube, Inc., 419 Philip St.,

Waterloo, Ontario, Canada) and PCMODEL (V5.13, Serena Software, Box 3076, Bloomington, Indiana, USA) has been used to calculate the PAH heat of formation (Yunker et al., 2000).

The PAH molecular ratios used in this study covered parent and alkylated PAH homologues, which are listed below:

Parent PAH

- Phenanthrene/anthracene
- Fluoranthene/pyrene
- Anthracene/(anthracene + phenanthrene)
- Fluoranthene/(fluoranthene + pyrene)
- Indeno1,2,3-(cd)pyrene/(indeno1,2,3-(cd) pyrene + benzo(ghi)perylene)
- Perylene abundance index: Perylene/Σ (5-ring PAH)
  (5-ring PAH = BbFl + BkFl + BaPi + BePi + DBAn. See Table 17-1 for codes.)

Alkylated PAH

- Phenanthrene + anthracene/(phenanthrene + anthracene + C1phenanthrenes)
- Pyrogenic index: Σ (other 3–6-ring PAHs)/ Σ (5 alkylated PAH series)
  (5 alkylated PAH series = naphthalenes, fluorenes, dibenzothiophenes, phenanthrenes, and chrysenes)

The cross-plots used in this study included the PAH data from the 2000 and 2003 campaigns together and included data from four different oils: a heavy fuel; an Arabian oil; a light oil; and a diesel fuel.

17.2.3.2 PAH Multivariate Statistical Analysis

The multivariate analysis of the PAH concentration data applied in this study was based on principal component analysis (PCA) (Statistica version 5.0) in which the sample projections were separately plotted for surveys 2000 and 2003. However, some preprocessing of the data was conducted before applying PCA, namely:

- A half-limit concentration was used for non-detected compounds (0.5 ng/g).
- Each compound concentration was normalized to the total PAH concentration in the sample in order to prevent the influence of the wide range of sample concentrations.
- The normalized concentration data for each compound were reduced by the average and standard deviation for that compound within all of the samples.

17.2.3.3 Biomarker Diagnostic Ratios

Biomarker analyses were performed by using a gas chromatograph coupled to a mass selective detector (MSD 5973N, Agilent), electron impact as ionization method, and selective ion monitoring as data acquisition mode for some diagnostic ions of these compounds [m/z 191, 177, 398, 412 for terpanes; m/z 217, 218 for steranes; m/z 231 for methylsteranes; m/z 259 for diasteranes and tetracyclic terpane polyprenoids (TPP); (Peters et al., 2005; Holba et al., 2000, 2003)]. GC conditions were similar to those reported in aromatic analyses, except for the oven temperature (55°C for 2 min, from 55 to 150°C at 15°C/min, from 150 to 320°C at 1.5°C/min, with a final time hold of 20 min). Quantification was performed by internal standard method using cholestane-d_4 or 5β-cholane. The analytical method was certified by analyzing the saturated hydrocarbon fraction of a reference crude oil and running blanks periodically in order to check contamination problems.

17.3 Hydrocarbon Results for Guanabara Bay Sediments

17.3.1 PAH Quantification and Distribution

The concentration of individual PAH compounds and the sum of 16 EPA priority PAHs (Σ 16 PAHs) and the total 38 PAHs (Σ PAH) in the sediments collected in Guanabara Bay in 2000 (just after the oil spill) and in 2003 are shown in Tables 17-1 and 17-2, respectively. For both sampling campaigns, the concentration of total PAH varied significantly throughout the bay, ranging from 559 to 58,439 ng/g dry weight (median concentration: 4,877 ng/g)

Table 17-1 Results for the Individual PAH (ng/g dry weight)* of Sediment Samples from Guanabara Bay — Campaign 2000 and Studied Oils

Compound	Code	Intertidal							
		T1	T7	T9	T15	T18	T22	T24	T28
Naphthalene	N	6	2	2	12	7	5	12	4
1-Methylnaphthalene	1MN	3	19	6	24	6	45	6	2
2-Methylnaphthalene	2MN	6	26	7	48	15	67	21	9
C_2Naphthalenes	C_2N	77	363	303	364	122	1,695	211	226
C_3Naphthalenes	C_3N	3	983	944	440	163	6,236	524	88
C_4Naphthalenes	C_4N	19	1,189	1,377	338	172	6,738	1,214	97
Acenaphthalene	Acl	2	2	2	9	12	12	7	6
Acenaphthene	Ace	1	18	13	17	6	60	9	3
Fluorene	F	4	26	30	28	17	93	10	8
C_1Fluorenes	C_1F	5	160	199	74	55	601	80	24
C_2Fluorenes	C_2F	28	536	636	115	101	1,602	682	57
C_3Fluorenes	C_3F	30	585	815	111	109	1,748	1,987	83
Dibenzothiophene	DBT	5	52	89	42	30	162	31	11
C_1Dibenzothiophenes	C_1DBT	5	206	438	94	75	757	301	42
C_2Dibenzothiophenes	C_2DBT	22	441	963	170	145	1,352	2,160	100
C_3Dibenzothiophenes	C_3DBT	31	425	884	152	135	1,157	5,109	111
Phenanthrene	Fe	17	148	277	120	76	486	54	47
C_1Phenanthrenes	C_1Fe	21	490	888	187	141	1,579	549	93
C_2Phenanthrenes	C_2Fe	30	781	1,759	277	230	2,294	5,033	163
C_3Phenanthrenes	C_3Fe	28	666	1,479	231	204	1,900	10,485	148
C_4Phenanthrenes	C_4Fe	1	321	538	120	110	839	6,494	85
Anthracene	An	4	26	46	20	21	122	38	23
Fluoranthene	Fl	27	37	45	108	91	56	72	79
Pyrene	Pi	22	57	110	99	91	247	499	96
C_1Pyrenes	C_1Pi	2	106	222	112	118	547	2,679	81
C_2Pyrenes	C_2Pi	1	161	311	108	99	868	5,114	93
Benz (a) anthracene	BaAn	14	35	55	65	59	151	854	69
Chrysene	C	16	46	78	74	61	226	543	61
C_1Chrysenes	C_1C	15.1	94	215	101	90	777	5,298	96
C_2Chrysenes	C_2C	9	95	234	124	99	1,070	6,695	107
Benz(b)fluoranthene	BbFl	25	37	52	88	89	88	76	117
Benz(k)fluoranthene	BkFl	11	13	14	28	20	23	157	37
Benz(a)pyrene	BePi	19	29	48	62	47	87	370	81
Benz(e)pyrene	BaPi	16	25	33	52	45	89	451	66
Perylene	Pe	30	62	18	50	79	62	159	33
Indeno (1,2,3-cd)pyrene	IPi	17	22	27	60	58	72	79	71
Dibenz(a,h)anthracene	DBAn	6	8	12	26	16	57	188	23
Benzo(ghi)perylene	BPe	17	21	23	50	49	82	186	72
Σ 16HPA		207	526	819	866	719	1,866	3,235	797
Σ Total HPA		691	8,259	13,191	4,198	3,058	34,048	58,439	2,614

* Surrogate recovery: 68%–117% (average = 98%); n.d.: not detected.

for the 2000 campaign and 400 to 52,384 ng/g dry weight (median concentration: 3,603 ng/g) for the 2003 campaign. However, based on nonparametric analysis (Mann–Whitney, Kolmogorov–Smirnov, Kruskal–Wallis), it was verified there was no statistical difference in total PAH concentrations between the 2000 and 2003 samples.

In order to compare the data of this study to previously reported data for Guanabara Bay, it is necessary to evaluate only the Σ 16 EPA priority PAH results, which varied from 207 to 13,425 ng/g and 184 to 5,110 ng/g for the 2000 and 2003 campaigns, respectively (Tables 17-1 and 17-2). The range of these hydrocarbons' contamination was the same observed in previous sediment surveys conducted by Lima (1996) (1,564 to 18,438 ng/g), Hamacher (1996) (554 to 1,894 ng/g), and Chalaux (1995) (1,051 to 5,861 ng/g).

	Subtidal												Oil			
T31	T32	T36	T46	T51	T54	T55	T56	T57	D34	D39	D43	D53	Ar.	A.L.	MF 380	D.M.
20	55	16	5	16	3	<1	45	8	8	83	3	71	1,452	115	478	304
10	39	9	8	10	<1	<1	28	2	6	21	<1	37	18,852	416	1,913	1,339
35	84	33	27	25	4	5	89	8	6	67	<1	93	10,558	426	3,349	1,983
234	629	294	166	294	9	100	315	16	328	218	189	166	61,032	2,166	11,304	7,237
92	1,458	55	148	56	7	56	163	10	55	132	24	91	55,182	2,303	12,259	8,397
84	1,536	50	109	59	9	41	156	<1	56	120	11	76	28,911	1,547	6,130	4,614
13	7	86	10	77	3	55	147	4	35	507	6	315	470	16	79	n.d.
4	15	7	3	8	<1	7	28	3	5	29	<1	10	1,886	4	316	55
13	38	21	18	18	2	18	55	5	10	82	6	59	1,405	40	300	200
47	192	31	60	10	2	23	87	6	17	88	11	66	3,725	158	787	474
73	618	43	90	114	4	33	255	8	41	128	28	118	5,530	322	1,203	887
122	971	67	103	211	10	45	326	9	108	217	22	163	3,934	477	860	934
25	85	13	34	13	<1	16	31	3	10	24	6	53	805	225	245	196
58	392	43	77	47	4	40	76	7	37	73	15	84	1,564	641	601	430
127	1,126	80	123	138	9	66	179	10	93	297	19	202	895	1,137	727	500
334	1,343	118	101	230	17	82	248	11	151	398	18	263	365	949	525	349
57	217	75	62	68	15	104	168	40	59	204	28	168	4,914	77	898	782
139	923	95	112	111	12	101	189	26	95	411	28	259	8,052	295	1,619	1,555
272	1,815	134	136	248	15	112	286	23	175	846	35	480	4,921	461	1,325	1,911
474	1,948	165	110	268	22	105	312	18	40	855	27	448	1,273	414	720	1,474
537	1,068	60	50	192	18	65	276	11	191	341	12	196	260	208	294	554
59	65	56	25	39	5	51	111	15	38	251	9	157	1,033	13	165	50
78	113	149	65	192	29	188	248	77	115	468	40	532	27	<0.5	6	4
138	225	218	72	227	34	221	629	84	171	921	51	791	125	9	67	44
280	453	270	57	337	21	192	667	53	210	1,772	33	1,209	84	39	160	145
544	661	224	48	306	20	133	561	29	225	1,505	25	879	23	94	211	172
126	126	198	57	170	23	199	281	65	102	1,025	36	927	1	n.d.	30	19
89	183	179	56	175	21	161	176	57	85	875	30	776	2	n.d.	54	63
457	517	291	57	264	21	187	383	40	165	1,452	25	1,105	2	34	158	110
764	673	241	42	191	18	119	231	21	179	690	14	620	1	46	172	9
133	107	278	76	286	44	323	528	55	184	2,153	53	1,246	n.d.	<0.5	5	4
33	101	114	38	129	17	109	198	23	70	838	20	478	n.d.	<0.5	n.d.	n.d.
122	88	286	66	272	34	274	434	51	128	1,161	38	1,382	n.d.	9	14	11
110	96	144	47	167	25	170	309	29	147	2,151	31	561	n.d.	n.d.	9	5
94	89	67	50	68	9	63	91	14	47	315	95	227	n.d.	n.d.	3	<1,5
127	73	298	77	228	36	256	464	51	148	1,660	35	1,120	n.d.	n.d.	n.d.	<1,5
110	43	119	38	70	10	77	144	21	43	551	8	485	n.d.	n.d.	1	<1,5
143	68	252	68	215	27	223	464	41	146	1,628	21	883	n.d.	n.d.	3	<1,5
1,264	1,524	2,352	735	2,189	303	2,267	4,119	600	1,366	13,425	381	9,399	11,314	282	2,415	1,534
6,174	18,240	4,877	2,487	5,398	559	4,019	9,816	952	3,730	24,555	1,048	16,793	217,282	12,638	46,990	34,807

The comparison of this study data with data from other estuarine and coastal regions in the world reported in the literature is presented in Table 17-3. The PAH concentrations in Guanabara Bay sediments collected at the time of the oil spill and 3 years after it are in a similar range as found at various international estuarine marine sites, with and without correlation to oil spills. These data suggest that the PAH concentrations in Guanabara sediments are not obviously related to the oil spill event, but rather to long-term anthropogenic input.

Considering the distribution of PAH of Guanabara Bay sediment samples for the 2000 and 2003 surveys, it was found, in general, a higher alkylated PAH contribution for intertidal samples from the 2000 campaign, where oil was observed. This can be observed in Figure 17-4, where all series of weathered alkylated

Table 17-2 Results for the Individual PAH (ng/g dry weight)* of Sediment Samples from Guanabara Bay — Campaign 2003 and Studied Oils

Compound	Code	Intertidal							
		T1	T7	T9	T15	T18	T22	T24	T28
Naphthalene	N	4	1	3	5	6	6	13	6
1-Methylnaphthalene	1MN	2	<1	2	8	3	4	5	20
2-Methylnaphthalene	2MN	3	1	4	7	5	11	16	10
C_2Naphthalenes	C_2N	32	5	18	64	35	55	100	133
C_3Naphthalenes	C_3N	12	12	20	102	42	91	113	206
C_4Naphthalenes	C_4N	7	32	33	120	60	143	219	319
Acenaphthalene	Acl	2	2	3	4	5	7	7	<2.5
Acenaphthene	Ace	<1	<1	2	21	2	2	3	3
Fluorene	F	2	1	4	13	6	8	8	9
C_1Fluorenes	C_1F	4	4	8	50	22	28	35	52
C_2Fluorenes	C_2F	5	23	20	81	46	75	212	184
C_3Fluorenes	C_3F	9	48	29	118	59	136	1,003	334
Dibenzothiophene	DBT	3	2	4	13	6	10	13	11
C_1Dibenzothiophenes	C_1DBT	5	7	5	41	18	44	117	47
C_2Dibenzothiophenes	C_2DBT	7	26	29	128	50	163	801	136
C_3Dibenzothiophenes	C_3DBT	7	56	37	173	75	403	3,350	180
Phenanthrene	Fe	12	7	25	42	21	27	30	40
C_1Phenanthrenes	C_1Fe	12	17	28	84	36	79	205	121
C_2Phenanthrenes	C_2Fe	11	60	58	194	75	249	1,723	318
C_3Phenanthrenes	C_3Fe	8	20	84	245	88	569	6,493	408
C_4Phenanthrenes	C_4Fe	3	72	56	169	63	588	8,491	257
Anthracene	An	3	2	7	10	5	17	31	6
Fluoranthene	Fl	23	28	65	99	58	52	56	35
Pyrene	Pi	25	48	57	52	49	84	625	69
C_1Pyrenes	C_1Pi	12	56	47	97	63	244	3,146	119
C_2Pyrenes	C_2Pi	7	71	43	72	49	496	7,638	173
Benz (a) anthracene	BaAn	17	20	38	41	24	48	622	10
Chrysene	C	13	19	32	42	30	55	467	35
C_1Chrysenes	C_1C	12	28	30	54	39	258	5,349	61
C_2Chrysenes	C_2C	5	51	34	60	43	488	8,711	61
Benz(b)fluoranthene	BbFl	23	22	46	78	48	82	279	21
Benz(k)fluoranthene	BkFl	10	8	15	21	13	20	41	5
Benz(a)pyrene	BePi	19	14	17	37	24	81	777	17
Benz(e)pyrene	BaPi	15	18	33	38	18	65	684	10
Perylene	Pe	32	24	13	41	51	72	215	167
Indeno (1,2,3-cd)pyrene	IPi	16	16	27	45	30	68	116	8
Dibenz(a,h)anthracene	DBAn	4	5	8	14	8	35	301	<2.5
Benzo(ghi)perylene	BPe	15	15	26	37	26	71	372	11
Σ 16PAH		184	212	388	560	349	647	3,653	269
Σ Total PAH		400	838	1,004	2,516	1,303	4,931	52,384	3,603

* Surrogate recovery: 61%–119% (average = 101%); n.d.: not detected.

PAHs are predominant in the 2000 survey for samples T7, T9, and T22.

On the other hand, for both campaigns, a predominance of four- and five-ring PAHs over other compounds for subtidal stations was observed, either inside the influence of the oil spill slick (T36, T51, D53) or beyond the influence (T55, T56, and T57), as can be seen in Figure 17-5(a) and (b), respectively. This feature is typical for estuarines near urban areas in which the PAHs in runoff are associated with combustion-derived particulate matter (Stout et al., 2004).

17.3.2 Hydrocarbon Source Identification

17.3.2.1 PAH Diagnostic Ratios

Comparisons of the set of PAH diagnostic ratios (above) were used to distinguish the

					Subtidal									Oil		
T31	T32	T36	T46	T51	T54	T55	T56	T57	D34	D39	D43	D53	Ar.	A.L.	MF 380	D.M.
7	13	20	19	7	20	24	45	<1	6	10	<1	49	1,452	115	478	304
6	40	22	10	11	11	8	17	1	5	13	n.d.	13	18,852	416	1,913	1,339
18	70	3	28	7	22	27	60	2	12	<1	n.d.	39	10,558	426	3,349	1,983
121	1,091	112	346	73	212	221	307	21	57	139	16	110	61,032	2,166	11,304	7,237
236	3,657	44	133	34	56	59	118	9	71	104	20	53	55,182	2,303	12,259	8,397
453	4,685	33	76	29	54	34	78	10	78	142	14	41	28,911	1,547	6,130	4,614
5	2	73	6	5	36	57	186	9	23	6	6	120	470	16	79	n.d.
3	16	6	2	80	9	10	32	2	4	3	<1	12	1,886	4	316	55
10	70	15	14	13	18	18	47	5	10	8	5	29	1,405	40	300	200
54	560	15	45	19	23	19	51	6	23	12	10	33	3,725	158	787	474
256	2,204	23	55	36	35	30	66	18	57	77	13	52	5,530	322	1,203	887
620	4,116	44	54	62	77	36	120	25	98	113	16	87	3,934	477	860	934
23	100	10	18	8	15	14	25	3	13	12	7	16	805	225	245	196
87	921	25	44	21	38	28	58	7	28	53	12	<1	1,564	641	601	430
447	2,087	59	71	84	94	56	116	24	88	140	18	131	895	1,137	727	500
872	2,170	82	55	165	155	59	140	49	147	154	15	224	365	949	525	349
62	300	70	49	51	103	99	154	39	53	52	24	102	4,914	77	898	782
223	1,657	85	83	97	109	100	181	32	88	104	26	119	8,052	295	1,619	1,555
189	3,905	119	96	60	129	111	209	51	146	199	4	233	4,921	461	1,325	1,911
1,216	2,191	120	71	270	168	96	197	35	187	185	3	295	1,273	414	720	1,474
964	2,376	73	29	138	90	48	106	27	138	121	<1	155	260	208	294	554
29	70	49	10	33	43	48	145	11	24	25	7	66	1,033	13	165	50
30	93	147	42	185	224	178	280	57	107	93	38	222	27	<0.5	6	4
126	393	181	49	224	248	262	885	85	117	131	42	383	125	9	67	44
268	921	211	35	329	217	259	755	80	125	118	24	609	84	39	160	145
547	1,529	165	26	298	168	155	435	62	157	149	16	516	23	94	211	172
75	94	170	29	219	155	227	392	62	71	86	28	351	1	n.d.	30	19
73	190	151	31	199	150	198	315	60	69	82	28	316	2	n.d.	54	63
130	435	225	31	246	175	216	506	53	112	138	25	533	2	34	158	110
486	757	153	25	159	132	134	256	29	159	175	17	316	1	46	172	9
69	46	305	55	411	200	311	825	84	126	86	55	373	n.d.	<0.5	5	4
15	9	118	23	144	81	110	335	29	46	36	19	136	n.d.	<0.5	n.d.	n.d.
49	36	284	44	267	178	270	686	69	91	73	38	355	n.d.	9	14	11
66	47	169	36	374	135	168	484	52	88	59	29	257	n.d.	n.d.	9	5
53	58	60	46	66	41	61	139	16	37	25	65	75	n.d.	n.d.	3	<1.5
43	48	205	34	310	184	177	455	55	128	60	40	334	n.d.	n.d.	n.d.	<1.5
19	nd	60	11	76	56	58	144	16	39	24	13	127	n.d.	n.d.	1	<1.5
44	58	153	32	281	169	147	386	48	117	57	36	301	n.d.	n.d.	3	<1.5
675	1,448	1,891	443	2,611	1,829	2,090	5,110	614	1,028	818	368	3,177	11,314	282	2,415	1,534
7,993	37,014	3,856	1,861	5,090	4,028	4,130	9,734	1,242	2,944	3,063	726	7,182	217,282	12,638	46,990	34,807

sources of the Guanabara Bay sediment samples including the parent PAH and the alkyl PAH ratios.

The double-ratio plot of phenanthrene/anthracene (Phe/An) versus fluoranthene/pyrene (Fl/Py) has been frequently used to distinguish a mixture of petrogenic and pyrolytic input for sediments (Baumard et al., 1998; Tam et al., 2001; Readman et al., 2002; Ke et al., 2002) despite the low discrimination capacity of Phe/An. Figure 17-6 depicts this parental ratio diagram for Guanabara Bay samples collected in both campaigns, 2000 and 2003, plotted together with the January 2000 oil spill sample (MF 380 derived from a crude oil from a Brazilian marginal basin), an Arabian crude oil (AL), frequently used in Brazilian refineries, a light oil (Ar), and a diesel fuel (DM) produced in a refinery from the south of the country. It was observed that this double-ratio

Table 17-3 Literature Data on PAH Concentration (ng/g dry weight) of Sediments from Various Coastal Sites in the World

Location	Number of PAH Analyzed	Concentration Range (ng/g)	References
Casco Bay, USA	23	16–20,748	Kennicutt et al., 1994
San Diego, USA	36	80–20,000	Anderson et al., 1996
San Francisco Bay, USA	17	2,653–27,680	Pereira et al., 1996
San Francisco Bay (Alameda), USA	43	751–11,059	Stout et al., 2004
Masan Bay, Korea	16	41–1,100	Khim et al., 1999
Gironde & Arcachon Bay, France	14	3.5–853	Sicre et al., 1987
Sarasota Bay, USA	11	17–26,771	Sherblom et al., 1995
Brisbane River Estuary, Australia	17	2,840–13,470	Kayal & Connell, 1989
Mersey Estuary, UK	13	5,310	Readman et al., 1986
Tamar Estuary, UK	13	8,630	Readman et al., 1986
Toulon Harbour, France	14	8,400	Baumard et al., 1998
Portland Harbor, USA	43	860–20,644	Stout et al., 2004
Eagle Harbor, WA, USA	43	8,524–80,913	Stout et al., 2004
Cerritos Channel, Los Angeles, USA	23	10,000	Brown et al., 1998
Jiulongjiang Estuary, China	21	515–1,522	Witt & Siegel, 2000
Baoshan, Shangai, China	7	61–7,618	Xu et al., 2001
Yalujiang Estuary, North China	10	79–1,500	Wu et al., 2003
Minjiang River Estuary, China	16	112–887	Zhang et al., 2004
Deep Bay (mudflat), China	16	238–416	Zhang et al., 2004
Masan Bay, Korea	24	207–2,670	Yim et al., 2005
Pearl River Estuary, China	16/47	156–10,810	Fu et al., 2001; Bixian et al., 2001
Pearl River Estuary, China	15	94–4,300	Fung et al., 2005
Rio de la Plata Estuary, Argentina	18	50–555,000	Colombo et al., 1989
Daya Bay, Hong Kong, China	16	115–1,134	Zhou & Maskaoui, 2003
Channel of Rio de la Plata (after oil spill), Argentina	16	10–70,000	Colombo, 2000
Guanabara Bay (campaign 2000)	16	207–13,425	This study
	38	559–58,439	
Guanabara Bay (campaign 2003)	16	184–3,653	
Brazil	38	400–52,384	

plot does not give a strong interpretation of PAH sources although some intertidal sediments (T1, T9, T15, and T18) showed pyrogenic characteristics, i.e., Fl/Py higher than 1 and Phe/An less than 10, as observed in the studies of Budizinski et al., 1997; Baumard et al., 1998; and Readman et al., 2002. All other samples of Guanabara Bay and even the studied oils (MF 380, the Arabian oil, the light oil, and the diesel fuel) presented mixed features in this double-ratio plot, fitting in its bottom left quadrant. This was expected since the pair of compounds phenanthrene and anthracene has small relative difference in thermodynamic stability between isomers, and this ratio is likely to be less effective in identifying PAH sources (Yunker et al., 2000).

A similar parental double ratio plot proposed by Yunker and collaborators (2000): anthracene/(anthracene + phenanthrene) versus fluoranthene/(fluoranthene + pyrene) was studied in order to verify the ratio's ability to distinguish between combustion and petroleum inputs. This can be seen in Figure 17-7. It must be highlighted that this proposed ratio Fl/(Fl + Py) has a more detailed boundary for combustion sources: it distinguishes fuel combustion from grass/wood/coal combustion. The boundaries for Fl/(Fl + Py) and other ratios that will be presented in this study were

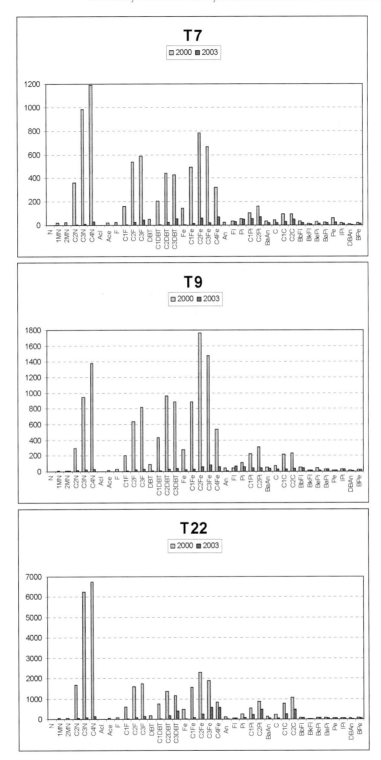

Figure 17-4 Predominance of alkylated PAH for intertidal samples from Guanabara Bay 2000 survey in relation to 2003.

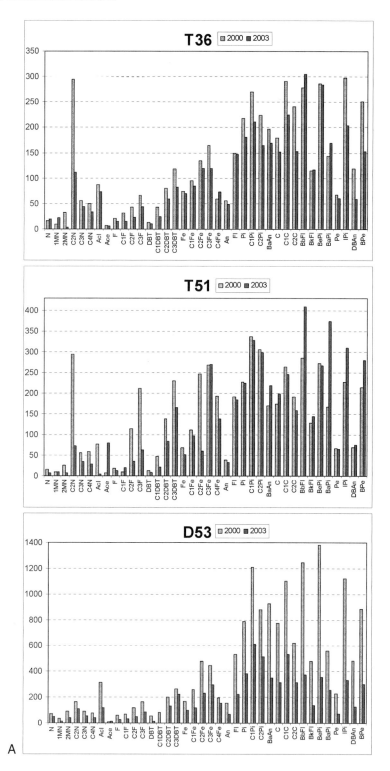

Figure 17-5 Predominance of four- and five-ring PAH for subtidal samples from Guanabara Bay for 2000 and 2003 campaigns. (a) Inside the influence of the oil spill slick; (b) outside the influence of the oil spill slick.

Case Study: Evaluation of Hydrocarbon Sources in Guanabara Bay, Brazil 519

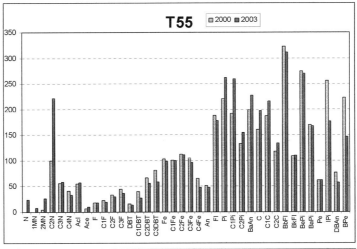

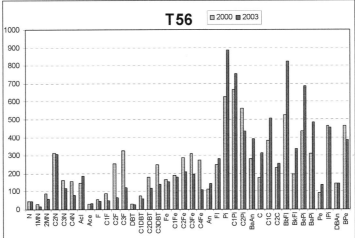

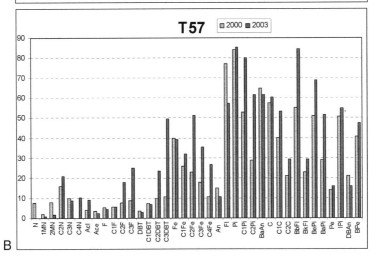

B

Figure 17-5, continued

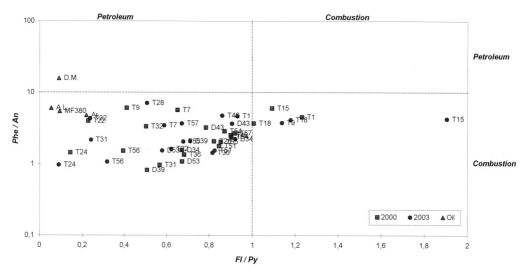

Figure 17-6 PAH cross-plot of phenanthrene/anthracene versus fluoranthene/pyrene for Guanabara Bay sediments and MF 380, Arabian oil (AL), light oil (Ar), and diesel oil (DM).

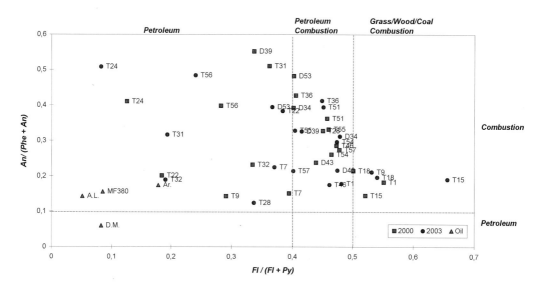

Figure 17-7 PAH cross-plot of anthracene/(anthracene + phenanthrenes) versus fluoranthene/(fluoranthene + pyrene) for Guanabara Bay sediments from campaigns 2000 and 2003 and MF 380, Arabian oil (AL), light oil (Ar), and diesel oil (DM).

established by Yunker and collaborators (2000) and were based on more than 40 studies from international literature, encompassing results of about 250 samples, including several types of petroleum, single sources of combustions (coal, wood, grass, oil), and environmental samples (bush fires, tunnels dust, road dust, urban air).

The low efficiency to determine PAH sources for anthracene and phenanthrene compounds was confirmed, which included all samples from Guanabara Bay in the category of pyrogenic characteristics (An/An + Phe higher than 0.1).

On the other hand, the ratio Fl/(Fl + Py) has shown a higher ability to distinguish combustion and petroleum inputs, separating the Guanabara Bay samples into two clusters: Fl/(Fl + Py) less than 0.4 for samples with petrogenic characteristics; and Fl/(Fl + Py) higher than 0.4 for those with combustion sources (Yunker et al., 2000, 2002). As mentioned before, the ratio showed a more sophisticated discrimination ability for the samples: the differentiation between wood/grass/coal combustion (ratio higher than 0.5) and petroleum combustion features (ratio range between 0.4 and 0.5) (Figure 17-7).

This was also observed in Fraser River samples (Yunker et al., 2000, 2002). The PAH pair fluoranthene and pyrene has a higher relative difference in thermodynamic stability between isomers than phenanthrene and anthracene, therefore being more effective to determine PAH sources. In general, the majority of samples from subtidal stations (from T31 on) presented combustion characteristics.

Other double-ratio plots reported in the literature have been analyzed for Guanabara Bay samples, and it was verified that the parental ratio indeno1,2,3-(cd)pyrene/(indeno1,2,3-(cd)pyrene + benzo(ghi)perylene) presented low efficiency to identify PAH sources (Figure 17-8). The majority of the samples corresponded to predominant combustion features, including the Arabian and light oils. Only the MF 380 was identified as a petroleum derivative sample.

In addition to the parental Fl/(Fl + Py) ratio, it was observed that the alkylated ratio phenanthrene + anthracene/(phenanthrene + anthracene + C1phenanthrene) (Yunker et al., 2000) has also exhibited high source discrimination capacity for Guanabara Bay samples. This can be seen in Figure 17-9, in which this ratio was plotted against Fl/(Fl + Py). The double-ratio plot of these two most promising ratios showed the highest ability to distinguish between pyrogenic and petrogenic sources in this study.

It could be observed that the ratio Phe + An/(Phe + An + C1Phe) separated samples into two clusters. Samples with a ratio higher than 0.5 would correspond to a combustion-dominant source; less than 0.5 could indicate petroleum or combustion characteristics,

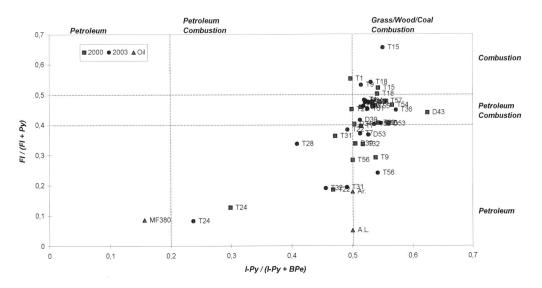

Figure 17-8 PAH cross-plot of indeno1,2,3-(cd)pyrene/(indeno1,2,3-(cd)pyrene + benzo(ghi)perylene) versus fluoranthene/(fluoranthene + pyrene) for Guanabara Bay sediments from campaigns 2000 and 2003 and MF 380, Arabian oil (AL), light oil (Ar), and diesel oil (DM).

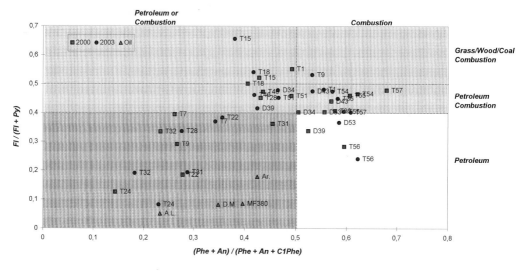

Figure 17-9 PAH cross-plot of fluoranthene/(fluoranthene + pyrene) versus phenanthrene + anthracene/(phenanthrene + anthracene + C1phenanthrenes) for Guanabara Bay sediments from campaigns 2000 and 2003 and MF 380, Arabian oil (AL), light oil (Ar), and diesel oil (DM). Interpretive guidelines based upon Yunker et al. (2000).

as claimed by Yunker and collaborators (2000).

As mentioned before, the ratio Fl/(Fl + Py) has shown the ability to distinguish between combustion and petroleum inputs, separating the samples into two clusters: ratio less than 0.4 for samples with petrogenic characteristics and ratio higher than 0.4 for those with combustion sources (petroleum combustion and grass/wood/coal combustion).

So, it was observed that combustion was the dominant source for samples with Fl/(Fl + Py) higher than 0.4 and Phe + An/(Phe + An + C1Phe) higher than 0.5 for Guanabara Bay samples. This corresponded to the top right quadrant of the diagram in Figure 17-9 (light shaded area). Samples with those characteristics were D34, T36, D43, D53, T54, T55, T57, for the 2000 survey, and T1, T9, T36, D43, T54, T55, T57, for the 2003 survey. This double-ratio source interpretation provided a good correlation with the whole PAH patterns for these samples, as can be seen in Figure 17-5(a) and (b), in which the PAH distribution of some samples is displayed, showing a strong pyrogenic signature.

Samples with Fl/(Fl + Py) less than 0.4 (petroleum input) and Phe + An/(Phe + An + C1Phe) less than 0.5 (could have petroleum or combustion input) would then correspond to a petroleum-dominant source. This corresponded to the bottom left quadrant of the diagram with a dark shaded area (Figure 17-9). For Guanabara Bay, the samples with those characteristics were T7, T9, T22, T24, T31, T32, for the 2000 survey, and T7, T22, T24, T28, T31, T32, for the 2003 survey. It should be emphasized that the analyzed oil samples (MF 380, Arabian oil, light oil, and diesel fuel) were clearly allocated in the bottom left quadrant of the diagram, confirming their petrogenic character. Once again, it could be verified that this source interpretation based on PAH double ratios comes to reflect the full PAH patterns for these samples (with petrogenic signature), as can be seen in Figure 17-4.

On the other hand, samples with Fl/(Fl + Py) higher than 0.4 (combustion input) and Phe + An/(Phe + An + C1Phe) less than 0.5 (petroleum or combustion input) would therefore correspond to a combustion-dominant source. This corresponded to the top left quadrant of the diagram of Figure 17-9 with the samples T1, T15, T18, T28, T51, for the 2000 survey, and samples T15, T18, T34, D39, T46, for the

2003 survey (light shaded area in the diagram). In this way, a great part of Guanabara Bay samples showed pyrolytic characteristics, allocated in both left and right top quadrants of the diagram. This covered samples from stations T1, T9, T15, T18, T28, D34, T36, D39, T43, T46, T51, T54, T55, and T57.

The exception for the double-ratio Fl/(Fl + Py) versus Phe + An/(Phe + An + C1Phe) to distinguish pyrogenic and petrogenic sources for Guanabara Bay samples was found in the bottom right quadrant of the diagram of Figure 17-9. The PAH source for stations D39 (survey 2000), D53 (survey 2003), and T56 (surveys 2000 and 2003) could not be clearly identified, probably due to a mixture of petrogenic and pyrolytic inputs and weathering processes.

A temporal investigation on the bay can be made comparing data from the 2000 and 2003 surveys obtained for the ratio Fl/(Fl + Py) versus Phe + An/(Phe + An + C1Phe). No significant changes were confirmed on the class of predominant source of the samples in this study. Only samples T9 and T28 have shown different PAH source contribution from one campaign to the other. In the 2000 survey, sample T9 has shown predominant petrogenic characteristics, corroborating the visual inspection of the intertidal region reached by the oil after the oil spill, where it was very affected.

Another compositional index used to differentiate the pyrogenic and petrogenic PAHs is the pyrogenic index reported by Wang et al. (1999b), which is defined as the ratio of the other EPA priority 3–6-ring PAHs to the total of 5 target alkylated PAH homologues [Σ (other 3–6-ring PAH)/Σ (5 alkylated PAH series)]. Based on more than 60 oils and petroleum products analyzed by Wang and collaborators, values up to 0.05 for the pyrogenic index unambiguously indicated the contribution of oil and refined products in the samples, while values greater than 0.5 (ratio increased tenfold) indicated combustion-derived sources for the samples. This ratio yielded high accuracy and consistency once the interference was minimized from concentration fluctuations from one compound to another.

For Guanabara Bay samples, this ratio (Figure 17-10) showed a good resolution, covering mostly subtidal samples and exceptionally T1 (from both the 2000 and 2003 surveys) and T9 (from the 2003 survey) with pyrolytic characteristics. This corroborated results from the cross-plot of Fl/(Fl + Py) versus Phe +

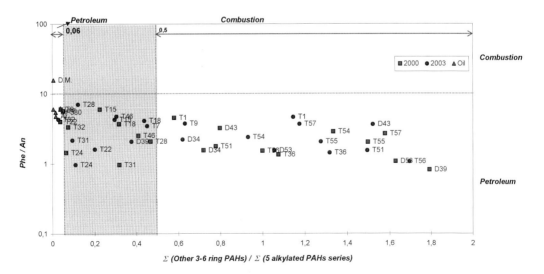

Figure 17-10 Plot of the relative ratios Σ (other 3–6-ring PAH)/Σ (5 alkylated PAH series) versus phenanthrene/anthracene for Guanabara Bay sediments from the campaigns 2000 and 2003 and MF 380, Arabian oil (AL), light oil (Ar), and diesel oil (DM). Interpretive guidelines based upon Wang et al. (1999b).

An/(Phe + An + C1Phe). Moreover, it can be seen in Figure 17-10 that the pyrogenic index included samples T7, T9, T22, T24 from the 2000 survey and T32 from the 2003 survey with clear petrogenic sources, along with analyzed oil samples. However, the ratio still showed a mixed feature for some samples within 0.06 to 0.5 values (gray-shaded area in the diagram), mainly from intertidal area.

As can be seen in Figure 17-11, a double-ratio plot of Σ (other 3–6-ring PAH)/Σ (5 alkylated PAH series) versus Fl/(Fl + Py) brought some resolution to the pyrogenic index in the range of 0.06 to 0.5 values. As shown previously, it would be expected that a predominance of petrogenic-derived sources for the samples with Fl/(Fl + Py) less than 0.4 while samples with Fl/(Fl + Py) values higher than 0.4 would show combustion characteristics. Therefore, the double-ratio Σ (other 3–6-ring PAH)/Σ (5 alkylated PAH series) versus Fl/(Fl + Py) would also confirm the contribution of petroleum in samples from the bottom left and middle quadrants of the diagram in Figure 17-11 [pyrogenic index less than 0.5 and Fl/(Fl + Py) less than 0.4: dark shaded area] and the combustion-derived sources for the samples from the top right and middle quadrants of the diagram [pyrogenic index higher than 0.06 and Fl/(Fl + Py) higher than 0.4: light shaded area]. Differently, some samples would present a mixed feature, located in the bottom right quadrant [samples T56 (surveys 2000 and 2003), D39 (survey 2000), and D53 (survey 2003)]. It can be observed that the final pattern of this corresponded reasonably to the pattern obtained from the double-ratio Fl/(Fl + Py) versus Phe + An/(Phe + An + C1Phe) of Figure 17-9, giving consistency to this study.

Another compositional index used to differentiate the natural and petrogenic PAHs was the relative perylene abundance reported by Baumard et al. (1998), which establishes the perylene abundance in relation to the 5-ring isomers, i.e., the ratio perylene/Σ (5-ring PAH = BbFl + BkFl + BaPi + BePi + DBAn) (Figure 17-12). It can be seen that the majority of samples from the intertidal region shows natural hydrocarbon contribution.

To sum up, the diagnostic ratios that exhibited a high ability to distinguish between combustion and petroleum inputs for Guanabara Bay sediments were Phe + An/(Phe + An + C1Phe) and Fl/(Fl + Py). In addition, the relative perylene abundance in relation to their 5-ring isomers also presented the ability to discriminate between natural and petrogenic

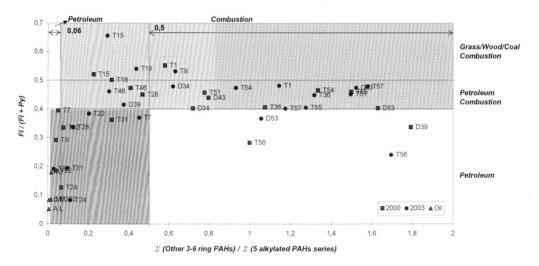

Figure 17-11 Plot of the relative ratios Σ (other 3–6-ring PAH)/Σ (5 alkylated PAH series) versus fluoranthene/(fluoranthene + pyrene) for Guanabara Bay sediments from the campaigns 2000 and 2003 and MF 380, Arabian oil (AL), light oil (Ar), and diesel oil (DM). Interpretive guidelines from Wang et al. (2002b).

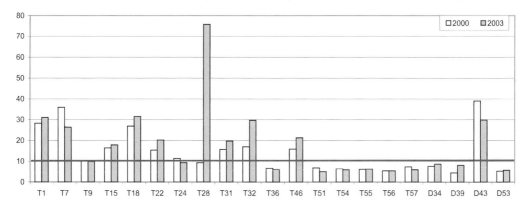

Figure 17-12 Relative perylene abundance for Guanabara Bay sediments from the campaigns 2000 and 2003.

PAH input. The PAH diagnostic ratios indicated that the Guanabara Bay sediments could be separated into three distinct groups: one with a dominant pattern and ratios consistent with petroleum-derived characteristics; another with a dominant pattern and ratios consistent with combustion-derived characteristics; and the third one with apparent mixed contributions of petrogenic and pyrolytic input.

17.3.2.2 PAH Principal Component Analysis

Figures 17-13 and 17-14 depict the factor score and factor loading results of PAH multivariate analysis of the 2000 and 2003 campaigns, respectively, in which the majority of the samples was separated into groups. In the plots, the distance and direction from the central axis have the same meaning for both samples and PAH variables.

For both campaigns, the first PCA defines two variable groups by separating alkylated PAH from parent PAH: on the left and right sides, respectively. The second PCA separates the PAHs into two groups: predominantly projected by all alkyl naphthalenes, C1 fluorenes, on the lower part; and alkyl dibenzothiophenes and alkyl phenanthrenes on the upper part (see Table 17-1 for codes).

In both datasets, the PCA model separated the sediment samples from Guanabara Bay and the four reference oils analyzed. For both campaigns, the oils projected on the lower left side of the y-axis with a high contribution of low-molecular-weight compounds, which are present in crudes but are usually weathered by the environment. Some samples (T7, T22, T24, T31, T32) projected on the upper left side of the y-axis, suggesting predominance of petrogenic input due to the contribution of alkylated phenanthrenes and dibenzothiophenes. On the other hand, other samples projected on the right side of the y-axis, including a group of samples with pyrolytic characteristics indicated by the dominance of high-molecular-weight compounds (T36, D43, T51, D53, T54, T55, T56, T57).

Therefore, the lower left side of the y-axis correlates to unweathered or less weathered oils; the upper left side correlates to samples with a more weathered petrogenic source predominance; and the lower right side correlates to samples with pyrolytic source predominance.

From one campaign to the other, it was confirmed that the class of predominant source of the samples had not significantly changed. The exceptions were T1, T7, T9, T28, T31, and D39. The intertidal samples T7 and T9 presented clear petrogenic features for the 2000 survey, corresponding to the visual inspection of the intertidal region reached by the oil after the oil spill. For the 2003 survey, the effect of the spill on these sites could no longer be an observed due mainly to the retention and cleaning effort carried out on the bay after the spill and to the natural washing capacity of the

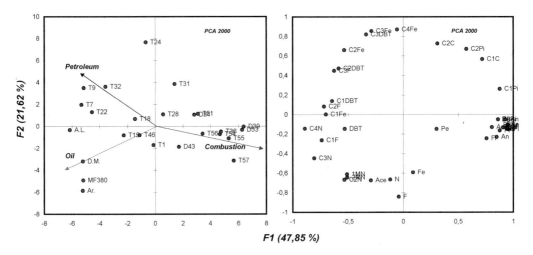

Figure 17-13 PCA projections of PAH variables and sediment samples from campaign 2000 and MF 380, Arabian oil (AL), light oil (Ar), and diesel oil (DM).

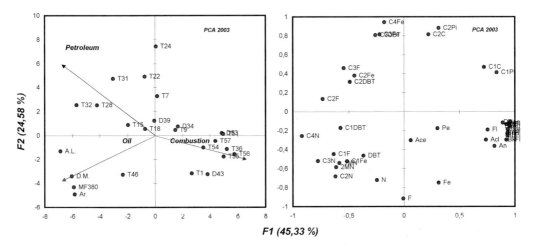

Figure 17-14 PCA projections of PAH variables and sediment samples from campaign 2003 and MF 380, Arabian oil (AL), light oil (Ar), and diesel oil (DM).

region. The full PAH pattern for these samples on the 2000 survey (Figure 17-4), with the predominance of weathered alkylated isomers, confirms this finding.

The intertidal sample T28, on the other hand, presented a clear petrogenic feature for the 2003 survey and could represent a minor independent event since its petrogenic characteristic was better revealed on the 2003 survey, having no relation to the January 2000 oil spill. The whole PAH pattern of sample T28 gives a good correlation with the PCA data, showing, in the 2003 survey, a higher concentration of weathered alkylated isomers and lower levels of four- and five-ring PAHs in relation to 2000.

For the subtidal samples T31 and D39, the change of predominant source to petrogenic features for the 2003 survey could be related to the heterogeneity of the bay sediment, also having no relation to the January 2000 oil spill.

Sample T1, on the other hand, showed a higher pyrolytic source predominance in the 2003 campaign in relation to 2000, despite its low PAH concentration. However, the whole

PAH pattern of the sample has not presented relevant differences between the two campaigns, except for an extraordinary high level of C2 naphthalenes in the 2000 survey, which could be responsible for the PCA results.

17.3.2.3 Biomarker Diagnostic Ratios

The biomarker distribution of the marine fuel MF 380 that was spilled in Guanabara Bay in January 2000 is shown in Figure 17-15. Both

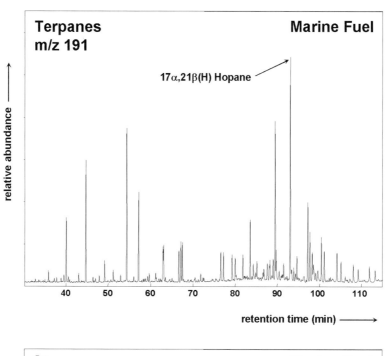

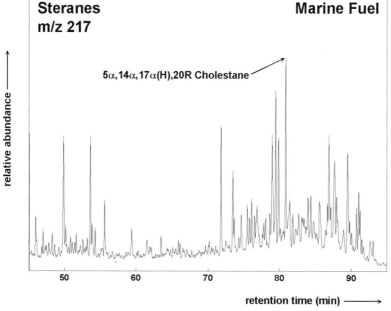

Figure 17-15 Terpane and sterane partial mass chromatograms of the spilled marine heavy fuel oil, January 2000.

Table 17-4 Results for Biomarkers, Terpanes, and Steranes of Sediment Samples from Guanabara Bay — Campaign 2000

Biomarker Ratios	Biomarker Diagnostic Ion (m/z)	Intertidal							
		T1	T7	T9	T15	T18	T22	T24	T28
Terpanes									
C24-tetracyclic/C26-tricyclic terpanes	191	0.73	0.49	0.46	0.79	0.80	0.69	0.57	0.78
gammacerane/C30-hopane	191	0.09	0.11	0.12	0.10	0.10	0.16	0.12	0.12
oleanane/(oleanane + C30-hopane)	191	0.03	0.00	0.00	0.03	0.04	0.03	0.01	0.02
(C29-hopane + C29-Ts)/C30-hopane	191	1.07	0.95	0.93	1.15	1.07	1.07	1.22	1.10
30-norhomohopane/(30-norhomohopane + C30-hopane)	191	0.097	0.066	0.068	0.104	0.088	0.120	0.089	0.103
C30-diahopane/C30-hopane	191	0.06	0.05	0.05	0.06	0.05	0.07	0.06	0.05
C35/C34-hopanes	191	1.00	0.86	0.93	0.93	0.91	0.87	0.88	0.94
C29/(C29 + C30-hopanes)	191	0.458	0.438	0.427	0.478	0.460	0.445	0.497	0.463
C29/C30-hopanes	191	0.85	0.78	0.75	0.91	0.85	0.80	0.99	0.86
hopanes/steranes	191, 217	7.86	6.41	6.10	6.17	5.63	5.36	7.85	6.19
tricyclic terpanes/hopanes	191	0.29	0.52	0.41	0.28	0.31	0.43	0.52	0.26
Ts/(Ts + Tm)	191	0.44	0.37	0.39	0.48	0.46	0.43	0.51	0.46
25-norhopane/C30-hopane	177	0.04	0.06	0.07	0.05	0.05	0.08	0.06	0.04
C29-hopane/C29-Ts	191	3.75	4.49	4.07	3.88	3.90	3.02	4.22	3.58
C30-35 hopanes/steranes	191	5.5	4.5	4.4	4.3	4.0	3.9	5.1	4.4
tetracyclic terpane polyprenoids/C27-diasteranes	259	0.66	0.69	0.68	0.63	0.59	0.69	0.79	0.62
Steranes									
% C27 5α,14β,17β(H),20S	218	32	40	39	31	31	34	36	33
% C28 5α,14β,17β(H),20S	218	24	23	25	27	28	27	23	25
% C29 5α,14β,17β(H),20S	218	44	37	37	42	41	39	41	42
C27/C29 5α,14β,17β(H),20S	218	0.72	1.09	1.05	0.74	0.75	0.87	0.86	0.77
C28/C29 5α,14β,17β(H),20S	218	0.55	0.63	0.67	0.63	0.69	0.71	0.55	0.59
C21 + C22/total steranes	217	0.09	0.09	0.07	0.09	0.10	0.10	0.08	0.07
C27 diasteranes/regular steranes	217	0.62	0.45	0.57	0.65	0.65	0.63	0.85	0.52
C29 5α,14α,17α(H) 20S/20S + 20R	217	0.51	0.53	0.54	0.50	0.50	0.53	0.64	0.48
C29 5α,14β,17β(H)/5α,14β,17β(H) + 5α,14α,17α(H)	217	0.53	0.46	0.48	0.54	0.53	0.47	0.41	0.53
C29/C27 5α,14α,17α(H),20S	217	0.55	0.54	0.53	0.55	0.56	0.61	0.74	0.54

terpane and sterane profiles are typical of a crude oil derived from a saline lacustrine depositional environment from the Brazilian marginal basins (Mello et al., 1988a, 1988b; Peters et al., 2005). The main molecular features are expressed by the following biomarker ratios: hopanes/steranes 6.3; tricyclic terpanes/hopanes 0.91; Ts/(Ts + Tm) 0.30; gammacerane/hopane 0.13; 30-norhopane/hopane 0.72; TPP/C27-diasteranes 0.70; C27-diasteranes/C27-regular steranes 0.72; 25-norhopane/C30-hopane 0.09; and C29-steranes 20S/20S + 20R 0.53.

Biomarker data for the sediment samples are presented in Tables 17-4 and 17-5. Based on selected biomarker ratios, it was observed that

					Subtidal							
T31	T32	T36	T46	T51	T54	T55	T56	T57	D34	D39	D43	D53
0.71	0.72	0.77	0.79	0.81	0.86	0.72	0.84	0.76	0.81	0.77	0.72	0.86
0.13	0.10	0.10	0.10	0.10	0.07	0.09	0.10	0.10	0.10	0.08	0.09	0.08
0.02	0.01	0.03	0.04	0.03	0.05	0.03	0.04	0.03	0.03	0.08	0.03	0.08
1.22	1.04	1.08	1.10	1.09	0.90	1.10	1.07	1.15	1.08	0.93	1.10	0.99
0.119	0.100	0.095	0.099	0.107	0.061	0.094	0.117	0.095	0.093	0.059	0.087	0.078
0.06	0.05	0.05	0.05	0.05	0.05	0.05	0.05	0.06	0.05	0.06	0.05	0.07
0.88	0.83	0.94	0.91	0.88	0.84	0.93	0.95	0.88	0.92	0.84	0.92	0.92
0.487	0.435	0.462	0.466	0.456	0.414	0.467	0.447	0.482	0.461	0.423	0.467	0.441
0.95	0.77	0.86	0.87	0.84	0.71	0.88	0.81	0.93	0.86	0.73	0.87	0.79
6.59	6.33	7.66	7.12	7.05	7.07	7.98	6.45	6.75	7.10	5.34	8.44	5.91
0.44	0.31	0.27	0.30	0.25	0.24	0.28	0.26	0.30	0.29	0.33	0.22	0.32
0.47	0.48	0.47	0.47	0.40	0.45	0.48	0.45	0.48	0.46	0.40	0.39	0.43
0.05	0.04	0.03	0.03	0.04	0.04	0.03	0.04	0.03	0.04	0.06	0.04	0.04
3.55	2.82	3.87	3.79	3.27	3.67	3.98	3.05	4.16	3.79	3.67	3.97	3.95
4.4	4.4	5.3	4.9	4.9	5.0	5.4	4.6	4.6	4.9	3.7	6.0	4.1
0.66	0.63	0.64	0.62	0.64	0.59	0.63	0.60	0.63	0.64	0.62	0.63	0.63
35	32	28	27	27	27	28	29	29	29	26	31	27
24	26	26	27	28	28	26	29	26	25	34	26	31
41	42	46	46	45	45	46	43	45	46	41	43	42
0.84	0.76	0.61	0.60	0.60	0.61	0.61	0.66	0.65	0.64	0.63	0.72	0.64
0.58	0.60	0.58	0.59	0.63	0.62	0.56	0.66	0.58	0.54	0.82	0.60	0.75
0.09	0.06	0.08	0.09	0.07	0.08	0.08	0.08	0.08	0.08	0.10	0.07	0.10
0.69	0.47	0.90	0.87	0.59	0.58	0.92	0.74	0.54	0.74	1.08	0.59	1.20
0.51	0.42	0.54	0.52	0.49	0.45	0.49	0.47	0.50	0.51	0.55	0.49	0.56
0.51	0.47	0.55	0.56	0.51	0.51	0.56	0.51	0.56	0.55	0.57	0.55	0.56
0.57	0.58	0.63	0.56	0.58	0.60	0.57	0.60	0.56	0.55	0.57	0.54	0.59

samples taken from the same site in both campaigns, 2000 and 2003, are very similar. An example is sample T24 sampled in both campaigns. Sample T24 presents a higher value for C29/C27 5α,14α,17α(H),20S sterane ratio compared to the other samples, 0.74 and 0.81, for campaigns 2000 and 2003, respectively (Figure 17-16). Such molecular features result in an uncommon sterane distribution showing a predominance of C27-sterane 5α,14β,17β(H),20S in sample T24.

In general, most of the samples from Guanabara Bay have not shown any correlation with the spilled marine fuel. There were a few exceptions such as two intertidal samples, T7 and T9, from the campaign 2000, that showed

Table 17-5 Results for Biomarkers, Terpanes, and Steranes of Sediment Samples from Guanabara Bay — Campaign 2003 the 30-Norhopane Series

Biomarker Ratios	Biomarker Diagnostic Ion (m/z)	Intertidal			Subtidal				
		T9	T15	T24	T31	T32	T51	T57	D39
Terpanes									
C24-tetracyclic/C26-tricyclic terpanes	191	0.57	0.73	0.57	0.83	0.86	0.75	0.81	0.81
Gammacerane/C30-hopane	191	0.12	0.11	0.19	0.11	0.09	0.11	0.10	0.12
Oleanane/(oleanane + C30-hopane)	191	0.01	0.02	0.00	0.01	0.01	0.04	0.04	0.01
(C29-hopane + C29-Ts)/C30-hopane	191	1.09	1.10	1.34	1.18	1.13	1.07	1.06	1.12
30-Norhomohopane/(30-norhomohopane + C30-hopane)	191	0.08	0.10	0.12	0.10	0.09	0.09	0.10	0.10
C30-diahopane/C30-hopane	191	0.07	0.05	0.10	0.05	0.05	0.06	0.06	0.06
C35/C34-hopanes	191	0.88	0.95	0.86	0.91	0.85	0.89	0.86	0.90
C29/(C29 + C30-hopanes)	191	0.46	0.46	0.51	0.48	0.47	0.46	0.45	0.46
C29/C30-hopanes	191	0.88	0.87	1.06	0.95	0.92	0.85	0.84	0.88
Hopanes/steranes	191, 217	8.26	7.42	7.62	7.30	6.89	8.01	7.96	7.09
Tricyclic terpanes/hopanes	191	0.35	0.34	0.73	0.31	0.28	0.29	0.25	0.27
Ts/(Ts + Tm)	191	0.45	0.47	0.53	0.48	0.49	0.48	0.48	0.50
25-norhopane/C30-hopane	177	0.05	0.04	0.09	0.03	0.03	0.03	0.03	0.03
C29-hopane/C29-Ts	191	4.14	3.76	3.69	4.21	4.35	3.90	3.81	3.73
C30-35 hopanes/steranes	191	5.9	5.2	4.8	5.0	4.7	5.7	5.7	4.98
Tetracyclic terpane polyprenoids/ C27-diasteranes	259	0.68	0.62	0.85	0.61	0.60	0.66	0.67	0.62
Steranes									
% C27 5α,14β,17β(H),20S	218	31	31	34	32	32	27	26	33
% C28 5α,14β,17β(H),20S	218	26	26	24	24	24	28	27	25
% C29 5α,14β,17β(H),20S	218	43	43	42	44	44	45	47	42
C27/C29 5α,14β,17β(H),20S	218	0.71	0.71	0.82	0.73	0.73	0.59	0.56	0.80
C28/C29 5α,14β,17β(H),20S	218	0.60	0.59	0.59	0.54	0.54	0.62	0.57	0.59
C21 + C22/total steranes	217	0.08	0.12	0.08	0.09	0.09	0.08	0.08	0.08
C27 diasteranes/regular steranes	217	0.96	0.92	1.27	0.75	0.67	1.12	0.91	0.80
C29 5α,14α,17α(H) 20S/20S + 20R	217	0.57	0.53	0.73	0.43	0.41	0.53	0.49	0.47
C29 5α,14β,17β(H)/5α,14β,17β(H) + 5α,14α,17α(H)	217	0.54	0.57	0.37	0.54	0.54	0.57	0.58	0.55
C29/C27 5α,14α,17α(H),20S	217	0.55	0.54	0.81	0.51	0.52	0.60	0.55	0.52

a slight correlation with the marine fuel oil spilled (Figure 17-17). Sample D43, from campaign 2000, presented a higher input of recent organic matter compared to other samples. Consequently, its mass chromatograms m/z 191 and 217 showed some interfering peaks.

On the other hand, there is a good biomarker correlation for most of the samples to petroleums derived from a marine carbonate depositional environment. The main features of this kind of crude oil are high values for the C29/C30-hopane and C35/C34-hopane ratios, ranging from 0.85 up to 1.10, the presence of oleanane, high values for the C24-tetracyclic/C26-tricyclic terpane ratio, ranging from 0.57 up to 0.86, values lower than 7.0 for the hopane/sterane ratio, and the presence of Table 17-5. On the other hand, the spilled fuel presents lower values for C29/C30-hopane and C35/C34-hopane ratios, ranging from 0.70 up to 0.86, the absence of oleanane, a lower value for the C24-tetracyclic/C26-tricyclic terpane ratio, 0.40, very low relative abundance or even absence of 30-norhopane series (Figure 17-18). Sterane distributions of sediment samples showed also a significant similarity to those of marine carbonate oils. Based on specific sterane ratios, such as C27/C29 5α,14β,17β(H),20S, C28/C29 5α,14β,17β(H),20S, and C29/C27 5α,

Figure 17-16 A plot of C27/C29-steranes 5α,14β,17β(H),20S against C29/C27-steranes 5α,14α,17α(H),20S showing the similarity between most of the samples and the correlation between both samples T24 collected in campaigns 2000 and 2003.

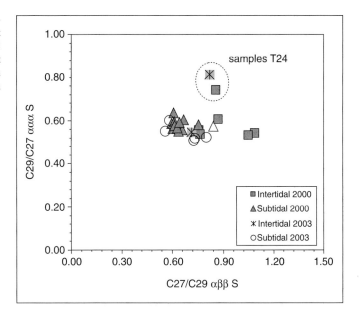

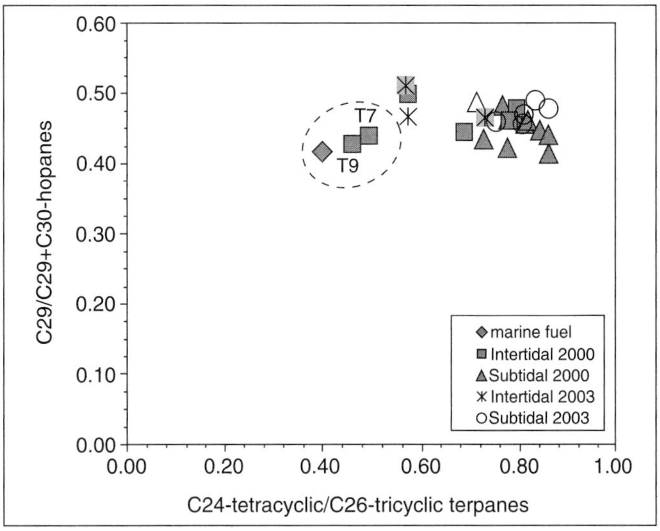

Figure 17-17 A plot of C24-tetracyclic/C26-tricyclic terpanes against C29/(C29 + C30-hopanes) highlighting the correlation between the marine fuel and samples T7 and T9 from the campaign 2000.

Figure 17-18 A plot of Ts/(Ts + Tm) against 30-norhomohopane/(30-norhomohopane + hopane) showing the correlation between samples and a marine carbonate-sourced crude oil.

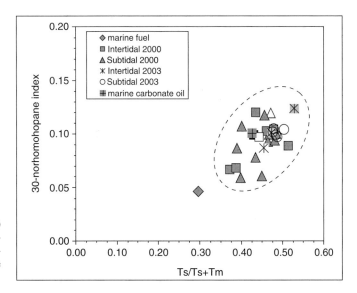

$14\alpha,17\alpha(H),20S$, this correlation can be observed (Figures 17-19 and 17-20).

17.4 Conclusions

Parent and alkyl PAH (a total of 38 compounds) and biomarker terpanes and steranes have been quantified in 21 sediment samples from a highly urban and industrialized ecosystem, the Guanabara Bay, Rio de Janeiro, Brazil. The study was carried out in two campaigns; the first immediately after the oil spill accident in January 2000, and the second, three years later. The sampling design included areas potentially affected and unaffected by the spilled oil, covering the entire ecosystem. The difference in PAH concentrations between the 2000 and 2003 samples was not statistically significant, reflecting the continued chronic anthropogenic pollution of the bay, which was similar to various international estuarine sites. The similarity of Guanabara Bay PAH levels could also be confirmed by comparing the data from both the 2000 and 2003 surveys with pre-

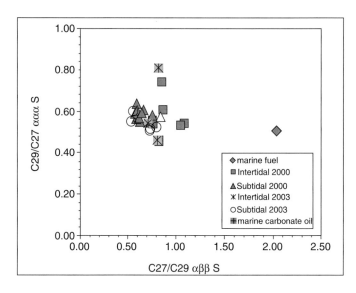

Figure 17-19 A plot of C27/C29-steranes $5\alpha,14\beta,17\beta(H),20S$ against C29/C27-steranes $5\alpha,14\alpha,17\alpha(H),20S$ showing the correlation between most of the samples and a marine carbonate-sourced crude oil.

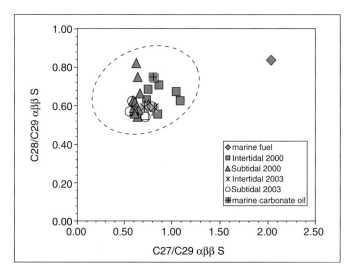

Figure 17-20 A plot of C27/C29-steranes $5\alpha,14\beta,17\beta(H),20S$ against C28/C29-steranes $5\alpha,14\beta,17\beta(H),20S$ showing the correlation between the sediment samples and a marine carbonate-sourced crude oil.

vious PAH data observed in the bay, which were in the same range.

The hydrocarbon source indications were made by using PAH ratios for the samples studied. Some diagnostic ratios exhibited a higher ability to distinguish combustion- and petroleum-derived PAH inputs for Guanabara Bay sediments, namely

Phenanthrene + anthracene/(phenanthrene + anthracene + C1phenanthrene)
Fluoranthene/(fluoranthene + pyrene)
Σ (other 3–6-ring PAHs)/Σ (5 alkylated PAH series)

In addition, the relative perylene abundance in relation to their 5-ring isomers also presented the ability to recognize a contribution from natural (biogenic) PAH input.

Furthermore, the PCA results also exhibited a promising capacity to separate the samples into groups, proving to be a helpful tool for PAH source identification in the environment, corroborating the diagnostic indexes. The results indicated clear patterns of dominantly petrogenic and dominantly pyrolytic hydrocarbons inflow to the bay.

Summarizing, the PAH sources to Guanabara Bay sediments can be separated into groups:

- samples with a clear pattern of petrogenic input — the majority located near the vicinity of the accident in January 2000;
- samples with combustion characteristics — those from the majority of subtidal stations;
- samples without a clear contribution of petrogenic or pyrolytic input, i.e., mixed sources.

A biogenetic contribution was also confirmed for most intertidal samples and for some subtidal samples.

In addition, no significant temporal changes were confirmed for the class of predominant sources in the samples in this study, revealing that the PAH concentrations in Guanabara Bay sediments are not obviously related to the oil spill event, but rather to long-term anthropogenic input. Furthermore, this result can also indicate that there was no significant impact of the oil spill of January 2000 into the Guanabara Bay sediments.

Based on biomarker data, it was observed that most of the sediments from Guanabara Bay have not shown any correlation with the spilled marine heavy fuel oil, except two samples that showed a slight correlation with the spilled fuel. Instead, the sediment samples collected in both campaigns showed a clear input of oils derived from a marine carbonate depositional environment, reflecting chronic anthropogenic pollution of the bay. The spilled MF 380 oil was produced from a Brazilian lacustrine oil that shows a different biomarker distribution pattern in comparison with the oils derived from a marine carbonate depositional environment.

Still more investigation seems to be necessary since Guanabara Bay is in a complex urban area with a significant anthropogenic hydrocarbon introduction process. Currently, a multidisciplinary project in Guanabara Bay is being carried out at the Petrobras Research Center, taking into account not only diagnostic hydrocarbon ratios but also geochemical biomarkers, individual compound carbon isotopic ratios, contaminants such as pesticides and coprostanol, contaminant normalized concentration to organic carbon or grain size distribution, ecotoxicity essays, and benthonic community evaluation, among others.

Acknowledgments

Field surveys were carried out by Federal Fluminense University and Petrobras SA, Brazil. Analyses of sediment samples for PAH were performed by PUC–Catholic University, and the biomarkers were performed by Petrobras SA, Brazil. The authors are grateful to Carlos German Massone of Gorceix Foundation, Dr. Arthur de Lemos Scofield of PUC–Catholic University, and Dr. Valter Jose Fernandes Junior of Federal University of Rio Grande do Norte, Brazil, for their participation in this study and would like to thank Andre Luiz dos Santos Brites and Rui Alexandre Oliveira da Fonseca for helping with the artwork for the chapter. The authors also wish to acknowledge

Dr. J. Michael Moldowan, Mauro Rocha Evangelho, and Irene Terezinha Gabardo for their constructive comments on the chapter.

References

Amador, E.S., Baia de Guanabara e ecossistemas periféricos: Homem e natureza, *Edição do Autor*, Rio de Janeiro, Brazil, 1997, pp. 148–159.

Anderson, J.W., F.C. Newton, J. Hardin, R.H. Tukey, and K.E. Richter, Chemistry and toxicity of sediments from San Diego Bay, including a biomarker (P450 RGS) response, In: *Environmental Toxicology and Risk Assessment: Biomarkers and Risk Assessment*, D.A. Bengtson and D.S. Henshel (eds.), Philadelphia, USA: American Society for Testing and Materials, 1996, **5**, 53–78.

Baumard, P., H. Budzinski, Q. Mchin, P. Garrigues, T. Burgeot, and J. Bellocq, Origin and bioavailability of PAH in Mediterranean Sea from mussel and sediment records, *Estuarine, Coastal and Shelf Sci.*, 1998, **47**, 77–90.

Bixian, M., F. Jiamo, Z. Gan, L. Zheng, M. Yushun, S. Guoying, and W. Xingm, Polycyclic aromatic hydrocarbons in sediments from the Pearl River and estuary, China: Spatial and temporal distribution and sources, *Appl. Geochem.*, 2001, **16**, 1429–1445.

Bentz, C.M. and F.P. Miranda, Application of remote sensing data for oil spill monitoring in the Guanabara Bay, Rio de Janeiro, Brazil, *Proc. Intl. Geosci. Remote Sensing Symp.*, 2001 Sydney, Australia.

Brown, D.W., B.B. McCain, B.H. Hornes, C.A. Sloan, K.L. Tilbury, S.M. Pierce, D.G. Burrows, S.-L. Chan, J.T. Landahl, and M.M. Krahn, Status, correlations and temporal trends of chemical contaminants in fish and sediment from selected sites on the Pacific Coast of the USA, *Mar. Poll. Bull.*, 1998, **37**(1–2), 67–85.

Budzinki, H., I. Jones, J. Bellocq, C. Piérard, and P. Garrigues, Evaluation of sediment contamination by polycyclic aromatic hydrocarbons in the Gironde Estuary, *Mar. Chem.*, 1997, **48**, 85–97.

Chalaux, N., Dinamica costanera de les aportations contaminants presents als efluents urbans, Universidade de Barcelona, Spain, 1995.

Colombo, J.C., Biogeochemical assessment of the 1999 Rio de la Plata oil spill, *Proc. Seventh Latin American Cong. Org. Geochem.*, Foz do Iguaçu, Brazil, 2000.

Colombo, J.C., E. Pelletier, C. Brochu, and M. Khalil, Determination of hydrocarbon sources using *n*-alkane and polyaromatic hydrocarbon indexes. Case Study: Rio de la Plata Estuary, Argentina, *Environ. Sci. Tech.*, 1989, **23**, 888–894.

Eglinton, G. and M. Calvin, Chemical fossils, *Sci. Amer.*, 1967, **261**, 32–43.

Eglinton, G., P.M. Scott, T. Besky, A.L. Burlingame, and M. Calvin, Hydrocarbons of biological origin from a one-billion-year-old sediment, *Science*, 1964, **145**, 263–264.

Elias, V.O., B.R.T. Simoneit, and J.N. Cardoso, Levoglucosan, a molecular fossil as indicator of biomass burning contribution in sediments of the Amazon Shelf, *Proc. Seventh Latin-American Congress on Org. Geochem.*, Foz do Iguaçu, Brazil, 2000, pp. 243–244.

EPA-3540, Soxhlet extraction, In: *Test Method for Evaluation Solid Waste Physical/Chemical Methods, Laboratory Manual*, Washington, DC: Environmental Protection Agency, 1986, v. I-B.

EPA-3630C, Silica gel clean-up, In: *Test Method for Evaluation Solid Waste Physical/Chemical Methods, Laboratory Manual*, Washington, DC: Environmental Protection Agency, 1986, Rev. 3.

EPA-8270C, Gas chromatography/mass spectrometry for semi-volatile organics capillary column technique. In: *Test Method for Evaluation Solid Waste Physical/Chemical Methods, Laboratory Manual*, Washington, DC: Environmental Protection Agency, 1986, v. I-B.

FEEMA, Qualidade de Água da Baía de Guanabara, Programa de Despoluição da Baía de Guanabara — Programas Ambientais Complementares 1998.

FEEMA, www.feema.rj.gov.br. Accessed on 08/01/2003.

Fu, J., Z. Wang, B. Mai, and Y. Kang, Field monitoring of toxic of organic pollution in sediments of Pearl River Estuary and its tributaries, *Water Sci. Tech.*, 2001, **43**, 83–89.

Fung, C.N., G.J. Zheng, D.W. Connell, X. Zhang, H.L. Wong, J.P. Giesy, Z. Fang, and P.K.S. Lam, Risks posed by trace organic contaminants in coastal sediments in the Pearl River Delta, China, *Mar. Poll. Bull.*, 2005, **50**(10), 1036–1049.

Gschwend, P.M. and R.A. Hites, Fluxes of polycyclic aromatic hydrocarbons to marine and lacustrine sediments in the Northeastern United States, *Geochimica et Cosmochimica Acta*, 1981, **45**, 2359–2367.

Hamacher, C., Determinação de hidrocarbonetos em amostras de água e sedimento da Baía de

Guanabara, M.Sc. Thesis, Catholic University PUC-Rio, Chemistry Department, Brazil, 1996.

Hites, R.A. and W.G. Biemann, Identification of specific organic compounds in a highly anoxid sediment by CG/MS and HRMS, *Adv. Chem. Ser.*, 1975, **147**, 188–201.

Holba, A.G., L. Ellis, E. Tegelaar, M.S. Singletary, and P. Albrecht, Tetracyclic polyprenoids: Indicators of fresh water (lacustrine) algal input, *Geol.*, 2000, **28**(3), 251–254.

Holba, A.G., L.I. Dzou, G.D. Wood, L. Ellis, P. Adam, P. Schaeffer, P. Albrecht, T. Greene, and W.B. Hughes, Application of tetracyclic polyprenoids as indicators of input from fresh-brackish water environments, *Org. Geochem.*, 2003, **34**(3), 441–469.

Kayal, S.I. and D.W. Connell, Occurrence and distribution of polycyclic aromatic hydrocarbons in surface sediments and water from the Brisbane River Estuary, Australia, *Estuarine Coastal and Shelf Sci.*, 1989, **29**, 473–487.

Ke, L., T.W.Y. Wong, Y.S. Wong, and N.F.Y. Tam, Fate of polycyclic aromatic hydrocarbon (PAH) contamination in a mangrove swamp in Hong Kong following an oil spill, *Mar. Poll. Bull.*, 2002, **45**, 339–347.

Kennicutt II, M.C., T.L. Wadw, B.J. Presley, A.G. Requejo, J.M. Brooks, and G.J. Denoux, Sediment contaminants in Casco Bay, Maine: Inventories, source and potential for biological impact, *Environ. Sci. Tech.*, 1994, **28**(1), 1–15.

Kennicutt II, M.C. (ed.), Gulf of Mexico offshore operations monitoring experiment; final report, U.S. Department of the Interior Minerals Management Service, Gulf of Mexico OCS Region, New Orleans; USA, 1995, p. 700.

Kjerfve, B., C.H.A. Ribeiro, G.T.M. Dias, A.M. Fillippo, and V.S. Quaresma, Oceanographic characteristics of an impacted coastal bay: Baía de Guanabara, *Continental Shelf Res.*, 1997, **17**(13), 1609–1643.

Khim, J.S., K. Kannan, D.L. Villeneuve, C.H. Koh, and J.P. Giesy, Characterization and distribution of trace organic contaminants in sediment from Masan Bay, Korea, 1. Instrumental analysis, *Environ. Sci. Tech.*, 1999, **33**, 4199–4205.

Lima, A.L.C., Geocronologia de hidrocarbonetos poliaromáticos (PAH) — Estudo de caso: Baía de Guanabara, M.Sc. Thesis, Catholic University PUC-Rio, Chemistry Department, Brazil, 1996.

Lipatou, E. and J. Albaigés, Atmospheric deposition of hydrophobic organic chemicals in the North-western Mediterrean Sea: Comparison with the Rhone River input, *Mar. Chem.*, 1994, **46**, 153–164.

Mayr, L.M., Avaliação ambiental da Baía de Guanabara com o suporte do geoprocessamento. Ph.D. Thesis, Rio de Janeiro Federal University — UFRJ, Geoscience Institute, Brazil, 1998.

Mello, M.R., P.C. Gaglianone, and J.R. Maxwell, Geochemical and biological marker assessment of depositional environments using Brazilian offshore oils, *Mar. Petrol. Geol.*, 1998a, **5**, 205–223.

Mello, M.R., N. Telnaes, and P.C. Gaglianone, Organic geochemical characterization of depositional paleoenvironments in Brazilian marginal basins, *Org. Geochem.*, 1998b, **13**, 31–46.

Meniconi, M.F.G., I.T. Gabardo, M.E.R. Carneiro, S.M. Barbanti, G.C. Silva, and C.G. Massone, Brazilian oil spills chemical characterization — case studies, *Environ. Forensics*, 2002, **3**(3/4), 303–321.

Neff, J., *Bioaccumulation in Marine Organisms*, 2nd ed., Elsevier, 1st ed., UK, 2002, pp. 269–277.

Pereira, W.E., F.D. Hostettler, and J.B. Rapp, Distributions and fate of chlorinated pesticides, biomarkers and polycyclic aromatic hydrocarbons in sediments along a contamination gradient from a point-source in San Francisco Bay, California, *Mar. Environ. Res.*, 1996, **41**, 299–314.

Peters, K.E., C.C. Walters, and J.M. Moldowan, *The Biomarker Guide*, 1st ed., Cambridge University Press, 2nd ed., UK, 2005.

Philp, R.P., Fossil fuel biomarkers: Application and spectra, *Methods in Geochem. Geophys.*, 1985, **23**.

Readman, J.W., M.R. Preston, and R.F.C. Mantoura, An integrated technique to quantify sewage, oil and PAH pollution in estuarine and coastal environments, *Mar. Poll. Bull.*, 1998, **17**, 298–308.

Readman, J.W., G. Fillman, I. Tolosa, J. Bartocci, J.P. Villeneuve, C. Catinni, and L.D. Mee, Petroleum and PAH contamination of the Black Sea, *Mar. Poll. Bull.*, 2002, **44**, 48–62.

Sherblom, P.M., D. Kelly, and R.H. Pierce, Baseline survey of pesticide and PAH concentrations from Sarasota Bay, Florida, USA, *Mar. Poll. Bull.*, 1995, **30**, 568–673.

Sicre, M.A., J.C. Marty, A. Saliot, X. Aparicio, J. Grimalt, and J. Albaige, Aliphatic and aromatic hydrocarbons in different sized aerosols over the Mediterranean Sea: Occurrence and origin, *Atmos. Environ.*, 1987, **21**, 2247–2259.

Stout, S.A., V.S. Mager, R.M. Uhler, J. Ickes, J. Abbott, and R. Brenne, Characterization of naturally occurring and anthropogenic PAHs in urban sediments — Wycoff/Eagle Harbour superfund site, *Environ. Forensics*, 2001, **2**, 287–300.

Stout, S.A., A.D. Uhler, and S.D. Emsbo-Mattingly, Comparative evaluation of background anthropogenic hydrocarbons in surficial sediments from nine urban waterways, *Environ. Sci. Tech.*, 2004, **38**, 2987–2994.

Tam, N.F.Y., L. Ke, X.H. Wang, and Y.S. Wong, Contamination of polycyclic aromatic hydrocarbons in surface sediments of mangroves swamps, *Environ. Poll.*, 2001, **114**, 255–2631.

Wang, Z., M. Fingas, and D.S. Page, Oil spill identification, *J. Chrom. A*, 1999a, **843**, 369–411.

Wang, Z., M. Fingas, Y.Y. Shu, L. Sigouin, M. Landriault, and P. Lambert, Quantitative characterization of PAHs in burn residue and soot samples and differentiation of pyrogenic PAHs from petrogenic PAHs — the 1994 mobile burn study, *Environ. Sci. Tech.*, 1999b, **33**, 3100–3109.

Witt, G. and H. Siegel, The consequences of the Oder Flood in 1997 on the distribution of polycyclic aromatic hydrocarbons (PAHs) in the Oder River Estuary, *Mar. Poll. Bull.*, 2000, **40**(12), 1124–1131.

Wu, Y., J. Zhang, and Z. Zhu, Polycyclic aromatic hydrocarbons in the sediments of the Yalujiang Estuary, North China, *Mar. Poll. Bull.*, 2003, **46**, 619–625.

Xu, S., X. Gao, M. Liu, and Chen, China's Yangtze Estuary II — phosphorus and polycyclic aromatic hydrocarbons in tidal flat sediments, *Geomorph.*, 2001, **41**, 207–217.

Yim, U.H., S.H. Hong, W.J. Shim, J.R. Oh, and M. Chang, Spatio-temporal distribution and characteristics of PAHs in sediments from Masan Bay, Korea, *Mar. Poll. Bull.*, 2005, **50**, 319–326.

Youngblood, W.W. and M. Blumer, Polycyclic aromatic hydrocarbons in Gulf of Maine sediments and Nova Scotia soils, *Geochimica et Cosmochimica Acta*, 1975, **38**, 303–1314.

Yunker, M.B., R.W. MacDonald, R. Brewer, S. Sylvestre, T. Tuominen, M. Sekela, R.H. Mitchell, D.W. Paton, B.R. Fowler, C. Gray, D. Goyette, and D. Sullivan, Assessment of natural and anthropogenic hydrocarbon inputs using PAHs as tracers, The Fraser River Basin and Strait of Georgia 1987–1997, *Report DOE FRAP*, Vancouver; BC: Environment Canada and Fisheries and Oceans Canada, 2000, p. 128.

Yunker, M.B., R.W. MacDonald, R. Brewer, R. Vingarzan, R.H. Mitchell, D. Goyette, and S. Sylvestre, PAHs in the Fraser River Basin: A critical appraisal of PAH ratios as indicators of PAH source and composition, *Org. Geochem.*, 2002, **33**, 489–515.

Zhang, Z.L., H.S. Homg, J.L. Zhou, and G. Yu, Phase association of polycyclic aromatic hydrocarbons in the Minjiang River Estuary, China, *The Science of the Total Environment*, 2004, **323**, 71–86.

Zhang, J., L. Cai, D. Yuan, and M. Chen, Distribution and sources of polynuclear aromatic hydrocarbons in mangrove surficial sediments of Deep Bay, China, *Mar. Poll. Bull.*, 2004, **49**, 479–486.

Zhou, J.L. and K. Maskaoui, Distribution of polycyclic aromatic hydrocarbons in water and surface sediments from Daya Bay, China, *Environ. Poll.*, 2003, **121**, 269–281.

Index

A

Abietane, 14
Aboveground storage tanks (ASTs), 31
Accidental oil spills, 1–2
Acenaphthalene, 514
Acenaphthene, 8, 267, 339, 514
Acenaphthylene, 5, 8, 267, 339
Acetate, 356–357
Acetogens, 357
Acid rain gases, 150
Acids, 360
Acinetobacter, 352
Acoustic systems, 435
Acyclic acids, 361
Acyclic isoprenoids, 14
Adamantanes, 114–117
   diagnostic ratios, 127
   empirical formula, 78
   mass spectra of, 88
   target ions, 78
Aegean Sea oil spill, 162
Aerobic biodegradation, 351–355
Aerobic respiration, 366, 367–368
Aerosol, 495
Air monitoring, 59–60
Air-oil interface, 429
Aircraft fuel, 2
Alaska North Slope (ANS) crude oil, 9, 16–17, 93–99, 111–113
Alberta Sweet Mix Blend (ASMB) crude oil, 9, 161, 434
Alcanivorax, 353
Aldehydes, 155
Algal debris, 38–39
Alicyclic compounds, 356
Aliphatic hydrocarbons, 14
Alkane alcohols, 352
Alkane carboxylase, 356
Alkane dioxygenase, 352–353
Alkanes, 14, 300, 353, 361, 496–497
Alkenes, 352
Alkylated benzenes, 362
Alkylated benzothiopenes, 152
Alkylated naphthalenes, 471
Alkylbenzenes, 132, 176–177
Alkylbiphenyls, 362
Alkylcyclohexanes, 24–26, 361

Alkyldiphenylmethanes, 362
Alkyltoluenes, 300
Alpha hydrogens, 80
Ameiurus nebulosus, 394–396
Amoco Cadiz oil spill, 5, 13
Anaerobic biodegradation, 355–357
Anaerobic respiration, 368–369
Anthracenes, 5, 8, 267, 339, 514
Anthracite, 40
Anthropogenic background hydrocarbons, 41–43
API gravity, 350, 354, 359–360
Arabian Light crude oil, 93–99, 111–113
Arcachon Bay, 516
Archaea, 198
Argo Merchant, 406, 407
Aromatic dioxygenases, 353
Aromatic hydrocarbons, 14–15
Aromatic steranes, 77, 104–109
Aromatic steroids, 240–244, 361, 362
Arrow oil spill, 133, 449
Artemia, 506
Aryl hydrocarbon nuclear translocator protein (ARNT), 388
Aryl hydrocarbon receptor (AhR), 388–389
Asphaltenes, 13, 211
Asphaltic bitumens, 216
Asphalts, 345–346, 494
ASTM 5623-94, 157
ASTM D3328, 5, 259–262, 330
ASTM D5739, 5, 259–262, 330–331
Asymmetric carbons, 80 81
Atmospheric deposition, 2
Atmospheric gas oils, 22
Atomic emission detection (AED), 157
Automotive gasoline, 22
Aviation gasoline, 22, 350
Azoarcus, 355–356

B

Baccharane, 90
Bacillus, 364
Background chemicals, 36–37
   anthropogenic, 41–43
      river runoff, 42–43
      urban, 42–43
   chemical fingerprinting of, 37

538 Index

Background chemicals (*Continued*)
  naturally-occurring, 37–41
    algal debris, 38–39
    oil seeps, 41
    particulate coal, 39–41
    vascular plants, 38–39
    wood charcoal, 39–41
  statistical analysis of, 37
  types of, 36
Bacteria, 350
Bacteriocides, 364
Baffin Island oil spill, 354
Baltic Carrier oil spill, 295, 315, 319
Barge vessel spills, 2
Base peak, 84
Baselines, 303–304
Battelle Memorial Institute, 230
Beach tars, 215–216
Beaufort Sea, 437
Benchtop quadropole GC/MS, 84
Benzanthracenes, 8, 339, 514
Benzenes, 5, 354–355
Benzo ring, opening of, 155
Benzoanthracene, 5, 267
Benzofluoranthenes, 5, 8, 267, 339, 514
Benzofluorenes, 237–239
Benzopyrelenes, 5, 8, 267, 339, 514
Benzopyrenes, 5, 8, 267, 339
Benzothiophenes, 27, 149–150, 152, 300
Benzylsuccinate synthase, 355
Beta-carotane, 124
Beta hydrogens, 80
Bicadinanes, 123–124
Bicyclic terpanes, 361
Bilge water, 28
Bioavailability assessment, 477–482
  biomarkers, 481–483
  passive sampling of PAH, 481
  uptake of PAH in biota, 477–482
Biodegradation, 349–370. *See also* Weathering
  aerobic, 351–355
  anaerobic, 355–357
  biochemistry of, 349–350
  and biomarker fingerprinting, 131
  of biomarkers, 360–361
  and chemical fingerprinting, 32–34
  and chemical transformation of crude oil, 153
  limiting factors, 362–364
    access of microbes to oil, 363–364
    lack of bacteriocides, 364
    nutrients, 363
    presence of microorganisms, 364
    temperature, 364
    terminal electron acceptors, 362–363
    water, 364
  microbial ecology of, 365–369
    deep subsurface ecology, 367–369
      aerobic respiration, 367–368
      anaerobic respiration, 368–369
    microbial communities, 365–367
  in reservoirs, 357–358
  subsurface, 357–362
  susceptibility to, 362
Biogenic hydrocarbons, 350, 451, 453
Biomarker fingerprinting, 73–74
  analysis methods for, 81–83
  application to oil spill studies, 117–121
  capillary GC/MS analysis in, 83–86
  effects of weathering on, 130–138
    biodegradation, 131
    dissolution, 131
    emulsification, 130
    evaporation, 130
    natural dispersion, 130–131
    oil-mineral aggregation, 132
    photooxidation, 131–132
    sedimentation, 132
  internal standards, 83
  source identification of spills, 134–138
    characterization of sesquiterpanes, 135
    evaluation of alkylated PAHs, 135–138
    evaluation of pentacyclic terpanes and steranes, 135–138
    product type-screening, 134–135
  surrogates, 83
Biomarkers, 9–10
  biodegradation of, 133–134, 360–361
  compound classes, 16, 74–77
    cyclic terpenoids, 75–77
    isoprenoids, 75
  cross-plots of, 128–130
    for oil and petroleum products, 129–130
    for spill source identification, 129
  in crude oils, 93–99
  definition of, 15
  diagnostic indices, 269–272
  diagnostic ratios of, 126–128, 239–240, 511, 514
  in heavy fuel oils (HFO), 336–338
  identification of, 90
  labeling and nomenclature of, 77–81
    asymmetric carbons, 80–81
    R and S stereoisomers, 81
    stereoisomers, 79–80
  in lubricating oils, 101–103
  mass fragments, 300

mass spectra of, 86–90
  in oil bioavailability assessment, 481–483
  in oil fractions, 104
  overview, 73–74
  in petroleum products, 99–101
  in qualitative fingerprinting, 264–265
  source-specific, 121–125
  weathered percentages, 134
Bioresistance, 360
BIOS oil spill, 133
Biosphere, 349
Biphenyl dioxygenase, 354
Biphenyls, 8, 178–180, 267, 300, 339
Biphytane, 198
Birds, effects of oil spills on, 216–217
Bishomohopanes, 242, 267
Bisnorhopane, 242, 267
Bitumens, 215–216
Bituminous coal, 40
Blue-sac disease, 394–395
Boil range indices, 271
Boiling points, 351
Bonn Agreement, 229
Boscan crude oil, 96–99
Botryococcane, 75, 123, 215
Botryococcus braunii, 121, 125
Bouchard 120, 191
Bouchard 65, 187–189
BP American Trader, 212–213
Brachymonas petroleovorans, 351
Branched alkanes, 362
Branched hydrocarbons, 352
Brevibacterium, 352
Brisbane River, 516
Bromochloroiodomethane, 81
Bulk isotope ratios, 210–213
Bunker C fuel oil, 101, 191–195, 329
Bunker fuel, 161, 327, 329
Buzzards Bay oil spill, 191–195

C

C-band radar, 427
Cadalene, 267
Cahn-Ingold-Prelog convention, 81
Calcite, 196
Calcium, 329
California crude oil, 96–99, 111–113
Cameras, 420–422
Capillary columns, 6
Carbon dioxide, 369
Carbon isotope ratios, 209–212
Carbon isotopes, 208
Carbon range analysis, 332

Carboxylated pyrrols, 361
Carboxylation, 356
Carboxylic acids, 155
Cargo washings, 2
Carotane, 124
Casco Bay, 516
Catabolic degradation, 387–393
Catalytic cracking, 24
Catechols, 352
CEN oil spill identification methodology, 229–254.
  See also Oil hydrocarbon fingerprinting
  case study, 251–254
  final evaluation and conclusions, 250–251
  overview, 230–231
  scope of, 231
  strategies, 231–233
  tiered levels of analysis and data treatment in, 233–251
    decision chart, 233–234
    GC/FID screening (level 1), 235–237
    GC/MS fingerprinting (level 2), 237–245
      biomarkers, 239–240
      diagnostic ratios, 237–238
      polycyclic aromatic compounds (PACs), 238–239
      sesquiterpanes, 240–245
    sample preparation and cleanup, 233–235
    treatment of results (level 3), 245–250
      criteria for diagnostic ratios, 245–246
      critical difference, 246
      duplicate analyses, 248–249
      multivariate statistics, 249–250
      oil sample comparison, 245
      repeatability limit, 246
      signal-to-noise (S/N) test, 246–248
    visual characterization of samples, 233–235
CERCLA, 36, 449
Cerritos Channel, 516
Cetane rating, 351
Chain-of-custody documentation, 70
Charcoal, 39–41
Charged-coupled devices (CCDs), 421, 439
Cheilanthane, 242
Chemical fingerprinting, 257–258
  control factors, 11–43
    background chemicals (tertiary controls), 36–43
    crude oil genesis (primary controls), 13–18
    petroleum refining (secondary controls), 18–26
    weathering (tertiary controls), 29–37
  of heavy fuel oils (HFO), 330–332
  of known-source spills, 3

Chemical fingerprinting (*Continued*)
  methods, 3–11
    GC/FID, 6–7
    GC/MS, 7–10
      petroleum biomarkers, 9–10
      polycyclic aromatic hydrocarbons (PAHs), 8–9
    historical perspective, 3–6
    quality assurance, 10–11
    quality control, 10
  of mystery spills, 2–3
  overview, 2–3
  qualitative, 258–263
  quantitative, 263–276
Chemical toxicity, 59–60
Chevron Biomarker Laboratory, 86
Chiral carbons, 80–81
Cholanes, 267
Cholestane, 14
  aerobic biodegradation of, 353
  diagnostic ratios, 243
  GC/FID analysis of, 267
  nomenclature, 81
Chromatograms, 174–175
ChromatTOF (software), 173
Chrysenes, 5
  biodegradation of, 361
  diagnostic indices, 271
  diagnostic ratios, 237–239
  GC/FID analysis of, 267
  in heavy fuel oils (HFO), 339
  mass fragments, 300
  in oil spills, 514
  response factor, 8
Chrysenothiophene, 149
Citrobacter freundii, 357
Climax communities, 366
Co-metabolism, 364
Coal, 39–41
Coal Oil Point, 196, 211
Coastal facility spills, 2
Coastal oil spills, 62–63
Coelution, 210
Coenzyme A, 352, 355
Coenzyme M, 352
COLD zone, 59
Columns, 82
Commercial ships, 328–329
Commet assay, 390
Compact airborne spectrographic scanner (CASI), 421
Compound-specific isotope analysis (CSIA), 208–209, 214–220

Comprehensive two-dimensional gas chromatography (GC × GC), 171–172
  chromatograms, 174–175
  data processing, 173–174
  detectors, 172–173
  modulation techniques, 172
  in oil spill fingerprinting, 181–201
    Buzzards Bay No. 6 spill, 191–195
    Mobile Bay marine diesel fuel spill, 181–184
    Santa Barbara oil seep, 195–201
    West Falmouth fuel spill, 184–187
    Winsor Cove fuel oil spill, 187–191
  in oil spill source identification, 158
  peak identity, 175–180
  petroleum applications, 180–181
Compressive waves, 437
Condensates, 350
Constitutional isomers, 79
Constrained least squares (CLS), 466–468
Continuous-flow instruments, 210
Conventional crude oils, 13
Conversion, 20–21
Correlation optimized warping (COW), 304–305
Crankcase oil, 494–495
Critical difference, 246
Cross-plots, 128–130
Crude oils, 13–18
  biodegradation of, 354
  biomarkers in, 93–99
  classification of, 13
  composition of, 13–15, 350
  density of, 350
  gas chromatograms of, 359
  genetically-similar, 260–261
  vs. heavy fuel oils (HFO), 343–346
  qualitatively-similar, 261
  refining, 18–26
    distillate fuels, 22–27
    gasoline, 21–22
    lubricating oils, 28
    residual fuels, 27–28
  source markers for, 160–162
  sulfur compounds in, 148–150
  ternary diagram for, *13*
  weathering, 29–37
Cryogenic trapping, 172
Currents, 411–412
Cyanobacteria, 365
Cyclic biomarkers, 81
Cyclic terpenoids, 75–77
Cyclododecanol, 352
Cyclohexane-monooxygenase, 351

Cyclohexanol, 352
Cytochrome P_{450}, 351, 388–390

D

Dalco Passage oil spill, 63
Data management, 67–71
Daughter ions, 85–86
Daya Bay, 516
Decahydronaphthalenes, 109, 187–191
Decalins, 109, 187–191, 267
Decision chart, 233
Deep Bay, 516
Deep subsurface ecology, 366–367
Delphi 4.0 (software), 296
Denitrification, 366
Density, 350
Density-driven current, 411–412
Desmethylsteranes, 198–200
Detectors, 172–173
Deuterium, 208
Diacholestane, 243, 267
Diagnostic indices, 269–272
Diagnostic power, 126, 313
Diagnostic ratios, 126–128
    biomarkers, 239–240
    comparison of, 249–254
    criteria for, 245–246, 251–254
    definition of, 232
    elimination of, 246–249
        duplicate analyses, 248–249
        signal-to-noise (S/N) test, 246–248
    from GC/MS fingerprinting, 237–238
    in oil hydrocarbon fingerprinting, 306–307
    oil sample comparison, 245
    polycyclic aromatic compounds (PACs), 238–239
    polycyclic aromatic hydrocarbons (PAHs), 510–511
    sesquiterpanes, 240–245
    signal-to-noise (S/N) test, 246–248
Diahopane, 122–124, 242, 528–530
Dialkylthiacyclopentanes, 149
Diamantanes, 78, 114–117, 127
Diamondoids, 78, 114–117
Diaromatics, 14
Diasteranes, 127, 300, 361, 362
Diastereomers, 79–80
Dibenzoanthracene, 5, 8, 267, 339, 514
Dibenzofuran, 8, 267, 339
Dibenzofuranes, 300
Dibenzothiophenes, 149–150
    biodegradation of, 355, 361
    in crude oils, 152, 161–162
    diagnostic ratios, 237–239
    in distillate fuels, 27
    in *Exxon Valdez* oil spill, 461
    GC/FID analysis of, 267
    in heavy fuel oils (HFO), 339
    mass fragments, 300
    in oil spills, 514
    response factor, 8
    source indices, 271
Dicarboxylic acid, 352
Dicholoromethane, 266
Dielectric constants, 437
Diesel fuel, 7, 22, 99–101, 152, 351
Diesel range organics (DRO), 38
Diethylbenzene, 353–355
Differential global positioning system (DGPS), 434
Dihydroxybenzoates, 352
Dimethylcyclohexane, 356
Dimethylcyclopentane, 356
Dimethyldibenzothiophenes, 152
Dimethylnaphthalenes, 361
Dimethylphenanthrenes, 361
Dinaphthothiophene, 149
Dinosterane, 123
Dioxin, 394
Dioxygenases, 351–353
Diphenyl disulfides, 155
Discrete biomarkers, 10
Dispersion, 415
Dissimilar match, 260
Dissolution, 32, 131, 382
Distillate fuels, 22–27
Distillation, 24, 350–351
Distribution coefficient, 384
Diterpanes, 14
Dodecahydrotriphenyliene, 198
Doppler shifts, 433
Drimane, 109–111, 244
Duplicate analyses, 248–249
Duque de Caxias refinery, 506
Dynamic time warping (DTW), 304

E

Eagle Harbor, 516
EC Oil Spill Research Laboratory, 85
Egyptian crude oil, 161
Emulsification, 32, 130, 415
Emulsions, 34
Enantiomers, 79
Environment Canada Oil Research Laboratory, 82–83
Environmental data, collection of, 63

ENVISAT satellite, 436
Enzyme-linked immunosorbent assay (ELISA), 390
EPA method 8015B, 5
Ergostanes, 81
Erika oil spill, 33, 217–218, 408
ERS-1 satellite, 436
ERS-2 satellite, 436
Ethane, 361
Ethene, 352
Ethoxyresorufin, 390
Ethyladamantanes, 114–117
Ethylbenzene, 5
Ethylbenzene dehydrogenase, 356
Ethylcholestane derivative, 243
Ethylmethylbiphenyls, 361
Ethylphenanthrenes, 361
Eudesmane, 14, 109
Eulerian approach, 408
Evaporation, 30–32, 415
Excimer laser, 424–425
Excitation-emission matrices (EEMs), 298, 307–308
Exposure limits, 59–60
Extracted ion plots (EIPs), 261
Extraction, 496
Extradiol dioxygenases, 352–353
Exxon Valdez oil spill, 449–450. *See also* Oil spills
   biodegradation in, 153
   biotic responses to, 396–398
   composition and chemistry of crude oil in, 453–457
   forensic investigations, 449–482
      hydrocarbon source allocation, 462–469
         qualitative allocation models, 464
         quantitative models, 464–469
         source allocation models, 462
      identification of hydrocarbon sources in, 450–453
         biogenic hydrocarbons, 453
         multiple sources, 450–451
         petrogenic hydrocarbons, 451–453
         pyrogenic hydrocarbons, 453
      investigation of hydrocarbons, 470–477
      oil bioavailability assessment, 477–482
         biomarkers, 481–483
         passive sampling of PAH, 481
         uptake of PAH in biota, 477–480
      PAH source allocation, 469–470
      PASHs as markers in, 162–164
      resolution of natural background, 458–462
   isotope ratios in, 211–212
   photooxidation in, 153
   phytane in, 387
   polycyclic aromatic hydrocarbons (PAHs) in, 387, 457–458
   presence of other hydrocarbons in, 13
   triplet ratios in, 128
   weathering products, 453–457
   weathering trends, 457–458
      data sources, 457
      major fractions, 457
      mass loss, 458
      polycyclic aromatic hydrocarbons (PAHs), 457–458

F

Factor analysis, 294
Fay, James, 413
Fermentative bacteria, 369
Finishing, 20–21
Flame ionization detectors (FID), 6–7
Flame photometric detection (FPD), 156–157
Flash point, 329
Flasher bottoms, 328
Florida oil spill, 184–187, 449
Fluoranthenes, 5
   GC/FID analysis of, 267
   in heavy fuel oils (HFO), 339
   mass fragments, 300
   in oil spills, 514
   response factor, 8
Fluorenes, 5
   GC/FID analysis of, 267
   in heavy fuel oils (HFO), 339
   mass fragments, 300
   in oil spills, 514
   response factor, 8
Fluorescence spectroscopy, 298–299, 307–308
Fluorescent aromatic contaminants (FACs), 481
Fluvio-deltaic oils, 214
Forecasting, 406–407
Forensic investigations, 55–71
   data collection in, 63–64
   data management, 67–71
   emergency response phase, 57–61
      chemical toxicity analysis, 59–60
      personal protective equipment (PPE), 60–61
      risk assessment and characterization, 58–59
      safety management, 57–58
      working environment safety, 60
   geographic boundary determination in, 61–63
      coastal oil spills, 62–63
      marine oil spills, 62–63
      terrestrial spills, 61–62

identification of hydrocarbon sources in, 450–453
  biogenic hydrocarbons, 453
  multiple sources, 450–451
  pyrogenic hydrocarbons, 453
petroleum PASH markers in, 159–164
reconnaissance survey in, 55–57
sampling plan and design, 64–67
  known-source spills, 64–66
  mystery spills, 66–67
site characterization in, 55–57
Fortner Software LLC, 173
Fractionation, 20–21, 496
Fraunhofer light discriminator, 424
Freighter spills, 2
Friedelene, 90
Fuel No. 5, 100–101
Fuel oil, 22, 338–343
Full scan mode, 84
Fumarate, 355
Fungi, 350

G

Gammacerane, 78, 120–121, 267, 528–530
Gas chromatography/mass spectrometry (GC/MS), 83–86
  baselines, 303–304
  benchtop quadropole, 84–85
  biomarkers, 9–10
  case study, 251
  in chemical fingerprinting, 7–10
  normalization, 306
  in oil hydrocarbon fingerprinting, 299–301
  overview, 4–6
  polycyclic aromatic hydrocarbons (PAHs), 8–9
  in quantitative fingerprinting, 268–269
  retention time, 304–305
  scan mode, 84
  selected ion monitoring (SIM), 84–85
  triple quadropole, 85–86
Gas oils, 22
Gases, 22
Gasoline, 21–22, 99, 351
GC/FID fingerprinting, 6–7
  case study, 251
  in CEN oil spill identification, 235–237
  in quantitative fingerprinting, 266–268
GC Image (software), 173
Gelbstoff, 424
Gene regulation, 364
Geobacter, 368
Geographic information system (GIS), 61, 68
Ghawar oil field, 357
Gironde Bay, 516
Global positioning system (GPS), 68, 420
Gram-negative bacteria, 356
Gravity, 350
Groundwater, 161
Guanara Bay, 506
Guanara Bay oil spill, 506–507
  chemical analysis, 508–510
    biomarkers, 509–510
    extract cleanup, 509
    PAH analysis, 508–510
    sediment sample extraction, 508–509
    source identification methods, 508–510
  hydrocarbon source identification, 514–532
    biomarker diagnostic ratios, 527–532
    PAH diagnostic ratios, 514–525
    PAH principal component analysis, 525–527
    PAH quantification, 511–514
  sampling design, 507–508
  source identification methods, 510–511
    biomarker diagnostic ratios, 511
    PAH diagnostic ratios, 510–511
    PAH multivariate analysis, 511
Gulf of Suez crude oil, 161
Gullfaks oilfied, 367

H

Halophilic bacteria, 364
Haven oil spill, 162, 435
Heavy crude oils, 13
Heavy fuel oils (HFO), 327–346. *See also* Crude oils; Petroleum
  boiling points, 332–333
  chemical fingerprinting, 330–332
  classification of, 329–330
  composition of, 333–336, 351
  vs. crude oil, 343–346
  historical perspective, 328
  ISO specifications for, 329
  isoalkane ratios, 331
  molecular variability in, 336–343
    biomarkers, 336–338
    polycyclic aromatic hydrocarbons (PAHs), 338–343
  overview, 327
  production of, 329–330
  quantitative fingerprinting of, 279–289
  samples and analytical methods, 332
  total hydrocarbons, 331
Heavy residual fuels, 101
Heavy straight-run distillates, 22

Heliscat, 428
Hematite, 368
Heptadecane, 236, 352
Herring, 474–475
Heterocyclic aromatic compounds, 300
Hexadecane, 352, 357
Hindcasting, 406–407
Holly oil, 196–201
Homodrimanes, 110, 244
Homohopanes, 201, 242, 267
Homopregnane derivative, 243
Hopanes, 80
    analysis of, 496–497
    biodegradation of, 133
        aerobic, 353
        subsurface, 360–361
        susceptibility, 362
    as biomarkers, 124, 199–201
    diagnostic indices, 271
    diagnostic ratios of, 242, 528–530
    GC/FID analysis of, 267
    as internal standards, 134
    isomers, 81
    mass spectra of, 87
    structures of, 14
HOT zone, 59
Houston Ship Channel, 211
Huntington Beach, 212
Hurricane Katrina, 31, 38–39
Hydrocarbons, 349–350
    accumulation by biota, 382–387
    aerobic biodegradation of, 351–355
    aliphatic, 14
    anaerobic biodegradation of, 355–357
    aromatic, 14
    biogenic, 350, 451, 453
    branching, 352
    catabolic degradation of, 387–393
    identification of, 381–382, 470–477
    labeling and nomenclature of, 77–81
    petrogenic, 350, 450–453
    pyrogenic, 451, 453
    structures of, 74–77
        cyclic terpenoids, 74–77
        isoprenoids, 74
    toxicity of, 393–396
Hydrodesulfurization, 27, 151
Hydrodynamic models, 412
Hydrogen isotopes, 208, 224
Hydrothermal vents, 365
HyperChrom (software), 173
Hyperspectral imaging, 421
Hyperthermophilic organism, 365, 368

I

Ice, 436–438
Inconclusive match, 260
Indan, 177
Indene, 177
Indenopyrene, 5, 8, 267, 339
Indeterminate match, 259
Industrial Materials Institute (NMI), 433
Infrared cameras, 438–439
Infrared sensors, 422–423
Integrated multivariate oil fingerprinting (IMOF), 296–297
Intentional operational discharges, 1–2
Intermediate fuel oil 180 (IFO 180), 100–101, 113, 330
Intermediate fuel oil 380 (IFO 380), 330
    biomarkers, 336–338
    chromatograms, 333–336
    polycyclic aromatic compounds (PACs), 338–342
Intermediate fuel oil (IFO), 279–289
Internal injection standard (IISTD), 497
Internal standards, 83
Intradiol dioxygenases, 352–353
Iron-reducing bacteria, 368
Iron reduction, 366
Iso-butane, 361
Iso-paraffins, 14
ISO specifications, 329–330
Isoalkane ratios, 331
Isoalkanes, 361
Isoprenes, 74
Isoprenoids, 7, 75
    biodegradation of, 361
    diagnostic indices, 271
    diagnostic ratios, 127, 253
    mass fragments, 300
    susceptibility to biodegradation, 362
Isopropyl benzene, 354–355
Isotope ratio mass spectrometer (IRMS), 208–210
Isotopes, 208
Isowake Experiments, 431

J

Jacob Luckenbach, 67
Jet fuel, 99
Jiulongjiang Estuary, 516
Johnston Creek crude oil, 460–461

K

Katalla Formation shale, 459–462
Kendrick mass defect, 154
Kendrick nominal mass, 154

Kerogens, 349, 467
Kerosene, 22, 113
Ketones, 352
Known-source spills, 3, 64–66
Knudsen Engineering Ltd., 437
Korea Diesel No. 2, 100–101, 110
Kulthieth Formation, 456–457
Kurdistan oil spill, 435
Kuskutz coals, 456, 459–461
Kuwait crude oil, 90

L

L-band radar, 427
La Luna crude oil, 359
Lactone, 352
LANDSAT satellite, 435
Langragian elements, 408
Laser fluorosensors, 419–420, 423–426, 438–439
Laser ultrasonic remote sensing of oil thickness (LURSOT), 433–435
Leco Corp., 173
Lewis acids, 114
Light distillates, 99
Light oils, 350
Light straight-run distillates, 22
Lignite, *40*
Lipid catabolic pathways, 352
Loading plots, 315–317
Low-density polyethylene strips (LDPEs), 481
Lowest effect level (LEL), 59
Lubricating oils, 28
  biomarkers, 101–103
  boiling point, 22
  carbon range, 22
  use in marine fuels, 328–329

M

Macrocyclic alkanes, 124–125
Malaysia, oil pollution sources in, 491–492
Manganese reduction, 366
Manmade petroleums, 12
Marine carbonate oils, 214
Marine diesel oil (MDO), 279, 284–286, 330
Marine Ecosystems Research Laboratory (MERL), 187
Marine fuel oils, 327–346. *See also* Crude oils; Petroleum
  boiling points, 332–333
  chemical fingerprinting, 330–332
  classification of, 329–330
  composition of, 333–336
  vs. crude oil, 343–346
  historical perspective, 328
  ISO specifications for, 329–330
  isoalkane ratios, 331
  molecular variability in, 336–343
    biomarkers, 336–338
    polycyclic aromatic hydrocarbons (PAHs), 338–343
  overview, 327
  production of, 329–330
  quantitative fingerprinting of, 279–289
  samples and analytical methods, 332
  total hydrocarbons, 331
Marine gas oil (MGO), 330
Marine oil spills, 57, 62–63, 181–184, 277–279, 327
Masan Bay, 516
Mass fragments, 300
Mass losses, 458
Mass-selective detection (MSD), 157
Mass selective detector (MSD), 85
Mass spectrometry, 4–6
Mass-to-charge ratio, 84–90
Match, 259
Math Works Inc., 173
Matlab (software), 173, 296
Matrices, 310–311
Maximum constructive interferences, 429
Maximum destructive interferences, 429
Mercaptans, 350
Mersey Estuary, 516
MESA oil, 105–107
Mesozoic petroleum system, 17
Methane, 361
Methanogenesis, 357, 366–367
Methanogens, 356–357, 369
Method detection limits (MDLs), 465–466
Methydiamantane index (MDI), 116
Methyl-ethyl benzene, 354–355
Methyl groups, 90, 155
Methyladamantane index (MAI), 116
Methyladamantanes, 114–117
Methylanthracene, 267
Methylation, 355
Methylbiphenyls, 361
Methylcholestanes, 243, 267
Methylchrysene, 267
Methyldibenzothiophenes, 152, 237–239, 267, 296
Methylfluoranthenes, 238
Methylfluorene, 295
Methylnaphthalenes, 222, 361, 514
Methylpentylsuccinate synthase, 355
Methylphenanthrenes, 237–239, 267, 271, 296, 361

Methylpyrenes, 237–239
Methylsteranes, 124
Metula oil spill, 133, 450
Microaerophilic bacteria, 368
Microbial communities, succession of, 365–367
Microbial ecology, 365
Microcarbon residues, 329
Microwave sensors, 426–428. *See also* Remote sensing
　radar, 426–428
　radiometers, 426
　scatterometers, 428
Mid-range distillates, 99–101
Midcontinent rift system, 367
Minas oil, 216
Minjiang Estuary, 516
Miocene Monterey Formation, 451
Mirror-image isomers, 81
Mixtures, separation of, 169–170
Mobile Bay marine diesel fuel spill, 181–184
Models and modeling, 405–416
　currents, 411–412
　oil spill movement, 406–407
　oil spill transport, 407–409
　overview, 405–406
　spreading, 413–414
　turbulence, 413
　weathering, 414–416
　wind, 409–411
Modulators, 172
Molecular fossils. *See* Biomarkers
Molecular markers, 501–502
Molybdopterin cofactor, 356
Monoaromatic steranes, 127
Monoaromatics, 14
Monocyclic acids, 361
Monomethylbenzothiophenes, 155
Monooxygenases, 351, 354
Monterey Formation, 128
Moretane, 242, 267, 271
Mousse, 34
Multi-detector electrooptical imaging scanner (MEIS), 421
Multi-linear models, 310–312
　higher-order arrays, 311–312
　two-way case, 310–311
Multi-spectral scanners, 438–439
Multidimensional separation, 170–171
Multivariate statistics, 308–314
　in comparison of diagnostic ratios, 249–250
　constrained least squares (CLS), 466–468
　multi-linear models, 310–312
　in oil spill fingerprinting, 294–296
　partial least squares (PLS), 468
　variable selection and scaling, 312–314
Mussels, 387, 391–393, 477–480
Mycobacterium vanbaalenii, 351
Mysidium gracile, 507
Mystery spills, 2–3, 66–67

N

N-alkanes, 90, 361, 362, 383
N-butane, 361
N-cyclohexyl-2-benzothiozolamine (NCBA), 497
N-way toolbox, 297
Naphthalene dioxygenase, 352
Naphthalenes, 339
Naphthenes, 14, 350
Naphthenic acids, 361
Naphthobenzothiophenes, 8, 267, 271, 300
Naphthothiophenes, 27, 149
Napthalene carboxylase, 356
Napthalenes, 5
　carbon isotope ratios, 222
　GC/FID analysis of, 267
　mass fragments, 300
　in oil spills, 8, 183, 514
　vapor pressure of, 382
National Petroleum Reserve, Alaska (NPRA), 367
Natural dispersion, 130–131
Natural hydrocarbons, 37–41
　algal debris, 38–39
　oil seeps, 41
　vascular plants, 38–39
　wood charcoal, 39–41
Natural resource damage assessment (NRDA), 265
Navy Fuel Oil (NFO), 67
Nd:YAG laser, 424, 433
NetCDF (software), 296
New Carisa, 68
Nigerian Bonnie Light crude oil, 16–17
Night vision cameras, 423
Nipisi oil spill, 90
NIST 1582 oil, 110
Nitrate-reducing bacteria, 366
Nitrogen isotope ratios, 211, 224
Nitrogen laser, 424
Nitrogen oxides, 150
Nomenclature, 77–81
Non-tanker spills, 2
Nonane, 353, 383
Nonaqueous phase liquids (NAPLs), 99
Nonmatch, 251, 259, 273, 319
Nordtest method, 229–230, 273–276
Norhomohopane, 528–530

Norhopanes, 360–361
  biomarker indices, 271
  diagnostic ratios, 242
  formation of, 360–361
  GC/FID analysis of, 267
  mass spectra of, 87
  susceptibility to biodegradation, 362
Normalization, 306
Normoretane, 242, 267
Norneohopane, 133, 242, 267
Norpristane, 267, 271

O

Ocean circulation, 411–412
Octadecane, 236, 352
Octane monooxygenase, 351
Octane rating, 351
Office of Solid Waste and Emergency Response, 36
Offshore rigs, discharges from, 2
Oil embargos, 328
Oil fractions, 104, 213, 457
Oil hydrocarbon fingerprinting, 293–294. *See also* Biomarker fingerprinting
  analytical methods, 298–300
    fluorescence spectroscopy, 298–299
    gas chromatography/mass spectrometry (GC/MS), 299–301
  data evaluation, 314–319
    numerical comparisons and statistical tests, 317–319
    score and loading plots, 315–317
  data preprocessing, 302–308
    diagnostic ratios, 306–307
    fluorescence spectra, 307–308
    partial GC-MS/SIM chromatograms, 303–306
      baseline removal, 303–304
      normalization, 306
      retention time alignment, 304–305
  integrated multivariate, 296–297
  multivariate statistics, 294–296, 308–314
    multi-linear models, 310–312
      higher-order arrays, 311–312
      two-way case, 310–311
    variable selection and scaling, 312–314
  quality assurance, 301–302
  quality control, 301–302
  sample preparation, 297
Oil-in-ice detectors, 436–438
Oil-mineral aggregation, 132
Oil production platform spills, 2
Oil seeps, 1, 41, 196–201

Oil spills, 1–3
  accidental, 1–2
  chemical toxicity of, 59–60
  coastal, 62–63
  evolution of, 413–416
    oil weathering, 414–416
    spreading, 413–414
  fingerprinting of, 181–201
    Buzzards Bay No. 6 fuel spill, 191–195
    Mobile Bay marine diesel fuel spill, 181–184
    Santa Barbara oil seep, 196–201
    West Falmouth No.2 fuel oil spill, 184–187
    Winsor Cove fuel oil spill, 187–191
  identification of, 231–233
  intentional operational discharges, 1–2
  known-source release, 3
  on land, 56
  marine, 62–63
  modeling of, 405–416
    forecasting and hindcasting, 406–407
    overview, 405–406
    transport, 407–413
      currents, 411–412
      turbulent diffusion, 413
      wind, 409–411
  multidimensional analysis of, 170–171
  mystery spills, 2–3
  remote sensing of, 419–439
    acoustic systems, 435
    detection of oil under ice, 436–438
    future trends in, 438–439
    laser fluorosensors, 423–426
    microwave sensors, 426–428
      radar, 426–428
      radiometers, 426
      scatterometers, 428
    optical sensors, 420–423
      infrared, 422–423
      night vision cameras, 423
      ultraviolet, 423
      visible, 420–422
      visible indications of oil, 420–422
    real-time displays and printers, 438
    satellite, 435–436
    slick thickness, 428–435
      remote sensors, 431–432
      specific sensors, 431–435
      visual indicators, 428–431
    visible indications of oil, 420
  sensitivity of birds to, 216–217
  source identification of, 125–130, 134–138
  sources of, 1–2

Oil spills (*Continued*)
  terrestrial, 61–62
  on water, 56
Oil-water interface, 429
Oily wastes, 28
Oleanane, 90
  diagnostic indices, 271
  diagnostic ratios, 242, 528–530
  GC/FID analysis of, 267
  identification of, 459
  in oil spills, 120–123
  structure of, 14
Operational discharges, 1–2
Optech Inc., 426
Optical frequency compensation device (OFCD), 433
Optical sensors, 420–423. *See also* Remote sensing
  infrared, 422–423
  night vision cameras, 423
  ultraviolet, 423
  visible, 420–422
Organo-sulfur compounds, 148–150
Orimulsion, 425
Orinoco crude oil, 96–99, 357
Oscillatoria, 365
Oserberg Field Centre oil, 315–317
Oserberg South East oil, 315–317
Oxidants, 365–366
Oxidation, 415–416
Oxidative element analyzers, 210
Oxidative metabolic gearing, 368
Oxygen, 351–352
Oxygen isotope ratios, 224

P

Paraffins, 14, 350, 359
Parallel factor analysis (PARAFAC), 307–312
Parent ions, 85–86
Partial least-squares regression, 295, 310, 468
Particles, 408–409
Particulate coal, 39–41
Pattern recognition, 294
Pearl River Estuary, 516
Pee Dee Belemnite (PBD), 208
Pentaaromatics, 14
Pentacyclic terpanes, 135–138
Pentacyclic triterpanes, 267
Pentakishomohopanes, 267
Pentanes, 361
PERF Method, 62
Perfluorotributylamine (PFTBA), 85, 268
Personal protective equipment (PPE), 60–61

Perylene, 40
  GC/FID analysis of, 267
  generation of, 453
  in heavy fuel oils (HFO), 339
  in oil spills, 8, 514
Petrobras, 506
Petrobras Geochemistry Laboratory, 85
Petrogenic hydrocarbons, 350, 450–453
Petroleum, 11–12. *See also* Crude oils
  aromatic steranes in, 104–109
  biomarkers in, 99–101
  chemical fingerprinting of, 258
  crude oil genesis (primary control), 13–18
  high-resolution separation of, 169
  manmade, 12
  natural, 11
  primary controls, 12
  refining, 18–26
    distillate fuels, 22–27
    gasoline, 21–22
    lubricating oils, 28
    residual fuels, 27–28
  secondary controls, 18–26
  sulfur compounds in, 148–150
  tertiary controls, 29–37
  weathering, 12, 29–37
Petroleum biodegradation. *See* Biodegradation
Petroleum biomarkers. *See* Biomarkers
Petroleum products, 350–351
Phenanthrenes, 5
  biodegradation of, 355, 361
  diagnostic ratios, 237–239
  in distillate fuels, 27
  in *Exxon Valdez* oil spill, 461
  in heavy fuel oils (HFO), 339
  mass fragments, 300
  in Mobile Bay marine fuel spill, 267
  in oil spills, 183–184, 514
  response factor, 8
Phenanthrothiophene, 149
Phenazines, 369
Phenols, 360, 361
Phosphorus, 329
Photochemical processes, 155
Photooxidation, 34, 131–132, 153, 382
Physical data, collection of, 63
Phytane, 75, 236, 267, 352
Pink salmon, 474–477
Pipeline spills, 2
Plant biomarkers, 90
Platform Elly crude oil, 96–99, 109
Point sources, 1
Point-to-point matching, 295

Polars, 15, 360
Polycyclic aromatic compounds (PACs), 160, 232–233, 238–239, 300
Polycyclic aromatic hydrocarbons (PAHs), 8–9. *See also* Biomarkers
  accumulation of, 390–393
  aerobic biodegradation of, 355
  analysis of, 497–498
  catabolic degradation of, 387–393
  depuration of, 390–393
  diagnostic indices, 269–272
  diagnostic ratios, 510–511, 514
  distribution coefficient of, 384
  in *Exxon Valdez* oil spill, 457–458
  in heavy fuel oils (HFO), 338–343
  in light oils and condensates, 350
  mass fragments, 300
  multivariate analysis of, 511
  persistence of, 390–393
  principal component analysis (PCA), 514
  pyrogenic index, 457–458
  in qualitative fingerprinting, 264–265
  quantitative evaluation of, 135–138
  in quantitative fingerprinting, 268, 464–465
  separation from PASHs, 157–158
  source allocation of, 469–470
  susceptibility to biodegradation, 362
  uptake in biota, 477–480
  in urban runoff, 42–43
Polycyclic aromatic sulfur heterocycles (PASHs), 147–164. *See also* Biomarkers
  analysis methods for, 155–159
    selective detection in gas chromatography, 156–157
      atomic emission detection (AED), 157
      flame photometric detection (FPD), 156–157
      mass-selective detection (MSD), 157
      sulfur chemiluminescence detection (SCD), 157
    separation from PAH, 157–158
    two-dimensional gas chromatography (GC x GC), 158
  in Egyptian crude oil, 152
  identification of, 150
  influence of refinery processes on, 150–152
  overview, 147–148
  quantification of, 158–159
  as source markers, 160–162
  stability of, 152–155
  structures of, 148–150
  as weathering markers, 162–163
Polydimethylsiloxane, 174

Polysilphenlylene, 174
Porphyrins, 125, 360, 361, 362
Portland Harbor, 516
Positive match, 250–251, 273, 319
Potentially responsible party (PRP), 230, 265
Poul Creek crude oil, 460–461
Pour point, 329
Precambrian period, 350
Pregnane derivative, 243
Prestige oil spill, 428
Primary controls, 13–18
Prince William Sound, 41, 128, 211–212, 295, 450–453
Principal component analysis (PCA), 249–250, 294–296, 310, 465–466
Principal component regression (PCR), 310
Printers, 438
Priority pollutants, 5
Pristane, 14
  biodegradation of, 352
  biomarker indices, 271
  diagnostic ratios, 236
  GC/FID analysis of, 267
Probable match, 251, 259, 273, 319
Procedural artifacts (PA), 471–474
Propane, 361
Propyl benzene, 354–355
Propylene, 352
Proteobacteria, 356–357
Prudhoe Bay oil, 115
Pseudomonads, 364
Pseudomonas, 153
Pseudomonas mendocina, 354
Pseudomonas putida, 351–354
Pseudomonas stutzeri, 353
Pyrenes, 5
  GC/FID analysis of, 267
  in heavy fuel oils (HFO), 339
  mass fragments, 300
  in oil spills, 514
  response factor, 8
Pyridines, 361
Pyrite, 364
Pyrogenic hydrocarbons, 451, 453
Pyrogenic index, 457–458
Pyrrols, 361

Q

Qualitative allocation models, 464
Qualitative fingerprinting, 258–263
  match criteria, 259–260
  shortcomings of, 260–263
    genetically similar oils, 260–261

Qualitative fingerprinting (*Continued*)
  mixing, 261–263
  qualitatively similar oils, 261
  weathered oils, 260
Quality Assurance Project Plan (QAPP), 70
Quality assurance (QA), 10–11, 70–71, 301–302
Quality control (QC), 10, 70–71, 301–302
Quality management plan (QMP), 10
Quantitative fingerprinting, 263–265
  in analysis of mixed source oil, 276–289
    fuel oil spills, 279–289
    marine oil spills, 277–279
    two-component mixing models, 276–277
  data generation for, 265–269
    data quality, 269
    GC/FID analysis, 266–268
    GC/MS analysis, 268–269
    sample collection, 265–266
    sample preparation, 266
  diagnostic indices, 269–272
  source identification protocols, 272–276
Quantitative models, 464–469
  PAH ratios, 464–465
  statistical models, 465–468
    constrained least squares (CLS), 466–468
    partial least squares (PLS), 468
  total organic carbon (TOC), 468–469
Queen Vein coals, 460–461
Quinones, 369

R
Radar, 420, 426–428, 438
Radar satellites, 436
Radarsat satellite, 436
Radiolysis, 352
Radiometers, 426, 438–439
Ralstonia pickettii, 351
Raman scattering, 298–299, 424
Rayleigh scattering, 298–299
Real-time displays, 438
Recent organic matter (ROM), 469
Recreational marine vessel, discharges from, 2
Reductive element analyzers, 210
Reference index (RI), 114
Refineries, 20–21, 350–351
Refractive index, 429, 459
Relative response factors (RRFs), 9, 266
Relative standard deviation (RSD), 249, 313
Remote sensing, 419–439
  acoustic systems, 435
  future trends in, 438–439
  microwave sensors, 426–428
    radar, 426–428
    radiometers, 426
    scatterometers, 428
  optical sensors, 420–423
    infrared, 422–423
    night vision cameras, 423
    ultraviolet, 423
  overview, 419–420
  real-time displays and printers, 438
  satellite, 435–436
  slick thickness, 428–435
    remote sensors, 431–432
    specific sensors, 431–435
    visual indicators, 428–431
  visible indications of oil, 420
Remotec Applications Ltd., 436
Repeatability limit, 246
Reservoir biodegradation, 357–358
Reservoir strata, 11
Residual fuel oils, 327–346. *See also* Crude oils; Petroleum
  boiling points, 332–333
  chemical fingerprinting, 330–332
  classification of, 329–330
  composition of, 333–336
  vs. crude oil, 343–346
  historical perspective, 328
  ISO specifications for, 329–330
  isoalkane ratios, 331
  molecular variability in, 336–343
    biomarkers, 336–338
    polycyclic aromatic hydrocarbons (PAHs), 338–343
  overview, 327
  production of, 329–330
  quantitative fingerprinting of, 279–289
  samples and analytical methods, 332
  total hydrocarbons, 331
Residual fuels, 27–28
Residuals, 328
Resolution power, 312
Resonance scattering, 437
Resorufin, 390
Retene, 237–239, 267, 271, 453
Retention time, 151, 304–305
Retention time markers, 10
% Ring number, 271
Rio de la Plata Estuary, 516
Risk assessment, 58–59
River runoff, 42–43
Rotational freedom, 310
Round Robin exercise, 230
Runoff, 42–43

S

Safety management, 57–58
Safety response zones, 58–59
Sampling, 64–67
    design, 65–66
    known-source spills, 64–66
    on land, 65
    oil source, 65
    water column, 64–65
San Francisco Bay, 516
Sandstones, 368
Santa Barbara oil seep, 196–201, 405
Sarasota Bay, 516
Satellite remote sensing, 435–436
Scanners, 421
Scatterometers, 428
Scores, 315–317
Scotia Light crude oil, 93–99, 111–113
Secondary biogenic methane, 369
Secondary controls, 18–35
Sedimentation, 132
Sediments, 492–494
Seepage, 195–201
Select ion monitoring (SIM), 268, 332
Selected ion monitoring (SIM), 7, 84–85
Semipermeable membrane devices (SPMDs), 220, 481
Separation methods, 169–170
Sequence of priority, 81
Sesquiterpanes, 109–114. *See also* Biomarkers
    as biomarkers, 24
    characterization of, 135
    cross-plots of, 129–130
    diagnostic ratios, 127, 137, 240–245
    labeling and nomenclature of, 78
    mass fragments, 300
    structures of, 14
Seychelles Islands, 218
Shear waves, 437
Shipborne radars, 428
Shoreline Cleanup Assessment Technique (SCAT), 62
Side-looking airborne radar (SLAR), 427
Signal-to-noise (S/N) test, 246–248
Silica gel, 82
Similar match, 259–260
Singular value decomposition (SVD), 310
SINTEF, 231
Site investigations, 55–71
    data collection in, 63–64
    data management, 67–71
    emergency response phase, 57–61
        chemical toxicity analysis, 59–60
        personal protective equipment (PPE), 60–61
        risk assessment and characterization, 58–59
        safety management, 57–58
        working environment safety, 60
    factors in, 57
    geographic boundary determination in, 61–63
        coastal oil spills, 62–63
        marine oil spills, 62–63
        terrestrial spills, 61–62
    petroleum PASH markers in, 159–164
    reconnaissance survey in, 55–57
    sampling plan and design, 64–67
        known-source spills, 64–66
        mystery spills, 66–67
    site characterization in, 55–57
Sockeye crude oil, 96–99
Sohxlet extraction, 219, 297
Source allocation models, 462
Source ratios, 160
Source-specific biomarkers, 121–125. *See also* Biomarkers
    4-methyl steranes, 124
    beta-carotane, 124
    bicadinanes, 124
    diahopane, 122–124
    hopanes, 124
    macrocyclic alkanes, 124–125
    polycyclic aromatic sulfur heterocycles (PASHs), 160–162
    porphyrins, 125
Source strata, 11
South Louisiana crude oil, 93–99
Southern California Monterey Formation, 212
Specific gravity, 350
Sphalerite, 364
Sphingomonas macrogolitabida, 353
Sphingomonas yanoikuyae, 354
Spills. *See also* Oil spills
Spreading, of oil spills, 413–414
Squalane, 75
St. Lawrence River, 218
Stable isotope ratios, 207–224
    basis of, 208
    bulk isotope ratios, 210–213
    compound-specific isotope analysis (CSIA), 214–220
    in geochemical materials, *209*
    instruments, 208–210
    measurement of, 207–210
    overview, 207
    standards, 208
    weathering, 220–224
Standard Mean Ocean Water (SMOW), 208

Steranes, 75–77. *See also* Biomarkers
  biodegradation of, 133, 353, 361–362
  diagnostic indices, 271
  diagnostic ratios, 127, 137, 528–530
  evaluation of, 135–138
  in *Exxon Valdez* oil spill, 461
  labeling and nomenclature of, 78
  mass fragments, 300
  mass spectra of, 86–90
  structures of, 14
Stereoisomers, 79–80
Steroids, 240
Stigmastanes, 81, 133–134
Still cameras, 438–439
Strait of Malacca, 489
  hydro-oceanographic conditions of, 489–491
  oil pollution sources in, 491–492
  oil spill events in, 491
  oil spill investigations, 492–502
    analytical procedure, 495–496
    biomarkers, 498–501
    instrumental analysis, 496–498
      alkanes, 496–497
      hopanes, 496–497
      N-cyclohexyl-2-benzothiozolamine (NCBA), 497
      polycyclic aromatic hydrocarbons (PAHs), 497–498
    methodology, 492–495
      aerosol samples, 495
      asphalt samples, 494
      extraction, 496
      fresh crankcase oil, 494
      sample collection, 492
      sediment samples, 492–494
      source petroleum, 492
      street dust samples, 493
      tar-ball samples, 492
      tire rubber, 495
      used crankcase oil, 495
    molecular markers, 501–502
    polycyclic aromatic hydrocarbons (PAHs), 502
  ship traffic in, 491
Street dust, 493
Subsurface biodegradation, 357–362
Subsurface circulation models, 412
Succinates, 355
Suez-Mediterranean pipelines, 163
Sulfate-reducing bacteria, 357, 366–367
Sulfate reduction, 366
Sulfides, 148
Sulfolobus solfataricus, 365

Sulfur, 148
Sulfur chemiluminescence detection (SCD), 157
Sulfur compounds, 26–27, 148–150, 214
Sulfur-free gasoline, 150
Sulfur isotope ratios, 211, 213, 224
Sulfur oxides, 150
Surrogates, 83
Synechococcus, 365
Synthetic aperture radar (SAR), 427
Syntrophic microorganisms, 356

T

Tamar Estuary, 516
Tanker spills, 2
Tar balls, 69, 120, 211–213, 492, 499–502
Tar sands, 357
Tensors, 311
Tenyo Maru oil spill, 422
Teresso industrial oil, 110, 113
Terminal electron acceptors, 362–363, 366, 369
Terpanes, 75
  biodegradation of, 133
  diagnostic ratios, 127, 137, 528–530
  evaluation of, 135–138
  in *Exxon Valdez* oil spill, 461
  labeling and nomenclature of, 78
  mass fragments, 300
Terpenes, 74
Terpenoids, 74–77
  acyclic, 75
  cyclic, 75–77
Terrestrial spills, 61–62
Tertiary controls, 29–43
  background chemicals, 37–43
  weathering, 29–36
Tertiary petroleum system, 17
Tetraaromatics, 14
Tetrakishomohopanes, 267
Tetralin, 353
Tetramethyladamantanes, 114–117
Tetramethylphenanthrenes, 361
TFE-fluorocarbon polymer, 265
Thaurea aromatica, 355
Thermal mapping, 432–433
Thermo Electron Corp., 173
Thermopiles, 364–365
Thermus, 364
Thiacyclohexane, 149
Thiamonocyclanes, 148
Thiapolycyclanes, 148
Thiopanes, 149
Thiopenes, 149–150

Three percent (3%) rule, 410
Tidal current, 411–412
Time-of-flight mass spectrometry (TOFMS), 173
Tire rubber, 495
Toluene, 5, 14, 353, 364
Toluene-3-monooxygenase, 351
Torrey Canyon, 405
Total ion chromatogram (TIC), 84
Total organic carbon (TOC), 459, 468–469
Total petroleum hydrocarbons (TPH), 38–39, 266
Toulon Harbor, 516
Toxicity, 59–60
Transform (software), 173
Transmittance, 424
Triaromatic steranes, 271, 300
Triaromatic steroids, 243, 267
Triaromatics, 14
Tricyclic terpanes, 361
Tricyclic triterpanes, 267
Trimethyl benzene, 354–355
Trimethylbiphenyls, 361
Trimethylnaphthalenes, 361
Triple quadropole GC-MS-MS, 85–86
Triplet ratios, 128
Trishomohopanes, 242, 267
Trisnorhopane, 242, 267
Trisnorneohopane, 242
Triterpanes, 14
   diagnostic indices, 271
   diagnostic ratios, 240–244, 253
   distribution of, 120
   GC/FID analysis of, 267
Troll crude oil, 93–99
Turbulence, 413
Two-component mixing models, 276–277

U

Ultraviolet cameras, 438
Ultraviolet sensors, 423
Uncertainty analysis, 409
Unresolved complex mixture (UCM), 359
   in GC/FID analysis, 7, 42–43
   in marine fuel oils, 333–334
   in oil spills, 184–186
Urban runoff, 42–43
Ursene, 90

V

Vacuum gas oils, 22
Van der Waals force, 382
Vanadium, 329
Vapor pressure, 382
Vapors, 59
Variable-outlier detection, 313
Variables, 312–314
Vascular plants, 38–39
Vegetation biomarkers, 90
Venezuela Fuel No. 6, 100–101
Vertical velocity sensor, 434
Vibrio fisheri, 507
Video cameras, 421, 438–439
Viscosity, 329, 351
Visible spectrum, 420–422
Volatile organic compounds, 60
Volatile solvents, 22

W

Wallach, Otto, 74
WARM zone, 59
Warping path, 304–305
Water-soluble fraction (WSF), 382, 471
Water-washing, 384
Waterborne oil spills, 62–63
Wavelengths, 429
Waxes, 34–35
Waxy bitumens, 215
Weathered percentages, 134
Weathering, 29–37
   biodegradation. *See* Biodegradation
   de-waxing, 34–35
   definition of, 12
   dissolution, 32, 131
   emulsification, 130
   evaporation, 30–32, 130
   GC/FID screening, 236–237
   and hydrogen accumulation by biota, 382–383
   markers, 162–163
   and mass loss, 458
   mousse formation, 34
   natural dispersion, 130–131
   oil-mineral aggregation, 132
   in oil spills, 414–416
   photooxidation, 34, 131–132
   and qualitative fingerprinting, 260
   sedimentation, 132
   and stable isotope ratios, 220–224
   wax enrichment, 34–35
Weathering ratios, 160
Weighted least-squares analysis, 295, 312–314
West Falmouth fuel spill, 184–187
West Texas crude oil, 111–113
Wind, 409–411
Wind drift factor, 410
Wind-driven current, 411–412

Winsor Cove fuel oil spill, 187–191
Working environment, safety of, 60

X
X-band radar, 427–428
XeCl excimer laser, 424–425
Xylene monooxygenase, 354
Xylenes, 5

Y
Yakataga Formation shale, 460–461
Yalujiang Estuary, 516

Z
Zinc, 329